AF692971

Plant Biotic and Abiotic Stresses —Volume II

Plant Biotic and Abiotic Stresses —Volume II

Editors

Hakim Manghwar
Wajid Zaman

MDPI

Basel • Beijing • Wuhan • Barcelona • Belgrade • Novi Sad • Cluj • Manchester

Editors
Hakim Manghwar
Chinese Academy of Sciences
Nanchang, China

Wajid Zaman
Yeungnam University
Gyeongsan, Republic of Korea

Editorial Office
MDPI
St. Alban-Anlage 66
4052 Basel, Switzerland

This is a reprint of articles from the Special Issue published online in the open access journal *Life* (ISSN 2075-1729) (available at: https://www.mdpi.com/journal/life/special_issues/_plant_stresses).

For citation purposes, cite each article independently as indicated on the article page online and as indicated below:

Lastname, A.A.; Lastname, B.B. Article Title. *Journal Name* **Year**, *Volume Number*, Page Range.

Volume II
ISBN 978-3-0365-8538-3 (Hbk)
ISBN 978-3-0365-8539-0 (PDF)
doi.org/10.3390/books978-3-0365-8539-0

Set
ISBN 978-3-0365-8534-5 (Hbk)
ISBN 978-3-0365-8535-2 (PDF)

© 2023 by the authors. Articles in this book are Open Access and distributed under the Creative Commons Attribution (CC BY) license. The book as a whole is distributed by MDPI under the terms and conditions of the Creative Commons Attribution-NonCommercial-NoDerivs (CC BY-NC-ND) license.

Contents

life

MDPI

Article

From Biodiversity to Musketry: Detection of Plant Diversity in Pre-Industrial Peloponnese during the *Flora Graeca* Expedition

Chrysanthi Chimona, Sophia Papadopoulou, Foteini Kolyva, Maria Mina and Sophia Rhizopoulou *

Section of Botany, Department of Biology, National and Kapodistrian University of Athens, 15784 Athens, Greece
* Correspondence: srhizop@biol.uoa.gr; Tel.: +30-210-7274513

Abstract: As the interest in natural, sustainable ecosystems arises in many fields, wild plant diversity is reconsidered. The present study is based on extant literature evidence from the journey of John Sibthorp (Professor of Botany, Oxford University) to Peloponnese (Greece) in pre-industrial time. In the year 1795, Peloponnese was a botanically unknown region, very dangerous for travellers and under civil unrest, in *conjuncture* with a pre-rebellion period. Our study reveals approximately 200 wild plant taxa that were collected from Peloponnese localities in 1795, transported to Oxford University (UK), and quoted in the magnificent edition *Flora Graeca Sibthorpiana* of the 19th century. Moreover, these plants currently constitute a living collection in Peloponnese, confirmed according to updated data on the vascular Flora of Greece. The presented lists constitute a source of information for plant biologists, linking the past to the present, shedding light on the study of adaptive traits of wild Mediterranean plants and revealing the temporal dimension of natural history. Nowadays, increasing and thorough understanding of the considered plants' functionality to abiotic and biotic environmental stimuli provides a new framework of sustainability and management options.

Keywords: archives; botanical collection; Greece; landscape; pre-rebellion period

Citation: Chimona, C.; Papadopoulou, S.; Kolyva, F.; Mina, M.; Rhizopoulou, S. From Biodiversity to Musketry: Detection of Plant Diversity in Pre-Industrial Peloponnese during the *Flora Graeca* Expedition. *Life* **2022**, *12*, 1957. https://doi.org/10.3390/life12121957

Academic Editors: Wajid Zaman and Hakim Manghwar

Received: 31 October 2022
Accepted: 19 November 2022
Published: 23 November 2022

Publisher's Note: MDPI stays neutral with regard to jurisdictional claims in published maps and institutional affiliations.

Copyright: © 2022 by the authors. Licensee MDPI, Basel, Switzerland. This article is an open access article distributed under the terms and conditions of the Creative Commons Attribution (CC BY) license (https://creativecommons.org/licenses/by/4.0/).

1. Introduction

In the 18th century, travelers' journey to Greece was also a journey through its history. The naturalists' travels were explorations, linked to searching for specimens of natural history. The travelers' observations became a way of identifying and revealing cultural and economic changes that have occurred over the last centuries. The botanical expeditions and the collections of specimens connected observations and descriptions with landscapes and environmental conditions; plants had been there for thousands of years, linked to the history and adapted to abiotic and biotic conditions of the localities [1–5].

John Sibthorp (1758–1796), Professor of Botany in the University of Oxford, decided to travel to unexplored areas of Greece, collecting and recording botanical specimens in the late 1780s and 1790s; at that time, Greece was an unknown region, very dangerous and difficult to visit owing to diseases, civil unrest, and bandit groups—known as armatoloi and klephts– that included illiterate peasants, artisans, and local clergy, together with the local notables and landowners in Peloponnese [6–8].

Sibthorp's main interest was linked to plants known since the classical antiquity and mainly quoted in the texts of Dioscorides (1st century AD) [9–14]. During the first exploration from 1786 to 1787, Sibthorp was accompanied by the Austrian painter Ferdinand Lucas Bauer (1760–1826) as his draughtsman [6,7]; this was a time when travelers were accompanied by a professional artist, whose work supplemented their discoveries with visual evidence [15–18]. Actually, the magnificent, illustrated edition *Flora Graeca Sibthorpiana* (hereafter *FGS*), published from 1806 to 1840, contains botanical hand-coloured engravings that are important icons of the Mediterranean flora [7,19,20].

John Sibthorp and his companion undertook a second botanical expedition to the Levant from 1794 to 1795. During this journey, they arrived in Peloponnese (Morea is the name used in their diaries and letters) on 26 February 1795 and visited numerous localities botanizing in a more or less largely unknown area, frequently hearing the firing of guns [6,15,21,22]. Those days, major parts of Peloponnese, electively ruled by semi-autonomous agas (persons of high rank or social position during the era of the Ottoman Empire [23]), were only nominally part of the Ottoman Empire [24,25].

Although substantial, revived research has been carried out on the content of *FGS* [7,8,20,26–30], the Peloponnese tour and the collected botanical specimens by Sibthorp in 1795 have received little attention [6] (pp. 164–169) [7] (pp. 144–146). The importance of studying local floras, historical and environmental conditions, distribution records, and species lists has been repeatedly stressed in the literature and awareness of this subject has recently been rising.

Plants collected during a pre-rebellion period (i.e., before the Greek Revolution of 1821) in Peloponnese correspond to "visual evidence" from a particular time (spring 1795), revealing regional plant species pool of this particular area, as well as physical, cultural, and aesthetic values of the natural environment. The main goal of this study was to study plants that have been recorded in Peloponnese in pre-industrial time, as functional components of a biodiversity, which, to the best of our knowledge, has not hitherto been published. A secondary goal of this study was to confirm the above-mentioned plant diversity in Peloponnese during the 21st century.

2. Materials and Methods

This research is based on our survey of written sources, i.e., books, travel reports, letters, diaries, plant catalogues, online published, and printed archives mainly linked to the "Flora Graeca" expedition in Peloponnese (Greece) in 1795 [6,7]. Two copies of *FGS*, i.e., a copy adorned the National Library of Greece since 1916 and another copy acquired by the Gennadius Library of Athens in 1967 were surveyed. Moreover, we studied the digitized published hand-coloured engravings and the original watercolours, together with the Mediterranean scenes that are freely available and accessible online via Digital Bodleian (https://digital.bodleian.ox.ac.uk/collections/flora-and-fauna-graeca/, accessed on 9 October 2022). In addition, rigorous research of the *Florae Graecae Prodromus* [30] (hereafter *Prodromus*) housed in the Department of Botany at National and Kapodistrian University of Athens in Greece was carried out; it has to be noted that the *Prodromus* contains indexes of modern Greek vernacular names of plants (Index Nominum Graecorum, pp. 383–391), ancient Greek names of plants quoted in Dioscorides' codex (Index Dioscoridem, pp. 392–404), and scientific names of plants (Index Generum et Synonymorum, pp. 405–422), as well as plant locality data [31]. Furthermore, two books were taken in consideration; the first by Robert Walpole (1781–1856, an English classical scholar with degrees from Trinity College at Cambridge in UK and Merton College at Oxford in UK, who travelled to Greece; his Memoirs including notes of various travelers' diaries, among them Sibthorp's and his companion [32] were first published in 1817) and the second by John Bacon Sawrey Morritt (1772–1843, who immediately after his BA degree from St. John's College at Cambridge in UK, started on the travels described in his book that was first published in 1914; Morritt travelled over a considerable part of Peloponnese in 1795 [33]). A plant taxon was included in the results if there was a record in *Prodromus* stating locality data from Peloponnese. Information linked to the currently accepted plant nomenclature and distribution was derived from the Flora of Greece web (https://portal.cybertaxonomy.org/flora-greece/, accessed on 21 October 2022).

3. Results

3.1. Peloponnese Tour

In Figure 1, the Peloponnese tours followed by Sibthorp and Morritt in 1795 are depicted in red and green lines, respectively. Sibthorp and his colleagues travelled from the island of Zakynthos to the port of Skaffidia (Ileia County); their route included Pyrgos, Lalla, and Tripolis, passing through several villages. The tour continued to Palaiepiscopi, ancient Tegea, and Arcadia. Next, they travelled to Argos and visited ancient Mycenae as well as Napoli di Roamin (Nafplion) in Argolida County. Then, they travelled to Korinthos and Patras, continued in Achaia County through villages, and proceeded to Ileia County again; from there, they followed different directions until they arrived in Kalamata (Messinia County). After Kalamata, they proceeded to Kutchuk Maina, Kardamili, Sparta (Laconia County), and Mystras; from there, they continued to Messini and Petallida and on 25 April 1795 they arrived at Zakynthos and, by ship, returned to England. Morritt's journey started from Kalamata; he visited Kutchukmaina, Palaeocastro and ancient Thuria (Messinia County), Corone, Abia, and Kitreés and, through various villages, went to Kardamili/Cardamyla; he arrived by boat at Platsa and then continued to Oetylos, Marathonisi (ancient Gythium), and Mystras (Laconia County).

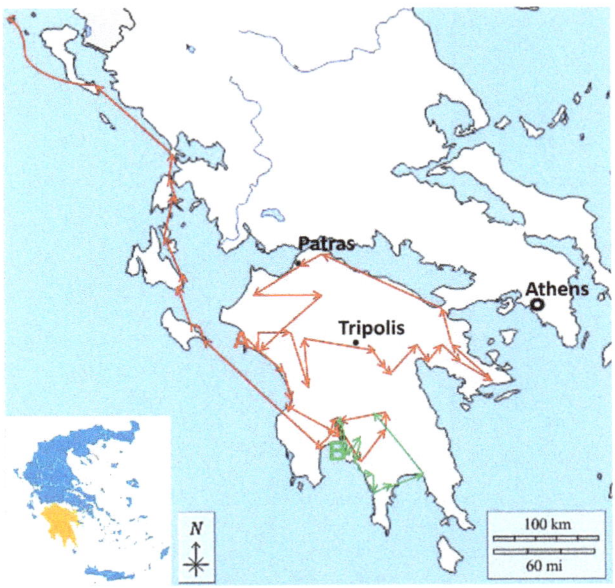

Figure 1. Map of Peloponnese (obtained by https://d-maps.com/ accessed on 10 October 2021 and modified accordingly), showing two tours, i.e., by Sibthorp (red line) and Morritt (green line) in 1795. The red symbol A indicates the start of Sibthorp's journey; red lines and arrows indicate locations and directions, respectively. In the insert, the map of Greece (blue) is presented and, in yellow, the Peloponnese peninsula is indicated. The green lines and arrows indicate locations and directions of Morritt's journey. The black-white dot indicates the capital of Greece, Athens (37.9838° N, 23.7275° E); the small black dots indicate the locations of cities: Patras (38.2466° N, 21.7346° E) and Tripolis (37.5101° N, 22.3726° E).

3.2. Plant Diversity in Pre-Industrial Peloponnese

Our study provides evidence for 183 plant taxa grown in pre-industrial Peloponnese, which had been collected during Sibthorp's expedition, drawn and cited in *FGS* (Table 1). Moreover, 21 plants quoted in *Prodromus* and linked to localities of Peloponnese, but neither drawn nor cited in *FGS*, were found (Table 2). Although citations for *prickly pear* [*Opuntia ficus-indica* (L.) Mill.], walnut (*Juglans regia* L.), and mulberries (*Morus nigra* L.) were found in the considered archival research concerning Peloponnese, these plants were neither drawn nor cited in both *FGS* and *Prodromus*. It should be mentioned that the botanist Sir James Edward Smith (1759–1828)—founder and first president of the Linnean Society of London—wrote the texts for the plants attested in *FGS* and *Prodromus* and excluded all species he regarded as not being part of the natural flora.

Table 1. List of plants found in Peloponnese and cited in *Flora Graeca Sibthorpiana* (*FGS*). First column: plant names quoted in the first edition of *FGS* (1806–1840). Second column: numerical register of hand-coloured engravings (plates) of plants cited in the first published edition of *FGS*. Third column: numerical register of the original watercolours by Ferdinand Bauer preserved at Oxford (MS. Sherard 241–245), digitized and electronically accessed via Digital Bodleian; whenever the picture of the original drawing was not digitally available, the digital hand-coloured engraving from the first printed edition is mentioned (Sherard 761 and 764). Fourth column: current scientific name.

Plant Name Cited in *FGS*	Engraving	Watercoulor	Scientific Name
Phillyrea latifolia	2	761: pl.2	*Phillyrea latifolia* L.
Olea europaea	3	761: pl.3	*Olea europaea* L.
Veronica glauca	7	244: f.66	*Veronica glauca* Sm.
Veronica triphyllos	10	244: f.69	*Veronica triphyllos* L.
Salvia triloba	17	244: f.158	*Salvia fruticosa* Mill.
Salvia ringens	18	244: f.159	*Salvia ringens* Sm.
Salvia sibthorpii	22	244: f.163	*Salvia virgata* Jacq.
Morina persica	28	761: pl.28	*Morina persica* L.
Crocus aureus	35	245: f.65	*Crocus flavus* Weston subsp. *flavus*
Iris florentina	39	245: f.69	*Iris albicans* Lange
Iris sisyrinchium	42	245: f.72	*Moraea sisyrinchium* (L.) Ker-Gawl.
Schoenus mucronatus	43	245: f.112	*Cyperus capitatus* Vand.
Saccharum ravennae	52	245: f.120	*Tripidium ravennae* (L.) H. Scholz
Panicum repens	61	245: f.130	*Panicum repens* L.
Briza minor	74	245: f.142	*Briza minor* L.
Festuga littoralis	80	245: f.148	*Aeluropus littoralis* (Gouan) Parl.
Bromus tectorum	82	245: f.150	*Bromus tectorum* L.
Bromus rubens	83	245: f.151	*Bromus rubens* L.
Stipa paleacea	86	245: f.154	*Stipa capensis* Thunb.
Triticum junceum	99	245: f.166	*Elytrigia juncea* (L.) Nevski
Valantia muralis	137	242: f.202	*Valantia muralis* L.
Crucianella latifolia	139	242: f.204	*Crucianella latifolia* L.
Plantago lagopus	144	244: f.182	*Plantago lagopus* L.
Hypecoum imberbe	156	241: f.30	*Hypecoum imberbe* Sm.
Anchusa tinctoria	166	244: f.33	*Anchusa tinctoria* L.
Cerinthe aspera	170	244: f.37	*Cerinthe major* L.
Cerinthe retorta	171	244: f.38	*Cerinthe retorta* Sm.
Asperugo procumbens	177	244: f.44	*Asperugo procumbens* L.
Lycopsis variegata	178	244: f.36	*Anchusella variegata* (L.) Bigazzi & al.
Primula vulgaris	184	244: f.175	*Primula vulgaris* Huds.
Lysimachia linum-stellatum	189	244: f.181	*Asterolinon linum-stellatum* (L.) Duby
Plumbago europaea	191	244: f.196	*Plumbago europaea* L.
Convolvulus siculus	196	244: f.15	*Convolvulus siculus* L.
Campanula rupestris	213	243: f.178	*Campanula rupestris* Sm.
Campanula drabifolia	215	243: f.180	*Campanula drabifolia* Sm.
Viola gracilis	222	241: f.85	*Viola gracilis* Sm.
Chironia maritima	237	244: f.9	*Centaurium maritimum* (L.) Fritsch
Chironia spicata	238	244: f.10	*Schenkia spicata* (L.) G. Mans.
Vitis vinifera	242	241: f.178	*Vitis vinifera* L.
Herniaria macrocarpa	252	242: f.125	*Herniaria incana* Lam.
Eryngium multifidum	259	242: f.148	*Eryngium amethystinum* L.
Bupleurum sibthorpianum	264	242: f.153	*Bupleurum falcatum* subsp. *cernuum* (Ten.) Arcang.
Echinophora spinosa	265	242: f.154	*Echinophora spinosa* L.
Echinophora tenuifolia	266	242: f.155	*Echinophora tenuifolia* L.
Artedia squamata	268	242: f.157	*Artedia squamata* L.
Peucedanum obtusifolium	277	242: f.175	*Selinum silaifolium* (Jacq.) Beck
Coriandrum sativum	283	242: f.170	*Coriandrum sativum* L.
Pastinaca opopanax	288	242: f.176	*Opopanax hispidus* (Friv.) Griseb.
Linum gallicum	303	241: f.160	*Linum trigynum* L.

Table 1. *Cont.*

Plant Name Cited in *FGS*	Engraving	Watercoulor	Scientific Name
Narcissus tazetta	308	245: f.73	*Narcissus tazetta* L.
Amaryllis lutea	310	245: f.75	*Sternbergia lutea* (L.) Spreng. subsp. *lutea*
Tulipa sibthorpiana	330	245: f.79	*Fritillaria sibthorpiana* (Sm.) Baker
Ornithogalum arvense	332	245: f.97	*Gagea villosa* (M. Bieb.) Sweet
Ornithogalum nanum	333	245: f.98	*Ornithogalum sibthorpii* Greuter
Asphodelus ramosus	334	245: f.99	*Asphodelus ramosus* L.
Anthericum graecum	336	245: f.101	*Gagea graeca* (L.) Irmisch
Asparagus acutifolius	337	245: f.102	*Asparagus acutifolius* L.
Hyacinthus romanus	340	245: f.105	*Bellevalia romana* (L.) Sweet
Frankenia hirsuta	343	241: f.88	*Frankenia hirsuta* L.
Erica arborea	351	243: f.190	*Erica arborea* L.
Arbutus unedo	373	243: f.191	*Arbutus unedo* L.
Arbutus andrachne	374	243: f.192	*Arbutus andrachne* L.
Saxifraga media	376	242: f.142	*Saxifraga sempervivum* K. Koch
Saxifraga rotundifolia	377	242: f.143	*Saxifraga rotundifolia* L.
Saxifraga cymbalaria	378	242: f.144	*Saxifraga sibthorpii* Boiss.
Dianthus cinnamomeus	400	241: f.110	*Dianthus cinnamomeus* Sm.
Silene nocturna	408	241: f.118	*Silene nocturna* L.
Silene behen	416	241: f.126	*Silene behen* L.
Silene italica	429	241: f.138	*Silene italica* (L.) Pers.
Silene staticifolia	434	241: f.144	*Silene bupleuroides* subsp. *staticifolia* (Sm.) Chowdhuri
Sedum tetraphyllum	448	242: f.135	*Sedum cepaea* L.
Oxalis corniculata	451	241: f.190	*Oxalis corniculata* L.
Cerastium pilosum	454	241: f.149	*Cerastium illyricum* Ard.
Cerastium tomentosum	455	241: f.150	*Cerastium candidissimum* Correns
Reseda alba	459	245: f.49	*Reseda alba* L.
Euphorbia spinosa	463	245: f.39	*Euphorbia acanthothamnos* Boiss.
Euphorbia leiosperma	465	245: f.41	*Euphorbia terracina* L.
Myrtus communis	475	242: f.120	*Myrtus communis* L.
Prunus prostrata	478	242: f. 109	*Prunus prostrata* Labill.
Pyrus aria	479	242: f.118	*Sorbus umbellata* (Desf.) Fritsch
Papaver somniferum	491	241: f.24	*Papaver somniferum* L.
Cistus monspeliensis	493	241: f.75	*Cistus monspeliensis* L.
Cistus incanus	494	241: f.74	*Cistus creticus* subsp. *eriocephalus* (Viv.) Greuter & Burdet
Cistus salviifolius	497	241: f.78	*Cistus salviifolius* L.
Cistus guttatus	498	241: f.79	*Tuberaria guttata* (L.) Fourr.
Cistus salicifolius	499	241: f.80	*Helianthemum salicifolium* (L.) Mill.
Delphinium consolida	504	241: f.7	*Consolida phrygia* (Boiss.) Soó
Anemone coronaria	514	241: f.17	*Anemone coronaria* L.
Ranunculus millefoliatus	521	241: f.4	*Ranunculus millefoliatus* Vahl
Satureja juliana	540	244: f.117	*Micromeria juliana* (L.) Rchb.
Satureja graeca	542	244: f.118	*Micromeria graeca* (L.) Rchb.
Satureja capitata	544	244: f.115	*Thymbra capitata* (L.) Cav.
Nepeta nuda	547	244: f.120	*Nepeta nuda* L.
Lamium maculatum	556	244: f.127	*Lamium maculatum* L.
Stachys orientalis	560	244: f.146	*Stachys obliqua* Waldst. & Kit.
Marrubium pseudodictamnus	562	244: f.147	*Ballota pseudodictamnus* (L.) Benth.
Prasium majus	584	244: f.155	*Prasium majus* L.
Bartsia latifolia	586	244: f.71	*Bellardia latifolia* (L.) Cuatrec.
Antirrhinum pelisserianum	591	244: f.76	*Linaria pelisseriana* (L.) Mill.
Antirrhinum chalepense	592	244: f.77	*Linaria chalepensis* (L.) Mill.
Antirrhinum reflexum	593	244: f.78	*Linaria triphylla* (L.) Mill.
Scrophularia canina	598	244: f.83	*Scrophularia canina* subsp. *bicolor* (Sm.) Greuter

Table 1. *Cont.*

Plant Name Cited in *FGS*	Engraving	Watercoulor	Scientific Name
Scrophularia caesia	604	244: f.89	*Scrophularia heterophylla* Willd.
Orobanche ramosa	608	244: f.93	*Phelipanche mutelii* (F.W. Schultz) Pomel
Acanthus spinosus	611	244: f.95	*Acanthus spinosus* L.
Bunias raphanifolia	612	241: f.33	*Rapistrum rugosum* (L.) All.
Aubrieta deltoidea	628	241: f.49	*Aubrieta deltoidea* (L.) DC.
Biscutella columnae	629	241: f.50	*Biscutella didyma* subsp. *apula* Nyman
Arabis verna	641	241: f.62	*Arabis verna* (L.) R. Br.
Erodium romanum	654	241: f.182	*Erodium acaule* (L.) Bech. & Thell.
Erodium gruinum	656	241: f.184	*Erodium gruinum* (L.) L'Hér.
Erodium malacoides	658	241: f.186	*Erodium malacoides* (L.) L'Hér.
Geranium tuberosum	659	241: f.187	*Geranium tuberosum* L.
Alcea ficifolia	663	241: f.166	*Alcea biennis* Winterl
Hibiscus trionum	666	241: f.169	*Hibiscus trionum* L.
Polygala venulosa	669	241: f.86	*Polygala venulosa* Sm.
Ononis antiquorum	675	242: f.11	*Ononis spinosa* subsp. *diacantha* (Rchb.) Greuter
Anthyllis tetraphylla	681	242: f.17	*Tripodion tetraphyllum* (L.) Fourr.
Orobus sessilifolius	692	242: f.27	*Lathyrus digitatus* (M. Bieb.) Fiori
Lathyrus sativus	695	242: f.31	*Lathyrus sativus* L.
Lathyrus grandiflorus	698	242: f.34	*Lathyrus grandiflorus* Sm.
Vicia polyphylla	699	242: f.35	*Vicia villosa* subsp. *varia* (Host) Corb.
Vicia melanops	701	242: f.37	*Vicia melanops* Sm.
Cytisus sessilifolius	705	242: f.41	*Podocytisus caramanicus* Boiss. & Heldr.
Coronilla emerus	710	242: f.46	*Hippocrepis emerus* (L.) Lassen
Coronilla securidaca	712	242: f.48	*Securigera securidaca* (L.) Degen & Dörfl.
Ornithopus compressus	714	242: f.50	*Ornithopus compressus* L.
Ornithopus scorpioides	715	242: f.51	*Coronilla scorpioides* (L.) W.D.J. Koch
Hippocrepis unisiliquosa	716	242: f.52	*Hippocrepis unisiliquosa* L.
Hedysarum caput-galli	723	242: f.59	*Onobrychis caput-galli* (L.) Lam.
Phaca baetica	727	242: f.63	*Erophaca baetica* (L.) Boiss.
Astragalus incanus	732	242: f.68	*Astragalus spruneri* Boiss.
Astragalus aristatus	735	242: f.71	*Astragalus thracicus* subsp. *parnassi* (Boiss.) Strid
Biserrula pelecinus	737	242: f.73	*Astragalus pelecinus* (L.) Barneby
Trifolium cherleri	745	242: f.81	*Trifolium cherleri* L.
Trifolium rotundifolium	747	764: pl.747	*Trigonella rotundifolia* (Sm.) Strid
Trifolium stellatum	750	242: f.86	*Trifolium stellatum* L.
Trifolium clypeatum	751	242: f.87	*Trifolium clypeatum* L.
Trifolium uniflorum	752	242: f.88	*Trifolium uniflorum* L.
Lotus tetragonolobus	755	242: f.91	*Tetragonolobus purpureus* Moench
Lotus edulis	756	242: f.92	*Lotus edulis* L.
Lotus creticus	758	242: f.94	*Lotus creticus* L.
Lotus hirsutus	759	242: f.95	*Dorycnium hirsutum* (L.) Ser.
Trigonella corniculata	761	242: f.97	*Trigonella corniculata* (L.) L.
Trigonella monspeliaca	765	242: f.101	*Medicago monspeliaca* (L.) Trautv.
Medicago marina	770	242: f.106	*Medicago marina* L.
Hypericum olympicum	772	241: f.171	*Hypericum olympicum* L.
Hypericum hircinum	773	241: f.172	*Hypericum hircinum* L.
Hypericum crispum	776	241: f.175	*Hypericum triquetrifolium* Turra
Scorzonera laciniata	788	243: f.144	*Podospermum laciniatum* (L.) DC.
Sonchus picroides	793	243: f.166	*Reichardia picroides* (L.) Roth
Crepis rubra	801	243: f.157	*Crepis rubra* L.
Hedypnois cretica	813	243: f.132	*Hedypnois rhagadioloides* (L.) F.W. Schmidt
Hypochoeris minima	816	243: f.123	*Hypochaeris arachnoides* Poir.
Lapsana stellata	817	243: f.126	*Rhagadiolus stellatus* (L.) Gaertn.
Catananche lutea	821	243: f.129	*Catananche lutea* L.
Carduus glycacanthus	826	243: f.96	*Jurinea glycacantha* DC.

Table 1. *Cont.*

Plant Name Cited in *FGS*	Engraving	Watercoulor	Scientific Name
Cnicus acarna	827	243: f.94	*Picnomon acarna* (L.) Cass.
Onopordum elatum	833	243: f.87	*Onopordum tauricum* Willd.
Cynara humilis	835	243: f.89	*Cynara cardunculus* L.
Carlina lanata	836	243: f.82	*Carlina lanata* L.
Carlina corymbosa	837	243: f.83	*Carlina corymbosa* subsp. *graeca* (Heldr. & Sartori) Nyman
Acarna cancellata	839	243: f.85	*Atractylis cancellata* L.
Carthamus lanatus	841	243: f.118	*Carthamus lanatus* L.
Carthamus caeruleus	843	243: f.120	*Carthamus caeruleus* L.
Staehelina chamaepeuce	847	243: f.90	*Ptilostemon chamaepeuce* (L.) Less.
Senecio trilobus	869	243: f.65	*Senecio trilobus* L.
Bellis annua	876	243: f.22	*Bellis annua* L.
Chrysanthemum coronarium	877	243: f.58	*Glebionis coronaria* (L.) Spach
Anthemis cota	880	243: f.35	*Anthemis altissima* L.
Anthemis altissima	881	243: f.36	*Anthemis altissima* L.
Achillea aegyptiaca	892	243: f.51	*Achillea taygetea* Boiss. & Heldr.
Centaurea benedicta	906	243: f.114	*Centaurea benedicta* (L.) L.
Centaurea aegyptiaca	907	243: f.102	*Centaurea aegyptiaca* Sm.
Centaurea melitensis	909	243: f.104	*Centaurea melitensis* L.
Centaurea collina	914	243: f.109	*Centaurea salonitana* Vis.
Centaurea galactites	919	243: f.115	*Galactites tomentosus* Moench
Filago pygmaea	921	243: f.28	*Filago pygmaea* L.
Orchis undulatifolia	927	245: f.58	*Orchis italica* Poir.
Orchis papilionacea	928	245: f.59	*Anacamptis papilionacea* subsp. *aegaea* (P. Delforge) L. Lewis & Kreutz
Ophrys fusca	930	245: f.61	*Ophrys fusca* Link
Pistacia terebinthus	956	242: f.4	*Pistacia terebinthus* L.
Atriplex halimus	962	245: f.8	*Atriplex halimus* L.

In 1795, in western Peloponnese, *Salicornia fruticosa* L. was observed growing near lake banks, *Asphodelus ramosus* L. near rivers, and *Bromus rubens* L. in between cultivated fields. Stands of *Phillyrea latifolia* L., *Erica arborea* L., *Arbutus unedo* L., *Pistacia lentiscus* L., *vernal* (spring) *Crocus flavus* Weston, and primroses (*Primula vulgaris* Huds.) in bloom—observed in early March 1795—were encountered. In the southern Peloponnese (county of Messinia), black mulberry trees (*Morus nigra* L.) and prickly pear surrounded many villages. Moreover, they depicted fig trees (*Ficus carica* L.), grapevines, cotton, grains, corn, olive trees, *Euphorbia exigua* L., *Euphorbia spinosa* L., *Lolium perenne* L., and *Orobanche ramosa* L. Some regions produced flax and tobacco. In the eastern Peloponnese, *Quercus* species, as well as corn, grains, grapevines, olive trees, fig trees, mulberry trees, and chestnut trees, had been detected. In the central Peloponnese (county of Arcadia), they visited oaks' forest; moreover, they observed a huge walnut tree (*Juglans regia* L.), *Hyacinthus romanus* L., and *Hyacinthus spicatus* Sm. in bloom. In addition, the presence of floating crystal-wort (*Riccia fluitans L.*) and *Boletus* (a genus of mushroom-producing fungi that comprises over 100 species) and the use of truffle were mentioned. Cultivation of pear trees with open blossoms (10 March 1795) and corns grown among the remains of cities and temples of the ancient Greek territories were detected.

John Sibthorp arrived in Peloponnese bearing a mode of seeing, endowing the professorship of "Agriculture and Rural Economy" in the University of Oxford, thus the state of the agriculture in Peloponnese attracted his attention in 1795; the cultivation of corn (*Zea mays* L.), cotton (*Gossypium hirsutum* L.), millet (*Panicum repens* L.), tobacco (*Nicotiana tabacum* L.), and wheat (*Triticum junceum* L. and *Aegilops comosa* Sm.) was detected.

The revived interest in *FGS* is partially due to recent publications [22,28,37–39], but mainly to biodiversity issues raised under the threat of climate change, which gives another dimension to the whole achievement. Moreover, exhibitions dedicated to the concept and the content of *Flora Graeca Sibthorpiana* contributed to public awareness, e.g., in Oxford entitled "Painting by numbers" (Bodleian Library, 29 – 9 July 2017, https://treasures.bodleian.ox.ac.uk/treasures/flora-graeca/ accessed on 9 May 2017) and Athens entitled "Flora Graeca" (Gennadius Library, 8 March–4 July 2016, https://www.ascsa.edu.gr/events/details/flora-graeca-exhibition, accessed on 8 March 2016).

In Table 1, we compiled a list of 183 wild plants cited in *FGS* and located in Peloponnese, which is indicative of the biodiversity, environmental physiology, phenology, and short flowering season in response to drought conditions, i.e., during the period of spring rainfall and the concomitantly active pollinators [40,41]. The later generations of plant biologists studied plant species grown in geographic locations visited by Sibthorp and his companion in Peloponnese, increasing the overall knowledge about distribution, ecophysiology, and taxonomy of plants quoted in *FGS* and *Prodromus* [42–49]. The mediterranean-type climate is characterized by a marked seasonality, typified by the alternation of a hot and dry period with a cold and wet period. For example, Sibthorp observed open flowers of *Anemone coronaria* L., *Oxalis corniculata* L., and *Asphodelus ramosus* L. on 27 February 1795, as well as of *Crocus flavus* Weston in early spring (cited as *Crocus aureus* in *FGS* and *Crocus vernus latifolius aureus* in *Prodromus*, vol. I, pp. 24–25); such observations are supported by recent publications [5,50,51]. Moreover, in the 21st century, it is known that seasonal blossom is related to adaptive floral traits; for example, the study of petals revealed a surface nano-sculpture that declines water droplet adhesion and enhances the water repellence of these fragile floral tissues, which are exposed to the rainy conditions of the early spring flowering season [52–54]. In *Anemone coronaria* L., the temperature plays a critical role in the onset of dormancy [55]. Other species possess deeply rooted systems that enhance drought resistance (e.g., *Myrtus communis* L., *Pistacia lentiscus* L., and *Quercus* species). In addition, recent research revealed leaf functional traits linked to hydrophobicity and water status, highlighting species' responses to drought conditions [56–58]; this may be critical for resilience in the face of increasing drought stress.

Moreover, Sibthorp noticed that oaks in Peloponnese were frequently infested with the mistletoe *Loranthus europaeus* Jacq. [59–61]; it is worth mentioning that he regarded the deciduous, yellow *Loranthus europaeus* Jacq. as the "true mistletoe of the ancients" [6] (p. 165).

Sibthorp and his companion visited a mountainy area, barren and stony beyond conception; it was hard work botanizing under harsh field conditions. The earth, washed by the rains and torrents from the higher parts, was supported on a plethora of terraces cultivated with wheat, cotton, maize, and millet, while olives and mulberry trees seemed to grow out of the rocky substrate itself. However, carpets of geophytes and numerous annual plants produced a spring flowering distinctive to the human eye. The results from this tour in the late 1790s, in pre-industrial landscapes, barely resembled the area we see today in Peloponnese, and brought information about numerous unknown to science (those days) wild plants, oak woodlands, pine forests, crops, cultivated areas, and arable lands of the monasteries [62]. Nowadays, several places of Peloponnese that Sibthorp visited in 1795 are included in the European network Natura 2000—i.e., the cornerstone of European Union nature conservation policy—of designated sites (https://eunis.eea.europa.eu/sites, accessed on 18 October 2022) relevant for flora and habitat protection [63–65], e.g., mountainy landscapes such as Parnonas: GR2520006, Mainalo (Arcadia): GR2520001, and Taygetos: GR2550006, as well as Folois plateau: GR2330002 and Olympia: GR2330004. Other progression was also recorded; that is, information linked to the current distribution of the considered plants, confirmed via the Flora of Greece web, contributed to our knowledge about natural stands of wild plants.

According to our study, on one hand, among the plants found in Peloponnese in 1795 and cited in *FGS* and *Prodromus*, there are species either widely distributed or grown in restricted areas, e.g., *Achillea taygetea* Boiss. & Heldr., *Erophaca baetica* Boiss., *Saxifraga sibthorpii* Boiss., and *Scilla messeniaca* Boiss. On the other hand, *Zea mays* L., originated from the Americas and found among the few cultivated species in isolated valleys in Peloponnese in pre-industrial time, might be attributed to the Venetian occupation of Peloponnese (1688–1715); during that period, when the area was dependent on the European market, plants might have been a product of cross-cultural communication between the conquerors and conquered [66–70].

Sibthorp's expedition in Peloponnese contributed to our understanding of botany in the field and revealed the diversity of plants grown in their habitats, in pre-industrial time. Historical time was linked to a gradually known plant diversity, as locations were explored and knowledge about the natural fertility of the land increased. However, anthropogenic pressure maintained by human activities, grazing, and fires in Peloponnese added to environmental stresses and caused profound transformation in the natural landscape, reducing the distribution of indigenous plants and enhancing a widespread concern about the extent of habitat and species loss [71–76]. This means that whatever effort can be made to study, maintain, and protect the diversity of ecosystems in this region is closely connected to a sustainable future, via the preservation of numerous plant taxa cited in the monumental *FGS* and *Prodromus*. Nowadays, Oxford Botanic Garden in UK (where visitors can enjoy the full sensory experience of walking through an aromatic Mediterranean landscape while learning about the work of Sibthorp and Bauer and its important botanical and horticultural legacy [35]) and Diomedes' Botanic Garden in Greece (due to the fact that administration of Diomedes' Botanic Garden is directly linked to the staff of the National and Kapodistrian University of Athens in Greece, this Garden has also been used for relevant, educational programs [37]) contain living collections of Mediterranean plants cited in *FGS*, which may be perceived as celebrations for Flora Graeca expeditions and *FGS* [35,37]. However, a larger number of plants quoted in *FGS* and *Prodromus* may be introduced and cultivated in the above-mentioned botanic gardens and/or the network of botanic gardens in Greece, in order to detect the diversity and the life-cycle of wild plants within the context of the seasons, floral colours in Mediterranean ecosystems, and collection and deposition of seeds in seed-banks. As such, botanic gardens can be used as common gardens, where researchers can conduct unmatched comparative research studies of plant ecophysiology, morphology, anatomy, and responses to climate change [77,78]. It is worth mentioning that Sibthorp introduced new species into English horticulture; moreover, he returned to Oxford from his eastern Mediterranean explorations with seeds, bulbs, and corms for the Botanic Garden, but few details of these collections have survived, and the plants and any knowledge about their propagation have been lost through many routes [7] (p. 180) and neglected [79] (p. 102).

This work provides a novel and valuable insight into the development of early plant environmental biology and is an important element of timelessness aspects of botany [80,81]. The study of plant diversity in Peloponnese peninsula, during the pre-rebellion period in Greece, tracing long-term changes in the region, is also a reminder that nature is often a repository at which nations look when crafting their identity.

5. Conclusions

The interest in archival material has been revived on account of research for a biodiversity threatened by climatic change. In this context, our research gives prominence to approximately 200 wild plant taxa found in Peloponnese (Greece)—most of them quoted in the magnificent edition *Flora Graeca Sibthorpiana* of the 19th century—and few cultivated introduced plants, all grown under ambient conditions and exposed to environmental stresses of the eastern Mediterranean during the pre-rebellion period, representing plant environmental issues in pre-industrial time, which have not hitherto been published.

Author Contributions: Conceptualization, S.R.; methodology, C.C., S.P., F.K., M.M. and S.R.; validation, C.C., S.P., F.K., M.M. and S.R.; investigation, C.C., S.P., F.K. and M.M.; resources, S.R.; data curation, C.C., S.P., F.K., M.M. and S.R.; writing—original draft preparation, C.C., S.P., F.K. and S.R.; writing—review and editing, C.C. and S.R.; supervision, S.R.; project administration, S.R.; funding acquisition, C.C., S.P. and F.K. All authors have read and agreed to the published version of the manuscript.

Funding: This research was implemented in the framework of the project entitled "Contribution of the National and Kapodistrian University of Athens to the research for the study of the history and memory of the Revolution of 1821", and funded by the National and Kapodistrian University of Athens, grant number 16614.

Institutional Review Board Statement: Not applicable.

Informed Consent Statement: Not applicable.

Data Availability Statement: The data are available from the authors upon request.

Acknowledgments: We would like to thank Stephen Harris (Department of Biology, University of Oxford, UK) for early discussions on the subject.

Conflicts of Interest: The authors declare no conflict of interest.

References

1. Raven, J.E. *Plants and Plant Lore in Ancient Greece*; Leopard's Head: Oxford, UK, 2000.
2. Rhizopoulou, S. Symbolic plant (s) of the Olympic Games. *J. Exp. Bot.* **2004**, *55*, 1601–1606. [CrossRef]
3. Rhizopoulou, S.; Marmarinos, M. Plants as an element of cultural heritage: What Oedipus does not see when he arrives at Colonus. *BIO* **2004**, *11*, 48–50.
4. Day, J. Botany meets archaeology: People and plants in the past. *J. Exp. Bot.* **2013**, *64*, 5805–5816. [CrossRef]
5. D'Agata, C.; Rhizopoulou, S. Cretan and Greek plants in Italian Renaissance gardens cited in archives. *Plant Biosyst.* **2022**, *156*, 598–605. [CrossRef]
6. Lack, H.W.; Mabberley, D.J. *The Flora Graeca Story, Sibthorp, Bauer and Hawkins in the Levant*; Oxford University Press: Oxford, UK, 1999.
7. Harris, S. *The magnificent Flora Graeca*; Bodleian Library: Oxford, UK, 2007.
8. Asdrachas, S. *Primitive Revolution, Amatoloi and Klephts (18–19th c.)*; Hellenic Open University: Athens, Greece, 2019.
9. Stearn, W.T. From Theophrastus and Dioscorides to Sibthorp and Smith: The background and origin of the Flora Graeca. *Biol. J. Linn. Soc.* **1976**, *8*, 285–298. [CrossRef]
10. Negbi, M. Theophrastus on geophytes. *Bot. J. Linn. Soc.* **1989**, *100*, 15–43. [CrossRef]
11. Scarborough, J. Theophrastus on herbals and herbal remedies. *J. His. Biol.* **1978**, *11*, 353–385. Available online: https://www.jstor.org/stable/4330714 (accessed on 7 September 2022). [CrossRef]
12. Weiher, E.; Van Der Werf, A.; Thompson, K.; Roderick, M.; Garnier, E.; Eriksson, O. Challenging Theophrastus: A common core list of plant traits for functional ecology. *J. Veg. Sci.* **1999**, *10*, 609–620. [CrossRef]
13. O'Neill, Y.V.; Infusino, M.; Medicina Antiqua. Codex Vindobonensis 93. Vienna, Österreichische National-bibliothek. *Bull. Hist. Med.* **2001**, *75*, 558–560. [CrossRef]
14. Irwin, M.E. Flower power in Medicine and Magic: Theophrastus' response to the rootcutters. *Mouseion* **2006**, *6*, 425–437. [CrossRef]
15. Greuter, W. The early botanical exploration of Greece. In *Progress in Botanical Research*; Tsekos, I., Moustakas, M., Eds.; Springer: Dordrecht, The Netherlands, 1998; pp. 9–20.
16. Krimbas, C.B. HW Lack with DJ Mabberley, The Flora Graeca Story—Sibthorp, Bauer and Hawkins in the Levant. *Hist. Rev.* **2004**, *1*, 275–285.
17. Nickelsen, K. Draughtsmen, botanists and nature: Constructing eighteenth-century botanical illustrations. *Stud. Hist. Philos. Biol. Biomed. Sci.* **2006**, *37*, 1–25. [CrossRef]
18. Riedl-Dorn, C.; Riedl, M. Ferdinand Bauer or Johann and Joseph Knapp? A rectification. *Gard. Bull.* **2019**, *71*, 123–142. [CrossRef]
19. Lack, H.W. The Sibthorpian herbarium at Oxford—guidelines for its use. *Taxon* **1997**, *46*, 253–263. [CrossRef]
20. Harris, S.A. Sibthorp, Bauer and the Flora Graeca. *OPS* **2008**, *15*, 7.
21. Harlan, D. Travel, Pictures, and a Victorian Gentleman in Greece. *Hesperia* **2009**, *78*, 421–453. Available online: https://www.jstor.org/stable/25622703 (accessed on 15 September 2022). [CrossRef]
22. Strid, A. The botanical exploration of Greece. *Plant Syst. Evol.* **2020**, *306*, 1–23. [CrossRef]
23. Kostantaras, D.J. Christian elites of the Peloponnese and the Ottoman state 1715–1821. *Eur. Hist. Q.* **2013**, *43*, 628–656. [CrossRef]
24. Andrews, K. *Castles of the Morea*; American School of Classical Studies at Athens: Princeton, NJ, USA, 2006.
25. Gündoğdu, B. Ottoman Constructions of Morea Rebellion, 1770s: A Comprehensive Study for Attitudes to the Greek Uprising. Unpublished. Ph.D. Thesis, University of Toronto, Toronto, ON, Canada, 2012.
26. Wise, R. A naturalist's paradise. *New Sci.* **1989**, *123*, 68. [CrossRef]
27. Mills, R. Flora Graeca online. *OPS* **2008**, *15*, 8.
28. Lack, H.W. Flora Graeca on the European continent. *Gard. Bull.* **2019**, *71*, 109–122. [CrossRef]
29. Sibthorp, J.; Smith, J.E. *Flora Graeca: Sive Plantarum Rariorum Historia, Quas in Provinciis Aut Insulis Graeciae*; Richard Taylor: London, UK, 1806–1840; ten volumes.
30. Sibthorp, J.; Smith, J.E. *Flora Graeca Prodromus*; Richard Taylor: London, UK, 1806, 1813; Volume 2.

31. Stearn, W.T. Sibthorp, Smith, the Flora Graeca and the Florae Graecae Prodromus. *Taxon* **1967**, *16*, 168–178. [CrossRef]
32. Walpole, R. *Memoirs Relating to European and Asiatic Turkey: And Other Countries of the East*; Cambridge University Press: Cambridge, UK, 2012.
33. Morritt, J.B. *The Letters of John BS Morritt of Rokeby: Descriptive of Journeys in Europe and Asia Minor in the Years 1794–1796*; Cambridge University Press: Cambridge, UK, 2011.
34. Lack, H.W. Lilac and horse-chestnut: Discovery and rediscovery. *Curtis's Bot. Mag.* **2000**, *17*, 109–141. Available online: https://www.jstor.org/stable/45065430 (accessed on 10 October 2022). [CrossRef]
35. Thorogood, C.J. The University of Oxford Botanic Garden: Sharing the scientific wonder and importance of plants with the world. *Curtis's Bot. Mag.* **2021**, *38*, 438–450. [CrossRef]
36. Turrill, W.B. Revision of Sibthorp's plants at Kew. *Bull. Misc. Inform. (R. Bot. Gard. Kew)* **1926**, *3*, 120–128. [CrossRef]
37. Rhizopoulou, S.; Lykos, A.; Delipetrou, P.; Vallianatou, I. Living Collection of Flora Graeca Sibthorpiana. *Sibbaldia* **2012**, *10*, 171–196. [CrossRef]
38. Mulholland, R. Ferdinand Bauer's *Flora Graeca* colour code. In *Technology and Practice: Studying Eighteenth Century Paintings and Works of Art oOn Paper*; Evens, H., Muir, K., Eds.; Archetype: London, UK, 2015; pp. 153–163.
39. Zografidis, A. Resurrection and typification of *Verbascum auriculatum* (Scrophulariaceae), a long-disused name in Flora Graeca Sibthorpiana. *Phytotaxa* **2018**, *361*, 233–243. [CrossRef]
40. Petanidou, T.; Lamborn, E. A land for flowers and bees: Studying pollination ecology in Mediterranean communities. *Plant Biosyst.* **2005**, *139*, 279–294. [CrossRef]
41. Rhizopoulou, S.; Pantazi, H. Constraints on floral water status of successively blossoming Mediterranean plants under natural conditions. *Acta Bot. Gallica* **2015**, *162*, 97–102. [CrossRef]
42. Giannopoulos, K.; Tan, K.; Vold, G. Contributions to the bulb flora of Ilias (NW Peloponnese, Greece): Amaryllidaceae, Araceae and Aristolochiaceae. *Phytol. Balcan.* **2021**, *27*, 97–106.
43. Atherden, M.; Hall, J.; Wright, J.C. A pollen diagram from the northeast Peloponnese, Greece: Implications for vegetation history and archaeology. *Holocene* **1993**, *3*, 351–356. [CrossRef]
44. Strid, A. Lost and found in the Greek flora. In Proceedings of the 3rd Global Botanic Gardens Congress, Wuhan, China, 16–20 April 2007; pp. 1–5.
45. Trigas, P.; Tsiftsis, S.; Tsiripidis, I.; Iatrou, G. Distribution patterns and conservation perspectives of the endemic flora of Peloponnese (Greece). *Folia Geobot.* **2012**, *47*, 421–439. [CrossRef]
46. Allen, H. *Mediterranean Ecogeography*; Routledge: London, UK, 2014.
47. Meletiou-Christou, M.S.; Rhizopoulou, S. Leaf functional traits of four evergreen species growing in Mediterranean environmental conditions. *Acta Physiol. Plant.* **2017**, *39*, 1–13. [CrossRef]
48. Chimona, C.; Rhizopoulou, S. Water economy through matching plant root elongation to Mediterranean landscapes. *World J. Res. Rev.* **2017**, *5*, 22–24. [CrossRef]
49. Cheminal, A.; Kokkoris, I.P.; Zotos, A.; Strid, A.; Dimopoulos, P. Assessing the Ecosystem Services Potential of Endemic Floras: A Systematic Review on the Greek Endemics of Peloponnese. *Sustainability* **2022**, *14*, 5926. [CrossRef]
50. Pastor-Férriz, T.; De-los-Mozos-Pascual, M.; Renau-Morata, B.; Nebauer, S.G.; Sanchis, E.; Busconi, M.; Fernádez, J.-A.; Kamenetsky, R.; Molina, R.V. Ongoing evolution in the genus *Crocus*: Diversity of flowering strategies on the way to hysteranthy. *Plants* **2021**, *10*, 477. [CrossRef]
51. Tan, K.; Giannopoulos, K. Contributions to the bulb flora of Ilias (NW Peloponnese, Greece): Iridaceae. *Phytol. Balcan.* **2022**, *28*, 85–101. [CrossRef]
52. Argiropoulos, A.; Rhizopoulou, S. Micromorphology of the petals of the invasive weed *Oxalis pes-caprae*. *Weed Biol. Manag.* **2012**, *12*, 47–52. [CrossRef]
53. Gkikas, D.; Argiropoulos, A.; Rhizopoulou, S. Epidermal focusing of light and modelling of reflectance in floral-petals with conically shaped epidermal cells. *Flora* **2015**, *212*, 38–45. [CrossRef]
54. Chimona, C.; Koukos, D.; Meletiou-Christou, M.S.; Spanakis, E.; Argiropoulos, A.; Rhizopoulou, S. Functional traits of floral and leaf surfaces of the early spring flowering *Asphodelus ramosus* in the Mediterranean region. *Flora* **2018**, *248*, 10–21. [CrossRef]
55. Ben-Hod, G.; Kigel, J.; Steinitz, B. Dormancy and flowering in *Anemone coronaria* L. as affected by photoperiod and temperature. *Ann. Bot.* **1988**, *61*, 623–633. [CrossRef]
56. Koukos, D.; Meletiou-Christou, M.S.; Rhizopoulou, S. Leaf surface wettability and fatty acid composition of *Arbutus unedo* and *Arbutus andrachne* grown under ambient conditions in a natural macchia. *Acta Bot. Gallica* **2015**, *162*, 225–232. [CrossRef]
57. Bertsouklis, K.F.; Papafotiou, M. Morphometric and molecular analysis of the three *Arbutus* species of Greece. *Not. Bot. Horti Agrobot.* **2016**, *44*, 423–430. [CrossRef]
58. Karatassiou, M.; Karaiskou, P.; Verykouki, E.; Rhizopoulou, S. Hydraulic Response of Deciduous and Evergreen Broadleaved Shrubs, Grown on Olympus Mountain in Greece, to Vapour Pressure Deficit. *Plants* **2022**, *11*, 1013. [CrossRef]
59. Glatzel, G. Mineral nutrition and water relations of hemiparasitic mistletoes: A question of partitioning. Experiments with *Loranthus europaeus* on *Quercus petraea* and *Quercus robur*. *Oecologia* **1983**, *53*, 193–201. [CrossRef]
60. Dimopoulos, P.; Bergmeier, E. Wood pasture in an ancient submediterranean oak forest (Peloponnese, Greece). *Ecol. Mediterr.* **2004**, *30*, 137–146. [CrossRef]

61. Katsarou, A.; Rhizopoulou, S.; Kefalas, P. Antioxidant potential of the aerial tissues of the mistletoe *Loranthus europaeus* Jacq. *Rec. Nat. Prod.* **2012**, *6*, 394–397.
62. Parveva, S. Agrarian land and harvest in South-West Peloponnese in the Early 18th Century. *Étud. Balk.* **2003**, *1*, 83–123.
63. Natura 2000. Available online: https://natura2000.eea.europa.eu/ (accessed on 24 October 2022).
64. Evans, D. Building the European union's Natura 2000 network. *Nat. Conserv.* **2012**, *1*, 11–26. [CrossRef]
65. Spiliopoulou, K.; Dimitrakopoulos, P.G.; Brooks, T.M.; Kelaidi, G.; Paragamian, K.; Kati, V.; Oikonomou, A.; Vavylis, D.; Trigas, P.; Lymberakis, P.; et al. The Natura 2000 network and the ranges of threatened species in Greece. *Biodivers. Conserv.* **2021**, *30*, 945–961. [CrossRef]
66. Harris, S. *What Have Plants ever Done for Us?* Bodleian Library, University of Oxford: Oxford, UK, 2015; pp. 155–159.
67. Stouraiti, A. Colonial encounters, local knowledge and the making of the cartographic archive in the Venetian Peloponnese. *Eur. Rev. Hist./Rev.* **2012**, *19*, 491–514. [CrossRef]
68. Goodman, M.M.; Galinat, W.C. The history and evolution of maize. *Crit. Rev. Plant Sci.* **1988**, *7*, 197–220. [CrossRef]
69. Janick, J.; Caneva, G. The first images of maize in Europe. *Maydica* **2005**, *50*, 71–80.
70. Ongaro, G. Maize diffusion in the Republic of Venice: The case of the Province of Vicenza (sixteenth-eighteenth century). In *Maize to the People! Cultivation, Consumption and Trade in the North-Eastern Mediterranean (Sixteenth-Nineteenth Century)*; Mocarelli, L., Panjek, A., Eds.; University of Primorska Press: Koper, Slovenia, 2020; pp. 25–46.
71. Jones-Walters, L.; Čivić, K.K. Wilderness and biodiversity. *J. Nat. Conserv.* **2010**, *18*, 338–339. [CrossRef]
72. Magurran, A.E.; Dornelas, M. Biological diversity in a changing world. *Philos. Trans. R. Soc. B* **2010**, *365*, 3593–3597. [CrossRef]
73. Paich, S.D. Where olive, lemon and laurel trees grow: A diachronic examination of cultural similarities under different names in greater Mediterranean history. *J. Intercult. Stud.* **2010**, *31*, 313–328. [CrossRef]
74. Rhizopoulou, S. Changing Mediterranean environment: Irrefutable evidence from pre-industrial, unpublished scenes contemporary with a mission (1786–1787) in the Levant. *Global Nest J.* **2012**, *14*, 516–524.
75. Paraskevopoulou, A.T.; Nektarios, P.A.; Kotsiris, G. Post-fire attitudes and perceptions of people towards the landscape character and development in the rural Peloponnese, a case study of the traditional village of Leontari, Arcadia, Greece. *J. Environ. Manag.* **2019**, *241*, 567–574. [CrossRef]
76. Gemitzi, A.; Koutsias, N. Assessment of properties of vegetation phenology in fire-affected areas from 2000 to 2015 in the Peloponnese, Greece. *RSASE* **2021**, *23*, 100535. [CrossRef]
77. Krishnan, S.; Novy, A. The role of botanic gardens in the twenty-first century. *CABI Rev.* **2016**, *11*, 1–10. [CrossRef]
78. Primack, R.B.; Ellwood, E.R.; Gallinat, A.S.; Miller-Rushing, A.J. The growing and vital role of botanical gardens in climate change research. *New Phytol.* **2021**, *231*, 917–932. [CrossRef]
79. Harris, S.A. *Oxford Botanic Garden & Arboretum. A Brief History*; Bodleian Library: Oxford, UK, 2017.
80. Harris, S.A. Sibthorp's Flora Graeca expedition and teaching Linnaean botany in Oxford physic garden. *Curtis's Bot. Mag.* **2021**, *38*, 451–471. [CrossRef]
81. Rhizopoulou, S.; Koukos, D.; Rhizopoulou, A.E. The botanical content of *Hypnerotomachia Poliphili* revisited. *Bot. Lett.* **2022**, 1–6. [CrossRef]

life

MDPI

Article

Effects of Salt Stress on the Antioxidant Activity and Malondialdehyde, Solution Protein, Proline, and Chlorophyll Contents of Three *Malus* Species

Dajiang Wang [1,2], Yuan Gao [2], Simiao Sun [2], Xiang Lu [1,2], Qingshan Li [1,2], Lianwen Li [2], Kun Wang [2,*] and Jihong Liu [1,3,*]

1 Xinjiang Production and Construction Corps Key Laboratory of Special Fruits and Vegetables Cultivation Physiology and Germplasm Resources Utilization, Agricultural College of Shihezi University, Shihezi 832003, China
2 National Repository of Apple Germplasm Resources, Key Laboratory of Horticulture Crops Germplasm Resources Utilization, Ministry of Agriculture and Rural Affairs of the People's Republic of China, Research Institute of Pomology, Chinese Academy of Agricultural Sciences (CAAS), Xingcheng 125100, China
3 Key Laboratory of Horticultural Plant Biology (MOE), College of Horticulture and Forestry Sciences, Huazhong Agricultural University, Wuhan 430070, China
* Correspondence: wangkun@caas.cn (K.W.); liujihong@mail.hzau.edu.cn (J.L.); Tel.: +86-429-3598120 (K.W.); +86-27-87282399 (J.L.)

Abstract: Understanding the different physiological responses of *Malus* species under salt stress in the seedling stages will be useful in breeding salt-tolerant dwarfing apple rootstocks. Seedlings of *Malus Zumi* (Mats.) Rehd. (*M. zumi*), *Malus sieversii* (Led.) Roem. (*M. sieversii*), and *Malus baccata* (L.) Borkh. (*M. baccata*) were treated with solution of 0, 0.20%, 0.40%, and 0.60% salinity. Physiological parameters of their leaves and roots were measured at 0 d, 4 d, 8 d and 12 d after salinity treatments. Superoxide dismutase (SOD), peroxidase (POD), catalase (CAT), malondialdehyde (MDA), solution protein (SP), and proline (PRO) initially increased and then decreased. The activities and contents of these parameters were higher in the 0.40% and 0.60% NaCl treatments than in the 0.20% treatment and in the 0% control. *M. zumi* was the most resistant to salt stress, showing the lowest content of MDA in the leaves and roots, which increased slightly under salt stress. *M. baccata* had the highest increase in both the content and proportion of MDA. High enzyme activity was shown to play an important role in the salt resistance of *M. zumi*. Moreover, it can be speculated that there are other substances that also play a major role. We found that osmotic regulation played a key role in response to salt stress for *M. baccata* even though it was sensitive to salt stress. For *M. sieversii*, both the osmotic regulation and enzymatic antioxidants were observed to play a major role in mitigating salt stress.

Keywords: *Malus* seedlings; NaCl treatments; enzyme activity; membrane damage; osmotic regulation

Citation: Wang, D.; Gao, Y.; Sun, S.; Lu, X.; Li, Q.; Li, L.; Wang, K.; Liu, J. Effects of Salt Stress on the Antioxidant Activity and Malondialdehyde, Solution Protein, Proline, and Chlorophyll Contents of Three *Malus* Species. *Life* **2022**, *12*, 1929. https://doi.org/10.3390/life12111929

Academic Editors: Wajid Zaman and Hakim Manghwar

Received: 4 November 2022
Accepted: 16 November 2022
Published: 18 November 2022

Publisher's Note: MDPI stays neutral with regard to jurisdictional claims in published maps and institutional affiliations.

Copyright: © 2022 by the authors. Licensee MDPI, Basel, Switzerland. This article is an open access article distributed under the terms and conditions of the Creative Commons Attribution (CC BY) license (https://creativecommons.org/licenses/by/4.0/).

1. Introduction

More than 800 million hectares of land and 32 million hectares of agricultural land are affected by salinity stress globally [1,2]. Moreover, it is estimated that soil salinization will cause deterioration of 50% of the land by the year 2050 [3]. Under salt stress, almost all plants exhibit adverse effects [4]. Salt stress causes water loss, iron ion absorption inhibition in roots, a reduction in the photosynthetic efficiency of leaves, and diminished tree growth, all of which seriously affect the healthy growth and yield formation of plants, including apple trees [5–7]. A salty environment produces two kinds of stress factors in plants: osmotic stress and ionic toxicity. The former obstructs water absorption in plants; the latter is toxic to the physiological function of plant metabolism. Moreover, both can lead to the production of reactive oxygen species (ROS), which damage the structure of cell membranes [8]. The changes in POD, SOD, and CAT activities can reflect the ability to scavenge ROS under stress in plants. SOD can dismutate O^{2-} to O_2 or H_2O_2, CAT can catalyze H_2O_2 to H_2O and O_2, and POD can direct oxidation of phenol or amine compounds with H_2O_2 as electron acceptor to eliminate the toxic H_2O_2 and phenol amine [9,10]. SP and PRO contents reflect the ability to overcome osmotic pressure. Plants synthesize PRO, SP, soluble sugars, and other osmolytes to promote osmotic balance at the cellular level; the biosynthesis of PRO is activated by stress [10,11]. MDA content, one of the most important products of membrane lipid peroxidation, reflects the degree of damage to the membrane system under biotic and abiotic stresses [12–16]. Salt stress was shown to cause lipid peroxidation as well as the accumulation of soluble sugars and PRO, and to increase the activity of antioxidant enzymes in both salt-resistant and salt-sensitive bread wheat [17]. Increasing NaCl was shown to increase the SOD and POD activities, as well as the PRO and MDA contents, in linseed [18]. MDA content and SOD, CAT, and POD activities were shown to increase with increasing salinity in lentils [19]. The enhancing and transporting of PRO in plant organs are important survival strategies against salt stress; exogenous PRO may enhance resistance to salt stress in lupine [20,21].

Apple, one of the most popular fruits globally, plays a major role in poverty alleviation and rural revitalization in north and northwest China. At present, the apple dwarfing rootstocks widely used in China and abroad—M26, MM106, M9, etc.—are not tolerant to salt [22]. The soil salt content in certain regions in north and northwest China exceeds 0.4% [23–25]. Northwest China is the new dominant producing area of apple. Salinity stress has, to a certain extent, restricted the development of the apple industry in these areas. Using salt land to grow apples is one of the ways to expand the apple industry in Northwest China. The breeding of salt-tolerant apple rootstock is an important guarantee to achieve this approach and is the theoretical basis for salt-tolerant breeding to understand the physiological mechanism of salt-tolerant *Malus* species.

There are approximately 55 species of *Malus* around the world [26]. Different species have been developed with special characteristics to adapt to the natural environment in the distribution center after a long period of natural selection. For example, *Malus xiaojinensis* Cheng et Jiang., *Malus toringoides* (Rehd.) Hughes., and *Malus kansuensis* (Batal.) Schneid. are tolerant to drought; *Malus hupehensis* (Pamp.) Rehd. and *Malus toringoides* (Rehd.) Hughes. are tolerant to waterlogging; *M. baccata* and *M. sieversii* are tolerant to cold; *Malus robusta* (Carr.) Rehd., *M. sieversii*, and *Malus sikkimensis* (Wenzig.) Koehne. are tolerant to salt; and even *M. zumi* is tolerant to a 0.60% salt content in soil [27,28]. There are three ways to avoid ionic toxicity in plants: salinity dilution, salinity regionalization, and salt rejection. Overcoming osmotic stress mainly depends on the content of osmotic regulating substances [29,30]. For *Malus* plants, the main mechanisms of salt stress resistance are salt rejection and ion regionalization. However, they have a long history of heredity and evolution, with different species exhibiting different tolerances to salt stress, and thus the adaptation mechanisms to salt stress are not all the same. Whether different *Malus* species take advantage of the same substance to scavenge ROS or regulate osmotic stress remains unknown. Therefore, understanding the physiological basis of salt resistance in different *Malus* species is very important for breeding salt-tolerant rootstocks.

Plants may be more tolerant to salt in the seedling stage than in the other growth stages [31]. *M. zumi*, *M. sieversii*, and *M. baccata* are high-resistance, medium-resistance, and salt-sensitive, respectively. The latter two are widely used as rootstocks in northwest and northeast China, but how they physiologically differ in terms of salt stress resistance remains unclear.

In the present study, the activities of POD, SOD, and CAT and the contents of MDA, SP, PRO, CHLa, and CHLb were compared in the three species under different NaCl treatments during the seedling period. We aimed to evaluate the effects on the physiological parameters of salt stress to elucidate the adaptive mechanisms of different *Malus* species to salinity stress. Our findings can be used as the basis for the breeding of salt-tolerant dwarfing rootstocks.

2. Materials and Methods

2.1. Plant Materials

M. zumi, *M. sieversii*, and *M. baccata* trees were planted in the National Repository of Apple Germplasm Resources (Xingcheng, Liaoning, China) in 2007. *M. zumi*, *M. sieversii*, and *M. baccata* are genotypes that are characterized as high-resistance, medium-resistance and salt-sensitive, respectively. The seeds of *M. zumi*, *M. sieversii*, and *M. baccata* were collected in the autumn of 2019 and were laminated for 60 days at 4 °C starting in late January of 2020. After germination, seeds were sown in a seedling tray in April and transplanted into plastic pots in June. One seedling was planted per pot.

2.2. Experimental Design

All experiments took place in a greenhouse. During the experiments, the average temperature was approximately 28 °C. The lowest temperature was approximately 16 °C and the highest temperature was 33 °C. The relative air humidity was 50–60%. A total of 200 seedlings that exhibited uniform growth, were 1 year old, and were approximately 30 cm tall were selected from each species, with 50 seedlings per group. Four groups were irrigated with either 0, 0.20%, 0.40%, or 0.60% NaCl solution, respectively, three times from 10 to 17 July 2020. Samples of roots and leaves were collected at 0, 4, 8, and 12 days after the last round of irrigation for each group. Leaves or roots of three plants were mixed for each group as one replicate with three replicates per group. The samples were rinsed with tap water to remove the soil and other surface debris and then washed with distilled water. All the samples were frozen in liquid nitrogen and stored at −80 °C.

2.3. Physiological Parameter Measurements

The nitroblue tetrazolium (NBT) method was used to determine the SOD activity [32]. The superoxide anion (O^{2-}) is produced by the xanthine and xanthine oxidase reaction system. O^{2-} reduces nitroblue tetrazole to generate blue formazan, for which the maximum absorption peak is 560 nm. SOD scavenges O^{2-}, which results in formazan being inhibited. The more blue the reaction liquid, the lower the SOD activity. The experimental steps used were those described in the kit instructions (kit series no.: SOD-2-Y, Comin Biotechnology, Suzhou, China; www.cominbio.com, accessed on 15 May 2020). A total of 1 mL of blank tube solution and measuring tube solution was absorbed into a glass colorimetric dish, and the absorbance value at 560 nm was recorded as Ab560 for the blank tube and Am560 for the measuring tube.

$$\text{PI (percentage of inhibition)} = \frac{\text{Ab560} - \text{Am560}}{\text{Ab560}} \times 100\%$$

$$\text{SOD activity (U/g, FW)} = 11.4 \times \frac{\text{PI}}{0.1 \times (1 - \text{PI})}$$

The guaiacol method was used to determine the POD activity [33]. POD catalyzes the oxidation of specific substrates with H_2O_2 and has a characteristic light absorption at 470 nm. The experimental steps used were those described in the kit instructions (kit series no.: POD-2-Y, Comin Biotechnology, Suzhou, China; www.cominbio.com, accessed on 15 May 2020). A total of 1 mL of supernatant was added into a glass colorimetric dish; the absorbance value at 470 nm was recorded as Ab470, and the value 1 min later was recorded as Al470.

$$\text{POD activity (U/g, FW)} = \frac{2000 \times (\text{Al470} - \text{Ab470})}{0.1}$$

The ultraviolet absorption method was utilized to determine the CAT activity [34]. H_2O_2 has a characteristic absorption peak at 240 nm, and CAT can decompose H_2O_2, so the absorbance of the reaction solution at 240 nm decreased with the reaction time. CAT activity could then be calculated according to the change rate of the absorbance. The experimental steps used were those described in the kit instructions (kit series no.: CAT-2-Y, Comin Biotechnology, Suzhou, China; www.cominbio.com, accessed on 17 May 2020). The absorbance value at 240 nm was recorded as Ab240, and the value 1min later was recorded as Al240.

$$\text{CAT activity (U/g, FW)} = \frac{687 \times (\text{Ab240} - \text{Al240})}{0.1}$$

The thiobarbituric acid (TBA) method was applied to measure the MDA content. MDA combined with thiobarbituric acid (TBA) to produce a red product with a maximum absorption peak at 532 nm. The content of lipid peroxide in the sample could be estimated after colorimetry; the MDA content was calculated as the difference between the absorbance values at 532 and 600 nm. The experimental steps used were those described in the kit instructions (kit series no.: MDA-2-Y, Comin Biotechnology, Suzhou, China; www.cominbio.com, accessed on 15 May 2020). A total of 1 mL of upper solution was absorbed into a glass colorimetric dish; the absorbance values at 532 and 600 nm were recorded as A532 and A600, respectively.

$$\text{MDA content (nmol/g, FW)} = \frac{25.8 \times (\text{A532} - \text{A600})}{0.1}$$

The bicinchoninc acid (BCA) method was performed to determine the SP content. Under alkaline conditions, cysteine, tryptophan, tyrosine, and peptide bonds in proteins can reduce Cu^{2+} to Cu^{+}. Two molecules of BCA combined with Cu^{+} to form a purple complex, which had an absorption peak at 562 nm. The experimental steps used were those described in the kit instructions (kit series no.: BCAP-2-W, Comin Biotechnology, Suzhou, China; www.cominbio.com, accessed on 15 May 2020). The absorbance values of a blank tube, standard tube, and measuring tube at 562 nm was recorded as Ab562 for the blank tube, As562 for the standard tube, and Am562 for measuring tube.

$$\text{SP content (mg/g, FW)} = \frac{0.5 \times (\text{Am562} - \text{Ab562})}{0.1 \times (\text{As562} - \text{Ab562})}$$

The sulfosalicylic acid (SSA) method was used to determine the PRO content. PRO was extracted with sulfosalicylic acid and reacted with an acidic ninhydrin solution to produce a red color after heating. The absorbance value was measured at 520 nm after extraction with methylbenzene. The experimental steps were those described in the kit instructions (kit series no.: PRO-2-Y, Comin Biotechnology, Suzhou, China; www.cominbio.com, accessed on 15 May 2020). The absorbance value was recorded as A520 at 520 nm.

$$\text{PRO content } (\mu\text{g/g, FW}) = \frac{19.2 \times (\text{A520} + 0.0021)}{0.1}$$

Determination of the CHLa and CHLb content was performed according to the experimental steps described in the kit instructions (kit series no.: CPL-2-G, Comin Biotechnology, Suzhou, China; www.cominbio.com, accessed on 15 May 2020). The absorbance values at 663 and 645 nm were determined and were denoted as A663 and A645, respectively.

$$\text{CHLa content (mg/g, FW)} = \frac{(12.7 \times \text{A663} - 2.69 \times \text{A645}) \times \text{Ve} \times \text{D}}{\text{m} \times 1000}$$

$$\text{CHLb content (mg/g, FW)} = \frac{(22.9 \times \text{A663} - 4.68 \times \text{A645}) \times \text{Ve} \times \text{D}}{\text{m} \times 1000}$$

where Ve is the extraction volume, D is the multiple dilution, and m is the sample weight.

2.4. Statistical Analysis

The experimental data collection and analysis were performed using Microsoft Excel 2010. The physiological parameters were subjected to an analysis of variance (ANOVA) using SPSS 19. The significant difference level, which was calculated using Tukey's test, was used to compare the differences among treatments and the control. The PCA was conducted using Origin 2019b software.

3. Results

3.1. Dynamic Effects on Physiological Parameters of Leaves of Malus Plants under Salinity Stress

The SOD, POD, and CAT activities of the three *Malus* species increased at first and then decreased in the leaves in the 0.20%, 0.40% and 0.60% NaCl treatment groups (Figures 1 and 2; Table S1). The activities of SOD, POD, and CAT were higher in the 0.40% and 0.60% NaCl treatments groups than in the 0.20% NaCl treatments group and in the 0% control. For all treatments, the highest activity of SOD, POD, and CAT was exhibited on the fourth day after NaCl treatment. At the peak time, the SOD activity increased by 49.33%, 38.54%, and 37.11% for *M. sieversii*, *M. baccata*, and *M. zumi*, respectively, in the 0.60% NaCl treatment group compared with the control treatment; moreover, the POD activity increased by 66.63%, 208.01%, and 164.22%, respectively; and the CAT activity increased by 250.68%, 86.67%, and 128.73%, respectively. POD increased the most for *M. baccata*, while SOD and CAT increased the most for *M. sieversii*. The increase in the SOD, POD, and CAT activities fell in the midrange for *M. zumi*. After 12 days of treatment, there was little difference between treatments and the control in terms of SOD, POD, and CAT activities in *M. zumi* and POD activity in *M. sieversii*, but the SOD, POD, and CAT activities remained obviously higher in *M. baccata* than in the control.

For all the NaCl treatments, the MDA content increased at first and then decreased in the three *Malus* species (Figure 2; Table S1). The MDA content was higher in the 0.40% and 0.60% NaCl treatments groups than in the 0.20% NaCl treatments group and in the control except for *M. zumi*. The times of peak occurrence were different, reaching a peak on the fourth day after NaCl treatment for *M. zumi* and *M. sieversii* and on the eighth day after NaCl treatment for *M. baccata*. At the time of peak MDA content for the 0.60% NaCl treatment group, the MDA content in *M. sieversii*, *M. baccata*, and *M. zumi* increased by 27.66%, 28.64%, and 18.17%, respectively, compared with the control treatment. After 12 days of treatment, there was little difference between treatments and the control for the three species, but *M. zumi* had an overall lower MDA content than the other two species both in the treatments and in the control.

The contents of SP and PRO increased at first and then decreased in all NaCl treatments for the three *Malus* species (Figure 3; Table S1). The SP and PRO contents were higher in the 0.60% NaCl treatment group than those in the 0.20% and 0.40% NaCl treatment groups and the control, with the exception of the PRO content in *M. sieversii*. The peak time was different for SP and PRO, with the former peaking on the fourth day after NaCl treatments and the latter on the eighth day after NaCl treatments. With the increasing NaCl concentration, the SP and PRO contents in the 0.20% NaCl treatment group and the control for the three *Malus* species was lower than in the 0.40% and 0.60% groups. Compared with the control treatment, in *M. sieversii*, *M. baccata*, and *M. zumi* subjected to 0.60% NaCl treatments, the content of SP increased by 76.58%, 38.26%, and 71.68%, respectively; and the content of PRO increased by 22.95%, 47.45%, and 61.07%, respectively, when considering their values at peak. After 12 days of treatment, there was little difference between treatments and the control for SP and PRO in the three species except for the SP content in *M. baccata*, which was slightly higher in the treatments than in the control.

Increasing the salinity concentration reduced the measured values of traits related to photosynthesis. The CHLa and CHLb content in the three *Malus* species decreased with prolongation of the NaCl treatment (Figure 4; Table S1). The values of all treatments were lower than those of the control in *M. baccata*, and there was no significant difference between treatments and control for *M. zumi* except on the twelfth day after NaCl treatment. Moreover, the CHLa and CHLb contents were lower in the 0.60% NaCl treatment groups than in the control for *M. sieversii*. In addition, after 12 days of NaCl treatment, the CHLa and CHLb contents were obviously higher in the control than in the treatments used for the three species.

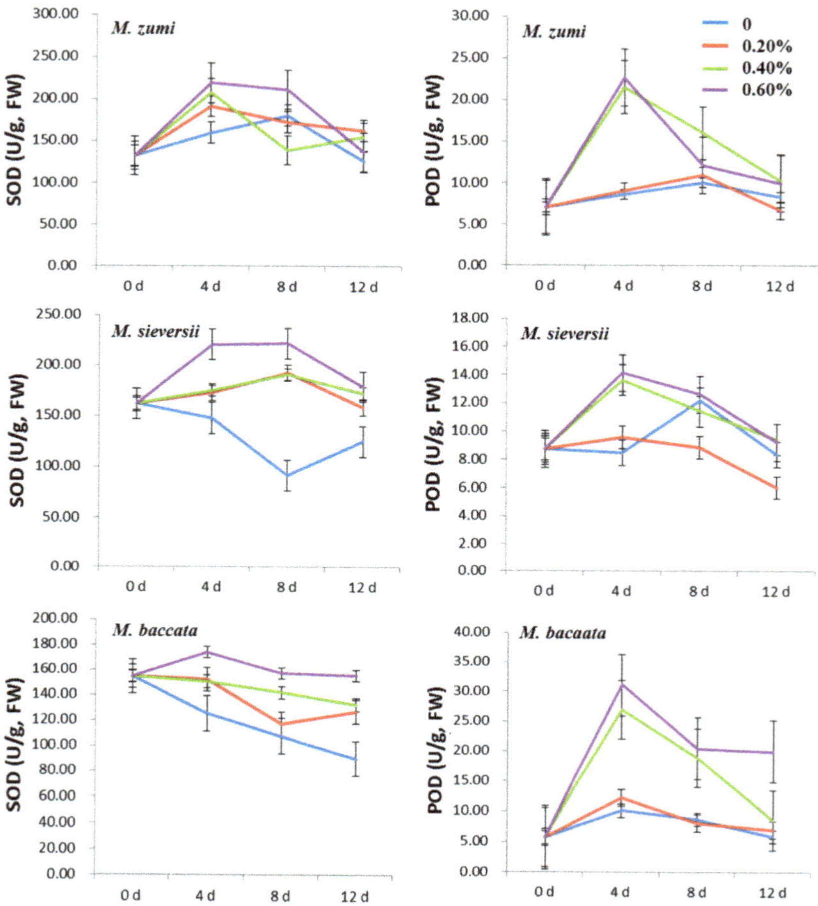

Figure 1. Change curve of leaf SOD and POD activities for the three *Malus* species under four levels of NaCl treatments.

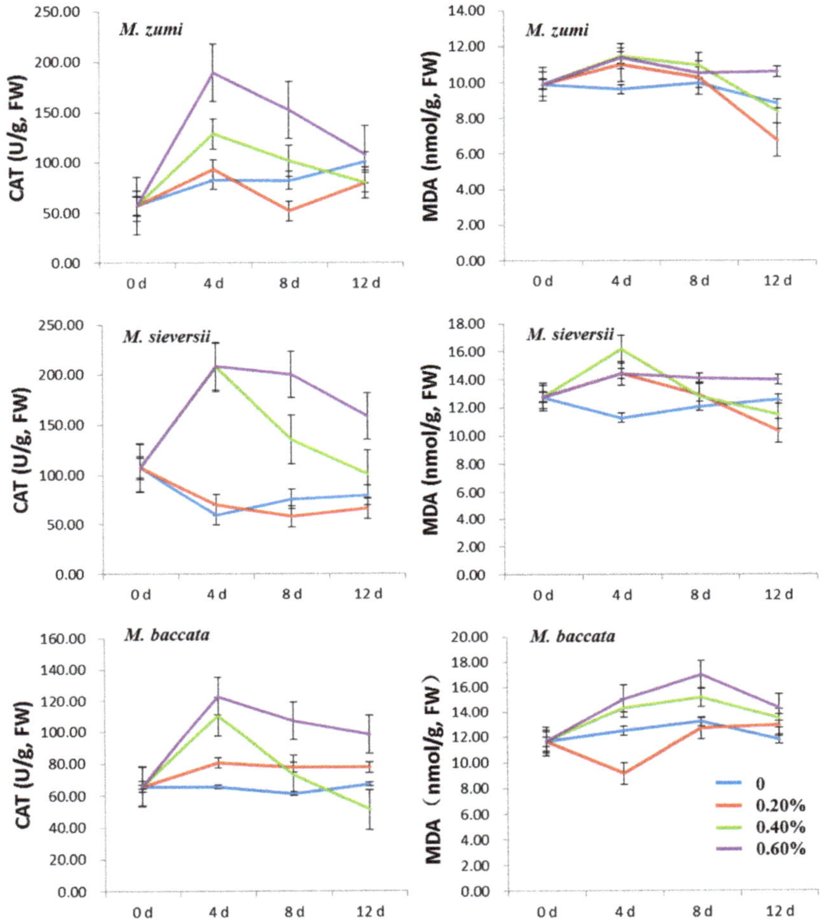

Figure 2. Change curve of leaf CAT activity and MDA content for three *Malus* species under four levels of NaCl treatments.

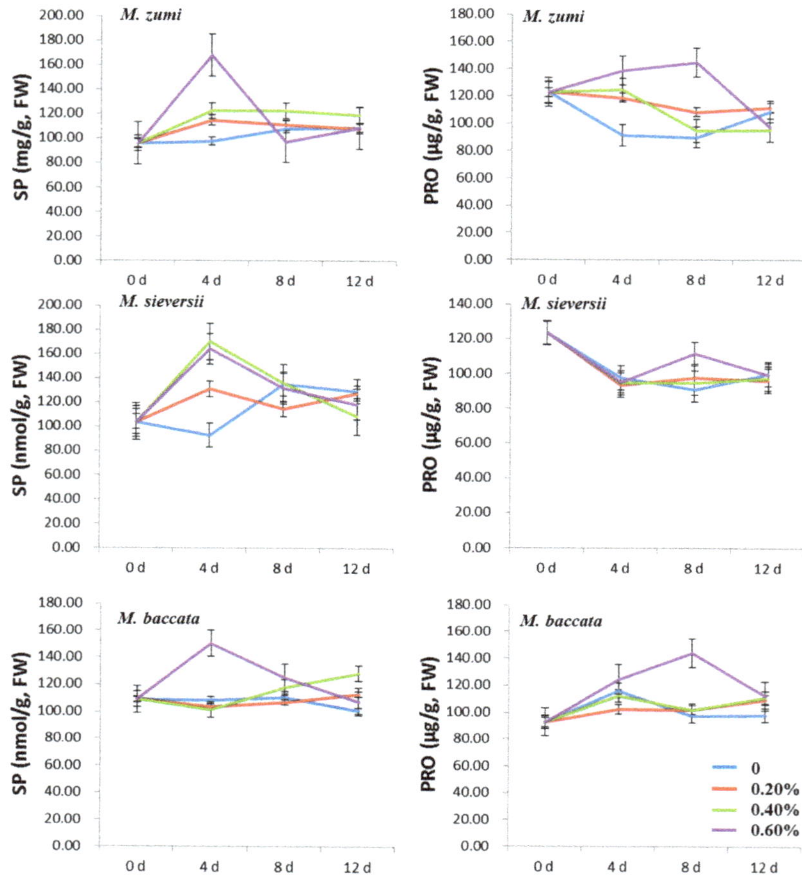

Figure 3. Change curve of leaf SP and PRO contents for the three *Malus* species under four levels of NaCl treatments.

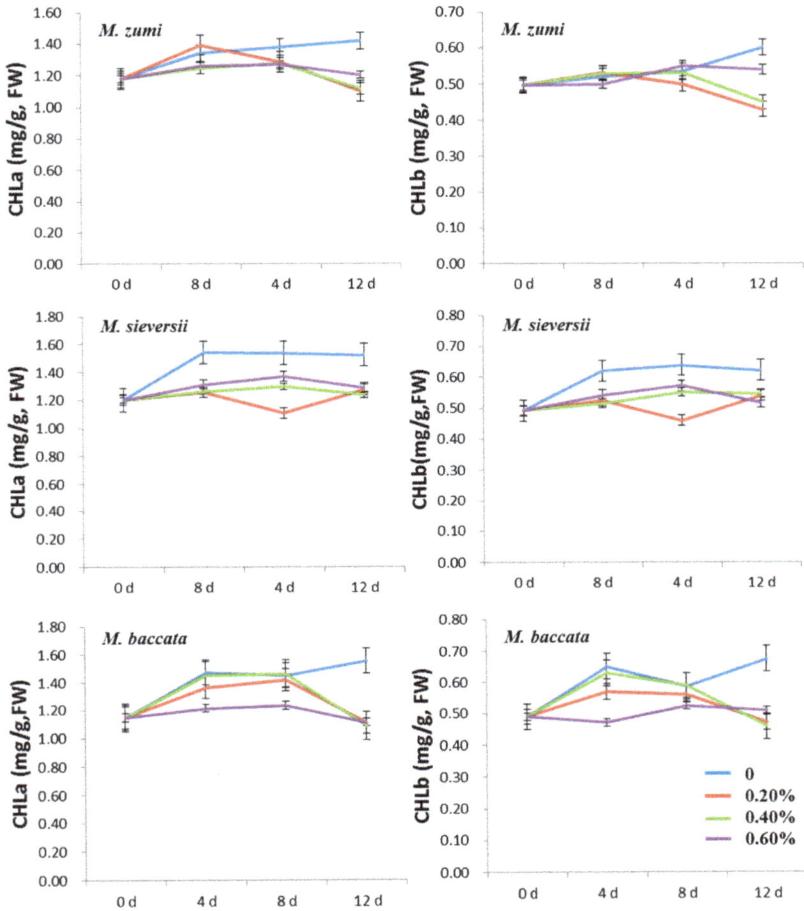

Figure 4. Change curve of leaf CHLa and CHLb contents in the three *Malus* species under four levels of NaCl treatments.

3.2. Dynamic Effects on Physiological Parameters of Malus Plant Roots under Salinity Stress

During the treatment periods, the SOD, POD, and CAT activities exhibited a similar trend in the roots as in leaves for the three *Malus* species (Figures 5 and 6; Table S2). With the increase in the NaCl concentration, the SOD, POD, and CAT activities were higher in the 0.40% and 0.60% NaCl treatment groups than in the 0.20% NaCl treatment group and control. At the time of peak of these enzyme activities in the 0.60% NaCl treatment group, the SOD activity increased by 6.70%, 51.35%, and 68.23% for *M. sieversii*, *M. baccata*, and *M. zumi*, respectively, compared with the control treatment; the POD activity increased by 243.32%, 1206.89%, and 285.30%, respectively; and the CAT activity increased by 409.05%, 49.98%, and 13.17%, respectively. After 12 days of treatment, there was little difference between treatments and the control for SOD in *M. zumi* or CAT in *M. baccata*, but the activity of POD remained higher than the control in the three species.

The MDA content exhibited a similar trend in the leaves as in the roots for the three *Malus* species (Figure 6; Table S2). With the increasing NaCl concentration, the MDA content was higher in the 0.40% and 0.60% NaCl treatment groups than in the 0.20% NaCl treatment group and in the control for *M. sieversii* and *M. baccata*, but there was no significant difference among treatments for *M. zumi*. At the time of peak in the 0.60% NaCl treatment group, the MDA content in *M. sieversii*, *M. baccata*, and *M. zumi* increased by 20.59%, 48.54%, and 34.77%, respectively, compared with the control. After 12 days of NaCl treatments, the MDA content tended to normal levels in *M. zumi* but were still higher in the treatments than in the control in *M. sieversii* and *M. baccata*.

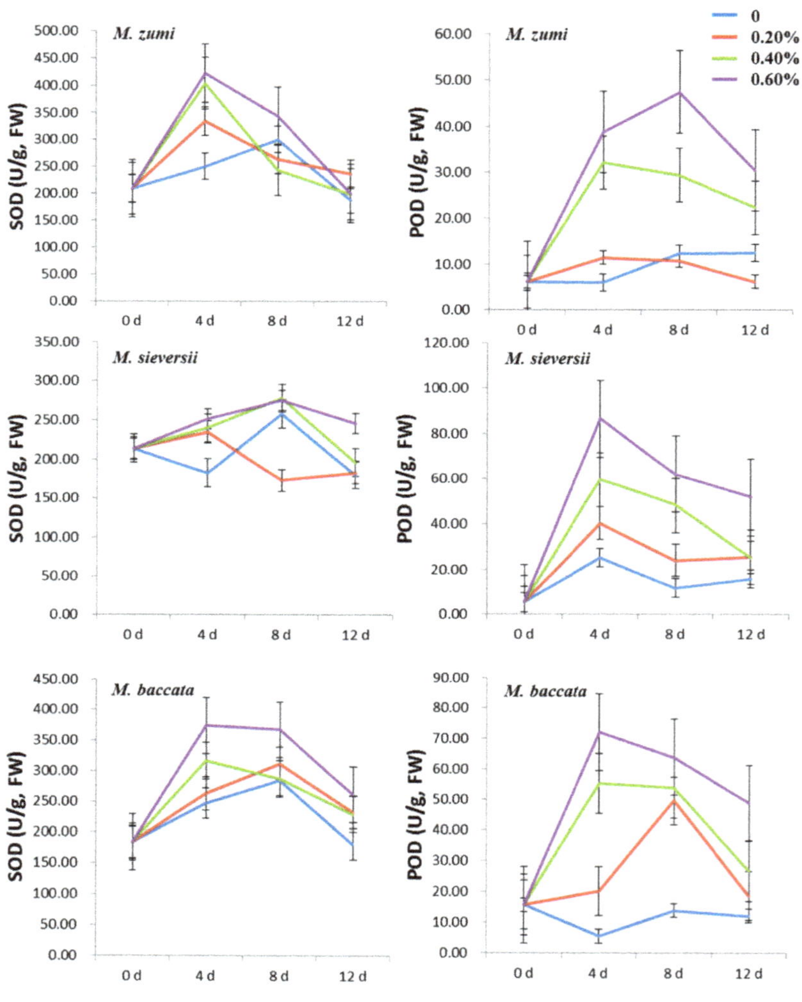

Figure 5. Change curve of root SOD and POD activities in the three *Malus* species under four levels of NaCl treatments.

The SP and PRO contents increased at first and then decreased in the three *Malus* species (Figure 7; Table S2). The SP and PRO contents were higher in the 0.40% and 0.60% NaCl treatment groups than in the 0.20% NaCl treatment group and the control. There was a difference in the peak time of SP; i.e., it was on the fourth day after NaCl treatments for *M. zumi* but on the eighth day after NaCl treatment for *M. sieversii* and *M. baccata*. With the increasing NaCl concentration, the SP and PRO contents were lower in the 0.20% NaCl treatment group and the control than in the 0.40% and 0.60% groups for the three *Malus* species. Considering the peak SP and PRO contents in the 0.60% NaCl treatment group, the

SP content increased by 40.22%, 46.32%, and 40.17% for *M. sieversii*, *M. baccata*, and *M. zumi*, respectively; and the PRO content increased by 34.03%, 52.26%, and 29.75%, respectively, compared with the control treatment. After 12 days of treatment, there was little difference between the treatments and control for SP and PRO in *M. zumi* and *M. sieversii*, while they remained higher in treatments than in the control in *M. baccata*.

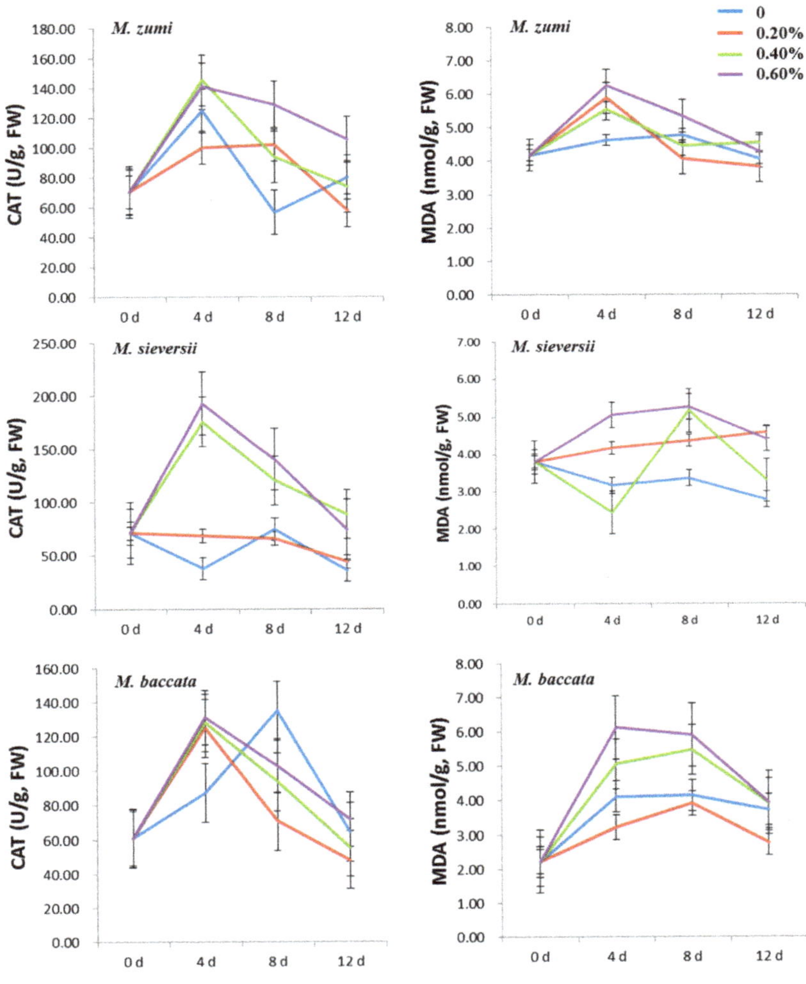

Figure 6. Change curve of root CAT activity and MDA in the three *Malus* species under four levels of NaCl treatments.

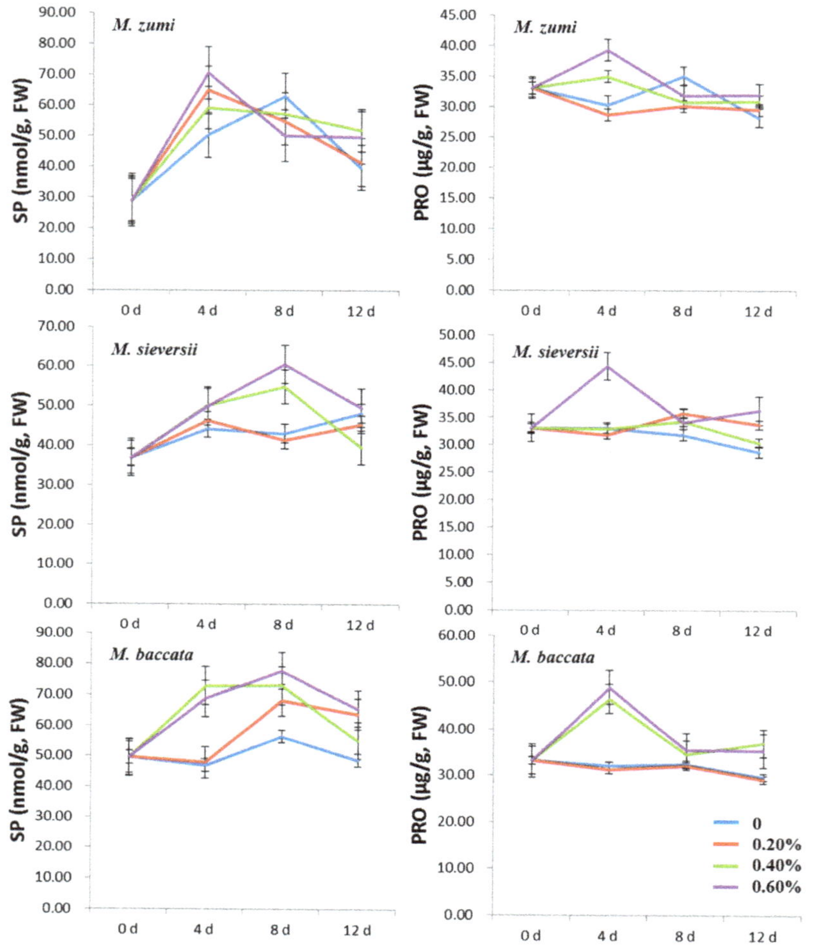

Figure 7. Change curve of root SP and PRO contents in the three *Malus* species under four levels of NaCl treatments.

3.3. Comparing the Effects on Physiological Parameters of Roots and Leaves of Malus Plants under Salinity Stress

The mean values of the physiological parameters were compared between the treatments and control (Tables S1 and S2), and significant differences at the 0.05 level were determined. The results showed that the SOD and POD activities were higher in the roots than in the leaves and that the MDA, SP, and PRO contents were higher in the leaves than in the roots, while there was no significant difference in CAT activity between leaves and roots. Under salt stress, the increase in SOD, POD, CAT, and MDA was higher in the roots than in the leaves, whereas the increases in SP and PRO were slightly higher in the leaves than in the roots.

The PCA, which included all physiological parameters from both the leaves and roots, showed that there was a significant difference between these tissues (Figure 8). Principal component 1 (PC1) and principal component 2 (PC2) could explain 67.64% and 19.64% of the variation, respectively, so the first two principle components could thus explain 87.28% of the variation. Leaves and roots could be distinguished by PC1, which mainly presented

CHLa and CHLb, and were positively correlated. Roots were separated from leaves mainly due to a lack of CHLa and CHLb contents, and most of the root physiological parameters were found on the left of the PCA plot. This indicated that the roots had higher SOD and POD activities because they were also found on the left. Thus, SOD and POD were other factors that distinguished roots from leaves.

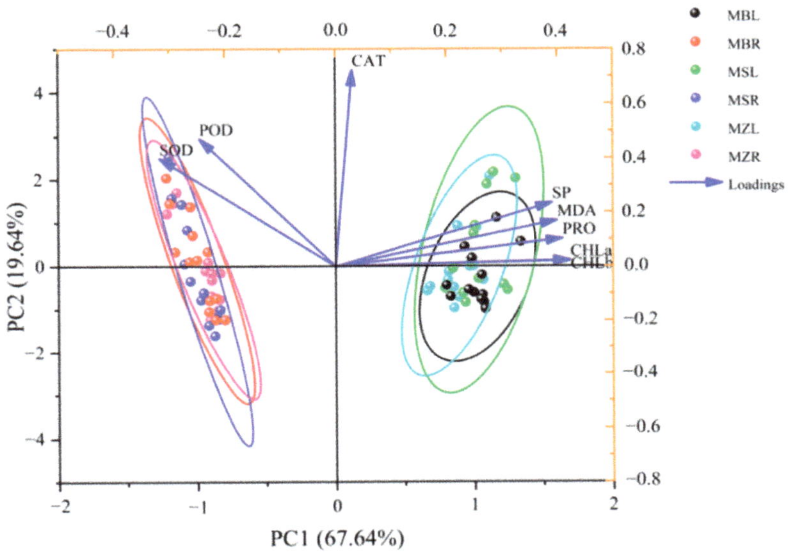

Figure 8. PCA of physiological parameters in leaves and roots for the three *Malus* species. Loading and scores plot of the first two principal components of the principal component analysis model. The left and bottom coordinates were the loading scores of the first two principle components, and the top and right coordinates were the scores of all the physiological parameters in the first two principle components. MZL—leaves of *M. zumi*; MSL—leaves of *M. sieversii*; MBL—leaves of *M. baccata*; MZR—roots of *M. zumi*; MSR—roots of *M. sieversii*; MBR—roots of *M. baccata*.

A PCA using the physiological parameters of the leaves was performed (Figure S1). PC1 and PC2 could explain 36.52% and 23.81% of the variation, respectively, so the first two principle components could thus explain 60.33% of the variation. *M. baccata* could be distinguished from other *Malus* species by PC2, which mainly presented CHLa and CHLb and had lower CHLa and CHLb contents. The other two species could not be distinguished by the PCA of the leaves. A PCA using the physiological parameters of the roots was applied (Figure S2). PC1 and PC2 could explain 59.19% and 14.75% of the variation, respectively, so the first two principle components could thus explain 73.94% of the variation, which was more than in the leaves. *M. zumi* and *M. sieversii* could be distinguished by PC2, which mainly presented POD, PRO, and CAT; *M. sieversii* had higher values for these parameters, while *M. baccata* could not be distinguished from the other two *Malus* species based on the comparison of roots.

4. Discussion

Salt ions cause little damage to the cell membranes of highly resistant varieties, and the MDA production is reduced. For the medium-resistance varieties, MDA causes damage to membranes via lipid peroxidation, which leads to the formation of ROS; however, they are able to protect cells from further oxidative damage via their own enzymatic defense systems [35]. The results of this study showed that *M. zumi* and *M. sieversii* had a lower MDA content than *M. baccata* in the NaCl treatment groups. This was possibly due to the higher SOD and CAT activities of *M. zumi* and *M. sieversii*. Moreover, the SOD, POD, and CAT activities were the highest in both the leaves and roots of *M. sieversii* in the control. In addition, *M. baccata* had the lowest SOD and CAT activities in the roots and the lowest POD activity in the leaves. The formation and elimination of ROS were in dynamic equilibrium under normal conditions in the plants. The SOD, POD, and CAT activities exhibited the ability to resist adversity without stress [36–38]. *M. zumi* is highly resistant to salt stress, but the enzyme activity in this species was not the highest. Aside from enzymatic antioxidants, non-enzymatic antioxidants also helped to scavenge these indigenously generated ROS [39]. The present results showed that *M. zumi* and *M. sieversii* had a higher PRO content in leaves of the control and that there was no significant difference in the root PRO content of the three species. It was interesting that *M. baccata* had the higher SP content in roots. The high content of osmotic substances in roots helped the plants to cope with water absorption disorders caused by salt stress. We therefore concluded that both enzymatic and non-enzymatic antioxidants played a major role in determining the medium resistance of *M. sieversii* due to its high activity of CAT in the leaves and roots and the content of SP in the leaves. We speculated that there were other endogenous substances that could bear primary responsibility for the high salt resistance of *M. zumi* due to its low contents of SP and PRO. Hannachi et al. [40] confirmed that salicylic acid could be involved in salt-stress tolerance; i.e., it was associated with the efficient antioxidant defense system for scavenging ROS. Various salinized plants showed high values of total soluble sugars and total free amino acids [41,42].

Salt damage was mitigated if a plant developed a series of responses to alleviate the associated stress [43–45]. In the present study, the POD, SOD, and CAT activities in the leaves and roots of *M. zumi*, *M. sieversii*, and *M. baccata* increased after NaCl treatments, reaching a peak on the fourth day in most cases. The SOD, POD, and CAT activities increased in order to rapidly scavenge ROS under salt stress [46]. The increasing activities of SOD, POD, and CAT in the first four days after treatments showed that these enzymes were the main substances that were scavenging ROS, while the enzymes that played a major role in the three species differed. In the 0.60% NaCl treatment groups, POD in *M. baccata* increased the most in the both leaves and roots after 4 days of NaCl treatments compared with the control; moreover, the SOD and CAT in *M. sieversii* increased the most in the leaves, and the SOD in *M. zumi* increased the most in the roots. Regarding osmotic regulation, *M. baccata* exhibited a significant increase in the proportions of SP and PRO in the roots. This indicated that *M. baccata* was sensitive to salt stress, and thus higher levels of SOD, POD, and CAT activities were needed to overcome the salt injury. The MDA content in *M. baccata* remained high after 12 days of treatment, by which point the enzyme activities and MDA content in *M. zumi* had returned to a normal level. ROS were continuously produced under salt stress, but the damage could be mitigated if they were eliminated in time [43]. It was demonstrated that *M. zumi* had a stronger recovery capability than *M. baccata*.

The SOD, POD, and CAT activities and the MDA, SP, and PRO contents increased with the increase in the NaCl concentration. Moreover, there was a greater accumulation of MDA for *M. baccata* in the different NaCl treatment groups over the same period. When comparing the different NaCl treatments, there was an obvious increase in the leaves of SOD and CAT for *M. sieversii* and in POD and PRO for *M. zumi* and *M. baccata*. In addition, SOD for *M. zumi*, POD and CAT for *M. sieversii* and *M. baccata*, and SP and PRO for *M. baccata* exhibited a significant increase in the roots. In a previous study on Chinese cabbage, the MDA content was shown to continually increase with the increase in the NaCl concentration [47]. In addition, sensitive cultivars were shown to accumulate more MDA in eggplant seedlings [48], and reduced MDA accumulation was a reflection of improved growth performance under salinity stress [49]. Various studies concluded that the level of PRO content should not be used as an indicator of salt resistance [44,45,50–52], while other studies reported that its increase under biological or abiotic stress was a type of victimization symptom. NaCl had little effect on the PRO content in salt-tolerant *Malus* plants, whereas the content in salt-sensitive species continued to increase significantly under a high salt concentration [53–55]. In our study, *M. baccata* suffered a more serious injury than the other species under the same NaCl treatments. This was mainly due to the lower CAT and SOD activities in both the roots and leaves under these treatments and to the lower SP content in the roots, although the activities of the aforementioned enzymes were high in the control. In addition, PRO accumulation occurred in both the leaves and roots of all three species when subjected to NaCl treatments, which supported the hypothesis that PRO accumulation was a symptom of salt damage and could be an indicator of resistance to salt in the *Malus* species.

Chlorophyll content is one of the most important indexes of photosynthesis. In this study, the CHLa and CHLb contents were slightly lower in the NaCl treatments than in the control for *M. zumi* and *M. sieversii*, while a significant difference between all NaCl treatments and the control was observed in *M. baccata*. Chlorophyll was suppressed with an increase in salinity [56,57]. Other reports have shown that chlorophyll contents were higher under salinity stress conditions [58]. The results of this study showed that salt stress may inhibit photosynthesis in *M. baccata*.

The main substances for scavenging ROS are different under salt stress in different plants. Responses to salt stress even differed in different tissues within the same plant. The SOD and POD activities were higher in the roots than in the leaves of the three *Malus* species. The SP and PRO contents were higher in the leaves than in the roots, but no significant differences in CAT were observed. Salt damage was directly harmful to plant roots, but the increasing proportion of MDA in the leaves was higher than in the roots in the three species for almost all periods of treatment. Both the leaves and roots responded to salt stress in plants. In a previous study, it was noted that root growth was positively correlated with aboveground growth and that the changes in physiological parameters of the leaves and roots varied for different *Malus* species [59]. We obtained similar results, which indicated that the roots were more tolerant to salt than the leaves. Moreover, the roots were shown to play a key role in eliminating ROS under salt stress in *Malus* plants, and the degree of damage under salt treatment was more serious to the leaves than the roots. Based on the PCA, we concluded that SOD and POD activities and CHLa and CHLb contents were the main factors that differentiated between the roots and leaves of *Malus* species under salt stress. Moreover, *M. zumi* and *M. baccata* were distinguished by POD, PRO, and CAT in the roots or MDA, SOD, and SP in the roots. *M. baccata* could be distinguished from *M. zumi* and *M. sieversii* by the CHLa, CHLb, and PRO contents in leaves.

5. Conclusions

The activity and content of the measured physiological parameters were higher in the 0.40% and 0.60% NaCl treatment groups than in the 0.20% group and the control. The SOD and POD activities were higher in the roots than in the leaves; while the MDA, SP, and PRO contents were higher in the leaves than in the roots. The resistance to salt stress of *M. zumi* was mainly due to the high SOD and POD activities under salt stress, while there were other substances that also may have played a major role in the salt stress response. Osmotic regulation was shown to play a greater role in the response to salt stress than enzymatic antioxidants in *M. baccata*, and both enzymatic antioxidants and osmotic regulation made a significant contribution to salt resistance in *M. sieversii*.

Supplementary Materials: The following supporting information can be downloaded at: https://www.mdpi.com/article/10.3390/life12111929/s1, Table S1: ANOVA analysis of physiological parameters in leaves for the three *Malus* species; Table S2: ANOVA analysis of physiological parameters in roots for the three *Malus* species; Figure S1: PCA of physiological parameters in leaves for the three *Malus* species; Figure S2: PCA of physiological parameters in roots for the three *Malus* species.

Author Contributions: D.W.; K.W. and J.L. planned and designed the experiments and were involved in the methodology; D.W.; Y.G. and S.S. performed the experiments and investigated and recorded data; X.L. and Q.L. were responsible for software; L.L. was responsible for field management; D.W.

analyzed the data and wrote the manuscript; K.W. and J.L. reviewed and edited the final manuscript. All authors have read and agreed to the published version of the manuscript.

Funding: This research was partly funded by the Agricultural Science and Technology Innovation Program (CAAS-ASTIP-2021-RIP-02) and the Fundamental Research Funds for Central Non-Profit Scientific Institutions (grant number 161018201614).

Institutional Review Board Statement: Not applicable.

Informed Consent Statement: Not applicable.

Data Availability Statement: All data generated or analyzed during this study are included in the published article.

Conflicts of Interest: The authors declare no conflict of interest.

References

1. Munns, R.; Tester, M. Mechanisms of salinity tolerance. *Annu. Rev. Plant Biol.* **2008**, *59*, 651–681. [CrossRef] [PubMed]
2. FAO. *Status of the World's Soil Resources (SWSR)–Main Report*; Food and Agriculture Organization of the United Nations and Intergovernmental Technical Panel on Soils; FAO: Rome, Italy, 2015; Volume 650.
3. Jamil, A.; Riaz, S.; Ashraf, M.; Foolad, M.R. Gene expression profiling of plants under salt stress. *Crit. Rev. Plant Sci.* **2011**, *30*, 435–458. [CrossRef]
4. Soltabayeva, A.; Ongaltay, A.; Omondi, J.O.; Srivastava, S. Morphological, physiological and molecular markers for salt-stressed plants. *Plants* **2021**, *10*, 243. [CrossRef] [PubMed]
5. Alasvandyari, F.; Mahdavi, B.; Hosseini, M.S. Glycine betaine affects the antioxidant system and ion accumulation and reduces salinity-induced damage in safflower seedlings. *Arch. Biol. Sci.* **2017**, *69*, 139–147. [CrossRef]
6. Richard, P.M.; Gennaro, F. Apple rootstocks: History, physiology, management, and breeding. *Hortic. Rev.* **2018**, *45*, 197–312. [CrossRef]
7. Chen, X.; Cheng, X.W.; Zhu, H.; Bañuelos, C.; Shutes, B.; Wu, H.T. Influence of salt stress on propagation, growth and nutrient uptake of typical aquatic plant species. *Nord. J. Bot.* **2019**, *37*, e02411. [CrossRef]
8. Slama, I.; Abdelly, C.; Bouchereau, A.; Flowers, T.; Savoure, A. Diversity, distribution and roles of osmoprotective compounds accumulated in halophytes under abiotic stress. *Ann. Bot.* **2015**, *115*, 433–447. [CrossRef]
9. Liang, W.J.; Ma, X.L.; Wan, P.; Liu, L.Y. Plant salt-tolerance mechanism: A review. *Biochem. Biophys. Res. Commun.* **2018**, *495*, 286–291. [CrossRef]
10. Akter, S.; Huang, J.J.; Waszczak, C.; Jacques, S.; Gevaert, K. Cysteines under ROS attack in plants: A proteomics view. *J. Exp. Bot.* **2015**, *66*, 2935–2941. [CrossRef]
11. Sharp, R.E.; Hsiao, T.C.; Silk, W.K. Growth of the maize primary root at low water potentials: II. Role of growth and deposition of hexoes and potassium in osmotic adjustment. *Plant Physiol.* **1990**, *93*, 1337–1346. [CrossRef]
12. Sudhakar, C.; Lakshmi, A.; Giridarakumar, S. Changes in the antioxidant enzyme efficacy in two high yielding genotypes of mulberry (*Morus alba* L.) under NaCl salinity. *Plant Sci.* **2001**, *141*, 613–619. [CrossRef]
13. Rahnama, H.; Ebrahimzadeh, H. The effect of NaCl on proline accumulation in potato seedlings and calli. *Acta Physiol. Plant* **2004**, *26*, 263–270. [CrossRef]
14. Neto, A.A.D.; Prisco, J.T.; Eneas-Filho, J.; Abreu, C.E.B.; Filho, G.E. Effect of salt stress on antioxidative enzymes and lipid peroxidation in leaves and roots of salt-tolerant and salt-sensitive maize genotypes. *Environ. Exp. Bot.* **2006**, *56*, 87–94. [CrossRef]
15. Ashraf, M.; Foolad, M.R. Role of glycine betaine and proline in improving plant abiotic stress resistance. *Environ. Exp. Bot.* **2007**, *59*, 206–216. [CrossRef]
16. Eraslan, F.; Inal, A.; Pilbeam, D.J.; Gunes, A. Interactive effects of salicylic acid and silicon on oxidative damage and antioxidant activity in spinach (*Spinacia oleracea* L. CV. Matador) grown under boron toxicity and salinity. *Plant Growth Regul.* **2008**, *55*, 207–219. [CrossRef]
17. Ibrahimova, U.; Suleymanova, Z.; Brestic, M.; Mammadov, A.; Ali, O.M.; Latef, A.A.H.A.; Hossain, A. Assessing the adaptive mechanisms of two bread wheat (*Triticum aestivum* L.) genotypes to salinity stress. *Agronomy* **2021**, *11*, 1979. [CrossRef]
18. Singh, A.; Shekhar, S.; Marker, S.; Ramteke, P.W. Changes in morpho-physiological attributes in nine genotypes of linseed (*Linum usitatissimum* L.) under different level of salt (NaCl) stress. *Vegetos* **2021**, *34*, 647–653. [CrossRef]
19. Yasir, T.A.; Khan, A.; Skalicky, M.; Wasaya, A.; Rehmani, M.I.A.; Sarar, N.; Mubeen, K.; Aziz, M.; Hassan, M.M.; Hassan, F.A.S.; et al. Exogenous sodium nitroprusside mitigates salt stress in lentil (*Lens culinaris* Medik.) by affecting the growth, yield, and biochemical properties. *Molecules* **2021**, *26*, 2576. [CrossRef]
20. Hossain, A.; Azeem, F.; Shahriar, S.M.; Islam, M.T. *Regulation of Proline Transporters in Salt Stress Response in Plants. Transporters and Plant Osmotic Stress*, 1st ed.; Elsevier: Amsterdam, The Netherlands, 2021; pp. 291–306. [CrossRef]
21. Rady, M.M.; Taha, R.S.; Mahdi, A.H.A. Proline enhances growth, productivity and anatomy of two varieties of *Lupinus termis* L. grown under salt stress. *S. Afr. J. Bot.* **2016**, *102*, 221–227. [CrossRef]

22. Du, Z.J.; Zhai, H.; Luo, X.S.; Cheng, S.H.; Pan, Z.Y. Field identification of salt tolerance of apple rootstocks. *China Fruits* **2001**, *2*, 1–4.
23. Fang, H.L.; Liu, G.H.; Kearney, M. Georelational analysis of soil type, soil salt content, landform, and land use in the Yellow River Delta, China. *Environ. Manag.* **2005**, *35*, 72–83. [CrossRef] [PubMed]
24. Wang, Y.G.; Li, Y.; Xiao, D.N. Catchment scale spatial variability of soil salt content in agricultural oasis, Northwest China. *Environ. Geol.* **2008**, *56*, 439–446. [CrossRef]
25. Li, F.Z.; Huang, Z.B.; Ma, Y.; Sun, Z.J. Improvement effects of different environmental materials on coastal saline-alkali soil in Yellow River Delta. *Mater. Sci. Forum* **2018**, *913*, 879–886. [CrossRef]
26. Wu, Z.Y.; Peter, H.R.; Hong, D.Y. *Flora of China*; Science Press: Beijing, China, 2003; Volume 9, pp. 179–189.
27. Gu, Y.L.; Zhao, H.X.; Ma, J.L.; Zhou, S.W. The adaption and application area on the salinity for *Malus zumi* Mats. *J. Tianjin Agric. Coll.* **1996**, *3*, 48–52.
28. Wang, Y.Z.; Feng, X.Z.; Luo, J.L. Multi-purpose and salt-tolerant economic tree-*Malus zumi*. *For. Sci. Technol.* **1999**, *24*, 53–55.
29. Zhao, K.F. Adaptation of plants to saline stress. *Bull. Biol.* **2002**, *51*, 7–10.
30. Beacham, A.M.; Hand, P.; Pink, D.A.; Monaghan, J.M. Analysis of *Brassica oleracea* early stage abiotic stress responses reveals tolerance in multiple crop types and for multiple sources of stress. *J. Sci. Food Agric.* **2017**, *97*, 5271–5277. [CrossRef]
31. Saeed, A.; Khan, A.A.; Saeed, N.; Saleem, M.F. Screening and evaluation of tomato germplasm for NaCl tolerance. *Acta Agric. Scand. Sect. B-Soil Plant Sci.* **2010**, *60*, 69–77. [CrossRef]
32. Zuo, Q. *Guidance of Plant Physiology and Biochemistry Experiment*; China Agriculture Press: Beijing, China, 1995; pp. 59–99.
33. Hammerschmidt, R.; Nuckles, E.M.; Kuc, J. Association of enhanced peroxidase activity with induced systemic resistance of cucumber to *Colletotrichum lagenarium*. *Physiol. Plant Pathol.* **1982**, *20*, 73–76. [CrossRef]
34. Rajeswari, V.; Paliwal, K. Peroxidase and catalase changes during in vitro adventitious shoot organogenesis from hypocotyls of *Albizia odoratissima* L.f. (Benth). *Acta Physiol. Plant.* **2008**, *30*, 825–832. [CrossRef]
35. Lu, J.J.; Duo, L.A.; Liu, X.J. Changes in SOD and POD activity and free proline content of *Lolium perenne* and *Festuca elata* leaves under different levels of salt stress. *Bull. Bot. Res.* **2004**, *24*, 115–119.
36. Kaur, H.; Sirhindi, G.; Bhardwaj, R.; Alyemeni, M.N.; Siddique, K.H.M.; Ahmad, P. 28-homobrassinolide regulates antioxidant enzyme activities and gene expression in response to salt- and temperature-induced oxidative stress in *Brassica juncea*. *Sci. Rep.* **2018**, *8*, 8735. [CrossRef]
37. Zhang, Z.H.; Cao, B.L.; Gao, S.; Xu, K. Grafting improves tomato drought tolerance through enhancing photosynthetic capacity and reducing ROS accumulation. *Protoplasma* **2019**, *256*, 1013–1024. [CrossRef]
38. Attia, H.; Al-Yasi, H.; Alamer, K.; Ali, E.; Hassan, F.; Elshazly, S.; Hessini, K. Induced anti-oxidation efficiency and others by salt stress in *Rosa damascena* Miller. *Sci. Hortic.* **2020**, *274*, 109681. [CrossRef]
39. Ahmad, P.; Jaleel, C.A.; Salem, M.A.; Nabi, G.; Sharma, S. Roles of enzymatic and nonenzymatic antioxidants in plants during abiotic stress. *Crit. Rev. Biotechnol.* **2010**, *30*, 161–175. [CrossRef]
40. Hassan, F.; Al-Yasi, H.; Ali, E.; Alamer, K.; Hessini, K.; Attia, H.; El-Shazly, S. Mitigation of salt-stress effects by moringa leaf extract or salicylic acid through motivating antioxidant machinery in damask rose. *Can. J. Plant Sci.* **2021**, *101*, 157–165. [CrossRef]
41. Khalafallah, A.A.; Tawfik, K.M.; Abd El-Gawad, Z.A. Tolerance of seven faba bean varieties to drought and salt stresses. *Res. J. Agric. Biol. Sci.* **2008**, *4*, 175–186.
42. Bornare, S.S.; Prasad, L.C.; Kumar, S. Comparative study of biochemical indicators of salinity tolerance of barley (*Hordeum vulgare* L.) with other crops: A review. *Can. J. Plant Breed.* **2013**, *1*, 97–102.
43. Bharathkumar, S.; Jena, P.P.; Kumar, J.; Baksh, S.; Reddy, J.N. Identification of new alleles in salt tolerant rice germplasm lines through phenotypic and genotypic screening. *Int. J. Agric. Biol.* **2016**, *18*, 441–448. [CrossRef]
44. Yuan, L.; Ali, K.; Zhang, L.Q. Effects of NaCl Stress on active oxygen metabolism and membrane stability in *Pistacia vera* seedlings. *Acta Phytoecol. Sinica* **2005**, *29*, 985–991.
45. Ma, H.X.; Meng, C.M.; Zhang, K.X.; Wang, K.Y.; Fan, H.; Li, Y.B. Study on physiological mechanism of using cottonseed meal to improve salt–alkali tolerance of cotton. *J. Plant Growth Regul.* **2020**, *40*, 126–136. [CrossRef]
46. Ahmad, P.; Latef, A.A.A.; Allah, A.E.F.; Hashem, A.; Sarwat, M.; Anjum, N.A.; Gucel, S. Calcium and potassium supplementation enhanced growth, osmolyte secondary metabolite production, and enzymatic antioxidant machinery in cadmium-exposed chickpea (*Cicer arietinum* L.). *Front Plant Sci.* **2016**, *7*, 513. [CrossRef] [PubMed]
47. Li, N.; Cao, B.L.; Chen, Z.J.; Xu, K. Root morphology ion absorption and antioxidative defense system of two Chinese cabbage cultivars (*Brassica rapa* L.) reveal the different adaptation mechanisms to salt and alkali stress. *Protoplasma* **2021**, *259*, 1–14. [CrossRef] [PubMed]
48. Hannachi, S.; Van Labeke, M.-C. Salt stress affects germination, seedling growth and physiological responses differentially in eggplant cultivars (*Solanum melongena* L.). *Sci. Hortic.* **2018**, *228*, 56–65. [CrossRef]
49. Ahanger, M.A.; Qin, C.; Qi, M.D.; Dong, X.X.; Ahmad, P.; Abd Allah, E.F.; Zhang, L. Spermine application alleviates salinity induced growth and photosynthetic inhibition in *Solanum lycopersicum* by modulating osmolyte and secondary metabolite accumulation and differentially regulating antioxidant metabolism. *Plant Physiol. Biochem.* **2019**, *144*, 1–13. [CrossRef]
50. Pan, R.Z. *Plant Physiology*, 4th ed.; Higher Education Press: Beijing, China, 2001; pp. 279–293.
51. Zhao, F.G.; Liu, Y.L.; Zhang, W.H. Proline metabolism in the leaves of barley seedlings and its relation to salt tolerance. *J. Nanjing Agric. Univ.* **2002**, *25*, 7–10.

52. Chen, Y.H.; Yan, Z.L.; Li, Y.H. Study on the characteristic of proline accumulation and active oxygen metabolism in *Rhizophora stylosa* under salt stress. *J. Xiamen Univ. (Nat. Sci. Edi.)* **2004**, *43*, 402–405.
53. Zhao, H.G.; Liu, Y.L. Advances in study on metabolism and regulation of proine in higher plants under stress. *Chin. Bull. Bot.* **1999**, *16*, 540–546.
54. Liu, E.E.; Zong, H.; Guo, Z.F. Effects of drought, salt and chilling stresses on proline accumulation shoot of rice seedlings. *J. Trop. Subtrop. Bot.* **2000**, *8*, 235–238.
55. Yang, H.B.; Han, Z.H.; Xu, X.F. Effects of NaCl and iso-osmotic polyethylene glycol on free proline content of *Malus*. *Plant Physiol. Commun.* **2005**, *41*, 157–162.
56. Dogan, M.; Tipirdamaz, R.U.K.I.Y.E.; Demir, Y. Salt resistance of tomato species grown in sand culture. *Plant Soil Environ.* **2010**, *56*, 499–507. [CrossRef]
57. Parvin, K.; Hasanuzzaman, M.; Bhuyan, M.H.; Nahar, K.; Mohsin, S.M.; Fujita, M. Comparative physiological and biochemical changes in tomato (*Solanum lycopersicum* L.) under salt stress and recovery: Role of antioxidant defense and glyoxalase systems. *Antioxidants* **2019**, *8*, 350. [CrossRef]
58. Wang, Y.; Nii, N. Changes in chlorophyll, ribulose bisphosphate carboxylase-oxygenase, glycine betaine content, photosynthesis and transpiration in *Amaranthus tricolor* leaves during salt stress. *J. Hortic. Sci. Biotechnol.* **2000**, *75*, 623–627. [CrossRef]
59. Gou, W.; Zheng, P.; Zheng, P.; Wang, K.; Zhang, L.; Akram, N.A. Salinity-induced callus browning and re-differentiation, root formation by plantlets and anatomical structures of plantlet leaves in two *Malus* species. *Pak. J. Bot.* **2016**, *48*, 1393–1398.

life

Article

Melatonin in Micro-Tom Tomato: Improved Drought Tolerance via the Regulation of the Photosynthetic Apparatus, Membrane Stability, Osmoprotectants, and Root System

Naveed Mushtaq [1,2], Shahid Iqbal [2], Faisal Hayat [3], Abdul Raziq [4], Asma Ayaz [5] and Wajid Zaman [6,*]

1 Jiangsu Key Laboratory for Horticultural Crop Genetic Improvement, Nanjing 210014, China
2 College of Horticulture, Nanjing Agricultural University, Nanjing 210095, China
3 College of Horticulture, Zhongkai University of Agriculture and Engineering, Guangzhou 510225, China
4 Institute of Crop Germplasm Resources (Institute of Biotechnology), Shandong Academy of Agricultural Sciences, Shandong Provincial Key Laboratory of Crop Genetic Improvement, Ecology and Physiology, Jinan 250100, China
5 State Key Laboratory of Biocatalysis and Enzyme Engineering, School of Life Sciences, Hubei University, Wuhan 430062, China
6 Department of Life Sciences, Yeungnam University, Gyeongsan 38541, Republic of Korea
* Correspondence: wajidzaman@yu.ac.kr

Abstract: Environmental variations caused by global climate change significantly affect plant yield and productivity. Because water scarcity is one of the most significant risks to agriculture's future, improving the performance of plants to cope with water stress is critical. Our research scrutinized the impact of melatonin application on the photosynthetic machinery, photosynthetic physiology, root system, osmoprotectant accumulation, and oxidative stress in tomato plants during drought. The results showed that melatonin-treated tomato plants had remarkably higher water levels, gas exchange activities, root system morphological parameters (average diameter, root activity, root forks, projected area, root crossings, root volume, root surface area, root length, root tips, and root numbers), osmoprotectant (proline, trehalose, fructose, sucrose, and GB) accumulation, and transcript levels of the photosynthetic genes SlPsb28, SlPetF, SlPsbP, SlPsbQ, SlPetE, and SlPsbW. In addition, melatonin effectively maintained the plants' photosynthetic physiology. Moreover, melatonin treatment maintained the soluble protein content and antioxidant capacity during drought. Melatonin application also resulted in membrane stability, evidenced by less electrolyte leakage and lower H_2O_2, MDA, and O_2^- levels in the drought-stress environment. Additionally, melatonin application enhanced the antioxidant defense enzymes and antioxidant-stress-resistance-related gene (SlCAT1, SlAPX, SlGR, SlDHAR, SlPOD, and SOD) transcript levels in plants. These outcomes imply that the impacts of melatonin treatment on improving drought resistance could be ascribed to the mitigation of photosynthetic function inhibition, the enhancement of the water status, and the alleviation of oxidative stress in tomato plants. Our study findings reveal new and incredible aspects of the response of melatonin-treated tomato plants to drought stress and provide a list of candidate targets for increasing plant tolerance to the drought-stress environment.

Keywords: abiotic stress; drought; tomato; reactive oxygen species; oxidative stress; melatonin; photosynthesis; climate changes; antioxidant system

Citation: Mushtaq, N.; Iqbal, S.; Hayat, F.; Raziq, A.; Ayaz, A.; Zaman, W. Melatonin in Micro-Tom Tomato: Improved Drought Tolerance via the Regulation of the Photosynthetic Apparatus, Membrane Stability, Osmoprotectants, and Root System. *Life* **2022**, *12*, 1922. https://doi.org/10.3390/life12111922

Academic Editor: Kousuke Hanada

Received: 4 November 2022
Accepted: 16 November 2022
Published: 18 November 2022

Publisher's Note: MDPI stays neutral with regard to jurisdictional claims in published maps and institutional affiliations.

Copyright: © 2022 by the authors. Licensee MDPI, Basel, Switzerland. This article is an open access article distributed under the terms and conditions of the Creative Commons Attribution (CC BY) license (https://creativecommons.org/licenses/by/4.0/).

1. Introduction

Due to variations in the global climate, the drought frequency, intensity, and interval are currently rising and have reached alarming levels [1]. Drought stress is the most important element among all environmental aspects linked to the predicted consequences of climate change on worldwide plant production [2]. Drought stress is caused by below-average rain along with warmer climates, resulting in extensive plant damage and reduced yield. Though a lack of water is the direct source of drought stress, increased evapotranspiration due to a warming environment is assumed to be the primary reason for severe drying due to global climate change [3]. Water scarcity hinders plant osmotic regulation, resulting in turgor loss, decreasing cell division, damaging membranes, disrupting the energy balance, and lowering plant growth overall [4]. Drought stress also exacerbates leaf aging and, even for a short duration, causes critical yearly yield losses [5].

Melatonin is a low-weight compound that has been found in most living organisms. Initially, melatonin was recognized in the conarium, which regulates the circadian clock, immune system, behavior, fertility, and the activities of antioxidants [6]. Melatonin may also serve an essential primary function in plant development (germination, vegetative phase, flowering, and delayed aging) and diverse stress responses as a naturally occurring antioxidant [7]. Melatonin has also enhanced resistance to various stresses, such as dehydration, salt, heavy metals, and cold and ambient temperatures [8]. Melatonin's effect on plant stress tolerance is due to its ability to improve the plant antioxidant defense system [9]. Drought stress leads to a decline in plant production mainly due to the impairment of photosynthetic efficiency [10]. Likewise, a decline in photosynthesis activity initiated due to drought stress can trigger a drop in the reaction center in the plant's PSII [11]. These changes can severely impair the photosynthetic apparatus if the plant cannot remove the additional energy [12]. Certainly, plants' captured energy could be used to initiate the photosynthesis process. Secondly, it could utilize the chlorophyll fluorescence process as a result of re-emitting. Thirdly, this additional energy could be dispersed in the form of heat from the plants. These functions happen in a competitive manner; thus, a relative rise in the productivity of one process will decrease the efficiency of the remaining two processes in the plant [13]. Consequently, the plant's capacity to disperse the extra energy can be analyzed by determining the chlorophyll fluorescence, especially under stress.

Plant root systems are important for adapting and surviving in a drought-stress environment [14]. It was previously revealed that melatonin treatment improved the root system in Arabidopsis and maize plants [15,16]. Drought also causes a rapid increase in ROS, inducing photo-inhibition, membrane injury, and oxidative stress damage in tomato plants [10]. In drought, excessive ROS accumulation can decrease the activity of chloroplasts, decrease photochemical efficiency, and ultimately decrease the photosynthesis and growth of tomato plants [11]. Plants possess antioxidant resistance mechanisms that include enzymes, such as SOD, APX, GR, POD, DHAR, and CAT, and non-enzymatic components, such as glutathione and ascorbate, all of which function together to counterbalance ROS and protect cells from oxidative injury [17]. In addition to these antioxidant functions, the tomato plant's tolerance to drought stress involves the accumulation of osmoprotectants such as proline, which help to maintain its osmotic balance. In this manner, the proline level can help to reduce water potential by allowing water movement into the cell interior, reducing the injury caused by excessive ions. Proline accumulation is triggered under water stress, retains higher water in cells, and protects against protein impairment [18]. Similarly, the induction of different osmoprotectants, such as GB and trehalose, is associated with drought stress resistance in plants. Trehalose and GB significantly maintain the membrane stability and water level and enhance the ROS scavenging process in plants, helping them survive under drought stress [19].

Climate change has quickly evolved into a climate concern, with unanticipated impacts on agricultural yield. Because water scarcity is one of the biggest concerns for plants and, importantly, the world population's future, assessing and discovering the ability of plants to grow with a limited water supply is critical [3]. Tomato (*Solanum lycopersicum*) is one of the most important vegetables grown globally and often encounters lower water availability because of climate change. In the current era, drought is the prime hindrance to tomato production in various areas around the world [20]. Improving drought resistance and maintaining tomato plants under water-limited conditions is a high-priority research focus [2]. However, several studies have revealed that the application of melatonin could improve drought tolerance in some plant species [7,21]. There is certainly no research focusing on the melatonin effects governing drought tolerance in Micro-Tom plants through osmoprotectant accumulation, membrane stability, and the protection of the photosynthetic apparatus. Furthermore, our research also determined the primary function of melatonin application in photosynthetic apparatus protection by employing the advancing field of plant phenotyping along with photosynthetic gene expression. We also explored the important role of melatonin application by analyzing the root system, ROS accumulation, water level, membrane damage, osmoprotectant genes, soluble proteins, polyphenol oxidase, the activation of antioxidants, and antioxidant capacity. This research provides details concerning melatonin's role in enhancing tomato drought tolerance.

2. Materials and Methods

Micro-Tom tomato seeds (Jiangsu Academy of Agricultural Sciences, China) were sown in vermiculate after sterilizing and washing with distilled water. After germination, the tomato plants were placed in a growth chamber under 600 μmol m^{-2} s^{-1} light, with a relative humidity of 70%, a temperature of 24 °C, and a 15 h day/9 h night photoperiod. After 12 days, the tomato plants were moved to pots (24 cm × 14 cm) and transferred to the greenhouse. The plants were watered daily with Hoagland's nutrient solution with a pH of 5.8. After 24 days, 96 tomato plants were separated into two groups. (1) Half (48) of the tomato plants were treated with 100 μM melatonin solution, and (2) the other 48 plants were treated with water only. The melatonin application was stopped after ten days. Next, four treatments were applied to the tomato plants in our study: (1) plants well-watered during the entire duration of the experiment, (2) well-watered plants pretreated with 100 μM melatonin application, (3) drought-stress-treated tomato plants that received the full water requirement for 12 days, followed by 10 days of water withholding without melatonin application, and (4) 10 days of water withholding in plants pretreated with 100 μM melatonin application (80 mL per tomato plant). The melatonin pretreatment was applied six times, with a 2-day interval, followed by up to 10 days of water withholding. Each stress condition included three replicates, using eight tomato plants per replicate. All stress treatments lasted for ten days; as in pilot experiments, tomato plants showed wilting on the ninth day of drought stress. The leaf samples were collected on the 10th day to conduct various analyses, including qPCR and biochemical and physiological examinations.

2.1. Total Antioxidant Capacity

The total antioxidant capacity (TAC) was estimated in tomato leaves using an analytical system (BioQuoChem). It employs a photochemiluminescence (PCL) approach, which enables the quantification of the antioxidant status and water-soluble antioxidant capacity (ACW). Extracts for estimating the TAC were made following a prior method, and quantification was performed as instructed by the protocol in [22].

2.2. Polyphenol Peroxidase, Protein, and ROS Fluorescence

The polyphenol peroxidase (PPO) activity was measured by dissolving the 2 mL crude sample in 4 mL of caffeic acid solution and 2 mL of potassium dihydrogen phosphate (KH_2PO_4) buffer. The samples were placed in the incubator at 32 °C for 15 min. In the final step, the PPO absorbance was measured with a spectrophotometer at 380 nm. In tomato leaves, soluble protein quantification was performed via the method described by Bradford [23]. For this method, the enzyme sample was mixed with 0.9 mL of water and 6 mL of brilliant blue G–250 solutions. The samples were placed on a shaker for 1 min, and the final absorbance was measured at 595 nm.

Tomato leaves were initially soaked in a PBS solution (0.02 mM) for at least 10–14 min. H_2DCFDA (15 um) was added after removing the PBS solution. Then, the leaves were vacuumed for 20 min and placed on slides to estimate the ROS fluorescence under a confocal microscope (Germany).

2.3. Endogenous Melatonin Content

In tomato plants, melatonin was estimated with a procedure illustrated in a previous study [24]. Two-gram leaf samples were crushed in liquid nitrogen, and 3 mL of acetone was added. Next, the tubes were placed into a shaker for 35 min at 28 °C. The tubes were centrifuged at nearly 3500× g (TGL-19 Benchtop High-Speed Multi-Functional Centrifuge, Noki, Zhengzhou, China) at 4 °C. After centrifugation, 3 mL of water was added to the tubes, and the samples were set up for measuring melatonin via high-performance liquid chromatography (HPLC, 1290 LC, Agilent, Santa Clara, CA, USA) equipped with a mass spectrometer (6470 LC-MS/MS, Agilent, Santa Clara, CA, USA).

2.4. H_2O_2, MDA, and O^{2-}

The levels of H_2O_2 were measured in tomato plants with a method explained in previous studies [25]. For this method, the level of peroxide in the solution was estimated by matching the absorbance to the standard curve. To estimate the MDA level, leaf samples were mixed with a PBS solution and placed in a water bath at 90 °C [21]. After that, the samples were cooled down at room temperature and centrifuged at 9000× g for about 6 min. After centrifugation, the supernatant was dissolved in TBA solution and vortexed for 2 min. The final reading of the solution was measured at 530 nm to estimate the MDA level.

The $O_2{}^-$ level in the tomato plant was calculated via the consequent protocol described by prior research [26]. Leaf samples (0.4 g) were dissolved in 5 mL of PBS buffer in the first step. Then, the samples were moved to ice for 18 min and centrifuged at 15,000× g for about 20 min. A total of 0.4 mL of the resultant solution was mixed with 0.4 mL of Na-PB solution, and then $NH_2OH.HCl$ solution was added to the tubes. The samples were then transferred for incubation at 28 °C for 40 min.

2.5. Osmoprotectants and Enzymatic Activities

Proline and GB contents were measured in tomato leaves according to the methodology described by previous studies [27]. Glucose, sucrose, and fructose contents were determined by following a previously described technique [28]. The enzymatic process was followed to measure the starch contents, as defined by earlier research [29]. Trehalose content was calculated in tomato leaves by employing the technique reported by prior research [30]. The proline enzyme SlP5CS and SlP5CR activities were analyzed by following the method illustrated by Rivero [31]. SlP5CS and SlP5CR activities were quantified at 630 nm [32]. SlBADH, SlSUS3, SlSPS, and SlT6PS activities were estimated via a previously illustrated approach [27].

2.6. Antioxidant Enzyme Activities

In tomato plants, the extraction of enzymes was performed through the procedure illustrated by [33]. After collecting leaf samples, the tissue was homogenized in cold PBS liquid. The samples were then centrifuged for 13 min at 7000× g and 5 °C. Furthermore, the enzyme activity was determined right away using the supernatant in the tubes.

The catalase (CAT) activity was finally measured by observing the reduction in absorbance of about 240 nm in the spectrophotometer [33]. The sample tubes contained 40µL of PBS liquid (pH 7.0) and 2 mL of 0.3% H_2O_2. Ninety microliters of enzyme extract was used in the tubes to start the process. The SOD analysis was performed using the protocol of previous studies [14]. First, 0.8 mL of plant extract was added to the tubes. Next, 2 mL of Na_2CO_3, 0.6 mL of NBT, and 0.2 mL of EDTA were added to the tubes. Then, 0.3 mL of $NH_2OH{\cdot}HCl$ was added to initiate the reaction in the tubes. In the final step, the reading was measured at 555 nm for about 2 min. The activity of APX was determined using a previously described method [34]. This protocol used a spectrophotometer to analyze the decrease in absorbance at 280 nm, which occurred during 5 min of the ascorbate oxidation process. The sample tubes contained 48 mM PBS liquid, 0.52 mM ascorbic acid, and 98 µL of crude enzyme. The reaction immediately started after the addition of 0.13 mM H_2O_2.

To analyze the DHAR activity in tomato plants, 0.12 mM EDTA, 0.18 mM DHA, 48 mM HEPES, and 2.6 mM GSH buffer were added to 19 μL of leaf extract. The extinction coefficient of 14 $mM^{-1}.cm^{-1}$ was used to analyze activity by determining the increase in the reaction at about 260 nm in the spectrophotometer [22]. The GR level was determined in tomato plants through the oxidation of NADPH according to a previously illustrated method [35]. In this process, the sample tubes contained 0.6 mM EDTA, 0.4 mM GSSG, 0.26 Mm NADPH, 48 mM HEPES buffer, and 98 μL of leaf extract. The reaction process was initiated by adding NADPH to the sample tubes. In the final step, the GR level was measured at 330 nm using the spectrophotometer. The POD level was analyzed in tomato leaves according to the method used in previous investigations [36]. First, the leaves were digested in 4 mL of PBS liquid. The tubes were centrifuged for 8 min at 25,000× *g*, and this liquid was used as the enzyme extract to measure the POD level. Next, 2 mM H_2O_2, 2.5 mL of PBS liquid, 8 mM guaiacol, and 48 mL of enzyme extract were added to the tubes. Finally, the POD level was monitored at 460 nm.

2.7. *Chlorophyll Fluorescence and Gas Exchange Parameters*

The following chlorophyll fluorescence parameters of tomato leaves were analyzed with the chlorophyll fluorescence apparatus ((Imaging PAM). This instrument measured the maximum quantum yield of PSII (Fv/Fm), effective PSII quantum yield (PSII), non-photochemical quenching (NPQ), and the electron transport rate using the Imaging Win application. The tomato leaves were dark-acclimated for 40 min before estimating the above parameters [10]. The photosynthesis-related parameters were analyzed in all tomato plants after ten days of treatment. These analyses were performed on the fully opened 3rd leaf about 4 h after the beginning of the light treatment [10]. The parameters include the photosynthesis activity rate, transpiration, stomatal conductance, and CO_2 assimilation with the use of a portable photosynthesis system (LI-COR Biosciences, Lincoln, NE, USA).

2.8. *Electrolyte Leakage and RWC*

Electrolyte leakage was measured to determine membrane stability using a previous technique [2]. Tomato leaf samples (about 110 mg) from different plants were obtained for every treatment. Next, the leaf samples were washed with distilled water for about 10 min. The leaves were then incubated at ambient temperature for 28 h in test tubes with 20 mL of water. In these leaf tissues, the initial conductivity (C1) was determined. After that, the leaf samples were boiled for 90 min at the boiling temperature and placed at 25 °C to cool down. The conductivity of these samples was determined as C2. Finally, electrolyte leakage was analyzed as the percentage proportion of C1–C2 in tomato leaves.

To calculate the fresh weight, fully extended leaves were taken after the stress treatments [27]. These leaves were then held for 14 h at a 23 °C temperature in water to determine the turgid weight (TW). After that, the leaves were placed at 85 °C for two days, and the dry weight (DW) was determined. Eventually, the relative water content (RWC) was calculated using the following method:

$$RWC = FW - DW/TW - DW \times 100$$

2.9. *The Root Activity, Morphology, and Fresh and Dry Weights*

After 10 days of drought-stress treatment, three plant roots (per treatment) were taken, washed with water, and scanned via a root scanner to analyze the root parameters. The root activity was measured by following the triphenyl tetrazolium chloride method explained in previous research [37]. In a solution of 10 mL of PBS puffer and TTC (0.45%), the root samples (600 mg) were immersed. Afterwards, these samples were moved to a dark place for 4 h at 35 °C, and 2.2 mL of H_2SO_4 was added to the solution. Lastly, the roots were blended and moved to a solution of 12 mL of ethyl acetate; the solution absorbance was estimated at 485 nm. After the drought-stress phase, the root samples were also separated to estimate the fresh and dried weights. First, the root fresh weight was measured, and the roots were moved to different paper bags and placed in the oven to remove the moisture. The root dry weight was determined after drying samples at 100 °C for 20 min and 80 °C for two days.

2.10. RNA Extraction and qPCR Analysis

After 10 days of treatment, the second leaves of the plants were utilized for RNA extraction. For this purpose, the RNeasy Plant Mini Kit was used following the protocol guidelines. Furthermore, cDNAs were produced with the PrimeScript™ RT Reagent Kit (TaKaRa) as per the company's directions. The primers used in this research are provided in Supplemental Table S1. For the qPCR analysis, the instructions of the SYBR Premix Ex Taq™ were followed, and the samples were placed in the Bio-Rad CFX-9 system. The transcript levels of genes were estimated by following the procedure of previous studies [27].

2.11. Statistical Analysis

All research parameter differences were distinguished by applying the ANOVA and LSD analysis program in the SPSS software (version 25.0). The research results are shown as the mean ± S.D. The heml program was utilized to create heat maps.

3. Results

To determine how the water status was affected in drought-stress environments, RWC was analyzed in plants. RWC revealed that melatonin-treated plants showed similar values to those found in untreated plants under well-watered conditions. RWC decreased in the drought-stressed plants compared to the control plants. Additionally, after 10 days of drought stress, the melatonin treatment maintained an elevated RWC compared to the untreated plants (Figure 1A). Furthermore, electrolyte leakage was also determined in tomato plants to evaluate the effect of drought stress on membrane stability. The melatonin-treated plants had the same electrolyte leakage values as the non-treated plants in a controlled environment. A significant increase in electrolyte leakage in the drought-stressed plants was observed compared to the control plants. Plants treated with melatonin revealed lower values of electrolyte leakage in comparison to the untreated plants in the drought environment (Figure 1B). These consequences suggest that the melatonin application supported membrane stability as an adaptation approach, which plays a primary role in increasing melatonin-treated plants' drought resistance by retaining relatively high water content.

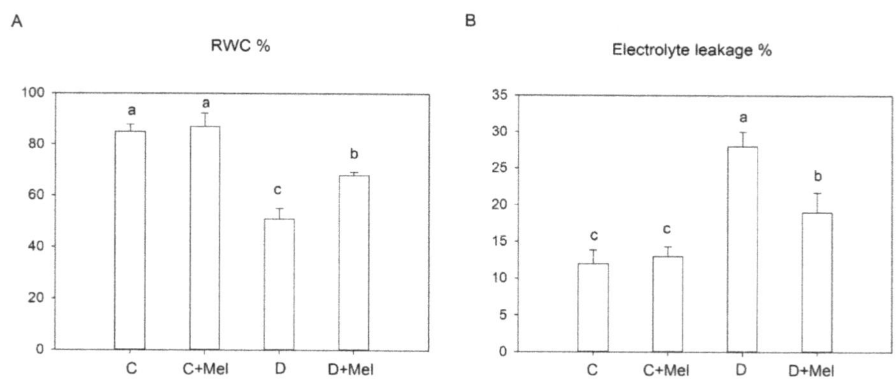

Figure 1. The physiological parameters relative water content (**A**) and electrolyte leakage (**B**) in tomato plant leaves under normal or 10-day drought-stress conditions with or without melatonin application. C: control; C+Mel: control with 100 μM melatonin pretreatment; D: drought, 10 days of withholding water; D+Mel: drought with 100 μM melatonin pretreatment. The values are the average of six replicates ± S.D. (n = 6). Significant differences among different treatments in the experiment were determined by LSD 0.05 test and are indicated by different letters.

Drought stress enhances the production of ROS, which impair the plant's function and membrane stability. The fluorescence determination of ROS accumulation in leaf discs of tomato plants after ten days of drought stress revealed that ROS levels were lower in melatonin-treated plants (Figure 2A). Melatonin is highly effective for ROS removal and scavenging under environmental stresses [22].

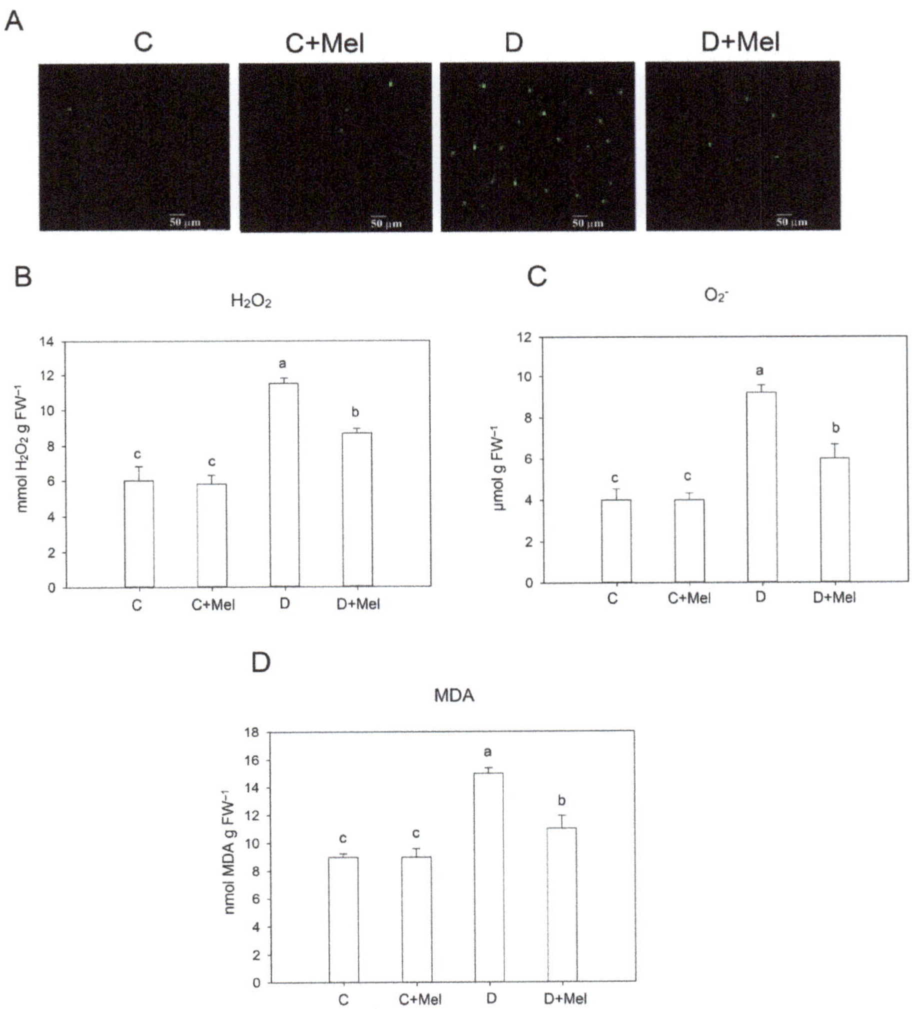

Figure 2. Impacts of melatonin application on the oxidative stress response in tomato plants. (**A**) ROS fluorescence and (**B**) H_2O_2, (**C**) O_2^-, and (**D**) MDA levels in tomato leaves after 10 days of drought stress. The green spots display the distribution of ROS. Bars, 50 µm. Data values are the means ± S.D. ($n = 6$). Significant differences among different treatments in the experiment were determined by LSD 0.05 test and are indicated by different letters.

H_2O_2 and O_2^- levels were analyzed in tomato plants exposed to a water deficit. Under well-watered conditions, the application of melatonin showed no considerable change in the tomato plants. Under dry conditions, considerable H_2O_2 and O_2^- accumulation occurred in the leaves; however, the melatonin-treated plants showed notably less H_2O_2

and O_2^- (ROS levels) (Figure 2B,C). Additionally, the malondialdehyde (MDA) level was analyzed in plants to determine the protective effect of exogenous melatonin on membrane stability during drought conditions. Drought impaired the integrity of the cell membrane, as evidenced by elevated MDA levels in drought-stressed tomato plants (Figure 2D). However, in drought treatments, melatonin application significantly reduced MDA levels in plants.

The gas exchange parameters were assessed to investigate the effects of melatonin application on photosynthetic machinery in tomato plants. Under the control condition, melatonin had no evident effect on the gas exchange parameters, including the photosynthetic rate (Pn), transpiration rate (tr), CO_2 assimilation (Ci), and stomatal conductance (St). The photosynthetic rate (Pn) and transpiration rate (Tr) decreased under the water-stress treatment compared to the control plants. The Ci and St parameters also showed the same decreasing trend under drought stress. We found that exogenous melatonin decreased these downtrends and improved the protective effect (Figure 3). These results confirm similar consequences, that is, that melatonin treatment protected the gas exchange parameters in plants under drought stress [9].

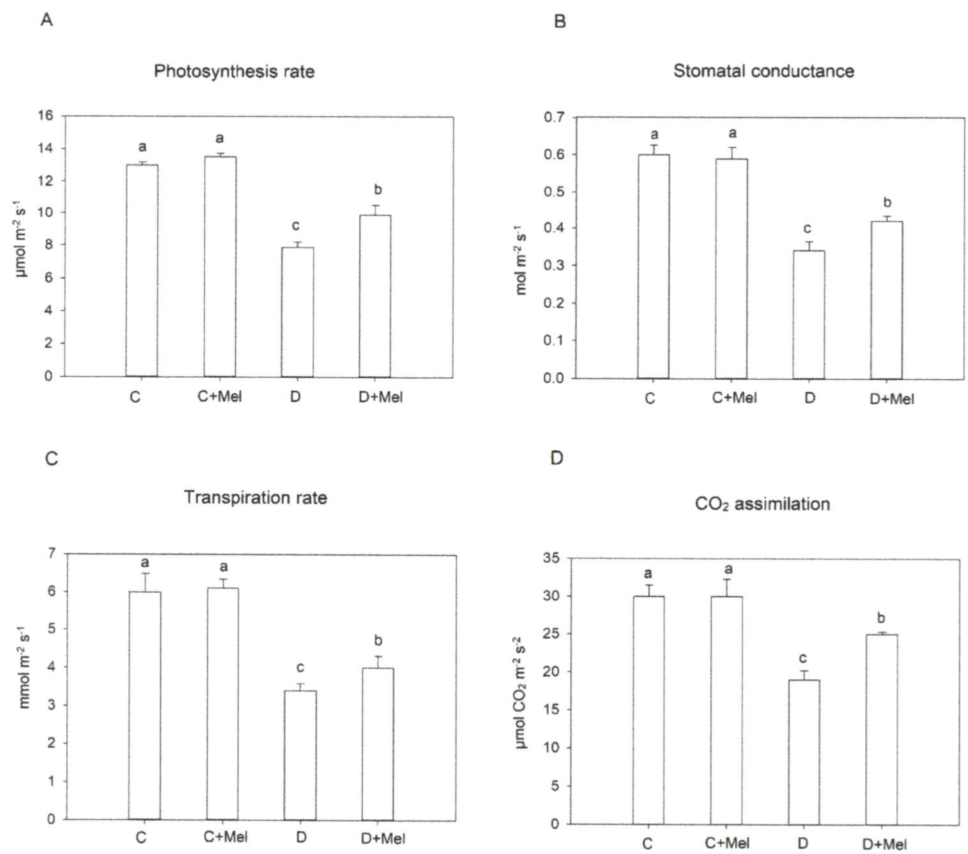

Figure 3. Melatonin application effect on tomato plants' photosynthetic parameters: (**A**) photosynthetic rate, (**B**) stomatal conductance, (**C**) transpiration rate, and (**D**) CO_2 assimilation rate after 10 days of drought stress. The values are the average of six replicates ± S.D. (n = 6). Significant differences among different treatments in the experiment were determined by LSD 0.05 test and are indicated by different letters.

Chlorophyll fluorescence is an effective technique for determining the photosynthetic physiology of plants. Fv/Fm demonstrates the probable photosynthetic capacity of tomato plants, which illustrates the portion of captured photons consuming the photochemistry in the leaves (Figure 4A,B). Fv/Fm was not significantly affected by the melatonin treatment under control conditions. Ten days of drought stress considerably reduced Fv/Fm in the drought-treated plants compared to melatonin-treated plants. At the start of the drought-stress treatment, the ETR reading was similar to that of the control plants. The ETR decreased considerably in the drought condition after 10 days (Figure 4C). However, NPQ increased in the tomato plants after 10 days of drought stress (Figure 4D). Interestingly, melatonin-treated plants maintained higher ETR and NPQ than drought-treated plants. The above outcomes suggest that melatonin application might be involved in protecting Micro-Tom tomato plants' photosynthetic physiology by maintaining normal PSII function and efficiency under drought stress.

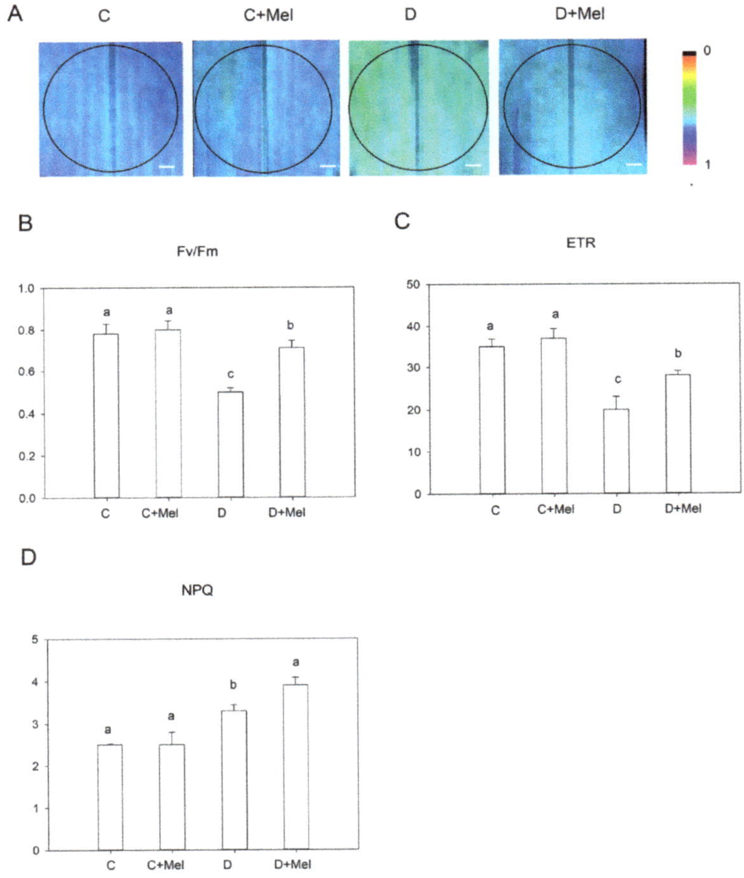

Figure 4. Chlorophyll fluorescence parameters in tomato plants after 10 days of drought stress. Fluorescence images of maximum PSII yield (Fv/Fm) (**A**), maximum PSII yield (Fv/Fm) values (**B**), electron transport rate (ETR) (**C**), and non-photochemical quenching (NPQ) (**D**). Scale bars represent 100 μm in the fluorescence images of maximum PSII yield. The values presented above are the average of six replicates ± S.D. (n = 6). Significant differences among different treatments in the experiment were determined by LSD 0.05 test and are indicated by different letters.

The effective efficiency of PSII (Fq/Fm) was analyzed to scrutinize the tomato plant's photosynthetic effectiveness under drought stress (Figure 5). The black color represents a lower value, and the magenta color illustrates a higher value of Fq/Fm (Figure 5A). Our results showed that drought stress decreased the Fq/Fm values in tomato plants. Conversely, melatonin treatment ameliorated the negative effect of drought, implying that exogenous melatonin could inhibit the effective efficiency of PSII. Compared to the control plants, the melatonin-treated plants displayed no observable differences in the Fq/Fm values under well-watered conditions.

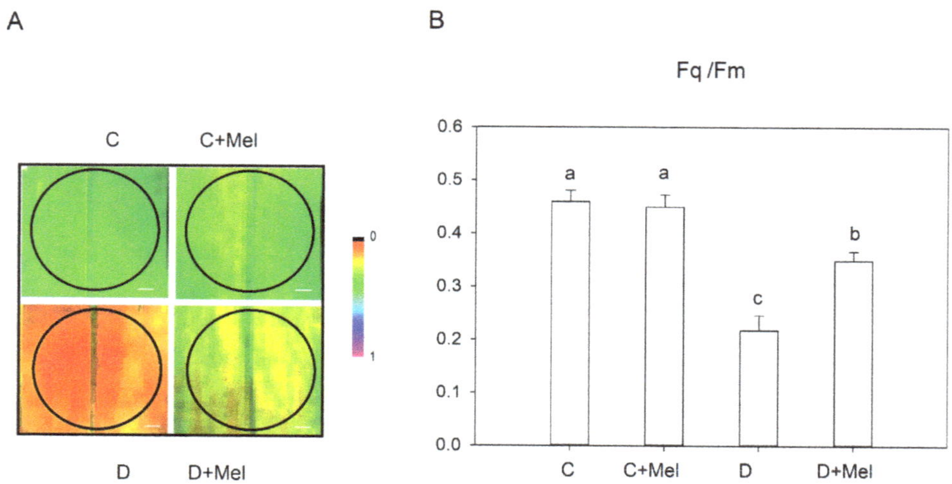

Figure 5. The effective efficiency of PSII (Fq/Fm) in tomato plants after 10 days of drought stress. Fluorescence images of (Fq/Fm) (**A**) and the values of (Fq/Fm) (**B**). Scale bars represent 100 μm in the effective efficiency of PSII. The values presented above are the average of six replicates ± S.D. ($n = 6$). Significant differences among different treatments in the experiment were determined by LSD 0.05 test and are indicated by different letters.

We examined the expression levels of several photosynthesis-related genes by qRT-PCR analysis during the stress treatment. Under control conditions, the expression of these genes was not different from that in the melatonin-treated plants. With the progression of water stress, the relative expression levels of photosynthetic genes were gradually downregulated. After 10 days of drought stress, the expression levels of SlPsb28, SlPetF, SlPsbP, SlPsbQ, SlPetE, and SlPsbW were significantly reduced compared to normal plants (Figure 6). We noticed that melatonin-treated plants exhibited high expression levels of these genes compared to drought-treated tomato plants. The results imply that melatonin application upregulates the expression of photosynthesis-related genes, thus retaining the PSII process in plants.

The above studies show that melatonin is indeed synthesized and found in tomato plant leaves, and its endogenous level depends upon the stress conditions and their duration. Endogenous melatonin levels were measured in the leaves to determine the result of 10 days of drought stress on melatonin production (Figure 7A). In the control plants, the leaf's melatonin content was about 19 pg/g^{-1} FW. The application of exogenous melatonin led to a significant difference in the melatonin level. Melatonin levels were significantly induced after plants were exposed to drought-stress conditions. Under drought stress, treatment with 100 μM melatonin increased endogenous melatonin levels by about 108 pg/g^{-1} FW. These findings reveal that melatonin is implicated in tomato plants' response to drought stress. Exogenous melatonin treatment might change endogenous melatonin levels during stress to alleviate drought stress.

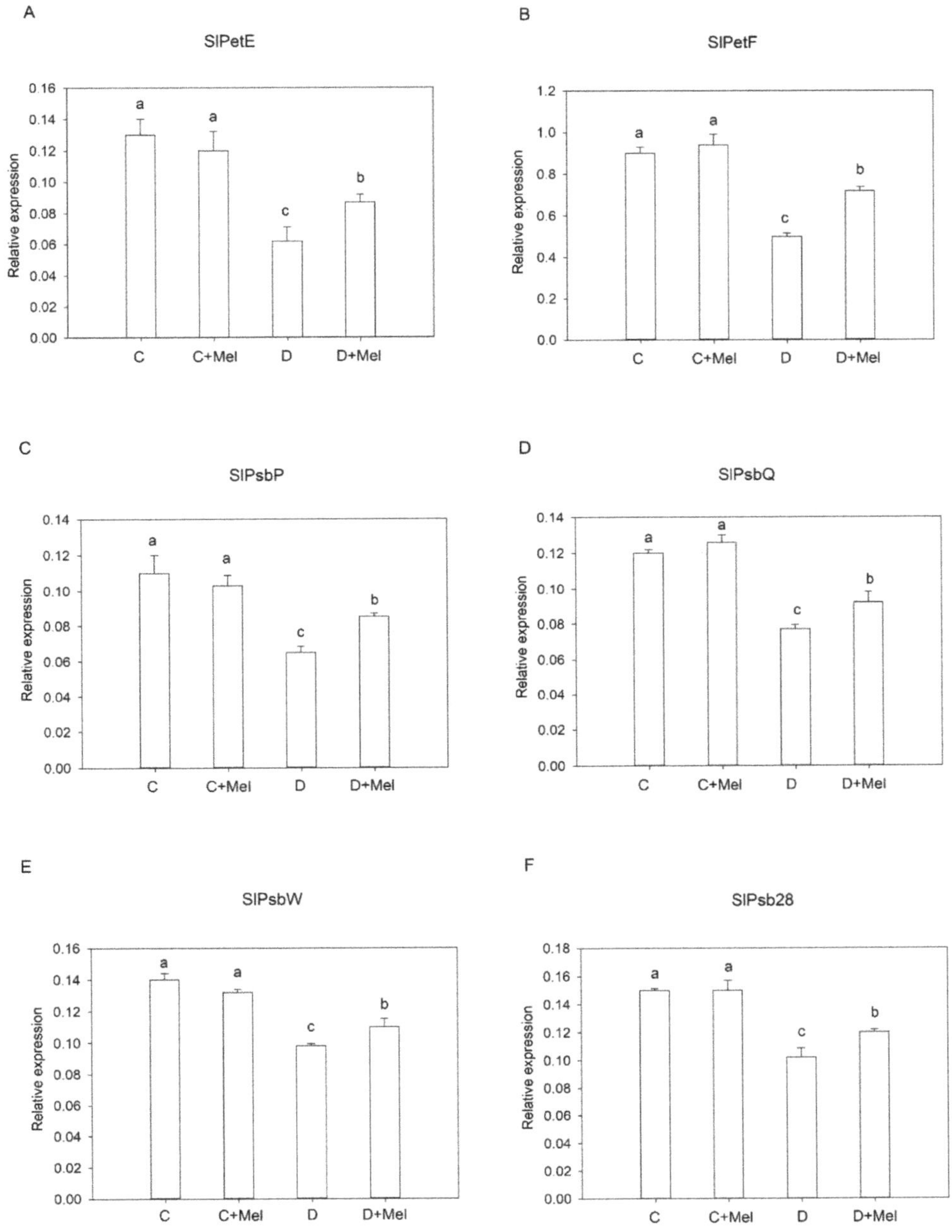

Figure 6. Effects of melatonin on the expression levels of photosynthetic-machinery-associated genes in tomato plants after drought stress. (**A**) SlPetE, (**B**) SlPetF, (**C**) SlPsbP, (**D**) SlPsbQ, (**E**) SlPsbW, and (**F**) SlPsb28 in melatonin-supplemented and non-supplemented plants under control or stress conditions. The values presented above are the average of six replicates ± S.D. (n = 6). Significant differences among different treatments in the experiment were determined by LSD 0.05 test and are indicated by different letters.

Table 2. List of plants found in Peloponnese and cited in *Prodromus*. First column: plant names alphabetically presented according to the name given in archives, which are quoted in *Prodromus*, but not referred in *FGS*. Second column: numerical register of volume and page, respectively, in *Prodromus*. Third column: current scientific name.

Plant Name Cited in *Prodromus*	Volume, Page	Scientific Name
Castanea sativa	2, 242	*Castanea sativa* Mill.
Corylus spp. (hazel)	2, 244	*Corylus avellana* L., *C. colurna* L.
Euphorbia apios	1, 326	*Euphorbia apios* L.
Ficus carica	2, 268	*Ficus carica* L.
Fraxinella	1, 271	*Dictamnus albus* L.
Globularia alypum	1, 78	*Globularia alypum* L.
Leontice altaica	1, 234	*Gymnospermium peloponnesiacum* (Phitos) Strid
Leontice chrysogonum	1, 234	*Bongardia chrysogonum* (L.) Spach
Leontice leontopetalum	1, 234	*Leontice leontopetalum* L.
Lolium	1, 70	*Lolium perenne* L., *L. subulatum* Vis., *L. temulentum* L.
Imperatoria	1, 199	*Imperatoria ostruthium* L.
Loranthus	1, 242	*Loranthus europaeus* Jacq.
Urtica	2, 233	*Urtica dioica* L., *U. pilulifera* L., *U. urens* L.
Quercus spp.	2, 239	*Quercus aegilops* L., *Q. coccifera* L., *Q. ilex* L., *Q. pubescens* Willd.
Pinus	2, 242	*Pinus pinea* L.
Rubus spp.	1, 349	*Rubus sanctus* Schreb., *R. canescens* DC.
Salicornia	1, 1	*Salicornia fruticosa* L., *S. perennans* Willd.
Satyrium	2, 215	*Satyrium* L., *Orchis* sp.
Scilla	1, 237	*Scilla nivalis* Boiss., *S. messeniaca* Boiss., *S. pneumonanthe* Speta
Viola	1, 145	*Viola scorpiuroides* Coss., *Viola graeca* (W. Becker) Halácsy
Nymphaea	1, 360–361	*Nymphaea alba* L.

4. Discussion

Professor John Sibthorp and his colleagues visited Greek territories twice in pre-industrial time, i.e., 1786–1787 and 1794–1795, and collected wild plants grown under natural conditions [7,16,34]. It was an outstanding achievement, considering the duration, the collections of specimens of plants from which "a legacy of 2462 pressed specimens are still preserved in the Sibthorpian Herbarium" [35] (Figure 2), and the geographical coverage, during the above-mentioned botanical expeditions. Moreover, a number of specimens found in Kew are of considerable importance as supplementing Sibthorp's collection at Oxford [36]; these specimens have been published [36] according to the sequence of plants cited in *Prodromus* [30].

Figure 2. Dried specimens of plants in the Sibthorpian herbarium at the University of Oxford, associated with the Flora Graeca expeditions and collected from the eastern Mediterranean in the 18th century. Courtesy of Stephen Harris, modified by Sophia Rhizopoulou.

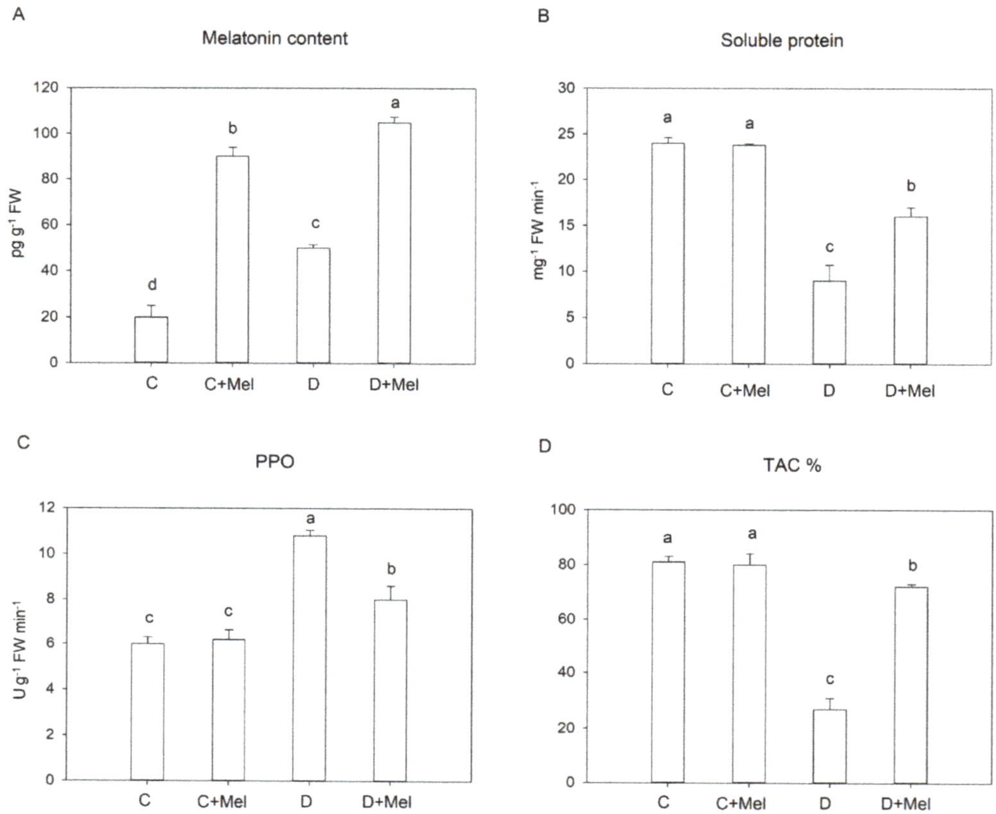

Figure 7. Changes in melatonin content (**A**), soluble proteins (**B**), PPO (**C**), and antioxidant capacity (**D**) in tomato leaves after melatonin treatment under 10 days of drought stress. The values presented above are the average of six replicates ± S.D. ($n = 6$). Significant differences among different treatments in the experiment were determined by LSD 0.05 test and are indicated by different letters.

The tomato plants' soluble protein contents were also measured under drought stress conditions. The soluble protein content was notably reduced in drought-stressed plants compared to well-watered plants. The melatonin application maintained the soluble protein content at higher levels than those in drought-treated plants (Figure 7B). Our outcome suggests that melatonin addition ameliorates the drought-stress effect through adjustments in PPO and soluble protein content in tomato plants. Drought stress effectively increased the activity of PPO in tomato plants, as perceived in Figure 7, compared to well-watered plants. The PPO activity decreased in melatonin-treated plants in comparison to drought-stressed plants. The melatonin-treated drought-stressed plants showed the most significant decrease in PPO activity compared to drought-treated plants (Figure 7C).

In this study, the total antioxidant capacity (TAC) was also analyzed in tomato plants to determine the effects of melatonin application and its role in antioxidant system activation during drought stress. As tomato plants grew under well-watered conditions, their TAC values were the same as those of melatonin-treated plants. The TAC values were significantly reduced in the drought-treated plants compared to control tomato plants. Importantly, plants treated with melatonin exhibited higher TAC activity in comparison to drought-stressed plants (Figure 7D). This study indicates that melatonin application might significantly activate the TAC in tomato plants to eliminate ROS during drought stress.

Root parameters, including average diameter (AD), root activity (RA), root forks (RF), projected area (PA), root crossings (RC), root volume (RV), root surface area (RSA), root length (RL), root tips (RT), root numbers (RN), fresh root weight (FRW), and dry root weight (DRW), were analyzed to determine the effect of drought stress on tomato plants (Table 1). The current analysis indicated that the drought environment significantly affected tomato plants' root-growth-related parameters. The results indicated that RA, RF, PA, RC, RSA, and RT were not improved by melatonin under control conditions. However, parameters such as RL, RN, FRW, RV, and DRW were significantly enhanced in melatonin-treated plants under control conditions. All root parameters were significantly decreased under drought-stress conditions, and, importantly, the application of melatonin alleviated these trends in melatonin-treated tomato plants.

Table 1. Effect of melatonin treatment on the root parameters of Micro-Tom plants under drought stress.

	C	C+Mel	D	D+Mel
RN	35 ± 0.05 [b]	40 ± 0.03 [a]	16 ± 0.01 [d]	26 ± 0.04 [c]
RL (cm)	38 ± 0.02 [b]	44 ± 0.07 [a]	21 ± 0.06 [d]	32 ± 0.03 [c]
RV (cm^3)	0.4 ± 0.001 [b]	0.49 ± 0.002 [a]	0.09 ± 0.003 [d]	0.25 ± 0.006 [c]
RSA (cm^2)	45 ± 0.3 [a]	46 ± 0.4 [a]	10 ± 0.1 [c]	22 ± 0.2 [b]
RC	181 ± 0.9 [a]	179 ± 0.6 [a]	42 ± 0.4 [c]	90 ± 0.8 [b]
RT	699 ± 6 [a]	696 ± 8 [a]	189 ± 3 [c]	380 ± 9 [b]
RF	598 ± 4 [a]	596 ± 9 [a]	150 ± 2 [c]	390 ± 5 [b]
AD (mm)	0.5 ± 0.002 [a]	0.47 ± 0.006 [a]	0.12 ± 0.009 [c]	0.35 ± 0.003 [b]
PA (cm^2)	13 ± 0.07 [a]	12 ± 0.02 [a]	4.2 ± 0.01 [c]	7 ± 0.04 [b]
RA (mg g^{-1} FW)	35 ± 0.05 [a]	37 ± 0.09 [a]	14 ± 0.02 [c]	25 ± 0.01 [b]
FRW (g)	2.7 ± 0.001 [b]	3.2 ± 0.007 [a]	0.5 ± 0.003 [d]	1.52 ± 0.004 [c]
DRW (g)	0.29 ± 0.002 [b]	0.36 ± 0.001 [a]	0.09 ± 0.06 [d]	0.18 ± 0.05 [c]

RN: root numbers; RL: root length; RSA: root surface area; RC: root crossings; RT: root tips; RF: root forks; AD: average diameter; PA: projected area; RA: root activity; FRW: fresh root weight; DRW: dry root weight. The values are the average of six replicates ± S.D (n = 6). Significant differences among different treatments in the experiment were determined by LSD 0.05 test and are indicated by different letters.

Plants develop antioxidant resistance mechanisms to remove accumulated ROS and protect the plants from damage. Consequently, the activation of the antioxidants CAT, SOD, APX, DHAR POD, and GR in tomato plants with and without the application of melatonin was evaluated. The melatonin application did not change the activities of these six antioxidants under control conditions in tomato plants. The antioxidants CAT, SOD, APX, DHAR POD, and GR were induced under drought stress. Melatonin application further enhanced the activities of the antioxidants CAT, SOD, APX, DHAR POD, and GR compared to drought-treated plants (Figure 8). Our study indicates that melatonin application enhances the activation of the antioxidant defense system to maintain ROS in tomato plants.

The expression of transcripts related to the antioxidant pathway was measured under the drought-stress conditions employed in our study to fill some gaps in the understanding of the explicit role of melatonin in the activation of some stress-resistance-related genes (SlCAT1, SlAPX, SlGR, SlDHAR1, SlPOD, and SOD) in a climate change environment (Figure 8). The use of melatonin under well-watered conditions resulted in the same expression of the transcripts in comparison to control tomato plants. Drought stress significantly induced the transcript levels of these genes compared to well-watered plants. Tomato plants treated with melatonin showed the highest expression of antioxidant genes compared to drought-stressed plants (Figure 9).

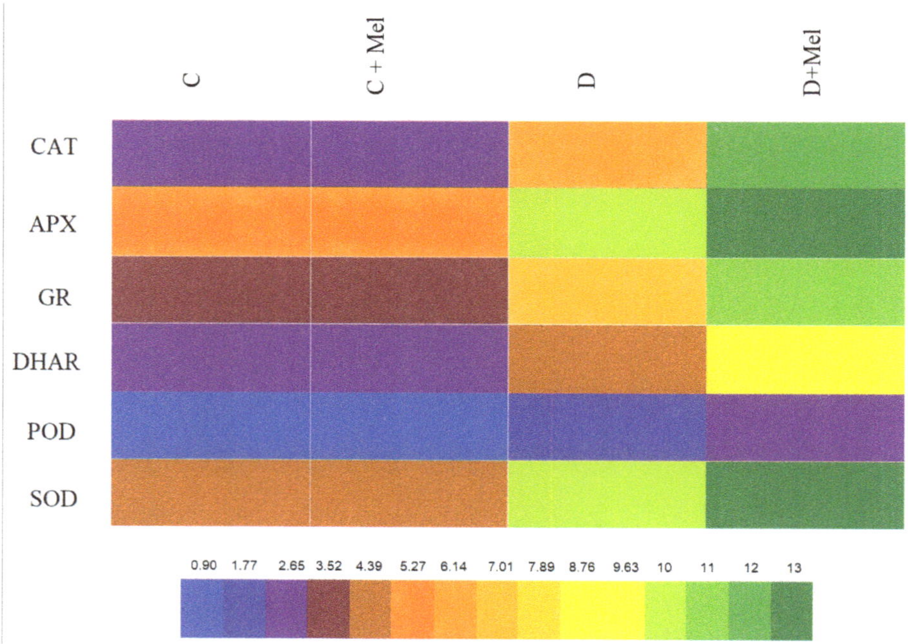

Figure 8. Effects of melatonin application on antioxidant defense enzyme activities. Catalase (CAT), ascorbate peroxidase (APX), glutathione reductase (GR), dehydroascorbate reductase (DHAR), peroxidase (POD), and superoxide dismutase (SOD) activities with or without melatonin in tomato leaves after 10 days of drought stress. The blue color illustrates lower values, and the green color illustrates higher values of antioxidant enzymes in the heat map. Significant differences among different treatments in the experiment were determined by LSD 0.05 test and are indicated by different letters.

To investigate the osmoprotectant accumulation response to drought stress, we analyzed the accumulation of proline, trehalose, GB, glucose, sucrose, fructose, and starch in melatonin-treated and non-treated plants (Figure 10). The color range shows the osmoprotectant values, from pink (minimum) to brown (maximum). No apparent osmoprotectant accumulation differences were found after melatonin application in normal water conditions. After 10 days of drought, the accumulation of proline, trehalose, GB, glucose, sucrose, and fructose was perceived in the plants compared to normal plants. On the other hand, drought-treated plants had reduced starch content. Notably, melatonin-treated plants presented a higher concentration of these osmoprotectants than drought-treated plants.

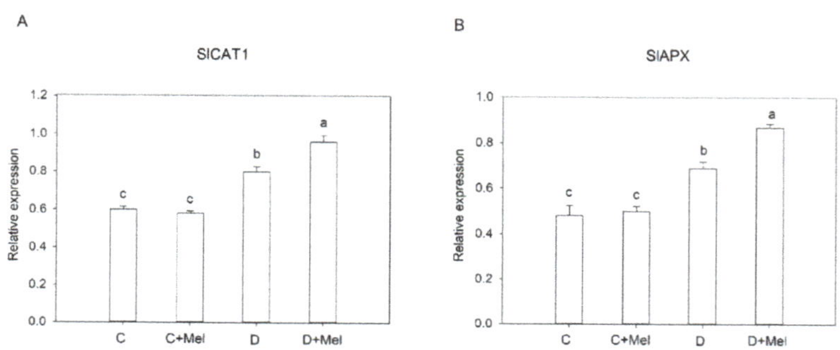

Figure 9. *Cont.*

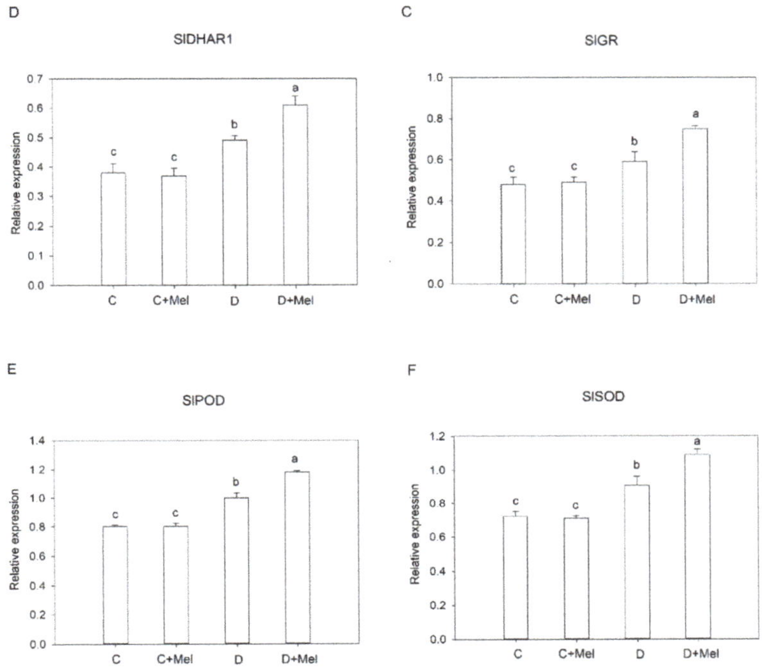

Figure 9. The expression levels of antioxidant-defense-related genes in tomato plants after 10 days of drought-stress treatment. SlCAT1 (**A**), SlAPX (**B**), SlGR (**C**), SlDHAR1 (**D**), SlPOD (**E**), and SOD (**F**) gene expression levels in tomato leaves with or without melatonin after 10 days of drought stress. The values presented above are the average of six replicates ± S.D. (n = 6). Significant differences among different treatments in the experiment were determined by LSD 0.05 test and are indicated by different letters.

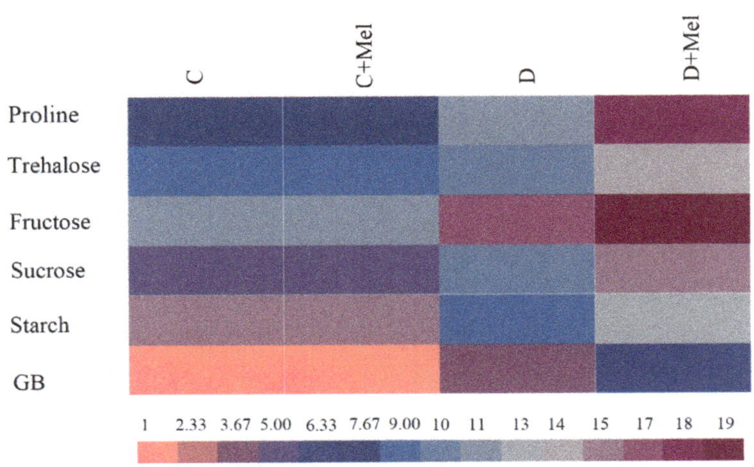

Figure 10. The activities of important osmoprotectants in tomato plants. Osmoprotectants (proline, trehalose, fructose, sucrose, starch, and GB) in melatonin-supplemented or non-supplemented tomato leaves after 10 days of drought stress. The pink color illustrates lower values, and the brown color illustrates higher values of osmoprotectants in the heat map. The values presented above are the average of six replicates ± S.D. (n = 6).

The transcript levels of osmoprotectant genes were analyzed to determine the role of the osmoprotectant biosynthesis pathway in coping with drought-stress environments (Figure 11). In control conditions, melatonin application did not affect these genes' transcript levels in tomato plants. Drought stress upregulated the expression of osmoprotectant genes compared to non-stressed tomato plants. Melatonin application further induced the expression of osmoprotectant genes such as SlP5CR, SlP5CS, SlSPS, SlBADH, SlT6PS, and SlSUS3 in tomato plants in the drought-stress phase.

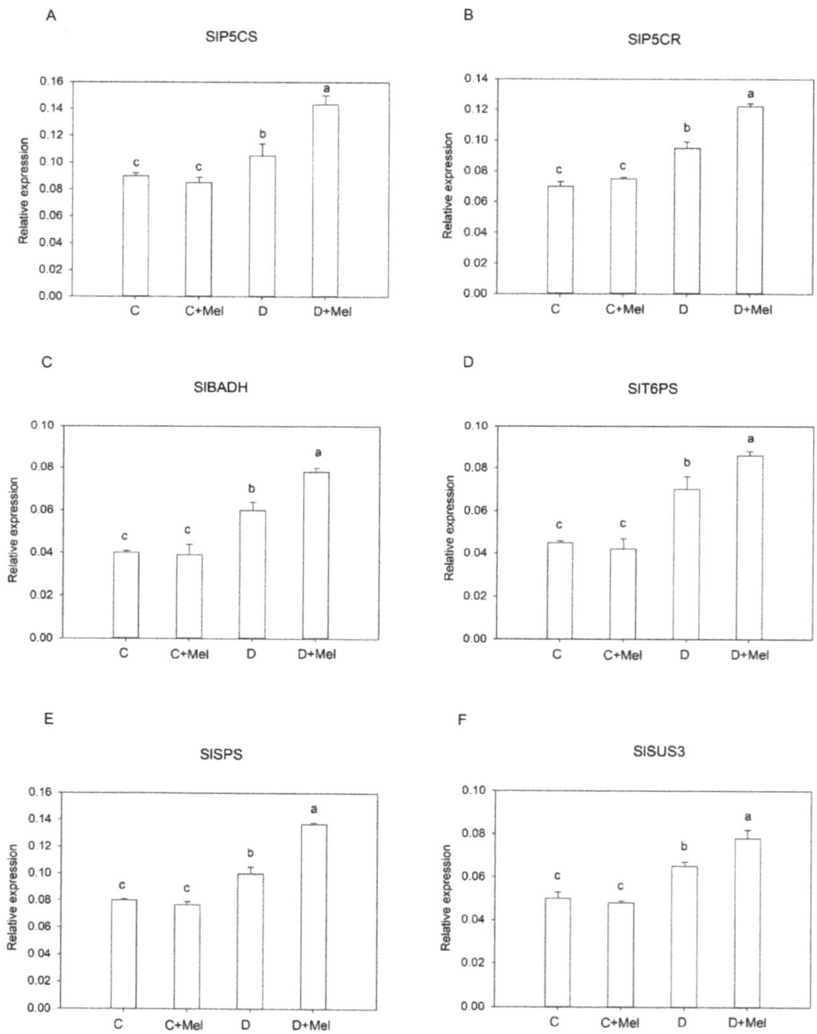

Figure 11. The expression levels of osmoprotectant genes in tomato plants. SlP5CS (**A**), SlP5CR (**B**), SlBADH (**C**), SlSPS (**D**), SlSUS3 (**E**), and SlT6PS (**F**) in tomato leaves with or without melatonin after 10 days of drought stress. The values presented above are the average of six replicates ± S.D. (n = 6). Significant differences among different treatments in the experiment were determined by LSD 0.05 test and are indicated by different letters.

The osmoprotectants' key enzyme activities were measured to further understand osmolyte accumulation in tomato plants. The color range shows the osmoprotectant values, from magenta (minimum) to brown (maximum). The accumulation of these enzymes was the same in melatonin-treated plants in well-watered conditions. The osmoprotectant enzymes were induced in drought-stressed plants compared to control plants. The application of melatonin significantly enhanced the activities of the osmoprotectant enzymes SlP5CR, SlP5CS, SlBADH, SlSPS, SlSUS3, and SlT6PS in the drought-stress environment (Figure 12).

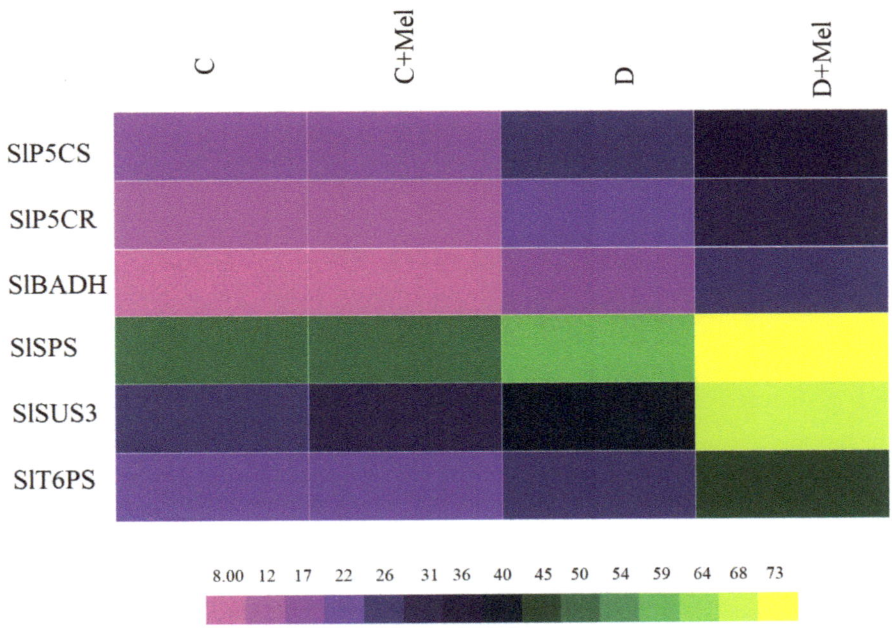

Figure 12. The enzymatic activity of osmoprotectants in tomato plants. SlP5CS, SlP5CS, SlBADH, SlSPS, SlSUS3, and SlT6PS were measured in melatonin-supplemented or non-supplemented tomato leaves after 10 days of drought stress. The magenta color illustrates lower values, and the yellow color illustrates higher values of osmoprotectants in the heat map. The values presented above are the average of six replicates ± S.D. ($n = 6$).

4. Discussion

Dehydration stress critically restricts plant growth, decreases its water status, and reduces crop production. At the same time, melatonin treatment enhances plant resistance to dehydration stress by alleviating stress effects [8,24]. This study provides strong evidence that melatonin treatment improves drought tolerance in tomato plants by protecting the photosynthetic apparatus, photosynthetic physiology, membrane stability, and root system and enhancing the activity of the antioxidant defense system and osmoprotectants. Furthermore, melatonin-treated plants presented higher endogenous melatonin content under drought stress, implying that endogenous melatonin could be associated with drought tolerance.

RWC is one indicator of the plant water level, and its adjustment is associated with the adaptation to drought stress [10]. Drought stress significantly reduces RWC in tomato plants, resulting in decreased water movement from the roots to the stem, decreased mesophyll turgidity, lower leaf water potential, or a decline in soil moisture [38]. In our research, RWC was maintained in melatonin-treated tomato plants, which, overall, improved the performance of these plants due to membrane protection, as supported by prior research [16].

Melatonin has been shown to increase the cuticle's thickness and ultimately prevent water loss through/from plants. Another study also demonstrated that melatonin treatment mediates tolerance to drought by maintaining the turgor and water ratio in plants [24].

MDA and electrolyte leakage are valuable indicators of cellular membrane damage induced by drought stress [33]. In the current research, drought-treated tomato plants showed a noticeable rise in electrolyte leakage. This outcome indicates that increased leakage could be caused by damage to membrane stability in plants due to 10 days of drought stress. Conversely, melatonin-treated plants showed decreased electrolyte leakage under the drought-stress scenario. In melatonin-treated plants, lower electrolyte leakage may be observed due to the protective mechanism's activation. Our study is consistent with previous research showing that drought-tolerant plants displayed an increased membrane strength mechanism [39]. Our research also suggests that tolerance to drought stress might be associated with low lipid peroxidation levels in plants. Furthermore, drought stress causes ROS generation. This mainly consists of O_2^- and H_2O_2, which function as signal transducers to activate cellular responses to stress [40]. The H_2O_2 and O^{2-} contents accumulated in melatonin-treated plants were considerably lower than those in plants not treated with melatonin. This outcome suggests that the enhanced drought resistance of plants could be due to effective ROS elimination. Our findings are highly consistent with previous research that demonstrated that lower H_2O_2 and O_2^- contents are associated with tomato plants' drought tolerance [41].

In plants, photosynthesis is the basis of plant growth and employs light energy to initiate the synthesis of organic compounds [10]. Drought is a major environmental stressor that prevents photosynthesis from taking place. The primary cause of the reduced photosynthetic rate during water stress is the limited ambient CO_2 distribution, mediated by stomatal closure [42]. The maintenance of relative water content through melatonin could function in mediating gas exchanges and thus biochemical processes because melatonin-treated tomato plants had greater stomatal conductance, a higher photosynthetic rate, and increased transpiration, allowing a better source of assimilation for leaf tissues. The application of melatonin could also play a role in maintaining carboxylation efficacy in tomato plants under drought stress. Previous studies revealed that plants treated with melatonin maintained Pn, Tr, and Gs compared to untreated plants during drought stress [24,43].

To scrutinize the regulatory mechanism of melatonin for photosynthesis during drought stress, we investigated photosynthetic gene expression in plants. In our study, melatonin-treated plants showed higher gene expression of crucial photosynthetic genes than non-treated tomato plants in drought stress. Ferredoxin (Fd) and Plastocyanin (Pc) are essential components of the photosynthetic electron transport chain because of their crucial role in electron transfer [44]. Consequently, our research suggests that the high expression of these genes may have contributed to the improved electron transport rate and PSII efficiency in melatonin-treated plants in the drought-stress treatment.

The use of chlorophyll fluorescence analysis to evaluate photosynthetic physiology in drought-stressed plants is a quick, sensitive, and non-invasive method [13]. The PSII photosynthetic machinery serves a critical function in the energy conversion process. Studies have shown that the drought-stress environment impairs the PSII reaction center in plants [12]. Our research suggests that melatonin treatment increases energy dissipation in tomato plants to improve the xanthophyll cycle. This process may increase the plant's ability to endure drought stress and ultimately reduce PSII machinery damage in plants. The plants' photosynthetic machinery functionality was observed under the drought-stress scenario by measuring Fv/Fm. Fv/Fm indicates the maximum photochemical efficacy of PSII and is utilized to reflect the level of injury to the photosynthetic apparatus in the drought environment [45]. In our study, Fv/Fm significantly decreased in non-melatonin-treated plants in the drought-stress environment. This could be due to the photosynthetic machinery being damaged. A previous study documented that melatonin can improve the efficiency of the electron transport rate in plants [46]. Similarly, the exogenous melatonin treatment efficiently improved the PSII reaction center response to drought stress in our study. This process maintained the electron transport rate and the efficiency of photochemical conversion in tolerant plants. Our research indicates that melatonin treatment may protect against the drought-induced impairment of the photosynthetic machinery.

Polyphenol oxidase is the primary enzyme supporting plants' defense system against stress by converting phenols to quinines, and PPO activity is induced under environmental stresses. A previous study revealed that plants under stress induce higher activities of PPO, which makes them more susceptible to these stress conditions [47]. In the current study, the PPO activity was significantly induced in drought-stressed plants compared to melatonin-treated plants. Our research outcomes are consistent with earlier research showing that melatonin application might alleviate oxidative stress damage in plants [48].

It is well recognized that the amount of soluble proteins in plants is a good indicator of their physiological status, especially under environmental stresses [49]. During water stress, the soluble protein content was considerably decreased compared to well-watered plants. Melatonin-treated drought-stressed plants maintained a higher soluble protein content than drought-treated plants. A previous study reported the same outcomes and proposed that drought stress decreased the soluble protein content by suppressing protein synthesis and triggering protein hydrolysis [50]. On the other hand, melatonin treatment may decrease protein breakdown while promoting new protein synthesis. Our studies suggest that exogenous melatonin may increase the generation of osmotic solutes, which play important roles in increasing plant-cell osmosis regulation and the water-holding capacity and controlling stomatal opening by removing ROS during drought stress.

Trehalose considerably improves and maintains the electron transport process in the PSII apparatus [27]. According to our findings, significant trehalose accumulation maintained the function of PSII in melatonin-treated plants. The maintained PSII function may be associated with the high electron transport rate, consistent with previous findings [51]. The plant's ability to cope with drought stress depends on the effective regulation of electrons. Trehalose has been shown in several recent studies to be effective for protective mechanisms of the plant's PSII machinery under stress conditions. Furthermore, treating buds with GB solutions significantly reduced H_2O_2 production, enhanced soluble sugar accumulation, and protected plant tissues from the impacts of environmental stress [52]. GB limits ROS formation and reduces lipid peroxidation in a stressful environment by protecting the photosynthetic machinery [53]. In our study, melatonin-treated plants significantly accumulated GB, the GB enzyme SlBADH, and key GB gene expression and decreased H_2O_2 and MDA levels after drought stress. This outcome suggests that GB could play an essential role in ROS scavenging and the activation of the antioxidant defense system to protect tomato plants.

Plants also initiate proline accumulation as a stress response to drought stress conditions. The activity of enzymes such as P5CR and P5CS is required to increase proline biosynthesis in plants [54]. Our research indicated that the proline level, the activity of the proline enzymes P5CR and P5CS, and the expression of the key proline genes SlP5CS and SlP5CR were induced in melatonin-treated plants. This outcome suggests that proline could have a role in stabilizing membrane structures, as evidenced by decreased electrolyte leakage in melatonin-treated plants, to cope with drought stress. In plants, high proline accumulation improves drought resistance [19,55]. A prior investigation indicated that the proline function stabilizes membranes and protein structures in plants [56]. Our research also observed significant sucrose accumulation in melatonin-treated plants under the drought-stress treatment. This process in melatonin-treated tomato plants might play a significant part in the increased energy observed in mitochondria under drought stress. Previous research showed that sucrose accumulation in plants protected cell mechanisms from the negative consequence of drought-stress treatment [57].

Climate-change-induced drought stress is frequently associated with elevated ROS levels in plants. Drought stress induces an increase in ROS generation, which affects plant growth, causes oxidative injury to the membrane, and, ultimately, plays a primary role in reducing plant efficiency in the drought-stress environment [10]. Plants have evolved a complex antioxidant defense system that includes non-enzymatic and enzymatic antioxidants to scavenge excess stress-generated ROS [17]. The important enzymatic antioxidants comprise CAT, GR, POD, APX, SOD, and DHAR, which perform crucial functions in protecting plants during drought stress [41]. In the current research, increased activities of APX, CAT, GR, POD SOD, and DHAR antioxidants was detected in melatonin-treated tomato plants under drought stress. This antioxidant defense system reduced drought-induced oxidative stress before it became harmful. Importantly, the potential of tomato plants to withstand 10 days of drought stress could be associated with the presence of the increased activity of these antioxidant enzymes, including POD, SOD, CAT, GR, and APX [58].

Root system enhancement regulates the ability of plants to attain nutrients and water. Root parameters directly affect plants' capacity to absorb and transport available water and nutrients [59]. Improved root growth and length could effectively help withstand the damage caused by drought-stress conditions in plants. In addition, the root surface area and fresh and dry weights can be employed as critical indicators to estimate the drought tolerance of plants. Previous studies have shown that drought stress affects tomato plants' root systems [60]. In our research, 10 days of drought-stress conditions affected the root system, including the average root diameter, root forks, projected area, root crossings, root volume, root surface area, root length, root tips, root numbers, fresh root weight, and dry root weight in Micro-Tom tomato plants. Nevertheless, tomato plants treated with melatonin showed protected roots in drought-stress scenarios. Previous studies reported that plants treated with melatonin enhanced their root system compared to untreated plants in the drought-stress phase [38,61]. Our study suggested that in Micro-Tom plants, melatonin treatment improved these plants' root systems in particular, which indicates the stringent association between nutrient use and water withholding in drought stress.

In the current research, a 100 μM melatonin treatment activated mechanisms that (1) maximally enhanced the ability to scavenge ROS in Micro-Tom tomato plants by initiating antioxidant metabolism and antioxidant defense genes, as well as by accumulating osmoprotectants that maintained the water status in leaves, which could lead to increased plant growth in the drought-stress phase. (2) The photosynthetic apparatus was protected to prevent ROS-induced oxidative stress damage by regulating photosynthetic activity, CO_2 assimilation, maximum PSII yield, and non-photochemical quenching. (3) Melatonin treatment induced osmotic adjustment, maintained membrane stability, and retained the water balance, improving light energy absorption by electron transport in the PSII system. (4) The root system architecture was significantly enhanced, which could enable and improve access to water. Our study is in line with previous studies that illustrated that antioxidant metabolism, the protection of PSII, and the enhancement of the root system and membrane stability is a crucial feature of plants that helps them to survive in drought-stress environments [54,60,61].

5. Conclusions

Our research demonstrated that melatonin application helps alleviate the negative effects of drought stress in Micro-Tom tomato plants. Our study highlighted that melatonin application protected plants' photosynthetic machinery and photosynthetic physiology. Furthermore, our study revealed the feasibility of treating Micro-Tom plants with exogenous melatonin, which improved the root system and decreased the sensitivity to drought stress. The ability of these tomato plants to increase the length, diameter, volume, and, importantly, the development of the root system is an imperative technological advance in horticulture crops. In addition, osmoprotectants and their related genes were evidently improved in melatonin-treated Micro-Tom tomato plants. In addition, melatonin-treated plants had less membrane damage after 10 days of drought stress, probably via the initiation of ROS scavenging by activating antioxidant enzymes and their related genes. This study unlocks a new research aspect for developing horticulture plants with enhanced tolerance to drought-stress environments. The overall intent is to minimize tomato plants' yield losses in climate change environments.

Supplementary Materials: The following supporting information can be downloaded at https://www.mdpi.com/article/10.3390/life12111922/s1, Table S1: Primers used in this study.

Author Contributions: Conceptualization, N.M.; methodology, N.M.; software, N.M., F.H. and S.I.; validation, N.M., S.I. and A.R.; formal analysis, N.M. and S.I.; investigation, N.M., F.H., W.Z. and A.A.; data curation, N.M.; writing—original draft preparation, N.M.; review and editing, N.M.; A.R., S.I., W.Z. and A.A.; visualization, N.M.; supervision, W.Z.; project administration, W.Z and N.M.; funding acquisition, W.Z. All authors have read and agreed to the published version of the manuscript.

Funding: This research received no external funding.

Institutional Review Board Statement: Not applicable.

Informed Consent Statement: Not applicable.

Data Availability Statement: Not applicable.

Conflicts of Interest: The authors declare no conflict of interest.

Abbreviations

APX, ascorbate peroxidase; KH_2PO_4, potassium dihydrogen phosphate; PBS, phosphate-buffered saline; HEPES, (4-(2-hydroxyethyl)-1-piperazineethanesulfonic acid); CAT, catalase; DHAR, dehydroascorbate; NBT, nitro blue tetrazolium; EDTA, ethylene diamine tetra-acetic acid reductase; GC-MS, gas chromatography–mass spectrometry; HPLC, high-performance liquid chromatography; GR, glutatione reductase; MDA, malondialdehyde; PPO, polyphenol peroxidase; POD, peroxidase; qRT-PCR, quantitative real-time PCR; ROS, reactive oxygen species; RWC, relative water content; SOD, superoxide dismutase; TAC, total antioxidant capacity; Na_2CO_3, sodium carbonate; $NH_2OH{\cdot}HCl$, hydroxylamine hydrochloride; ACW, water-soluble antioxidant capacity; PCL, photochemiluminescence.

References

1. Liu, D.; Zhang, C.; Ogaya, R.; Fernández-Martínez, M.; Pugh, T.A.M.; Peñuelas, J. Increasing climatic sensitivity of global grassland vegetation biomass and species diversity correlates with water availability. *New Phytol.* **2021**, *230*, 1761–1771. [CrossRef]
2. Mushtaq, N.; Wang, Y.; Fan, J.; Li, Y.; Ding, J. Down-Regulation of Cytokinin Receptor Gene SlHK2 Improves Plant Tolerance to Drought, Heat, and Combined Stresses in Tomato. *Plants* **2022**, *11*, 154. [CrossRef]
3. Zandalinas, S.I.; Balfagón, D.; Gómez-Cadenas, A.; Mittler, R. Responses of plants to climate change: Metabolic changes during abiotic stress combination in plants. *J. Exp. Bot.* **2022**, *73*, 3335–3354. [CrossRef]
4. Yang, X.; Lu, M.; Wang, Y.; Wang, Y.; Liu, Z.; Chen, S. Response mechanism of plants to drought stress. *Horticulturae* **2021**, *7*, 50. [CrossRef]
5. Sperdouli, I.; Mellidou, I.; Moustakas, M. Harnessing chlorophyll fluorescence for phenotyping analysis of wild and cultivated tomato for high photochemical efficiency under water deficit for climate change resilience. *Climate* **2021**, *9*, 154. [CrossRef]
6. Kuwabara, W.M.T.; Gomes, P.R.L.; Andrade-Silva, J.; Júnior, J.M.S.; Amaral, F.G.; Cipolla-Neto, J. Melatonin and its ubiquitous effects on cell function and survival: A review. *Melatonin Res.* **2022**, *5*, 192–208. [CrossRef]
7. Sun, C.; Liu, L.; Wang, L.; Li, B.; Jin, C.; Lin, X. Melatonin: A master regulator of plant development and stress responses. *J. Integr. Plant Biol.* **2021**, *63*, 126–145. [CrossRef]
8. Khan, M.; Ali, S.; Manghwar, H.; Saqib, S.; Ullah, F.; Ayaz, A.; Zaman, W. Melatonin function and crosstalk with other phytohormones under normal and stressful conditions. *Genes* **2022**, *13*, 1699. [CrossRef]
9. Hardeland, R. Melatonin in plants–diversity of levels and multiplicity of functions. *Front. Plant Sci.* **2016**, *7*, 198. [CrossRef]
10. Liang, G.; Liu, J.; Zhang, J.; Guo, J. Effects of drought stress on photosynthetic and physiological parameters of tomato. *J. Am. Soc. Hortic. Sci.* **2020**, *145*, 12–17. [CrossRef]
11. Yang, X.; Li, Y.; Chen, H.; Huang, J.; Zhang, Y.; Qi, M.; Liu, Y.; Li, T. Photosynthetic response mechanism of soil salinity-induced cross-tolerance to subsequent drought stress in tomato plants. *Plants* **2020**, *9*, 363. [CrossRef] [PubMed]
12. Hussain, M.I.; Reigosa, M.J. Allelochemical stress inhibits growth, leaf water relations, PSII photochemistry, non-photochemical fluorescence quenching, and heat energy dissipation in three C3 perennial species. *J. Exp. Bot.* **2011**, *62*, 4533–4545. [CrossRef] [PubMed]
13. Harbinson, J. Improving the accuracy of chlorophyll fluorescence measurements. *Plant Cell Environ.* **2013**, *36*, 1751–1754. [CrossRef] [PubMed]
14. Gapińska, M.; Skłodowska, M.; Gabara, B. Effect of short-and long-term salinity on the activities of antioxidative enzymes and lipid peroxidation in tomato roots. *Acta Physiol. Plant.* **2008**, *30*, 11–18. [CrossRef]

15. Yang, L.; Sun, Q.; Wang, Y.; Chan, Z. Global transcriptomic network of melatonin regulated root growth in Arabidopsis. *Gene* **2021**, *764*, 145082. [CrossRef] [PubMed]
16. Su, X.; Fan, X.; Shao, R.; Guo, J.; Wang, Y.; Yang, J.; Yang, Q.; Guo, L. Physiological and iTRAQ-based proteomic analyses reveal that melatonin alleviates oxidative damage in maize leaves exposed to drought stress. *Plant Physiol. Biochem.* **2019**, *142*, 263–274. [CrossRef]
17. Hasanuzzaman, M.; Bhuyan, M.B.; Zulfiqar, F.; Raza, A.; Mohsin, S.M.; Mahmud, J.A.; Fujita, M.; Fotopoulos, V. Reactive oxygen species and antioxidant defense in plants under abiotic stress: Revisiting the crucial role of a universal defense regulator. *Antioxidants* **2020**, *9*, 681. [CrossRef]
18. Burritt, D.J. Proline and the cryopreservation of plant tissues: Functions and practical applications. In *Current Frontiers in Cryopreservation*; BoD–Books on Demand: Norderstedt, Germany, 2012; pp. 415–426.
19. Semida, W.M.; Abdelkhalik, A.; Rady, M.O.A.; Marey, R.A.; Abd El-Mageed, T.A. Exogenously applied proline enhances growth and productivity of drought stressed onion by improving photosynthetic efficiency, water use efficiency and up-regulating osmoprotectants. *Sci. Hortic.* **2020**, *272*, 109580. [CrossRef]
20. Jangid, K.K.; Dwivedi, P. Physiological responses of drought stress in tomato: A review. *Int. J. Agric. Environ. Biotechnol.* **2016**, *9*, 53. [CrossRef]
21. Li, C.; Tan, D.-X.; Liang, D.; Chang, C.; Jia, D.; Ma, F. Melatonin mediates the regulation of ABA metabolism, free-radical scavenging, and stomatal behaviour in two Malus species under drought stress. *J. Exp. Bot.* **2015**, *66*, 669–680. [CrossRef]
22. Mestre, T.C.; Garcia-Sanchez, F.; Rubio, F.; Martinez, V.; Rivero, R.M. Glutathione homeostasis as an important and novel factor controlling blossom-end rot development in calcium-deficient tomato fruits. *J. Plant Physiol.* **2012**, *169*, 1719–1727. [CrossRef]
23. Nicoli, M.C.; Elizalde, B.E.; Pitotti, A.; Lerici, C.R. Effect of sugars and maillard reaction products on polyphenol oxidase and peroxidase activity in food. *J. Food Biochem.* **1991**, *15*, 169–184. [CrossRef]
24. Liu, J.; Wang, W.; Wang, L.; Sun, Y. Exogenous melatonin improves seedling health index and drought tolerance in tomato. *Plant Growth Regul.* **2015**, *77*, 317–326. [CrossRef]
25. Brennan, T.; Frenkel, C. Involvement of hydrogen peroxide in the regulation of senescence in pear. *Plant Physiology* **1977**, *59*, 411–416. [CrossRef]
26. Elstner, E.F.; Heupel, A. Inhibition of nitrite formation from hydroxylammoniumchloride: A simple assay for superoxide dismutase. *Anal. Biochem.* **1976**, *70*, 616–620. [CrossRef]
27. Rivero, R.M.; Mestre, T.C.; Mittler, R.O.N.; Rubio, F.; Garcia-Sanchez, F.; Martinez, V. The combined effect of salinity and heat reveals a specific physiological, biochemical and molecular response in tomato plants. *Plant Cell Environ.* **2014**, *37*, 1059–1073. [CrossRef]
28. Blunden, C.A.; Wilson, M.F. A specific method for the determination of soluble sugars in plant extracts using enzymatic analysis and its application to the sugar content of developing pear fruit buds. *Anal. Biochem.* **1985**, *151*, 403–408. [CrossRef]
29. Rasmussen, T.S.; Henry, R.J. Starch determination in horticultural plant material by an enzymic-colorimetric procedure. *J. Sci. Food Agric.* **1990**, *52*, 159–170. [CrossRef]
30. Schulze, U.; Larsen, M.E.; Villadsen, J. Determination of intracellular trehalose and glycogen in *Saccharomyces cerevisiae*. *Anal. Biochem.* **1995**, *228*, 143–149. [CrossRef]
31. Rivero, R.M.; Ruiz, J.M.; Romero, L.M. Importance of N source on heat stress tolerance due to the accumulation of proline and quaternary ammonium compounds in tomato plants. *Plant Biol.* **2004**, *6*, 702–707. [CrossRef]
32. Geladopoulos, T.P.; Sotiroudis, T.G.; Evangelopoulos, A.E. A malachite green colorimetric assay for protein phosphatase activity. *Anal. Biochem.* **1991**, *192*, 112–116. [CrossRef]
33. Camejo, D.; Jiménez, A.; Alarcón, J.J.; Torres, W.; Gómez, J.M.; Sevilla, F. Changes in photosynthetic parameters and antioxidant activities following heat-shock treatment in tomato plants. *Funct. Plant Biol.* **2006**, *33*, 177–187. [CrossRef]
34. Nakano, Y.; Asada, K. Hydrogen peroxide is scavenged by ascorbate-specific peroxidase in spinach chloroplasts. *Plant Cell Physiol.* **1981**, *22*, 867–880.
35. Zafar, S.A.; Hameed, A.; Nawaz, M.A.; Ma, W.; Noor, M.A.; Hussain, M.; Mehboob ur, R. Mechanisms and molecular approaches for heat tolerance in rice (*Oryza sativa* L.) under climate change scenario. *J. Integr. Agric.* **2018**, *17*, 726–738. [CrossRef]
36. Shah, K.; Kumar, R.G.; Verma, S.; Dubey, R.S. Effect of cadmium on lipid peroxidation, superoxide anion generation and activities of antioxidant enzymes in growing rice seedlings. *Plant Sci.* **2001**, *161*, 1135–1144. [CrossRef]
37. Comas, L.H.; Eissenstat, D.M.; Lakso, A.N. Assessing root death and root system dynamics in a study of grape canopy pruning. *New Phytol.* **2000**, *147*, 171–178. [CrossRef]
38. Altaf, M.A.; Shahid, R.; Ren, M.-X.; Naz, S.; Altaf, M.M.; Khan, L.U.; Tiwari, R.K.; Lal, M.K.; Shahid, M.A.; Kumar, R. Melatonin Improves Drought Stress Tolerance of Tomato by Modulation Plant Growth, Root Architecture, Photosynthesis, and Antioxidant Defense System. *Antioxidants* **2022**, *11*, 309. [CrossRef] [PubMed]
39. Farooq, M.; Wahid, A.; Kobayashi, N.; Fujita, D.; Basra, S.M.A. Plant drought stress: Effects, mechanisms and management. *Agron. Sustain. Dev.* **2009**, *29*, 185–212. [CrossRef]
40. Rai, G.K.; Parveen, A.; Jamwal, G.; Basu, U.; Kumar, R.R.; Rai, P.K.; Sharma, J.P.; Alalawy, A.I.; Al-Duais, M.A.; Hossain, M.A.J.A. Leaf Proteome Response to Drought Stress and Antioxidant Potential in Tomato (*Solanum lycopersicum* L.). *Atmosphere* **2021**, *12*, 1021. [CrossRef]

41. Ijaz, R.; Ejaz, J.; Gao, S.; Liu, T.; Imtiaz, M.; Ye, Z.; Wang, T. Overexpression of annexin gene AnnSp2, enhances drought and salt tolerance through modulation of ABA synthesis and scavenging ROS in tomato. *Sci. Rep.* **2017**, *7*, 1–14. [CrossRef]
42. Liu, D.; Wu, L.; Naeem, M.S.; Liu, H.; Deng, X.; Xu, L.; Zhang, F.; Zhou, W. 5-Aminolevulinic acid enhances photosynthetic gas exchange, chlorophyll fluorescence and antioxidant system in oilseed rape under drought stress. *Acta Physiol. Plant.* **2013**, *35*, 2747–2759. [CrossRef]
43. Nguyen, T.T.; Fuentes, S.; Marschner, P. Effects of compost on water availability and gas exchange in tomato during drought and recovery. *Plant Soil Environ.* **2012**, *58*, 495–502. [CrossRef]
44. Zhang, Y.; Yu, S.H.I.; Gong, H.-J.; Zhao, H.-L.; Li, H.-L.; Hu, Y.-H.; Wang, Y.-C. Beneficial effects of silicon on photosynthesis of tomato seedlings under water stress. *J. Integr. Agric.* **2018**, *17*, 2151–2159. [CrossRef]
45. Kusaba, M.; Ito, H.; Morita, R.; Iida, S.; Sato, Y.; Fujimoto, M.; Kawasaki, S.; Tanaka, R.; Hirochika, H.; Nishimura, M. Rice NON-YELLOW COLORING1 is involved in light-harvesting complex II and grana degradation during leaf senescence. *Plant Cell* **2007**, *19*, 1362–1375. [CrossRef]
46. Zhao, H.; Su, T.; Huo, L.; Wei, H.; Jiang, Y.; Xu, L.; Ma, F. Unveiling the mechanism of melatonin impacts on maize seedling growth: Sugar metabolism as a case. *J. Pineal Res.* **2015**, *59*, 255–266. [CrossRef]
47. He, F.; Shi, Y.-J.; Zhao, Q.; Zhao, K.-J.; Cui, X.-L.; Chen, L.-H.; Yang, H.-B.; Zhang, F.; Mi, J.-X.; Huang, J.-L. Genome-wide investigation and expression profiling of polyphenol oxidase (PPO) family genes uncover likely functions in organ development and stress responses in *Populus trichocarpa*. *BMC Genom.* **2021**, *22*, 1–15. [CrossRef]
48. Ahmad, S.; Wang, G.-Y.; Muhammad, I.; Chi, Y.-X.; Zeeshan, M.; Nasar, J.; Zhou, X.-B. Interactive Effects of Melatonin and Nitrogen Improve Drought Tolerance of Maize Seedlings by Regulating Growth and Physiochemical Attributes. *Antioxidants* **2022**, *11*, 359. [CrossRef]
49. Liang, D.; Ni, Z.; Xia, H.; Xie, Y.; Lv, X.; Wang, J.; Lin, L.; Deng, Q.; Luo, X. Exogenous melatonin promotes biomass accumulation and photosynthesis of kiwifruit seedlings under drought stress. *Sci. Hortic.* **2019**, *246*, 34–43. [CrossRef]
50. Meng, J.F.; Xu, T.F.; Wang, Z.Z.; Fang, Y.L.; Xi, Z.M.; Zhang, Z.W. The ameliorative effects of exogenous melatonin on grape cuttings under water-deficient stress: Antioxidant metabolites, leaf anatomy, and chloroplast morphology. *J. Pineal Res.* **2014**, *57*, 200–212. [CrossRef]
51. Luo, Y.; Wang, W.; Fan, Y.Z.; Gao, Y.M.; Wang, D. Exogenously-supplied trehalose provides better protection for D1 protein in winter wheat under heat stress. *Russ. J. Plant Physiol.* **2018**, *65*, 115–122. [CrossRef]
52. Zhang, G.; Liu, Y.; Ni, Y.; Meng, Z.; Lu, T.; Li, T. Exogenous calcium alleviates low night temperature stress on the photosynthetic apparatus of tomato leaves. *PLoS ONE* **2014**, *9*, e97322. [CrossRef] [PubMed]
53. Chen, T.H.H.; Murata, N. Glycinebetaine: An effective protectant against abiotic stress in plants. *Trends Plant Sci.* **2008**, *13*, 499–505. [CrossRef] [PubMed]
54. Nounjan, N.; Nghia, P.T.; Theerakulpisut, P. Exogenous proline and trehalose promote recovery of rice seedlings from salt-stress and differentially modulate antioxidant enzymes and expression of related genes. *J. Plant Physiol.* **2012**, *169*, 596–604. [CrossRef] [PubMed]
55. Montesinos-Pereira, D.; Barrameda-Medina, Y.; Romero, L.; Ruiz, J.M.; Sánchez-Rodríguez, E. Genotype differences in the metabolism of proline and polyamines under moderate drought in tomato plants. *Plant Biol.* **2014**, *16*, 1050–1057. [CrossRef] [PubMed]
56. Evers, D.; Lefevre, I.; Legay, S.; Lamoureux, D.; Hausman, J.-F.; Rosales, R.O.G.; Marca, L.R.T.; Hoffmann, L.; Bonierbale, M.; Schafleitner, R. Identification of drought-responsive compounds in potato through a combined transcriptomic and targeted metabolite approach. *J. Exp. Bot.* **2010**, *61*, 2327–2343. [CrossRef]
57. Jing, F.; Miao, Y.; Zhang, P.; Chen, T.; Liu, Y.; Ma, J.; Li, M.; Yang, D. Characterization of TaSPP-5A gene associated with sucrose content in wheat (*Triticum aestivum* L.). *BMC Plant Biol.* **2022**, *22*, 1–11. [CrossRef]
58. Moustakas, M. Plant Photochemistry, Reactive Oxygen Species, and Photoprotection. *Photochem* **2021**, *2*, 5–8. [CrossRef]
59. Wang, W.; Ding, G.-D.; White, P.J.; Wang, X.-H.; Jin, K.-M.; Xu, F.-S.; Shi, L. Mapping and cloning of quantitative trait loci for phosphorus efficiency in crops: Opportunities and challenges. *Plant Soil* **2019**, *439*, 91–112. [CrossRef]
60. Xiong, R.; Liu, S.; Considine, M.J.; Siddique, K.H.M.; Lam, H.M.; Chen, Y. Root system architecture, physiological and transcriptional traits of soybean (*Glycine max* L.) in response to water deficit: A review. *Physiol. Plant.* **2021**, *172*, 405–418. [CrossRef]
61. Tiwari, R.K.; Lal, M.K.; Kumar, R.; Chourasia, K.N.; Naga, K.C.; Kumar, D.; Das, S.K.; Zinta, G. Mechanistic insights on melatonin-mediated drought stress mitigation in plants. *Physiol. Plant.* **2021**, *172*, 1212–1226. [CrossRef]

life

MDPI

Review

Chitosan: An Autocidal Molecule of Plant Pathogenic Fungus

Debanjana Debnath [1], Ipsita Samal [2], Chinmayee Mohapatra [1], Snehasish Routray [2], Mahipal Singh Kesawat [3] and Rini Labanya [4,*]

1 Department of Plant Pathology, Faculty of Agriculture, Sri Sri University, Cuttack 754006, Odisha, India
2 Department of Entomology, Faculty of Agriculture, Sri Sri University, Cuttack 754006, Odisha, India
3 Department of Genetics and Plant Breeding, Faculty of Agriculture, Sri Sri University, Cuttack 754006, Odisha, India
4 Department of Soil Science & Agricultural Chemistry, Faculty of Agriculture, Sri Sri University, Cuttack 754006, Odisha, India
* Correspondence: rini.l@srisriuniversity.edu.in

Abstract: The rise in the world's food demand with the increasing population threatens the existence of civilization with two equally valuable concerns: increase in global food production and sustainability in the ecosystem. Furthermore, biotic and abiotic stresses are adversely affecting agricultural production. Among them, losses caused by insect pests and pathogens have been shown to be more destructive to agricultural production. However, for winning the battle against the abundance of insect pests and pathogens and their nature of resistance development, the team of researchers is searching for an alternative way to minimize losses caused by them. Chitosan, a natural biopolymer, coupled with a proper application method and effective dose could be an integral part of sustainable alternatives in the safer agricultural sector. In this review, we have integrated the insight knowledge of chitin-chitosan interaction, successful and efficient use of chitosan, recommended and practical methods of use with well-defined doses, and last but not least the dual but contrast mode of action of the chitosan in hosts and as well as in pathogens.

Keywords: chitosan; pathogen; sustainable; plant protection

Citation: Debnath, D.; Samal, I.; Mohapatra, C.; Routray, S.; Kesawat, M.S.; Labanya, R. Chitosan: An Autocidal Molecule of Plant Pathogenic Fungus. *Life* **2022**, *12*, 1908. https://doi.org/10.3390/life12111908

Academic Editors: Wajid Zaman and Hakim Manghwar

Received: 12 September 2022
Accepted: 9 November 2022
Published: 16 November 2022

Publisher's Note: MDPI stays neutral with regard to jurisdictional claims in published maps and institutional affiliations.

Copyright: © 2022 by the authors. Licensee MDPI, Basel, Switzerland. This article is an open access article distributed under the terms and conditions of the Creative Commons Attribution (CC BY) license (https://creativecommons.org/licenses/by/4.0/).

1. Introduction

Global climatic changes are posing a threat to the security of the world's food supply by adversely affecting plant growth, development, and yield in multiple directions, such as by causing abiotic stresses to plants and as well as encouraging and strengthening the biotic populations by increasing their resistance against conventional chemical management procedures. In this situation, the researchers are trying to identify some natural compounds or their derivatives which can be effectively established themselves as an ideal one to replace the chemicals against which the pathogens are growing resistance gradually. Chitosan, a chemically and physically diverse compound and long-chain polymer of *N*-acetyl-glucosamine and d-glucosamine derivative of chitin (second most prevalent polysaccharide after cellulose), was first discovered in 1859 by Rouget [1]. Due to its potential for usage in antiviral, antifungal, and antibacterial products, chitosan and its derivatives have gained attention in recent years. Other than chitosan, some other chitin-related compounds and chitin derivatives have also been identified as possible plant protection agents [2]. Chitosan, an aminoglucopyranan made up of N-acetyl-D-glucosamine (GlcNAc) and glucosamine (GlcN) residues, is currently being deemed essential due to its appealing features and biological activities [3,4]. An amino group and two hydroxyl groups make up the three reactive functional groups that make up chitosan.

According to Amine et al. 2021 chitin and its derivative chitosan are effective at boosting plant defence against pathogens in monocotyledons and dicotyledons [5]. Cell wall lignification, cytoplasmic acidification, membrane depolarization, changes in ion flux and protein phosphorylation, phytoalexin biosynthesis, production of reactive oxygen species, jasmonic acid biosynthesis, and activation of chitinase and glucanase are all targets of this increased plant defense [6].In addition to this, many dicot species have been observed to produce callose, proteinase inhibitors, and phytoalexin in response to chitosan that also have significant roles in plant defense [7]. In the current situation, the rising incidences of resistance, residue, and resurgence (3Rs) have facilitated the increased use of naturalytes in disease and pest control [8]. Consequently, chitosan, a byproduct of fungal and insect chitin, may be used as an autocidal agent to kill invasive disease causing pathogen.

2. Structure and Formation of Chitin

Chitin, a β (1,4)-linked homopolymer of N-acetylglucosamine, is a simple polysaccharide that is an important component of fungal cell wall [9]. It is an amino sugar biopolymer that forms elaborate structures such as insect cuticles and peritrophic membranes when combined with a range of proteins. This polymer is mostly used as a structural component, and it is similar to cellulose and collagen in plants and vertebrates, respectively [10]. 75% of the overall weight of shellfish including crab and shrimp generally discarded as waste among which 20-58% is made up of chitin [11].

Broadly, chitin is defined as a β-(1–4) linked linear cationic heteropolymer consisting of 2-acetamide-2-deoxy-D-glucopyranose (N–acetyl–D–glucosamine, GlcNAc) and randomly distributed units of 2-amino-2-deoxy-D-glucopyranose (D–glucosamine, GlcN) [12]. Due to the presence of the acetoamide groups in the *trans* position, chitin exhibits two important properties: a high degree of crystallinity and a lack of solubility in water and organic solvents [13].

X-ray diffraction analysis revealed that chitin generally occurs in three different crystalline forms, termed α-, β-, and γ-chitin [10], which mainly differ from each other in the degree of hydration, size of the unit cell, and the number of chitin chains per unit cell [14]. All the chains exhibit an anti-parallel orientation in the α form, whereas in the β form the chains are arranged in a parallel manner and in the γ form sets of two parallel strands alternate with single anti-parallel strands. All three crystalline forms are primarily found in the chitinous structures of insects. The α form is most prevalent in chitinous cuticles, whereas the β and γ forms are frequently found in cocoons [15,16].

Due to the properties such as biodegradability, biocompatibility, non-toxicity, physiological inertness, and gel-forming properties, chitin has been found to have countless applications in different industries, e.g., food, cosmetic, pharmaceutical, manufacturing of synthetic materials, agriculture, and even electronics for the production of biosensors [17,18].

In the first step, glycogen is converted by glycogen phosphorylase to glucose-1-P, which is either fed into glycolysis or used for trehalose synthesis [19]. Additionally, the enzyme trehalase can mobilise trehalose by hydrolyzing it to glucose. This is followed by the enzymes hexokinase, phosphoglucomutase, and glucose-6-P isomerase converting glucose to fructose-6-P. Finally, from fructose-6-P the chitin biosynthetic pathway branches off, with the first enzyme catalyzing this branch being glutamine-fructose-6-phosphate amidotransferase [10]. The reaction catalyzed by GFAT converts fructose-6-P into glucosamine-6-phosphate by transferring the ammonia from the co-substrate l-glutamine and isomerizing the resulting fructosimine-6-phosphate. Next, an acetyl group from coenzyme A is added by glucosamine-6-P acetyltransferase to obtain N-acetylglucosamine (GlcNAc)-6-P [20] (Figure 1), whose phosphate is then transferred from the C-6 to the C-1 position catalyzed by phosphoacetylglucosamine mutase. The resulting GlcNAc-1-P, finally, is uridinylated by UDP-GlcNAc pyrophosphorylase yielding UDP-GlcNAc which serves as a substrate for the chitin synthase, transferring the sugar moiety of UDP-GlcNAc to the growing chitin chain. Chitin is degraded by chitinases and N-acetylglucosaminidases yielding GlcNAc which can be reused for chitin biosynthesis [21].

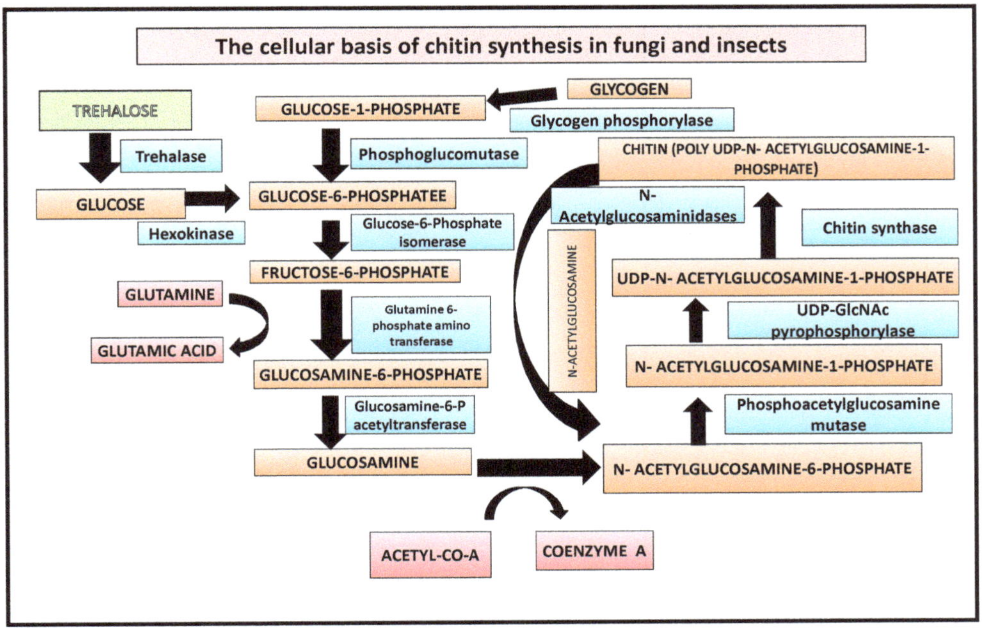

Figure 1. Biochemical basis of chitin synthesis in fungi and insects.

Deacetylation and Hydrolysis of Chitin/Chitosan

Unfortunately, despite having so many advantages, there is very limited use of chitin due to resistance to different physical and chemical agents because of its highly ordered crystalline structure and its lack of solubility in water or any organic solvents. Chitosan, an N-deacetylated derivative of chitin that is soluble in aqueous solutions of both organic and inorganic acids, is frequently employed to get around this restriction [22]. This cationic polymer resembles chitin and also consists of β-(1–4) linked N–acetyl–D–glucosamine and the D–glucosamine residues [23]. Chitosan, which is produced industrially by hydrolyzing the amino acetyl groups of chitins, is less frequent in nature than chitin. It is also naturally present in the cell walls of filamentous fungi primarily belonging to the *Zygomycetes* class [24]. Carboxymethyl-chitin (CM-chitin), as a water-soluble anionic polymer, is the second most studied derivative of chitin after chitosan. Most of the chitin and chitosan biological properties are directly related to their physicochemical characteristics. These characteristics include degree of deacetylation, molecular mass, and the amount of moisture content [25].

Enzymatic degradation of chitin can be achieved by two different paths:

1. Chitin can be degraded by first being solubilized by deacetylation (Figure 2). This process is carried out by chitin deacetylases, and the derived substrate (chitosan) is hydrolysed by chitosanases [18,26].
2. The chitinolytic process requires direct hydrolysis of the beta-1,4 glycosidic bonds between the GlcNAc units by chitinases. Chitinases are produced by higher plants, which use the enzymes to defend themselves against pathogenic attacks by degrading chitin in the cell walls of fungi and bacteria [27]. Plant chitinases have molecular weights ranging from 25 to 40 kD and can be acidic or basic. Endochitinases and exochitinases are the two types of chitinases [28]. Chitinase genes from biocontrol fungi such as *Trichoderma* have significantly higher antifungal activity than comparable plant genes. These fungal genes encode for chitinolytic enzymes, which have higher antifungal activity similar to chemical fungicides [29].

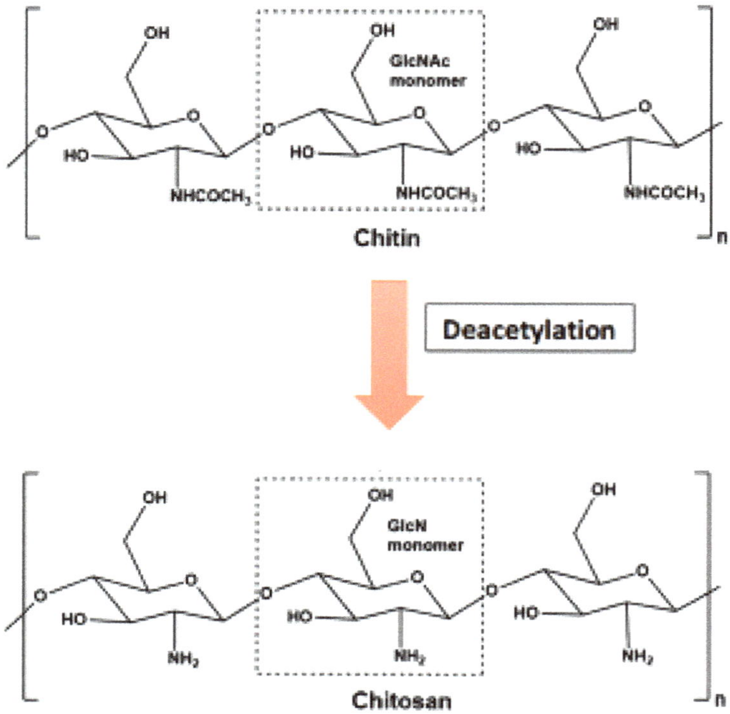

Figure 2. Conversion of chitosan from chitin through deacetylation.

Despite having many significant similarities in the molecular structures of chitin and chitosan, the physicochemical characteristics of both the biopolymers and the reactions are often surprisingly distinct [18]. Both polymers have the same reactive hydroxyl and amino groups in different molecular ratios, but the lower crystallinity of chitosan makes it more accessible to reagents [12]. Perhaps the most important and crucial difference between chitin and chitosan in terms of their applications is based on their degree of deacetylation and their solubility. Unlike chitin, chitosan has a below pKa (pH~6.5) in most acidic aqueous solutions such as acetic acid, formic acid, lactic acid, citric acid, and other solvents such as dimethylsulfoxide and p-toluenesulfonic acid [30].

3. Successful Use of Chitin/Chitosan against Plant Pathogen with Special Reference of Pathogenic Fungus

With changing times and increasing knowledge, producers have begun to identify alternatives to toxic chemicals that are harmful not only to consumers but also to the environment and ecosystems. In these circumstances, use of chitosan against plant pathogens is receiving popularity and wider acceptance due to its eco-friendly nature and abundance of its source [7]. The irony is that chitin and chitosan are obtained from the fungus or the insect and this molecule has unique characteristics to cause damage to multiple plant pathogens, even including fungi itself singly or in combination which make them an autocidal component for fungi (Table 1). A combination of allicin (5% allicin ME 100–time dilution liquid) + chitosan (100–time dilution liquid) showed 85.97% control effect against powdery mildew which was significantly ($p < 0.01$) higher than allicin (76.70% of) and chitosan (70.93%) alone [31]. Some recent research suggesting the use of chitin/chitosan has been mentioned below.

Table 1. Efficacy of chitosan and chitin derivatives against different microorganisms.

Sl.No	Chitin Derivative	Target Pathogen	Remarks	References
1	Chitosan supplemented with 0.05% boron and 0.05%	*Pseudomonas syringae pv. actinidiae*	Inhibition of the growth of bacteriumin in vitro condition	[32]
2	Chitin (Shrimp shell)	*Aspergillus fumigatus*, *Aspergillus flavus* and *Aspergillus niger*	In in vitro condition, highest inhibition (19.5 mm) in case of *Aspergillus fumigatus*	[33]
3	Chitin methanol extract	*Aspergillus fumigatus*, *Aspergillus flavus* and *Aspergillus niger*	Highest inhibition (16.5 mm) in case of *A. fumigatus*	[33]
4	Chitosan	*Aspergillus fumigates*, *Aspergillus flavus* and *Aspergillus niger*	Highest inhibition (14 mm) in case of *A. fumigatus*	[33]
5	Chitosan	*Alternaria solani*	Complete inhibition in in vitro condition at 5.0 g/lit	[1]
6	Chitin (CT), 6-amino-chitin (NCT) and 3,6-diamino-chitin (DNCT)	*F. oxysporum f.* sp. *cucumerium*, *B. cinerea*, *C. lagenarium*, *P. asparagi*, *F. oxysporum f. niveum*, and *G. zeae*	In in vitro condition, DNCT showed Highest inhibition zone (11.4–20.4 mm) > NCT > CT	[34]
7	Chitosan	*Aspergillus flavus*, *Rhizoctonia solani* and *Alterneria alterneta*	Growth inhibition was highest in case of *Aspergillus flavus* (10.66 mm.)	[35]
8	Chitosan-polyacrylic acid nanoparticles	*Aspergillus flavus*, *Fusarium oxysporum*, *Fusarium solani*, *Aspergillus terreus*, *Alternaria tenuis*, *Aspergillus niger* and *Sclerotium rolfsii*	Inhibition percentage was highest in case of *Aspergillus flavus* (60%),	[36]
9	Chitosan	*F. proliferatum* and *F. verticillioides*	Reduce deoxynivalenol (DON) and fumonisin (FBs) production on irradiated maize and wheat grains and growth rates of both the pathogens decreased.	[37]
10	Chitosan	*Colletotrichum capsici*	7.67% post-emergence seedling mortality where Seeds were treated with 1% chitosan.	[38]
11	Chitosan	*Fusarium oxysporum radicisycopersici*, *F. oxysporum lycopersici*, *F. solani*, *Rhizoctonia solani*, *Sclerotium rolfsii*, *Macrophomina phaseolinae*, *Pythium* sp. and *Phytophthora* sp.	In 5 g/lit concentration, 100% inhibition can occur against every tested pathogen	[39]
12	Chitosan	*Fusarium oxysporum f.* sp. *radicislycopersici*	16.60% and 42.8% reduction of disease severity in application of chitosan 1 g/lit and *T. harzianum* + Chitosan 1.0 g/lit	[40]

4. Application of Chitosan

After discovering chitosan's antipathogenic activity, especially its antifungal activity, the second major concern of researchers was how to use chitosan effectively. The appropriate way of application is directly proportional to the rate of success and higher efficiency. On the other hand, application methods should be easy and cheap in nature.

4.1. Seed Treatment

Use of chitosan in various dose dependent manners has been proved to be very effective against different pathogens (Table 2). Chitosan not only creates a barrier between pathogens and healthy embryos, but it also helps retain moisture which increases germination rate and also affects plant vitality. effectively combats pathogens and also induces defense in seedlings grown from treated seeds [41].

Table 2. Chitosan seed treatment in different crops and their efficacy.

Sl. No	Crop Name	Respective Dose	Efficiency on Target Pathogen	Reference
1	Fenugreek	2.0 g/lit	In pot and field studies, seeds treated with chitosan greatly reduced root rot disease severity of *Fusarium solani*	[1]
2	Potato	4.0 g/lit of acetic acid-distilled water solution	Reduced dry rot severity observed in case of *F. oxysporum* (60.0%)and *F. sambucinum* (48.2%) by chitosan treatment.	[42]
3	Chilli	1%	100% mycelium growth inhibition and the lowest (7.67%) post-emergence seedling mortality was observed against *Colletotrichum capsici*	[38]
5	Cucumber	500 ppm	500 ppm chitosan seed treatment showed 100% disease resistance against damping off caused by *Phytophthora capsici*	[43]

4.2. Chitosan Used for Soil Amendment

The use of chitosan as a soil conditioner or soil treatment has enormous direct or indirect effects on plant pathogens. There are many experiments performed by researchers who noted different activities of chitosan against different pathogens.Corsee et al. (2015) demonstrated a new concept of chitosan-induced plant defense mechanism, in which kiwi plant immunity was enhanced by applying chitosan to the growth medium by acting as a trigger to increase the activity of guaiacol peroxidase. (G-POD), ascorbate peroxidase (APX), phenylalanine ammonia lyase (PAL) and polyphenol oxidase (PPO) that regulate plant defense [44]. Moreover, adding chitosan to soil promotes the growth and abundance of beneficial microorganisms such as Pseudomonas fluorescens, actinomycetes, mycorrhizal fungi and rhizobia [45,46]. Both chitin and chitosan are taken up by soil at different rates, and chitosan acts as chitin when parasitic on tomato fields [47] and neither chitosan nor chitin showed phytotoxicity to the host.

4.3. Chitosan Used as Foliar Spray

Foliar sprays are known for their ease of use and direct contact with infected pathogens and the symptoms they cause. They also create a protective barrier between pathogen and host, kill contact-associated pathogens, impede fungal sporulation, and most importantly help the host to induce defense mechanisms [40,48]. Some of the examples has been cited below in Table 3.

Table 3. Foliar spray of chitosan and their efficacy.

Sl.No	Crop	Effective Dose	Pathogen	Activity	Reference
1	Tomato	0.5 g/lit	*Rhizoctonia solani*	58.8% disease reduction in Pre-emergence damping off after 10 days	[40]
2	Cucumber	0.05–0.1%	*Colletotrichum* sp.	Disease control and reduction of lesion than the untreated one	[49]
3	Tea	0.01%	*Exobasidium vexans*	67.73% less disease incidence than control	[50]

Table 3. *Cont.*

Sl.No	Crop	Effective Dose	Pathogen	Activity	Reference
4	Turmeric	0.1%	*Pythium aphanidermatum*	Reduced disease severity and increased chitinase activity.	[51]
5	Grape	0.8%	*Plasmopara vitccola*	81% disease reduction	[52]

4.4. Chitosan Used as Post Harvest Fruit Treatment

Chitosan shows a dual mode of action in post-harvest disease control where it reduces the growth of decay-causing fungi and foodborne pathogens and induces resistance responses in the host tissues (Table 4). Chitosan coating forms a semipermeable film on the surface of fruit and vegetables, thereby delaying the rate of respiration, decreasing weight loss, maintaining the overall quality, and prolonging the shelf life.

Table 4. Postharvest disease control by chitosan effect in different crops.

Sl.No	Crop	Effective Dose	Pathogen	Activity	Reference
1	Pomegranate	0.1–10 g/lit of chitosan	*Botrytis* sp., *Penicillium* sp. and *Pilidiella granati*	Reduced rot incidence by 18–66%	[53]
2	Jujube fruit.	20 mg/ml	*Penicillium expansum*	More than 80% inhibition of incidence one day after treatment.	[54]
3	Kiwi fruit	5 gm/lit	*Gray mold* *(B. cinerea)* and *blue mold* *(P. expansum)*	Disease Incidence was 46% (gray mold) and 65% (blue mold) comparing to untreated control.	[55]
4	Peach	Chitosan and oligochitosan 5 g/lit	*Brown rot* *(Monilinia fruiticola)*	Disease incidence drastically reduced and in both the cases only 20% DI occurred.	[56]
5	Rose Apple	2%	*Penecillium expansum*	Disease incidence was only 14% which was 24% less than control	[57]

4.5. Effect of Chitosan in Plant Disease Control

In combating infectious diseases, chitosan exhibits different mechanisms of action against various pathogens (Figure 3). Some of them are directly related to inhibiting pathogen growth and multiplication, while others are involved in activating or enhancing plant defenses to combat pathogens and achieve sustainable yields. Thus, it is possible to produce visible positive changes in the host by improving yield-related parameters and to enhance the plant's own defenses so that the host can resist attack by plant pathogens, is the dual nature of chitosan's action. On the other hand, chitosan adversely affects or causes negative changes in pathogen growth and fertility.

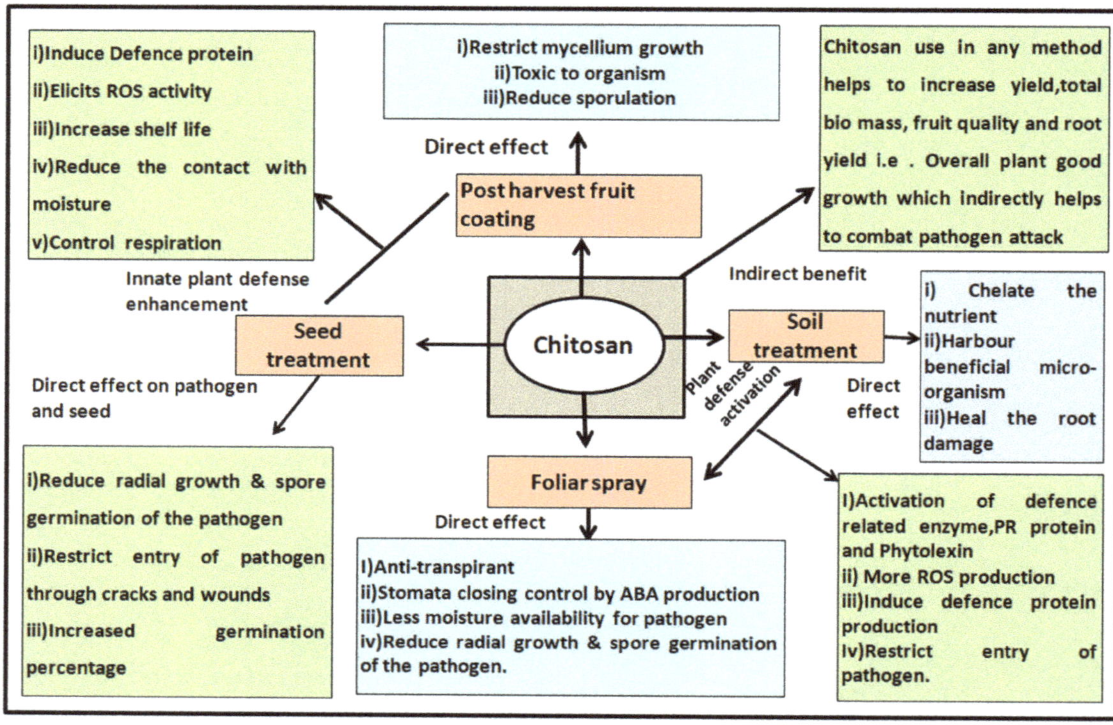

Figure 3. Methods of use of chitosan and their impact on host and pathogen both directly and indirectly in inducing defence mechanism or direct killing.

1. Indirect treatment of disease by activating defenses and improving plant health

Different methods of chitosan treatment enhance plant defenses by producing various defense enzymes, proteins and phytorexins. Sidia et al. In 2018, mentioned broader ideas regarding the defense-related activities of seed treatments with chitosan nanoparticles (CNP) [58]. CNP treatment expressed high levels of pathogenesis-related proteins PR1 and PR5. Seedlings treated with chitosan nanoparticles increased maximal phenylalanine ammonia lyase (PAL), polyphenol oxidase (PPO) and peroxidase (POX) activities by 1.08-fold, 1.10-fold and 1.10-fold, respectively. with normal chitosan treatment. It induced both systemic and permanent tolerance and showed significant protection against downy mildew. Treatment with chitosan also increased plant vigor, recording 89% seed germination. All this leads to a lower downy mildew incidence in the treated plants. H. 18.1% and 19.6% for CNP and chitosan treatment, respectively. Zen et al. (2010) also noted increases in peroxidase (POD) and superoxide dismutase (SOD) activities, as well as increases in glutathione (GSH) and hydrogen peroxide (H2O2) levels following post-harvest treatment of fruit with the application of chitosan on Navel Orange (Citrus sinensis L. Osbeck)[59]. Peroxidase and poly-phenol oxidase are responsible for eliciting plant defence. Peroxidase activity and peroxidase gene expression both were increased by chitosan treatment in many folds comparable to the control [56,60]. Chitosan significantly increases polyphenol oxidase activity by catalysing the phenolic substances responsible for lignin synthesis, that gives strength to the host cell wall and promotes prevention of pathogen entry [61,62]. Catalase enzyme responsible for degradation of H_2O into H_2O_2 and O_2 increased by the chitosan treatment in peach indicates that it plays a distinct responsibility in increasing defense, controlling aging and senescence [56]. In contrast, chitosan treatment can cause some visible impact on plants. Seed germination rate, i.e., 94.45% was increased by application of 5% chitosan by creating a semi-permeable membrane on the seed surface that helps maintain seed moisture and improve germination [63]. A combination of seed treatment and foliar application also helps increase plant height and yield [38]. With more vigorous disposition and improved yield parameters, these seeds help plants maintain yields and fend off attack by pathogens.

2. Direct method of pathogen control.

a. Chitosan mediated detrimental changes in plant pathogen

Chitin of fungal cell wall hydrolyses by chitinase enzyme and decomposes in the fungal cell wall. The presence of the chitins generally helps in inducing the activation of pathogenesis related (PR) protein. Activation of this PR protein helps in two ways. First, it catalyzes the hydrolysis process, and second, it activates the phenylalanine ammonia-lyase enzyme involved in triggering and controlling defense mechanisms through the phenylpropanoid pathway [64]. Chitosan concentrations can cause rapid potassium efflux, leading to a decrease in H+-ATPase that accumulates protons inside the cell, causing negative changes in H+/K+ exchange transport across the membrane of *Rhizopus stolonifera* [65]. Chitosan's mechanism of action has been brilliantly explained by various researchers. With combined treatment (1% seed treatment + 0.5% foliar application) with chitosan, the investigator recorded the lowest DI (10.08%) at his 90 DAT (days after treatment) [38]. Chitosan application significantly reduced the release of zoospores from the sporangia of *Phytophthora capsici* and also restricted the movement of the zoospores and transformed them into round cystospores [43]. Chitosan affects germination and hyphal morphology of economically important post-harvest fungal pathogens (e.g., *Rhizopus stolonifer* and *Botrytis cinerea*) [66,67]. This polymer also hampers the growth of different plant pathogenic and mycoparasitic fungi including *Alternaria* spp., *Colletotrichium* spp., or *Trichoderma* spp.). Chitosan was recently found to permeabilize the plasma membrane of N.crass and flow cytometry was performed. This triggers intracellular production of reactive oxygen species (ROS) and cell death [68]. RNAseq data and gene ontology (GO) analysis revealed oxidoreductase activity, plasma membrane, and transport as the main categories induced by chitosan [69]. Chitosan also enhances oxidative metabolism, respiration, and GO transport functions in the plasma membrane of the model yeast Saccharomyces cerevisiae, and the stress response and cell wall integrity genes were also identified to be induced by chitosan [70].

b. Molecular mechanism of chitosan on sensitive and tolerant fungi

Chitosan generally exhibits different modes of action in host plants and fungi. In fungi, plasma membrane fluidity generally determines the susceptibility of fungi to chitosan. High plasma membrane fluidity is caused by the presence of polyunsaturated fatty acids (FFAs) such as linolenic acid and is observed in chitosan-sensitive fungi such as *Neurospora crassa* [71]. Lopez-Moya et al. (2019) described that the opposite mechanism of action can be observed in the plasma membrane of chitosan-resistant fungi with low fluidity and saturated FFAs [72]. This fluidity effect is the single key factor that triggers ROS production in fungal cells, leading to cell death [73]. In addition, genes encoding lipase class III, monosaccharide transport, and glutathione transferase play a role in targeting chitosan in fungi, ensuring plasma membrane repair and buffering of ROS, leading to chitosan damage to cells and modulates the antifungal activity of fungi [69]. In chitosan-resistant fungi, chitosan is degraded by the action of chitin deactylase or chitosanase [18]. Chitosan has been shown to be self-defeating against pathogens and, conversely, to support biocontrol mechanism fungi [74].

A summary of the chitosan mode of action can be visualized by Figure 4.

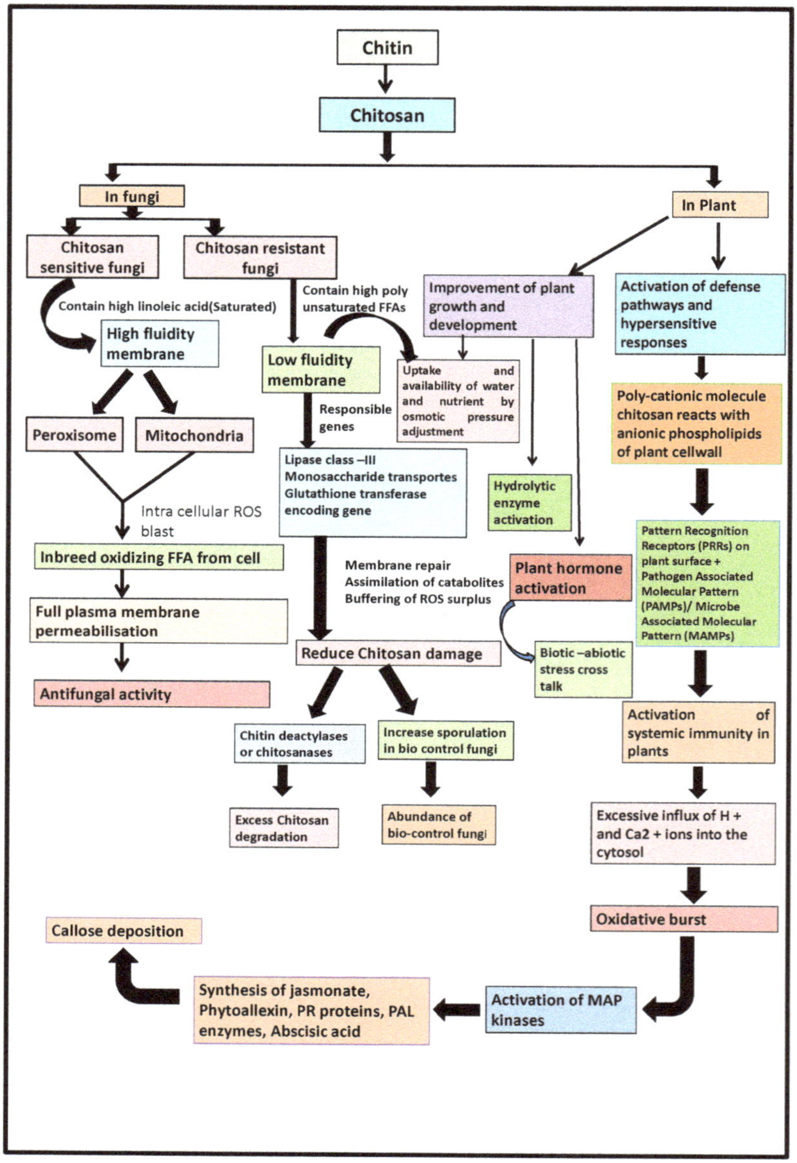

Figure 4. A summary of the chitosan mode of action.

5. Conclusions

Since the discovery of chitosan, researchers have conducted extensive research to understand its new properties, new application methods and its efficacy. Study of chitosan has repeatedly demonstrated its versatile properties for combating plant diseases through its broad-spectrum anti phytopathogenic activity and by inducing the plant's own defense activity. For this reason, in this era in chitosan is being highlighted as a legitimate alternative to chemicals and greatly ensuring food and environmental safety.

6. Future Aspect

Although extensive research has been conducted, further research is needed to identify the exact mechanism of action of chitosan in combating pathogens. Studies of the efficacy of chitosan to combat viral and prokaryotic plant pathogenic diseases need to be more comprehensive. The persistence of induced defense in the host, gene expression after chitosan application and inherited changes in defense are all awaiting exploitation and staining by researchers.

Author Contributions: Conceptualization, D.D. and I.S.; methodology, C.M.; software, M.S.K.; validation, S.R., R.L. and D.D.; formal analysis, I.S.; investigation, D.D. and I.S.; resources, D.D. and I.S.; data curation, D.D., M.S.K. and C.M.; writing—original draft preparation, D.D., I.S. and C.M.; writing—review and editing, R.L. and M.S.K.; visualization, D.D.; supervision, R.L. and M.S.K. All authors have read and agreed to the published version of the manuscript.

Funding: The research has no funding from any sources.

Institutional Review Board Statement: Not applicable.

Informed Consent Statement: Not applicable.

Data Availability Statement: Not applicable.

Conflicts of Interest: The authors declare no conflict of interest.

References

1. Ghule, M.R.; Ramteke, P.K.; Ramteke, S.D.; Kodre, P.S.; Langote, A.; Gaikwad, A.V.; Holkar, S.K.; Jambhekar, H. Impact of chitosan seed treatment of fenugreek for management of root rot disease caused by Fusarium solani under in vitro and in vivo conditions. *3 Biotech* **2021**, *11*, 290. [CrossRef] [PubMed]
2. Sun, R.; Liu, C.; Zhang, H.; Wang, Q. Benzoylurea Chitin Synthesis Inhibitors. *J. Agric. Food Chem.* **2015**, *63*, 6847–6865. [CrossRef] [PubMed]
3. Jayakumar, R.; Prabaharan, M.; Kumar, P.T.S.; Nair, S.V.; Tamura, H. Biomaterials based on chitin and chitosan in wound dressing applications. *Biotechnol. Adv.* **2011**, *29*, 322–337. [CrossRef]
4. Prasad, R.; Chandrika, K.; Godbole, V. A novel chitosan biopolymer based Trichoderma delivery system: Storage stability, persistence and bio efficacy against seed and soil borne diseases of oilseed crops. *Microbiol. Res.* **2020**, *237*, 126487. [CrossRef] [PubMed]
5. Amine, R.; Tarek, C.; Hassane, E.; Noureddine, E.H.; Khadija, O. Chemical Proprieties of Biopolymers (Chitin/Chitosan) and their Synergic Effects with Endophytic *Bacillus* Species: Unlimited Applications in Agriculture. *Molecules* **2021**, *26*, 1117. [CrossRef]
6. Kean, T.; Thanou, M. Biodegradation, biodistribution and toxicity of chitosan. *Adv. Drug Deliv. Rev.* **2010**, *62*, 3–11. [CrossRef]
7. Hassan, O.; Chang, T. Chitosan for Eco-friendly Control of Plant Disease. *Asian J. Plant Pathol.* **2017**, *11*, 53–70. [CrossRef]
8. Jones, M.; Kujundzic, M.; John, S.; Bismarck, A. Crab vs. Mushroom: A Review of Crustacean and Fungal Chitin in Wound Treatment. *Mar. Drugs* **2020**, *18*, 64. [CrossRef]
9. Chen, J.-K.; Shen, C.-R.; Liu, C.-L. N-Acetylglucosamine: Production and Applications. *Mar. Drugs* **2010**, *8*, 2493–2516. [CrossRef]
10. Merzendorfer, H. The cellular basis of chitin synthesis in fungi and insects: Common principles and differences. *Eur. J. Cell Biol.* **2011**, *90*, 759–769. [CrossRef]
11. Wang, S.L.; Chang, W.T. Purification and characterization of two bifunctional chitinases/lysozymes extracellularly produced by *Pseudomonas aeruginosa* K-187 in a shrimp and crab shell powder medium. *Appl. Environ. Microbiol.* **1997**, *63*, 380–386. [CrossRef]
12. Mourya, V.; Inamdar, N.N. Chitosan-modifications and applications: Opportunities galore. *React. Funct. Polym.* **2008**, *68*, 1013–1051. [CrossRef]
13. Synowiecki, J.; Al-Khateeb, N.A. Production, Properties, and Some New Applications of Chitin and Its Derivatives. *Crit. Rev. Food Sci. Nutr.* **2003**, *43*, 145–171. [CrossRef]
14. Kramer, K.J.; Koga, D. Insect chitin: Physical state, synthesis, degradation and metabolic regulation. *Insect Biochem.* **1986**, *16*, 851–877. [CrossRef]
15. Merzendorfer, H.; Zimoch, L. Chitin metabolism in insects: Structure, function and regulation of chitin synthases and chitinases. *J. Exp. Biol.* **2003**, *206*, 4393–4412. [CrossRef]
16. Hu, X.; Chen, L.; Xiang, X.; Yang, R.; Yu, S.; Wu, X. Proteomic analysis of peritrophic membrane (PM) from the midgut of fifth-instar larvae, Bombyx mori. *Mol. Biol. Rep.* **2012**, *39*, 3427–3434. [CrossRef]
17. Rinaudo, M. Chitin and chitosan: Properties and applications. *Prog. Polym. Sci.* **2006**, *31*, 603–632. [CrossRef]
18. Kaczmarek, M.B.; Struszczyk-Swita, K.; Li, X.; Szczęsna-Antczak, M.; Daroch, M. Enzymatic Modifications of Chitin, Chitosan, and Chitooligosaccharides. *Front. Bioeng. Biotechnol.* **2019**, *7*, 243. [CrossRef]
19. Arrese, E.L.; Soulages, J.L. Insect Fat Body: Energy, Metabolism, and Regulation. *Annu. Rev. Èntomol.* **2010**, *55*, 207–225. [CrossRef]

20. Salama, D.M.; El-Aziz, M.A.; Rizk, F.A.; Elwahed, M.A. Applications of nanotechnology on vegetable crops. *Chemosphere* **2021**, *266*, 129026. [CrossRef]
21. Suresh, P.V.; Kumar, P.K.A. Enhanced degradation of α-chitin materials prepared from shrimp processing byproduct and production of N-acetyl-d-glucosamine by thermoactive chitinases from soil mesophilic fungi. *Biogeochemistry* **2012**, *23*, 597–607. [CrossRef] [PubMed]
22. Brasselet, C.; Pierre, G.; Dubessay, P.; Dols-Lafargue, M.; Coulon, J.; Maupeu, J.; Vallet-Courbin, A.; de Baynast, H.; Doco, T.; Michaud, P.; et al. Modification of Chitosan for the Generation of Functional Derivatives. *Appl. Sci.* **2019**, *9*, 1321. [CrossRef]
23. Mourya, V.K.; Inamdar, N.N.; Choudhari, Y.M. Chitooligosaccharides: Synthesis, characterization and applications. *Polym. Sci. Ser. A* **2011**, *53*, 583–612. [CrossRef]
24. Chatterjee, S.; Adhya, M.; Guha, A.; Chatterjee, B. Chitosan from *Mucor rouxii*: Production and physico-chemical characterization. *Process Biochem.* **2005**, *40*, 395–400. [CrossRef]
25. LogithKumar, R.; KeshavNarayan, A.; Dhivya, S.; Chawla, A.; Saravanan, S.; Selvamurugan, N. A review of chitosan and its derivatives in bone tissue engineering. *Carbohydr. Polym.* **2016**, *151*, 172–188. [CrossRef] [PubMed]
26. Arnold, N.D.; Brück, W.M.; Garbe, D.; Brück, T.B. Enzymatic Modification of Native Chitin and Conversion to Specialty Chemical Products. *Mar. Drugs* **2020**, *18*, 93. [CrossRef] [PubMed]
27. Sharma, N.; Sharma, K.; Gaur, R.; Gupta, V. Role of Chitinase in Plant Defense. *Asian J. Biochem.* **2011**, *6*, 29–37. [CrossRef]
28. Rathore, A.S.; Gupta, R.D. Chitinases from Bacteria to Human: Properties, Applications, and Future Perspectives. *Enzym. Res.* **2015**, *2015*, 791907. [CrossRef]
29. Loc, N.H.; Huy, N.D.; Quang, H.T.; Lan, T.T.; Ha, T.T.T. Characterisation and antifungal activity of extracellular chitinase from a biocontrol fungus, *Trichoderma asperellum* PQ34. *Mycology* **2019**, *11*, 38–48. [CrossRef]
30. Roy, J.C.; Salaun, F.; Giraud, S.; Ferri, A.; Chen, G.; Guan, J. Solubility of chitin: Solvents, solution behaviors and their related mechanisms. In *Solubility of Polysaccharides*; Xu, Z., Ed.; Intech Open: London, UK, 2017. [CrossRef]
31. Li, J.; Li, R.; Zhang, C.; Guo, Z.; Wu, X.; An, H. Co-Application of Allicin and Chitosan Increases Resistance of *Rosa roxburghii* against Powdery Mildew and Enhances Its Yield and Quality. *Antibiotics* **2021**, *10*, 1449. [CrossRef]
32. Ferrante, P.; Scortichini, M. Molecular and phenotypic features of *Pseudomonas syringae* pv. *actinidiae isolated during recent epidemics of bacterial canker on yellow kiwifruit (Actinidia chinensis) in central Italy. Plant Pathol.* **2010**, *59*, 954–962. [CrossRef]
33. Huda, M.S. Antifungal and Antioxidant Activities of methanol extract of Chitin, Chitosan and Shrimp shell waste. *Int. J. Pharm. Phytopharm. Res.* **2018**, *8*, 25–30. [CrossRef]
34. Zhang, J.; Luan, F.; Li, Q.; Gu, G.; Dong, F.; Guo, Z. Synthesis of Novel Chitin Derivatives Bearing Amino Groups and Evaluation of Their Antifungal Activity. *Mar. Drugs* **2018**, *16*, 380. [CrossRef]
35. Shakeel, Q.; Lyu, A.; Zhang, J.; Wu, M.; Li, G.; Hsiang, T.; Yang, L. Biocontrol of *Aspergillus flavus* on Peanut Kernels Using *Streptomyces yanglinensis* 3-10. *Front. Microbiol.* **2018**, *9*, 1049. [CrossRef]
36. Xing, K.; Shen, X.; Zhu, X.; Ju, X.; Miao, X.; Tian, J.; Feng, Z.; Peng, X.; Jiang, J.; Qin, S. Synthesis and in vitro antifungal efficacy of oleoyl-chitosan nanoparticles against plant pathogenic fungi. *Int. J. Biol. Macromol.* **2016**, *82*, 830–836. [CrossRef]
37. Zachetti, V.G.L.; Cendoya, E.; Nichea, M.J.; Chulze, S.N.; Ramirez, M.L. Preliminary Study on the Use of Chitosan as an Eco-Friendly Alternative to Control Fusarium Growth and Mycotoxin Production on Maize and Wheat. *Pathogens* **2019**, *8*, 29. [CrossRef]
38. Akter, J.; Jannat, R.; Hossain, M.; Ahmed, J.U.; Rubayet, T. Chitosan for Plant Growth Promotion and Disease Suppression against Anthracnose in Chilli. *Int. J. Environ. Agric. Biotechnol.* **2018**, *3*, 806–817. [CrossRef]
39. Riad, S.R.; El-Mohamedy, M.M.; Abdel-Kader, F.; Abd-El-Kareem, M.M.; El-Mougy, N.S. Inhibitory effect of antagonistic bio-agents and chitosan on the growth of tomato root rot pathogens In vitro. *J. Agric. Technol.* **2013**, *9*, 1521–1533.
40. El-Mohamedy, S.R.R.; Shafeek, M.R.; Abd El-Samad, E.E.-D.H.; Salama, D.M.; Rizk, F.A. Field application of plant resistance inducers (PRIs) to control important root rot diseases and improvement growth and yield of green bean (*Phaseolus vulgaris* L.). *Aust. J. Crop Sci.* **2017**, *11*, 496–505. [CrossRef]
41. Li, R.; He, J.; Xie, H.; Wang, W.; Bose, S.K.; Sun, Y.; Hu, J.; Yin, H. Effects of chitosan nanoparticles on seed germination and seedling growth of wheat (*Triticum aestivum* L.). *Int. J. Biol. Macromol.* **2019**, *126*, 91–100. [CrossRef]
42. Xue, H.; Yang, Z. *Potato Dry Rot Caused by Fusarium spp. and Mycotoxins Accumulation and Management*; Intech Open: London, UK, 2021. [CrossRef]
43. Zohara, F.; Surovy, M.Z.; Khatun, A.; Prince, F.R.K.; Akanda, A.M.; Rahman, M.; Islam, T. Chitosan biostimulant controls infection of cucumber by Phytophthora capsici through suppression of asexual reproduction of the pathogen. *Acta Agrobot.* **2019**, *72*, 1763. [CrossRef]
44. Corsi, B.; Riccioni, L.; Forni, C. In vitro cultures of Actinidia deliciosa (A. Chev) C.F. Liang & A.R. Ferguson: A tool to study the SAR induction of chitosan treatment. *Org. Agric.* **2015**, *5*, 189–198. [CrossRef]
45. Murphy, J.G.; Rafferty, S.M.; Cassells, A.C. Stimulation of wild strawberry (Fragaria vesca) arbuscular mycorrhizas by addition of shellfish waste to the growth substrate: Interaction between mycorrhization, substrate amendment and susceptibility to red core (*Phytophthora fragariae*). *Appl. Soil Ecol.* **2000**, *15*, 153–158. [CrossRef]
46. Mendes, R.; Kruijt, M.; de Bruijn, I.; Dekkers, E.; Van Der Voort, M.; Schneider, J.H.; Piceno, Y.M.; DeSantis, T.Z.; Andersen, G.L.; Bakker, P.A.; et al. Deciphering the Rhizosphere Microbiome for Disease-Suppressive Bacteria. *Science* **2011**, *332*, 1097–1100. [CrossRef] [PubMed]

47. Radwan, M.A.; Farrag, S.A.A.; Abu-Elamayem, M.M.; Ahmed, N.S. Extraction, characterization, and nematicidal activity of chitin and chitosan derived from shrimp shell wastes. *Biol. Fertil. Soils* **2012**, *48*, 463–468. [CrossRef]
48. Rahman, H.; Shovan, L.R.; Hjeljord, L.G.; Aam, B.B.; Eijsink, V.G.H.; Sørlie, M.; Tronsmo, A. Inhibition of Fungal Plant Pathogens by Synergistic Action of Chito-Oligosaccharides and Commercially Available Fungicides. *PLoS ONE* **2014**, *9*, e93192. [CrossRef]
49. Dodgson, J.; Dodgson, W. Comparison of Effects of Chitin and Chitosan for Control of Colletotrichum sp. on Cucumbers. *J. Pure Appl. Microbiol.* **2017**, *11*, 87–93. [CrossRef]
50. Chandra, S.; Chakraborty, N.; Panda, K.; Acharya, K. Chitosan-induced immunity in *Camellia sinensis* (L.) O. Kuntze against blister blight disease is mediated by nitric-oxide. *Plant Physiol. Biochem.* **2017**, *115*, 298–307. [CrossRef]
51. Anusuya, S.; Sathiyabama, M. Effect of Chitosan on Rhizome Rot Disease of Turmeric Caused by *Pythium aphanidermatum*. *ISRN Biotechnol.* **2014**, *2014*, 305349. [CrossRef]
52. Romanazzi, G.; Feliziani, E.; Baños, S.B.; Sivakumar, D. Shelf life extension of fresh fruit and vegetables by chitosan treatment. *Crit. Rev. Food Sci. Nutr.* **2017**, *57*, 579–601. [CrossRef]
53. Munhuweyi, K.; Lennox, C.L.; Meitz-Hopkins, J.C.; Caleb, O.J.; Sigge, G.O.; Opara, U.L. Investigating the effects of crab shell chitosan on fungal mycelial growth and postharvest quality attributes of pomegranate whole fruit and arils. *Sci. Hortic.* **2017**, *220*, 78–89. [CrossRef]
54. Wang, L.; Wu, H.; Qin, G.; Meng, X. Chitosan disrupts *Penicillium expansum* and controls postharvest blue mold of jujube fruit. *Food Control* **2014**, *41*, 56–62. [CrossRef]
55. Zheng, F.; Zheng, W.; Li, L.; Pan, S.; Liu, M.; Zhang, W.; Liu, H.; Zhu, C. Chitosan Controls Postharvest Decay and Elicits Defense Response in Kiwifruit. *Food Bioprocess Technol.* **2017**, *10*, 1937–1945. [CrossRef]
56. Ma, Z.; Yang, L.; Yan, H.; Kennedy, J.F.; Meng, X. Chitosan and oligochitosan enhance the resistance of peach fruit to brown rot. *Carbohydr. Polym.* **2013**, *94*, 272–277. [CrossRef]
57. Plainsirichai, M.; Leelaphatthanapanich, S.; Wongsachai, N. Effect of chitosan on the quality of rose apples (*Syzygium agueum* Alston) cv. tabtimchan stored at an ambient temperature. *APCBEE Procedia* **2014**, *8*, 317–322. [CrossRef]
58. Siddaiah, C.N.; Prasanth, K.V.H.; Satyanarayana, N.R.; Mudili, V.; Gupta, V.K.; Kalagatur, N.K.; Satyavati, T.; Dai, X.-F.; Chen, J.-Y.; Mocan, A.; et al. Chitosan nanoparticles having higher degree of acetylation induce resistance against pearl millet downy mildew through nitric oxide generation. *Sci. Rep.* **2018**, *8*, 2485. [CrossRef]
59. Zeng, K.; Deng, Y.; Ming, J.; Deng, L. Induction of disease resistance and ROS metabolism in navel oranges by chitosan. *Sci. Hortic.* **2010**, *126*, 223–228. [CrossRef]
60. Meng, X.; Yang, L.; Kennedy, J.F.; Tian, S. Effects of chitosan and oligochitosan on growth of two fungal pathogens and physiological properties in pear fruit. *Carbohydr. Polym.* **2010**, *81*, 70–75. [CrossRef]
61. Li, S.-J.; Zhu, T.-H. Biochemical response and induced resistance against anthracnose (*Colletotrichum camelliae*) of camellia (*Camellia pitardii*) by chitosan oligosaccharide application. *For. Pathol.* **2013**, *43*, 67–76. [CrossRef]
62. Li, B.; Liu, B.; Shan, C.; Ibrahim, M.; Lou, Y.; Wang, Y.; Xie, G.; Li, H.-Y.; Sun, G. Antibacterial activity of two chitosan solutions and their effect on rice bacterial leaf blight and leaf streak. *Pest Manag. Sci.* **2013**, *69*, 312–320. [CrossRef]
63. Zeng, D.; Luo, X.; Tu, R. Application of Bioactive Coatings Based on Chitosan for Soybean Seed Protection. *Int. J. Carbohydr. Chem.* **2012**, *2012*, 104565. [CrossRef]
64. Hadwiger, L.A. Multiple effects of chitosan on plant systems: Solid science or hype. *Plant Sci.* **2013**, *208*, 42–49. [CrossRef] [PubMed]
65. Garcia-Rincona, J.; Vega-Perezb, J.; Guerra-Sanchezb, M.G.; Hernandez-Lauzardoa, A.N.; Pena-Diazc, A.; Vallea, M.G.V.-D. Effect of chitosan on growth and plasma membrane properties of *Rhizopus stolonifer* (Ehrenb.:Fr.) Vuill. *Pestic. Biochem. Physiol.* **2010**, *97*, 275–278. [CrossRef]
66. Ghaouth, A.; Arul, J.; Ponnampalam, R.; Boulet, M. Chitosan Coating Effect on Storability and Quality of Fresh Strawberries. *J. Food Sci.* **1991**, *56*, 1618–1620. [CrossRef]
67. Verleet, A.; Mincke, S.; Stevens, C.V. Recent developments in antibacterial and antifungal chitosan and its derivatives. *Carbohydr. Polym.* **2017**, *164*, 268–283. [CrossRef] [PubMed]
68. Lopez-Moya, F.; Colom-Valiente, M.F.; Martinez-Peinado, P.; Martinez-Lopez, J.E.; Puelles, E.; Sempere-Ortells, J.M.; Lopez-Llorca, L.V. Carbon and nitrogen limitation increase chitosan antifungal activity in *Neurospora crassa* and fungal human pathogens. *Fungal Biol.* **2015**, *119*, 154–169. [CrossRef]
69. Lopez-Moya, F.; Kowbel, D.; Nueda, M.J.; Palma-Guerrero, J.; Glass, N.L.; Lopez-Llorca, L.V. *Neurospora crassa* transcriptomics reveals oxidative stress and plasma membrane homeostasis biology genes as key targets in response to chitosan. *Mol. BioSyst.* **2016**, *12*, 391–403. [CrossRef]
70. Jaime, M.D.; Lopez-Llorca, L.V.; Conesa, A.; Lee, A.Y.; Proctor, M.; Heisler, L.E.; Gebbia, M.; Giaever, G.; Westwood, J.T.; Nislow, C. Identification of yeast genes that confer resistance to chitosan oligosaccharide (COS) using chemogenomics. *BMC Genom.* **2012**, *13*, 267. [CrossRef]
71. Palma-Guerrero, J.; Huang, I.-C.; Jansson, H.-B.; Salinas, J.; Lopez-Llorca, L.V.; Read, N. Chitosan permeabilizes the plasma membrane and kills cells of *Neurospora crassa* in an energy dependent manner. *Fungal Genet. Biol.* **2009**, *46*, 585–594. [CrossRef]
72. Lopez-Moya, F.; Suarez-Fernandez, M.; Lopez-Llorca, L.V. Molecular Mechanisms of Chitosan Interactions with Fungi and Plants. *Int. J. Mol. Sci.* **2019**, *20*, 332. [CrossRef]

73. Ke, C.-L.; Deng, F.-S.; Chuang, C.-Y.; Lin, C.-H. Antimicrobial Actions and Applications of Chitosan. *Polymers* **2021**, *13*, 904. [CrossRef]
74. Singh, T.; Chittenden, C. Synergistic Ability of Chitosan and *Trichoderma harzianum* to Control the Growth and Discolouration of Common Sapstain Fungi of *Pinus radiata*. *Forests* **2021**, *12*, 542. [CrossRef]

life

Review

Molecular Breeding and Drought Tolerance in Chickpea

Ruchi Asati [1], Manoj Kumar Tripathi [1,2,*], Sushma Tiwari [1,2], Rakesh Kumar Yadav [1] and Niraj Tripathi [3,*]

[1] Department of Genetics & Plant Breeding, College of Agriculture, Rajmata Vijayaraje Scindia Krishi Vishwa Vidyalaya, Gwalior 474002, India
[2] Department of Plant Molecular Biology & Biotechnology, College of Agriculture, Rajmata Vijayaraje Scindia Krishi Vishwa Vidyalaya, Gwalior 474002, India
[3] Directorate of Research Services, Jawaharlal Nehru Agricultural University, Jabalpur 482004, India
* Correspondence: drmanojtripathi64@gmail.com (M.K.T.); tripathi.niraj@gmail.com (N.T.)

Abstract: *Cicer arietinum* L. is the third greatest widely planted imperative pulse crop worldwide, and it belongs to the Leguminosae family. Drought is the utmost common abiotic factor on plants, distressing their water status and limiting their growth and development. Chickpea genotypes have the natural ability to fight drought stress using certain strategies viz., escape, avoidance and tolerance. Assorted breeding methods, including hybridization, mutation, and marker-aided breeding, genome sequencing along with omics approaches, could be used to improve the chickpea germplasm lines(s) against drought stress. Root features, for instance depth and root biomass, have been recognized as the greatest beneficial morphological factors for managing terminal drought tolerance in the chickpea. Marker-aided selection, for example, is a genomics-assisted breeding (GAB) strategy that can considerably increase crop breeding accuracy and competence. These breeding technologies, notably marker-assisted breeding, omics, and plant physiology knowledge, underlined the importance of chickpea breeding and can be used in future crop improvement programmes to generate drought-tolerant cultivars(s).

Keywords: abiotic stress; candidate genes; drought tolerance; crop improvement; climate change

Citation: Asati, R.; Tripathi, M.K.; Tiwari, S.; Yadav, R.K.; Tripathi, N. Molecular Breeding and Drought Tolerance in Chickpea. *Life* **2022**, *12*, 1846. https://doi.org/10.3390/life12111846

Academic Editors: Hakim Manghwar and Wajid Zaman

Received: 4 October 2022
Accepted: 7 November 2022
Published: 11 November 2022

Publisher's Note: MDPI stays neutral with regard to jurisdictional claims in published maps and institutional affiliations.

Copyright: © 2022 by the authors. Licensee MDPI, Basel, Switzerland. This article is an open access article distributed under the terms and conditions of the Creative Commons Attribution (CC BY) license (https://creativecommons.org/licenses/by/4.0/).

1. Introduction

Chickpea is a diploid annual crop that is extremely self-pollinated [1]. After the faba bean and field pea, it is the world's third most significant pules crop [2]. It is a popular cool-season legume crop with a 738-megabyte genome size [3]. With an annual production of 10.13 million tonnes from a land area of 9.44 million hectares and a productivity of 1073 kg ha^{-1}, India is the greatest producer of chickpeas in the world [4]. Chickpeas are grown in 52 countries, together with Africa, Asia, Australia, and South Europe [5]. Mexico, Turkey, Canada, Iran, Australia, Tanzania, Ethiopia, Spain, and Burma are also notable producers of chickpea. Its seeds come in two varieties. The 'desi' chickpea is hardy in character, while the Kabuli chickpea has a delicate seed coat and appears to have evolved from the desi varieties [6,7]. In semi-arid zones, chickpea is cultivated in the form of a dry weather crop [8]; however, in cold climatic zones, it is grown as a rainfed crop [9,10]. In actuality, about 90% of the chickpea crop is cultivated in a rainfed environment [11–13]. Without irrigation, the crop is affected [14] at vegetative as well as reproductive phases. After illnesses, drought is the second most significant constraint to the yield of chickpea crop [15]. Drought has been reported as a factor of 40–50 percent yield reduction in chickpea [11,12,16–18].

The chickpea is also termed as the "poor man's meat" [19], since it is important for supplying protein sources [20]. Nutritionists have also highlighted its importance due to high nutritional contents in it [21]. Chickpea is high in lysine and arginine [22], but low in methionine and cystine [23]. In general, the Kabuli type contains more protein than the desi kinds. It contains more calcium and phosphorus than most other pulse crops [24,25].

Chickpea seeds comprise 23% protein, 64% total carbohydrates (47% starch, 6% soluble sugar), 5% fat, 6% crude fibre, and 2% ash on average, as well as micronutrients, for example phosphorus, calcium, magnesium, iron, and zinc [26]. Recently, Singh et al. [21] also reported chickpea as good source of Fe and Zn. Consequently, Samineni et al. [27] examined the effects of drought stress on nutritional parameters of chickpea and observed significant differences in the nutritional contents due to stress.

Chickpea is mostly cultivated in the post-rainy season [28], using soil moisture that has been retained from the previous rainy season [29]. As a result, the crop is frequently subjected to severe heat and drought pressure [12,13,22]. Drought, among other abiotic factors, has a significant impact on chickpea output [30]. Drought and heat stress have been reported to have reduced chickpea yields by about 50% due to the damaging effects of the membrane and reduced photosynthesis [31].

The four climatic elements that are changing will have an impact on how much water plants consume [32]. These elements include rising CO_2 concentrations and temperatures, more erratic precipitation, and changes in humidity. Due to the increased variability in precipitation during the growing season and more so in soils with low water holding capacity, these climate changes may result in an increase in the atmospheric water demand by crops and an increase in the potential for limitations in the availability of water in the soil. In the long run, breeding cultivars with high water use efficiency (WUE) is a more realistic and cost-effective strategy for raising yields in drought-prone locations. WUE promotes modest water absorption while maintaining elevated WUE, which is a crucial component of breeding programmes because of its yields in drought-prone areas. Any WUE is impacted by changes above the soil surface because they have an impact on the soil water balance by the evaporation and penetration of soil water. The majority of the chickpea crop is grown on residual moisture; however, additional irrigation can increase yields. At some sites in India, irrigation during the pre-flowering stage and at the beginning of the pod fill led to an increase in yield. Chickpeas' reproductive cycle was prolonged by irrigation, which also increased plant biomass and increased the number of pods per plant.

The greatest sustained surface winds of tropical storms range from 39 to 73 mph, and they are fast rotating storm systems with an organized centre over warm tropical oceans. These storms have a wide range in size and can cause a variety of dangers for the impacted areas, including tornadoes, catastrophic winds, coastal floods, and inland flooding. The effects of tropical cyclones on drought have been extensively studied, but less research has been conducted on how smaller tropical storms affect the severity of drought. According to research, rainfall is not necessarily inversely correlated with the strength of a tropical cyclone; therefore, tropical storms can sometimes provide more rain than expected. The question of whether tropical storms can help to lessen and mitigate drought conditions is now being researched. Water deficit and surplus are related to drought and tropical storms (TS), respectively. When it comes to monitoring dryness, soil moisture is a crucial element of the hydrological cycle, since it reflects the water that TS rainfall has penetrated or stored. Soil moisture data can be used to determine whether TS can alleviate extremely severe drought situations [33]. The authors calculated the frequency of TS afflicted places in the US, including the ratio of droughts that TS exacerbated and alleviated, and the regions where TS have a significant impact on the offset of drought. Based on a high-resolution data set, the findings demonstrate extensive spatial information about the offset of drought conditions and offer potential guidance for future drought and TS mitigation.

Drought has a substantial influence on crop growth and photosynthesis, both of which are directly related to production [34,35]. Drought researchers must assess growth as well as physiological responses such as chlorophyll index [36], relative water content [37], membrane stability index, and biomass when determining the influence of drought on various crop metrics. The quantitative character of attributes and the prevalence of linkage between desired and undesired genes make developing drought-tolerant agricultural variants difficult [38]. Many experiments on the effects of drought on numerous chickpea features, such as root attributes, shoot biomass, and early maturity, have been conducted [39]. In this crop, various experiments have been performed successfully and published with specific conclusions on different aspects, such as morphological, physiological, biochemical, and molecular characteristics [40–42].

Advances genomics has made it possible to tag genes [43] associated to agronomic qualities, as well as the tolerance/resistance to abiotic and biotic challenges [44]. It is playing a significant role in the transfer of labelled genes through molecular breeding [45,46], quickly and accurately. In chickpeas, microsatellite and sequence-tagged microsatellite site markers have been found to be more beneficial [47,48]. As stress resistance/tolerance is governed by numerous genes [49], quantitative trait loci (QTL) mapping has proven to be effective in identifying and tagging the genes [50] involved for disease resistance/tolerance in plants. Foreground, recombinant and background selection are all examples of marker-assisted backcrossing. Linkage disequilibrium (LD) and association mapping are also determined using markers [51].

Next-generation sequencing (NGS) is a segment of revolutionary biology being a frontier area in crop science and produces correct data, with the results of significant throughput [52] and reduction in the need for fragment-cloning processes, which were the initial requirement for Sanger sequencing. NGS is used for the identification and mapping of mutations in targeted genotype [53,54]. Aside from whole genome sequencing (WGS), NGS also provides a platform for whole transcriptome shotgun sequencing, which is also termed as RNA sequencing (RNA-seq) [55,56] and whole-exome sequencing [57], which exhibits for functional variations [58], targeted or candidate gene sequencing [59].In the examination of large numbers of samples, RNA-seq enables a more precise and sensitive measurement of gene expression levels than microarrays [60].

Transcriptomics is the technology used to study the transcriptome of an organism [61]. Transcriptome is the complete set of genes [62] expressed under specific conditions by the genome of the targeted organism. MicroRNA (miRNA), transfer RNA (tRNA), messenger RNA (mRNA), ribosomal RNA (rRNA) and other non-coding RNA are all found in the transcriptome (ncRNA). Transcriptomics of chickpea [63] has provided insight into mechanisms of drought tolerance/avoidance, as well as pathogenesis-related and developmental processes [64]. Transcriptomics may undoubtedly have a greater impact on chickpea breeding in the future, including the use of microarrays.

Proteomics is the study of whole protein complement in a cell, tissue or organism in detail [65]. Mass spectrometry and protein microarrays can be used to analyse the proteome [66].

Role of various genes of plants under drought stress conditions have been recognized clearly [67]. Drought responsive mechanisms are activated in response to drought stress, which is a regular occurrence in plants. Morphological and structural changes [68], drought-resistant gene expression, hormonal and other biochemical changes are among these pathways [69]. Environmental stresses have the ability to change the developmental behaviour of plants. These alterations in plant growth and development [70,71] mostly resulted in lower yields [72]. We attempted to review the status and progress based on the existing literature on drought stress tests conducted in the chickpea.

2. Drought in Chickpea

Chickpea productivity has been found to be around 995 kg ha^{-1} on a global scale, which is quite low [73]. Drought, terminal heat [74], excessive salt, and cold are abiotic variables [75], whereas *Ascochyta* blight, *Fusarium* wilt, and *Helicoverpa* are biotic factors that have been recognised as key drivers of yield reduction in chickpea [76]. Drought stress was identified as a major cause in around 50% of chickpea output losses worldwide.

Several factors are responsible for complexity of drought stress (Table 1), including severity of drought, stage of crop, and duration of drought stress [77]. Two types of drought stresses, i.e., terminal and intermittent, have been reported with their impacts on crop plants [78]. During terminal drought, soil water availability diminishes over time, potentially leading to severe drought stress later in crop development. Intermittent drought is defined as a series of short episodes of insufficient rain or irrigation that occur at different times during the growing season [79]. Due to its limited cultivation on marginal terrain, chickpea is suffering from terminal drought stress. Intermittent and terminal drought stress is caused by breaks in rainfall combined with less moisture in terminal growth stages [80]. Apart from morpho-physiological factors various genes and proteins are also responsible for drought tolerance in chickpea crop (Table 2).

Table 1. Relevance of various physiological traits contributing to drought adaptation in chickpea.

Physiological Traits	Related with	References
Early phenology (early flowering, early podding)	Drought escape/conservative water-use strategy	[81–83]
Crop growth rate	High water harvest	[47]
Shoot biomass	High shoot biomass at maturity contribute to a higher grain yield under drought	[84]
Pod abortion and seed filling	High seed/grain yield could help in drought and heat stress tolerance	[85]
Biomass partitioning	Greater biomass partitioning to grain helps in drought and heat stress tolerance	[46,47,86]
Pod number; high pod number	Grain yield and contributes to heat, drought tolerance	[87]
Pod production	Number of pods/plants is more affected at early stage than late stage under drought stress	[88]
Specific leaf area	SLA has a positive effect on grain yield at reproductive stage	[89]
Cell membrane stability	Related to drought, heat, and cold tolerance	[30,90–92]
Canopy temperature depression	Cooler canopy contributes to drought avoidance and has a positive association with seed yield under drought stress, and it also contributes to heat stress tolerance	[93–95]
Canopy conductance	Associated to both heat and drought stress tolerance	[96]
Carbon isotope Discrimination	Transpiration efficiency	[97]
Recycling of CO_2 inside the pod	Maintain seed filling	[98]
Antioxidants enzymes, proline, anthocyanin content, trehalose, sucrose, and nonreducing sugars	Increase in antioxidant enzymes, proline, trehalose and anthocyanin content during vegetative stage causes drought and cold stress tolerance	[99]
Relative water content	Increase in relative water content causes drought stress tolerance	[100,101]
Chlorophyll content; carotenoid content	Higher chlorophyll content and carotenoid content helps in heat stress tolerance	[55,96]
(Na+ and K+) ion uptake	(Na+ and K+) ion uptake cause drought tolerance	[102]
Chlorophyll a fluorescence FO, FM, PSII, ETR, FV/FM	Enable preventing PSII photochemistry from damage and helps in both drought and heat stress tolerance	[102,103]
Plant transpiration rate	Low plant transpiration rate helps in conserving soil water	[104,105]
Transpiration efficiency	It decides ultimate yield	[106,107]
Early vigour	Associated to both heat and drought stress tolerance	[108]
Pollen traits (pollen viability, fertility, and pollentube germination)	High pollen viability and fertility under heat stress are associated to heat stress tolerance	[109]
Abscisic acid (ABA)	Under drought increase in ABA causes closure of stomata, thus reducing assimilate production that leads to the inhibition of seed set	[108]
Root architectural trait prolific root system, root branch, root density root depth, root area, and root volume	Prolific root system is associated to grain yield	[47]
	Deep rooting helps in using conserved soil moisture from subsoil and helps in avoiding terminal drought stress	[108]

Due to the overabundance of wheat in irrigated areas in India, chickpea growth is primarily limited to rainfed areas. Crops in rainfed areas are experiencing water shortages, particularly during the sowing and terminal growth periods.

Soil and plant management are important for minimizing water stress. For this purpose, various experiments have been conducted with the applications of different agents. Gypsum can enhance overall plant growth, since it is a moderately soluble source of the crucial plant nutrients, calcium and sulphur. Gypsum supplements can also enhance the physical and chemical characteristics of soils, hence lowering nutrient concentrations in surface water runoff and reducing soil erosion losses. The most often used addition for reclaiming sodic soil is gypsum, which can also be found in synthetic soils used in nursery, greenhouse, and landscaping applications. Gypsum can be used for a variety of purposes in agriculture and horticulture, which could be advantageous to users. There are currently no recognized standards that outline the broad best management practices for using gypsum in agricultural applications.

Table 2. List of some genes conferring adaptation to drought and other abiotic stresses in chickpea.

Treatment	Traits	Gene	References
Drought	Abiotic stress response	CarERF116	[110]
Drought	Biotic and abiotic stresses	Aquaporins gene family (*CaAQPs*)	[111]
Drought	Drought stress response	DEGs	[112]
Drought, heat and cold stress	Process of plant development	CarLEA4	[113]
Drought and heat stress	Root traits, plat morphology, transpiration, and yield traits	Marker–trait association	[47]

Drought tolerance is a complex phenomenon that involves defence mechanisms as well as stress-induced signal responses [114,115]. Drought stress triggers a number of physiological, biochemical, and molecular responses (Figure 1) that can be classified into six categories: drought escape [116], avoidance [117], tolerance [118], resistance [119], abandonment [120], and drought adaptation [12,121]. Some chickpea genotypes have been identified as drought sensitive [122] and others as drought tolerant [123,124]. Plant breeders apply different ways of selection and development of drought tolerant crop genotypes. Different strategies are important to protect plants from harmful effects of drought.

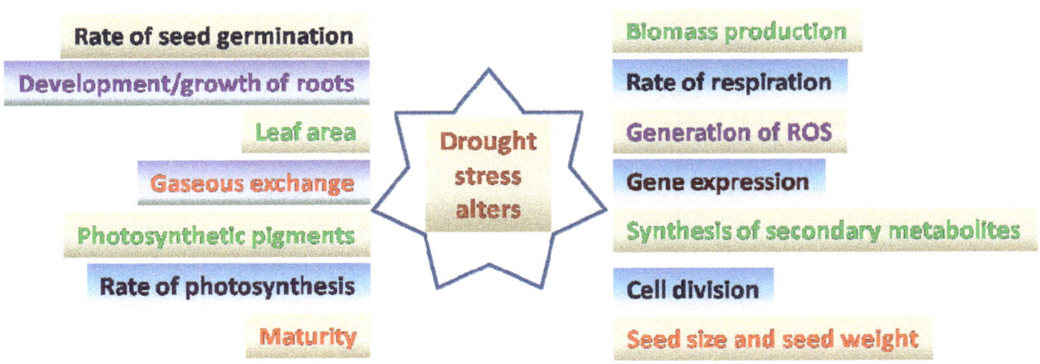

Figure 1. Diagrammatic representation of effects of drought stress on chickpea.

Drought escape is the capacity of the plant to complete its life cycle before experiencing a major water deficit. Drought escape causes early flowering and maturity, as well as better yield potential, allowing plants to finish reproduction before drought strikes [125]. Crop longevity is governed in part by genotype and in part by the environment, and it impacts the crop's ability to withstand climatic conditions such as drought. To achieve large seed yields, it is necessary to match the plant growth time to soil moisture availability. The genotypes with early maturity have the capacity to escape the terminal drought stress, whereas the genotypes with late maturity generally needs well-watered environments. The length of the growing phase and yield potential are positively associated with each other. In this consequence, the development of shorter duration crop is important for the reliable management of drought stress in the chickpea. The timing of flowering is a key feature of a plant's response to extreme drought and high temperatures [126]. Early maturity lets the crop circumvent the passé of stress, hence short duration cultivars may be produced to minimise production loss from terminal dryness. However, under ideal growth conditions, the yield is often associated with crop duration, and any decrease in crop interval underneath the optimal will tax yield [29,52].

Drought escape is a critical strategy for preventing chickpea crop from drought [12]. Water supply is harmonised with phenological development in drought escape. Early maturity aids in escaping terminal dryness and is a key feature in germplasm screening. However, growers are frequently incapable to reorganize for early planting owing to climatic factors [52].

Drought avoidance is explained as a plant's aptitude to retain a high tissue water potential contempt in a lack of soil moisture [125]. Processes involved in the enhancement of water intake, its storage in plant cells, and limiting water loss are associated with drought avoidance. Other mechanisms, including deep rooting, increased level of hydraulic conductance, reduced level of epidermal conductance, radiation absorption and reduced leaf area have also been reported to be linked with drought avoidance in plants. Deep rooting promotes water intake, which is helpful in reducing water losses. In the chickpea, the stomata remains closed during the day to minimise water loss during drought, and as a result, the carbon assimilation is impeded, lowering production [126,127].

Root biomass plays major role in absorbing water [128], as it is advantageous even in the condition of less moisture in the soil. It means there is a linkage between the root system and water stress tolerance [129,130], thus, in the current scenario, breeders are focused in the development of cultivars with larger root systems [131]. From integrating large root features, cultivars have been developed by chickpea breeders with increased drought tolerance [43,132]. Because root size is governed by intrinsic genetic variables [133,134] and modified by multiple environmental signals, such as nutrition and moisture accessibility in the soil, it is a complicated feature [135]. During the vegetative growth stage, susceptible genotypes absorb more water than tolerant genotypes, whereas tolerant genotypes absorb more water during the reproductive stage [136]. The intake of water during the vegetative as well as reproductive stages of plants has a direct relation with seed yield [137]. The importance of roots, rather than just root growth, is determined by their temporal water intake [138]. The best method for screening the germplasm for water usage competence (WUE) is carbon isotope discernment (13C), and this method has also been adopted in the chickpea [139,140].

One of the important impacts of drought stress is stomatal closure. Drought stress reduces the stomatal conductance and transpiration rate. This declines the CO_2 fixation and photosynthesis due to the reduction in the internal CO_2 concentration of the leaf (Ci). All of these factors have their role on the reduction in yield due to the reduced rate of photosynthesis [141].

The reduced rate of photosynthesis is directly related to extreme drought stress, and it is a result of the decreased chlorophyll content. Because of the lowered chlorophyll content, continuous poor moisture availability reduces light collecting capacity, triggering the generation of reactive oxygen species due to excessive energy absorption [142]. This is also a cause of damaged photosynthetic machinery. The principal cause of chlorophyll depletion is reactive oxygen species [143]. Reduction in photosynthetic activities under drought stress, have been experimented in chickpea genotypes [55] and this reduction was found to be linked to reduced ATP synthesis [144,145]. The yield reduction in chickpea genotypes due to the flower and pod drop under heat and drought stress circumstances was also noticed [146,147].

The leaf surface is also an imperative characteristic of plants in relation to drought stress. As tiny leaf surfaces lose less water [148], waxy leaves have high water preservation potential. Waxy leaves have the ability of a reflectance of irradiation and the reduction of water loss. This helps in the reduction of leaf temperature and provides tolerance against drought condition. The preservation of water in leaves with a reduced leaf temperature are directly related to the drought tolerant behaviour of plants. Drought stress raises leaf temperature in a variable manner, as tolerance genotypes have lower leaf temperatures than sensitive genotypes [149]. One drought tolerant chickpea variety 'Gokce' has been developed by ICARDA through the gene pyramiding method, which can be survived under severe drought conditions. This variety possess some other important features, such as early maturity, resistance to *Ascochyta* blight, increased seed size, and suitability for mechanised harvesting [150].

The drought tolerance refers to a plant's ability to maintain its metabolism in a water shortage [35] condition with low tissue water potential [58]. Two types of traits are responsible for the drought tolerance in plants, i.e., constitutive characters and acquired behaviours. The constitutive traits affect the yield at mild to moderate levels of drought stress, whereas the acquired traits affect the yield at severe levels of drought stress. Drought tolerance features are largely concerned with cellular structural protection against the effects of cellular dehydration. Due to a reduction in the plant tissue water content, dehydrins and late embryogenesis of abundant (LEA) proteins accumulate [151]. These proteins act as chaperones [152].

In recent years, the role of reactive oxygen species (ROS) in stress signalling has been widely researched and evaluated [153,154]. The extreme creation of ROS causes oxidative damage and, lastly, cell death [155]. The role of ROS as a signalling molecule or in the oxidative damage depends upon the equilibrium between production and the scavenging of them [156]. The scavenging of ROS under drought stress depends upon the action of antioxidants in the cell [157,158].

Pushpavalli et al. [159] emphasised the need of selecting chickpea genotypes that can withstand various shocks rather than simply one. High temperature stress, in addition to drought stress, is a new threat to chickpea production [33,160]. According to Kalra et al. [161], a temperature increase of 18 °C above a particular threshold causes a significant loss in chickpea output. Furthermore, it is predicted that a global temperature increase of 2–38 °C, along with erratic rainfall patterns, would pose a threat to chickpea yield. In agriculture, the yield is the most important parameter for crops, and a reduction in yield cannot be compromised at any level. There is a strong association between drought tolerance with yield in a crop [122,162]. This is because yield-related traits of crops have been found to be sensitive under drought stress [163].

3. Antioxidant Defence

Plants have multi-level systems of antioxidant defence [124] with main function to maintain homeostasis inside the cell. This system counteracts ROS and protects the cell from oxidative damage. In the absence of the sufficient quantity of an antioxidant to neutralize ROS, reactions such as biomolecule oxidation, lipid peroxidation, and protein damage, as well as nucleic acid (DNA, RNA) oxidation and apoptosis activation, may occur [90].

The antioxidant defence system has both enzymatic and non-enzymatic components. The enzymatic component involves superoxide dismutase, catalase, peroxidase, ascorbate peroxidase, and glutathione reductase. However, the non-enzymatic component involves cysteine, reduced glutathione, and ascorbic acid [164,165].

In plants, ascorbate peroxidase is an important antioxidant enzyme, and glutathione reductase is important for sustaining the reduced glutathione pool during stress [166]. In various plants, two glutathione reductase corresponding to deoxyribonucleic acids have been recognized, one producing cytosolic isoforms and the other encoding glutathione reductase proteins, which dually embattle chloroplasts and mitochondria [127]. Superoxide dismutase is a key enzyme that catalyses the detachment of two superoxide molecules into O_2 and H_2O_2 [167]. The drought tolerance of a particular plant species can be linked to increased antioxidant enzyme activity [168].

Proline appears to play a variety of activities under stress situations as a multifunctional amino acid, including stabilizing proteins, membranes, and subcellular structures as well as defending cellular functioning by scavenging reactive oxygen species (ROS). The functional diversification of proline metabolism is more complicated as a result of the compartmentalization of proline production and degradation in the cytosol, chloroplast, and mitochondria. When the electron transport chain is saturated under stressful circumstances, the increased rate of proline production in the chloroplast can help to stabilize the redox balance and maintain cellular homeostasis by dissipating the excess reducing potential. Proline is one of the most widely dispersed suitable solutes and a key component of plant stress resistance that increases in plants under adverse environmental conditions. Proline serves as a superb osmolyte and also has important functions as a metal chelator, antioxidant defence molecule, and stress signalling molecule. By regulating mitochondrial activity, affecting cell growth, inducing certain gene expression, and stabilising membranes, it promotes stress tolerance by reducing electrolyte leakage, bringing ROS concentrations back into normal levels, and promoting stress recovery.

One of the elements driving drought resistance in the chickpea is proline build up in different plant sections due to the increased activity of proline synthesising enzymes [169]. Drought tolerant genotypes of chickpea had higher proline contents than sensitive genotypes [130]. Earlier, an increase in the leaf proline concentration under water-deprived conditions indicates an efficient osmotic regulating system in the chickpea. To modify the osmotic potential, proline, glycine betaine, and soluble carbohydrates are accumulated in response to drought stress [170].

4. Plant Growth Regulators

Plant growth regulators, basically known as phytohormones, can be administered externally or synthesised inside the plant [171]. Auxins, cytokinins, gibberellins, ethylene, and abscisic acid have all been referred to as plant growth regulators. However, in recent studies, brassinosteroids (BRs) and various compounds of jasmonic acid, cytokinin, salicylic acid, strigolactones, and some peptides have been identified as plant hormones. The concentrations of Auxins, gibberellins, and cytokinin are negatively related to drought, but abscisic acid and ethylene have a positive association [172]. Drought stress inhibits the formation of endogenous auxins, which is frequently accompanied by a rise in the levels of abscisic acid and ethylene [173].

Abscisic acid and cytokinin are thought to play opposing functions in drought stress. Under water stress, an upsurge in abscisic acid and a diminution in cytokinin levels favour stomatal closure and minimise water loss by transpiration [174]. Abscisic acid affects the relative growth rates of different plant parts, such as the root-to-shoot dry weight ratio, leaf area development inhibition, and the generation of prolific and deeper roots. It can influence the rate of transpiration by closing the stomata, and it may be implicated in the machinery providing drought tolerance in plants. Under drought conditions, ABA formation inhibits the lateral root growth [175].

Ethylene is a growth inhibitory hormone that acts as a part in both inhibiting and stimulating growth in response to environmental factors. The plants can maximise growth and resist abiotic challenges such as drought to avoid this adversity, and this response also requires ethylene synthesis [134].

Plant growth and development are known to be affected by polyamines. There has been an increasing interest in the role of polyamines in the plant defence against environmental stressors, and substantial research energies have been conducted in the last twenty years. The overexpression of the apple spermidine synthase gene, for example, results in high amounts of spermidine synthase, which enhances abiotic stress resistance, for instance, drought tolerance [176].

5. Role of Conventional Breeding

Breeders commonly employ traditional breeding techniques such as introduction, selection, hybridization, and mutation [106]. Hybridization is used to blend the desired characteristics from different parents into a single cultivar [85]. Any hybridization program's success hinges on the selection of proper parents. Single, multiple, and three-way crosses [107] have all been employed for the hybridisations in the chickpea crop [177].

Among the different branches of breeding technology, the mutation breeding has been found as a powerful strategy [178] for the creation of genetic variability in crops [179]. It is considered under advanced breeding technologies [180]. According to Kumar et al. [181], fifteen chickpea varieties have been developed through mutation breeding and most of them are under the cultivation chain. The first chickpea variety developed in the year 1984 through mutation breeding in India was Kiran (RSG-2), which was the mutant form of RSG-10. This variety possess higher numbers of pods, early maturity, high yield, and tolerance to salinity stress [182].

The evaluation of different genotypes of a plant species in response to the drought controlled condition is needed [183]. In different studies, the phenopsis has been used as a non-automated control-guided drought screening method [145,184,185] to examine the performance of several *Arabidopsis* ecotypes [186]. In this regard, the *ERECTA* gene [147] responsible for the growth and development of the plant as well as the stomatal development, the *ESKIMO1* gene [187], governs the plant water relation in *Arabidopsis,* which have been examined well. Similarly, some biosynthesis genes in the chickpea [188] have been well studied using controlled drought.

New cultivars, landraces, wild relatives, or a new crop species for the region could all be introduced. By using this technique, it is feasible to find a desirable genotype with a higher yield and better environmental tolerance, while also increasing the genetic variety. Through the international exchange of the germplasm and the inclusion of the crop of wild relatives and landraces, significant progress has been made in recent decades in enhancing the genetic diversity of the cultivated chickpea. Landraces are an important resource of novel genes in crop breeding. Landraces may possess genes for resistance against various biotic as well as abiotic stresses. For the identification of drought tolerant chickpea landraces, a field study was conducted by Kumar et al. [189]. The experiment included 37 chickpea landraces collected from ICARDA. Based on various morphological as well as physiological parameters, two landraces viz.,IG5856 (Jordan) and IG5904 (Iraq), were identified as drought tolerant.

6. Role of Molecular Breeding

Genetic Diversity

Complex abiotic stress such as drought requires a large group of genetic resources [190] to study the genetics of these stresses authentically. To fulfil the objective of molecular breeding in crop improvement, it is important to characterize the plant genetic resources. In molecular breeding, the characterization of plant genetic resources depends on the availability of DNA-based markers [191]. The molecular markers may be hybridization-based or PCR-based, depending on the technique of the detection of nucleotide variation [192]. The restriction fragment length of polymorphism (RFLP)is one of the molecular markers based on hybridization. Random-amplified polymorphic DNA (RAPD), amplified fragment length polymorphism (AFLP), simple sequence repeats (SSR) or microsatellite, sequence-tagged sites (STS), and cleaved amplified polymorphic sequence (CAPS) are all PCR-dependent molecular markers [158]. In the advanced category, single nucleotide polymorphism (SNP), single feature polymorphism (SFP), and diversity array technology (DArT) have been included [193]. In the field of crop improvement, molecular markers are categorized into dominant and co-dominant. The multi-locus markers (RAPD, ISSR, AFLP, etc.) come under the dominant category, while the single locus markers (SSR, STS, etc.) come under the co-dominant. This categorization of markers is basically based on their efficiency to discriminate homozygous and heterozygous genotypes. Dominant markers cannot differentiate homozygous and heterozygous genotypes. However, co-dominant markers have the ability to differentiate them.

In comparison to molecular markers, i.e., the hybridisation-based RFLP, the PCR-based RAPD and ISSR, SSR, and biochemical markers, i.e., isozyme, have a low polymorphic ability, which could be related to the decreased polymorphism in structural genes in the chickpea genome [194].

In some of the previous studies, the RAPD markers were utilised to detect genetic relationships amongst *Cicer* species [195]. The non-reproducibility nature of this dominant character is the main reason of the limited applicability of it [196]. However, the sequence characterised the amplified region (SCAR) markers developed with the use of RAPD markers, which are more suitable for the detection of the desired gene in crops, including the chickpea [197]. Amplification fragment length polymorphisms (AFLPs) were also reported to be uncommon in *Cicer arietinum* [198]. However, the availability of some reports on use of these markers in chickpea for diversity analysis [199] and the screening of abiotic as well as biotic resistant genotypes [200] proved their importance.

Microsatellite and STMS (sequence-tagged microsatellite site) markers are numerous, scattered throughout the genome, and highly polymorphic [198]. STMS raises the likelihood of finding polymorphism by a factor of ten. As a result, any genetic enhancement initiative should begin with a study of genetic variability. These markers become an important part of molecular breeding in the chickpea [201].

More than 3000 microsatellites [202,203], 15,000 DArT arrays [204,205], and SNPs [206] markers have been developed in the last few years in the chickpea. Because of a few of the specific characteristics, including co-dominance, abundance, repeatability, higher polymorphism and large genome coverage, SSR markers have proven their efficiency in the field of molecular breeding [207]. Subsequently, the data obtained after the use of SSR markers for molecular characterization or fingerprinting can be used to determine the genotypic identity of an individual. The applications of ISSR markers for the genotypic identification of chickpea genotypes in association with the seed germination and flowering time reported recently by Yadav et al. [208]. In this sequence, SNPs/InDels were also used recently for the gene identification and analysis in chickpea [209]. Basu et al. [210] identified SNPs linked with seed yield and Rajkumar et al. [211] identified SNPs linked with seed size and seed weight in the chickpea.

7. QTLs and Their Relevance with Drought Tolerance in Chickpea

A crucial requirement for identifying and integrating genes in linkage maps for marker–aided selection (MAS) is the knowledge of the agronomic trait inheritance [212]. The marker-assisted selection [54] and mapping of QTL (Quantitative Trait Loci) have been proposed to improve chickpea productivity [213]. Linkage map construction [214] and attribute mapping [215] were both conducted with available markers in the chickpea. Many research groups have focused their studies on abiotic stresses [216,217]. After completion of the sequencing of the desi and kabuli chickpea genomes [218], a genome-wide physical map was also generated. Furthermore, QTL studies have also been carried out in the chickpea (Table 3) to better understand the genetics of drought tolerance [217,219] and salt tolerance [220]. Varshney et al. [221] identified 'QTL-hotspot' regions that contain QTLs for a number of drought-related characteristics in the chickpea. They also reported the linkage between QTL hotspots and SSR markers. In the marker-assisted selection, the chickpea genotype ICC-4958 is used as a control for root studies under drought condition due to the large root character. This character makes this genotype suitable to use as a parent for transferring drought tolerance QTL-hotspot regions into the desired genotype. Recently, Muriuki et al. [222] also evaluated the root traits of some chickpea genotypes under drought stress and found that some of the *desi* genotypes (ICC4958, ICCV 00108, ICCV 92944 and ICCV 92318) performed fine.

Table 3. List of QTLs identified for drought tolerance in chickpea.

Mapping Approach	Numbers of QTLs	Markers Used	Statistical Method Used	References
Biparental	15 QTLs	SSR		[151]
Biparental	93 QTLs	SSR	Composite interval mapping-epistatic mapping (ICIM-EPI)	[213]
Biparental and backcross	QTL-hotspot	SSR, AFLP		[13]
Biparental	QTL-hotspot	SSR	Composite interval Mapping	[214]
GWAS	312 significant model MTAs	DArT, SNP	Mixed linear	[58]
Biparental	164 main-effect QTLs	SNP, CAPS	Composite interval mapping	[215]
Biparental	QTL-hotspot_a(15genes)	SNP	ICIM-ADD mapping method	[216]
Biparental	3 candidates Genes	SNP		[217]
Biparental	12 QTLs	SNP		[218]
Biparental	21 QTLs	SNP	Composite interval mapping	[223]
GWAS	Several MTAs	SNP		[224,225]

The advanced genomics involves a genome-wide association study (GWAS) that helps researchers in the screening of a wide range of genotypes [226,227] with different phenotypic or agronomic characters, and it also helps in the identification of the variability present among them. GWAS also helps in the identification of the association between the marker and a specific trait of interest [28,228]. The majority of these connotation investigations used either GWAS or candidate gene sequencing. Recently, the GWAS-based association mapping for the drought tolerance in the chickpea and for salinity tolerance have been performed. Apart from these, other examples of the association mapping in the chickpea are also available, i.e., for iron and zinc concentration in seeds [229] and *Fusarium* wilt resistance [230,231].

In some of the earlier studies, a combined analysis of the GWAS and sequencing of candidate gene [223] has been found to be more suitable in crop improvement. The GWAS study on two sets of chickpea genotypes with a different degree of their response in drought conditions helped in the discrimination of these genotypes on the basis of the single nucleotide polymorphisms generated.

Scientific efforts made on the improvement of the chickpea crop made it possible to generate not only a bi-parental plant population but also multi-parent populations [224]. The need of a multi-parental population was due to issues such as narrow genetic variability and limited efficiency of the bi-parental population [225] during the multiple trait analysis [232]. Multi-parent advanced generation inter-cross (MAGIC) populations for the chickpea are being established [233] to create diverse patterns of recombination [234]. The purpose of creating multi-parent populations is to advance the precision of QTL mapping [235] and discover specific loci regulating to the trait of interest [236]. ICRISAT and ICARDA played a major role in the development of a few MAGIC populations in the chickpea. One example of the MAGIC population developed at ICRISAT is the results of crossing eight varieties and advance breeding lines (ICC 4958, ICCV 10, JAKI 9218, JG 11, JG 130, JG 16, ICCV 97105, and ICCV 00108) with eight different founder parents [6,237]. Similar to this one, the MAGIC population developed at ICARDA was the result of crossing 12 different parents [238]. These plant populations accelerate the detection, isolation, and transfer of critical candidate genes to help in the development of chickpea varieties with superior agronomic traits [237]. One more approach (target-induced local lesions in genome -TILLING) [239] was adopted in the validation of the drought responsive gene in chickpea [58]. The selection and further use of agronomically superior genotypes of chickpea for the development of new varieties are the basic objectives of breeders involved in the chickpea crop improvement [240].

8. Attempts to Develop Drought Tolerant Varieties

One of the drought tolerant high yielding Ethiopian chickpea varieties,'Geletu', was developed and released in the year 2019 through the marker-assisted back-crossing after multi-location trials. During the development of this variety, the'QTL-hotspot' linked to drought tolerance was introgressed into an Indian chickpea cultivar JG11 from ICC4958 (gldc.cgiar.org). Recurrent selection is a crucial breeding technique used to increase crop plant populations. It is a productive method used in plant breeding to enhance the quantitative traits through repeated crossing and selection. Among the genomics-assisted selection methods, the marker-assisted back crossing has been found better for the introgression of the targeted region of the genome into a desired genotype. Consequently, the introgression of the 'QTL hotspot' region for the development of the drought-tolerant chickpea through molecular breeding has been found effective. Similar to this, numerous drought tolerance characters were introgressed into three elite Indian chickpea varieties: Pusa 372, Pusa 362, and DCP 92-3from ICC 4958. Recently, drought tolerant root traits have been introgressed into Kenyan chickpea varieties using the marker-assisted backcrossing approach [241].

Initially, the Pusa 372, chickpea variety was released as a drought tolerant variety for cultivation in the central, north-east, and north-west plains zones. However, under drought conditions, this variety's output has decreased in recent years. To enhance drought tolerance in this variety, the MABC approach was adopted to introgress the 'QTL-hotspot' region from ICC 4958 into 'Pusa 372'. Recently, the Pusa 372 was released with improved drought tolerance under the name 'Pusa 10216'. This improved chickpea variety is the example of the first enhanced drought tolerant chickpea variety developed through the MABC approach. ICRISAT in collaboration with other research institutes in India is in the process of the development and release of drought tolerant chickpea varieties, i.e., IPC L4-14 and BGM 4005.Both of these varieties were developed by transferring a 'QTL-hotspot' from ICC4958 into DCP92-3 and Pusa362, respectively (https://www.icrisat.org/new-climate-resilient-disease-resistant-chickpea-varieties-coming-farmers-way/, accessed on 20 October 2022).

9. Whole-Genome Re-Sequencing

Whole-genome sequencing is the most thorough NGS technology [242], allowing for the complete genome sequencing and identification of variations in both exonic and non-coding areas, as well as the structural variant detection. Due to a paucity of genetic knowledge, the chickpea was formerly referred to as an orphan crop well adopted to suboptimal growing environments [243] However, researchers published the first draft genomes of the desi and kabuli chickpeas in 2013. The chickpea genome sequencing was based on advances in high-throughput sequencing and next-generation approaches. The BAC end genetic map and DArT markers were used to offer information on SSR and SNP molecular markers [244]. Both the kabuli and desi chickpea genomes have been updated, as well as a comparative examination of the two varieties. The QTLs associated to drought tolerance were reported by Jaganathan et al. [243]. The drought-responsive genomic areas were identified and employed in breeding approaches such as the marker-assisted gene interrogation and genetic gain to improve production in harsh climatic circumstances.

After publication of the draft genome sequence of the chickpea, the sequencing-based technique for the improvement of this crop has open multiple windows [244]. Furthermore, re-sequencing of a large number of chickpea assents collected from 45 nations enabled the identification of various candidate genes with their associations to a large number of agronomical characters [245]. The results of these experiments revealed the origin and migration routes of chickpea in the world. Re-sequencing data helped in the identification of 50,590 SNPs, and this data was used to develop the 'Axiom®CicerSNP Array' [246]. This SNP platform is being employed in the character mapping and identification of QTLs. Recently, Rajkumar et al. [247] reported the re-sequencing of large and small seed chickpea genotypes and 266 SNPs associated with seed size and seed weight. The findings of the study may help in the selection and categorization of chickpea genotypes on the basis of the size and weight of their seeds.

Next generation sequencing technology [248] has made possible the development of new markers for the improvement of the chickpea [249]. One of the important concepts, 'The 3000 Chickpea Genome Sequencing Initiative' [250], is an important step in the field of chickpea improvement. This initiative is helpful in the identification of variations in genomic sequences (rare alleles, markers) and their role in the determination of various agronomic characteristics, including yield and resistance/tolerance, against biotic and abiotic stresses. A thorough map of variation in 3171 farmed and 195 wild accessions was recently published in a project to give publicly accessible tools for chickpea genomics research and breeding [250]. This study also demonstrated the variations among the cultivated and wild progenitors of chickpea.

10. Pangenome and Super-Pangenome

The availability of the whole genome sequences of multiple individuals makes it feasible to compare them for the identification of diversity among them [251]. This approach may be termed as a comparative genomics [252] analysis, which allows for the identification of bio-markers linked with taxonomic as well as morphological and functional characteristics [253]. In this sequence, the pangenome concept was arisen, which allows for the accurate and efficient comparison of the genomes of a wide range of individuals [254]. A recently proposed revolutionary approach known as super-pangenome, allows the construction of the pangenomes of many species within a specified genus [255]. These concepts facilitate the identification of novel variations among individuals from different sources. These advanced technologies have their importance in crop breeding due to their accuracy and efficiency. The construction of pangenomes has their advantages in the identification of signature genomic areas relevant to crop domestication and evolution. Chickpea landraces and varieties have been sequenced to build the pangenome. These pangenome may be coupled with phenotypic traits and alleles associated with various characteristics and may also help in the identification of abiotic stress tolerance [256]. Pangenomes also provide a platform for the accurate identification of target genes for genome editing using CRISPR-clustered, regularly interspaced, short palindromic repeat technology [257].

11. Omics Approaches

Complex genetic traits, including drought tolerance, need advanced tools for their dissection, along with crop improvement [258]. Multiple omics approaches have revolutionized the identification of genes [259] as well as the metabolic database [260].Investigations have been carried out on transcriptomic analysis with the applications of the NGS technology in chickpea [261]. Multiple examples on transcriptomics' evaluation in the chickpea are available as developing seeds [262], development and function [263], tissue specificity [264], and salinity tolerance [265] as well as root transcriptomics for drought tolerance. About 20,162 ESTs in the chickpea under salt and drought stress circumstances have been reported (Table 4). Recently, Kaashyap et al. [266] performed a comparative flower transcriptomic analysis to analyse the reproductive success under the salinity stress in the chickpea.

Table 4. Advanced technologies adopted to identify drought responsive differentially expressed genes/ESTs in chickpea.

Differentially Expressed Genes/ESTs	Technique/Platform Used	References
1562 genes, 2592 genes	Illumina HiSeq 3000	[267]
1624 differentially expressed genes	Illumina platform	[103]
20,162 ESTs	-	[266]
53 ESTs	cDNA library	[268]
3062 unigenes	Suppression subtraction hybridization	[258]
44,639 differentially expressed sequences	Roche/454 and Illumina/Solexa	[269,270]
7532 unitags and 880 unitags	SuperSAGE	[267]
4053 and 1330	Illumina HiSeq 2000 platform	[271,272]
261 (shoot) and 169 (root)	Illumina TrueSeq RNA	[273]
15,947 differentially expressed genes	Illumina HiSeq 2000	[274]

The RNA-Seq technique has also been used to analyse differential regulation of genes under drought stress in the chickpea [112]. Kumar et al. [113] analysed the gene expression of polyethylene glycol-stimulated drought stress in the chickpea, and thousands of differentially expressed genes (DEGs) were identified. Earlier, DEGs were detected in the kabuli chickpea under drought conditions [275]. Using RNA-Seq technique, it was used for the development of an inclusive *C. arietinum* Gene Expression Atlas (CaGEA) based on a drought tolerant ICC 4958 cultivar [269]. The findings of this study also validated the 'QTL hotspot' for drought tolerance in the chickpea.

The regulation of metabolic activities plays a major role in maintaining the osmotic potential of the cell under drought stress [276]. With the applications of the metabolomics method, many important metabolites were identified with a different regulation pattern during a drought [277] as well as in the salinity [278] in the chickpea. Similar to this, in an earlier investigation conducted on *Arabidopsis thaliana*, various genes were identified with their similar contributions under both salinity and drought stresses. The production of similar metabolites under both abiotic stresses indicates a common tolerance mechanism for drought as well as in the salinity in plants.

The proteomics' analysis has also been performed in the chickpea for the identification of changes at a protein level under abiotic stresses. Earlier, a comparative proteomics analysis was conducted on one chickpea cultivar (JG-62), and novel dehydration-responsive proteins were detected [279]. In this sequence, Jaiswal et al. [280] reported the role of Sad1/UNC-84 in dehydration signalling. Recently, Vessal et al. [281] analysed the proteomic responses of drought sensitive and tolerant chickpea cultivars and identified changes in terms of the requirement of relative leaf water content for tolerant and susceptible cultivars. Drought responsive root proteins were also analysed recently by Gupta et al. [282].

Phenomics is an emerging tool in plant research, and is used to describe the use of genomics in phenotyping [283]. Phenomics studies for different phenotypic traits as well as seed yield have been conducted for the drought tolerance in chickpea [32]. The drought tolerance in chickpea is determined through phenotyping, and for this purpose, high-throughput screening technologies [32,44,284] have been adopted.

12. Role of Candidate Genes

A contender or candidate gene is thought to be linked to a specific disease or phenotypic character [285], and the biological function(s) of that has been derived either directly or indirectly from other investigations, including, for instance, the genome-wide association studies [286], the traditional map-based positional cloning technique, and the more recent next-generation sequencing (NGS) method [287]. Candidate gene studies are low-cost and rapid to conduct, and they focus on finding genes that have already been linked to the disease and hence have a prior knowledge of gene function [288]. Few of the important candidate genes detected in chickpea for the abiotic stress tolerance are Snf-1-related kinase (*AKIN*), *DREB2A*, dehydrin (*DHN*), *CAP2*, and *Myb* transcription factor (MYB) [289], Table 5. Despite the fact that multiple genes have been linked to drought resistance, the association study based on candidate gene sequencing has received little attention.

Table 5. List of various genes/transcription factors and their roles in response to drought and other abiotic stresses in chickpea.

S.No.	Gene/Transcription Factor	Function	References
1	DREB	Dehydration responsive element binding proteins	[290]
2	Dehydrin (DHN)	Response to water stress	[291,292]
3	STPK	Drought stress	[293]
4	CAD	Response to abiotic stress	[294]
5	AMADH	Wound healing, abiotic stress responsive	[295,296]
6	TCS	Abiotic stresses tolerance	[297]
7	EREBP	Ethylene responsive	[298]
8	LEA Gene	Response to water stress	[299]
9	AKIN	Positive regulator of drought tolerance	[300]
10	Myb transcription factor	Stress	[301]
11	ASR	Abscisic acid stress and ripening gene	[302]
12	SuSy	Sucrose synthase	[303]
13	CAP2	Promoter of DREB2A	[297]
14	ERECTA	Transpiration efficiency regulator	[298]
15	SPS	Sucrose phosphate synthase	[300]
16	CAMTA	Salinity and drought tolerance	[304]
17	CarNAC4	Salt and Drought tolerance	[305]
18	CaNAC	Drought tolerance	[306]
19	CarERF	Drought stress	[109]
20	CaSWEET	Abiotic stress tolerance	[307]

13. Transcription Factors and Their Role in Drought Tolerance in Chickpea

Transcription factors induce the cis-elements in the promoter provinces of different stress-responsive genes to accelerate the countenance of various downstream genes, which have their part in the stress tolerance [308] of plants. According to Riechmann et al. [309], nearly 1500 transcription factors have been reported in the *Arabidopsis thaliana* genome, which have their part in stress-responsive gene expression.

14. Dehydration Responsive Element Binding Proteins (*DREBs*)

DREBs (dehydration responsive element binding proteins) are key plant transcription factors [12]. They are responsible for the regulation of many stresses' responsive genes. One of the transcription factors, AtDREB1a, was identified from *Arabidopsis thaliana* [310], with its role under abiotic stress. Recently, Das et al. [290] reported a better performance of *AtDREB1a* transgenic chickpea lines under water stress conditions.

15. Dehydrin (*DHN*)

Dehydrin (*DHN*) are stress-responsive proteins that are found when the temperature is low or when the body is dehydrated [311]. Protein dehydrin protects the embryo and seed tissues under water scarcity [312]. A better performance of transgenic plants overexpressing

DHN than wild-type plants [313] have been identified. The role of the *DHN* gene in the Pusa1103 and Pusa362 genotypes of the chickpea has been found to be linked with the drought tolerance. Furthermore, in comparison to other genotypes, these genotypes were recognised as drought tolerant due to their better response [291].

16. Serine/Threonine Protein Kinase (*STPK*) Gene

Serine/threonine protein kinase (*STPK*), tyrosine protein kinase (*TPK*), and histidine protein kinase (*HPK*) are the three types of eukaryotic protein kinases [314]. In Arabidopsis, chickpea, and rice, the *STPK* family gene *AtSnRK2.8* is reported with an enhanced degree of drought tolerance [293].

17. Cinnamyl Alcohol Dehydrogenase (*CAD*)

CAD is thought to be important in plant defence against a variety of biotic and abiotic stressors. When employing primers developed for the contig exhibiting match with the *CAD* gene of *Arabidopsis thaliana*, a homologue form of this gene was recovered from eight chickpea genotypes [294].

18. Ethylene-Responsive Element Binding Protein (*EREBP*) Gene

Ethylene-responsive element binding factors (ERFs) are a new type of transcription factor that are only found in plants. The ERF domain, a highly conserved DNA binding domain, is the protein family's distinguishing trait. Primers for the chickpea were constructed using a contig sequence that was found to be identical to the *Arabidopsis thaliana* ethylene-responsive transcription factor. The amplification of eight chickpea genotypes yielded amplicons of around 400 bp. In plants, the *AP2/EREBP* genes have a variety of roles in developmental processes and stress responses [315].

19. Amino-Aldehyde Dehydrogenase (*AMADH*)

In some crops, an *AMADH* gene has been reported with its association to osmotic stress tolerance [316] by detoxifying hazardous aminoaldehydes; this gene has a function in physiological as well as metabolic responses under abiotic stresses [295]. On the basis of functional characterisation of the *AMADH* gene in *Arabidopsis* [317], the role of this gene should be examined in the chickpea.

20. ERECTA Gene

The *ERECTA* gene has a part in leaf organogenesis, lowering the density of the stomata on the leaf underside and thereby lowering evapotranspiration. It can also control the transpiration by the alteration of the leaf epidermal cell expansion, proliferation of mesophyll cells, and cell–cell interactions. The *ERECTA* gene has been demonstrated to regulate the growth and development of the organ and flower in *Arabidopsis* via encouraging cell proliferation [318]. Complementation tests on the wilting mutant *Arabidopsis* plants confirmed the involvement of the *ERECTA* gene to the water usage competence [319]. Pioneer Hi-Bred International, Inc. has patented the *ZmERECTA* genes from maize, which were implicated in the drought tolerance in crop plants.

21. Late-Embryogenesis Abundant (LEA) Proteins

The attainment of dehydration tolerance and the behaviour of plants to drought have been linked to late-embryogenesis abundant (LEA) proteins. Increased LEA and Dehydrin expression in genotypes during the vegetative, flowering, and podding stages could represent an adaptation to assist the plant survival by supplying the energy for growth and survival [29]. Leaf age inhibited arLEA4 expression, which changed during seed and pod development, including during germination. Drought, salt, heat, cold, ABA, IAA, GA_3, and MeJA all significantly increased the *CarLEA4* expression. *CarLEA4* is a LEA assembly 4 protein that may participate in a variety of plant developmental processes as well as abiotic stress responses [299].

22. Myeloblastosis (MYB) Gene

Plants have a big transcription factor (TF) family, called the myeloblastosis (MYB) gene [320]. It plays a role in the secondary metabolism regulation, hormonal and climatic condition response, cell differentiation, and resistance to drought and other abiotic stimuli. Under drought stress, arrays of MYB-transcription factors are involved in the generation of epicuticular waxes [321]. These waxes seal the plant's aerial component and reduce water loss through the leaf surfaces [322]. In an experiment, the root tissue of ICC 4958 (drought tolerant), ICC 1882 (drought sensitive), JG 11 (elite), and JG 11+ (introgression line) were employed to recognize the role of the 1R-MYB gene in the machinery of the drought tolerance in the chickpea. The findings of this experiment were suggested to conduct more experiments on this aspect in the chickpea. Recently, Caballo et al. [301] observed that CaRAX1/2a codes a MYB transcription factor that is exactly articulated in the meristem of chickpea. These results disclosed that the single flower gene (*SFL*) encodes for MYB, which works as a central factor responsible for the regulation of the numbers of flowers in chickpea inflorescence.

23. *S-Adenosylmethionine Synthetase* Gene

S-adenosylmethionine (*SAM*) is a precursor in the production of polyamines and ethylene [323]. In plants, the action of 1- aminocyclopropane-1-carboxylate (ACC) synthase and ACO (ACC oxidase) is responsible for ethylene biosynthesis, while the activity of *SAM* decarboxylase is responsible for spermidine and spermine production. The exogenous polyamine administration or overexpression of polyamine production genes has been demonstrated to improve abiotic stress tolerance. Primers were developed using a contig sequence that was comparable to the S-adenosylmethionine synthetase 1 (*SAM1*) gene of *Arabidopsis thaliana* for the isolation of the S-adenosylmethionine synthetase 1 gene homologue in the chickpea. The PCR amplification revealed amplicons of roughly 300 bp in eight chickpea genotypes [324].

Expression of *SAM* gene in pigeon pea (*Cajanus cajan* L.) was evaluated under drought, heavy metal ($CdCl_2$), and cold stresses. The enhanced up-regulation of *SAM* gene in the leaves were recorded after three days [325].

24. Abscisic Acid Stress and Ripening Gene

Among many other genes, the abscisic acid stress and ripening (*ASR*) gene plays a critical role in controlling various plant stresses. The *ASR* gene has been reported in plants, and is induced by abscisic acid and different abiotic stresses during the process of fruit ripening [268]. Reports on *ASR* genes with their responses in different plant species under drought, salt, and cold stresses [326] confirm their role. Transgenic *Arabidopsis* demonstrated the over-expression of the *ASR* gene in response to drought and salt stresses [302]. Genotypes of rice also presented the association of the *ASR* gene expression [327]. Similarly, Cortés et al. [328] reported the potential significance of the *ASR1* gene in the common bean. In a recent study conducted on the chickpea under drought stress, increased *ASR* gene expression was observed. The increased expression may have helped the drought-tolerant chickpea genotypes function better under stress. This hypothetical *ASR* protein could have boosted the activity of the *ASR* gene as a transcription factor mediating drought responses in chickpeas.

25. ABRE-Binding Protein (*AREB*)

Various genes that are activated by abscisic acid (ABA) have been discovered to be drought stress-inducible. Such ABA-regulated genes have conserved cis-elements in their promoter regions known as ABA responsive elements (*ABREs*), which use bZIP-type *AREB/ABF* transcription factors to regulate the gene expression. ABA and water stress upregulate the expression of the *AREB/ABF* gene. Expression of *AREB* gene under drought stress has been reported by Yoshida et al. [329] in *Arabidopsis thanliana.*

26. Sucrose Synthase (*SuSy*) Gene

Sucrose synthase (*SuSy*) is a crucial enzyme that hydrolyzes sucrose directly to provide substrates for plant metabolism. It is also used as a bio-marker for plant sink strength [330]. Plant sink strength improvement could contribute to increased plant growth and yield [331]). In an experiment, cultivars and treatments had a strong and positive association between the seed dry weight at maturity and peak sucrose synthase movement. Sucrose synthase is a decent physiological indication to employ in chickpea breeding for larger seeds.

Sucrose synthase activity has a major role in chickpea seed growth. The supremacy of the sink, as measured by the sucrose synthase movement most of the time, hinges upon genetic features of a genotype along with the accessibility of water obtainable at seed filling [332]. In both the large-seeded kabuli and the small-seeded desi varieties, the water shortage reduced the enzyme action and seed size, but the higher enzyme action in the large-seeded kabuli, pre-dominantly at the late seed filling stage, seemed to persuade a better remobilization of the integrates from the pod wall and seed coat. The cotyledons' greater sucrose synthase action is taken into account. The strong association between sucrose synthase activity during rapid seed filling and final seed dry weight accumulation, and therefore seed size, advises that the sink strength is an important element of the seed size in chickpea. The tight link between the sucrose synthase activity during rapid seed filling and final seed dry weight build up, and thus seed size, implies that the sink strength is a key factor in the chickpea seed growth. Higher cotyledon sucrose synthase activity is vital in breeding for better seed size in chickpeas, regardless of the growth environment [303].

27. *CAP2* Gene

The AP_2 subgroup of proteins has two copies of the DNA-binding domain (BD), detached by an insertion province [333]. *CAP_2* is a C-Repeat binding factor (CBF) that muddles to the DRE/CRT (dehydration responsive element/C-repeat element)(CCGAC) found in the promoters of abiotic-stress responsive genes. Dehydration, excessive salinity, and exogenous ABA treatment all enhanced *CAP_2* gene expression. The incidence of roughly 60-amino-acid long AP_2/ERF DNA-binding realms in these transcription regulators allows them to connect directly with GC-rich cis-acting elements (GCC box/C-repeat) in the promoter of their target genes. Ectopic expression of CAP_2 in tobacco resulted in increased drought, salinity, and heat tolerance, as well as improved transgenic plant growth [334]. The enhanced accumulation of the CaZF transcript was caused by the transient expression of *CAP_2* in chickpea leaves. *CAP_2* activates the CaZF promoter through interacting with C-repeat elements (CRTs) in CaZF promoter, according to the gel mobility shift and transient promoter-reporter tests. The *CAP_2* protein interacts with the CaZF promoter in vivo, according to a chromatin immunoprecipitation (ChIP) test [335].

28. Sucrose Phosphate Synthase (*SPS*) Gene

The sucrose phosphate synthase (*SPS*) gene has an important function in the sucrose production in different plant species. It regulates sucrose metabolism in drought sensitive and tolerant genotypes [300]. Sucrose biosynthesis and sucrose degradation determines the level of sucrose in a genotype, and an optimum level of it is important for growth and development under environmental stress in plants [336]. Some reports are available on the unchanged or decreased level of the *SPS* activity in maize, potato, soybean, and some other crops, however, some reports advocate the increment in the *SPS* activity in rice, wheat crops, and *Arabidopsis* [336]. The significance of the *SPS* gene has been studied in chickpea under a low temperature by Sharma et al. [300].

29. Genome Editing Options

The genome editing approach, specially CRISPR-Cas9, has proved their efficiency in the development of climate resilient cultivars of different crops [337]. Two genes, *namely* RVE7 and 4CL, have been identified in the chickpea and their association with drought tolerance.The CRISPR/Cas9-mediated editing of the chickpea protoplast was reported for the first time by Badhan et al. [338], where they reported knock-outs of 4CL andRVE7 genes, which are associated with drought tolerance mechanisms. This report laid down a foundation for future genome editing options in the chickpea [339]. Genome editing approaches with the applications of CRISPR-Cas9 (Figure 2) may be helpful in the development of abiotic stress tolerance in chickpea genotypes including drought.

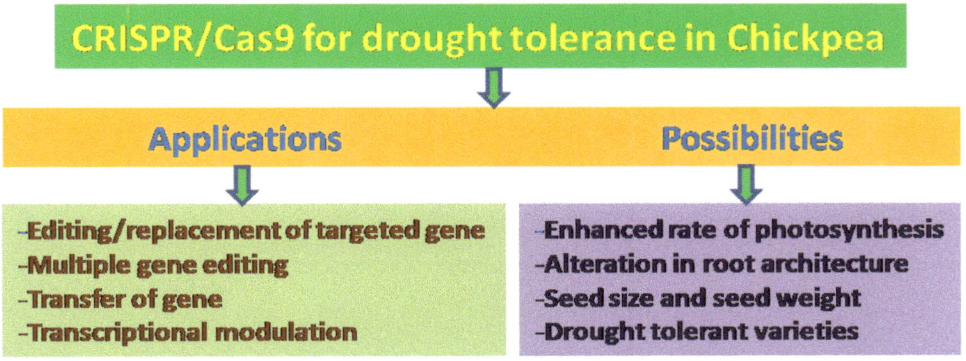

Figure 2. Diagrammatic representation of scopes of CRISPR/Cas9 applications and possibilities in chickpea improvement.

30. Conclusions

As previously stated, the changes in the plant shape and internal biochemical characteristics during drought stress have been extensively characterised in previous studies. Plant drought stress techniques can help us better use scientific means to improve plant tolerance to water shortage environments and increase crop yields, allowing us to play a larger role. As a result, by thoroughly examining and summarising the mechanisms of the chickpea plant response to drought, this study provides essential background knowledge and theoretical framework for selective breeding, cross breeding, and molecular breeding of the chickpea in the future. Drought tolerant land races/germplasm lines may be employed in classical as well as molecular breeding programmes to breed drought tolerant cultivars in future by using the available scientific data.

Author Contributions: Writing: R.A., S.T., R.K.Y. and N.T.; Editing: M.K.T. All authors have read and agreed to the published version of the manuscript.

Funding: This research received no external funding.

Institutional Review Board Statement: Not applicable.

Informed Consent Statement: Not applicable.

Data Availability Statement: Not applicable.

Conflicts of Interest: The authors declare no conflict of interest.

References

1. Madurapperumage, A.; Tang, L.; Thavarajah, P.; Bridges, W.; Shipe, E.; Vandemark, G.; Thavarajah, D. Chickpea (*Cicer arietinum* L.) as a Source of Essential Fatty Acids–A Biofortification Approach. *Front. Plant Sci.* **2021**, *12*, 734980. [CrossRef] [PubMed]
2. Singh, A.; Nath, O.; Singh, S.; Kumar, S.; Singh, I.K. Genome-wide identification of the MAPK gene family in chickpea and expression analysis during development and stress response. *Plant Gene* **2018**, *13*, 25–35. [CrossRef]
3. Varshney, R.K.; Gaur, P.M.; Chamarthi, S.K.; Krishnamurthy, L.; Tripathi, S.; Kashiwagi, J.; Samineni, S.; Singh, V.K.; Thudi, M.; Jaganathan, D. Fast-track introgression of 'QTL-hotspot' for root traits and other drought tolerance traits in JG 11, an elite and leading variety of chickpea. *Plant Genome* **2013**, *6*, 3. [CrossRef]

4. Directorate of Economics and Statistics. 2019. Available online: https://eands.dacnet.nic.in/ (accessed on 20 October 2022).
5. Ganjeali, A.; Porsa, H.; Bagheri, A. Assessment of Iranian chickpea (*Cicer arietinum* L.) germplasms for drought tolerance. *Agric. Water Manag.* **2011**, *98*, 1477–1484. [CrossRef]
6. Gaur, P.M.; Samineni, S.; Thudi, M.; Tripathi, S.; Sajja, S.B.; Jayalakshmi, V.; Mannur, D.M.; Vijayakumar, A.G.; Ganga Rao, N.V.; Ojiewo, C.; et al. Integrated breeding approaches for improving drought and heat adaptation in chickpea (*Cicer arietinum* L.). *Plant Breed.* **2019**, *138*, 389–400. [CrossRef]
7. Gaur, R.; Verma, S.; Pradhan, S.; Ambreen, H.; Bhatia, S. A high-density SNP-based linkage map using genotyping-by-sequencing and its utilization for improved genome assembly of chickpea (*Cicer arietinum* L.). *Funct. Integr. Genom.* **2020**, *20*, 763–773. [CrossRef] [PubMed]
8. Sahu, V.K.; Tiwari, S.; Gupta, N.; Tripathi, M.K.; Yasin, M. Evaluation of physiological and biochemical contents in desi and Kabuli chickpea. *Legume Res.* **2020**, *45*, 1197–1208. [CrossRef]
9. Kumar, T.; Hamwieh, A.; Swain, N.; Sarker, A. Identification and morphological characterization of promising kabuli chickpea genotypes for short-season environment in central India. *J. Genet.* **2021**, *100*, 33. [CrossRef]
10. Tahir, N.A.R.; Karim, H.F.H. Impact of magnetic application on the parameters related to growth of chickpea (*Cicer arietinum* L.). *Jordan J. Biol. Sci.* **2010**, *3*, 175–184.
11. Mohammed, A.; Tana, T.; Singh, P.; Korecha, D.; Mollad, A. Management options for rainfed chickpea (*Cicer arietinum* L.) in northeast Ethiopia under climate change condition. *Clim. Risk Manag.* **2017**, *16*, 222–233. [CrossRef]
12. Rani, A.; Devi, P.; Jha, U.C.; Sharma, K.D.; Siddique, K.H.M.; Nayyar, H. Developing Climate-Resilient Chickpea Involving Physiological and Molecular Approaches with a Focus on Temperature and Drought Stresses. *Front. Plant Sci.* **2020**, *10*, 1759. [CrossRef] [PubMed]
13. Arif, A.; Parveen, N.; Waheed, M.Q.; Atif, R.M.; Waqar, I.; Shah, T.M. A Comparative Study for Assessing the Drought-Tolerance of Chickpea Under Varying Natural Growth Environments. *Front. Plant Sci.* **2021**, *11*, 607869. [CrossRef]
14. Kumar, J.; Abbo, S. Genetics of flowering time in chickpea and its bearing on productivity in semiarid environments. *Adv. Agron.* **2001**, *72*, 107–138.
15. Singh, K.B.; Malhotra, R.S.; Halila, M.H.; Knights, E.J.; Verma, M.M. Current status and future strategy in breeding chickpea for resistance to biotic and abiotic stresses. *Euphytica* **1994**, *73*, 137–149. [CrossRef]
16. Millan, T.; Clarke, H.J.; Siddique, K.H.; Buhariwalla, H.K.; Gaur, P.M.; Kumar, J.; Gil, J.; Kahl, G.; Winter, P. Chickpea molecular breeding: New tools and concepts. *Euphytica* **2006**, *147*, 81–103. [CrossRef]
17. Jameel, S.; Hameed, A.; Shah, T.M. Investigation of Distinctive Morpho-Physio and Biochemical Alterations in Desi Chickpea at Seedling Stage Under Irrigation, Heat, and Combined Stress. *Front. Plant Sci.* **2021**, *12*, 692745. [CrossRef]
18. Sachdeva, S.; Bharadwaj, C.; Patil, B.S.; Pal, M.; Roorkiwal, M.; Varshney, R.K. Agronomic Performance of Chickpea Affected by Drought Stress at Different Growth Stages. *Agronomy* **2022**, *12*, 995. [CrossRef]
19. Grewal, S.K.; Sharma, K.P.; Bharadwaj, R.D.; Hegde, V.; Tripathi, S.; Singh, S.; Jain, P.K.; Agrawal, P.K.; Mondal, B. Understanding genotypic variation and identification of promising genotypes for iron and zinc content in chickpea (*Cicer arietinum* L.). *J. Food Compos. Anal.* **2020**, *88*, 103458. [CrossRef]
20. Sahu, V.K.; Tiwari, S.; Tripathi, M.K.; Gupta, N.; Tomar, R.S.; Yasin, M. Morpho-physiological and biochemical traits analysis for *Fusarium* wilt disease using gene-basedmarkers in desi and Kabuli genotypes of chickpea (*Cicerarietinum* L.). *Indian J. Genet.* **2020**, *80*, 163–172.
21. Singh, S.; Babu, K.S.; Arora, A.; Panwar, R.K.; Verma, S.K. Genetic studies for biofortification traits in chickpea. *J. Food Legumes* **2021**, *34*, 17–20.
22. Wallace, T.C.; Murray, R.; Zelman, K.M. The Nutritional Value and Health Benefits of Chickpeas and Hummus. *Nutrients* **2016**, *8*, 766. [CrossRef] [PubMed]
23. Iqbal, A.; Ateeq, N.; Khalil, I.A.; Perveen, S.; Saleemullah, S. Physicochemical characteristics and amino acid profile of chickpea cultivars grown in Pakistan. *J. FoodSer.* **2006**, *17*, 94–101. [CrossRef]
24. Singh, S.; Singh, D.; Rao, V.U.M. Seedling establishment of chickpea cultivars in varying sowing environments under field conditions. *J. Agrometeorol.* **2009**, *11*, 148–151. [CrossRef]
25. Hirdyani, H. Nutritional composition of Chickpea (*Cicer arietinum*-L) and value added products. *Indian J. Community Health Haryana J. Agron.* **2014**, *4*, 116–118.
26. Jukanti, A.K.; Gaur, P.M.; Gowda, C.L.L.; Chibbar, R.N. Chickpea: Nutritional properties and its benefits. *Br. J. Nutr.* **2012**, *108*, S11–S26. [CrossRef]
27. Samineni, S.; Mahendrakar, M.D.; Shankar, N.; Hotti, A.; Chand, U.; Rathore, A.; Gaur, P.M. Impact of heat and drought stresses on grain nutrient content in chickpea: Genome-wide marker-trait associations for protein, Fe and Zn. *Environ. Exp. Bot.* **2022**, *194*, 104688. [CrossRef]
28. Korbu, L.; Fikre, A.; Tesfaye, K.; Funga, A.; Bekele, D.; Ojiewo, C.O. Response of chickpea to varying moisture stress conditions in Ethiopia. *Agrosyst. Geosci. Environ.* **2022**, *5*, e20234. [CrossRef]
29. Marjani, A.; Farsi, M.; Hervan, E.M.; Ganjeali, A. Comparative analysis of LEA and Dehydrin genes in response to drought stress in chickpea phonological different. *Int. J. Biosci.* **2014**, *4*, 49–57.
30. Krishnamurthy, L.; Kashiwagi, J.; Gaur, P.M.; Upadhyaya, H.D.; Vadez, V. Sources of tolerance to terminal drought in the chickpea (*Cicer arietinum* L.) minicore germplasm. *Field Crops Res.* **2010**, *119*, 322–330. [CrossRef]

31. Upadhyaya, H.D.; Kashiwagi, J.; Varshney, R.K.; Gaur, P.M.; Saxena, K.B.; Krishnamurthy, L.; Gowda, C.L.L.; Pundir, R.P.S.; Chaturvedi, S.K.; Basu, P.S.; et al. Phenotyping chickpeas and pigeonpeas for adaptation to drought. *Front. Physiol.* **2012**, *3*, 179. [CrossRef]
32. Awasthi, R.; Gaur, P.; Turner, N.C.; Vadez, V.; Siddique, K.H.M.; Nayyar, H. Effects of individual and combined heat and drought stress during seed filling on the oxidative metabolism and yield of chickpea (*Cicer arietinum*) genotypes differing in heat and drought tolerance. *Crop Pasture Sci.* **2017**, *68*, 823–841. [CrossRef]
33. Devasirvatham, V.; Tan, D.K.Y. Impact of High Temperature and Drought Stresses on Chickpea Production. *Agronomy* **2018**, *8*, 145. [CrossRef]
34. Choudhary, M.L.; Tripathi, M.K.; Gupta, N.; Tiwari, S.; Tripathi, N.; Parihar, P.; Pandya, R.K. Screening of pearl millet [Pennisetum glaucum [L] R Br] germplasm lines against drought tolerance based on biochemical traits. *Curr. J. Appl. Sci. Technol.* **2021**, *40*, 1–12. [CrossRef]
35. Choudhary, M.L.; Tripathi, M.K.; Tiwari, S.; Pandya, R.K.; Gupta, N.; Tripathi, N.; Parihar, P. Screening of pearl millet [*Pennisetum glaucum* (L) R Br] germplasm lines for drought tolerance based on morpho-physiological traits and SSR markers. *Curr. J. Appl. Sci. Technol.* **2021**, *40*, 46–63. [CrossRef]
36. Raddi, S.; Giannetti, F.; Martini, S.; Farinella, F.; Chirici, G.; Tani, A.; Maltoni, A.; Mariotti, B. Monitoring drought response and chlorophyll content in Quercus by consumer-grade, near-infrared (NIR) camera: A comparison with reflectance spectroscopy. *New For.* **2022**, *53*, 241–265. [CrossRef]
37. Sapes, A.; Sala, G. Relative water content consistently predicts drought mortality risk in seedling populations with different morphology, physiology and times to death. *Plant Cell Environ.* **2021**, *44*, 3322–3335. [CrossRef]
38. Maqbool, M.A.; Aslam, M.; Ali, H.; Shah, T.M.; Farid, B.; Zaman, Q.U. Drought tolerance indices-based evaluation of chickpea advanced lines under different water treatments. *Res. Crops* **2015**, *16*, 336–344. [CrossRef]
39. Kashiwagi, J.; Krishnamurthy, L.; Upadhyaya, H.D.; Krishna, H.; Chandra, S.; Vadez, V.; Serraj, R. Genetic variability of drought-avoidance root traits in the mini-core germplasm collection of chickpeas (*Cicer arietinum* L). *Euphytica* **2005**, *146*, 213–222. [CrossRef]
40. Kashiwagi, J.; Krishnamurthy, L.; Upadhyaya, H.D.; Gaur, P.M. Rapid screening technique for canopy temperature status and its relevance to drought tolerance improvement in chickpea. *J. SAT Agric. Res.* **2008**, *6*, 1–4.
41. Krishnamurthy, L.; Kashiwagi, J.; Upadhyaya, H.D.; Serraj, R. Genetic diversity of drought-avoidance root traits in the mini-core germplasm collection of chickpeas. *Int. Chick. Pigeonpea Newslett.* **2003**, *10*, 21–24.
42. Ramamoorthy, P.; Lakshmanan, K.; Upadhyaya, H.D.; Vadez, V.; Varshney, R.K. Root traits confer grain yield advantages under terminal drought in chickpea (*Cicer arietinum* L.). *Field Crops Res.* **2017**, *201*, 146–161. [CrossRef] [PubMed]
43. Kumar, N.; Soren, K.R.; Bharadwaj, C.; Sneha Priya, P.R.; Shrivastava, A.K.; Pal, M.; Roorkiwal, M.; Kumar, K.; Patil, B.S.; Soni, A.; et al. Genome-wide transcriptome analysis and physiological variation modulates gene regulatory networks acclimating salinity tolerance in chickpea. *Environ. Exp. Bot.* **2021**, *187*, 104478. [CrossRef]
44. Hosseinzadeh, S.R.; Amiri, H.; Ismaili, A. Evaluation of photosynthesis, physiological, and biochemical responses of chickpea (*Cicer arietinum* L. cv. Pirouz) under water deficit stress and use of vermicompost fertilizer. *J. Integrat. Agric.* **2018**. [CrossRef]
45. Hussain, T.; Akram, Z.; Shabbir, G.; Manaf, A.; Ahmed, M. Identification of drought tolerant Chickpea genotypes through multi trait stability index. *Saudi J. Biol. Sci.* **2021**, *28*, 6818–6828. [CrossRef]
46. Salahvarzi, M.; Nasr Esfahani, M.; Shirzadi, N.; Burritt, D.J.; Tran, L.P. Genotype- and tissue-specific physiological and biochemical changes of two chickpea (*Cicer arietinum*) varieties following a rapid dehydration. *Physiol. Plant* **2021**, *172*, 1822–1834. [CrossRef]
47. Thudi, M.; Upadhyaya, H.D.; Rathore, A.; Gaur, P.M.; Krishnamurthy, L.; Roorkiwal, M.; Nayak, S.N.; Chaturvedi, S.K.; Basu, P.S.; Gangarao, N.V.P.R.; et al. Genetic dissection of drought and heat tolerance in chickpea through genomewide and candidate gene-based association mapping approaches. *PLoS ONE* **2014**, *9*, e96758. [CrossRef]
48. Palit, P.; Ghosh, R.; Tolani, P.; Tarafdar, A.; Chitikineni, A.; Bajaj, P.; Sharma, M.; Kudapa, H.; Varshney, R.K. Molecular and Physiological Alterations in Chickpea under Elevated CO2 Concentrations. *Plant Cell Physiol.* **2020**, *61*, 1449–1463. [CrossRef] [PubMed]
49. Kanca, O.; Bellen, H.J.; Schnorrer, F. Gene Tagging Strategies to Assess Protein Expression, Localization, and Function in Drosophila. *Genetics* **2017**, *207*, 389–412, Erratum in *Genetics* **2017**, *207*, 1711. [CrossRef]
50. Santiago, C.R.d.N.; Assis, R.d.A.B.; Moreira, L.M.; Digiampietri, L.A. Gene Tags Assessment by Comparative Genomics (GTACG): A User-Friendly Framework for Bacterial Comparative Genomics. *Front. Genet.* **2019**, *10*, 725. [CrossRef]
51. Wanga, M.A.; Shimelis, H.; Mashilo, J.; Laing, M.D. Opportunities and challenges of speed breeding: A review. *Plant Breed.* **2021**, *140*, 185–194. [CrossRef]
52. Ali, Z.; Merrium, S.; Habib-ur-Rahman, M.; Hakeem, S.; Saddique, M.A.B.; Sher, M.A. Wetting mechanism and morphological adaptation; leaf rolling enhancing atmospheric water acquisition in wheat crop—A review. *Environ. Sci. Pollut. Res.* **2022**, *29*, 30967–30985. [CrossRef] [PubMed]
53. Ray, S.; Satya, P. Next generation sequencing technologies for next generation plant breeding. *Front. Plant Sci.* **2014**, *5*, 367. [CrossRef] [PubMed]
54. Verma, S.; Gupta, S.; Bandhiwal, N.; Kumar, T.; Bharadwaj, C.; Bhatia, S. High-density linkage map construction and mapping of seed trait QTLs in chickpea (*Cicer arietinum* L.) using Genotyping-by-Sequencing (GBS). *Sci. Rep.* **2015**, *5*, 17512. [CrossRef] [PubMed]

55. Sahu, P.K.; Sao, R.; Mondal, S.; Vishwakarma, G.; Gupta, S.K.; Kumar, V.; Singh, S.; Sharma, D.; Das, B.K. Next Generation Sequencing Based Forward Genetic Approaches for Identification and Mapping of Causal Mutations in Crop Plants: A Comprehensive Review. *Plants* **2020**, *9*, 1355. [CrossRef]
56. Wang, Z.; Gerstein, M.; Snyder, M. RNA-Seq: A revolutionary tool for transcriptomics. *Nat. Rev. Genet.* **2009**, *10*, 57–63. [CrossRef] [PubMed]
57. Kaur, P.; Gaikwad, K. From Genomes to GENE-omes: Exome Sequencing Concept and Applications in Crop Improvement. *Front. Plant Sci.* **2017**, *8*, 2164. [CrossRef]
58. Leo, V.C.; Morgan, N.V.; Bern, D.; Jones, M.L.; Lowe, G.C.; Lordkipanidzé, M.; Drake, S.; Simpson, M.A.; Gissen, P.; Mumford, A.; et al. Use of next-generation sequencing and candidate gene analysis to identify underlying defects in patients with inherited platelet function disorders. *J. Thromb. Haemost.* **2015**, *13*, 643–650. [CrossRef] [PubMed]
59. Kulski, J.K.; Suzuki, S.; Ozaki, Y.; Mitsunaga, S.; Inoko, H.; Shiina, T. Phase HLA genotyping by next generation sequencing—A comparison between two massively parallel sequencing bench-top systems, the Roche GS Junior and Ion Torrent PGM. In *HLA and Associated Important Diseases*; Xi, Y., Ed.; Intech.: Rijeka, Croatia, 2014; pp. 141–181.
60. Rao, M.S.; Van Vleet, T.R.; Ciurlionis, R.; Buck, W.R.; Mittelstadt, S.W.; Blomme, E.A.G.; Liguori, M.J. Comparison of RNA-Seq and Microarray Gene Expression Platforms for the Toxicogenomic Evaluation of Liver from Short-Term Rat Toxicity Studies. *Front. Genet.* **2019**, *9*, 636. [CrossRef] [PubMed]
61. Lowe, R.; Shirley, N.; Bleackley, M.; Dolan, S.; Shafee, T. Transcriptomics technologies. *PLoS Comput. Biol.* **2017**, *13*, e1005457. [CrossRef]
62. Wang, B.; Kumar, V.; Olson, A.; Ware, D. Reviving the Transcriptome Studies: An Insight into the Emergence of Single-Molecule Transcriptome Sequencing. *Front. Genet.* **2019**, *10*, 384. [CrossRef]
63. Bhaskarla, V.; Zinta, G.; Ford, R.; Jain, M.; Varshney, R.K.; Mantri, N. Comparative Root Transcriptomics Provide Insights into Drought Adaptation Strategies in Chickpea (*Cicer arietinum* L.). *Int. J. Mol. Sci.* **2020**, *21*, 1781. [CrossRef] [PubMed]
64. Upasani, M.L.; Limaye, B.M.; Gurjar, G.S.; Kasibhatla, S.M.; Joshi, R.R.; Kadoo, N.Y.; Gupta, V.S. Chickpea-Fusarium oxysporum interaction transcriptome reveals differential modulation of plant defense strategies. *Sci. Rep.* **2017**, *7*, 7746. [CrossRef] [PubMed]
65. Aslam, B.; Basit, M.; Nisar, M.A.; Khurshid, M.; Rasool, M.H. Proteomics: Technologies and Their Applications. *J. Chromatogr. Sci.* **2017**, *55*, 182–196. [CrossRef] [PubMed]
66. Ahrens, C.H.; Brunner, E.; Qeli, E.; Basler, K.; Aebersold, R. Generating and navigating proteome maps using mass spectrometry. *Nat. Rev. Mol. Cell Biol.* **2010**, *11*, 789–801. [CrossRef]
67. Yang, X.; Lu, M.; Wang, Y.; Wang, Y.; Liu, Z.; Chen, S. Response Mechanism of Plants to Drought Stress. *Horticulturae* **2021**, *7*, 50. [CrossRef]
68. Mishra, N.; Tripathi, M.K.; Tiwari, S.; Tripathi, N.; Gupta, N.; Sharma, A. Morphological and physiological performance of Indian soybean [*Glycine max* (L.) Merrill] genotypes in respect to drought. *Legume Res. Int. J.* **2021**. [CrossRef]
69. Mishra, N.; Tripathi, M.K.; Tripathi, N.; Tiwari, S.; Gupta, N.; Sharma, A.; Shrivastav, M.K. Changes in biochemical and antioxidant enzymes activities play significant role in drought tolerance in soybean. *Int. J. Agric. Technol.* **2021**, *17*, 1425–1446.
70. Atta, K.; Singh, A.P.; Adhikary, S.; Mondal, S.; Dewanjee, S. Drought Stress: Manifestation and Mechanisms of Alleviation in Plants. In *Drought [Working Title] [Internet]*; Eyvaz, A.P.M., Albahnasawi, A., Tekbaş, M.M., Gürbulak, M.E., Eds.; IntechOpen: London, UK, 2022. Available online: https://www.intechopen.com/online-first/80834 (accessed on 30 April 2022).
71. Mishra, N.; Tripathi, M.K.; Tiwari, S.; Tripathi, N.; Ahuja, A.; Sapre, S.; Tiwari, S. Cell suspension culture and in vitro screening for drought tolerance in soybean using poly-ethylene glycol. *Plants* **2021**, *10*, 517. [CrossRef]
72. Tiwari, S.; Lata, C.; Chauhan, P.S.; Nautiyal, C.S. Pseudomonas putida attunes morphophysiological, biochemical and molecular responses in Cicer arietinum L. during drought stress and recovery. *Plant Physiol. Biochem.* **2016**, *99*, 108–117. [CrossRef]
73. Toker, C.; Canci, H. Selection for drought and heat resistance in chickpea under terminal drought conditions. In *Food Legumes for Nutritional Security and Sustainable Agriculture: 4th International Food Legumes Research Conference*; Kharkwal, M.C., Ed.; Indian Agricultural Research Institute: New Delhi, India, 2006; pp. 18–22.
74. Varshney, R.K.; Thudi, M.; Pandey, M.K.; Tardieu, F.; Ojiewo, C.; Vadez, V.; Whitbread, A.M.; Siddique, K.H.M.; Nguyen, H.T.; Carberry, P.S.; et al. Accelerating genetic gains in legumes for the development of prosperous smallholder agriculture: Integrating genomics, phenotyping, systems modelling and agronomy. *J. Exp. Bot.* **2018**, *69*, 3293–3312. [CrossRef] [PubMed]
75. Basu, S.; Ramegowda, V.; Kumar, A.; Pereira, A. Plant adaptation to drought stress. *F1000Res* **2016**, *5*, F1000 Faculty Rev-1554. [CrossRef]
76. Comas, L.H.; Becker, S.R.; Cruz, V.M.V.; Byrne, P.F.; Dierig, D.A. Root traits contributing to plant productivity under drought. *Front. Plant Sci.* **2013**, *4*, 442. [CrossRef]
77. Martignago, D.; Rico-Medina, A.; Blasco-Escámez, D.; Fontanet-Manzaneque, J.B.; Caño-Delgado, A.I. Drought Resistance by Engineering Plant Tissue-Specific Responses. *Front. Plant Sci.* **2020**, *10*, 1676. [CrossRef] [PubMed]
78. Xu, Z.; Zhou, G.; Shimizu, H. Plant responses to drought and rewatering. *Plant Signal. Behav.* **2010**, *5*, 649–654. [CrossRef] [PubMed]
79. Bhargava, S.; Sawant, K. Drought stress adaptation: Metabolic adjustment and regulation of gene expression. *Plant Breed.* **2012**, *132*, 21–32. [CrossRef]
80. Keskin, M.E.; Terzi, Ö.; Taylan, E.D.; Küçükyaman, D. Meteorological drought analysis using artificial neural networks. *Acad. J.* **2011**, *6*, 4469–4477.

81. Berger, J.; Palta, J.; Vadez, V. An integrated framework for crop adaptation to dry environments: Responses to transient and terminal drought. *Plant Sci.* **2016**, *253*, 58–67. [CrossRef] [PubMed]
82. Maliro, M.F.A.; MacNeil, D.; Redden, B.; Kollmorgen, J.F.; Pittock, C. Sampling strategies and screening of chickpea (*Cicerarietinum* L.) germplasm for salt tolerance. *Genet. Resour. Crop. Evol.* **2008**, *55*, 53–63. [CrossRef]
83. Kashiwagi, J.; Krishnamurthy, L.; Purushothaman, R.; Upadhyaya, H.D.; Gaur, P.M.; Gowda, C.L.L.; Ito, O.; Varshney, R.K. Scope for improvement of yield under drought through the root traits in chickpea (*Cicer arietinum* L.). *Field Crops Res.* **2015**, *170*, 47–54. [CrossRef]
84. Serraj, R.; Sinclair, T.R. Osmolyte Accumulation: Can It Really Help Increase Crop Yield under Drought Conditions? *Plant Cell Environ.* **2002**, *25*, 333–341. [CrossRef] [PubMed]
85. Goulet, B.E.; Roda, F.; Hopkins, R. Hybridization in Plants: Old Ideas, New Techniques. *Plant Physiol.* **2017**, *173*, 65–78. [CrossRef]
86. Purushothaman, R.; Krishnamurthy, L.; Upadhyaya, H.D.; Vadez, V.; Varshney, R.K. Shoot traits and their relevance in terminal drought tolerance of chickpea (*Cicer arietinum* L.). *Field Crop. Res.* **2016**, *197*, 10–27. [CrossRef]
87. Leport, L.; Turner, N.C.; Davies, S.L.; Siddique, K.H.M. Variation in pod production and abortion among chickpea cultivars under terminal drought. *Eur. J. Agron.* **2006**, *24*, 236–246. [CrossRef]
88. Nayyar, H.; Bains, T.; Kumar, S. Low temperature induced floral abortion in chickpea: Relationship to abscisic acid and cryoprotectants in reproductive organs. *Environ. Exp. Bot.* **2005**, *53*, 39–47. [CrossRef]
89. Rahbarian, R.; Nejad, R.K.; Ganjeali, A.; Bagheri, A.; Najafi, F. Drought stress effects on photosynthesis, chlorophyll fluorescence and water relations in tolerant and susceptible chickpea (*Cicer arietınum* L.) genotypes. *Acta Biol. Crac. Ser. Bot.* **2011**, *53*, 47–56. [CrossRef]
90. Gill, S.S.; Tuteja, N. Reactive oxygen species and antioxidant machinery in abiotic stress tolerance in crop plants. *Plant Physiol. Biochem.* **2010**, *48*, 909–930. [CrossRef] [PubMed]
91. Kashiwagi, J.; Krishnamurty, L.; Gaur, P.M.; Chandra, S.; Upadhyaya, H.D. Estimation of gene effects of the drought avoidance root characteristicsin chickpea (*C. arietinum* L.). *Field Crops Res.* **2008**, *105*, 64–69. [CrossRef]
92. Purushothaman, R.; Thudi, M.; Krishnamurthy, L.; Upadhyaya, H.D.; Kashiwagi, J.; Gowda, C.L.L.; Varshney, R.K. Association of mid-reproductive stage canopy temperature depression with the molecular markers and grain yields of chickpea (*Cicer arietinum* L.) germplasm under terminal drought. *Field Crops Res.* **2015**, *174*, 1–11. [CrossRef]
93. Sofi, P.A.; Ara, A.; Gull, M.; Rehman, K. Canopy Temperature Depression as an Effective Physiological Trait for Drought Screening. In *Drought-Detection and Solutions*; IntechOpen: London, UK, 2019. [CrossRef]
94. Karimzadeh, S.H.; Nezami, A.; Nabati, J.; Oskoueian, E.; Ahmadi-Lahijani, M.J. The physiological, biochemical, and molecular modifications of chickpea (*Cicer arietinum* L.) seedlings under freezing stress. *J. Plant Growth Regul.* **2021**, *41*, 1109–1124. [CrossRef]
95. Sivasakthi, K.; Tharanya, M.; Kholová, J.; Wangari Muriuki, R.; Thirunalasundari, T.; Vadez, V. Chickpea Genotypes Contrasting for Vigor and Canopy Conductance Also Differ in Their Dependence on Different Water Transport Pathways. *Front. Plant Sci.* **2017**, *8*, 1663. [CrossRef]
96. Krishnamurthy, L.; Kashiwagi, J.; Tobita, S.; Ito, O.; Upadhyaya, H.D.; Gowda, C.L.L.; Gaur, P.M.; Sheshshayee, M.S.; Singh, S.; Vadez, V.; et al. Variation in carbon isotope discrimination and its relationship with harvest index in the reference collection of chickpea germplasm. *Funct. Plant Biol.* **2013**, *14*, 1350–1361. [CrossRef] [PubMed]
97. Ma, Q.; Behboudian, M.H.; Turner, N.C.; Palta, J.A. Gas exchange by pods and subtending leaves and internal recycling of internal CO2 by pods of chickpea (*Cicer arietinum* L.) subjected to water stess. *J. Exp. Bot.* **2001**, *52*, 123–131. [CrossRef] [PubMed]
98. Macar, T.K.; Ekmekci, Y. Alterations in Photochemical and Physiological Activities of Chickpea (*Cicer arietinum* L.) Cultivars under Drought Stress. *J. Agron. Crop. Sci.* **2009**, *195*, 335–346. [CrossRef]
99. Shayla, B.; Inderjit, S.; Satinder, S.; Ashutosh, K.S.G.B.; Sonia, S.; Karan, K.; Kaur, G.S.C.B.; Harsh, N.; Sarvjeet, S. Use of morpho-physiological and biochemical traits to identify sources of drought and heat tolerance in chickpea (*Cicer arietinum*). *Crop. Pasture Sci.* **2021**, *72*, 801–814.
100. Talebi, R.; Ensafi, M.H.; Baghebani, N.; Karami, E.; Mohammadi, K. Physiological responses of chickpea (*Cicer arietinum*) genotypes to drought stress. *Environ. Exp. Biol.* **2013**, *11*, 9–15.
101. Jangpromma, N.; Songsri, P.; Thammasirirak, S.; Jaisil, P. Rapid assessment of chlorophyll content in sugarcane using a SPAD chlorophyll meter across different water stress conditions. *Asian J. Plant Sci.* **2010**, *9*, 368–374. [CrossRef]
102. Lamaoui, M.; Jemo, M.; Datla, R.; Bekkaoui, F. Heat and Drought Stresses in Crops and Approaches for Their Mitigation. *Front. Chem.* **2018**, *6*, 26. [CrossRef] [PubMed]
103. Devasirvatham, V.; Tan, D.K.Y.; Gaur, P.M.; Raju, T.N.; Trethowan, R.M. High temperature tolerance in chickpea and its implications for plant improvement. *Crop. Pasture Sci.* **2012**, *63*, 419–428. [CrossRef]
104. Chen, Y.; Ghanem, M.E.; Siddique, K.H. Characterising root trait variability in chickpea (*Cicer arietinum* L.) germplasm. *J. Exp. Bot.* **2017**, *68*, 1987–1999. [CrossRef]
105. Badhan, S.; Kole, P.; Ball, A.; Mantri, N. RNA sequencing of leaf tissues from two contrasting chickpea genotypes reveals mechanisms for drought tolerance. *Plant Physiol. Biochem.* **2018**, *129*, 295–304. [CrossRef]
106. Ahmar, S.; Gill, R.A.; Jung, K.H.; Faheem, A.; Qasim, M.U.; Mubeen, M.; Zhou, W. Conventional and Molecular Techniques from Simple Breeding to Speed Breeding in Crop Plants: Recent Advances and Future Outlook. *Int. J. Mol. Sci.* **2020**, *21*, 2590. [CrossRef]
107. Stefaniak, T.; McPhee, K. Comparison of Hybridization Techniques in Chickpea. *Crop. Sci.* **2017**, *57*, 843–846. [CrossRef]

108. Deokar, A.A.; Tar'an, B. Genome-wide analysis of the aquaporin gene family in chickpea (*Cicer arietinum* L.). *Front. Plant Sci.* **2016**, *7*, 1802. [CrossRef]
109. Deokar, A.A.; Kondawar, V.; Kohli, D.; Aslam, M.; Jain, P.K.; Karuppayil, S.M.; Varshney, R.K.; Srinivasan, R. The CarERF genes in chickpea (*Cicerarietinum* L.) and the identification of CarERF116 as abiotic stress responsive transcription factor. *Funct. Integr. Genom.* **2015**, *15*, 27–46. [CrossRef] [PubMed]
110. Hamwieh, A.; Imtiaz, M.; Malhotra, R.S. Multi-environment QTL analyses for drought-related traits in a recombinant inbred population of chickpea (*Cicer arietinum* L.). *Theor. Appl. Genet.* **2013**, *126*, 1025–1038. [CrossRef] [PubMed]
111. Singh, V.K.; Khan, A.W.; Jaganathan, D.; Thudi, M.; Roorkiwal, M.; Takagi, H.; Garg, V.; Kumar, V.; Chitikineni, A.; Gaur, P.M.; et al. QTL-seq for rapid identification of candidate genes for 100-seed weight and root/total plant dry weight ratio under rainfed conditions in chickpea. *Plant. Biotechnol. J.* **2016**, *14*, 2110–2119. [CrossRef]
112. Kudapa, H.; Garg, V.; Chitikineni, A.; Varshney, R.K. The RNA-Seqbased high resolution gene expression atlas of chickpea (*Cicer arietinum* L.) reveals dynamic spatiotemporal changes associated with growth and development. *Plant Cell Environ.* **2018**, *41*, 2209–2225. [CrossRef]
113. Kumar, M.; Chauhan, A.S.; Yusuf, M.A.; Sanyal, I.; Chauhan, P.S. Transcriptome sequencing of chickpea (*Cicer arietinum* L.) genotypes for identifcation of drought-responsive genes under drought stress condition. *Plant Mol. Biol. Rep.* **2019**, *37*, 186–203. [CrossRef]
114. Gupta, S.; Singh, A.; Singh, P.; Kewat, R.N. Effect of drought stress or carbohydrate content in drought tolerant and susceptible chickpea genotypes. *J. Crop. Sci. Biotechnol.* **2015**, *4*, 35–38.
115. Sachdeva, S.; Bharadwaj, C.; Singh, R.K.; Jain, P.K.; Patil, B.S.; Roorkiwal, M.; Varshney, R. Characterization of ASR gene and its role in drought tolerance in chickpea (*Cicer arietinum* L.). *PLoS ONE* **2020**, *15*, e0234550. [CrossRef]
116. Kooyers, N.J. The evolution of drought escape and avoidance in natural herbaceous populations. *Plant Sci.* **2015**, *234*, 155–162. [CrossRef] [PubMed]
117. Siddique, K.H.M.; Loss, S.P.; Thomson, B.D. Cool season grain legumes in dryland Mediterranean environments of Western Australia: Significance of early flowering. In *Management of Agricultural Drought*; Saxena, N.P., Ed.; Science Publishers: Enfield, NH, USA, 2003; pp. 151–161.
118. Saeed, M.; Francis, C.A. Yield stability in relation to maturity in grain sorghum. *Crop. Sci.* **1983**, *23*, 683–687. [CrossRef]
119. Richards, M.F.; Maphosa, L.; Preston, A.L. Impact of Sowing Time on Chickpea (*Cicer arietinum* L.) Biomass Accumulation and Yield. *Agronomy* **2022**, *12*, 160. [CrossRef]
120. Shimray, P.U.; Bajaj, D.; Shrivastava, R.; Daware, A.; Upadhyaya, H.D.; Kumar, R.; Bhardwaj, C.; Tyagi, A.K.; Parida, S.K. Identifying Transcription Factor Genes Associated with Yield Traits in Chickpea. *Plant Mol. Biol. Rep.* **2017**, *35*, 562–574. [CrossRef]
121. Sabaghpour, S.H.; Kumar, J.; Rao, T.N. Inheritance of growth vigor and its association with other characters in chickpea. *Plant Breed.* **2003**, *122*, 542–544. [CrossRef]
122. Sun, F.; Chen, Q.; Chen, Q.; Jiang, M.; Gao, W.; Qu, Y. Screening of Key Drought Tolerance Indices for Cotton at the Flowering and Boll Setting Stage Using the Dimension Reduction Method. *Front. Plant Sci.* **2021**, *12*, 619926. [CrossRef]
123. Sabaghpour, S.H.; Mahmodi, A.A.; Saeed, A.; Kamel, M.; Malhotra, R.S. Study on chickpea drought tolerance lines under dryland condition of Iran. *Indian J. Crop. Sci.* **2006**, *1*, 70–73.
124. Dumanović, J.; Nepovimova, E.; Natić, M.; Kuča, K.; Jaćević, V. The Significance of Reactive Oxygen Species and Antioxidant Defense System in Plants: A Concise Overview. *Front. Plant Sci.* **2021**, *11*, 552969. [CrossRef]
125. Levitt, J. *Responses of Plants to Environmental Stresses. Volume II. Water, Radiation, Salt, and Other Stresses*; Academic Press: Cambridge, MA, USA, 1980.
126. Kamanga, R.M.; Mbega, E.; Ndakidemi, P. Drought Tolerance Mechanisms in Plants: Physiological Responses Associated with Water Deficit Stress in *Solanum lycopersicum*. *Adv. Crop. Sci. Technol.* **2018**, *6*, 362. [CrossRef]
127. Pastori, G.; Foyer, C.H.; Mullineaux, P. Low temperature-induced changes in the distribution of H_2O_2 and antioxidants between the bundle sheath and mesophyll cells of maize leaves. *J. Exp. Bot.* **2000**, *51*, 107–113. [CrossRef]
128. Li, H.; Ma, X.; Lu, Y.; Ren, R.; Cui, B.; Si, B. Growing deep roots has opposing impacts on the transpiration of apple trees planted in subhumid loess region. *Agric. Water Manag.* **2021**, *258*, 107207. [CrossRef]
129. Maqbool, M.A.; Aslam, M.; Ali, H.; Shah, T.M.; Atta, B.M. GGE biplot analysis-based selection of superior chickpea (*Cicerarietinum* L.) inbred lines under variable water environments. *Pak. J. Bot.* **2015**, *47*, 1901–1908.
130. Sohrabi, Y.; Heidari, G.; Weisany, W.; Ghasemi Golezani, K.; Mohammadi, K. Some physiological responses of chickpea cultivars to arbuscular mycorrhiza under drought stress. *Russ. J. Plant Physiol.* **2012**, *59*, 708–716. [CrossRef]
131. Kashiwagi, J.; Krishnamurthy, L.; Crouch, J.H.; Serraj, R. Variability of root length density and its contributions to seed yield in chickpea (*Cicer arietinum* L) under terminal drought stress. *Field Crops Res.* **2006**, *95*, 171–181. [CrossRef]
132. Gaur, P.M.; Krishnamurthy, L.; Kashiwagi, J. Improving drought avoidance root traits in chickpea (*Cicer arietinum* L.) current status of research at ICRISAT. *Plant Prod. Sci.* **2008**, *11*, 3–11. [CrossRef]
133. Davies, W.; Zhang, J. Root signals and the regulation of growth and development of plants in dry soil. *Ann. Rev. Plant Physiol. Plant Mol. Biol.* **1991**, *42*, 55–76. [CrossRef]
134. Pierik, R.; Sasidharan, R.; Voesenek, L.A.C.J. Growth control by ethylene: Adjusting phenotypes to the environment. *J. Plant Growth Regul.* **2007**, *26*, 188–200. [CrossRef]

135. Zaman-Allah, M.; Jenkinson, D.M.; Vandez, V. Chickpea genotypes contrasting for seed yield under terminal drought stress in the field differ for traits related to the control of water use. *Funct. Plant Biol.* **2011**, *38*, 270–281. [CrossRef]
136. Pang, J.; Turner, N.C.; Du, Y.-L.; Colmer, T.D.; Siddique, K.H.M. Pattern of Water Use and Seed Yield under Terminal Drought in Chickpea Genotypes. *Front. Plant Sci.* **2017**, *8*, 1375. [CrossRef]
137. Zaman-Allah, M.; Jenkinson, D.M.; Vandez, V. A conservative pattern of water use, rather than deep or profuse rooting, is critical for terminal drought tolerance of chickpea. *J. Exp. Bot.* **2011**, *38*, 270–281. [CrossRef]
138. Kashiwagi, J.; Krishnamurthy, L.; Singh Sube Gaur, P.M.; Upadhyaya, H.D.; Panwar, J.D.S.; Basu, P.S.; Ito, O.; Tobita, S. Relationship between transpiration efficiency and carbon isotope discrimination in chickpea (*Cicer arietinum* L.). *J. SAT Agric. Res.* **2006**, *2*, 1–3.
139. Nadal-Sala, D.; Grote, R.; Birami, B.; Knüver, T.; Rehschuh, R.; Schwarz, S.; Ruehr, N.K. Leaf Shedding and Non-Stomatal Limitations of Photosynthesis Mitigate Hydraulic Conductance Losses in Scots Pine Saplings During Severe Drought Stress. *Front. Plant Sci.* **2021**, *12*, 715127. [CrossRef] [PubMed]
140. Li, Y.; Li, H.; Li Zhang, S. Improving water-use efficiency by decreasing stomatal conductance and transpiration rate to maintain higher ear photosynthetic rate in drought-resistant wheat. *Crop. J.* **2017**, *5*, 231–239. [CrossRef]
141. Flexas, J.; Medrano, H. Drought-inhibition of photosynthesis in C3 plants: Stomatal and non-stomatal limitations revisited. *Ann. Bot.* **2002**, *89*, 183–189. [CrossRef] [PubMed]
142. Cornic, G. Drought stress inhibits photosynthesis by decreasing stomatal aperture not by affecting ATP synthesis. *Trends Plant Sci.* **2000**, *5*, 187–188. [CrossRef]
143. Flexas, J.; Galmes, J.; Ribas-Carbo, M.; Medrano, H. The effects of water stress on plant respiration. In *Plant Respiration: From Cell to Ecosystem*; Lambers, H., Ribas-Carbo, M., Eds.; Advances in Photosynthesis and Respiration Series; Springer: Dordrecht, The Netherlands, 2005; Chapter 6; Volume 18, pp. 85–94.
144. Farooq, M.; Gogoi, N.; Barthakur, S.; Baroowa, B.; Bharadwaj, N.; Alghamdi, S.S.; Siddique, K.H.M. Drought stress in grain legumes during reproduction and grain filling. *J. Agron. Crop. Sci.* **2017**, *203*, 81–102. [CrossRef]
145. Xiao, Q.; Bai, X.; Zhang, C.; He, Y. Advanced high-throughput plant phenotyping techniques for genome-wide association studies: A review. *J. Adv. Res.* **2022**, *35*, 215–230. [CrossRef]
146. Nayyar, H.; Singh, S.; Kaur, S.; Kumar, S.; Upadhyaya, H.D. Differential sensitivity of macrocarpa and microcarpa types of chickpea (*Cicer arietinum* L.) to water stress: Association of contrasting stress response with oxidative injury. *J. Integr. Plant Biol.* **2006**, *48*, 1318–1329. [CrossRef]
147. Masle, J.; Gilmore, S.R.; Farquhar, G.D. The ERECTA gene regulates plant transpiration efficiency in Arabidopsis. *Nature* **2005**, *436*, 866–870. [CrossRef]
148. Saxena, N.P. Management of drought in chickpea-holistic approach. In *Management of Agricultural Drought-Agronomic and Genetic Options*; Saxena, N.P., Ed.; Oxford and IBH Publishing Co. Pvt. Ltd.: New Delhi, India, 2003; pp. 103–122.
149. Hemati, A.; Mofidi-Chelan, M.; Amirifar, A.; Moghiseh, E.; Asgari Lajayer, B. Drought Tolerance Mechanisms in Crop Plants. In *Plant Stress Mitigators*; Vaishnav, A., Arya, S., Choudhary, D.K., Eds.; Springer: Singapore, 2022. [CrossRef]
150. Imtiaz, M.; Malhotra, R.S. Reduce stress: Breed for drought tolerance. *ICARDA Caravan Sci. Food Sec.* **2009**, *26*, 34–36.
151. Close, T.J. Dehydrins: Emergence of a biochemical role of a family of plant dehydration proteins. *Physiol. Plant* **1996**, *97*, 795–803. [CrossRef]
152. Hara, M.; Terashima, S.; Kuboi, T. Characterization and cryoprotective activity of cold-responsive dehydrin from Citrus unshiu. *J. Plant Physiol.* **2001**, *158*, 1333–1339. [CrossRef]
153. Hung, S.H.; Yu, C.W. Hydrogen peroxide functions as a stress signal in plants. *Bot. Bull. Acad. Sinica* **2005**, *41*, 1–10.
154. Chen, Z.; Gallie, D.R. The ascorbic acid redox state controls guard cell signaling and stomatal movement. *Plant Cell.* **2004**, *16*, 1143–1162. [CrossRef] [PubMed]
155. Sharma, P.; Jha, A.B.; Dubey, R.S.; Pessarakli, M. Reactive Oxygen Species, Oxidative Damage, and Antioxidative Defense Mechanism in Plants under Stressful Conditions. *J. Bot.* **2012**, *2012*, 217037. [CrossRef]
156. Das, K.; Roychoudhury, A. Reactive oxygen species (ROS) and response of antioxidants as ROS-scavengers during environmental stress in plants. *Front. Environ. Sci.* **2014**, *2*, 53. [CrossRef]
157. Huang, H.; Ullah, F.; Zhou, D.-X.; Yi, M.; Zhao, Y. Mechanisms of ROS Regulation of Plant Development and Stress Responses. *Front. Plant Sci.* **2019**, *10*, 800. [CrossRef]
158. Gupta, P.K.; Varshney, R.K.; Prasad, M. Molecular markers: Principles and methodology. In *Molecular Techniques in Crop Improvement*; Jain, S.M., Ahloowalia, B.S., Brar, D.S., Eds.; Kluwer Academic Publishers: Dordrecht, The Netherlands, 2002; pp. 9–54.
159. Pushpavalli, R.; Krishnamurthy, L.; Thudi, M.; Gaur, P.M.; Rao, M.V.; Siddique, K.H.M.; Colmer, T.D.; Turner, N.C.; Varshney, R.K.; Vadez, V. Two key genomic regions harbour QTLs for salinity tolerance in ICCV 2 JG 11 derived chickpea (*Cicerarietinum* L.) recombinant inbred lines. *BMC Plant Biol.* **2015**, *15*, 124. [CrossRef]
160. Li, Y.; Ruperao, P.; Batley, J.; Edwards, D.; Khan, T.; Colmer, T.D.; Pang, J.; Siddique, K.H.M.; Sutton, T. Investigating Drought Tolerance in Chickpea Using Genome-Wide Association Mapping and Genomic Selection Based on Whole-Genome Resequencing Data. *Front. Plant Sci.* **2018**, *9*, 190. [CrossRef]
161. Kalra, N.; Chakraborty, D.; Sharma, A.; Rai, H.K.; Jolly, M.; Chander, S.; Kumar, P.R.; Bhadraray Barman, D.; Mittal, R.B.; Lal, M.; et al. Effect of increasing temperature on yield of some winter crops in northwest India. *Curr. Sci.* **2008**, *94*, 82–88.

162. Monneveux, P.; Ribaut, J.-M. Secondary traits for drought tolerance improvement in cereals. In *Drought Adaptation in Cereals*; Ribaut, J.-M., Ed.; The Haworth Press Inc.: Binghamton, NY, USA, 2006; pp. 97–143.
163. Shah, T.M.; Imran, M.; Atta, B.M.; Ashraf, M.Y.; Hameed, A.; Waqar, I.; Shafiq, M.; Hussain, K.; Naveed, M.; Aslam, M.; et al. Selection and screening of drought tolerant high yielding chickpea genotypes based on physio-biochemical indices and multi-environmental yield trials. *BMC Plant Biol.* **2020**, *20*, 171. [CrossRef] [PubMed]
164. Gong, H.; Zhu, X.; Chen, K.; Wang, S.; Zhang, C. Silicon alleviates oxidative damage of wheat plants in pots under drought. *Plant Sci.* **2005**, *169*, 313–321. [CrossRef]
165. Hasegawa, P.M.; Bressan, R.A.; Zhu, J.K.; Bohnert, H.J. Plant cellular and molecular responses to high salinity. *Annu. Rev. Plant. Biol.* **2000**, *51*, 463–499. [CrossRef]
166. Fazeli, F.; Ghorbanli, M.; Niknam, V. Effect of drought on biomass, protein content, lipid peroxidation and antioxidant enzymes in two sesame cultivars. *Biol. Plant.* **2007**, *51*, 98–103. [CrossRef]
167. Lima, A.L.S.; DaMatta, F.M.; Pinheiro, H.A.; Totola, M.R.; Loureiro, M.E. Photo- chemical responses and oxidative stress in two clones of Coffea canephora under water deficit conditions. *Environ. Exp. Bot.* **2002**, *47*, 239–247. [CrossRef]
168. Chew, O.; Whelan, J.; Miller, A.H. Molecular definition of the ascorbate-gluta- thione cycle in Arabidopsis mitochondria reveals dual targeting of antioxidant defences in plants. *J. Biol. Chem.* **2003**, *278*, 46869–46877. [CrossRef]
169. Kaur, D.; Grewal, S.K.; Kaur, J.; Singh, S. Differential proline metabolism in vegetative and reproductive tissues determines drought tolerance in chickpea. *Biol. Plant.* **2017**, *61*, 359–366. [CrossRef]
170. Gurrieri, L.; Merico, M.; Trost, P.; Forlani, G.; Sparla, F. Impact of drought on soluble sugars and free proline content in selected Ara bidopsis mutants. *Biology* **2020**, *9*, 367. [CrossRef]
171. Iqbal, S.; Wang, X.; Mubeen, I.; Kamran, M.; Kanwal, I.; Díaz, G.A.; Abbas, A.; Parveen, A.; Atiq, M.N.; Alshaya, H.; et al. Phytohormones Trigger Drought Tolerance in Crop Plants: Outlook and Future Perspectives. *Front. Plant Sci.* **2022**, *12*, 799318. [CrossRef]
172. Taiz, L.; Zeiger, E. *Plant Physiology*, 4th ed.; Sinauer Associates Inc. Publishers: Sunderland, MA, USA, 2006.
173. Dubois, M.; Van den Broeck, L.; Inzé, D. The Pivotal Role of Ethylene in Plant Growth. *Trends Plant Sci.* **2018**, *23*, 311–323. [CrossRef]
174. Nilsen, E.T.; Orcutte, D.M. Phytohormones and plant responses to stress. In *Physiology of Plant under Stress: Abiotic Factors*; John Wiley and Sons: New York, NY, USA, 1996; pp. 183–198.
175. Taylor, I.B. Genetics of ABA synthesis. In *Abscisic Acid: Physiology and Biochemistry*; Davies, W.J., Jones, H.G., Eds.; Bios Scientific Publishers Ltd.: Oxford, UK, 1991; pp. 23–38.
176. Kasukabe, Y.; He, L.; Nada, K.; Misawa, S.; Ihara, I.; Tachibana, S. Overexpression of spermidine synthase enhances tolerance to multiple environmental stresses and up- regulates the expression of various stress-regulated genes in transgenic Arabidopsis thaliana. *Plant Cell. Physiol.* **2004**, *45*, 712–722. [CrossRef] [PubMed]
177. Salimath, P.M.; Toker, C.; Sandhu, J.S.; Kumar, J.; Suma, B.; Yadav, S.S.; Bahl, P.N. Conventional breeding methods. In *Chickpea Breeding and Management*; Yadav, S.S., Redden, R.J., Chen, W., Sharma, B., Eds.; CABI: Wallingford, UK, 2007; pp. 369–390.
178. Holme, I.B.; Gregersen, P.L.; Brinch-Pedersen, H. Induced Genetic Variation in Crop Plants by Random or Targeted Mutagenesis: Convergence and Differences. *Front. Plant Sci.* **2019**, *10*, 1468. [CrossRef] [PubMed]
179. Micke, A. Genetic improvement of food legumes in developing countries by mutation induction. In *World Crops: Cool Season Food Legumes*; Summerfield, R.J., Ed.; Kluwer Academic: Dordrecht, The Netherlands, 1988; pp. 1031–1047.
180. Amri-Tiliouine, W.; Laouar, M.; Abdelguerfi, A.; Jankowicz-Cieslak, J.; Jankuloski, L.; Till, B.J. Genetic Variability Induced by Gamma Rays and Preliminary Results of Low-Cost TILLING on M2 Generation of Chickpea (*Cicer arietinum* L.). *Front. Plant Sci.* **2018**, *9*, 1568. [CrossRef] [PubMed]
181. Kumar, S.; Katna, G.; Sharma, N. Mutation breeding in chickpea. *Adv. Plants Agric. Res.* **2019**, *9*, 355–362. [CrossRef]
182. Dua, R.P.; Chaturvedi, S.K.; Shiv, S. *Reference Varieties of chickpea for IPR Regime*; IIPR: Kanpur, India, 2001.
183. Ma, P.; Zhang, X.; Chen, L.; Zhao, Q.; Zhang, Q.; Wang, Z.; Tang, H.; Yu, Q.; Zhang, M.; Ming, R.; et al. Comparative analysis of sucrose phosphate synthase (SPS) gene family between Saccharum officinarum and Saccharum spontaneum. *BMC Plant Biol.* **2020**, *20*, 422. [CrossRef]
184. Harb, A.; Krishnan, A.; Ambavaram, M.M.; Pereira, A. Molecular and physiological analysis of drought stress in Arabidopsis reveals early responses leading to acclimation in plant growth. *Plant Physiol.* **2010**, *154*, 1254–1271. [CrossRef]
185. Yao, J.; Sun, D.; Cen, H.; Xu, H.; Weng, H.; Yuan, F.; He, Y. Phenotyping of Arabidopsis Drought Stress Response Using Kinetic Chlorophyll Fluorescence and Multicolor Fluorescence Imaging. *Front. Plant Sci.* **2018**, *9*, 603. [CrossRef]
186. Granier, C.; Aguirrezabal, L.; Chenu, K.; Cookson, S.J.; Dauzat, M.; Hamard, P.; Thioux, J.J.; Rolland, G.; Bouchier-Combaud, S.; Lebaudy, A. PHENOPSIS, an automated platform for reproducible phenotyping of plant responses to soil water deficit in Arabidopsis thaliana permitted the identification of an accession with low sensitivity to soil water deficit. *New Phytol.* **2006**, *169*, 623–635. [CrossRef]
187. Bouchabke-Coussa, O.; Quashie, M.L.; Seoane-Redondo, J.; Fortabat, M.N.; Gery, C.; Yu, A.; Linderme, D.; Trouverie, J.; Granier, F.; Téoulé, E.; et al. ESKIMO1 is a key gene involved in water economy as well as cold acclimation and salt tolerance. *BMC Plant Biol.* **2008**, *8*, 125. [CrossRef]
188. Bhatnagar-Mathur, P.; Vadez, V.; Devi, M.; Lavanya, M.; Vani, G.; Sharma, K. Genetic engineering of chickpea (*Cicer arietinum* L.) with the P5CSF129A gene for osmoregulation with implications on drought tolerance. *Mol. Breed.* **2009**, *23*, 591–606. [CrossRef]

189. Kumar, T.; Bharadwaj, C.; Rizvi, A.H.; Sarker, A.; Tripathi, S.; Alam, A.; Chauhan, S.K. Chickpea Landraces: A Valuable and Divergent Source for Drought Tolerance. *Int. J. Trop. Agric.* **2015**, *33*, 1–6.
190. Amiteye, S. Basic concepts and methodologies of DNA marker systems in plant molecular breeding. *Heliyon* **2021**, *7*, e08093. [CrossRef]
191. Tiwari, S.; Tripathi, N.; Tsuji, K.; Tantwai, K. Genetic diversity and population structure of Indian soybean (*Glycine max* (L.) Merr.) as revealed by microsatellite markers. *Physiol. Mol. Biol. Plants* **2019**, *25*, 953–964. [CrossRef] [PubMed]
192. Kachare, S.; Tiwari, S.; Tripathi, N.; Thakur, V.V. Assessment of genetic diversity of soybean (*Glycine max* (L.) Merr.) genotypes using qualitative traits and microsatellite makers. *Agric. Res.* **2019**, *9*, 23–34. [CrossRef]
193. Gupta, P.K.; Rustgi, S.; Mir, R.R. Array-based high-throughput DNA markers for crop improvement. *Heredity* **2008**, *101*, 5–18. [CrossRef] [PubMed]
194. Serret, M.D.; Udupa, S.M.; Weigand, F. Assessment of genetic diversity of cultivated chickpea using microsatellite-derived RFLP markers: Implications for origin. *Plant Breed.* **2006**, *116*, 573–578. [CrossRef]
195. Iruela, M.; Rubio, J.; Cubero, J.I.; Gil, J.; Mill´an, T. Phylogenetic analysis in the genus Cicer and cultivated chickpea using RAPD and ISSR markers. *Theor. Appl. Genet.* **2002**, *104*, 643–651. [CrossRef]
196. Yadav, P.; Tomar, V.S.; Mishra, J.K.; Chauhan, A.K.S. Relationship among annual Cicer species using random amplified polymorphic DNA markers. *Indian J. Life Sci. Gale Acad.* 2012. Available online: link.gale.com/apps/doc/A357968924/AONE?u=anon~{}91e4fa6e&sid=googleScholar&xid=cbcb2ca8 (accessed on 20 October 2022).
197. Tripathi, N.; Chouhan, D.S.; Saini, N.; Tiwari, S. Assessment of genetic variations among highly endangered medicinal plant *Bacopa monnieri* (L.) from Central India using RAPD and ISSR analysis. *3 Biotech* **2012**, *2*, 327–336. [CrossRef]
198. Singh, R.; Singhal, V.; Randhawa, G.J. Molecular Analysis of Chickpea (*Cicer arietinum* L) Cultivars Using AFLP and STMS Markers. *J. Plant Biochem. Biotechnol.* **2008**, *17*, 167–171. [CrossRef]
199. Hajibarat, Z.; Saidi, A.; Hajibarat, Z.; Talebi, R. Characterization of genetic diversity in chickpea using SSR markers, Start Codon Targeted Polymorphism (SCoT) and Conserved DNA-Derived Polymorphism (CDDP). *Physiol. Mol. Biol. Plants* **2015**, *21*, 365–373. [CrossRef]
200. Gajbhiye, P.; Yadav, A.S. Molecular markers in chickpea improvement: A review. *Plant Arch.* **2012**, *12*, 589–597.
201. Bharadwaj, C.; Chauhan, S.K.; Rajguru, G.; Srivastava, R.; Satyavathi, C.T.; Yadav, S. Diversity analysis of chickpea (*Cicer arietinum* L.) using STMS markers. *Indian J. Agric. Sci.* **2010**, *80*, 947–951.
202. Varshney, R.K.; Kudapa, H.; Roorkiwal, M.; Thudi, M.; Pandey, M.K.; Saxena, R.K.; Chamarthi, S.K.; Mohan, S.M.; Mallikarjuna, N.; Upadhyaya, H.; et al. Advances in genetics and molecular breeding of three legume crops of semi-arid tropics using next generation sequencing and high-throughput genotyping technologies. *J. Biosci.* **2012**, *37*, 811–820. [CrossRef] [PubMed]
203. Nayak, S.N.; Zhu, H.; Varghese, N.; Datta, S.; Choi, H.K.; Horres, R.; Jüngling, R.; Singh, J.; Kavi Kishor, P.B.; Sivaramakrishnan, S.; et al. Integration of novel SSR and gene-based SNP marker loci in the chickpea genetic map and establishment of new anchor points with Medicago truncatula genome. *Theor. Appl. Genet.* **2010**, *120*, 1415–1441. [CrossRef]
204. Seyedimoradi, H.; Talebi, R.; Kanouni, H.; Naji, A.M.; Karami, E. Genetic diversity and population structure analysis of chickpea (*Cicer arietinum* L.) advanced breeding lines using whole-genome DArTseq-generated SilicoDArT markers. *Braz. J. Bot.* **2020**, *43*, 541–549. [CrossRef]
205. Ahmed, S.M.; Alsamman, A.M.; Jighly, A.; Mubarak, M.H.; Al-Shamaa, K.; Istanbuli, T.; Momtaz, O.A.; Allali, A.E.; Hamwieh, A. Genome-wide association analysis of chickpea germplasms differing for salinity tolerance based on DArTseq markers. *PLoS ONE* **2021**, *16*, e0260709. [CrossRef]
206. Hiremath, P.J.; Kumar, A.; Penmetsa, R.V.; Farmer, A.; Schlueter, J.A.; Chamarthi, S.K.; Whaley, A.M.; Carrasquilla-Garcia, N.; Gaur, P.M.; Upadhyaya, H.D.; et al. Large-scale development of cost-effective SNP marker assays for diversity assessment and genetic mapping in chickpea and comparative mapping in legumes. *Plant Biotechnol. J.* **2012**, *10*, 716–732. [CrossRef]
207. Varshney, R.K. Exciting journey of 10 years from genomes to felds and markets: Some success stories of genomics-assisted breeding in chickpea, pigeonpea and groundnut. *Plant Sci.* **2016**, *242*, 98–107. [CrossRef]
208. Yadav, G.; Jayaswal, D.; Jayaswall, K.; Bhandawat, A.; Singh, A.N.; Tilgam, J.; Rai, A.K.; Chaturvedi, R.; Kumar, A.; Kumar, S.; et al. Identification and characterization of chickpea genotypes for early flowering and higher seed germination through molecular markers. *Mol. Biol. Rep.* **2022**, *49*, 6181–6188. [CrossRef]
209. Rajkumar, M.S.; Garg, R.; Jain, M. Genome-wide discovery of DNA polymorphisms via resequencing of chickpea cultivars with contrasting response to drought stress. *Physiol. Plant* **2022**, *174*, e13611. [CrossRef]
210. Basu, U.; Srivastava, R.; Bajaj, D.; Thakro, V.; Daware, A.; Malik, N.; Upadhyaya, H.D.; Parida, S.K. Genome-wide generation and genotyping of informative SNPs to scan molecular signatures for seed yield in chickpea. *Sci. Rep.* **2018**, *8*, 13240. [CrossRef] [PubMed]
211. Rajkumar, M.S.; Garg, R.; Jain, M. Genome-wide discovery of DNA polymorphisms among chickpea cultivars with contrasting seed size/weight and their functional relevance. *Sci. Rep.* **2018**, *8*, 16795. [CrossRef] [PubMed]
212. Manchikatla, P.K.; Kalavikatte, D.; Mallikarjuna, B.P.; Palakurthi, R.; Khan, A.W.; Jha, U.C.; Bajaj, P.; Singam, P.; Chitikineni, A.; Varshney, R.K.; et al. MutMap Approach Enables Rapid Identification of Candidate Genes and Development of Markers Associated with Early Flowering and Enhanced Seed Size in Chickpea (*Cicer arietinum* L.). *Front. Plant Sci.* **2021**, *12*, 688694. [CrossRef] [PubMed]

213. Srivastava, R.; Singh, M.; Bajaj, D.; Parida, S.K. A High-Resolution In Del (Insertion–Deletion) Markers-Anchored Consensus Genetic Map Identifies Major QTLs Governing Pod Number and Seed Yield in Chickpea. Front. *Plant Sci.* **2016**, *7*, 1362. [CrossRef]
214. Mallikarjuna, B.P.; Samineni, S.; Thudi, M.; Sajja, S.B.; Khan, A.W.; Patil, A.; Viswanatha, K.P.; Varshney, R.K.; Gaur, P.M. Molecular Mapping of Flowering Time Major Genes and QTLs in Chickpea (*Cicer arietinum* L.). *Front. Plant Sci.* **2017**, *8*, 1140. [CrossRef]
215. Barmukh, R.; Soren, K.R.; Madugula, P.; Gangwar, P.; Shanmugavadivel, P.S.; Bharadwaj, C.; Konda, A.K.; Chaturvedi, S.K.; Bhandari, A.; Rajain, K.; et al. Construction of a high-density genetic map and QTL analysis for yield, yield components and agronomic traits in chickpea (*Cicer arietinum* L.). *PLoS ONE* **2021**, *16*, e0251669. [CrossRef]
216. Rezaei, M.K.; Deokar, A.A.; Arganosa, G.; Roorkiwal, M.; Pandey, S.K.; Warkentin, T.D.; Varshney, R.K.; Tar An, B. Mapping Quantitative Trait Loci for Carotenoid Concentration in Three F2 Populations of Chickpea. *Plant Genome* **2019**, *12*, 67. [CrossRef]
217. Varshney, R.K.; Thudi, M.; Nayak, S.N.; Gaur, P.M.; Kashiwagi, J.; Krishnamurthy, L.; Jaganathan, D.; Koppolu, J.; Bohra, A.; Tripathi, S.; et al. Genetic dissection of drought tolerance in chickpea (*Cicer arietinum* L.). *Theor. Appl. Genet.* **2014**, *127*, 445–462. [CrossRef]
218. Varshney, R.K.; Song, C.; Saxena, R.K.; Azam, S.; Yu, S.; Sharpe, A.G.; Cannon, S.; Baek, J.; Rosen, B.D.; Tar'an, B.; et al. Draft genome sequence of chickpea (*Cicer arietinum*) provides a resource for trait improvement. *Nat. Biotechnol.* **2013**, *31*, 240–246. [CrossRef]
219. Jain, D.; Chattopadhyay, D. Analysis of gene expression in response to water deficit of chickpea (*Cicer arietinum* L.) varieties differing in drought tolerance. *BMC Plant Biol.* **2010**, *10*, 24. [CrossRef]
220. Roorkiwal, M.; Sharma, P.C. Sequence similarity-based identification of abiotic stress responsive genes in chickpea. *Bioinformation* **2012**, *8*, 92–97. [CrossRef] [PubMed]
221. Varshney, R.K.; Mir, R.R.; Bhatia, S.; Thudi, M.; Hu, Y.; Azam, S.; Zhang, Y.; Jagannathan, D. Integrated physical map with the genetic maps and reference genome sequence for chickpea (*Cicer arietinum* L.) improvement. *Funct. Integr. Genom.* **2014**, *14*, 59–73. [CrossRef] [PubMed]
222. Muriuki, R.; Kimurto, P.K.; Towett, B.K.; Vadez, V.; Gangarao, R. Study of root traits of chickpea (*Cicer arietinum* L.) under drought stress. *Afr. J. Plant Sci.* **2020**, *14*, 420–435.
223. Upadhyaya, H.D.; Dwivedi, S.L.; Baum, M.; Varshney, R.K.; Udupa, S.M.; Gowda, C.L.L.; Hoisington, D.; Singh, S. Genetic structure, diversity, and allelic richness in composite collection and reference set in chickpea (*Cicer arietinum* L.). *BMC Plant Biol.* **2008**, *8*, 106. [CrossRef]
224. Roorkiwal, M.; Nayak, S.N.; Thudi, M.; Upadhyaya, H.D.; Brunel, D.; Mournet, P.; This, D.; Sharma, P.C.; Varshney, R.K. Allele diversity for abiotic stress responsive candidate genes in chickpea reference set using gene-based SNP markers. *Front. Plant Sci.* **2014**, *5*, 248. [CrossRef]
225. Roorkiwal, M.; Bharadwaj, C.; Barmukh, R.; Dixit, G.P.; Thudi, M.; Gaur, P.M.; Chaturvedi, S.K.; Fikre, A.; Hamwieh, A.; Kumar, S.; et al. Integrating genomics for chickpea improvement: Achievements and opportunities. *Theor. Appl. Genet.* **2020**, *133*, 1703–1720. [CrossRef]
226. Upadhyaya, H.D.; Bajaj, D.; Das, S.; Kumar, V.; Gowda, C.L.L.; Sharma, S.; Tyagi, A.K.; Parida, S.K. Genetic dissection of seed-iron and zinc concentrations in chickpea. *Sci. Rep.* **2016**, *6*, 24050. [CrossRef]
227. Zhao, K.; Tung, C.W.; Eizenga, G.C.; Wright, M.H.; Ali, M.L.; Price, A.H.; Norton, G.J.; Islam, M.R.; Reynolds, A.; Mezey, J.; et al. Genome wide association mapping reveals a rich genetic architecture of complex traits in Oryza sativa. *Nat. Commun.* **2011**, *2*, 467. [CrossRef]
228. Pasam, R.K.; Sharma, R.; Malosetti, M.; van Eeuwijk, F.A.; Haseneyer, G.; Kilian, B.; Graner, A. Genome-wide association studies for agronomical traits in a worldwide spring barley collection. *BMC Plant Biol.* **2012**, *12*, 16. [CrossRef]
229. Morris, G.P.; Ramu, P.; Deshpande, S.P.; Hash, C.T.; Shah, T.; Upadhyaya, H.D.; Riera-Lizarazu, O.; Brown, P.J.; Acharya, C.B.; Mitchell, S.E.; et al. Population genomic and genome-wide association studies of agroclimatic traits in sorghum. *Proc. Natl. Acad. Sci. USA* **2013**, *110*, 453–458. [CrossRef]
230. Karaca, N.; Ates, D.; Nemli, S.; Ozkuru, E.; Yilmaz, H.; Yagmur, B.; Kartal, C.; Tosun, M.; Ozdestan, O.; Otles, S.; et al. Identification of SNP Markers Associated with Iron and Zinc Concentrations in Cicer Seeds. *Curr. Genom.* **2020**, *21*, 212–223. [CrossRef] [PubMed]
231. Jha, U.; Jha, R.; Bohra, A.; Manjunatha, L.; Saabale, P.; Parida, S.; Singh, N. Association mapping of genomic loci linked with Fusarium wilt resistance (Foc2) in chickpea. *Plant Genet. Resour. Charact. Util.* **2021**, *19*, 195–202. [CrossRef]
232. Varshney, R.K.; Pandey, M.K.; Bohra, A.; Singh, V.K.; Thudi, M.; Saxena, R.K. Toward the sequence-based breeding in legumes in the post-genome sequencing era. *Theor. Appl. Genet.* **2019**, *132*, 797–816. [CrossRef]
233. Samineni, S.; Sajja, S.B.; Mondal, B.; Chand, U.; Thudi, M.; Varshney, R.K.; Gaur, P.M. MAGIC lines in chickpea: Development and exploitation of genetic diversity. *Euphytica (TSI)* **2021**, *217*, 1–12. [CrossRef]
234. Gangurde, S.S.; Wang, H.; Yaduru, S.; Pandey, M.K.; Fountain, J.C.; Chu, Y.; Isleib, T.; Holbrook, C.C.; Xavier, A.; Culbreath, A.K.; et al. Nested-association mapping (NAM)-based genetic dissection uncovers candidate genes for seed and pod weights in peanut (*Arachis hypogaea*). *Plant Biotechnol. J.* **2020**, *18*, 1457–1471. [CrossRef]
235. McMullen, M.D.; Kresovich, S.; Villeda, H.S.; Bradbury, P.; Li, H.; Sun, Q.; Flint-Garcia, S.; Thornsberry, J.; Acharya, C.; Bottoms, C.; et al. Genetic properties of the maize nested association mapping population. *Science* **2009**, *325*, 737–740. [CrossRef]

236. Bajaj, D.; Srivastava, R.; Nath, M.; Tripathi, S.; Bharadwaj, C.; Upadhyaya, H.D.; Tyagi, A.K.; Parida, S.K. EcoTILLING-Based Association Mapping Efficiently Delineates Functionally Relevant Natural Allelic Variants of Candidate Genes Governing Agronomic Traits in Chickpea. *Front. Plant Sci.* **2016**, *7*, 450. [CrossRef]
237. Palakurthi, R.; Jayalakshmi, V.; Kumar, Y.; Kulwal, P.; Yasin, M.; Kute, N.S.; Laxuman, C.; Yeri, S.; Vemula, A.; Rathore, A.; et al. Translational Chickpea Genomics Consortium to Accelerate Genetic Gains in Chickpea (*Cicer arietinum* L.). *Plants* **2021**, *10*, 2583. [CrossRef]
238. Varshney, R.K.; Saxena, R.K.; Upadhyaya, H.D.; Khan, A.W.; Yu, Y.; Kim, C.; Rathore, A.; Kim, D.; Kim, J.; An, S.; et al. Whole-genome resequencing of 292 pigeonpea accessions identifes genomic regions associated with domestication and agronomic traits. *Nat. Genet.* **2017**, *49*, 1082–1088. [CrossRef]
239. Arriagada, O.; Cacciuttolo, F.; Cabeza, R.A.; Carrasco, B.; Schwember, A.R. A Comprehensive Review on Chickpea (*Cicer arietinum* L.) Breeding for Abiotic Stress Tolerance and Climate Change Resilience. *Int. J. Mol. Sci.* **2022**, *23*, 6794. [CrossRef]
240. Kanzi, A.M.; San, J.E.; Chimukangara, B.; Wilkinson, E.; Fish, M.; Ramsuran, V.; de Oliveira, T. Next Generation Sequencing and Bioinformatics Analysis of Family Genetic Inheritance. *Front. Genet.* **2020**, *11*, 544162. [CrossRef] [PubMed]
241. Kosgei, A.J.; PKKimurtoPMGaurMAYeboahSKOffei, E.Y. Danquah, Introgression of drought tolerance root traits into Kenyan commercial chickpea varieties using marker assisted backcrossing. *Afr. Crop. Sci. J.* **2022**, *30*, 31–50. [CrossRef]
242. Jain, A.; Govindaraj, G.M.; Edavazhippurath, A.; Faisal, N.; Bhoyar, R.C.; Gupta, V.; Uppuluri, R.; Manakkad, S.P.; Kashyap, A.; Kumar, A.; et al. Whole genome sequencing identifies novel structural variant in a large Indian family affected with X-linked agammaglobulinemia. *PLoS ONE* **2021**, *16*, e0254407. [CrossRef] [PubMed]
243. Jaganathan, D.; Thudi, M.; Kale, S.; Azam, S.; Roorkiwal, M.; Gaur, P.M.; Kishor, P.B.; Nguyen, H.; Sutton, T.; Varshney, R.K. Genotyping-by-sequencing based intra-specific genetic map refines a "QTL-hotspot" region for drought tolerance in chickpea. *Mol. Genet. Genom.* **2015**, *290*, 559–571. [CrossRef]
244. Thudi, M.; Chitikineni, A.; Liu, X.; He, W.; Roorkiwal, M.; Yang, W.; Jian, J.; Doddamani, D.; Gaur, P.M.; Rathore, A.; et al. Recent breeding programs enhanced genetic diversity in both desi and kabuli varieties of chickpea (*Cicer arietinum* L.). *Sci. Rep.* **2016**, *6*, 38636. [CrossRef]
245. Varshney, R.; Thudi, M.; Upadhyaya, H.; Dwivedi, S.; Udupa, S.; Furman, B.; Baum, M.; Hoisington, D. A SSR kit to study genetic diversity in chickpea (*Cicer arietinum* L.). *Plant Genet. Resour.* **2014**, *12*, S118–S120. [CrossRef]
246. Roorkiwal, M.; Jain, A.; Kale, S.M.; Doddamani, D.; Chitikineni, A.; Thudi, M.; Varshney, R.K. Development and evaluation of highdensityAxiom® CicerSNP Array for high-resolution genetic mapping and breeding applications in chickpea. *Plant Biotechnol. J.* **2018**, *16*, 890–901. [CrossRef]
247. Rajkumar, M.S.; Garg, R.; Jain, M. Genome resequencing reveals DNA polymorphisms associated with seed size/weight determination in chickpea. *Genomics* **2021**, *113*, 1458–1468. [CrossRef]
248. Varshney, R.K.; Bohra, A.; Yu, J.; Graner, A.; Zhang, Q.; Sorrells, M.E. Designing Future Crops: Genomics-Assisted Breeding Comes of Age. *Trends Plant Sci.* **2021**, *26*, 631–649. [CrossRef]
249. Varshney, R.K.; Graner, A.; Sorrells, M.E. Genic microsatellite markers in plants: Features and applications. *Trends Biotechnol.* **2005**, *23*, 48–55. [CrossRef]
250. Varshney, R.K.; Roorkiwal, M.; Sun, S.; Bajaj, P.; Chitikineni, A.; Thudi, M.; Singh, N.P.; Du, X.; Upadhyaya, H.D.; Khan, A.W.; et al. A chickpea genetic variation map based on the sequencing of 3366 genomes. *Nature* **2021**, *599*, 622–627. [CrossRef] [PubMed]
251. Coletta, D.R.; Qiu, Y.; Ou, S.; Hufford, M.B.; Hirsch, C.N. How the pan-genome is changing crop genomics and improvement. *Genome Biol.* **2021**, *22*, 3. [CrossRef]
252. Hardison, R.C. Comparative genomics. *PLoS Biol.* **2003**, *1*, E58. [CrossRef] [PubMed]
253. Boutte, J.; Maillet, L.; Chaussepied, T.; Letort, S.; Aury, J.M.; Belser, C.; Boideau, F.; Brunet, A.; Coriton, O.; Deniot, G.; et al. Genome Size Variation and Comparative Genomics Reveal Intraspecific Diversity in Brassica rapa. *Front. Plant Sci.* **2020**, *11*, 577536. [CrossRef] [PubMed]
254. Ebler, J.; Ebert, P.; Clarke, W.E.; Rausch, T.; Audano, P.A.; Houwaart, T.; Mao, Y.; Korbel, J.O.; Eichler, E.E.; Zody, M.C.; et al. Pangenome-based genome inference allows efficient and accurate genotyping across a wide spectrum of variant classes. *Nat. Genet.* **2022**, *54*, 518–525. [CrossRef] [PubMed]
255. Khan, A.W.; Garg, V.; Roorkiwal, M.; Golicz, A.A.; Edwards, D.; Varshney, R.K. Super-Pangenome by Integrating the Wild Side of a Species for Accelerated Crop Improvement. *Trends Plant Sci.* **2020**, *25*, 148–158. [CrossRef]
256. Lei, L.; Goltsman, E.; Goodstein, D.; Wu, G.A.; Rokhsar, D.S.; Vogel, P.J. Plant Pan-Genomics Comes of Age. *Annu. Rev. Plant Biol.* **2021**, *72*, 411–435. [CrossRef]
257. Tay Fernandez, C.G.; Nestor, B.J.; Danilevicz, M.F.; Gill, M.; Petereit, J.; Bayer, P.E.; Finnegan, P.M.; Batley, J.; Edwards, D. Pangenomes as a Resource to Accelerate Breeding of Under-Utilised Crop Species. *Int. J. Mol. Sci.* **2022**, *23*, 2671. [CrossRef]
258. Jha, U.C.; Bohra, A.; Nayyar, H. Advances in "omics" approaches to tackle drought stress in grain legumes. *Plant Breed.* **2019**, *139*, 1–27. [CrossRef]
259. Yang, Y.; Saand, M.A.; Huang, L.; Abdelaal, W.B.; Zhang, J.; Wu, Y.; Li, J.; Sirohi, M.H.; Wang, F. Applications of Multi-Omics Technologies for Crop Improvement. *Front. Plant Sci.* **2021**, *12*, 563953. [CrossRef]
260. Kumar, R.; Sharma, V.; Suresh, S.; Ramrao, D.P.; Veershetty, A.; Kumar, S.; Priscilla, K.; Hangargi, B.; Narasanna, R.; Pandey, M.K.; et al. Understanding Omics Driven Plant Improvement and de novo Crop Domestication: Some Examples. *Front. Genet.* **2021**, *12*, 637141. [CrossRef] [PubMed]

261. Sreenivasulu, N.; Kishor, P.B.K.; Varshney, R.K.; Altschmied, L. Mining functional information from cereal genomics–the utility of expressed sequence tags. *Curr. Sci.* **2002**, *83*, 965–973.
262. Pradhan, S.; Bandhiwal, N.; Shah, N.; Kant, C.; Gaur, R.; Bhatia, S. Global transcriptome analysis of developing chickpea (*Cicer arietinum* L.) seeds. *Front. Plant Sci.* **2014**, *5*, 698. [CrossRef] [PubMed]
263. Agarwal, G.; Jhanwar, S.; Priya, P.; Singh, V.K.; Saxena, M.S.; Parida, S.K.; Garg, R.; Tyagi, A.K.; Jain, M. Comparative Analysis of Kabuli Chickpea Transcriptome with Desi and Wild Chickpea Provides a Rich Resource for Development of Functional Markers. *PLoS ONE* **2012**, *7*, e52443. [CrossRef]
264. Garg, R.; Patel, R.K.; Jhanwar, S.; Priya, P.; Bhattacharjee, A.; Yadav, G.; Bhatia, S.; Chattopadhyay, D.; Tyagi, A.K.; Jain , M. Gene discovery and tissue-specific transcriptome analysis in chickpea with massively parallel pyrosequencing and web resource development. *Plant Physiol.* **2011**, *1564*, 1661–1678. [CrossRef]
265. Kumar, N.; Singh, S.S.; Ghosh, P.K.; Hazra, K.K.; Venkatesh, M.S.; Praharaj, C.S.; Singh, M.K.; Kumar MSenthilBasu, P.S.; Yadav, A.; Yadav, S.L.; et al. Improving chickpea productivity in rice-fallow of Indo-Gangetic Plain with soil moisture conservation and cultivar selection. *J. Food Legumes* **2020**, *33*, 28–35.
266. Kaashyap, M.; Ford, R.; Mann, A.; Varshney, R.K.; Siddique, K.H.M.; Mantri, N. Comparative Flower Transcriptome Network Analysis Reveals DEGs Involved in Chickpea Reproductive Success during Salinity. *Plants* **2022**, *11*, 434. [CrossRef]
267. Molina, C.; Rotter, B.; Horres, R.; Udupa, S.M.; Besser, B.; Bellarmino, L.; Baum, M.; Matsumura, H.; Terauchi, R.; Kahl, G.; et al. SuperSAGE: The drought stress-responsive transcriptome of chickpea roots. *BMC Genom.* **2008**, *9*, 553. [CrossRef]
268. Carrari, F.; Fernie, A.R.; Iusem, N.D. Heard it through the grapevine? ABA and sugar cross-talk: The ASR story. *Trends Plant Sci.* **2004**, *9*, 57–59. [CrossRef]
269. Kumar, M.; Kumar Patel, M.; Kumar, N.; Bajpai, A.B.; Siddique, K.H.M. Metabolomics and Molecular Approaches Reveal Drought Stress Tolerance in Plants. *Int. J. Mol. Sci.* **2021**, *22*, 9108. [CrossRef]
270. Deokar, A.A.; Kondawar, V.; Jain, P.K.; Karuppayil, S.M.; Raju, N.L.; Vadez, V.; Varshney, R.K.; Srinivasan, R. Comparative analysis of expressed sequence tags (ESTs) between drought-tolerant and -susceptible genotypes of chickpea under terminal drought stress. *BMC Plant Biol.* **2011**, *11*, 70. [CrossRef] [PubMed]
271. Garg, R.; Shankar, R.; Thakkar, B.; Kudapa, H.; Krishnamurthy, L.; Mantri, N.; Varshney, R.K.; Bhatia, S.; Jain, M. Transcriptome analyses reveal genotype- and developmental stage-specific molecular responses to drought and salinity stresses in chickpea. *Sci. Rep.* **2016**, *6*, 19228. [CrossRef] [PubMed]
272. Raes, J.; Rohde, A.; Christensen, J.H.; Van de Peer, Y.; Boerjan, W. Genome-wide characterization of the lignification toolbox in Arabidopsis. *Plant Physiol.* **2003**, *133*, 1051–1071. [CrossRef] [PubMed]
273. Kumar, N.; Bharadwaj, C.; Soni, A.; Sachdeva, S.; Yadav, M.C.; Pal, M.; Soren, K.R.; Meena, M.C.; Roorkiwal, M.; Varshney, R.K.; et al. Physio-morphological and molecular analysis for salt tolerance in chickpea (*Cicer arietinum*). *Indian J. Agric. Sci.* **2020**, *90*, 804–808.
274. Ahmad, B.; Azeem, F.; Ali, M.A.; Nawaz, M.A.; Nadeem, H.; Abbas, A.; Batool, R.; Atif, R.M.; Ijaz, U.; Nieves-Cordones, M.; et al. Genome-wide identification and expression analysis of two component system genes in Cicer arietinum. *Genomics* **2020**, *112*, 1371–1383. [CrossRef] [PubMed]
275. Mashaki, K.M.; Garg, V.; Ghomi, A.A.N.; Kudapa, H.; Chitikineni, A.; Nezhad, K.Z.; Yamchi, A.; Soltanloo, H.; Varshney, R.K.; Thudi, M. RNA-Seq analysis revealed genes associated with drought stress response in kabuli chickpea (*Cicer arietinum* L.). *PLoS ONE* **2018**, *13*, e0199774. [CrossRef]
276. Khan, N.; Bano, A.; Rahman, M.A.; Rathinasabapathi, B.; Babar, M.A. UPLC-HRMS-based untargeted metabolic profiling reveals changes in chickpea (*Cicer arietinum*) metabolome following long-term drought stress. *Plant Cell Environ.* **2019**, *42*, 115–132. [CrossRef]
277. Dias, D.A.; Hill, C.B.; Jayasinghe, N.S.; Atieno, J.; Sutton, T.; Roessner, U. Quantitative profling of polar primary metabolites of two chickpea cultivars with contrasting responses to salinity. *J. Chromatogr. B Anal. Technol. Biomed. Life Sci* **2015**, *1000*, 1. [CrossRef]
278. Shariatipour, N.; Heidari, B. Investigation of Drought and Salinity Tolerance Related Genes and their Regulatory Mechanisms in Arabidopsis (*Arabidopsis thaliana*). *Open Bioinform. J.* **2018**, *11*, 12–28. [CrossRef]
279. Bhushan, D.; Pandey, A.; Choudhary, M.K.; Datta, A.; Chakraborty, S.; Chakraborty, N. Comparative proteomics analysis of differentially expressed proteins in chickpea extracellular matrix during dehydration stress. *Mol. Cell Prot.* **2007**, *6*, 1868–1884. [CrossRef]
280. Jaiswal, D.; Mishra, P.; Subba, P.; Rathi, D.; Chakraborty, S.; Chakraborty, N. Membrane-associated proteomics of chickpea identifies Sad1/UNC-84 protein (CaSUN1), a novel component of dehydration signaling. *Sci. Rep.* **2014**, *4*, 4177. [CrossRef]
281. Vessal, S.; Arefian, M.; Siddique, K.H.M. Proteomic responses to progressive dehydration stress in leaves of chickpea seedlings. *BMC Genom.* **2020**, *21*, 523. [CrossRef] [PubMed]
282. Gupta, S.; Mishra, S.K.; Misra, S.; Pandey, V.; Agrawal, L.; Nautiyal, C.S.; Chauhan, P.S. Revealing the complexity of protein abundance in chickpea root under drought-stress using a comparative proteomics approach. *Plant Physiol. Biochem.* **2020**, *151*, 88–102. [CrossRef] [PubMed]
283. Azimi, S.; Kaur, T.; Gandhi, T.K. Water Stress Identification in Chickpea Plant Shoot Images using Deep Learning. In Proceedings of the 2020 IEEE 17th India Council International Conference (INDICON), New Delhi, India, 10–13 December 2020; pp. 1–7. [CrossRef]

284. Clark, R.T.; MacCurdy, R.B.; Jung, J.K.; Shaff, J.E.; McCouch, S.R.; Aneshansley, D.J.; Kochian, L.V. Three-dimensional root phenotyping with a novel imaging and software platform. *Plant Physiol.* **2011**, *156*, 455–465. [CrossRef] [PubMed]
285. Zhu, M.; Zhao, S. Candidate gene identification approach: Progress and challenges. *Int. J. Biol. Sci.* **2007**, *3*, 420–427. [CrossRef] [PubMed]
286. Cano-Gamez, E.; Trynka, G. From GWAS to Function: Using Functional Genomics to Identify the Mechanisms Underlying Complex Diseases. *Front. Genet.* **2020**, *11*, 424. [CrossRef]
287. Zhong, C.; Sun, S.; Li, Y.; Duan, C.; Zhu, Z. Next-generation sequencing to identify candidate genes and develop diagnostic markers for a novel Phytophthora resistance gene, RpsHC18, in soybean. *Theor. Appl. Genet.* **2018**, *131*, 525–538. [CrossRef]
288. Dossa, K.; Li, D.; Zhou, R.; Yu, J.; Wang, L.; Zhang, Y.; You, J.; Liu, A.; Mmadi, M.A.; Fonceka, D.; et al. The genetic basis of drought tolerance in the high oil crop Sesamum indicum. *Plant Biotechnol. J.* **2019**, *17*, 1788–1803. [CrossRef]
289. Bhattarai, G.; Shi, A.; Feng, C.; Dhillon, B.; Mou, B.; Correll, J.C. Genome Wide Association Studies in Multiple Spinach Breeding Populations Refine Downy Mildew Race 13 Resistance Genes. *Front. Plant Sci.* **2020**, *11*, 563187. [CrossRef]
290. Das, A.; Basu, P.S.; Kumar, M.; Ansari, J.; Shukla, A.; Thakur, S.; Singh, P.; Datta, S.; Chaturvedi, S.K.; Sheshshayee, M.S.; et al. Transgenic chickpea (*Cicer arietinum* L.) harbouring *AtDREB1a* are physiologically better adapted to water deficit. *BMC Plant Biol.* **2021**, *21*, 39. [CrossRef]
291. Puhakainen, T.; Hess, M.W.; Makela, P.; Svensson, J.; Heino, P.; Palva, E.T. Over expression of multiple dehydrin genes enhances tolerance to freezing stress in Arabidopsis. *Plant Mol. Biol.* **2004**, *54*, 743–753. [CrossRef] [PubMed]
292. Allagulova, C.R.; Gimalov, F.R.; Shakirova, F.M.; Vakhitov, V.A. The plant dehydrins: Structure and putative functions. *Biochemistry* **2003**, *68*, 1157–1165. [CrossRef]
293. Kumar, T.; Tiwari, N.; Bharadwaj, C.; Sarker, A.; Pappula, S.P.R.; Singh, S.; Singh, M. Identification of Allelic Variation in Drought Responsive Dehydrin Gene Based on Sequence Similarity in Chickpea (*Cicer arietinum* L.). *Front. Genet.* **2020**, *11*, 584527. [CrossRef] [PubMed]
294. Ichimura, K.; Mizoguchi, T.; Yoshida, R.; Yuasa, T.; Shinozaki, K. Various abiotic stresses rapidly activate Arabidopsis MAP kinases AtMPK4 and AtMPK6. *Plant J.* **2000**, *24*, 655–665. [CrossRef]
295. Stiti, N.; Missihoun, T.D.; Kotchoni, S.O.; Kirch, H.H.; Bartels, D. Aldehyde dehydrogenases in *Arabidopsis thaliana*: Biochemical requirements, metabolic pathways, and functional analysis. *Front. Plant Sci.* **2011**, *2*, 65. [CrossRef]
296. Park, J.M.; Park, C.J.; Lee, S.B.; Ham, B.K.; Shin, R.; Paek, K.H. Overexpression of the tobacco Tsi1 gene encoding an EREBP/AP2-type transcription factor enhances resistance against pathogen attack and osmotic stress in tobacco. *Plant Cell* **2001**, *13*, 1035–1046. [CrossRef]
297. Shukla, R.K.; Tripathi, V.; Jain, D.; Yadav, R.K.; Chattopadhyay, D. CAP2 enhances germination of transgenic tobacco seeds at high temperature and promotes heat stress tolerance in yeast. *FEBS J.* **2009**, *276*, 5252–5262. [CrossRef]
298. Shukla, U.C.; Yadav, O.P. Effect of phosphorus and zinc on nodulation and nitrogen fixation in chickpea (*Cicer arietinum* L.). *J. Plant Soil* **2004**, *65*, 239–248. [CrossRef]
299. Gu, H.; Jia, Y.; Wang, X.; Chen, Q.; Shi, S.; Ma, L.; Zhang, J.; Zhang, H.; Ma, H. Identification and characterization of a *LEA* family gene *CarLEA4* from chickpea (*Cicer arietinum* L.). *Mol. Biol. Rep.* **2012**, *39*, 3565–3572. [CrossRef]
300. Sharma, K.D.; Patil, D.; Kiran, A. Characterization and differential expression of sucrose and starch metabolism genes in contrasting chickpea (*Cicer arietinum* L.) genotypes under low temperature. *J. Genet.* **2021**, *100*, 71. [CrossRef]
301. Caballo, C.; Berbel, A.; Ortega, R.; Gil, J.; Millán, T.; Rubio, J.; Madueño, F. The SINGLE FLOWER (SFL) gene encodes a MYB transcription factor that regulates the number of flowers produced by the inflorescence of chickpea. *New Phytol.* **2022**, *234*, 827–836. [CrossRef] [PubMed]
302. Yang, C.Y.; Chen, Y.C.; Jauh, G.Y.; Wang, C.S. A Lily ASR protein involves abscisic acid signaling and confers drought and salt resistance in Arabidopsis. *Plant Physiol.* **2005**, *139*, 836–846. [CrossRef] [PubMed]
303. Li, W.; Huang, D.; Wang, B.; Hou, X.; Zhang, R.; Yan, M.; Liao, W. Changes of starch and sucrose content and related gene expression during the growth and development of Lanzhou lily bulb. *PLoS ONE* **2022**, *17*, e0262506. [CrossRef]
304. Meenakshi Kumar, A.; Kumar, V.; Dubey, A.K.; Narayan, S.; Sawant, M.V.; Pande, V.; Sirke, P.; Sanyal, I. CAMTA transcription factor enhances salinity and drought tolerance in chickpea (*Cicer arietinum* L.). *Plant Cell Tissue Organ Cult.* **2022**, *148*, 319–330. [CrossRef]
305. Yu, X.; Liu, Y.; Wang, S.; Tao, Y.; Wang, Z.; Shu, Y.; Peng, H.; Mijiti, A.; Wang, Z.; Zhang, H.; et al. CarNAC4, a NAC-type chickpea transcription factor conferring enhanced drought and salt stress tolerances in Arabidopsis. *Plant Cell Rep.* **2016**, *35*, 613–627. [CrossRef] [PubMed]
306. Nguyen, K.H.; Ha, C.V.; Watanabe, Y.; Tran, U.T.; Nasr Esfahani, M.; Nguyen, D.V.; Tran, L.-S.P. Correlation between differential drought tolerability of two contrasting drought-responsive chickpea cultivars and differential expression of a subset of CaNAC genes under normal and dehydration conditions. *Front. Plant Sci.* **2015**, *6*, 449. [CrossRef] [PubMed]
307. La, H.V.; Chu, H.D.; Tran, C.D.; Nguyen, K.H.; Le, Q.T.N.; Hoang, C.M.; Cao, B.P.; Pham, A.T.C.; Nguyen, B.D.; Nguyen, T.Q.; et al. Insights into the gene and protein structures of the CaSWEET family members in chickpea (*Cicer arietinum*), and their gene expression patterns in different organs under various stress and abscisic acid treatments. *Gene* **2022**, *819*, 146210. [CrossRef]
308. Agarwal, P.K.; Jha, B. Transcription factors in plants and ABA dependent and independent abiotic stress signalling. *Biol. Plant* **2010**, *54*, 201–212. [CrossRef]

309. Riechmann, J.L.; Heard, J.; Martin, G.; Reuber, L.; Jiang, C.; Keddie, J.; Adam, L.; Pineda, O.; Ratcliffe, O.J.; Samaha, R.R.; et al. Arabidopsis transcription factor: Genome wide comparative analysis among eukaryotes. *Science* **2000**, *290*, 2105–2110. [CrossRef]
310. Lata, C.; Prasad, M. Role of DREBs in regulation of abiotic stress responses in plants. *J. Exp. Bot.* **2011**, *62*, 4731–4748. [CrossRef]
311. Anbazhagan, K.; Bhatnagar-Mathur, P.; Vadez, V.; Dumbala, S.R.; Kishor, P.B.; Sharma, K.K. DREB1A overexpression in transgenic chickpea alters key traits influencing plant water budget across water regimes. *Plant Cell Rep.* **2015**, *34*, 199–210. [CrossRef] [PubMed]
312. Hanin, M.; Brini, F.; Ebel, C.; Toda, Y.; Takeda, S.; Masmoudi, K. Plant dehydrins and stress tolerance: Versatile proteins for complex mechanisms. *Plant Signal. Behav.* **2011**, *6*, 1503–1509. [CrossRef] [PubMed]
313. Wise, M.J.; Tunnacliffe, A. POPP the question: What do LEA protein do? *Trends Plant Sci.* **2004**, *9*, 13–17. [CrossRef] [PubMed]
314. Puhakainen, T.; Li, C.; Boije-Malm, M.; Kangasjarvi, J.; Heino, P.; Palva, E.T. Short-day potentiation of low temperature-induced gene expression of a C-repeat-binding factor-controlled gene during cold acclimation in silver birch. *Plant Physiol.* **2004**, *136*, 4299–4307. [CrossRef]
315. Rudrabhatla, P.; Reddy, M.M.; Rajasekharan, R. Genome-wide analysis and experimentation of plant serine/threonine/tyrosine-specific protein kinases. *Plant Mol. Biol.* **2006**, *60*, 293–319. [CrossRef]
316. Petrivalský, M.; Brauner, F.; Luhová, L.; Gagneul, D.; Sebela, M. Aminoaldehyde dehydrogenase activity during wound healing of mechanically injured pea seedlings. *J. Plant Physiol.* **2007**, *164*, 1410–1418. [CrossRef]
317. Skibbe, D.S.; Liu, F.; Wen, T.J.; Yandeau, M.D.; Cui, X.; Cao, J.; Simmons, C.R.; Schnable, P.S. Characterization of the aldehyde dehydrogenase gene families of *Zea mays* and *Arabidopsis*. *Plant Mol. Biol.* **2002**, *48*, 751–764. [CrossRef]
318. Shpak, E.D.; Berthiaume, C.T.; Hill, E.J.; Torii, K.U. Synergistic interaction of three ERECTA-family receptor-like kinases controls Arabidopsis organ growth and flower development by promoting cell proliferation. *Development* **2004**, *131*, 1491–1501. [CrossRef]
319. Mandel, T.; Moreau, F.; Kutsher, Y.; Fletcher, J.C.; Carles, C.C.; Eshed Williams, L. The ERECTA receptor kinase regulates Arabidopsis shoot apical meristem size, phyllotaxy and floral meristem identity. *Development* **2014**, *141*, 830–841. [CrossRef] [PubMed]
320. Zhu, S.; Shi, W.; Jie, Y.; Zhou, Q.; Song, C. A MYB transcription factor, BnMYB2, cloned from ramie (*Boehmeria nivea*) is involved in cadmium tolerance and accumulation. *PLoS ONE* **2020**, *15*, e0233375. [CrossRef]
321. Wei, Q.; Chen, R.; Wei, X.; Liu, Y.; Zhao, S.; Yin, X.; Xie, T. Genome-wide identification of R2R3-MYB family in wheat and functional characteristics of the abiotic stress responsive gene TaMYB344. *BMC Genom.* **2020**, *21*, 792. [CrossRef] [PubMed]
322. Islam, K.; Rawoof, A.; Ahmad, I.; Dubey, M.; Momo, J.; Ramchiary, N. Capsicum chinense MYB Transcription Factor Genes: Identification, Expression Analysis, and Their Conservation and Diversification with Other Solanaceae Genomes. *Front. Plant Sci.* **2021**, *12*, 721265. [CrossRef] [PubMed]
323. Heidari, P.; Mazloomi, F.; Nussbaumer, T.; Barcaccia, G. Insights into the SAM Synthetase Gene Family and Its Roles in Tomato Seedlings under Abiotic Stresses and Hormone Treatments. *Plants* **2020**, *9*, 586. [CrossRef]
324. Sun, Y.; Locasale, J.W. Rethinking the bioavailability and cellular transport properties of S-adenosylmethionine. *Cell Stress* **2021**, *6*, 1–5. [CrossRef] [PubMed]
325. Radadiya, N.; Parekh, V.P.; Dobariya, B.; Mahatma, L.; Mahatma, M.K. Abiotic stresses alter expression of S-Adenosylmethionine synthetase gene, polyamines and antioxidant activity in pigeon pea (*Cajanus cajan* L.). *Legume Res.* **2016**, *2016*, 905–913.
326. Joo, J.; Lee, Y.H.; Kim, Y.K.; Nahm, B.H.; Song, S.I. Abiotic stress responsive rice ASR1 and ASR3 exhibit different tissue-dependent sugar and hormone-sensitivities. *Mol. Cells* **2013**, *35*, 421–435. [CrossRef]
327. Philippe, R.; Courtois, B.; McNally, K.L.; Mournet, P.; El-Malki, R.; Paslier, M.C.; Fabre, D.; Billot, C.; Brunel, D.; Glaszmann, J.C.; et al. Structure, allelic diversity and selection of Asr genes, candidate for drought tolerance, in Oryza sativa L. and wild relatives. *Theor. Appl. Genet.* **2010**, *121*, 769–787. [CrossRef]
328. Cortés, A.J.; Monserrate, F.A.; Ramírez-Villegas, J.; Madriñán, S.; Blair, M.W. Drought tolerance in wild plant populations: The case of common beans (*Phaseolus vulgaris* L.). *PLoS ONE* **2013**, *8*, e62898. [CrossRef]
329. Yoshida, T.; Fujita, Y.; Sayama, H.; Kidokoro, S.; Maruyama, K.; Mizoi, J.; Shinozaki, K.; Yamaguchi-Shinozaki, K. AREB1, AREB2 and AFB2 are master transcription factors that cooperatively regulate ABRE dependent ABA signalling involved in drought stress tolerance and require ABA for full activation. *Plant J.* **2010**, *61*, 672–685. [CrossRef]
330. Stein, O.; Granot, D. An Overview of Sucrose Synthases in Plants. *Front. Plant Sci.* **2019**, *10*, 95. [CrossRef]
331. Yao, D.; Gonzales-Vigil, E.; Mansfield, S.D. Arabidopsis sucrose synthase localization indicates a primary role in sucrose translocation in phloem. *J. Exp. Bot.* **2020**, *71*, 1858–1869. [CrossRef] [PubMed]
332. Kumar, A.; Turner, N. Growth and sucrose synthase activity of developing chickpea (*Cicer arietinum* L.) seeds under field conditions. *Aust. J. Crop. Sci.* **2009**, *3*, 20–27.
333. Okamuro, J.K.; Caster, B.; Villarroel, R.; Montagu, M.V.; Jofuku, K.D. The AP2 domain of APETALA2 defines a large new family of DNA binding proteins in Arabidopsis. *Proc. Natl. Acad. Sci. USA* **1997**, *94*, 7076–7081. [CrossRef] [PubMed]
334. Shukla, R.K.; Raha, S.; Tripathi, V.; Chattopadhyay, D. Expression of CAP2, an APETALA2-family transcription factor from chickpea, enhances growth and tolerance to dehydration and salt stress in transgenic tobacco. *Plant Physiol.* **2006**, *142*, 113–123. [CrossRef] [PubMed]
335. Jain, D.; Chattopadhyay, D. Promoter of *CaZF*, a Chickpea Gene That Positively Regulates Growth and Stress Tolerance, Is Activated by an AP2-Family Transcription Factor CAP2. *PLoS ONE* **2013**, *8*, e56737. [CrossRef] [PubMed]

336. Nemati, F.; Ghanati, F.; Ahmadi Gavlighi, H.; Sharifi, M. Comparison of sucrose metabolism in wheat seedlings during drought stress and subsequent recovery. *Biol. Plant* **2018**, *62*, 595–599. [CrossRef]
337. Bhowmik, P.; Konkin, D.; Polowick, P.; Hodgins, C.L.; Subedi, M.; Xiang, D.; Yu, B.; Patterson, N.; Rajagopalan, N.; Babic, V.; et al. CRISPR/Cas9 gene editing in legume crops: Opportunities and challenges. *Legume Sci.* **2021**, *3*, e96. [CrossRef]
338. Badhan, S.; Ball, A.S.; Mantri, N. First Report of CRISPR/Cas9 Mediated DNA-Free Editing of 4CL and RVE7 Genes in Chickpea Protoplasts. *Int. J. Mol. Sci.* **2021**, *22*, 396. [CrossRef]
339. Razzaq, M.K.; Akhter, M.; Ahmad, R.M.; Cheema, K.L.; Hima, A.; Karikari, B.; Raza, G.; Xing, G.; Gai, J.; Khurshid, M. CRISPR-Cas9 based stress tolerance: New hope for abiotic stress tolerance in chickpea (*Cicer arietinum*). *Mol. Biol. Rep.* **2022**, *49*, 8977–8985. [CrossRef]

life

Article

Adaptation of Some Quinoa Genotypes (*Chenopodium quinoa* Willd.), Grown in a Saharan Climate in Algeria

Kelthoum Maamri [1,2], Ouiza Djerroudi Zidane [2], Ahmed Chaabena [2], Gabriele Fiene [3,4] and Didier Bazile [3,4,*]

1 Research Laboratory on Phoeniciculture, Faculty Science of Nature and Life, Kasdi Merbah Ouargla University, Ouargla 30000, Algeria
2 Saharan Bio-Resources Laboratory, Safeguarding and Valorization, Kasdi Merbah Ouargla University, Ouargla 30000, Algeria
3 CIRAD, UMR SENS, F-34398 Montpellier, France
4 SENS, Univ Montpellier, CIRAD, F-34398 Montpellier, France
* Correspondence: didier.bazile@cirad.fr

Abstract: Agriculture in southern Algeria faces several challenges that hinder its development, including drought, high temperatures and the excessive salinity of soil and groundwater. The introduction of crops resistant to these factors is one of the solutions chosen to address these abiotic constraints. This research aimed to evaluate the behavior of quinoa (*Chenopodium Quinoa* Willd.) grown in the Ouargla region of southeastern Algeria. Five varieties of quinoa (*Santa maria, Giza1, Amarilla Sacaca, Blanca de Junin* and *Kancolla)* were tested at two sites that differed in terms of soil salinity (9.95 mS/cm and 0.85 mS/cm) during 2019 and 2020. A complete random block experimental design with four repetitions was used for the agronomic tests. Our results clearly show that higher grain yields were obtained at the high salinity site (site 1) compared to the low salinity site (site 2). However, plant height, grain yield per plant and harvest index differed between varieties and sites. In contrast, stem diameter was not greatly affected by salinity. The varieties that seem to be best adapted to the growing conditions of the Ouargla region are, in descending order: *Santa Maria, Giza1, Amarilla Sacaca* and *Blanca de Junin*. When testing quinoa in new environments, it is critical to adapt the cropping cycle of varieties to avoid very high temperatures. The choice to switch to winter cultivation instead of spring cultivation can be an essential criterion for success. The biogeographical approach conducted in this research opens up new perspectives for the adaptation and cultivation of quinoa outside its region of origin to satisfy the food security of the people of North Africa.

Keywords: adaptation; *Chenopodium quinoa* Willd.; genotypes; salinity; Sahara; Algeria

Citation: Maamri, K.; Zidane, O.D.; Chaabena, A.; Fiene, G.; Bazile, D. Adaptation of Some Quinoa Genotypes (*Chenopodium quinoa* Willd.), Grown in a Saharan Climate in Algeria. *Life* **2022**, *12*, 1854. https://doi.org/10.3390/life12111854

Academic Editors: Hakim Manghwar and Wajid Zaman

Received: 21 October 2022
Accepted: 2 November 2022
Published: 11 November 2022

Publisher's Note: MDPI stays neutral with regard to jurisdictional claims in published maps and institutional affiliations.

Copyright: © 2022 by the authors. Licensee MDPI, Basel, Switzerland. This article is an open access article distributed under the terms and conditions of the Creative Commons Attribution (CC BY) license (https://creativecommons.org/licenses/by/4.0/).

1. Introduction

Chenopodium quinoa Willd., a plant native to the Andean highlands, was first domesticated around Lake Titicaca, which lies at an altitude of 3800 m along the Peruvian-Bolivian border [1]. The domestication of quinoa began about 7000 years ago [2], and the process is considered to be ongoing today as the crop continues to be adapted to new environments. For centuries, quinoa was a staple food for the people of the Andes [3–5]. Today, its use is mainly based on human consumption of the grains, such as with cereals.

Quinoa is an optional halophyte plant [6]. A dicotyledonous herbaceous belonging to the Amaranthaceae, the fruit is a tiny achene whose seed color varies between white, yellow, purple and black [7]. Quinoa is one of the most nutritious food crops currently known in the world. The seeds contain high-quality proteins, as they possess all nine essential amino acids, including lysine, methionine and threonine, which are rare and often a limiting factor in cereals and legumes [8,9].

Worldwide, there are over 6000 accessions of quinoa grown by farmers [10]. The genetic diversity of the quinoa species can be classified into five major ecotypes [11] (highlands, inter-Andean valley, *salares* (salt lakes), *yungas* (subtropical forests) and coastal lowlands) according to their adaptation to the specific agro-ecological conditions of the main production areas [2,12,13]. The high genetic diversity of the species offers opportunities to take advantage of its hardiness and promote its wide adaptation [14]. The needs of the crop vary considerably depending on the local variety or cultivar [15].

The hardiness of the species allows quinoa to thrive in a wide range of climatic conditions, including desert-like, hot, dry, cold and temperate and rainy, and hot with high humidity [1,16–18]. Several scientific studies have confirmed that thanks to physiological mechanisms, quinoa can tolerate very dry conditions and drought [19].

The ideal average temperature for quinoa development and growth is around 15 °C to 20 °C, but some varieties can also withstand extreme temperatures of −8 °C to +38 °C [20].

Periods of temperature sensitivity have been recorded, mainly when seed germination occurs at cold temperatures (frost) and when flowering occurs at high temperatures [21,22]. There are varieties adapted to short days or long days, and there are others insensitive to photoperiod [23,24]. Depending on the photoperiod sensitivity of each variety, the duration of growth can be modified according to the length of days and temperatures [25,26]. Photoperiod sensitivity is a key factor in the adaptation of this crop to new latitudes [24], and should be considered along with the analysis of the distribution of daily temperatures during the crop development cycle.

Quinoa has exceptional nutritional properties, with a high protein content compared to cereals, combined with a good balance and sufficient content of all essential amino acids [27–30]. Quinoa thus represents an opportunity for farmers exposed to an increasingly drier climate [31]. It is also one solution for the rehabilitation of salt-affected land, as quinoa is considered one of the most promising food crops for sustainable agriculture in regions affected by soil and water salinization [32]. This is why its status as a facultative halophyte makes it an alternative cash crop for land and water unsuitable for conventional crops in arid and semi-arid regions [33,34]. Quinoa has the ability to grow and complete its life cycle under high salinity levels that are almost similar to those found in seawater [35–37].

Quinoa is a viable alternative in areas limited by climate change and soil salinization. These constraints affect the conditions under which crops can grow and influence the nutritional quality of the grains. Soil and water salinity is ubiquitous, with about one billion hectares affected worldwide in 2021 [38]. Against this backdrop, quinoa appears to be a hardy crop with interesting agronomic and physiological traits. It can grow under different stress conditions such as soil salinity and acidity, and can under certain conditions tolerate episodes of drought and frost [1,39]. This ability of quinoa to grow under extreme stress conditions has encouraged researchers to take it out of the Andes and attempt to adapt it to other parts of the world [20]. Today, various studies are investigating its adaptation in Europe [40], but also and above all in marginal arid and semi-arid zones [41].

Given that the Earth's population will reach nine billion within the next few decades, global food security is becoming an increasingly urgent concern. Today, 870 million people already suffer from hunger in underdeveloped countries [7], and two billion people are estimated to be undernourished [42].

Faced with these challenges, the hardiness of quinoa, linked to the species' very high genetic diversity, and it's incredible nutritional richness mean that it is increasingly appreciated by producers and researchers. Today the plant is cultivated or under testing in over 125 countries [43] and its cultivation continues to develop rapidly [41,44–46].

The spatial and temporal expansion of quinoa around the world went through several stages [47] over time. During phase 1 (before the 1900s), quinoa was limited to the Andean countries (Colombia, Ecuador, Peru, Bolivia, Argentina and Chile). It was considered a local food crop and a staple food for Andean populations [46,48,49]. In phase 2 (between 1901 and 1969), quinoa was imported into Africa as an experiment. The first known trial outside the Andes took place in 1935 in Kenya, and other trials on quinoa's response to nutrient deficiencies and tolerance to abiotic stresses (salinity and temperature) were conducted between 1950 and 1968 [47,50]. In phase 3 (between 1970 and 1989), quinoa was introduced into northern continents, North America (Colorado (USA) [50,51]), Europe

(England, Denmark and the Netherlands [1,50,52]), and Asia (India and China [46,50,53]). During this period, quinoa was also tested in Brazil and Cuba [20,50]. At the end of the 1980s, quinoa was present in 11 countries outside the Andes. During phase 4 (between 1990 and 2012), quinoa spread to 30 new countries, propelled by the project "American and European Test of Quinoa" between 1996 and 1998 [1,47], which gave birth to the first variety (Atlas) and cultivars (Carmen) in Europe [1,46]. In 2012, quinoa appeared in a few countries in the Mediterranean region [47,50]. In phase 5 (between 2013 and 2018), following the declaration of the International Year of Quinoa in 2013, it was tested in 76 countries: 31 in Africa, 24 in Asia, and 15 in Europe [46]. An FAO regional project entitled "Technical Assistance for Strengthening the Food System Associated with Quinoa", was also launched in 2013–2015, implementing the distribution of quinoa accessions among national research institutions in eight countries of North Africa and the Middle East (Algeria, Egypt, Iraq, Iran, Lebanon, Mauritania, Sudan and Yemen) [41] to evaluate these genotypes under semi-arid and arid conditions.

Algeria is one of the countries that has benefited from the expansion of quinoa thanks to the scientific and technical expertise provided by the FAO to assess the behavior of this crop when it was first introduced into the country in 2013–2014. During this first experiment, eight trial sites were chosen to represent the different agro-ecological regions of the country. These were Baïnem (Algiers), Setif, Tiaret, Relizane, Guelma, Biskra, El Oued and Adrar. International cooperation under the aegis of FAO made it possible to evaluate 16 quinoa genotypes (Q21, Q12, Q29, Q18, Q26, Q22, Q27, Giza1, Giza2, *Sajama, Santamaria, Amarilla Marangani, Amarille Sacaca, Blanca de Junin, Kancolla* and *Salcedo Inea*) under arid and semi-arid conditions in order to characterize the phenological development of plants and determine the yield components according to the selected varieties and sites. The first trials were carried out in the autumn of 2014 at seven sites, namely Baïnem (Algiers), Setif, Tiaret, Biskra, El Oued, Adrar and Relizane, and trials were conducted at the other two sites, Guelma and Relizane, the following spring of 2015 (Figure 1). The yield of these trials ranged from 0 to 2.62 t/ha. Despite the various experimental trials carried out on these experimental stations, some with good agronomic results, the cultivation of quinoa has remained at an elementary stage and has not yet met the conditions to be generalized, or better known, across Algeria. New research is currently being carried out on the morphological characterization of certain varieties of quinoa, as well as the effect of saline stress on the physiological performance of these plants. This article presents the first results of the most recent research conducted in Algeria in the arid region of Ouargla. The primary objective of the introduction of quinoa in Algeria is to find alternative species in order to continue to exploit marginal lands affected by salinity, drought and very high temperatures. The aim is to determine whether quinoa is hardy enough to cope with the current and future challenges of the Saharan agrosystem and withstand desert conditions that continue to deteriorate.

To gain a better mastery of quinoa cultivation techniques in Algeria, multiple studies in the different agro-ecological regions of the country are required. The present study was conducted during 2019 and 2020 to assess the behavior of quinoa and its growth under Saharan conditions.

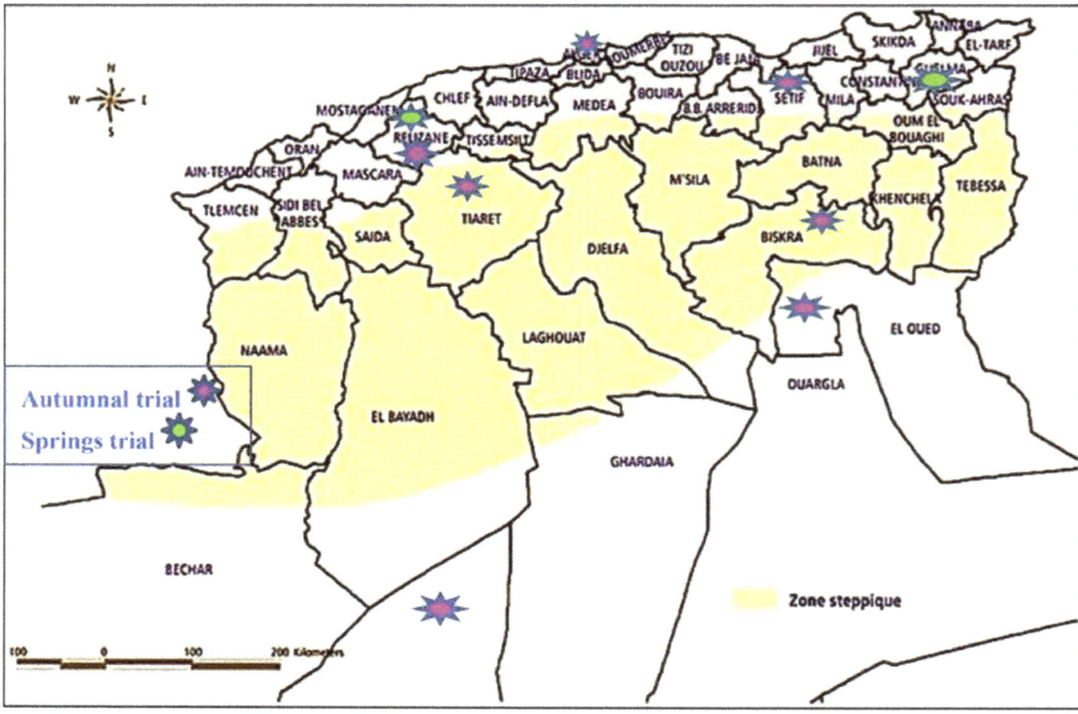

Figure 1. Location of the first quinoa cultivation trials in Algeria. *Own elaboration adapted from a personal communication from the Technical Institute for the Development of Saharan Agriculture (ITDAS).*

2. Materials and Methods

2.1. General Presentation of Agriculture in Algeria

Algeria is the largest country in Africa—since the splitting of Sudan—with an area of 2,381,741 km^2. Due to its vast surface area and distinctive geographical position, extending from the shores of the Mediterranean to the Sahara, the country has a wide range of climates. The northern part, which extends from the Mediterranean coast and includes the Tell Atlas, has a Mediterranean climate, while the rest of the country has a predominantly desert (Saharan) climate (Table 1). Between these two major climatic types, there are many transitional climates in the space between the Tell Atlas and the Saharan Atlas mountain chains, including a semi-arid climate, which corresponds to a Mediterranean climate with a persistent drought over a large part of the year [54]. As a whole, and despite its northern facade and 1200 km long Mediterranean coastline, Algeria is much more semi-arid and arid than humid, with a dominant climate that is hot and dry for most of the year. The desert part (Sahara) covers more than 89% of the country, or about 2 million km^2, while the utilized agricultural area (UAA) covers 8.5 million ha, representing about 19.7% of the country's surface, of which 15% is irrigated. The ratio of hectares per capita is also the lowest in the Maghreb region; it is estimated at 0.19 ha/inhabitant, compared to 0.27 ha/inhabitant for Morocco and 0.45 ha/inhabitant for Tunisia [55].

In Algeria, field crops, particularly cereals, occupy more than half of the UAA and are mainly found in semi-arid areas, highlands and sub-humid areas (Figure 2). Arboriculture occupies just over 10% of the UAA and is represented by olive, date palm and other fruit trees. Vegetable crops cover about 5% of the UAA [56].

Table 1. Table of bioclimatic zones in Algeria.

Bioclimatic Zones	Annual Rainfall (mm)	Percentage of Total Area (%)
Humid	900–1800	0.4
Sub-humid	600–900	1.42
Semi-arid	300–600	4.12
Arid	300–100	4.78
Saharan	<100	89.5

Own elaboration adapted from the Commissariat for the development of agriculture in the Saharan regions (CDARS/Ministry of Agriculture and Rural Development, 2017) [54].

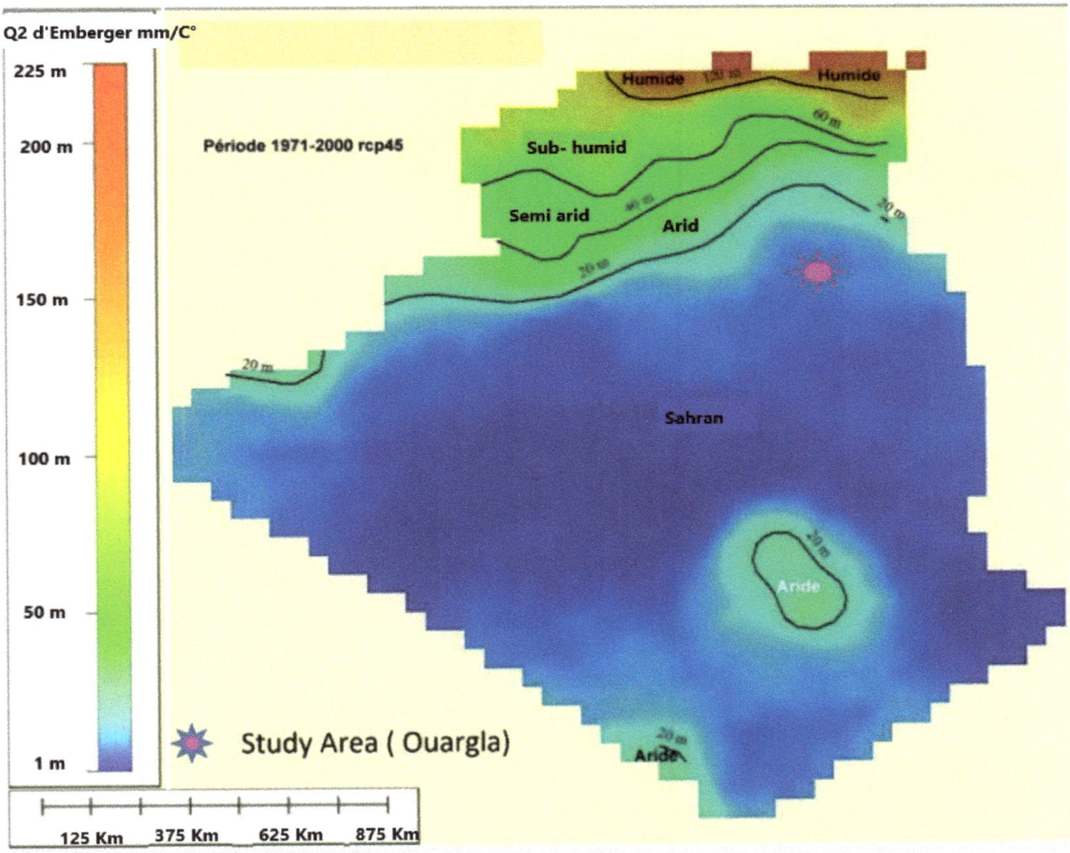

Figure 2. Distribution of bioclimatic zones in Algeria according to Emberger's Q2 measurement [54]. *Own elaboration adapted from the Commissariat for the development of agriculture in the Sa-haran regions (CDARS from the study on the improvement of livestock conditions in Saharan rangelands (CDARS/Ministry of Agriculture and Rural Development, 2017).*

The agriculture of the oases in the south is organized around gardens planted with date palms, associated with market gardening and fruit trees, irrigated by traditional techniques (submersion, seguia (Open-air water supply pipes for irrigation, usually made of earth) and foggaras (Underground pipe (draining gallery), to bring water from upstream to downstream, for agricultural and other needs, located in the region of Touat, Gourara and Tidikelt)). The extension of agriculture in the south outside the oases is progressing in the form of modern land development schemes created under the Law on Access to Agricultural Land Ownership of 1983 (APFA). Most are near traditional palm groves [57]. These are mainly oriented towards pivot cereal farming and date palms irrigated by drip systems (Figure 3).

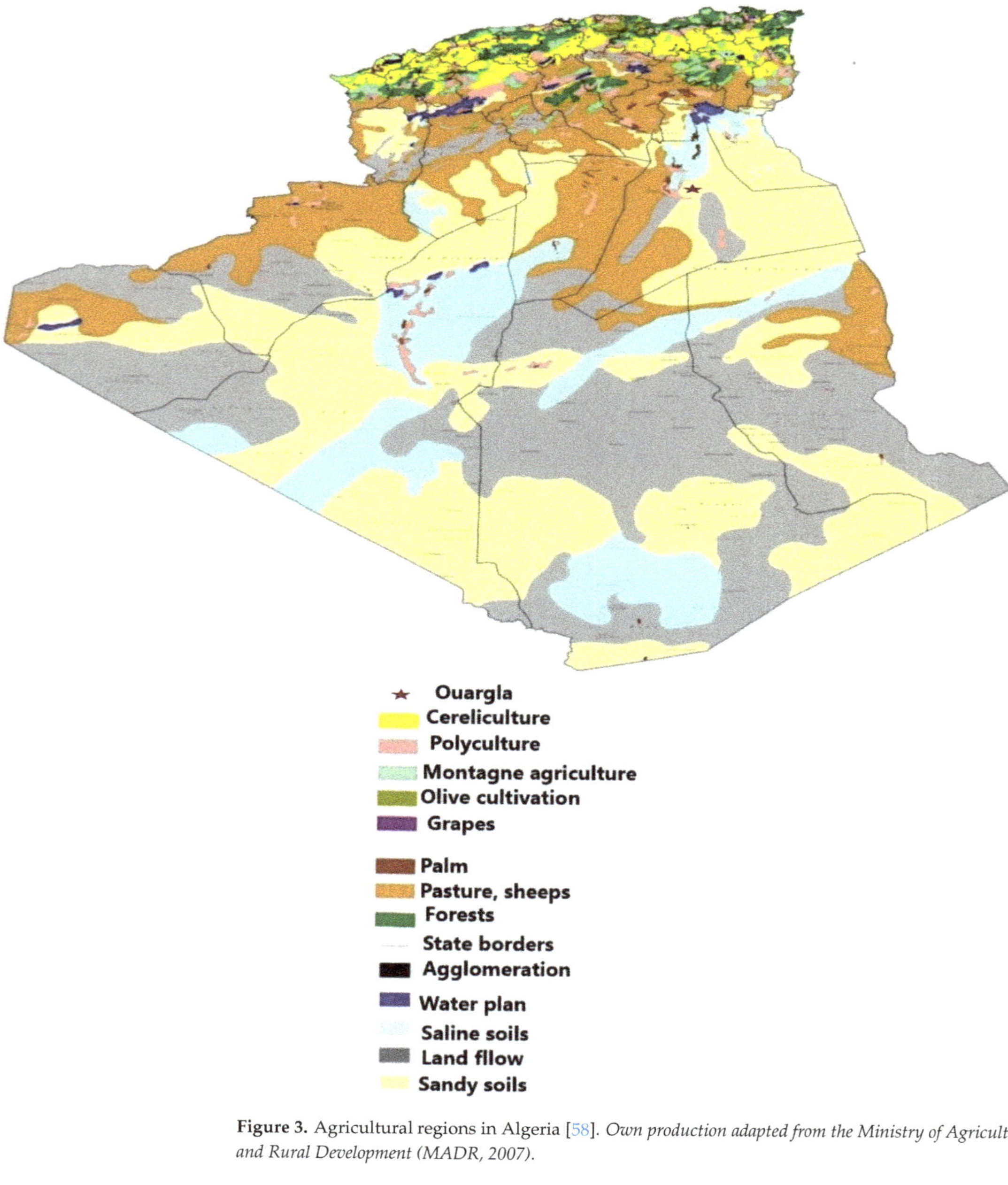

Figure 3. Agricultural regions in Algeria [58]. *Own production adapted from the Ministry of Agriculture and Rural Development (MADR, 2007).*

2.2. Challenges and Constraints of Saharan Agriculture

The Algerian Sahara is a huge biogeographical entity covering 2,000,000 km^2. This natural area faces climatic, soil and anthropogenic challenges, with consequences for the degradation of Saharan agriculture and existing cropping systems, and consequently on the food and nutrition security of the population. The rainfall regime of the Sahara is characterized by low rainfall of about 150 mm per year north of the Sahara, but less than 50 mm in most other Saharan regions along with very high temperatures (over 40 °C), which accentuate the effects of drought [59,60]. Winds also are an aggravating factor, and they are challenging due to the transport of sand that they cause. They are relatively frequent and their speeds are important from April to July, which causes the *siroco* (or sand wind) responsible for silting phenomena with the formation and displacement of dunes [60]. All of these negative conditions make it impossible to grow crops without irrigation in the Saharan zone.

Other constraining aspects of the Saharan climate include both the very high daily thermal amplitude and the annual thermal amplitude. The very low temperatures recorded during the first three months of the year cause frosts and are a limiting factor to be taken into account for crop cycles [59]. Sahara soils are generally composed of sandy mineral substrates, devoid of organic matter, with a coarse texture, low water retention capacity, and limited depth. The Saharan waters are generally chlorinated, chlorinated-sulphated, or sulphated-chlorinated. The chlorine concentration of irrigation water is generally greater than 10 mEq/L [59]. The quality of irrigation water is most often poor because of this primary salinity of geological waters. This is further increased by poor water resource management, which is called secondary salinization [61]. The salinity of the waters is probably one of the main reasons for the low yields obtained in certain irrigated areas of Algeria. The magnesium concentration of irrigation water also is sometimes high. Finally, it appears that these waters have a high ion concentration, which gives them a high risk of salinization and generates risks of toxicity by Na and Cl ions [59].

2.3. Study Sites

The wilaya of Ouargla is located northeast of the northern Sahara. It covers an area of 163,233 km^2. It is characterized by an arid climate, with an average monthly temperature of 42.8 °C in July and a minimum average of 4 °C in January. The average annual rainfall is 50 mm. The texture of the soil is usually sandy or sandy-silty.

The trials conducted for this research were conducted during the winter period of the 2019/2020 campaign to compare two different sites in the Ouargla region, both in open fields and under irrigation. The first site (1) is located in the experimental farm of the Faculty of Natural Sciences and Life of Kasdi Merbah University in the municipality of Ouargla (31°56′20.82″ N latitude, 5°17′33.71″ E longitude, altitude 246 m). The second site (2) is located at the Technical Institute for the Development of Saharan Agriculture, which is in the municipality of Hassi Ben Abdellah (ITDAS), with coordinates 32°0′25.59″ N latitude, 5°27′48.63″ E longitude, altitude 446 m (Figure 4).

Groundwater is the main source of irrigation water used by farmers in the region. The first site is irrigated by the Mio-Pliocene aquifer with an EC = 3.03 mS/cm and a pH = 7.71, while the second site is irrigated by the Intercalary continental (CI) aquifer or the Albian (EC = 2.45 mS/cm and pH = 7.87).

The soil is sandy in texture with an alkaline pH (pH = 7.89 and 7.54) for both sites, but they differ in terms of EC; the first site has highly saline soil, while the second site has low salinity.

The same experimental set-up was adopted in both sites, consisting of a randomized complete block design with four replications. Each treatment (genotype) was represented only once in each block and the distribution of treatments was randomized. The area of each elementary plot was 10 m^2. Each plot comprised five rows spaced 40 cm apart; the inter-plant spacing was 20 cm. Measurements were carried out on the plants in the middle row of each plot. The trials were carried out in the open field and under drip irrigation, with an organic fertilization of 40 t/ha. Sowing was done on 17 and 26 October 2019 for sites 1 and 2, respectively.

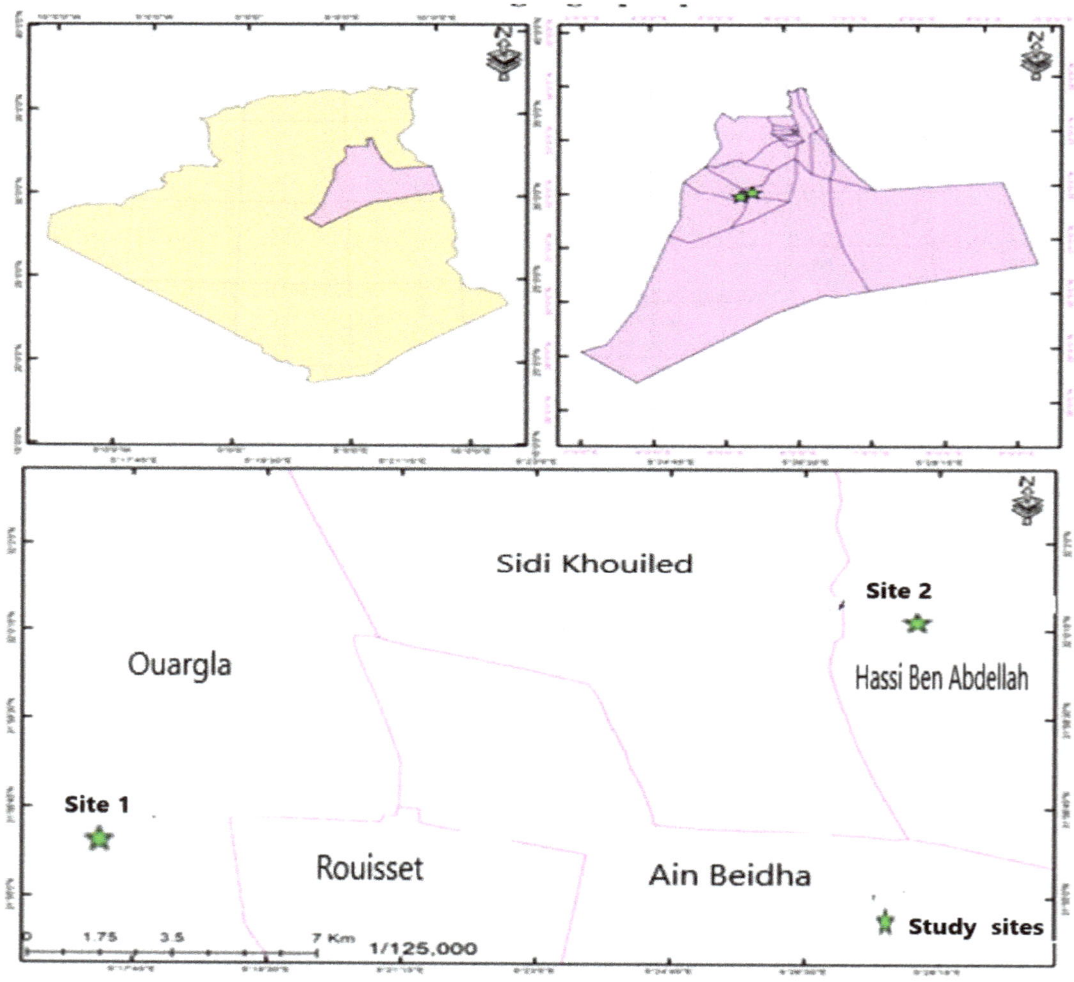

Figure 4. Location map of study sites.

The Ouargla region is characterized by a Saharan climate, with very low rainfall, high temperatures, and high evaporation (Table 2). The annual average maximum and minimum temperatures measured in our study were 31.2 °C and 16.5 °C, respectively; the highest temperature recorded was 48.2 °C during the month of July and the lowest −1 °C in January. Rainfall is rare and irregular, and the total annual rainfall is 13.21 mm.

Table 2. Climate data for the study area (October–April) during the crop cycle of the 2019/2020 crop year.

Month	Max Temperature (°C)	Min Temperature (°C)	Relative Humidity (%)
October 2019	31	17.2	35.8
November 2019	23.3	9.3	37.3
December 2019	21.1	7.1	46.1
January 2020	19	3.2	46.1
February 2020	23.4	6.7	35.5
March 2020	25.8	11.3	33.3
April 2020	30.7	16.4	29

Data source: National Meteorological Office (O.N.M) Ouargla.

2.4. Plant Material Used

The plant material used in this study included five genotypes of quinoa (*Chenopodium quinoa* Willd.). These were *Santa Maria*, Giza1, *Amarilla Sacaca, Blanca de Junin* and *Kancolla*, which were provided by ITDAS and whose seeds were produced in the FAO trial plots. Table 3 shows some characteristics of the seeds used.

Table 3. Characteristics of quinoa genotypes used.

Genotypes	Origin	Institution	Seed Color
Santa Maria	Cultivar of Bolivia	ITDAS	White and brown
Giza1	Cultivar of Egypt	ITDAS	Beige
Amarilla Sacaca (Q102)	Variety from Peru	ITDAS	Orange
Kancolla (Q104)	Variety from Peru	ITDAS	Yellow, brown and beige
Q103	Variety from Peru	ITDAS	Yellow, brown and beige

2.5. Morphological and Agronomic Measurements

The choice of indicators for characterization and monitoring of plant development was made using the book "Descriptors for Quinoa (*Chenopodium quinoa* Willd.) and its Wild Relatives" (Bioversity International, FAO, PROINPA, INIAF and IFAD). The indicators selected to be measured at harvest (physiological maturity) were: plant height (PHT), stem diameter (SD), grain yield per plant (GYP), harvest index (HI) calculated as the ratio of GYP to total shoot dry matter, and the number of days from sowing to maturity.

Considering the BBCH Method applied to quinoa, a field is at stage when 50% of the plant has the corresponding development level [62,63]. For harvesting, a growth stage above 95 was considered as physiological maturity and for plant measuring.

2.6. Statistical Analyses

For morphological and agronomic measurements, an analysis of variance (ANOVA) was conducted using XL-STAT software (2014), and parameter means were compared using a Tukey's test ($p \leq 0.05$).

3. Results

3.1. Number of Days of Genotype Growth

The statistical analysis revealed a significant difference in the number of days until grain maturity between the varieties studied, and this in the two study sites (site 1 $p < 0.02$ and site 2 $p < 0.03$) (Figure 5). Genotype Q102 had the longest maturity time at both sites, with 162 and 164 days for sites 1 and 2, respectively, while the shortest maturity time was 136 days for Giza1 in site 1 and 152 days for Q104 in site 2. All of the varieties reached maturity proportionally earlier under the conditions of the first site compared to those of the second site.

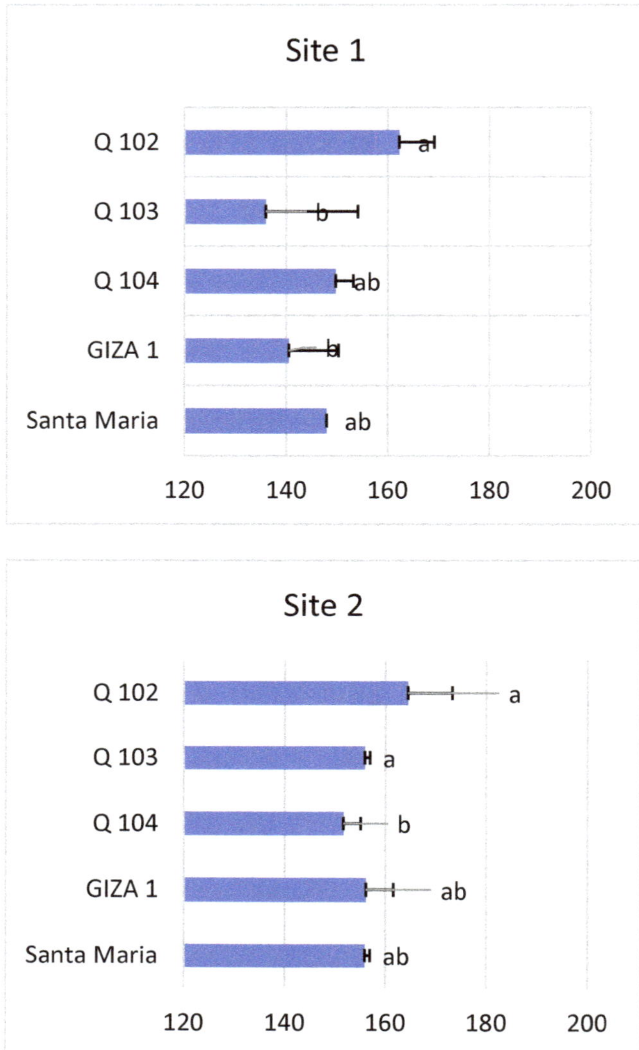

Figure 5. Total duration of the growth cycle of quinoa varieties at both sites. *Means with different letters showed statistical difference ($p \leq 0.02$ for site 1 and $p \leq 0.03$ for site 2).*

This difference can be explained by the difference between the date of sowing and the altitude. Temperatures decrease at higher altitudes, which lengthens the growing cycle for the same photoperiod.

3.2. Plant Height

The plant height of the quinoa varieties showed a significant difference in each of the two sites (site 1 $p < 0.01$ and site 2 $p < 0.02$). Genotype Q102 recorded the greatest height, with an average of 55.73 and 50.97 cm in sites 1 and 2, respectively (Figure 6).

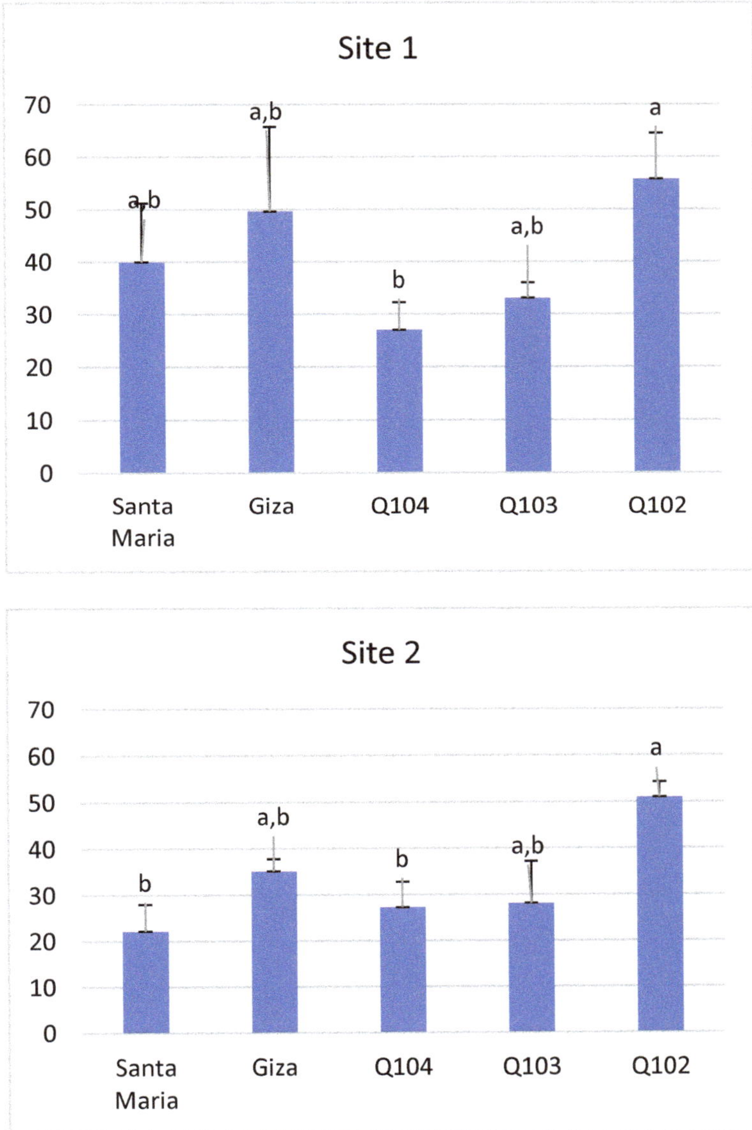

Figure 6. Plant height (cm) at both study sites. *Means with different letters showed statistical difference ($p \leq 0.01$ for site 1 and $p \leq 0.02$ for site 2).*

The lowest height was noted for genotype Q104, with an average of 27.01 cm in site 1, and for Santa Maria with an average of 22.31 cm in site 2.

We observed that the height of the quinoa plant for all of the varieties studied, with the exception of Q104, was always higher under the saline conditions of site 1 (EC = 9.95 mS/cm) than under the non-saline conditions of site 2 (EC = 0.85 mS/cm).

3.3. Stem Diameter

For stem diameter, differences between genotypes were not significant at the chosen threshold at each of the two sites ($p < 0.08$ and $p < 0.83$). Locality variation and salinity did not significantly affect stem diameter, which varied between 4.66 mm and 6.71 mm (Figure 7). Genotypes Q102 and Q104 showed the highest diameters on site 1 (6.71 mm) and site 2 (5.67 mm), respectively. This criterion is a possible selection criterion for good stability of large panicle plants, and it gives them a resistance factor when the plants are exposed to wind. In the Saharan region of Ouargla in Algeria, the fields in the desert are very exposed to climatic phenomenons. The wind velocity is one of them and the resistance of quinoa through the diameter of the stems is of importance for crop adaptation.

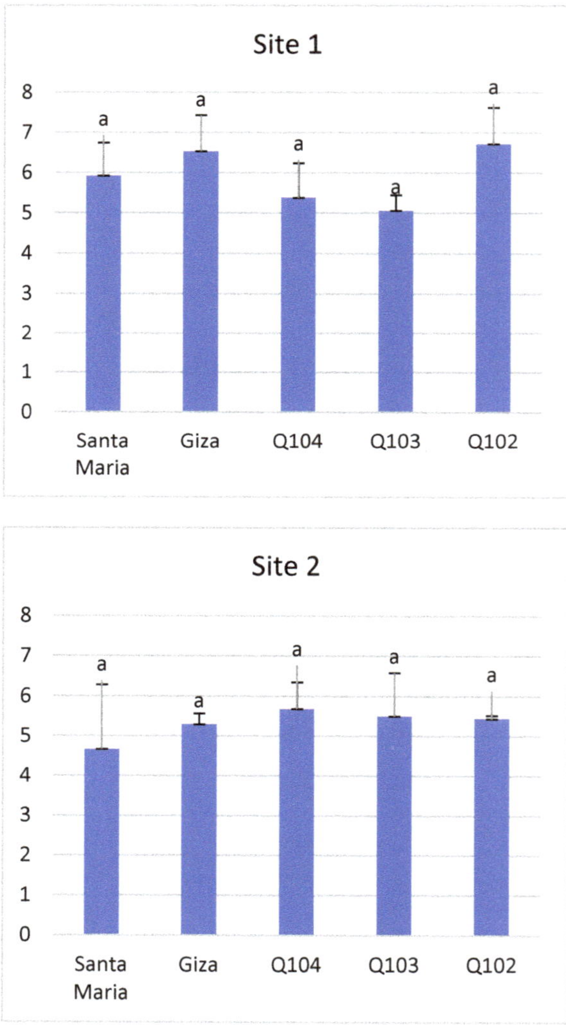

Figure 7. Stem diameter (mm) at both study sites. *Means with different letters showed statistical difference ($p \leq 0.08$ for site 1 and $p \leq 0.83$ for site 2).*

3.4. Grain Yield

The study of grain yield presented in Figure 8 showed a significant difference ($p < 0.01$) in site 1. The highest yields are recorded for the Bolivian variety *Santa Maria* (12.47 g/plant), followed by the Peruvian Q102 (10.24 g/plant), and then by the varieties Giza1 and Q103. The variety with the lowest grain yield was Q104 (4.06 g/plant).

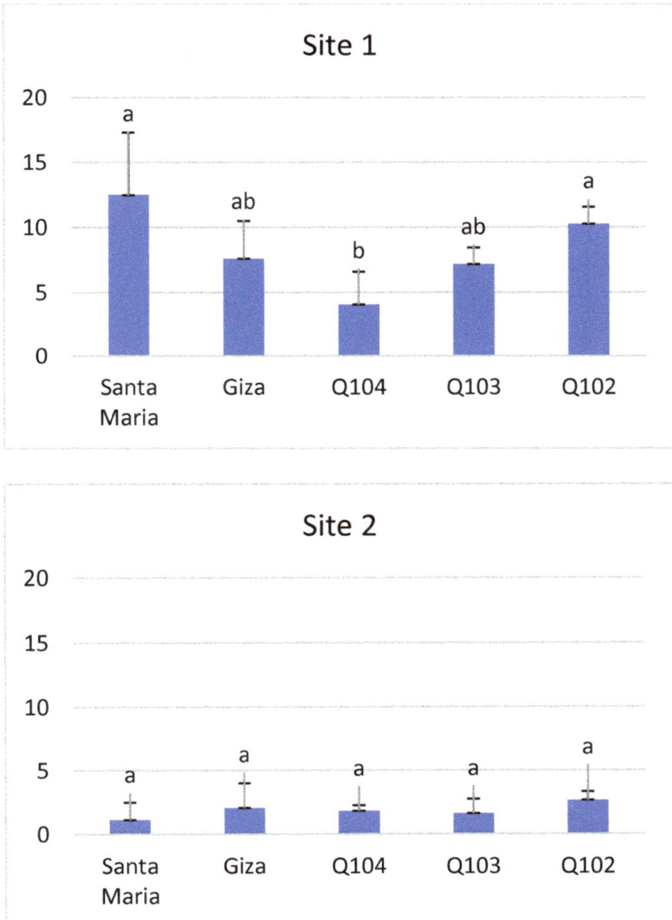

Figure 8. Seed/plant yield (g) at the two study sites. *Means with different letters showed statistical difference ($p \leq 0.01$ for site 1 and $p \leq 0.78$ for site 2).*

In contrast, the statistical differences were not significant ($p < 0.78$) for site 2. *Santa Maria* scored the lowest yield, with an average of 1.13 g/plant, and the highest yield was obtained in the variety Q102, with an average of 2.67 g/plant, and in Giza1 (2.08 g/plant) (Figure 8).

3.5. Harvest Index

Bhargava et al. [64] and Bertero et al. [65] reported that differences are generally significant in the quinoa harvest index by variety and locality, which is consistent with our results. The harvest index (Figure 9) varied considerably between the two sites. The highest indices were recorded in site 1, where Giza1 and Q103 had the highest indexes (0.82 and 0.80), with non-significant differences ($p < 0.43$) between the genotypes.

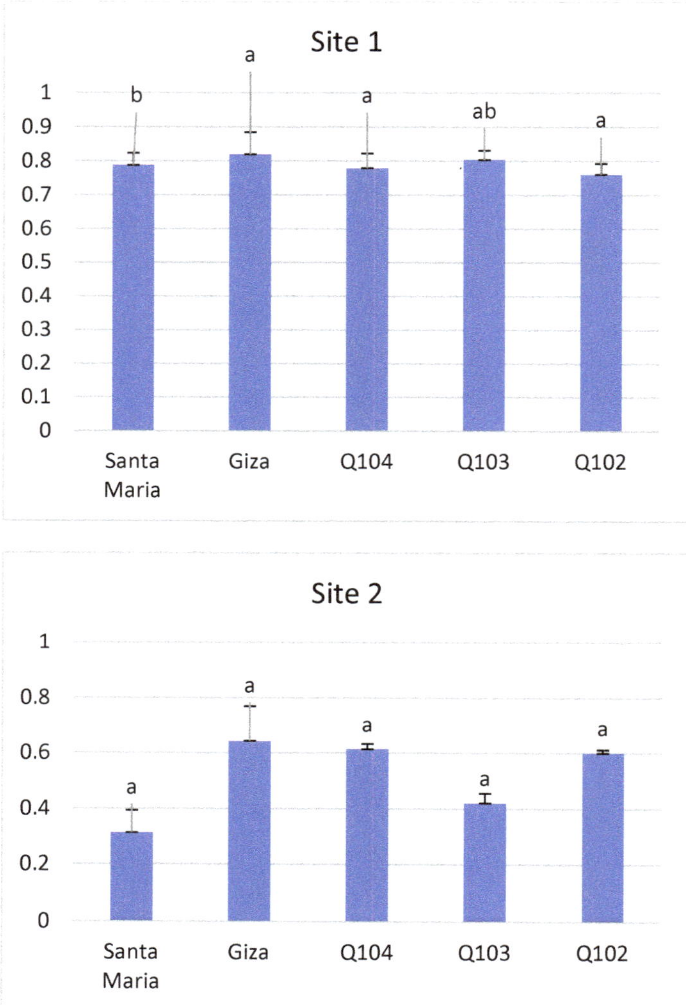

Figure 9. Harvest index at the two study sites. *Means with different letters showed statistical difference ($p \leq 0.43$ for site 1 and $p \leq 0.42$ for site 2). No significant differences.*

The values of the harvest indexes of site 2 were lower compared to those of site 1, varying between 0.64 and 0.31, with the most important again observed at the level of genotypes Giza1 and Q104 (0.64 and 0.61).

4. Discussion

The results of this research clearly show the capacity of quinoa to adapt and tolerate the extreme agro-climatic factors of the Ouargla region (southern Algeria), which is characterized by its aridity, drought and soil salinity. The ripening time of all varieties was shorter at site 1 compared to site 2.

The estimate of the total growth duration of the five varieties studied in southern Algeria was between 136 and 164 days. The total growth time of all varieties was short at site 1 (136–162 days) compared to site 2 (152–164 days) and this variation may be due to differences in temperatures, which are always strongly influenced by altitude.

Jacobsen and Stolen [26] reported that the total growth duration in South America was between 110 and 190 days, while in northern Europe the total duration was somewhat shorter (109–182 days) [66]. In northern India, Bhargava et al. [64] reported a total duration of between 109 and 163 days. In the latter cases, it was a spring crop (sowing in November for harvest in February), while in our case, we tested quinoa as a winter crop (sowing in September and harvest in March).

In addition, the differences between the temperature requirements of the varieties justify the contrast between their number of days of ripening. These results are similar to those provided by Szilagyi and Jornsgard [67] and Tan and Temel [68], who reported that quinoa genotypes require different daylight hours and temperatures, while their maturation phases are also different in Romania and Turkey. In site 2, where sowing was late, the duration of growth was also longer compared to site 1, contrary to results reported by Tan and Temel [69], who conducted experiments in the provinces of Erzurum and Iÿdÿr in Eastern Anatolia, where quinoa was then a summer crop (April–September). Quinoa genotypes reached maturity earlier under Erzurum's conditions, where sowing was late, compared to Iÿdÿr's conditions, where quinoa matured later when planted earlier.

The highest plant height was that of the late-ripening Peruvian variety Q102 in both study sites. These results corroborate those of Tan and Temel [69], who revealed in their trials in Turkey that late-ripening varieties, such as Oro de Valle and Mint Vanilla, grew higher than those that matured early, such as Q-52 and Moqu Arrochilla. These authors pointed to both genetic differences between varieties and variations in the environment as reasons for their results.

Our results revealed that plant heights differed between varieties within the same site, and this is explained by intrinsic genetic differences. Similar results were found in different geographic regions (Pulvento et al. [70], Bhargava et al. [64], Tan et al. [68]). These authors observed differences in plant heights between varieties in different regions, namely southern Italy in a Mediterranean region with a sub-humid climate (summer crop-sowing in May), in Eastern Turkey (Anatolia province) as a summer crop (sowing in April) and in northern India as a spring crop (sowing mid-November).

With the exception of Q104, the plant height of all of the varieties studied was greater under the saline conditions of site 1 (EC = 9.95 mS/cm) than the non-saline conditions of site 2 (EC = 0.85 mS/cm). This result is not in line with those of Hirich et al. [71] and Hirich et al. [72], according to which the salt tolerance threshold is equal to 9 mS/cm, which normally should lead to increased stress on plants, and, as a consequence, reduced growth. These authors showed in their trials that the increase in salinity (EC = 8 dS/m) negatively affected plant height and led to a severe reduction of 73% compared to EC = 1 dS/m.

The Santa Maria, Q102 and Giza 1 genotypes can be considered to be high-yielding varieties. This reflects a greater adaptability of these quinoa varieties to the agro-climatic conditions of southern Algeria. The differences observed in grain yield between varieties can probably be explained by the intrinsic performance of the varieties and their tolerance to salinity. It was noted that variability was not related to geographical origin. This is illustrated by the Peruvian varieties, of which Q104 had the lowest yields. However, high and medium grain yields were found for varieties Q102 and Q103, which corroborates the results of Bhargava et al. [64] in their trials in India where they found strong significant differences in yield between Bolivian varieties. This variation is very marked when analyzing the harvest index between the two sites, which can be explained by the late sowing in site 2. This led to a proportional increase in temperatures during flowering that was especially detrimental to the late-ripening variety, with the corollary of a decrease in yield and harvest index [73].

5. Perspectives

Quinoa is a crop that has attracted attention in recent decades and has been the subject of extensive research recently carried out in all regions of the world. This research has confirmed that quinoa can tolerate various abiotic stresses, including salinity, and that it is an example of an alternative crop in regions that are characterized by a harsh climate, with excessive heat, severe drought and high salinity [71,72,74,75]. Our research is among the first studies in Algeria that can be compared with previous studies conducted on the adaptation of the quinoa species around the world. Due to the increasing problems of salinity in the world, especially in arid areas, and the need for new alternative crops that are more adapted to difficult conditions (saline, drought, high temperature), the results of this research validate the potential of quinoa to be introduced into the cropping systems of the Ouargla area, which are based on phoeniciculture and various associated crops including alfalfa and barley, but also potatoes, fodder corn and cereal cultivation under pivot with durum wheat and soft wheat. Unfortunately, salinity is becoming a major constraint affecting cereal production in the arid zone of Algeria. This constraint is responsible for the drop in yield and is becoming a major evaluation criterion for agricultural development in these regions [76]. Production is approximately 3.6 t/ha [77], the thresholds for a 100% reduction in yields have now been reached, and there will be no more cereal production in 48 to 70 years due to the constant increase in salinity [76].

Nevertheless, after the introduction of quinoa in the Algerian Sahara, the results of this research and of other demonstration trials at research stations in the drylands of Algeria corroborate studies conducted in Morocco and Egypt on the suitability of quinoa for adaptation in drylands. This study indicates that quinoa could be proposed for crop diversification, integrating it into existing cropping systems, as an under-crop in palm groves (fodder crop) and in rotation with field crops (wheat and maize), in order to enrich current cropping systems with alternative species and increase their sustainability in this region.

Like all of the lower Sahara, the Ouargla region is characterized by a desert climate, with large thermal amplitudes between the minima and maxima, and very low rainfall of about 50 mm per year [60]. It has already been pointed out that very high temperatures present an important constraint for the choice of crops [22]. In our case, high (maximum) temperatures, especially in summer, can exceed 50 °C (July), and the average monthly minimum temperature is 4.5 °C in winter (January). The seasonal distribution of low and high temperatures is an essential criterion to properly position the cultivation cycle with an optimal sowing date that avoids the risk of frosts for seedlings and allows plants to develop until flowering before the onset of very high temperatures.

Quinoa is a hardy halophyte plant that can adapt to different geographical areas and abiotic stresses, including drought, frost and heat stress [6,23,71,78]. Quinoa can tolerate a wide range of temperatures (−8 °C to 35 °C) depending on genotypic characteristics and the phenological stage [79]. Despite its adaptation outside its geographical area and its resistance to various abiotic stresses including heat stress, high temperatures during the germination and flowering phase significantly affect the plant and the grain yield in particular. In this context, several studies have focused on the tolerance of quinoa to heat stress (Pulvento et al. [70]; Peterson and Murphy [80]; Yang et al. [81]; Lesjak and Calderini. [82]; Alvar-Beltrán et al. [83]). These studies analyzed the effect of heat stress on different stages of quinoa development including germination and flowering. Lesjak and Calderini [82] in Chile reported a decrease in seed yield when the flowering temperature reached 34 °C. In Italy, it has been shown that the Titicaca variety responded negatively to high temperatures, with a decrease in seed yields to half when the flowering period occurs, around July [70]. Other studies have shown that higher temperatures (20 °C to 25 °C) may promote quinoa growth compared to lower temperatures (8 °C to 18 °C) [81]. In addition, based on tests conducted in Burkina Faso and in the Sahelian, MENA and Mediterranean regions, Alvar-Beltrán et al. [83] found that high temperatures (between 34 °C and 38 °C) on cv. Titicaca causes seed yield losses (25% reduction). For this reason, 38 °C was considered the maximum temperature threshold at flowering. The same authors also found that at temperatures above 34 °C, there was a decrease of over 50% for seed germination.

As quinoa has been introduced in Algeria only recently, farmers' knowledge and know-how regarding this plant remain very limited. This work therefore proposes a new approach to the optimal period for growing quinoa in the region of Ouargla and Oued Righ to avoid extreme heat. Based on meteorological data (monthly mean temperature-MMT) provided by CRU-TS4.03 [84] downscaled with World Clim2.1 [85], over a period of 28 years (1990–2018) we produced 12 maps to assess the monthly climate risk of high temperatures in the two Saharan regions of Ouargla and Oued Righ using ArcGis10.9 software (ESRI-France: 92195 Meudon, France).

Based on previous studies, it appeared essential to represent the distribution of the maximum average temperature over the entire year in the study area (Ouargla and Oued Righ region) in order to identify the appropriate period for quinoa cultivation in these regions outside the risk periods (Figure 10).

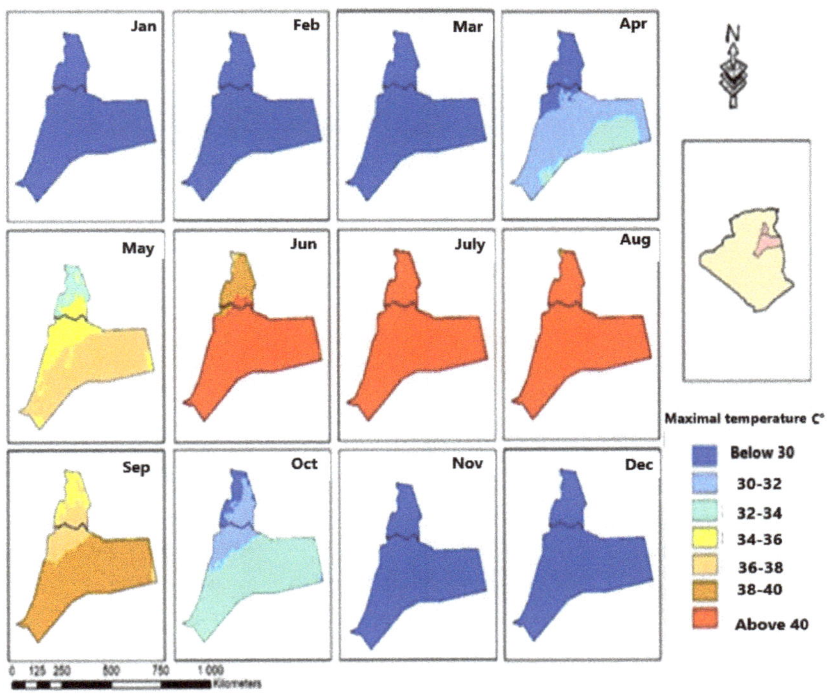

Figure 10. Mean monthly temperatures in two regions in southeastern Algeria (1990–2018).

This figure shows the automatically generated monthly average temperature classes. The first blue class groups MMTs below 34 °C, which are very adequate for the growth of quinoa and are very close to the ideal temperature (optimal between 15 °C and 20 °C) [12,21,66,83,86,87], and correspond to the October-May period for the northern part of the Oued Righ region, and October–April for the Ouargla region. In the second yellow class, the MMTs vary between 34 °C and 38 °C, which are the limit temperatures according to Mamedi et al. [87]. The highest seed germination percentages are between 0 °C and 35 °C, which can regress to 40 °C, thus Alvar-Beltrán et al. [83] reported that most seed yield losses (25% reduction) occurred between temperature levels of 34 °C and 38 °C. For this reason, 38 °C was considered the maximum temperature threshold at flowering, which corresponds to the period of September and May for the region of Ouargla and the south of Oued Righ.

In the brown and red third class, MMTs are above 38 °C, the critical MMT threshold to avoid the coincidence with the period of flowering and germination, and they correspond to the period from June to August for the northern part of the region of Ouargla and Oued Righ, and to September for the southern part.

To avoid high temperatures during the flowering period in the region of Ouargla and Oued Righ in southeast Algeria, we suggest the September–May growing period. It is preferable to sow in September for long-cycle varieties and in October for short-cycle varieties in order for flowering to occur between the months of December–January, when MMTs are adequate (below 30 °C), and also to avoid the frequent, high-speed winds (ranging from 19 to 28 m/s) that occur between March and June.

These high MMTs to be avoided for the cultivation of quinoa give us a first indication for the positioning of the cultivation cycle according to the seasons. However, more detailed work on a daily scale is still necessary to better take into account the real risk incurred by quinoa plants.

6. Conclusions

A great variability between genotypes with the same geographical origin was observed, in particular between Peruvian varieties. Q102 showed very high yields and morphological traits at above-average values, whereas values for Q103 were low, as well as for most traits of Q104. This confirms that the variability was not only related to the geographical origin of the varieties but to genetic factors intrinsic to each variety tested.

All of the varieties selected for the trials in this study performed well on the site where saline levels were very high. It can be concluded that the cultivation of quinoa is possible in the environments affected, even strongly, by salinity. The development of the quinoa plant under these conditions manifests acceptable morphological and agronomic traits. This study shows that varieties Q102, Giza1, Santa Maria and Q103 seem to be best adapted to conditions in southern Algeria. The limitations mentioned in the discussion encourage us to extend the evaluation of this crop in other agro-ecological conditions to better assess its adaptation potential in the arid zones of Algeria.

Author Contributions: Conceptualization, D.B., O.D.Z. and K.M.; methodology, D.B., O.D.Z. and K.M.; investigation, K.M.; data curation, K.M.; statistical analysis, A.C.; writing—original draft preparation, K.M.; writing—review and editing, D.B., O.D.Z., G.F., A.C. and K.M.; supervision, D.B. and O.D.Z. All authors have read and agreed to the published version of the manuscript.

Funding: This research received no external funding.

Informed Consent Statement: Not applicable.

Data Availability Statement: Data available only on demand and after the PhD thesis presentation (K.M.).

Acknowledgments: The authors wish to thank Ismail Guettaa from the Commission for the Development of Agriculture in the Saharan Regions (CDARS, Ouargla, Algeria) for helping us for elaborating the maps. The authors thank also Grace Delobel, English native speaker (USA) and professional translator, for the English edition of our paper.

Conflicts of Interest: The authors declare that they have no conflict of interest.

References

1. Jacobsen, S.E. The worldwide potential for quinoa (*Chenopodium quinoa* Willd.). *Food Rev. Int.* **2003**, *19*, 167–177. [CrossRef]
2. Bazile, D.; Fuentes, F.; Mujica, A. Historical perspectives and domestication. In *Quinoa: Botany, Production and Uses*; Bhargava, A., Srivastava, S., Eds.; CABI: Wallingford, UK, 2013; pp. 16–35.
3. Tagle, M.B.; Planella, M.T. *La Quínoa en la Zona Central de Chile, Supervivencia de una Tradición Prehispánica*; Editorial IKU: Santiago, Chile, 2002.
4. Planella, M.T.; Scherson, R.; McRostie, V. Sitio El Plomo y nuevos registros de cultígenos iniciales en cazadores del Arcaico IV en alto Maipo, Chile central. *Chungara Rev. Antropol. Chil.* **2011**, *43*, 189–202. [CrossRef]

5. Martinez, E.A.; Fuentes, F.; Bazile, D. History of Quinoa: Its origin, domestication, diversification and cultivation with particular reference to the chilean context. In *Quinoa: Improvement and Sustainable Production*; Murphy, K., Matanguihan, J., Eds.; Wiley-Blackwell: Hoboken, NJ, USA, 2015; pp. 19–24. [CrossRef]
6. Adolf, V.I.; Shabala, S.; Andersen, M.N.; Razzaghi, F.; Jacobsen, S.E. Varietal differences of quinoa's tolerance to saline conditions. *Plant Soil* **2012**, *357*, 117–129. [CrossRef]
7. Ruiz, K.B.; Biondi, S.; Oses, R.; Acuña-Rodríguez, I.S.; Antognoni, F.; Martinez-Mosqueira, E.A.; Coulibaly, A.; Canahua-Murillo, A.; Pinto, M.; Zurita-Silva, A.; et al. Quinoa biodiversity and sustainability for food security under climate change: A review. *Agron. Sustain. Dev.* **2014**, *34*, 349–359. [CrossRef]
8. Wright, K.H.; Huber, K.C.; Fairbanks, D.J.; Huber, C.S. Isolation and characterization of *Atriplex hortensin* and sweet *Chenopodium quinoa* starches. *Cereal Chem.* **2002**, *79*, 715–719. [CrossRef]
9. Repo-Carrasco, R.; Espinoza, C.; Jacobsen, S.E. Nutritional value and use of the andean crops quinoa (*Chenopodium quinoa*) and Kañiwa (*Chenopodium pallidicaule*). *Food Rev. Int.* **2003**, *19*, 179–189. [CrossRef]
10. Rojas, W.; Pinto, M.; Alanoca, C.; Gómez-Pando, L.; León-Lobos, P.; Alercia, A. Quinoa genetic resources and ex situ conservation. In *State of the Art Report on Quinoa Around the World*; Bazile, D., Bertero, D., Nieto, C., Eds.; FAO: Rome, Italy; CIRAD: Montpellier, France, 2013; pp. 56–82.
11. Fuentes, F.; Bazile, D.; Bhargava, A.; Martinez, E.A. Implications of farmers' seed exchanges for on-farm conservation of quinoa, as revealed by its genetic diversity in Chile. *J. Agric. Sci.* **2012**, *150*, 702–716. [CrossRef]
12. Wood, J.F.; Winkel, T.; Lhomme, J.P.; Raffaillac, J.P.; Rocheteau, A. Response of some Andean cultivars of quinoa (*Chenopodium quinoa* Willd.) to temperature: Effects on germination, phenology, growth and freezing. *Eur. J. Agron.* **2006**, *25*, 299–308.
13. Bazile, D.; Martinez, E.A.; Fuentes, F. Diversity of quinoa in a biogeographical Island: A review of constraints and potential from arid to temperate regions of Chile. *Not. Bot. Horti Agrobot. Cluj Napoca* **2014**, *42*, 289–298. [CrossRef]
14. Louafi, S.; Bazile, D.; Noyer, J.L. Conserving and cultivating agricultural genetic diversity: Transcending established divides. In *Cultivating Biodiversity to Transform Agriculture*; Hainzelin, E., Ed.; Springer: Heidelberg, Germany, 2013; pp. 181–230.
15. Chevassus-au-Louis, B.; Bazile, D. Cultivating diversity. *Cah. Agric.* **2008**, *17*, 77–78. [CrossRef]
16. Jacobsen, S.E. Adaptation of quinoa (*Chenopodium quinoa*) to Northern European agriculture: Studies on developmental pattern. *Euphytica* **1997**, *96*, 41–48. [CrossRef]
17. Bosque, H.; Lemeur, R.; Van Damme, P.; Jacobsen, S.E. Ecophysiological analysis of drought and salinity stress of Quinoa (*Chenopodium quinoa* Willd.). *Food Rev. Int.* **2003**, *19*, 111–119. [CrossRef]
18. Gesinski, K. Evaluation of the development and yielding potential of *Chenopodium quinoa* Willd. Under the climatic conditions of Europe. *Acta Agrobot.* **2008**, *61*, 185–189. [CrossRef]
19. Cocozza, C.; Pulvento, C.; Lavini, A.; Riccardi, M.; d'Andria, R.; Tognetti, R. Effects of increasing salinity stress and decreasing water availability on ecophysiological traits of Quinoa (*Chenopodium quinoa* Willd) grown in a Mediterranean-type agroecosystem. *J. Agron. Crop Sci.* **2013**, *199*, 229–240. [CrossRef]
20. Bazile, D.; Bertero, H.D.; Nieto, C. *State of the Art Report on Quinoa around the World in 2013*; FAO: Santiago, Chile; CIRAD: Montpellier, France, 2015; p. 603.
21. FAO. *Quinoa an Ancient Crop to Contribute to World Food Security*; FAO Regional Office for Latin America and the Caribbean: Santiago, Chile, 2011.
22. García-Parra, M.A.; Roa-Acosta, D.F.; Stechauner-Rohringer, R.; García-Molano, F.; Bazile, D.; Plazas-Leguizamón, N. Effect of temperature on the growth and development of quinoa plants (*Chenopodium quinoa* Willd.): A review on a global scale. *Sylwan* **2020**, *164*, 411–423. Available online: https://sylwan.lasy.gov.pl/ (accessed on 1 October 2022).
23. Bertero, H.D.; King, R.W.; Hall, A.J. Modelling photoperiod and temperature responses of flowering in quinoa (*Chenopodium quinoa* Willd.). *Field Crop. Res.* **1999**, *63*, 19–34. [CrossRef]
24. Bertero, H.D. Effects of photoperiod, temperature and radiation on the rate of leaf appearance in quinoa (*Chenopodium quinoa* Willd.) under field conditions. *Ann. Bot.* **2001**, *87*, 495–502. [CrossRef]
25. Risi, J.; Galwey, N.W. Genotype X Environment interaction in the Andean grain crop quinoa (*Chenopodium quinoa* Willd.) in temperate environments. *Plant Breed.* **1991**, *107*, 141–147. [CrossRef]
26. Jacobsen, S.E.; Stølen, O. Quinoa—Morphology and phenology and prospects for its production as a new crop in Europe. *Eur. J. Agron.* **1993**, *2*, 19–29. [CrossRef]
27. Vega-Gálvez, A.; Miranda, M.; Vergara, J.; Uribe, E.; Puente, L.; Martínez, E.A. Nutrition facts and functional potential of quinoa (*Chenopodium quinoa* Willd.), an ancient Andean grain: A review. *J. Sci. Food Agric.* **2010**, *90*, 2541–2547. [CrossRef]
28. Maureira, H.; Martínez, E.A. Nutritional aspects of six quinoa (*Chenopodium quinoa* Willd.) ecotypes from three geographical areas of Chile. *Chil. J. Agric. Res.* **2012**, *72*, 175–181. [CrossRef]
29. Pathan, S.; Siddiqui, R.A. Nutritional Composition and Bioactive Components in Quinoa (*Chenopodium quinoa* Willd.) Greens: A Review. *Nutrients* **2022**, *14*, 558. [CrossRef]
30. González, J.A.; Konishi, Y.; Bruno, M.; Valoy, M.; Prado, F.E. Interrelationships among seed yield, total protein and amino acid composition of ten quinoa (*Chenopodium quinoa*) cultivars from two different agroecological regions. *J. Sci. Food Agric.* **2012**, *92*, 1222–1229. [CrossRef] [PubMed]
31. Martínez, E.A.; Veas, E.; Jorquera, C.; San Martín, R.; Jara, P. Reintroduction of *Chenopodium quinoa* Willd. into arid Chile: Cultivation of two lowland races under extremely low irrigation. *J. Agro. Crop Sci.* **2009**, *195*, 1–10. [CrossRef]

32. Eisa, S.; Eid, M.A.; Abd El-Samad, E.H.; Hussin, S.A.; Abdel-Ati, A.A.; El-Bordeny, N.E.; Ali, S.H.; Al-Sayed Hanan, M.A.; Lotfy, M.E.; Masoud, A.M.; et al. *Chenopodium quinoa* Willd. A new cash crop halophyte for saline regions of Egypt. *Aust. J. Crop Sci.* **2017**, *11*, 343–351. [CrossRef]
33. Eisa, S.; Koyro, H.W.; Kogel, K.H.; Imani, J. Induction of somatic embryogenesis in cultured cells of *Chenopodium quinoa*. *Plant Cell Tissue Organ. Cult.* **2005**, *81*, 243–246. [CrossRef]
34. González, J.A.; Hinojosa, L.; Mercado, M.I.; Fernández-Turiel, J.L.; Bazile, D.; Ponessa, G.I.; Eisa, S.; González, D.A.; Rejas, M.; Hussin, S.; et al. A long journey of CICA-17 quinoa variety to salinity conditions in Egypt: Mineral concentration in the seeds. *Plants* **2021**, *10*, 407. [CrossRef]
35. Koyro, H.W.; Eisa, S.S. Effect of salinity on composition, viability and germination of seeds of *Chenopodium quinoa* Willd. *Plant Soil* **2008**, *302*, 79–90. [CrossRef]
36. Shabala, S.; Hariadi, Y.; Jacobsen, S.E. Genotypic difference in salinity tolerance in quinoa is determined by differential control of xylem Na^+ loading and stomatal density. *J. Plant Physiol.* **2013**, *17*, 906–914. [CrossRef]
37. Panuccio, M.R.; Jacobsen, S.E.; Akhtar, S.S.; Musscolo, A. Effect of saline water on seed germination and early seedling growth of halophyte quinoa. *Aob. Plants* **2014**, *6*, 1–18. [CrossRef]
38. Benedetti, F.; Caon, L. *Global Soil Laboratory Assessment 2020—Laboratories' Capacities and Needs*; FAO: Rome, Italy, 2021.
39. Bhargava, A.; Shukla, S.; Ohri, D. *Chenopodium quinoa*: An Indian perspective. *Ind. Crop. Prod.* **2006**, *23*, 73–87. [CrossRef]
40. Jacobsen, S.E. The scope for adaptation of quinoa in northern latitudes of Europe. *J. Agron. Crop Sci.* **2017**, *203*, 603–613. [CrossRef]
41. Bazile, D.; Pulvento, C.; Verniau, A.; Al-Nusairi, M.S.; Ba, D.; Breidy, J.; Hassan, L.; Mohammed, M.I.; Mambetov, O.; Otambekova, M.; et al. Worldwide evaluations of quinoa: Preliminary results from post international year of quinoa FAO projects in nine countries. *Forehead. Plant Sci.* **2016**, *7*, 850. [CrossRef]
42. Jacobsen, S.J.; Sørensen, M.; Pedersen, S.M.; Weiner, J. Feeding the world: Genetically modified crops versus agricultural biodiversity. *Agron Sustain. Dev.* **2013**, *33*, 651–662. [CrossRef]
43. Katwal, T.B.; Bazile, D. First adaptation of quinoa in the Bhutanese mountain agriculture systems. *PLoS ONE* **2020**, *15*, e0219804. [CrossRef] [PubMed]
44. Bazile, D. *Quinoa, Les Enjeux D'une Conquête*; Editions Quae: Versailles, rance, 2015.
45. Andreotti, F.; Bazile, D.; Biaggi, M.C.; Callo-Concha, D.; Jacquet, J.; Jemal, O.M.; King, I.O.; Mbosso, C.; Padulosi, S.; Speelman, E.N.; et al. When neglected species gain global interest: Lessons learned from quinoa's boom and bust for teff and minor millet. *Glob. Food Secur.* **2022**, *32*, 100613. [CrossRef]
46. Alandia, L.; Rodriguez, J.P.; Jacobsen, S.E.; Bazile, D.; Condori, B. Global expansion of quinoa and challenges for the Andean region. *Glob. Food Secur.* **2020**, *26*, 100429. [CrossRef]
47. Bazile, D.; Baudron, F. The dynamics of the global expansion of quinoa growing in view of its high biodiversity. In *State of the Art Report on Quinoa around the World in 2013*; Bazile, D., Bertero, H.D., Nieto, C., Eds.; FAO: Santiago, Chile, 2015; pp. 42–55. Available online: http://www.fao.org/quinoa-2013/publications/detail/en/item/278923/icode/?no_mobile=1 (accessed on 1 September 2022).
48. Tapia, M.; Gandarillas, H.; Alandia, S.; Cardozo, A.; Mujica, A.; Ortiz, R.; Otazu, V.; Rea, J.; Salas, B.; Zanabria, E. *Quinuay Kaniwa Cultivos Andinos*; Mario, T., Gandarillas, H., Alandia, S., Cardozo, A., Mujica, A., Eds.; CIID; IICA: Bogota, Colombia, 1979; Volume 40, p. 228.
49. Maughan, P.J.; Bonifacio, A.; Coleman, E.C.; Jellen, E.N.; Stevens, M.R.; Fairbanks, D.J. Chapter 9: Quinoa (*Chenopodium quinoa*). In *Pulses, Sugar and Tuber Crops*; Springer: Berlin/Heidelberg, Germany, 2007; pp. 147–158. [CrossRef]
50. Bazile, D.; Jacobsen, S.E.; Verniau, A. The global expansion of quinoa: Trends and limits. *Forehead. Plant Sci.* **2016**, *7*, 622. [CrossRef]
51. Alandia, L.; Blajos, J.; Rojas, W. Chapter 7: Expansion of quinoa cultivation to countries outside the Andean region. In *Quinoa: An Ancient Crop to Contribute to World Food Security*; Food and Agriculture Organization of the United Nations, Ed.; FAO—Regional Office for the Americas and the Caribbean: Santiago, Chile, 2011; pp. 43–45. Available online: http://www.fao.org/3/aq287e/aq287e.pdf (accessed on 1 September 2022).
52. Galwey, N.W. The potential of quinoa as a multi-purpose crop for agricultural diversification: A review. *Ind. Crop. Prod.* **1992**, *1*, 101–106. [CrossRef]
53. Bhargava, A.; Ohri, D. Chapter 6.2. Quinoa in the Indian subcontinent. In *State of the Art Report on Quinoa around the World in 2013*; Bazile, D., Bertero, D., Nieto, C., Eds.; FAO: Rome, Italy; CIRAD: Montpellier, France, 2014; pp. 511–523. Available online: http://www.fao.org/3/a-i4042e.pdf (accessed on 1 October 2022).
54. CDARS. Phase II: Study of plant species and mapping of rangelands. In *Study on the Improvement of Breeding Conditions in Saharan rangelands*; Commissariat au Développement de l'Agriculture dans les Régions Sahariennes: Alger, Algeria, 2017; p. 43.
55. MADR. Database—Directorate of Agricultural Statistics and Information Systems. Available online: www.minagri.dz (accessed on 1 September 2022).
56. Bessaoud, O.; Pellissier, J.P.; Rolland, J.P.; Khechimi, W. *Synthesis report on agriculture in Algeria*; Research Report; CIHEAM-IAMM: Montpellier, France, 2019; p. 82. ffhal-02137632e.
57. Coast, M. From oases to development areas, the astonishing revival of Saharan agriculture. *Mediterranean* **2002**, *99*, 5–14.
58. MADR. *Agricultural Map of Algeria*; Ministry of Agriculture and Rural Development: Alger, Algeria, 2007; p. 37.

59. CDARS. Phase 5: Analysis of constraints: Ways and means to overcome them. Commissariat for the Development of Agriculture in the Saharan Regions. In *Study of the General Master Plan for the Development of the Saharan Regions*; CDARS: Alger, Algeria, 1999; p. 56.
60. Khadhraoui, A. *Water and Environmental Impact in the Algerian Sahara Palm Grove of Temacine (Oued Rhir)*; CDARS: Alger, Algeria, 2007; p. 299, ISBN 978-9961-797-05-1.
61. Khadhraoui, A. *Water and Soil in Algeria*; Minsitry of Water Resources: Alger, Algeria, 2006; p. 391, ISBN 9947-0-1193-3.
62. Sosa-Zuniga, V.; Brito, V.; Fuentes, F.; Steinfort, U. Phenological growth stages of quinoa (*Chenopodium quinoa*) based on the BBCH scale. *Ann. Appl. Biol.* **2017**, *171*, 117–124. [CrossRef]
63. Stanschewski, C.S.; Rey, E.; Fiene, G.; Craine, E.B.; Wellman, G.; Melino, V.J.; Patiranage, D.S.R.; Johansen, K.; Schmöckel, S.M.; Bertero, D.; et al. Quinoa Phenotyping Methodologies: An International Consensus. *Plants* **2021**, *10*, 1759. [CrossRef] [PubMed]
64. Bhargava, A.; Shukla, S.; Ohri, D. Genetic variability and interrelationship among various morphological and quality traits in quinoa (*Chenopodium quinoa* Willd.). *Field Crop. Res.* **2007**, *101*, 104–116. [CrossRef]
65. Bertero, H.D.; De La Vega, A.J.; Correa, G.; Jacobsen, S.E.; Mujica, A. Genotype and genotype by environment interaction effects for grain yield and grain size of quinoa (*Chenopodium quinoa* Willd.) as revealed by pattern analysis of international multi environment trials. *Field Crop. Res.* **2004**, *89*, 299–318. [CrossRef]
66. Jacobsen, S.E.; Bach, A.P. The influence of temperature on seed ger-mination rate in quinoa quinoa (*Chenopodium quinoa* Willd.). *Seed Sci. Technol.* **2004**, *26*, 515–523.
67. Szilagyi, L.; Jornsgard, B. *Preliminary Agronomic Evaluation of Chenopodium quinoa Willd. under Climatic Conditions of Romania, Scientific Papers. Series A.*; Agronomy, University of Agronomic Sciences and Veterinary Medicine of Bucharest, Faculty of Agriculture: Buchares, Romania, 2014; pp. 339–343.
68. Tan, M.; Temel, S. Studies on the adaptation of quinoa (*Chenopodium quinoa* Willd.) to eastern anatolia region of turkey. *Agrofor* **2018**, *2*, 33–39. [CrossRef]
69. Tan, M.; Temel, S. Performance of some quinoa (*Chenopodium quinoa* Willd.) genotypes grown in different climate conditions. *Turk. J. Field Crop.* **2018**, *23*, 180–186. [CrossRef]
70. Pulvento, C.; Riccardi, M.; Lavini, A.; D'Andria, R.; Iafelice, G.; Marconi, E. Field Trial Evaluation of Two *Chenopodium quinoa* Genotypes Grown Under Rain-Fed Conditions in a Typical Mediterranean Environment in South Italy. *J. Agron. Crop Sci.* **2010**, *196*, 407–411. [CrossRef]
71. Hirich, A.; Jelloul, A.; Choukr-Allah, R.; Jacobsen, S.-E. Saline Water Irrigation of Quinoa and Chickpea: Seedling Rate, Stomatal Conductance and Yield Responses. *J. Agron. Crop Sci.* **2014**, *200*, 378–389. [CrossRef]
72. Hirich, A.; Choukr-Allah, R.; Ezzaiar, R.; Shabbir, S.A.; Lyamani, A. Introduction of alternative crops as a solution to groundwater and soil salinization in the Laayoune area, South Morocco. *Euro-Mediterr. J. Environ. Integr.* **2021**, *6*, 1–13. [CrossRef]
73. Rodriguez, J.P.; Ono, E.; Abdullah, A.M.S.; Choukr-Allah, R.; Abdelaziz, H. *Cultivation of Quinoa (Chenopodium quinoa) in Desert Ecoregion*; Springer: Cham, Switzerland, 2020; pp. 145–161. [CrossRef]
74. Choukr-Allah, R.; Rao, N.K.; Hirich, A.; Shahid, M.; Alshankiti, A.; Toderich, K.; Gill, S.; Butt, K.U.R. Quinoa for Marginal Environments: Toward Future Food and Nutritional Security in MENA and Central Asia Regions. *Front. Plant Sci.* **2016**, *7*, 346. [CrossRef] [PubMed]
75. Iqbal, S.; Basra, S.M.A.; Afzal, I.; Wahid, A.; Saddiq, M.S.; Hafeez, M.B.; Jacobsen, S.-E. Yield potential and salt tolerance of quinoa on salt-degraded soils of Pakistan. *J. Agron. Crop Sci.* **2019**, *205*, 13–21. [CrossRef]
76. Benbrahim, F.; Mohamed, B.; Kemassi, A.; Darem, S.; Imade, H.; CHikhi, F.; Halilat, M. Evaluating the sustainability of grain production under pivot through the study of soil salinization in the region of Ouargla. *Cienc. E Tec. Vitivinic.* **2016**, *31*, 107–123.
77. Dsa. *Agricultural Services Branch*; Statistics Services: Ouargla, Algeria, 2018.
78. Jacobsen, S.E.; Mujica, A.; Jensen, C.R. The resistance of quinoa (*Chenopodium quinoa* Willd.) to adverse abiotic factors. *Food Rev. Int.* **2003**, *19*, 99–109. [CrossRef]
79. Jacobsen, S.E.; Monteros, C.; Christiansen, J.L.; Bravo, L.A.; Corcuera, L.J.; Mujica, A. Plant responses of quinoa (*Chenopodium quinoa* Willd.) to frost at various phenological stages. *Eur. J. Agron.* **2005**, *22*, 131–139. [CrossRef]
80. Peterson, A.; Murphy, K.M. Quinoa cultivation for temperate North America: Considerations and areas for investigation. In *Quinoa: Improvement and Sustainable Production*; Murphy, K., Matanguihan, J., Eds.; John Wiley & Sons, Inc.: Hoboken, NJ, USA, 2015; pp. 173–192, ISBN 978-1-118-62804-1.
81. Yang, A.; Akhtar, S.S.; Amjad, M.; Iqbal, S.; Jacobsen, S.-E. Growth and Physiological Responses of Quinoa to Drought and Temperature Stress. *J. Agron. Crop Sci.* **2016**, *202*, 445–453. [CrossRef]
82. Lesjak, J.; Calderini, D.F. Increased Night Temperature Negatively Affects Grain Yield, Biomass and Grain Number in Chilean Quinoa. *Front. Plant Sci.* **2017**, *8*, 352. [CrossRef]
83. Alvar-Beltrán, J.; Verdi, L.; Marta, A.D.; Dao, A.; Vivoli, R.; Sanou, J.; Orlandini, S. The effect of heat stress on quinoa (cv. Titicaca) under controlled climatic conditions. *J. Agric. Sci.* **2020**, *158*, 255–261. [CrossRef]
84. Harris IP, D.J.; Jones, P.D.; Osborn, T.J.; Lister, D.H. Updated high-resolution grids of monthly climatic observations–the CRU TS3. 10 Dataset. *Int. J. Climatol.* **2014**, *34*, 623–642. [CrossRef]
85. Fick, S.E.; Hijmans, R.J. WorldClim 2: New 1-km spatial resolution climate surfaces for global land areas. *Int. J. Climatol.* **2017**, *37*, 4302–4315. [CrossRef]

86. González, J.A.; Buedo, S.E.; Bruno, M.; Prado, F.E. Quantifying Cardinal Temperatures in Quinoa (*Chenopodium quinoa*) Cultivars. *Lilloa* **2017**, *54*, 179–194. [CrossRef]
87. Mamedi, A.; Afshari, R.T.; Oveisi, M. Cardinal temperatures for seed germination of three Quinoa (*Chenopodium quinoa* Willd.) cultivars. *Iran. J. Field Crop Sci.* **2017**, *48*, 89–100. [CrossRef]

life

MDPI

Article

Evaluation of the Ecological Environment Affected by *Cry1Ah1* in Poplar

Ali Movahedi [1,†], Hui Wei [2,†], Abdul Razak Alhassan [1], Raphael Dzinyela [1], Pu Wang [1], Weibo Sun [1], Qiang Zhuge [1,*] and Chen Xu [3,*]

1 College of Biology and the Environment, Nanjing Forestry University, Nanjing 210037, China
2 Key Laboratory of Landscape Plant Genetics and Breeding, School of Life Sciences, Nantong University, Nantong 226019, China
3 Nanjing Key Laboratory of Quality and Safety of Agricultural Product, Nanjing Xiaozhuang University, Nanjing 211171, China
* Correspondence: qzhuge@njfu.edu.cn (Q.Z.); xuchen@njxzc.edu.cn (C.X.)
† These authors contributed equally to this work.

Abstract: *Populus* is a genus of globally significant plantation trees used widely in industrial and agricultural production. Poplars are easily damaged by *Micromelalopha troglodyta* and *Hyphantria cunea*, resulting in decreasing quality. *Bt* toxin-encoded by the *Cry* gene has been widely adopted in poplar breeding because of its strong insect resistance. There is still no comprehensive and sufficient information about the effects of *Cry1Ah1*-modified (CM) poplars on the ecological environment. Here, we sampled the rhizosphere soils of field-grown CM and non-transgenic (NT) poplars and applied 16S rRNA and internal transcribed spacer amplicon Illumina MiSeq sequencing to determine the bacterial community associated with the CM and NT poplars. Based on the high-throughput sequencing of samples, we found that the predominant taxa included *Proteobacteria* (about 40% of the total bacteria), *Acidobacteria* (about 20% of the total bacteria), and *Actinobacteria* (about 20% of the total bacteria) collected from the natural rhizosphere of NT and CM poplars. In addition, studies on the microbial diversity of poplar showed that *Cry1Ah1* expression has no significant influence on rhizosphere soil alkaline nitrogen, but significantly affects soil phosphorus, soil microbial biomass nitrogen, and carbon. The results exhibited a similar bacterial community structure between CM varieties affected by the expression of *Cry1Ah1* and non-transgenic poplars. In addition, *Cry1Ah1* expression revealed no significant influence on the composition of rhizosphere microbiomes. These results broadly reflect the effect of the *Bt* toxin-encoded by *Cry1Ah1* on the ecology and environment and provide a clear path for researchers to continue research in this field in the future.

Keywords: *Bt* toxins; *Cry1Ah1* transgenic poplar; ecology; environment; rhizosphere

Citation: Movahedi, A.; Wei, H.; Alhassan, A.R.; Dzinyela, R.; Wang, P.; Sun, W.; Zhuge, Q.; Xu, C. Evaluation of the Ecological Environment Affected by *Cry1Ah1* in Poplar. *Life* **2022**, *12*, 1830. https://doi.org/10.3390/life12111830

Academic Editors: Hakim Manghwar and Wajid Zaman

Received: 27 September 2022
Accepted: 7 November 2022
Published: 9 November 2022

Publisher's Note: MDPI stays neutral with regard to jurisdictional claims in published maps and institutional affiliations.

Copyright: © 2022 by the authors. Licensee MDPI, Basel, Switzerland. This article is an open access article distributed under the terms and conditions of the Creative Commons Attribution (CC BY) license (https://creativecommons.org/licenses/by/4.0/).

1. Introduction

Poplar (*Populus*) is a genus of globally important plantation trees used widely in industrial and agricultural production [1]. However, with the deterioration of the global environment, characterized by increasing salt, drought, pest, and disease stresses, the global production of poplar is becoming challenging. One approach to address this challenge is genetic modification. The manipulation of critical genes has been applied to alter poplar characteristics in transgenic lines, resulting in improved traits for better growth in adverse environments [2–4].

Insecticide resistance based on *Bacillus thuringiensis* (*Bt*) has allowed the development of a variety of insect resistance proteins for commercial genetically modified (GM) crops [5]. In addition, *Bt* toxin-encoded *Cry* genes have been widely applied in commercial GM crops, improving plant resistance to insect pests [6,7]. Despite the benefits of *Bt*-modified plants, a significant potential disadvantage is their effect on soil chemical properties and the structure and diversity of rhizosphere microorganisms, including bacteria and fungi [6].

Therefore, rhizosphere microorganisms associated with *Bt*-modified plants are highly interesting [8–10]. However, regarding the complexity of field conditions and the lack of a unified approach to microorganism analysis, studies exploring the influence of *Bt*-modified plants on soil microorganisms sometimes produce conflicting results [11–14]. For example, one study showed that soil microorganisms are not adversely affected by the cultivation of *Bt*-modified cotton plants [15].

In contrast, another study suggested that exogenous gene products expressed by Bt-modified transgenic crops may interact with soil microorganisms and affect their activities and functions [16]. *Bt*-modified plants have been found to alter bacterial population diversity compared with non-transgenic (NT) plants [17] and may enhance the population size of soil microbial communities [18,19]. Compared with NT rice, *Bt*-modified rice has been found to have a short-term effect on rhizosphere microbial community function [20]. Thus, there are some differences among studies on the impact of *Bt*-modified plant varieties on the soil microbial community. Further research is needed to evaluate the safety of *Bt*-modified crop plants. Since the commercialization of GM plants, the global planting area of GM crops has been overgrown, and new varieties of GM plants have been emerging continually [21]. GM plants have provided great economic and environmental benefits worldwide, such as increased plant yield and reduced chemical fertilizer and pesticide application. However, the potential impacts of GM plants on the ecological environment have raised concerns [22]. Soil microorganisms are essential to the soil ecosystem and involve various biochemical processes, including organic matter accumulation, mineralization, nutrient transformation, and circulation [23]. In addition, GM plants communicate with soil microorganisms during the growth process; therefore, research on the potential effects of GM plants on the soil microbial community is of great significance in evaluating their potential risks [24–26]. The study aimed to identify the effects of *Cry1Ah1*-modified (CM) poplar plantation on soil chemical properties and the diversity and structure of the soil microorganism community after three years of growth under field conditions. We applied high-throughput sequencing of 16S rRNA and internal transcribed spacer 1 (ITS1) to evaluate the diversity and composition differences in rhizosphere microorganisms between NT and CM varieties. The results contribute to our knowledge of how CM varieties affect rhizosphere soil microbial community function and provide reference information for evaluating the safety of CM varieties.

2. Materials and Methods

2.1. Plant Material and Experimental Field Design

In the previous study, the *Cry1Ah1* gene was cloned into the destination vector pH35GS. CM poplars, 'Nanlin 895' (*Populus deltoides* × *Populus euramericana*), were regenerated by inoculating poplar leaf discs with Agrobacterium tumefaciens strain LBA4404, including recombinant plasmid pH35GS-*Cry1Ah1* [2]. To study the effects of CM poplars on a natural soil ecosystem, we designed a field test in Sihong, Jiangsu Province (118°68 N, 33°72 E). To identify the influence of CM varieties on the rhizosphere soil microbiome, we planted NT poplars in the experimental field. Three-year-old NT and CM varieties marked as A5-0, A4-6, Z1-3, A5-23, and A3-4 were selected, and six plots were established with four replicates per clone (Figure S1). An additional 3 m wide isolation zone was established between communities, with about 676 m^2 (26 m × 26 m). The poplars were cultivated by cutting in March 2017 with permission from the State Forestry Administration. Moreover, the topography, physiognomy, soil, air temperature, vegetation, cultivation management, and other natural conditions were consistent. In addition, the experimental field was managed conventionally without chemical fertilizers or pesticides. The experimental field also confirmed similar soil characteristics and microenvironment.

There are four poplars in each small line area, including NT and CM varieties (lines A5-0, A4-6, Z1-3, A5-23, and A3-4) (Figure S1). The weeds and leaves were removed from the soil surface and a soil extractor was used to take out a soil column about 50 cm (diameter 8 cm) around every poplar rhizosphere. The fine roots in the 10–30 cm soil column were carefully taken out and the soil within 3 mm of fine roots was considered the rhizosphere soil. The rhizosphere soil of every four poplar trees in the small area was collected and mixed into one sample. The rhizosphere soil samples were taken out from the entire experimental field through this sampling method, and there were six duplicate rhizosphere soil samples for NT and each CM variety. All rhizosphere soil samples were passed through a 10-mesh sieve, mixed thoroughly, added to sterile centrifuge tubes, and then placed in a liquid nitrogen tank for transport to the laboratory.

2.2. Identification of Cry1Ah1 Expression Level and Insecticidal Activity

The fully expanded poplar leaves were collected from NT and CM varieties, and the collected samples were placed in a liquid nitrogen tank for transport to the laboratory. All the collected leaves, including NT and CM varieties, were assayed to detect *Cry1Ah1* expression levels using an ELISA kit (EnviroLogix, Portland, ME, USA). In addition, pupae of Micromelalopha troglodyta were collected from poplars cultivated in the field and hatched in a culture room at 27 ± 2 °C and 74% humidity with a 14 h light/10 h dark photoperiod. The eggs were collected from female adults and instar larvae were fed with NT and CM varieties. Larval mortality of M. troglodyta was counted on the 6th and 12th days. Three independent biological samples were performed.

2.3. Determination of Rhizosphere Soil Physical and Chemical Indices

Fumigation with chloroform was used to obtain the microbial biomass of nitrogen (MBN), microbial biomass of carbon (MBC), and microbial biomass of phosphorus (MBP) from the soil in the rhizosphere [12,27–29]. The rhizosphere soil samples treated with chloroform fumigation and non-chloroform fumigation were extracted with K2SO4 solution. The rhizosphere MBC, MBN, and MBP were measured using a Vario TOC cube analyzer (Elementar, Langenselbold, Germany). Briefly, fresh rhizosphere soil samples were dissolved in chloroform and the mixtures were boiled for 5 min in a vacuum. Then, 0.5 mol/L K2SO4 was added to the mixtures and subjected to fumigation in the dark at 25 °C for 24 h. The resulting mixture was then filtered with a quantitative filter paper. Simultaneously, the control groups were performed similarly, except that the rhizosphere soil samples were added to the reaction mixture for detection.

The rhizosphere soil alkaline nitrogen was determined using the Conway method (i.e., the alkali hydrolysis diffusion method). Briefly, 10 mL of NaOH was used to dissolve air-dried soil samples set in the outer chamber of a diffusion dish and 2 mL of boric acid (an indicator solution) was placed in the inner chamber. After incubation at 40 °C for 24 h, NH3 in the inner chamber absorption solution was titrated with 0.005 mol/L H2SO4 as a standard solution. Simultaneously, the control groups were performed similarly, except that the rhizosphere soil samples were added to the reaction mixture for detection.

The rhizosphere soil phosphorus was identified by molybdenum–antimony colorimetry. Briefly, air-dried soil samples were mixed with 0.5 mol/L NaHCO3 and activated carbon, shaken for 30 min, and filtered immediately with phosphate-free filter paper. Then, 1–5 mL of filtrate was extracted and the absorbance value was determined. Soil pH was also measured using the glass electrode method. Finally, the rhizosphere soil samples were mixed with 2.5× water volume and the suspension pH was determined using a PP-25 Professional Meter electrode (Sartorius, Germany). Simultaneously, the control groups were performed similarly, except that the rhizosphere soil samples were added to the reaction mixture for detection.

2.4. Rhizosphere Soil DNA Extraction and High-Throughput Sequencing

Using the Fast DNA Spin kit for soil (MP Biomedicals, Santa Ana, CA, USA), 36 independent rhizosphere soil samples were obtained from the NT poplars and five CM varieties (lines A5-0, A4-6, Z1-3, A5-23, and A3-4). Triplicate DNA extractions from each replicate of rhizosphere soil samples were mixed and composited into one DNA sample to overcome the heterogeneity. The quality and integrity of the DNA were determined by electrophoresis on 0.8% agarose gel and the extracted DNA samples were diluted 10-fold and stored at −80 °C for further molecular analyses. Using the extracted genomic DNA as a template, the V3-V4 region (515f/907r) of the 16S rRNA gene and ITS1 region (1737f/2043r) of the ITS1 rRNA gene were amplified to identify the composition and diversity of the microbiological community [30,31]. We then performed high-throughput sequencing using the Illumina novaseq platform. The raw data were filtered using the Trimmatic software to obtain high-quality clean paired-end reads, spliced using the FLASH software. The minimum overlap length was set to 10 bp and the maximum mismatch ratio of the splicing sequence was 0.1. After filtering, influential splicing segment clean tags were obtained. All clean tags were clustered using the VSEARCH software (https://github.com/torognes/vsearch, accessed on 12 April 2021). The clean tags were denoised in amplicon sequence variants (ASVs) and chimeras were filtered with UNOISE3. Taxonomic assignment of ASVs was performed in QIIME 2 v2018.2 [32] (https://qiime2.org, accessed on 12 April 2021) using the QIIME 2 feature classifier plugin [33].

All data analyses were performed using SPSS 19.0 software (IBM, Armonk, NY, USA). Differences in the physical and chemical properties between NT and CM varieties were evaluated using one-way analysis of variance (ANOVA) and Tukey's post hoc comparison. The alpha diversity analyses were performed using the Chao1. It showed species indices to determine the community diversity of rhizosphere bacteria and the phylogenetic diversity (PD) whole-tree and Shannon indices to determine community richness and evenness [34]. According to ASVs' clustering results, alpha diversity was calculated in Mothur v.1.30.1 with rarefaction analysis after subsampling the libraries to an exact size [35]. UniFrac was conducted with beta diversity analysis and phylogenetic analysis. Bray–Curtis distances between NT and CM varieties were calculated and visualized with principal component analysis (PCA).

3. Results

3.1. Effects of CM Varieties on M. Troglodyta

The Bt-Cry1A ELISA kit was used to identify the *Cry1Ah1* expression level in NT and CM varieties. The results showed that *Cry1Ah1* was expressed in CM varieties, and lines A4-6 and A5-0 had a higher *Cry1Ah1* expression level. Conversely, line A3-4 had a lower *Cry1Ah1* expression level (Table S1). In addition, the insecticidal activities of CM varieties were identified, and the results showed that the CM varieties had higher insecticidal activity to *M. troglodyta* than NT poplars (Figure S2). Significantly, lines A4-6 and A5-0 with higher *Cry1Ah1* expression levels exhibited relatively more substantial insecticidal activity than *M. troglodyta*.

3.2. Effects of CM Poplars on Rhizosphere Soil Chemistry Patterns

During the first three years of poplar establishment, the mean soil pH ranged from 7.73 to 8.23 in rhizosphere soil (Figure 1A). Moreover, there was no significant change in rhizosphere soil pH between NT and CM varieties by the sampling date. For the CM varieties (lines A5-0, A4-6, Z1-3, A5-23, and A3-4), rhizosphere soil alkaline nitrogen ranged from 64.54 to 83.15 mg/kg, with similar values observed for NT poplars, and no significant difference between NT and CM varieties (Figure 1B). However, the CM varieties had significantly lower rhizosphere soil available phosphorus in the field-grown stage than NT poplars (Figure 1C). The rhizosphere MBC contents of NT poplars ranged from 160 to 172 mg/kg and differed significantly from those of the CM varieties (Figure 1D). In addition, CM varieties had significantly lower MBN and MBP contents than NT poplars (Figure 1E,F).

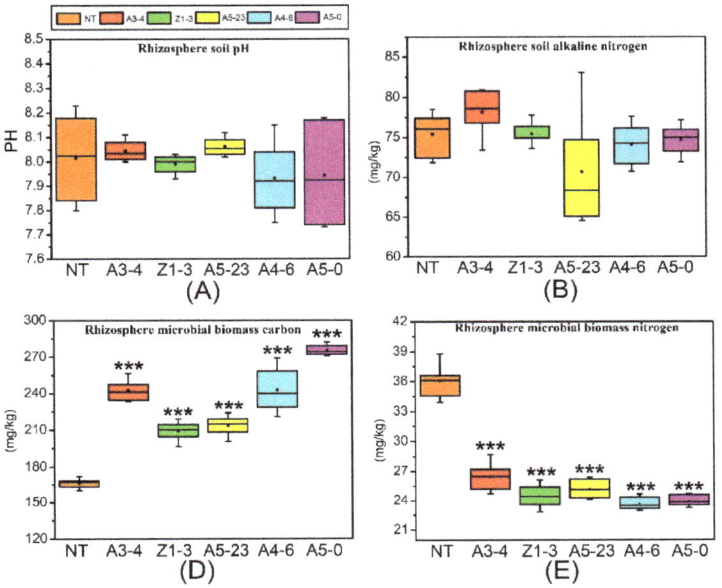

Figure 1. Evaluation of rhizosphere soil physical and chemical properties in non-transgenic (NT) and *Cry1Ah1*-modified (CM) poplar varieties. Analysis of rhizosphere soil pH (**A**), rhizosphere soil alkaline nitrogen (**B**), rhizosphere soil phosphorus (**C**), rhizosphere microbial biomass carbon (**D**), rhizosphere microbial biomass nitrogen (**E**), and rhizosphere microbial biomass phosphorus (**F**) in NT and CM poplar varieties. Data were analyzed using one-way analysis of variance (ANOVA) and Tukey's post hoc comparison. *** $p < 0.001$.

3.3. Data Quality Control and ASVs' Analysis

Using the Illumina hiseq, an average of 53,567 and 68,783 16S rDNA tags and 33,750 and 87,922 ITS1 tags were generated from the rhizosphere microbiome. Chimeras and short tag sequences were removed to obtain high-quality clean tags comprising an average of 33,390 and 86,787 16S rDNA tags and 21,523 and 22,555 ITS1 tags (Table S2). Moreover, clean tag distributions of rhizosphere bacteria were visualized. The results showed that clean tags ranged from 200 to 440 bp and clean tags with lengths of 420 to 440 bp occupied the largest proportion (Figure S3A). In contrast, clean tag distributions of rhizosphere fungi ranged from 200 to 360 bp and clean tags with lengths of 200 to 260 bp occupied the largest proportion (Figure S3B). Using Qiime ver. 2.0 and Vsearch 2.7.1, the chimeric and organelle sequences were removed to produce 10,787 rhizosphere bacterial community sequencing ASVs and 7732 fungal community sequencing ASVs (Table S3).

3.4. Rhizosphere Bacterial Diversity

To construct alpha rarefaction curves and evaluate the putative differences in the alpha diversity, the mothur was applied to perform ASV rarefaction analysis based on ASV clustering results. The sample rarefaction curves of rhizosphere bacteria illustrated that most NT and CM varieties saturate around 6500–7000 ASVs, suggesting slight differences in the diversity of the rhizosphere bacterial community between NT and CM varieties (Figure 2A). The Shannon–Wiener curves were also constructed to evaluate the rhizosphere bacterial diversity. The results showed that the Shannon curves are flat when the number of reads reaches 10,000, illustrating that the amount of sequencing data is sufficient to reflect the vast majority of rhizosphere bacterial information in the 36 samples. In addition,

Shannon curves of the 36 samples fitted together (Figure 2B) suggested that rhizosphere bacterial communities in different sequencing depths share similar diversity.

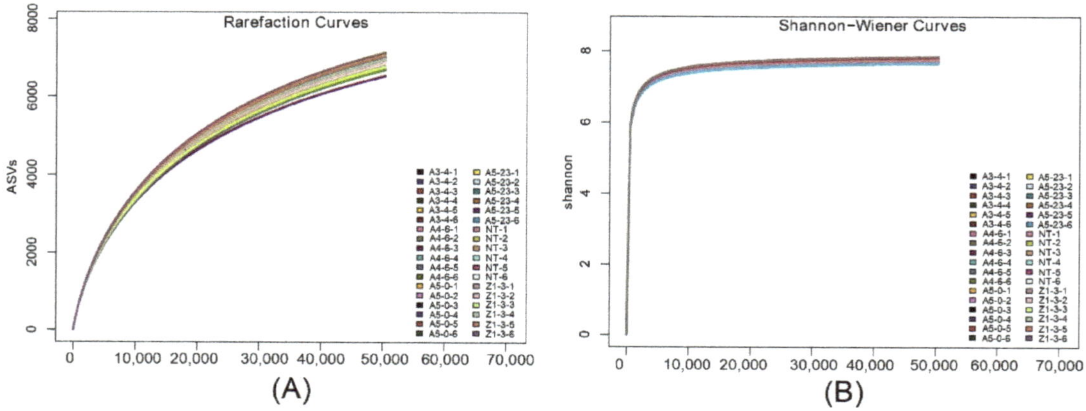

Figure 2. Based on the results of ASV clustering, the rarefaction of ASV has been analyzed. (**A**), Differences in the diversity of the rhizosphere bacterial community between NT and CM varieties have been shown. (**B**), The diversity of the rhizosphere bacteria from NT and CM has been compared by Shannon-Wiener curves.

The alpha diversity analysis was used to reflect the richness and diversity of rhizosphere bacteria. The Chao1 indexes in NT poplars had highly similar results in CM varieties, including A5-23, Z1-3, and A3-4. At the same time, the *Cry1Ah1* expression might increase the community richness of rhizosphere bacteria, for which the analysis of Chao1 showed no dominant difference between NT and CM varieties (Figure 3A and Table S4). Moreover, the observed species in NT poplars had no dominant differences compared with CM varieties, except for line A4-6 (Figure 3B and Table S4). The PD whole tree in NT and CM varieties shared similar features with the observed species (Figure 3C and Table S4). There are no significant differences in the analysis of the Shannon curves (Figure 3C,D and Table S4), which suggests that the community diversity of rhizosphere bacteria was similar between NT and CM varieties. Based on the alpha diversity, we concluded that *Cry1Ah1* expression slightly influences the rhizosphere bacterial richness, but does not affect the community diversity of rhizosphere bacteria.

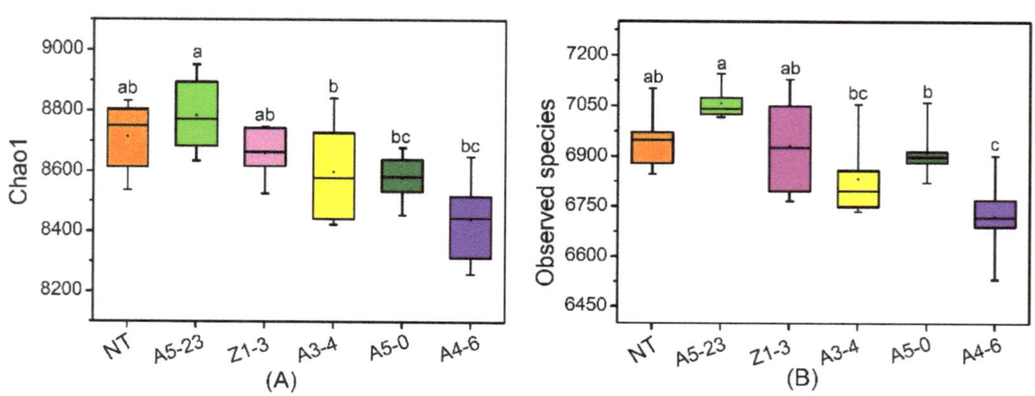

Figure 3. *Cont.*

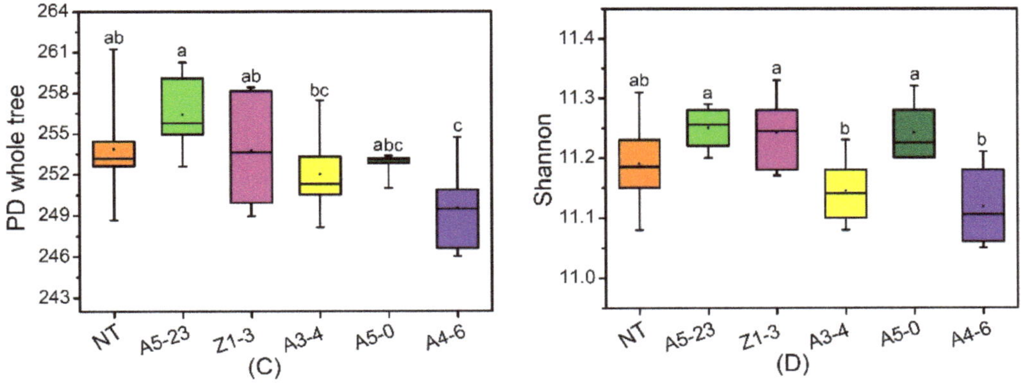

Figure 3. The analysis of the taxonomic distinctiveness of rhizosphere bacteria has been presented based on the alpha diversity as (**A**) the Chao1 index, (**B**) the observed species index, (**C**) the phylogenetic diversity (PD) whole-tree index, and (D) the Shannon index. Data were analyzed using one-way ANOVA and Tukey's post hoc comparison. Significant differences ($p < 0.05$) are indicated in lowercase letters. The NT, A5-23, Z1-3, A3-4, A5-0, and A4-6 lines were represented by orange, green, fusica, yellow, juniper, and violet rectangles, respectively.

A Bray–Curtis dissimilarity matrix was calculated on normalized and square-root-transformed read abundance data to compare the composition of rhizosphere bacterial members between NT and CM varieties. Based on weighted UniFrac, beta diversity analysis with PCA was applied to analyze the bacterial community structures among NT, A5-0, A4-6, Z1-3, and A5-23, and A5-0, A4-6, Z1-3, A5-23, A3-4, and NT overlapped each other and could not be separated (Figure 4), which indicated that the community structures of NT and CM varieties were similar. The *Cry1Ah1* expression did not affect the bacterial community structures.

3.5. The Higher Cry1Ah1 Expression Level May Have Marginal Effects on Rhizosphere Bacteria of Field-Grown Poplars

We investigated the taxonomic distinctiveness of poplar rhizosphere bacteria to determine the effect of *Cry1Ah1* expression on the rhizosphere bacteria. In addition, the DeSeq2 was used to select the putative statistically differential rhizosphere bacteria. The relative abundances of NT and CM varieties of rhizosphere bacteria at the phylum, class, order, family, and genus levels were identified. At the phylum level, Firmicutes, Myxococcota, Nitrospirota, Sva0485, Fibrobacterota, Latescibacterota, Desulfobacterota, and Proteobacteria, considered the dominant bacteria, were found in the rhizosphere bacterial community (Figure 5A). The DeSeq2 analysis found that the dominant bacteria share similar abundances between NT and CM varieties. In contrast, the relative abundances of Cyanobacteria and Methylomirabilota showed a significant difference between NT and line A3-4 (Figure 6A). Moreover, Methylomirabilota, Proteobacteria, Zixibacteria, MBNT15, Dadabacteria, Thermoplasmatota, Cyanobacteria, Chloroflexi, Bacteroidota, Acidobacteriota, Firmicutes, and Myxococcota were present at different abundances between NT and line A4-6 (Figure 6B). In addition, a few rhizosphere bacteria abundances were the difference between NT and A5-0, A5-23, or Z1-3 (Figure 6C–E). According to the above evidence, *Cry1Ah1* expression does not influence most rhizosphere bacteria abundances and only changes a small part of rhizosphere bacteria abundances.

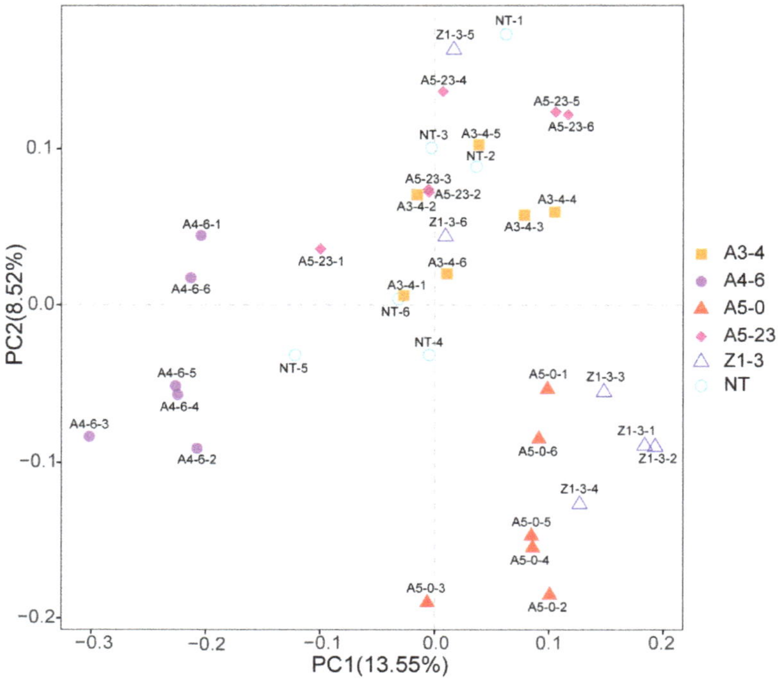

Figure 4. Principal component analysis (PCA) of rhizosphere bacterial communities at the ASVs level. ASVs were defined at a 97% sequence similarity cut-off in mothur. The differences and distances among NT, A5-0, A4-6, Z1-3, A5-23, and A3-4 can be visualized based on an analysis of ASVs' composition.

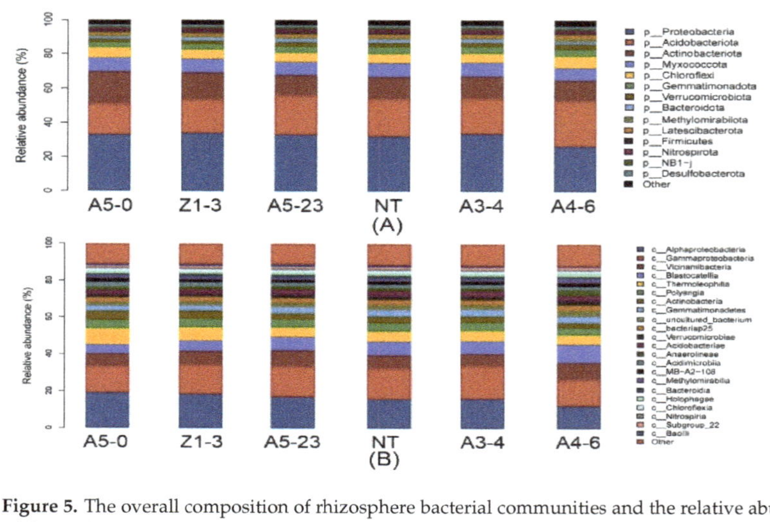

Figure 5. The overall composition of rhizosphere bacterial communities and the relative abundances of rhizosphere bacteria at the phylum or class level between NT and CM varieties. Phylum-level (**A**) and class-level (**B**) taxonomic analysis of bacterial distribution in rhizosphere soil samples of the NT and CM varieties based on 16S amplicon data.

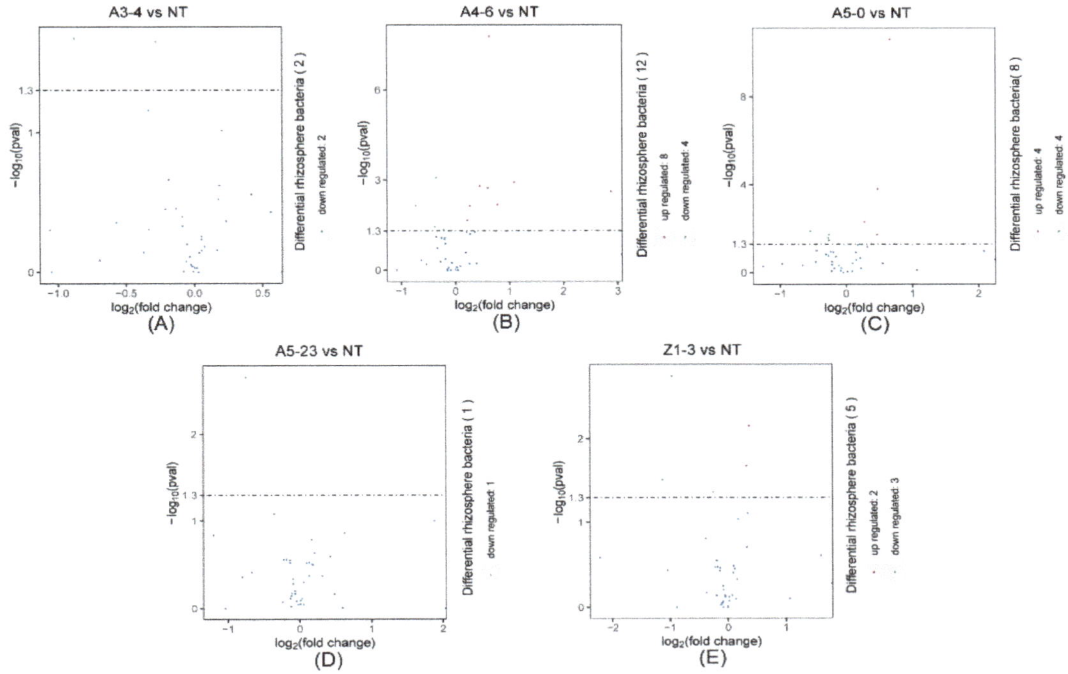

Figure 6. The DeSeq2 analysis for the selection of statistically differential rhizosphere bacteria at the phylum level between NT and lines A3-4 (**A**), A4-6 (**B**), A5-0 (**C**), A5-23 (**D**), and Z1-3 (**E**).

At the class level, the rhizosphere bacterial community composition of NT and CM varieties was similar (Figure 5B). The relative abundances of rhizosphere bacteria had no significant difference between NT and line A3-4, except for Cyanobacteria and Methylomirabilia (Figure 7A).

In native fields, Alphaproteobacteria, Deltaproteobacteria, Betaproteobacteria, Subgroup6, Blastocatellia, and Thermoleophilia accounted for approximately 60% of the total rhizosphere bacteria and were present at similar relative abundances in NT and CM varieties (Figure 5B). In addition, rhizosphere bacteria with lower relative abundances, such as Actinobacteria, Gemmatimonadetes, Nitrospira, Holophagae, and Acidimicrobiia, were significantly different between NT and line A4-6 (Figure 7B). Compared with NT, the relative abundances of rhizosphere bacteria in lines A5-0, A5-23, and Z1-3 were similar and only small parts of rhizosphere bacteria abundances displayed differences (Figure 7C–E). At the species level, a minor part of rhizosphere bacteria abundances from NT was slightly lower or higher than the CM varieties, indicating that the CM varieties had little influence on rhizosphere bacteria with lower relative abundances (Figure 8). We investigated the taxonomic distinctiveness of poplar rhizosphere soil fungi to determine whether *Cry1Ah1* expression affected rhizosphere fungus communities. The Chao1 analysis showed that the community richness of rhizosphere fungi in CM varieties shares a similar community richness to NT, except for line A5-0 (Figure S4A). In addition, no significant difference was present in the observed species, PD whole tree, and Shannon between NT and CM varieties. However, the observed species and PD whole tree in line A5-23 had no slight difference in line Z1-3 (Figure S4B–D). The alpha diversity showed that *Cry1Ah1* expression may slightly improve the fungal community richness, but does not influence the diversity of rhizosphere fungi. PCA was used to evaluate the fungal community structures among NT and CM varieties based on the Bray–Curtis dissimilarity matrix. The results showed that NT and CM varieties are gathered together and the *Cry1Ah1* expression does not affect the fungal community structures (Figure S5). The dominant fungal phyla in poplar rhizosphere soils included Ascomycota, Basidiomycota, and Mortierellomycota (Figure S6A); the sequence load of the six dominant phyla, represented by high sequence numbers, represented more than approximately 80% of the total sequence, whereas that of low-abundance phyla comprised less than 20% of the entire sequence (Figure S6A). Except for Ascomycota, there was no significant difference among rhizosphere fungi between NT and CM varieties (Table 1). Based on the analysis of rhizosphere fungal abundance in NT and CM varieties, we concluded that *Cry1Ah1* expression has no significant influence on the relative abundances of most rhizosphere fungi and only affects a few rhizosphere fungal abundances. We filtered extremely rare ASVs from the dataset to determine relative abundances at the class, order, family, and genus levels.

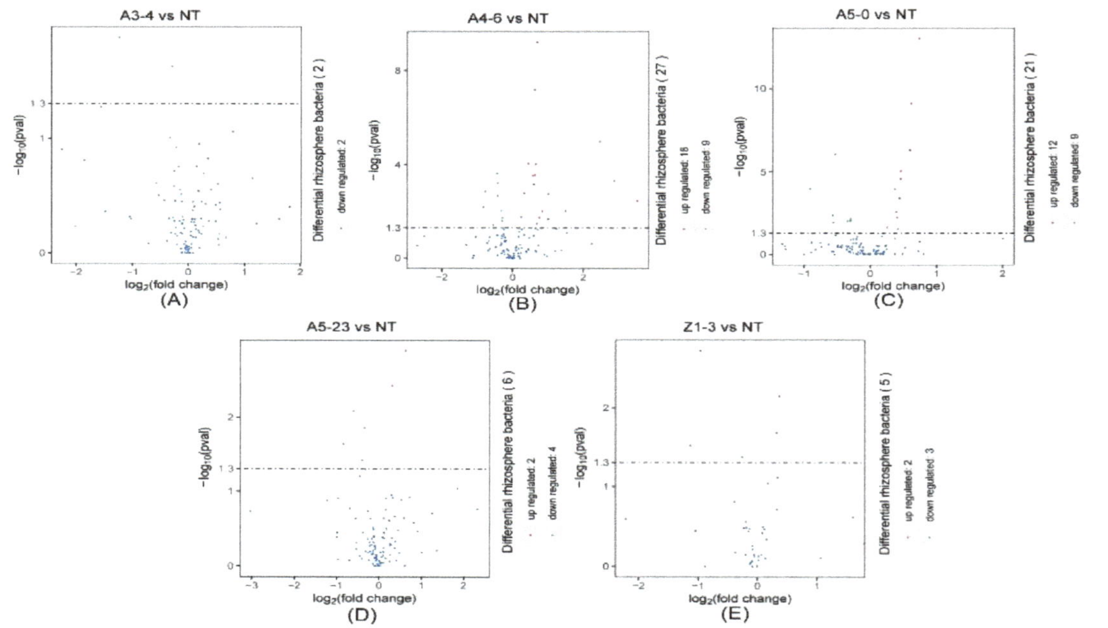

Figure 7. The DeSeq2 analysis for the selection of statistically differential rhizosphere bacteria at a class level between NT and lines A3-4 (**A**), A4-6 (**B**), A5-0 (**C**), A5-23 (**D**), and Z1-3 (**E**).

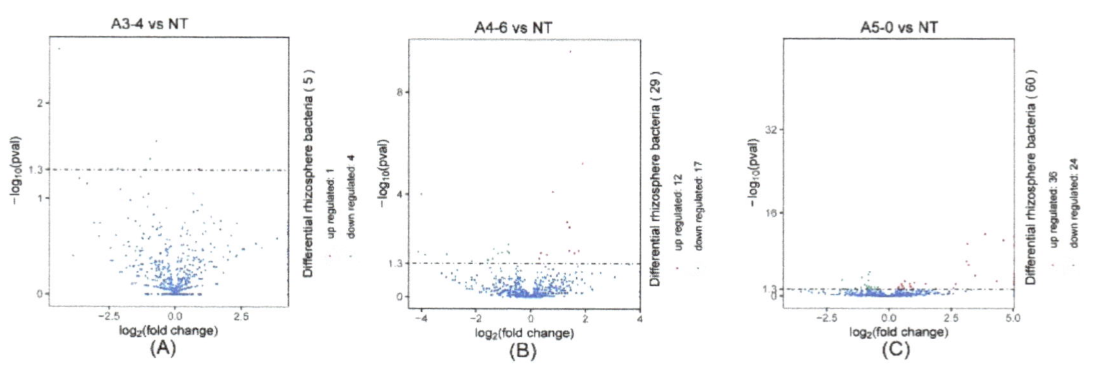

Figure 8. *Cont.*

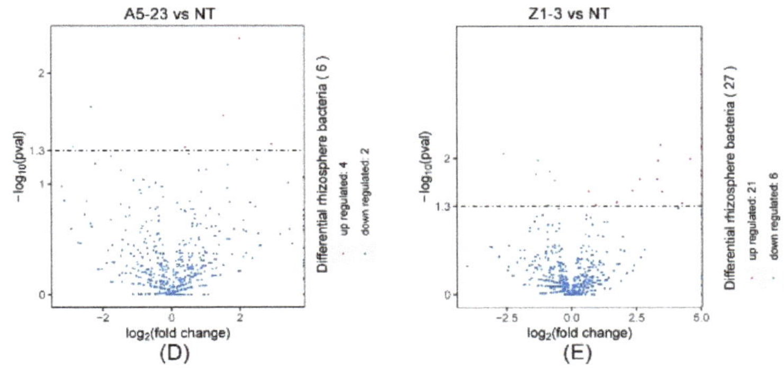

Figure 8. The DeSeq2 analysis for the selection of statistically differential rhizosphere bacteria at the species level between NT and lines A3-4 (**A**), A4-6 (**B**), A5-0 (**C**), A5-23 (**D**), and Z1-3 (**E**).

Table 1. The relative abundances of rhizosphere fungi at the phylum level between NT and CM varieties. Data were analyzed using the Kruskal–Wallis comparison. "$p < 0.05$" indicates a significant difference between NT and CM varieties.

ASV	Test-Statistic	p	FDR P	Bonferroni P	A5-0 Mean	NT Mean	A5-23 Mean	A4-6 Mean	A3-4 Mean	Z1-3 Mean
p__Ascomycota	13.56156156	0.018647881	0.354309737	0.354309737	0.38735852	0.61138326	0.61042142	0.243904621	0.484990829	0.392447323
p__Basidiomycota	10.16216216	0.070768701	0.391712711	1	0.227967163	0.135322999	0.253847358	0.417125218	0.331415246	0.198798819
p__Chlorophyta	10.15615616	0.070929628	0.391712711	1	0.01805686	0.011899969	0.006559522	0.009774974	0.01853778	0.021434483
p__Chytridiomycota	9.092451776	0.105432904	0.391712711	1	0.001884535	0.003696372	0.001263812	0.000542433	0.000810853	0.002152955
p__Cercozoa	8.697065259	0.121774576	0.391712711	1	0.045888695	0.017676598	0.006710509	0.007348007	0.007040442	0.00904241
p__unidentified	8.156156156	0.147836947	0.391712711	1	0.159939382	0.07572809	0.076952758	0.25286315	0.111400036	0.232972084
p__Entomophthoromycota	7.806407963	0.167232549	0.391712711	1	0.000117434	0.000849998	0.000995392	6.15×10^{-5}	6.15×10^{-5}	3.36×10^{-5}
p__Rozellomycota	7.579196217	0.181002706	0.391712711	1	0.000380262	0.001314141	0.000665459	0.000687827	0.00061513	0.001101642
p__Mortierellomycota	7.507507508	0.185548126	0.391712711	1	0.13757102	0.115996734	0.026478549	0.028184136	0.020472643	0.116136536
p__Blastocladiomycota	6.846918489	0.232276395	0.44132515	1	0.003265781	0.001213484	4.47×10^{-5}	0.003282557	0.00246052	0.001302957
p__Olpidiomycota	5.728450555	0.333544833	0.537634093	1	0.000866774	5.03×10^{-5}	7.83×10^{-5}	0.000111842	0.000117434	0.000123026
p__Zoopagomycota	5.322944896	0.377751263	0.537634093	1	0.000715788	0.002024337	0.000419407	0.000726972	0.000726972	0.001588154
p__Glomeromycota	5.162162162	0.396412153	0.537634093	1	0.01409766	0.021809153	0.014824632	0.034631817	0.017508836	0.021311457
p__Entorrhizomycota	5	0.415880187	0.537634093	1	0	0	1.12×10^{-5}	0	0	0
p__Monoblepharomycota	4.930281072	0.424447968	0.537634093	1	0.000352302	0.000335525	6.71×10^{-5}	4.47×10^{-5}	0.000173355	0.000313157
p__Ciliophora	3.628726784	0.60400532	0.61504672	1	0.000363486	1.12×10^{-5}	4.47×10^{-5}	5.59×10^{-6}	0.002829598	0.000184539
p__GS19	3.627922155	0.604125841	0.61504672	1	2.80×10^{-5}	0	2.24×10^{-5}	0	3.91×10^{-5}	5.59×10^{-6}
p__Kickxellomycota	3.615658975	0.605963711	0.61504672	1	0.00096184	0.000548025	0.00048092	0.000570393	0.000609538	0.00072138
p__Mucoromycota	3.555235853	0.61504672	0.61504672	1	0.000184539	0.000139802	0.000111842	0.00013421	0.000190131	0.000329933

Similar relative abundances of most rhizosphere fungi were observed between NT and CM varieties (Figure S6B–E). However, the relative abundances of Archaeosporomycetes, Agaricostilbomycetes, Lobulomycetes, Cystobasidiomycetes, Schizosaccharomycetes, Sordariomycetes, and Ascomycota Incertae sedis at the class level were different between NT and CM varieties (Table S5). Compared with NT poplars, only 13 kinds of 148 rhizosphere fungi relative abundances in CM varieties were different (Table S6). Moreover, the major rhizosphere fungi at the family level showed similar abundances between NT and CM varieties. In contrast, only 7.9% of rhizosphere fungi with a lower abundance were present at differences between NT and CM varieties (Table S7). In addition, most rhizosphere fungi abundances were found to have no differences between NT and CM varieties at the genus level (Table S8). The rhizosphere fungal abundance analysis indicated that most rhizosphere fungi are low and not significantly different between NT and CM varieties. Only relative abundances of a few rhizosphere fungi were different in NT poplars and CM varieties.

4. Discussion

Biological diversity comprises community composition, structure, and function [36]. Interactions between soil microorganisms and other organisms influence nutrient cycling, which is essential in soil condition, quality, and health [12,37,38]. The rhizosphere is a functional interface for material exchange between plants and soil ecosystems. Plants assimilate CO_2 during photosynthesis and transport some photosynthetic products to their underground parts, promoting the growth and metabolism of soil microorganisms, which transform organic nutrients into inorganic forms for plant absorption and utilization [39]. With the recent emergence of transgenic plants, the impact of their cultivation on the structure and function of the rhizosphere microbial community has become a concern [16,40,41]. Therefore, the structural diversity of the rhizosphere microbial community is an essential index for evaluating the effects of GM on the soil ecological environment.

4.1. Changes in Rhizosphere Soil MBC, MBN, and MBP Content in NT and CM Varieties

The rhizosphere soil microbial biomass is essential for assessing active soil nutrients and a sensitive indicator of environmental change in terrestrial ecosystems [42–44]. In addition, MBC, MBN, and MBP participate in the ecosystem cycling of carbon, nitrogen, and phosphorus [45,46]. However, the effects of GM plants on MBC, MBN, and MBP have not been reported. Therefore, in the present study, we examined the impact of field-cultivated CM plants on MBC, MBN, and MBP contents to study the relationship between CM plant growth and carbon, nitrogen, and phosphorus transformation in natural soil. As a part of active soil carbon, MBC is the driving force of soil organic matter decomposition, closely related to the cycling of soil elements. We found that rhizosphere MBC content is significantly higher in CM varieties than in NT poplars.

Conversely, rhizosphere MBN content decreased in CM varieties grown in the field, affecting rhizosphere microorganisms' growth, metabolism, and structure. As an essential source of active soil nitrogen, MBN plays a vital role in regulating soil nitrogen supply [47]. In addition, MBP is the most active component of soil organic phosphorus, which governs the mineralization and fixation of soil phosphorus. Thus, MBP is an essential source of soil phosphorus, reflecting functional capacity and turnover intensity [48,49]. Our results showed that CM varieties alter the rhizosphere MBN and MBP contents, at least during the study period, affecting the capacity of soil microorganisms to metabolize carbon, nitrogen, and phosphorus.

The rhizosphere MBC/MBN ratio can reflect the rhizosphere microbial community structure. The MBC/MBN ratio is about 5:1 for bacteria, 6:1 for actinomycetes, and 10:1 for fungi [50–54]. Based on our results, the MBC/MBN ratio for NT poplars in our study site was about 4.6 ± 0.3, indicating that rhizosphere bacteria may play a dominant role in determining rhizosphere MBC and MBN contents. However, the MBC/MBN ratio for CM varieties was about 9.2 ± 1.7, suggesting that rhizosphere microbial fungi in CM varieties participate widely in rhizosphere soil microenvironment regulation. Xu et al. [30] systematically analyzed MBC, MBN, and MBP in the global terrestrial ecosystem. They reported mean values of MBC/MBN, MBC/MBP, and MBN/MBP ratios of 7.6, 42.4, and 5.6. In the present study, the rhizosphere MBC/MBP and rhizosphere MBN/MBP ratios for NT poplars were 63.2 ± 3.1 and 13.6 ± 0.9, respectively, higher than those reported for the global terrestrial ecosystem. This discrepancy may be because of the low nitrogen and phosphorus content in the experimental field, resulting in lower rhizosphere MBN and rhizosphere MBP contents and lower rhizosphere MBC/MBP and rhizosphere MBN/MBP ratios. Compared with NT poplars, CM varieties showed higher rhizosphere MBC/MBN, rhizosphere MBC/MBP, and rhizosphere MBN/MBP ratios; thus, *Cry1Ah1* transformation may directly affect the growth of rhizosphere microorganisms or inhibit rhizosphere microbial activity, thereby affecting the metabolism of MBC, MBN, and MBP.

4.2. Effects of *Cry1Ah1* Expression on Native Rhizosphere Communities

Bt protein confers strong insecticide resistance as a dominant trait of GM crops; it has been widely used in transgenic breeding to achieve insecticide-resistant plants. Xu et al. [2] showed that field-planted CM poplars had strong insecticide resistance. With increasing GM crops worldwide, GM crop cultivations' environmental and ecological impact has raised concerns globally. Some studies have shown that GM crops seriously affect biodiversity and threaten the environment [55]. Whether GM crops affect rhizosphere microbial composition, structure, and function has become a widely studied question in oncology and food safety [56]. The plant rhizosphere is a dynamic microenvironment in which many factors, such as plant species, soil type, and root location [57–59], affect the composition and structure of microbial communities around plant roots [60]. Therefore, to avoid the interference of these factors in examining rhizosphere microbial communities in the present study, we selected poplar trees planted in a single location and collected samples simultaneously. Such a design can effectively avoid the influence of other factors on the results, focusing only on the effects of poplar type (NT or CM) on the rhizosphere microbial community.

Many studies have explored the relationship between soil biodiversity and the ecological safety of transgenic plants. The high insect resistance of Bt-maize makes it an important transgenic crop and it has been found not to affect soil microbial communities [61] or rhizosphere communities [6]. Field-cultivated *Bt* transgenic cotton has no significant effect on rhizosphere communities other than WT cotton [62]. In the present study, the community diversity and rhizosphere bacterial abundances in NT and CM varieties showed no significant difference. Similarly, Cry1Ac-sugarcane was found to have no impact on rhizosphere microbial diversity or enzyme activity compared with NT sugarcane within a single crop season [63]. Furthermore, NT and Bt-modified rice detected no persistent or adverse effects on the rhizosphere bacterial community population [64]. In addition, Li et al. [15] reported that Bt transgenic rice could change bacterial community composition but not fungal abundance or community structure. Some communities contained a few dominant taxa, whereas others contained many low-abundance taxa. According to the alpha diversity, we found the observed species and PD whole tree differ between NT and line A4-6, which suggested a difference in species richness and community diversity between NT and line A4-6. Concerning the subsequent analysis of relative abundances of rhizosphere bacteria in NT and CM varieties, we concluded that the difference in species richness and community diversity might have originated from the rhizosphere bacteria with low abundance. Weinhold et al. [65] performed a similarity percentage analysis to identify major differences in abundance within groups and found that highly abundant families contributed significantly to dissimilarities. Therefore, according to the alpha diversity, *Cry1Ah1* expression had no significant influence on the rhizosphere bacterial richness and community diversity of rhizosphere bacteria. Based on identifying relative abundances of rhizosphere bacteria in NT and CM varieties, we found no significant difference in the abundances of major rhizosphere bacteria between NT and CM varieties, while the only differences were found in the minor rhizosphere bacteria between NT and CM varieties. We also concluded that the *Cry1Ah1* expression does not affect the relative abundances of major rhizosphere bacteria in native fields and *Cry1Ah1* expression had no impact on most rhizosphere fungal abundances. Moreover, the taxonomic diversity and structure of rhizosphere fungal communities and the relative abundances of most rhizosphere fungi were similar among NT and CM varieties. However, a small fraction of rhizosphere fungal abundances in special CM varieties differed from those in NT poplars. Based on these findings, we concluded that *Cry1Ah1* expression has no effect on the rhizosphere microbial community composition and large numbers of microbial abundances.

5. Conclusions

This study revealed no significant effects of NT and CM cultivation on microbial population and community structure; meanwhile, most rhizosphere bacteria shared similar relative abundances between the NT and CM varieties, suggesting that *Cry1Ah1* expression has no effects on the major rhizosphere bacteria abundances. In conclusion, *Cry1Ah1* expression has no change in microbial population and community structure and does not impact most rhizosphere bacterial abundances.

Supplementary Materials: The following supporting information can be downloaded at https://www.mdpi.com/article/10.3390/life12111830/s1. Figure S1: Poplar growth and experimental design for poplar field trials. Diagram of the experimental design for poplar field trials. The NT and CM varieties were arranged randomly. The CM varieties are A5-0, A4-6, Z1-3, A5-23, and A3-4. Figure S2: Insecticidal activity of NT and CM varieties against Micromelalopha troglodyta. The larval mortality of *M. troglodyta* on days 6 (A) and 12 (B) of feeding on NT and CM varieties. Data were analyzed using one-way ANOVA and Tukey's post hoc comparison. * $p < 0.05$ and ** $p < 0.01$, respectively. Figure S3: The distribution of rhizosphere microbe clean tags. The clean tags' distribution of rhizosphere bacteria (A) and rhizosphere fungi (B). Figure S4: Analysis of the taxonomic distinctiveness of rhizosphere fungi based on alpha diversity. (A) Chao1 index. (B) Observed species index. (C) Phylogenetic diversity whole-tree index. (D) Shannon index. Data were analyzed using one-way ANOVA and Tukey's post hoc comparison. Significant differences ($p < 0.05$) are indicated with lowercase letters. Figure S5: PCA of rhizosphere fungal communities at the level of ASVs. ASVs were defined at a 97% sequence similarity cut-off in mothur. The differences and distances among NT, A5-0, A4-6, Z1-3, A5-23, and A3-4 can be visualized based on an analysis of ASVs' composition. Figure S6: Overall composition of rhizosphere fungal communities and the relative abundances of rhizosphere bacteria between NT and CM varieties. Phylum-level (A), class-level (B), order-level (C), family-level (D), and genus-level (E) taxonomic analysis of fungal distribution in rhizosphere soil samples of the NT and CM varieties based on ITS1 amplicon data. Table S1: Analysis of *Cry1Ah1* expression level in NT and CM varieties. Table S2: The 16S rDNA tags and ITS1 tags generated from the rhizosphere microbiome. Table S3: ASVs for bacterial and fungal community sequencing. Table S4: The alpha diversity analysis of rhizosphere bacteria in NT and CM varieties. Table S5: The relative abundances of rhizosphere fungi at the class level between NT and CM varieties. Data were analyzed using the Kruskal–Wallis comparison. "$p < 0.05$" indicates a significant difference between NT and CM varieties. Table S6: The relative abundances of rhizosphere fungi at the order level between NT and CM varieties. Data were analyzed using the Kruskal–Wallis comparison. "$p < 0.05$" indicates a significant difference between NT and CM varieties. Table S7: The relative abundances of rhizosphere fungi between NT and CM varieties at the family level. Data were analyzed using a Kruskal–Wallis comparison. "$p < 0.05$" indicates a significant difference between NT and CM varieties. Table S8: The relative abundances of rhizosphere fungi at the genus level between NT and CM varieties. Data were analyzed using a Kruskal–Wallis comparison. "$p < 0.05$" indicates a significant difference between NT and CM varieties.

Author Contributions: A.M. and H.W. wrote the first draft of the manuscript. A.M. and H.W. revised the manuscript. H.W., A.R.A., R.D., P.W. and W.S. performed the experiments. Q.Z. and C.X. funded and finalized. All authors have read and agreed to the published version of the manuscript.

Funding: This work was supported by the National Key Program on Transgenic Research (2018ZX08020002), Nanjing Key Laboratory of Quality and safety of agricultural products (NJGS2021-16), and The National Natural Science Foundation of China (31400347).

Institutional Review Board Statement: Not applicable.

Informed Consent Statement: Not applicable.

Data Availability Statement: Not applicable.

Acknowledgments: We thank all students and researchers who helped us analyze and calculate data during SPSS calculations.

Conflicts of Interest: The authors declare no conflict of interest.

References

1. Boerjan, W. Biotechnology and the domestication of forest trees. *Curr. Opin. Biotechnol.* **2005**, *16*, 159–166. [CrossRef] [PubMed]
2. Xu, C.; Wei, H.; Wang, L.; Yin, T.; Zhuge, Q. Optimization of the cry1Ah1 Sequence Enhances the Hyper-Resistance of Transgenic Poplars to Hyphantria cunea. *Front. Plant Sci.* **2019**, *10*, 335. [CrossRef] [PubMed]
3. He, F.; Niu, M.-X.; Feng, C.-H.; Li, H.-G.; Su, Y.; Su, W.-L.; Pang, H.; Yang, Y.; Yu, X.; Wang, H.-L.; et al. PeSTZ1 confers salt stress tolerance by scavenging the accumulation of ROS through regulating the expression of PeZAT12 and PeAPX2 in Populus. *Tree Physiol.* **2020**, *40*, 1292–1311. [CrossRef]
4. Wang, S.; Liu, J.; Dong, Y.; Li, Y.; Huang, Y.; Ren, M.; Yang, M.; Wang, J. Dynamic monitoring of the impact of insect-resistant transgenic poplar field stands on arthropod communities. *Forest Ecol. Manag.* **2022**, *505*, 119921. [CrossRef]
5. Gassmann, A.J.; Reisig, D.D. Management of Insect Pests with Bt Crops in the United States. *Annu. Rev. Entomol.* **2022**, *68*, 31–49. [CrossRef] [PubMed]
6. Dohrmann, A.B.; Küting, M.; Jünemann, S.; Jaenicke, S.; Schlüter, A.; Tebbe, C.C. Importance of rare taxa for bacterial diversity in the rhizosphere of Bt- and conventional maize varieties. *ISME J.* **2013**, *7*, 37–49. [CrossRef]
7. Deng, J.; Wang, Y.; Yang, F.; Liu, Y.; Liu, B. Persistence of insecticidal Cry toxins in Bt rice residues under field conditions estimated by biological and immunological assays. *Sci. Total Environ.* **2019**, *679*, 45–51. [CrossRef]
8. Zhaolei, L.; Naishun, B.; Jun, C.; Xueping, C.; Manqiu, X.; Feng, W.; Zhiping, S.; Changming, F. Effects of long-term cultivation of transgenic Bt rice (Kefeng-6) on soil microbial functioning and C cycling. *Sci. Rep.* **2017**, *7*, 4647. [CrossRef]
9. Zhaolei, L.; Naishun, B.; Xueping, C.; Jun, C.; Manqiu, X.; Zhiping, S.; Ming, N.; Changming, F. Soil incubation studies with Cry1Ac protein indicate no adverse effect of Bt crops on soil microbial communities. *Ecotoxicol. Environ. Saf.* **2018**, *152*, 33–41. [CrossRef]
10. Cotta, S.R.; Dias, A.C.F.; Marriel, I.E.; Andreote, F.D.; Seldin, L.; van Elsas, J.D. Different effects of transgenic maize and non-transgenic maize on nitrogen-transforming archaea and bacteria in tropical soils. *Appl. Environ. Microbiol.* **2014**, *80*, 6437–6445. [CrossRef]
11. Shu, Y.; Du, Y.; Wang, J. Presence of Cry1Ab in the Bt maize—Aphid (Rhopalosiphum maidis)—Ladybeetle (Propylea japonica) system has no adverse effects on insect biological parameters. *Entomol. Exp. Appl.* **2019**, *167*, 553–560. [CrossRef]
12. Čerevková, A.; Miklisová, D.; Szoboszlay, M.; Tebbe, C.C.; Cagáň, L'. The responses of soil nematode communities to Bt maize cultivation at four field sites across Europe. *Soil Biol. Biochem.* **2018**, *119*, 194–202. [CrossRef]
13. Li, P.; Li, Y.; Shi, J.; Yu, Z.; Pan, A.; Tang, X.; Ming, F. Impact of transgenic Cry1Ac+pTI cotton on diversity and dynamics of rhizosphere bacterial community of different root environments. *Sci. Total Environ.* **2018**, *637*, 233–243. [CrossRef]
14. Liu, Y.; Li, J.; Neal Stewart, C.; Luo, Z.; Xiao, N. The effects of the presence of Bt-transgenic oilseed rape in wild mustard populations on the rhizosphere nematode and microbial communities. *Sci. Total Environ.* **2015**, *530–531*, 263–270. [CrossRef] [PubMed]
15. Li, P.; Ye, S.; Liu, H.; Pan, A.; Ming, F.; Tang, X. Cultivation of Drought-Tolerant and Insect-Resistant Rice Affects Soil Bacterial, but Not Fungal, Abundances and Community Structures. *Front. Microbiol.* **2018**, *9*, 1390. [CrossRef]
16. Jansson, J.K.; Hofmockel, K.S. Soil microbiomes and climate change. *Nat. Rev. Microbiol.* **2020**, *18*, 35–46. [CrossRef]
17. Singh, A.K.; Rai, G.K.; Singh, M.; Dubey, S.K. Bacterial Community Structure in the Rhizosphere of a Cry1Ac Bt-Brinjal Crop and Comparison to Its Non-transgenic Counterpart in the Tropical Soil. *Microb. Ecol.* **2013**, *66*, 927–939. [CrossRef] [PubMed]
18. Hu, H.; Xie, M.; Yu, Y.; Zhang, Q. Transgenic Bt cotton tissues have no apparent impact on soil microorganisms. *Plant Soil Environ.* **2018**, *59*, 366–371. [CrossRef]
19. Velmourougane, K.; Sahu, A. Impact of transgenic cottons expressing cry1Ac on soil biological attributes. *Plant Soil Environ.* **2013**, *59*, 108–114. [CrossRef]
20. Wei, M.; Tan, F.; Hong, Z.; Cheng, K.; Xiao, W.; Lingxi, J.; Zhao, K.; Tang, X. Impact of Bt-transgenic rice (SHK601) on soil ecosystems in the rhizosphere during crop development. *Plant Soil Environ.* **2018**, *58*, 217–223. [CrossRef]
21. Wu, N.; Shi, W.; Liu, W.; Gao, Z.; Han, L.; Wang, X. Differential impact of Bt-transgenic rice plantings on bacterial com-munity in three niches over consecutive years. *Ecotox. Environ. Safe* **2021**, *223*, 112569. [CrossRef]
22. Mina, U. An approach for impact assessment of transgenic plants on soil ecosystem. *Appl. Ecol. Env. Res.* **2008**, *6*, 1–19. [CrossRef]
23. Jacoby, R.; Peukert, M.; Succurro, A.; Koprivova, A.; Kopriva, S. The Role of Soil Microorganisms in Plant Mineral Nutrition-Current Knowledge and Future Directions. *Front. Plant Sci.* **2017**, *8*, 1617. [CrossRef] [PubMed]
24. Liu, M.; Sui, X.; Hu, Y.; Feng, F. Microbial community structure and the relationship with soil carbon and nitrogen in an original Korean pine forest of Changbai Mountain, China. *BMC Microbiol.* **2019**, *19*, 218. [CrossRef]
25. DeBruyn, J.M.; Bevard, D.A.; Essington, M.E.; McKnight, J.Y.; Schaeffer, S.M.; Baxter, H.L.; Mazarei, M.; Mann, D.G.J.; Dixon, R.A.; Chen, F. Field-grown transgenic switchgrass (*Panicum virgatum* L.) with altered lignin does not affect soil chemistry, microbiology, and carbon storage potential. *Glob. Change Biol. Bioenergy* **2017**, *9*, 1100–1109. [CrossRef]
26. Abdul Rahman, N.S.N.; Abdul Hamid, N.W.; Nadarajah, K. Effects of Abiotic Stress on Soil Microbiome. *Int. J. Mol. Sci.* **2021**, *22*, 9036. [CrossRef] [PubMed]
27. Cui, Y.; Wang, X.; Zhang, X.; Ju, W.; Duan, C.; Guo, X.; Wang, Y.; Fang, L. Soil moisture mediates microbial carbon and phosphorus metabolism during vegetation succession in a semiarid region. *Soil Biol. Biochem.* **2020**, *147*, 107814. [CrossRef]
28. Gao, D.; Bai, E.; Li, M.; Zhao, C.; Yu, K.; Hagedorn, F. Responses of soil nitrogen and phosphorus cycling to drying and rewetting cycles: A meta-analysis. *Soil Biol. Biochem.* **2020**, *148*, 107896. [CrossRef]

29. Dong, H.; Zhang, S.; Lin, J.; Zhu, B. Responses of soil microbial biomass carbon and dissolved organic carbon to dry-ing-rewetting cycles: A meta-analysis. *Catena* **2021**, *207*, 105610. [CrossRef]
30. Xu, N.; Tan, G.; Wang, H.; Gai, X. Effect of biochar additions to soil on nitrogen leaching, microbial biomass and bacterial community structure. *Eur. J. Soil Biol.* **2016**, *74*, 1–8. [CrossRef]
31. Dennis, K.L.; Wang, Y.; Blatner, N.R.; Wang, S.; Saadalla, A.; Trudeau, E.; Roers, A.; Weaver, C.T.; Lee, J.J.; Gilbert, J.A.; et al. Adenomatous polyps are driven by microbe-instigated focal inflammation and are controlled by IL-10-producing T cells. *Cancer Res.* **2013**, *73*, 5905–5913. [CrossRef] [PubMed]
32. Caporaso, J.G.; Kuczynski, J.; Stombaugh, J.; Bittinger, K.; Bushman, F.D.; Costello, E.K.; Fierer, N.; Peña, A.G.; Goodrich, J.K.; Gordon, J.I.; et al. QIIME allows analysis of high-throughput community sequencing data. *Nat. Methods* **2010**, *7*, 335–336. [CrossRef]
33. Bokulich, N.A.; Dillon, M.R.; Zhang, Y.; Rideout, J.R.; Bolyen, E.; Li, H.; Albert, P.S.; Caporaso, J.G. q2-longitudinal: Longitudinal and Paired-Sample Analyses of Microbiome Data. *mSystems* **2018**, *3*, e00219-18. [CrossRef] [PubMed]
34. Lozupone, C.; Knight, R. UniFrac: A new phylogenetic method for comparing microbial communities. *Appl. Environ. Microbiol.* **2005**, *71*, 8228–8235. [CrossRef]
35. Amato, S.M.; Orman, M.A.; Brynildsen, M.P. Metabolic control of persister formation in Escherichia coli. *Mol. Cell* **2013**, *50*, 475–487. [CrossRef]
36. Librán-Embid, F.; Klaus, F.; Tscharntke, T.; Grass, I. Unmanned aerial vehicles for biodiversity-friendly agricultural land-scapes-A systematic review. *Sci. Total Environ.* **2020**, *732*, 139204. [CrossRef] [PubMed]
37. Ito, T.; Araki, M.; Higashi, T.; Komatsuzaki, M.; Kaneko, N.; Ohta, H. Responses of soil nematode community structure to soil carbon changes due to different tillage and cover crop management practices over a nine-year period in Kanto, Japan. *Appl. Soil Ecol.* **2015**, *89*, 50–58. [CrossRef]
38. Hu, J.; Chen, G.; Hassan, W.M.; Lan, J.; Si, W.; Wang, W.; Li, G.; Du, G. The impact of fertilization intensity on soil nematode communities in a Tibetan Plateau grassland ecosystem. *Appl. Soil Ecol.* **2022**, *170*, 104258. [CrossRef]
39. Mahnert, A.; Moissl-Eichinger, C.; Berg, G. Microbiome interplay: Plants alter microbial abundance and diversity within the built environment. *Front. Microbiol.* **2015**, *6*, 887. [CrossRef]
40. Pickett, J.A. The essential need for GM crops. *Nat. Plants* **2016**, *2*, 16078. [CrossRef] [PubMed]
41. Schmid, M.W.; Hahl, T.; van Moorsel, S.J.; Wagg, C.; de Deyn, G.B.; Schmid, B. Feedbacks of plant identity and diversity on the diversity and community composition of rhizosphere microbiomes from a long-term biodiversity experiment. *Mol. Ecol.* **2019**, *28*, 863–878. [CrossRef]
42. Jat, H.S.; Datta, A.; Choudhary, M.; Sharma, P.C.; Yadav, A.K.; Choudhary, V.; Gathala, M.K.; Jat, M.L.; McDonald, A. Climate Smart Agriculture practices improve soil organic carbon pools, biological properties and crop productivity in cereal-based systems of North-West India. *Catena* **2019**, *181*, 104059. [CrossRef]
43. Zhang, H.; Hobbie, E.A.; Feng, P.; Niu, L.; Hu, K. Can conservation agriculture mitigate climate change and reduce environmental impacts for intensive cropping systems in North China Plain? *Sci. Total Environ.* **2022**, *806*, 151194. [CrossRef] [PubMed]
44. Singh, J.S.; Gupta, V.K. Soil microbial biomass: A key soil driver in management of ecosystem functioning. *Sci. Total Environ.* **2018**, *634*, 497–500. [CrossRef]
45. Chen, H.; Zhao, X.; Chen, X.; Lin, Q.; Li, G. Seasonal changes of soil microbial C, N, P and associated nutrient dynamics in a semiarid grassland of north China. *Appl. Soil Ecol.* **2018**, *128*, 89–97. [CrossRef]
46. Li, P.; Yang, Y.; Han, W.; Fang, J. Global patterns of soil microbial nitrogen and phosphorus stoichiometry in forest ecosystems. *Glob. Ecol. Biogeogr.* **2014**, *23*, 979–987. [CrossRef]
47. Xu, X.; Thornton, P.E.; Post, W.M. A global analysis of soil microbial biomass carbon, nitrogen and phosphorus in terrestrial ecosystems: Concepts and applications. *Glob. Ecol. Biogeogr.* **2013**, *22*, 737–749. [CrossRef]
48. Xu, Y.; Du, A.; Wang, Z.; Zhu, W.; Li, C.; Wu, L. Effects of different rotation periods of Eucalyptus plantations on soil physiochemical properties, enzyme activities, microbial biomass and microbial community structure and diversity. *Forest. Ecol. Manag.* **2020**, *456*, 117683. [CrossRef]
49. Gao, G.; Tuo, D.; Han, X.; Jiao, L.; Li, J.; Fu, B. Effects of land-use patterns on soil carbon and nitrogen variations along revegetated hillslopes in the Chinese Loess Plateau. *Sci. Total Environ.* **2020**, *746*, 141156. [CrossRef] [PubMed]
50. Lopes, L.D.; Fernandes, M.F. Changes in microbial community structure and physiological profile in a kaolinitic tropical soil under different conservation agricultural practices. *Appl. Soil Ecol.* **2020**, *152*, 103545. [CrossRef]
51. Liu, L.; Wu, L.; Knauth, S.; Eickhorst, T. Degradation of transgenic Bt-rice straw incorporated with two different paddy soils. *J. Environ. Manag.* **2019**, *244*, 415–421. [CrossRef] [PubMed]
52. Wu, H.; Zeng, G.; Liang, J.; Zhang, J.; Cai, Q.; Huang, L.; Li, X.; Zhu, H.; Hu, C.; Shen, S. Changes of soil microbial biomass and bacterial community structure in Dongting Lake: Impacts of 50,000 dams of Yangtze River. *Ecol. Eng.* **2013**, *54*, 72–78. [CrossRef]
53. Wang, Z.; Zhao, M.; Yan, Z.; Yang, Y.; Niklas, K.J.; Huang, H.; Mipam, T.D.; He, X.; Hu, H.; Wright, S.J. Global patterns and predictors of soil microbial biomass carbon, nitrogen, and phosphorus in terrestrial ecosystems. *Catena* **2022**, *211*, 106037. [CrossRef]
54. Hallama, M.; Pekrun, C.; Lambers, H.; Kandeler, E. *Hidden Miners: The Roles of Cover Crops and Soil Microorganisms in Phosphorus Cycling through Agroecosystems*; Universität Hohenheim: Hohenheim, Germany, 2019.

55. Stange, M.; Barrett, R.D.H.; Hendry, A.P. The importance of genomic variation for biodiversity, ecosystems and people. *Nat. Rev. Genet.* **2021**, *22*, 89–105. [CrossRef] [PubMed]
56. Breed, M.F.; Harrison, P.A.; Blyth, C.; Byrne, M.; Gaget, V.; Gellie, N.J.C.; Groom, S.V.C.; Hodgson, R.; Mills, J.G.; Prowse, T.A.A.; et al. The potential of genomics for restoring ecosystems and biodiversity. *Nat. Rev. Genet.* **2019**, *20*, 615–628. [CrossRef]
57. Wang, X.; Whalley, W.R.; Miller, A.J.; White, P.J.; Zhang, F.; Shen, J. Sustainable cropping requires adaptation to a hetero-geneous rhizosphere. *Trends. Plant Sci.* **2020**, *25*, 1194–1202. [CrossRef]
58. Buee, M.; de Boer, W.F.; Martin, F.; van Overbeek, L.S.; Jurkevitch, E. The rhizosphere zoo: An overview of plant-associated communities of microorganisms, including phages, bacteria, archaea, and fungi, and of some of their structuring factors. *Plant Soil* **2009**, *321*, 189–212. [CrossRef]
59. Ling, N.; Wang, T.; Kuzyakov, Y. Rhizosphere bacteriome structure and functions. *Nat. Commun.* **2022**, *13*, 836. [CrossRef]
60. Wu, Y.; Sun, J.; Yu, P.; Zhang, W.; Lin, Y.; Ma, D. The rhizosphere bacterial community contributes to the nutritional competitive advantage of weedy rice over cultivated rice in paddy soil. *BMC Microbiol.* **2022**, *22*, 232. [CrossRef]
61. Oliveira, A.P.; Pampulha, M.E.; Bennett, J.P. A two-year field study with transgenic Bacillus thuringiensis maize: Effects on soil microorganisms. *Sci. Total Environ.* **2008**, *405*, 351–357. [CrossRef]
62. Zhang, Y.; Xie, M.; Wu, G.; Peng, D.; Yu, W. A 3-year field investigation of impacts of Monsanto's transgenic Bt-cotton NC 33B on rhizosphere microbial communities in northern China. *Appl. Soil Ecol.* **2015**, *89*, 18–24. [CrossRef]
63. Zhou, D.; Xu, L.; Gao, S.; Guo, J.; Luo, J.; You, Q.; Que, Y. Cry1Ac transgenic sugarcane does not affect the diversity of microbial communities and has no significant effect on enzyme activities in rhizosphere soil within one crop season. *Front. Plant Sci.* **2016**, *7*, 265. [CrossRef]
64. Wu, W.X.; Liu, W.; Lu, H.H.; Chen, Y.X.; Medha, D.; Janice, T. Use of 13C labeling to assess carbon partitioning in transgenic and non-transgenic (parental) rice and their rhizosphere soil microbial communities. *FEMS Microbiol. Ecol.* **2009**, *67*, 93–102. [CrossRef] [PubMed]
65. Weinhold, A.; Karimi Dorcheh, E.; Li, R.; Rameshkumar, N.; Baldwin, I.T. Antimicrobial peptide expression in a wild tobacco plant reveals the limits of host-microbe-manipulations in the field. *eLife* **2018**, *7*, e28715. [CrossRef] [PubMed]

life

MDPI

Article

Catabolism of Glucosinolates into Nitriles Revealed by RNA Sequencing of *Arabidopsis thaliana* Seedlings after Non-Thermal Plasma-Seed Treatment

Alexandra Waskow [1,*,†], Anthony Guihur [2,†], Alan Howling [1] and Ivo Furno [1]

1 Swiss Plasma Center (SPC), École Polytechnique Fédérale de Lausanne (EPFL), CH-1015 Lausanne, Switzerland
2 Department of Plant Molecular Biology, Faculty of Biology and Medicine, University of Lausanne (UNIL), CH-1015 Lausanne, Switzerland
* Correspondence: alexandra.waskow@epfl.ch
† These authors contributed equally to this work.

Abstract: Non-thermal plasma-seed treatments could be an environmentally friendly method to modulate plant properties. Since it remains unclear how plasmas affect seeds, RNA sequencing was used here to analyze gene transcription changes in 7-day-old *Arabidopsis thaliana* (L.) Heynh. seedlings grown from surface dielectric barrier discharge plasma-treated seeds. In a previous study, seeds were analyzed 6 days after plasma exposure and a plant stress and defense response was observed. Here, we performed a pathway analysis on differentially expressed genes and our results revealed again an increased expression of plant stress and defense, specifically glucosinolate pathway-related compounds. The main difference was that a different part of the plant defense response changed at 7 days, which was not previously observed at 6 days. With a 24-h delayed extraction time point, the glucosinolates were selectively broken down into nitriles among all of the glucosinolates catabolic products. Although information about nitriles is limited, it protects plants against biotic stresses and has variable toxicity depending on the interacting organism. More work needs to be performed to better understand which plasma seed treatment parameters affect plant defense; however, these preliminary findings suggest that an optimized plasma treatment could be used to elicit a plant defense response.

Keywords: non-thermal plasma; plant defense; glucosinolates; nitriles; RNA sequencing; *Arabidopsis thaliana*

Citation: Waskow, A.; Guihur, A.; Howling, A.; Furno, I. Catabolism of Glucosinolates into Nitriles Revealed by RNA Sequencing of *Arabidopsis thaliana* Seedlings after Non-Thermal Plasma-Seed Treatment. *Life* **2022**, *12*, 1822. https://doi.org/10.3390/life12111822

Academic Editors: Hakim Manghwar and Wajid Zaman

Received: 7 October 2022
Accepted: 4 November 2022
Published: 8 November 2022

Publisher's Note: MDPI stays neutral with regard to jurisdictional claims in published maps and institutional affiliations.

Copyright: © 2022 by the authors. Licensee MDPI, Basel, Switzerland. This article is an open access article distributed under the terms and conditions of the Creative Commons Attribution (CC BY) license (https://creativecommons.org/licenses/by/4.0/).

1. Introduction

The demand in agriculture to minimize or replace current chemical practices has been increasing in recent years and now, biologicals, soil health, and traditional farming practices are gaining traction. Among these approaches, investigations into cold, non-thermal plasma applications on seeds and plants are increasing.

There are a variety of stressors which can elicit a plant defense response, such as heat, chemical, or mechanical stress, and now, the potential of plasma is being explored as a non-toxic, soft chemical treatment. Plasma could potentially avoid additional mechanical and heat damage and due to its multiple components and synergies, it can potentially trigger unique defense responses and outcomes. In theory, an optimized plasma treatment should not produce any toxic residues, and to date, multiple studies have shown that plasma treatments can support germination, growth, disease and stress resistance, delay senescence, and improve crop yield [1–7].

Plasma is produced when a gas is ionized; it is a combination of UV photons, electric field, electrons, ions, heat, and reactive oxygen and nitrogen species (RONS). Biological applications of non-thermal plasmas are possible because a high-temperature chemistry can be attained at a low gas temperature [1]. There are multiple plasma device configurations but the most common are, by far, the Dielectric barrier discharges (DBDs) at atmospheric pressure. The dielectric layer, unique to this configuration, is used as an insulation barrier to prevent sparking which can eventually lead to arcing at high voltages. Moreover, treatment time and duty cycle are a few examples of variables which can be adjusted to ignite a plasma at a sufficiently low gas temperature, a requirement for biological substrates sensitive to heat, such as seeds.

As cited previously, successful results have been obtained globally, yet there are no clear guidelines which outline the relevant plasma treatment parameters for plasma-induced plant effects. Moreover, due to the limited body of knowledge, these changes are currently unpredictable. Novel information concerning the mechanisms can be discovered by analyzing changes in gene expression, methylation patterns, or protein expression in plants grown from plasma-treated seeds using high throughput methods [8–11]. Among these methods, studies have mostly resorted to using a quantitative PCR (qPCR) to measure the expression of specific genes of interest, and more recently micro-arrays or RNA sequencing (RNA-seq), although the latter is currently limited [12–31].

Based on our previous studies [32,33], we observed accelerated germination by modifying the plasma treatment times and voltage values. Here, we used RNA sequencing to study the mechanisms behind this plasma-induced phenotype, accelerated germination. *Arabidopsis thaliana* (L.) Heynh. seeds were treated with a dry air plasma, which ignites at the edges of high voltage electrodes in a surface dielectric barrier discharge (SDBD). *A. thaliana* is a plant model organism with an entirely sequenced genome and therefore, it is more feasible to probe the underlying mechanisms and effects of plasma–plant interactions. We first decided to analyze whole seedlings to capture a global overview of the main processes and to avoid influencing the results with additional mechanical stress as a result of separating and isolating different tissue types. However, future experiments should explore tissue-specific changes by analyzing the roots and shoots separately. Furthermore, dry seeds with a low moisture content of 7.66% were used in this study since moisture can influence the plasma seed treatment results [34].

This study is a follow-up of our previous study. RNA sequencing was previously used to analyze 6-day-old seedlings from plasma-treated seeds using two different plasma treatment times of 60 s and 80 s at 7.75 kV $\pm$ 3% [32]. Here, RNA sequencing was used once again as a pioneering, preliminary study with similar plasma conditions using a 60 s treatment time at 7.75 kV $\pm$ 3% except with a different sampling time point: 7-day-old seedlings [33]. Although we compared our two transcriptome datasets, we are mindful of the minor, albeit relevant, differences. The plasma seed treatments across both studies were performed on the same day and grown in parallel. Furthermore, the same plasma treatment time of 60 s was used again. Minor differences mainly arose from the different sampling time points and voltage values (inherent to plasma treatments). Although 8 kV was measured in the previous time series study, we assumed that 7.5 kV would produce a similar plasma since the voltage was 7.75 kV $\pm$ 3%. These two voltages differ only by 6%, which is within the experimental error of the voltage supply to the DBD electrodes.

Specifically in this paper, 6-day-old and 7-day-old untreated, control seedlings were first analyzed and compared. This was followed by a comparison between the untreated, control seedlings at days 6 and 7 with the corresponding plasma-treated seedlings to determine whether the changes in gene expression were due to the plant age or plasma treatment. Finally, we concluded with the differentially expressed genes and pathways of 7-day-old plasma-treated seedlings relative to the untreated 7-day-old seedlings. Increased transcription of glucosinolate-related enzymes such as two genes encoding myrosinases (AT1G51470, AT1G47600) were observed again, suggesting the breakdown of glucosinolates. However, the most striking finding from this study was that the glucosinolates were specifically broken down into nitriles, as indicated by the increased expression of a nitrile specifier protein (AT3G16390), which seems to promote only the production of simple nitriles (Table S3). To our knowledge, this has not been previously reported in the literature for plasma agriculture.

2. Materials and Methods

2.1. Seed Material

Arabidopsis thaliana Col-0 seeds were cultivated at the Department of Plant Molecular Biology at the University of Lausanne. They were grown in a plant chamber room and harvested in May 2019. A thermogravimetric analysis (TGA 4000, Perkin Elmer, Waltham, MA, USA) was completed to verify the low moisture content of the seeds, which was 7.66%. These seeds were stored in Eppendorf or Falcon tubes and kept at room temperature in the dark until used for experiments [32,33]. At the time of the experiments performed in this study, the seeds were 18–20 months old.

2.2. Surface Dielectric Barrier Discharge Description

The materials and design of the SDBD device (Sihon Electronics) used alumina as a dielectric and a printed striped pattern for the high voltage electrodes. The SDBD was housed in a closed stainless steel reactor chamber, 18 cm diameter and 11 cm high. For the plasma-seed treatment, the seeds were situated underneath the SDBD device, and were resting on Teflon cylinders approximately 3.7 mm away from the plasma. The seeds were not overlapping. Each individual *Arabidopsis* seed had an area of approximately 0.1 mm^2 [35]. The seed-plasma treatment was static. Additional details can be found in previous studies [32,33].

2.3. RNA Isolation, Library Construction and RNA Sequencing

After the seeds were treated with plasma, they were sown on water agar plates (within hours). The 48-h time point after sowing was used to measure the germination rate. The samples were incubated for another 5 days. After 7 days from the time of sowing, the total RNA (up to 100 mg) was extracted from three biological replicates using a lysing kit with 1.4 mm zirconium beads in 0.5 mL tubes in a Precellys machine (Bertin, Montingy-le-Bretonneux, France). A custom program in Precellys was used entirely at 4 °C as follows: 30 s at 6000 rpm, 10 s at 0 rpm (break), and 30 s at 6000 rpm. To isolate the RNA, InnuPREP Plant RNA kit (Analytic Jena, Jena, Germany) was used and the extracted RNA was quantified using a nanodrop (DS-11 Microvolume Spectrophotometer).

The RNA quality was determined using a Fragment Analyzer (Agilent Technologies, Santa Clara, CA, USA) and all samples used in this study had an RNA quality number (RQN) above 8.3. The Lausanne Genomic Technologies Facility at the University of Lausanne prepared the library and the RNA sequencing (https://www.unil.ch/gtf, accessed on 15 June 2021). For the RNA-seq libraries, 400 ng of total RNA was used in combination with the Illumina TruSeq Stranded mRNA reagents (Illumina) using a unique dual indexing strategy, and following the official protocol automated on a Sciclone liquid handling robot (PerkinElmer, Waltham, MA, USA). A fluorimetric method (QubIT, Life Technologies, Carlsbad, CA, USA) was used to quantify the libraries and a Fragment Analyzer determined the quality (Agilent Technologies, Santa Clara, CA, USA).

Cluster generation was performed with 2 nM of an equimolar pool from the resulting libraries using the Illumina HiSeq 3000/4000 SR Cluster Kit reagents. It was then sequenced on the Illumina HiSeq 4000 SR platform (single end) using HiSeq 3000/4000 SBS Kit reagents for 150 cycles (single end). The sequencing data were demultiplexed using the bcl2fastq2 Conversion Software (version 2.20, Illumina, San Diego, CA, USA). This produced 31–37 million of 150 bp long single-end reads for each library independently, which were then sequenced (Table S1).

Quality control (phred score > 20) and adapter trimming with FastQC (0.11.976), and BBDuk were performed on the raw reads. Any matches to ribosomal RNA were eliminated with fastq_screen (v. 0.9.3). The *Arabidopsis* reference genome sequence (Araport11) and the default parameters in STAR v2.7.5 were used for read alignment. The count matrix was generated with FeatureCounts v1.6.2 in order to calculate gene expression values as raw read counts. This was used to obtain RPKM to make heatmaps with an in-house script.

The gene expression profiles are of 7-day-old seedlings (NCBI project number PRJNA800224). The seeds were treated with a 7.5 kV plasma for 60 s and grown until the 7th day in the same agar plate under continuous light to reduce biological variability. A pool of 30 seedlings represents one biological replicate. An average of ~47 million raw reads of 150 base pairs (bp) were produced. After filtering, ~46 million clean reads per library were retained (Figure S1, Table S1) and ~96% were mapped to the *A. thaliana* reference genome (Table S2).

DESeq2 package from R software v1.30.1 [36] after rlog transformation and Wald test with the *p*-value adjusted using the Benjamini and Hochberg method (FDR), were used to analyze count read values in order to identify the differentially expressed genes (DEGs) between untreated and treated samples. ShinyGO v.0.76 software was used to find GO categories of differentially expressed genes with *p*-value cut-off set at <0.05 [37]. The results were based on customized background genes from our RNA-seq, which yield more accurate results for enrichment analysis [38]. The transcriptome data are available in NCBI Bioproject Code: PRJNA800224 [32].

3. Results

3.1. Global RNA-seq Analysis of Young Seedlings Grown from Plasma-Treated Seeds

The normalized gene expression values were used in the Principal Component Analysis (PCA). There was clustering among the three replicates of the untreated 6-day-old seedlings and untreated 7-day-old seedlings (Figures 1 and 2). However, there was little clustering among the three replicates of the 7-day-old plasma-treated seedlings and the untreated 7-day-old seedlings. Nevertheless, there was clustering when using two replicates. Therefore, for further data analysis, one replicate of the control and plasma-treated sample was removed to reduce the variability (Figure 3, Figures 4 and S2–S4). In total, 75% of the variance was explained by the first two principal components (59% by PC1 and 26% by PC2). From the 32,833 genes across four samples, 21,168 genes passed the selected threshold; each biological replicate had more than two reads (see Methods for more details).

A false discovery rate (FDR) < 0.15 and a log2foldchange (FC) > 1 were used for the analysis. There were 27 upregulated genes, and 29 downregulated genes for a total of 56 differentially expressed genes (DEGs). It should be noted that only a few enriched genes were identified in this study and therefore, our statements are made tentatively. The main focus of this manuscript is to highlight the similarities and differences between 6-day-old and 7-day-old seedlings treated with a similar plasma. Both the 6-day-old and 7-day-old seedlings had transcriptional changes related to the glucosinolate metabolism, yet only the transcription of a nitrile specifier protein was detected in the 7-day-old seedlings.

3.2. Comparison of Gene Expression between 6-Day-Old and 7-Day-Old Untreated, Control Seedlings

To ensure that the observed changes in secondary metabolism were caused by the plasma treatment and not plant age, we cross-referenced our data and analyzed the gene expression profiles for 6-day-old untreated, control seedlings (data taken from our previous study) and 7-day-old untreated seedlings (data obtained during this study). PCA analysis and hierarchical heat map clustering revealed significant differences between 6- and 7-day-old seedlings (Figure 1A,B). However, based on pathway enrichment analysis, the main differences were linked to plant development (Figure 1C,D).

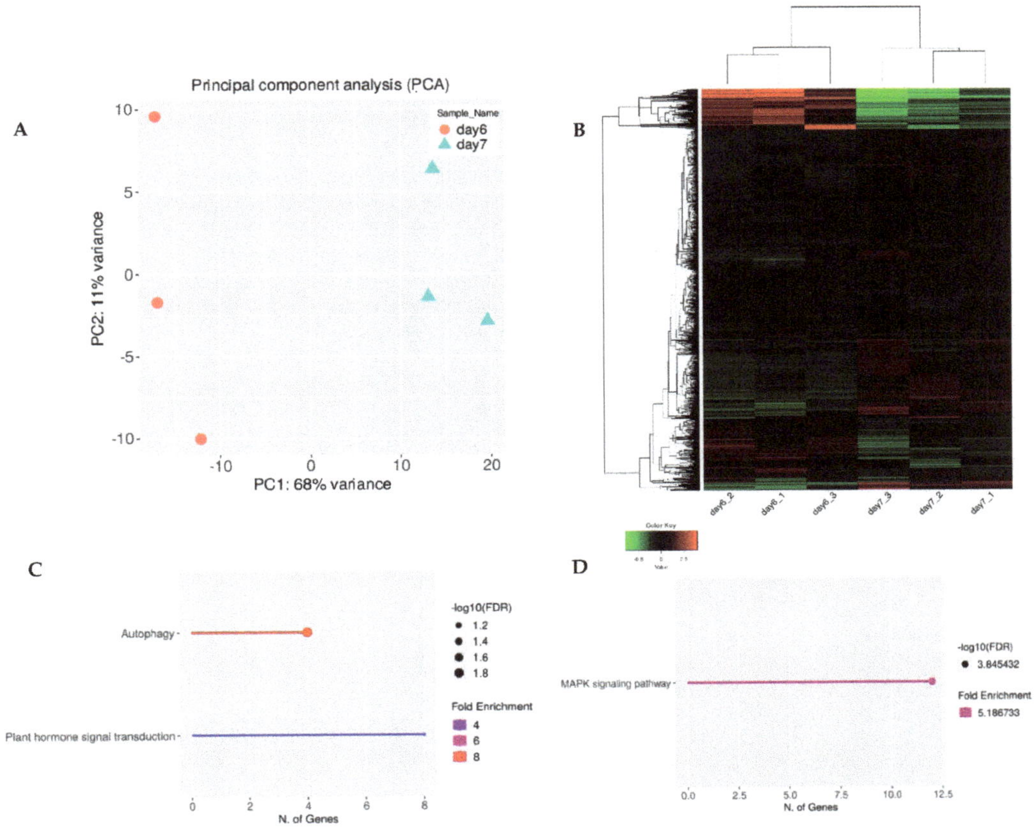

Figure 1. (**A**) Normalized gene expression values used in Principal Component Analysis (PCA) for 6-day-old and 7-day-old samples. The two components, PC1 and PC2, are shown on the X- and Y-axes with 68% and 11% variance, respectively. Orange circles represent triplicates of 6-day-old untreated *A. thaliana* seeds grown into seedlings and blue triangles represent the same as the orange circles except that they were 7-day-old seedlings. Each point in the plot represents a biological replicate, representing 30 seedlings, with a total of 6 biological replicates in the plot. (**B**) The full transcriptome for 6-day-old and 7-day-old untreated seedlings represented as a heat map (Z-scaled reads per kilobase of exon per million reads mapped (RPKM)). The relative expression profile of the top 2000 variable genes were selected based on the lowest standard deviation using Euclidean distance and are shown as hierarchical clustering. The columns represent individual samples, and the rows represent genes. The color scale represents the relative read count of genes: green indicates low relative read counts; red indicates high relative read counts; black indicates zero (no change). (**C**) Pathway enrichment analysis of upregulated genes using KEGG category. (**D**) Pathway enrichment analysis of downregulated genes using KEGG category. Significant differences between untreated 6-day-old and 7-day-old untreated, control seedlings were due to plant development.

3.3. Comparison of DEGs between 6-Day-Old and 7-Day-Old Untreated, Control Seedlings to DEGs in Plasma-Treated Seedlings

We then compared the DEGs of untreated, control seedlings and seedlings grown from plasma-treated seeds to check the similarities and differences in the gene expression profiles. To make it possible to produce a Venn diagram, the DEGs from 6-day-old and 7-day-old, untreated seedlings needed to be identified and then compared. By using the changing genes between the 6th and 7th days, a comparison could then be made with the DEGs of plasma-treated seedlings. More details about the DEGs in the 60 s and 80 s plasma-treated samples can be found in a previous study [32]. In Figure 2A,C, there is a minor overlap of significantly DEGs for both days 6 and 7. The untreated and 60 s plasma-treated seedlings showed 88 genes in common between the two conditions; however, there were 5869 DEGs and 181 DEGs in the untreated and plasma-treated samples, respectively. A similar pattern was observed with 80 s plasma-treated seedlings. There were 5838 DEGs and 303 DEGs for untreated and plasma-treated samples, respectively, of which 119 genes overlapped between the two conditions. In both instances, the overlapping genes were related to the primary metabolism (Figure 2B,D), which is involved in growth and development. These genes were found in pathways related to photosynthesis and oxidative phosphorylation, which are known to produce energy. This provided more confidence to ascribe the changes in secondary metabolism to the plasma treatment.

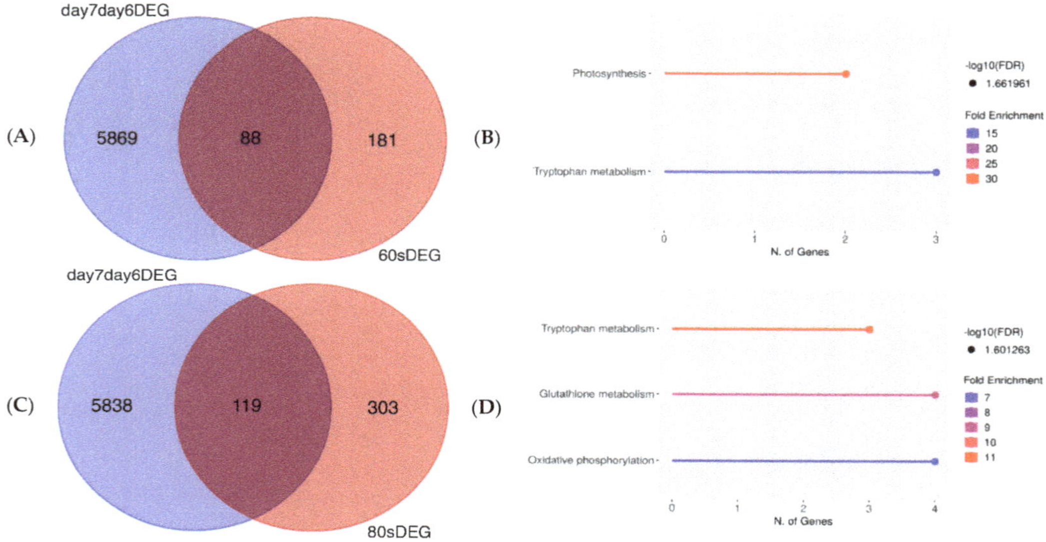

Figure 2. (**A**) Venn diagram showing number of DEGs which overlapped or differed between DEGs in 60 s plasma treatment triplicates from [32] (red) and DEGs shared between 6-day-old and 7-day-old untreated seedlings (blue). (**B**) Pathway enrichment analysis using KEGG for DEGs in (**A**). (**C**) Venn diagram showing number of DEGs which overlapped or differed between 80 s plasma treatment triplicates from [32] (red) compared to DEGs shared between 6-day-old and 7-day-old untreated seedlings (blue). (**D**) Pathway enrichment analysis using KEGG for genes in (**C**). Venn diagrams demonstrate very few genes in common between untreated and plasma-treated seeds grown into seedlings. Related genes are involved in primary metabolism and growth.

3.4. Gene Expression of Plasma-Treated Seeds Grown into 7-Day-Old Seedlings

Gene ontology (GO) analysis of specific DEG groups was used in ShinyGO v0.76 software for the pathway enrichment analysis on two replicates of the 7-day-old seedlings grown from plasma-treated seeds [37]. In Figure 3A–C, the upregulated genes after 60 s plasma treatment at 7.5 kV are organized into biological process, cellular component, and molecular function categories, respectively. The number of genes and fold enrichment in the pathway are shown in the lollipop diagrams, whereas the hierarchical tree clustering is shown in the supplemental section (Figure S3). The individual genes are listed in Table S3. Overall, gene expression increased in the secondary metabolic pathways, mainly for products from the glucosinolate metabolism. Specifically, gene expression in nitrile biosynthesis and metabolism was highly upregulated only in 7-day-old seedlings

(Figure 3A, Table S3). Within the cellular component category, components concerning the cell periphery were upregulated and equally had the highest number of upregulated genes (Figure 3B). The upregulated molecular functions were enzymatic reactions related to glucosinolates, glucohydrolase activity, and other enzymes involved in glucosinolate metabolism (Figure 3C).

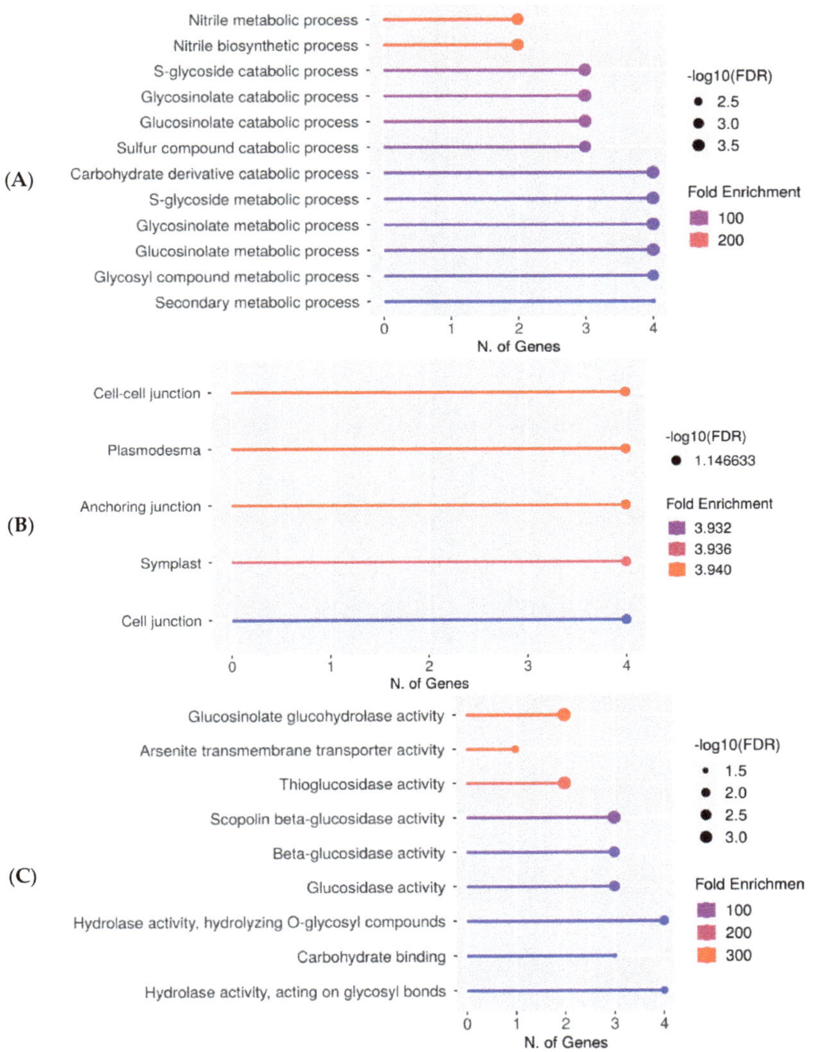

Figure 3. Upregulated gene enrichment analysis after 60 s plasma treatment at 7.5 kV. GO fold enrichment, significance (FDR in log10), and number of genes in each pathway are given in the lollipop diagrams in the following categories: (**A**) biological process, (**B**) cellular component, and (**C**) molecular function.

The lollipop diagrams for the downregulated genes based on two replicates are shown in Figure 4, whereas the hierarchical clustering trees can be found in the supplemental section (Figure S4). The individual genes are listed in Table S3. Overall, gene expression decreased across the diverse pathways related to response to stress or chemical stimulus.

Specifically, the cellular responses to iron ion starvation and reactive oxygen species were highly downregulated, whereas responses to the oxygen-containing compound had the highest number of enriched genes (Figure 4A). Within the cellular component category, lysosome and lytic vacuole were the most downregulated and the extracellular region had the highest number of downregulated genes (Figure 4B). The downregulated molecular functions were enzymatic reactions involved in nitrate transmembrane transporter activity or oxidative response (Figure 4C).

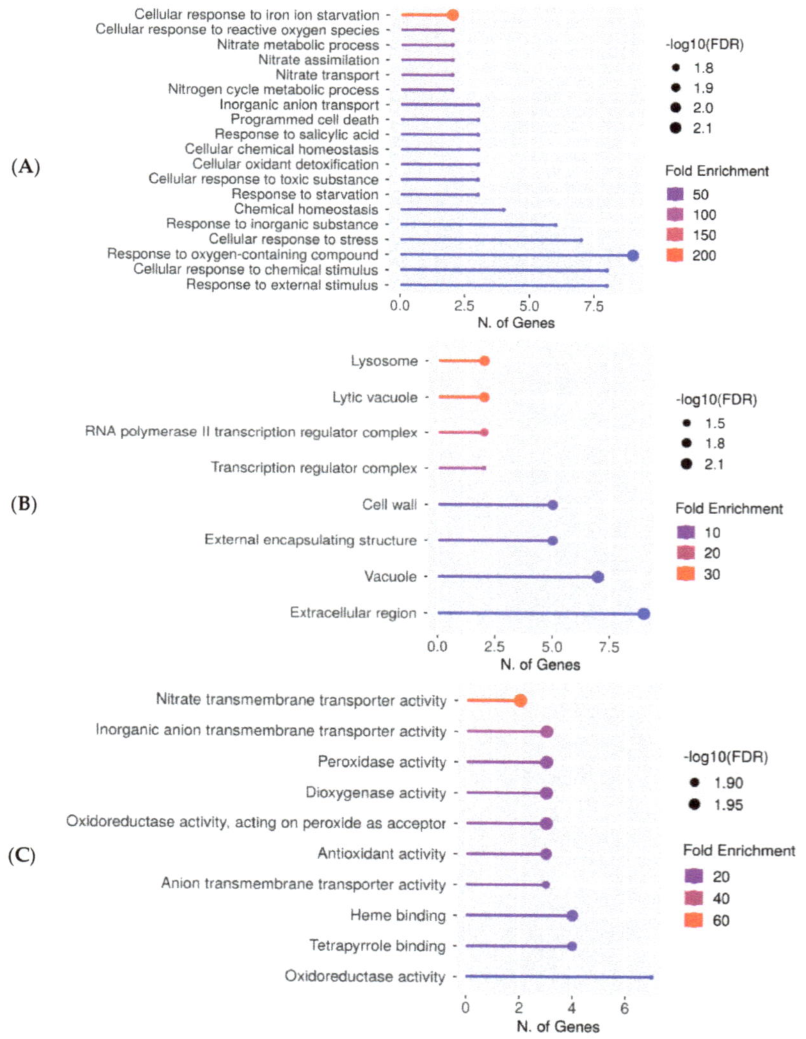

Figure 4. Downregulated gene enrichment analysis after 60 s plasma treatment at 7.5 kV. GO fold enrichment, significance (FDR in log10), and number of genes in each pathway are given in the lollipop diagrams in the following categories: (**A**) biological process, (**B**) cellular component, and (**C**) molecular function.

4. Discussion

4.1. Comparison between Our Studies

The aim of our study was to investigate how plasma-seed treatments affect the subsequent seed development on a molecular level. Approximately one-week-old seedlings were used after the root and shoot emergence for the following reasons: first, to ensure that transcriptional changes would be detected and second, because seedlings have increased sensitivity to stress. Limited treatment times and voltages were used to minimize additional stresses such as heat. Moreover, only the two formerly mentioned parameters among five (voltage, time, gas flow rate, plasma-seed gap distance, and frequency) resulted in accelerated germination, which increased our confidence that there would be detectable molecular changes [32,33]. Our previous findings are supported by a study performed by Šerá et al. [39], which showed how the pools of hormones change depending on a short or long plasma treatment time. However, this has not yet been shown using transcriptomics and therefore, this was completed using different plasma treatment times [32], and in this study, using different sampling time points.

The time series study [32] used two plasma treatment times of 60 and 80 s at 8 kV and the RNA was extracted 6 days after plasma-seed treatment. In this study, a single plasma treatment time of 60 s at 7.5 kV was used but the RNA was extracted 24 h later, 7 days after plasma-seed treatment. Previously, we demonstrated that a brief dry synthetic air plasma-seed treatment had a long-term memory effect since it modulated the primary and secondary metabolisms of 6-day-old seedlings. First, the transcription of genes belonging to the phenylpropanoid pathway were upregulated after a 60 s treatment. This pathway is responsible for lignin cell wall reinforcement and the production of antimicrobial compounds, such as phytoalexins. We tentatively interpreted this as a bacterial or fungal plant pathogen defense response. Second, the transcription of genes belonging to the glucosinolate pathway was upregulated after an 80 s treatment. We hypothesized and interpreted this response as a feeding deterrent and thus, as an insect and herbivore defense response. In both instances, it seems that plasma behaved as an oxidative stress and possibly as a wounding. Both the 6-day-old and 7-day-old seedlings had transcriptional changes related to glucosinolate metabolism, yet only the transcription of a nitrile specifier protein was detected only in the 7-day-old seedlings.

Concerning the dataset presented here, we rationalize that the lack of clear clusters between the triplicates of the untreated and plasma-treated samples could be due to the inherent seed variability, plasma-seed treatment variability, or the different sampling time point. For example, fewer DEGs were identified in this study, likely due to the response dampening over time. The latter is possible because this was demonstrated in another study with *Andrographis* where the earliest and latest time points had fewer DEGs [26]. Nevertheless, the results of the data analysis were coherent with previous observations and therefore, they were further analyzed and interpreted.

Prior to the DEG analysis of the 7-day-old seedlings, the gene profiles were compared between the untreated and plasma-treated samples to ensure that the gene expression changes were in fact due to the plasma treatment and not the plant physiology. Indeed, the genes which were common between the two sample types were involved in developmental processes (Figures 1 and 2), indicating that the DEGs were a result of plasma treatment.

Since these changes could be attributed to the plasma treatment, a comparison could then be made more confidently. When comparing the results from this study to our previous study, the gene expression trend in the first study showed a few upregulated genes and vastly more downregulated genes, whereas here, it was an equal ratio of up- and downregulated genes. The list of genes specifically induced after plasma treatment is shown in Table S3. It was initially expected that the upregulation of the phenylpropanoid pathway would be observed again when comparing 60 s at 8 kV to 60 s at 7.5 kV despite the plant age difference. However, the 60 s at 7.5 kV mimicked more closely the 80 s at 8 kV since there was an increased gene expression in glucosinolate-related production and enzymatic activity (Figure 3). Upon further thought, it would be reasonable to observe this since there was an additional 24 h prior to sampling or in other words, extraction. We hypothesize that the phenylpropanoid response shifted towards a glucosinolate response. Perhaps over time, the plant runs through a sequence of pathways, led by gene expression changes, and the same events could be observed with a less intense plasma, with a later sampling time point. In other words, each of these plasma-treatments might have sequentially undergone a phenylpropanoid biosynthesis response, followed by a glucosinolate biosynthesis response, and then nitrile biosynthesis. Depending on the plasma intensity and elapsed time, a different response might have been observed. It is entirely plausible that gene transcription could have changed within 24 h; it is the case that some heat shock proteins change within only 30 min [40].

Regarding the downregulated genes, there were subtle differences in the gene expression of hormones, where only auxin catabolism was observed in the previous study but here, salicylic acid (SA) was observed for the first time (Figure 4A). Auxin was reasonable to observe since aldoxime is a precursor to indole glucosinolates, camalexin, or auxin [41]. It is known that indole glucosinolates are blocked by high levels of auxin and it is likely the same inversely. However, SA is involved in systemic acquired resistance, which would be complementary to the upregulation of the secondary metabolism.

There are similarities between the two studies, which remain with the organelles, especially lysosomes being the most downregulated. The same rationale as before applies again here. The oxidized proteins as a result of plasma treatment could have been cleared before the extraction. Furthermore, oxidation played a role again in eliciting a response since many functions related to oxidation, detoxification, or chemical stress were observed (Figure 4C).

4.2. Data Support the Hypothesis about Wounding and Oxidative Stress as a Plant Response to Plasma

If the upregulated genes are analyzed closely, the data supports the proposed hypothesis where plasma could be interpreted as a wounding from an insect or a penetrating fungus or bacterium. We observed here an increase in pathway enrichment for cell wall biogenesis. We hypothesize that this was either related to growth or the plant was repairing damage and reinforcing the cell wall. It remains unknown which plasma components caused these transcriptional responses. To answer this question, it would require a detailed quantification of all the relevant RONS and their spatial distribution as a function of discharge parameters in the presence of the seeds. This would require investigations well beyond the scope of the present paper. However, preliminary studies in this direction have been undertaken, so we speculate that the cascade of transcriptional changes could be due to the diffusion of low concentration short-lived RONS, such as NO, which somehow does not affect the seed surface substantially. Based on our previous findings [27], we detected the presence of NO amongst other species with preliminary LIF studies. However, we observed no changes in the concentrations of carbon, oxygen, nitrogen, or other elements at the seed surface after plasma treatment using XPS (Figure S5). We assume that it may not be necessary for the plasma to interact with the entire seed surface to have changes on a molecular level. It has previously been shown that the seed surface facing the plasma was the only surface to experience any surface changes [1]. If the observed effects occur with only partial exposure of the seed to plasma, then the effects when the whole surface is exposed (for example, in a fluidized bed plasma reactor) can be assumed to be even stronger. In our study, seeds were checked before and after the treatment for any obvious changes in seed positioning and we can confirm that it was a static treatment with no seed movement. Since the seeds were not overlapping and an individual *Arabidopsis* seed is roughly 0.1 mm^2, probably half of each seed surface (0.05 mm^2) was in contact with plasma-derived components. Moreover, an indirect treatment using a 3.7 mm plasma-seed gap is unlikely to affect the seed with ions, electrons, or electric fields and thus, this leads us to believe it was RONS diffusion.

Nevertheless, there were transcriptional changes concerning the cell wall; specifically, extensin for cell wall protection (AT1G26240) and chitinase family protein (AT2G43610) were also upregulated, suggesting again cell wall reinforcement and protection against invasion. This could be because mechanical stimulus was detected, which triggers plant defense against wounding since the gene expression of mechanosensitive channel of small conductance-like 9 (AT5G19520) was upregulated. In case this is a response to the plasma-generated RONS, the plasma component could have travelled through an aquaporin, based on the DEGs in this study, which is known to be involved in hydrogen peroxide transport (AT4G19030). Alternatively, it may be due to a few UV photons since a gene involved in DNA repair and toleration (AT3G12610) was upregulated and this would be a typical response to UV.

From this mechanical stimulus, the plant might have responded with cell wall loosening using expansin, which has been mentioned before in other studies and one of these genes (AT5G02260) was upregulated in our dataset. It is interesting to see that the listed effects persisted for 7 days, even though they would be expected to occur shortly after the plasma-seed treatment and with the onset of germination (within the first 48 h in this case). It is often mentioned that abscisic acid decreases and gibberellic acid increases prior to germination and in our dataset, a gene (AT5G15230) which promotes gibberellic acid and exhibits redox activity was upregulated. Since the extraction took place 6 or 7 days after the plasma treatment, the response to the initial stimulus might have evolved into a glucosinolate response, which has been observed in both datasets, specifically with the upregulation of two genes encoding myrosinases (AT1G51470, AT1G47600). The novel aspect of this work was that these glucosinolates were further broken down into nitriles, as indicated by the increased expression of a nitrile specifier protein (AT3G16390), which seems to promote only the production of simple nitriles. There were no major transcriptional changes in the genes involved in thiocyanate formation.

4.3. Plasma Defense Activated with Increased Nitrile Synthesis

There are several breakdown products when glucosinolates are in contact with myrosinases, enzymes which are typically stored in different compartments. Once in contact, it results in a defense response with the production of thiocyanates, isocyanates, or nitriles (see [41], Figure 1). It was shown in another study where young mustard greens after plasma treatment had an increase or even doubling of isothiocynates [42,43], which is coherent with the activation of the glucosinolate pathway here.

These processes are regulated by MYB transcription factors, which were not strongly observed in our list, but there was a strong presence of nitrile-related genes. Although the breakdown of glucosinolates into nitriles is not novel [44], to our knowledge, it is the first time that this has been observed after plasma treatment. We cannot be certain that all glucosinolates are broken down into only nitriles without additional metabolic studies, so we propose these dynamic changes based on the obtained RNA-seq data. Considering that there is only one other study plasma-treating *Arabidopsis*, it will become clearer in time whether this speculation is true [27]. It is not yet clear why nitriles are favored over other forms. Ultimately, nitriles can be broken down further into cyanogenic glucosides, so it appears that a very potent response might be elicited from the plant after plasma treatment. In some instances, nitriles are less toxic than isothiocyanates [45]. However, certain organisms have evolved to consume breakdown products through coevolution [46]. In other instances, it can be more toxic to some herbivores or particularly to insects, so it seems to depend on what is attacking the plant [46]. The plant needs to carefully manage its resources, especially for defense responses. It is generally costly for the plant to defend itself, especially at the expense of growth. Therefore, the plant needs to choose when it would be best to mount a response and to which extent. There is a diverse combination of glucosinolate plant responses which are dependent on the concentration and type of glucosinolate, as well as the type of glucosinolate hydrolysis or breakdown products. This intricacy is a reflection of the complex interplay between microbial pathogens, insects, herbivores, and plants. If only glucosinolate biosynthesis is taken into consideration, there are 120 different glucosinolates to date, which is an impressive number to achieve such diversity from just a few amino acids [47]. The variation in glucosinolates is further amplified after glucosinolate breakdown; products can be broken down into a single glucosinolate or into products with diverse physicochemical and biological behavior. Therefore, it is difficult to reach a conclusion about the nitrile synthesis and plasmas without understanding the fundamental biology [48].

5. Conclusions

Overall, our findings here support the previously proposed hypothesis: the plant interprets plasma as an oxidative stress as well as a wounding, seemingly tuned to the plasma intensity [32]. Yet this time, another dimension was added by varying a biological parameter (sampling time point) instead of a physical parameter (plasma treatment time). Specifically, what was observed here was that the breakdown of glucosinolates led to the production of nitriles and not isothiocyanates. Although plasma can elicit a plant defense response and assuming that the biosynthesis of particular compounds can be increased and beneficial, it is important to understand at what cost and under which context it would remain so. For example, the biosynthesis of sweeteners increased at the expense of other secondary metabolites [49]. In our study, nitriles were not as poisonous as other glucosinolate breakdown products to certain biota so it may not be problematic but it might be better tolerated than other forms. Therefore, it is very difficult to foresee what effect this would have on plant–biota interactions without multiple bioassays reflecting a more natural environment. Caution should be exercised to not make the plant more susceptible to attacks.

So far, it seems that different pathways can be activated when using different combinations of plants and plasma treatments. For example, plants belonging to the *Brassica* family, such as our studies with *A. thaliana*, can activate the glucosinolate pathway after plasma treatment. However, this is absent in other plant families so other plants such as basil and pea might activate the phenylpropanoid pathway to increase essential oil production or increase lignification, simply due to the plant characteristics rather than the plasma. It thus would be valuable to understand how these changes occur between different plants. In the event that similar pathways are activated, there might be a preference of one pathway over the other. Therefore, one of the next aims should be to understand under which conditions a pathway is activated meanwhile following the genetic changes in parallel [50,51].

Furthermore, it would be important to identify the limit of plasma treatment before it becomes deleterious or activates apoptosis, programmed cell death. In certain contexts, the biosynthesis of particular compounds could be desired and therefore, it might not matter that the plants would die a few days later, given that they are harvested beforehand. However, this would be critical to understand if the plants are grown over the long-term. Therefore, experiments monitoring changes over time after sowing using multiple time points would expedite our understanding. As an example, an extended version of a study performed by the authors of [52] where the authors monitored and observed increased heat shock proteins in the first two days after sowing in plasma-treated corn should certainly be considered when designing experiments.

Lastly, to echo what was stated previously, a multi-omics approach will significantly advance the pace of this research field. Overall, plasma duration was studied before [32], and now, sampling/extraction time has been studied, which are both among the first, pioneering transcriptomics studies performed using *Arabidopsis* and plasma and hopefully, more will be completed in the future. Since there are presently only two *Arabidopsis*-plasma studies [27,33], this study was used to first detect any major changes in gene expression to be able to improve on the experimental design for future studies. Now that we have observed changes primarily in the glucosinolate biosynthesis and breakdown, future experiments will include mass spectrometry metabolomics where the glucosinolate and their breakdown products are extracted and quantified to directly validate these observations at a transcriptomic level. Furthermore, the concentrations of these defense metabolites should be studied indirectly using bioassays with predators, such as caterpillars or fungal pathogens, to determine if these plasma-induced effects do, in fact, protect the plant. In order to improve our general understanding, this will require the investigation of multiple layers, such as genomics, epigenomics, proteomics, and metabolomics. Since the data interpretation from one of these layers cannot be easily transposed to another, the implications remain unpredictable, i.e., gene expression does not always correlate with protein expression. Moreover, genes and proteins are subtly regulated through methylation or phosphorylation, respectively, and therefore, analyzing each and every layer can reveal information about the subtle fine-tuning of plasma treatments. Specifically, it would be interesting to look at the epigenomics and the modification of wrapped DNA since changes in methylation are presently observed but are not yet correlated with the phenotypic changes induced by plasma treatment [53].

In summary, it appears that regardless of minor changes in the plasma-seed experiment, a similar sequence of events might be observed: phenylpropanoid response is triggered first, followed by glucosinolate biosynthesis, and then catabolism into nitriles. Future studies will include more variables in the parameter space, such as multiple sampling time points and variables in the plasma treatment. This will help to determine whether the same response is observed regardless of minor plasma treatment changes or whether minor changes in plasma can trigger exclusively different responses. Nevertheless, our pioneering study brings new facts and clues in the field of a non-thermal plasma treatment with possible agronomic interests, which could serve as the foundation for future studies to build upon.

Supplementary Materials: The following supporting information can be downloaded at: https://www.mdpi.com/article/10.3390/life12111822/s1. Figure S1: Quality control for 7-day-old, untreated, control seedlings and 60 s, 7.5 kV plasma-treated seeds grown into seedlings (A) Genes retained in each sample (B) normalization of samples (C) density plot to demonstrate that profiles are similar to proceed with analysis.; Figure S2: (Left) Principal Component Analysis (PCA) conducted on the normalized gene expression values of the 7-day-old samples at 7.5 kV. X- and Y-axes show PC1 and PC2, respectively, with the amount of variance contained in each component, which 59% and 26%. Each point in the plot represents a biological replicate, representing 30 seedlings, with a total of 4 biological replicates in the plot. Symbols of the same colors are replicates of the same experimental group where orange represents the control which are untreated A.thaliana seeds grown into seedlings and blue represents 7.5 kV plasma-treated A.thaliana seeds grown into seedlings. (Right) Heat map of the expression patterns (Z-scaled reads per kilobase of exon per million reads mapped (RPKM)) of the full transcriptome for 7-day-old samples at 7.5 kV. Hierarchical clustering of the relative expression profile of the top 2000 variable genes selected based on the lowest standard deviation using Euclidean distance. Individual samples are shown in columns, and genes in rows. The left vertical axis shows clusters of genes. The color scale represents the relative read count of genes: green indicates low relative read counts; red indicates high relative read counts; black indicates zero (no change); Figure S3: A hierarchical clustering tree for the upregulated genes after 7.5 kV plasma treatment in 7-day-old seedlings, summarizing the correlation among significant pathways within GO categories (A) biological process (B) cellular component (C) molecular function. Pathways with many shared genes are clustered together. The blue, bigger dots followed by the FDR values indicate more significant P-values.; Figure S4: A hierarchical clustering tree for the downregulated genes after 7.5 kV plasma treatment in 7-day-old seedlings, summarizing the correlation among significant pathways within GO categories (A) biological process (B) cellular component, and (C) molecular function. Pathways with many shared genes are clustered together. The blue, bigger dots followed by the FDR values indicate more significant P-values.; Figure S5: X-ray photoelectron spectroscopy (XPS) analysis with atomic concentration table of carbon, nitrogen, oxygen, silicon, and potassium in untreated, and plasma-treated Arabidopsis seeds. From top to bottom, plasma treatment used different times, voltages, and plasma-seed gap distances.; Table S1: Number of NGS-RNA-seq reads before and after quality check on the raw sequencing data for 7-day-old seedlings treated with 7.5 kV plasma.; Table S2: Number of clean reads mapped against A. thaliana genome for 7-day-old seedlings treated with 7.5 kV plasma.; Table S3: List of DEGs using a 60 s, 7.5 kV plasma treatment with (left) upregulated genes and (right) downregulated genes with their corresponding fold change.

Author Contributions: Conceptualization, A.W.; methodology, A.W. and A.G.; software, A.G.; formal analysis, A.W. and A.G.; investigation, A.W. and A.G.; data curation, A.W. and A.G.; writing—original draft preparation, A.W. and A.G.; writing—review and editing, A.W., A.G., A.H., I.F.; visualization, A.W. and A.G.; funding acquisition, I.F. All authors have read and agreed to the published version of the manuscript.

Funding: This work was supported by an ad-hoc grant of the Swiss Federal budgetary framework 2017–2020 for the ETH Domain, and by the University of Lausanne and the Swiss National Fund grant n°CRSK-3_196689 to AG.

Institutional Review Board Statement: Not applicable.

Informed Consent Statement: Not applicable.

Data Availability Statement: The data presented in this study are available in this article, supplementary material and the transcriptome data are available in NCBI Bioproject Code: PRJNA800224.

Acknowledgments: The authors would like to thank the Lausanne Genomic Technologies Facility for performing the RNA sequencing.

Conflicts of Interest: The authors declare no conflict of interest. The funders had no role in the design of the study; in the collection, analyses, or interpretation of data; in the writing of the manuscript, or in the decision to publish the results.

References

1. Waskow, A.; Howling, A.; Furno, I. Mechanisms of Plasma-Seed Treatments as a Potential Seed Processing Technology. *Front. Phys.* **2021**, *9*, 174. [CrossRef]
2. FAO. News Article: Pollutants from Agriculture a Serious Threat to World's Water. Available online: https://www.fao.org/news/story/en/item/1141534/icode/ (accessed on 8 August 2022).
3. Land Is a Critical Resource, IPCC Report Says—IPCC. Available online: https://www.ipcc.ch/2019/08/08/land-is-a-critical-resource_srccl/ (accessed on 8 August 2022).
4. Guragain, R.P.; Baniya, H.B.; Dhungana, S.; Chhetri, G.K.; Sedhai, B.; Basnet, N.; Shakya, A.; Pandey, B.P.; Pradhan, S.P.; Joshi, U.M.; et al. Effect of plasma treatment on the seed germination and seedling growth of radish (*Raphanus sativus*). *Plasma Sci. Technol.* **2022**, *24*, 015502. [CrossRef]
5. Li, K.; Zhong, C.; Shi, Q.; Bi, H.; Gong, B. Cold plasma seed treatment improves chilling resistance of tomato plants through hydrogen peroxide and abscisic acid signaling pathway. *Free Radic. Biol. Med.* **2021**, *172*, 286–297. [CrossRef] [PubMed]
6. Renáta, Š.; Nicolette, V.; Monika, B.; Stanislav, K.; Eliška, G.; Veronika, M.; L'udmila, S. Enhanced In situ Activity of Peroxidases and Lignification of Root Tissues after Exposure to Non-Thermal Plasma Increases the Resistance of Pea Seedlings. *Plasma Chem. Plasma Process.* **2021**, *41*, 903–922. [CrossRef]
7. Ghasemzadeh, N.; Iranbakhsh, A.; Oraghi-Ardebili, Z.; Saadatmand, S.; Jahanbakhsh-Godehkahriz, S. Cold plasma can alleviate cadmium stress by optimizing growth and yield of wheat (*Triticum aestivum* L.) through changes in physio-biochemical properties and fatty acid profile. *Environ. Sci. Pollut. Res.* **2022**, *29*, 35897–35907. [CrossRef] [PubMed]
8. Suriyasak, C.; Hatanaka, K.; Tanaka, H.; Okumura, T.; Yamashita, D.; Attri, P.; Koga, K.; Shiratani, M.; Hamaoka, N.; Ishibashi, Y. Alterations of DNA Methylation Caused by Cold Plasma Treatment Restore Delayed Germination of Heat-Stressed Rice (*Oryza sativa* L.) Seeds. *ACS Agric. Sci. Technol.* **2021**, *1*, 5–10. [CrossRef]
9. Tamošiūnė, I.; Gelvonauskienė, D.; Haimi, P.; Mildažienė, V.; Koga, K.; Shiratani, M.; Baniulis, D. Cold Plasma Treatment of Sunflower Seeds Modulates Plant-Associated Microbiome and Stimulates Root and Lateral Organ Growth. *Front. Plant Sci.* **2020**, *11*, 1347. [CrossRef]
10. Tamošiūnė, I.; Gelvonauskienė, D.; Ragauskaitė, L.; Koga, K.; Shiratani, M.; Baniulis, D. Cold plasma treatment of *Arabidopsis thaliana* (L.) seeds modulates plant-associated microbiome composition. *Appl. Phys. Express* **2020**, *13*, 076001. [CrossRef]
11. Mildaziene, V.; Aleknavičiūtė, V.; Žūkienė, R.; Paužaitė, G.; Naučienė, Z.; Filatova, I.; Lyushkevich, V.; Haimi, P.; Tamošiūnė, I.; Baniulis, D. Treatment of Common Sunflower (*Helianthus annus* L.) Seeds with Radio-frequency Electromagnetic Field and Cold Plasma Induces Changes in Seed Phytohormone Balance, Seedling Development and Leaf Protein Expression. *Sci. Rep.* **2019**, *9*, 6437. [CrossRef]
12. Ji, S.H.; Ki, S.H.; Kang, M.H.; Choi, J.S.; Park, Y.; Oh, J.; Kim, S.B.; Yoo, S.J.; Choi, E.H.; Park, G. Characterization of physical and biochemical changes in plasma treated spinach seed during germination. *J. Phys. D Appl. Phys.* **2018**, *51*, 145205. [CrossRef]
13. Guo, Q.; Wang, Y.; Zhang, H.; Qu, G.; Wang, T.; Sun, Q.; Liang, D. Alleviation of adverse effects of drought stress on wheat seed germination using atmospheric dielectric barrier discharge plasma treatment. *Sci. Rep.* **2017**, *7*, 16680. [CrossRef]

14. Rahman, M.; Sajib, S.A.; Rahi, S.; Tahura, S.; Roy, N.C.; Parvez, S.; Reza, A.; Talukder, M.R.; Kabir, A.H. Mechanisms and Signaling Associated with LPDBD Plasma Mediated Growth Improvement in Wheat. *Sci. Rep.* **2018**, *8*, 10498. [CrossRef] [PubMed]
15. Adhikari, B.; Adhikari, M.; Ghimire, B.; Park, G.; Choi, E.H. Cold Atmospheric Plasma-Activated Water Irrigation Induces Defense Hormone and Gene expression in Tomato seedlings. *Sci. Rep.* **2019**, *9*, 16080. [CrossRef] [PubMed]
16. Adhikari, B.; Adhikari, M.; Ghimire, B.; Adhikari, B.C.; Park, G.; Choi, E.H. Cold plasma seed priming modulates growth, redox homeostasis and stress response by inducing reactive species in tomato (*Solanum lycopersicum*). *Free Radic. Biol. Med.* **2020**, *156*, 57–69. [CrossRef] [PubMed]
17. Li, K.; Zhang, L.; Shao, C.; Zhong, C.; Cao, B.; Shi, Q.; Gong, B. Utilising cold plasma seed treatment technologies to delay cotyledon senescence in tomato seedlings. *Sci. Hortic.* **2021**, *281*, 109911. [CrossRef]
18. Ghasempour, M.; Iranbakhsh, A.; Ebadi, M.; Ardebili, Z.O. Seed priming with cold plasma improved seedling performance, secondary metabolism, and expression of deacetylvindoline O-acetyltransferase gene in *Catharanthus roseus*. *Contrib. Plasma Phys.* **2020**, *60*, e201900159. [CrossRef]
19. Islam, S.; Omar, F.B.; Sajib, S.A.; Roy, N.C.; Reza, A.; Hasan, M.; Talukder, M.R.; Kabir, A.H. Effects of LPDBD Plasma and Plasma Activated Water on Germination and Growth in Rapeseed (*Brassica napus*). *Gesunde Pflanz.* **2019**, *71*, 175–185. [CrossRef]
20. Iranbakhsh, A.; Ardebili, Z.O.; Molaei, H.; Ardebili, N.O.; Amini, M. Cold Plasma Up-Regulated Expressions of WRKY1 Transcription Factor and Genes Involved in Biosynthesis of Cannabinoids in Hemp (*Cannabis sativa* L.). *Plasma Chem. Plasma Process.* **2020**, *40*, 527–537. [CrossRef]
21. Ghaemi, M.; Majd, A.; Iranbakhsh, A. Transcriptional responses following seed priming with cold plasma and electromagnetic field in *Salvia nemorosa* L. *J. Theor. Appl. Phys.* **2020**, *14*, 323–328. [CrossRef]
22. Ebrahimibasabi, E.; Ebrahimi, A.; Momeni, M.; Amerian, M.R. Elevated expression of diosgenin-related genes and stimulation of the defense system in *Trigonella foenum-graecum* (Fenugreek) by cold plasma treatment. *Sci. Hortic.* **2020**, *271*, 109494. [CrossRef]
23. Sajib, S.A.; Billah, M.; Mahmud, S.; Miah, M.; Hossain, F.; Omar, F.B.; Roy, N.C.; Hoque, K.M.F.; Talukder, M.R.; Kabir, A.H.; et al. Plasma activated water: The next generation eco-friendly stimulant for enhancing plant seed germination, vigor and increased enzyme activity, a study on black gram (*Vigna mungo* L.). *Plasma Chem. Plasma Process.* **2019**, *40*, 119–143. [CrossRef]
24. Perez, S.M.; Biondi, E.; Laurita, R.; Proto, M.; Sarti, F.; Gherardi, M.; Bertaccini, A.; Colombo, V. Plasma activated water as resistance inducer against bacterial leaf spot of tomato. *PLoS ONE* **2019**, *14*, e0217788. [CrossRef]
25. Ka, D.H.; Priatama, R.A.; Park, J.Y.; Park, S.J.; Kim, S.B.; Lee, I.A.; Lee, Y.K. Plasma-Activated Water Modulates Root Hair Cell Density via Root Developmental Genes in *Arabidopsis thaliana* L. *Appl. Sci.* **2021**, *11*, 2240. [CrossRef]
26. Tong, J.; He, R.; Tang, X.; Li, M.; Wan, J. Transcriptomic analysis of seed germination improvement of Andrographis paniculata responding to air plasma treatment. *PLoS ONE* **2020**, *15*, e0240939. [CrossRef] [PubMed]
27. Cui, D.; Yin, Y.; Li, H.; Hu, X.; Zhuang, J.; Ma, R.; Jiao, Z. Comparative transcriptome analysis of atmospheric pressure cold plasma enhanced early seedling growth in *Arabidopsis thaliana*. *Plasma Sci. Technol.* **2021**, *23*, 085502. [CrossRef]
28. Han, B.; Yu, N.-N.; Zheng, W.; Zhang, L.-N.; Liu, Y.; Yu, J.-B.; Zhang, Y.-Q.; Park, G.; Sun, H.-N.; Kwon, T. Effect of non-thermal plasma (NTP) on common sunflower (*Helianthus annus* L.) seed growth via upregulation of antioxidant activity and energy metabolism-related gene expression. *Plant Growth Regul.* **2021**, *95*, 271–281. [CrossRef]
29. Wang, P.; Chen, F.; Huang, Y.; Sun, H.; Fan, M.; Ning, D.; Wang, T.; Wang, H.; Liu, M. New insights of low-temperature plasma effects on seeds germination of *Platycodon grandiflorum*. *Isr. J. Plant Sci.* **2021**, *68*, 306–318. [CrossRef]
30. Hasan, M.; Sohan, S.R.; Sajib, S.A.; Hossain, F.; Miah, M.; Maruf, M.H.; Khalid-Bin-Ferdaus, K.; Kabir, A.H.; Talukder, M.R.; Rashid, M.; et al. The Effect of Low-Pressure Dielectric Barrier Discharge (LPDBD) Plasma in Boosting Germination, Growth, and Nutritional Properties in Wheat. *Plasma Chem. Plasma Process.* **2021**, *42*, 339–362. [CrossRef]
31. Hossain, F.; Sohan, S.R.; Hasan, M.; Miah, M.; Sajib, S.A.; Karmakar, S.; Khalid-Bin-Ferdaus, K.M.; Kabir, A.H.; Rashid, M.; Talukder, M.R.; et al. Enhancement of Seed Germination Rate and Growth of Maize (*Zea mays* L.) Through LPDBD Ar/Air Plasma. *J. Soil Sci. Plant Nutr.* **2022**, *22*, 1778–1791. [CrossRef]
32. Waskow, A.; Guihur, A.; Howling, A.; Furno, I. RNA Sequencing of *Arabidopsis thaliana* Seedlings after Non-Thermal Plasma-Seed Treatment Reveals Upregulation in Plant Stress and Defense Pathways. *Int. J. Mol. Sci.* **2022**, *23*, 3070. [CrossRef]
33. Waskow, A.; Ibba, L.; Leftley, M.; Howling, A.; Ambrico, P.F.; Furno, I. An In Situ FTIR Study of DBD Plasma Parameters for Accelerated Germination of *Arabidopsis thaliana* Seeds. *Int. J. Mol. Sci.* **2021**, *22*, 11540. [CrossRef] [PubMed]
34. Waskow, A.; Avino, F.; Howling, A.; Furno, I. Entering the plasma agriculture field: An attempt to standardize protocols for plasma treatment of seeds. *Plasma Process. Polym.* **2021**, *19*, e2100152. [CrossRef]
35. Herridge, R.P.; Day, R.C.; Baldwin, S.; Macknight, R.C. Rapid analysis of seed size in *Arabidopsis* for mutant and QTL discovery. *Plant Methods* **2011**, *7*, 3. [CrossRef]
36. Love, M.I.; Huber, W.; Anders, S. Moderated estimation of fold change and dispersion for RNA-seq data with DESeq2. *Genome Biol.* **2014**, *15*, 550. [CrossRef] [PubMed]
37. Ge, S.X.; Jung, D.; Yao, R. ShinyGO: A graphical gene-set enrichment tool for animals and plants. *Bioinformatics* **2020**, *36*, 2628–2629. [CrossRef] [PubMed]
38. Wijesooriya, K.; Jadaan, S.A.; Perera, K.L.; Kaur, T.; Ziemann, M. Guidelines for reliable and reproducible functional enrichment analysis. *bioRxiv.* **2021**. [CrossRef]

39. Šerá, B.; Vanková, R.; Roháček, K.; Šerý, M. Gliding Arc Plasma Treatment of Maize (*Zea mays* L.) Grains Promotes Seed Germination and Early Growth, Affecting Hormone Pools, but Not Significantly Photosynthetic Parameters. *Agronomy* **2021**, *11*, 2066. [CrossRef]
40. Guihur, A.; Rebeaud, M.E.; Goloubinoff, P. How do plants feel the heat and survive? *Trends Biochem. Sci.* **2022**, *47*, 824–838. [CrossRef]
41. Zhao, Y.; Wang, J.; Liu, Y.; Miao, H.; Cai, C.; Shao, Z.; Guo, R.; Sun, B.; Jiansheng, W.; Zhang, L.; et al. Classic myrosinase-dependent degradation of indole glucosinolate attenuates fumonisin B1-induced programmed cell death in *Arabidopsis*. *Plant J.* **2015**, *81*, 920–933. [CrossRef]
42. Saengha, W.; Karirat, T.; Buranrat, B.; Matra, K.; Deeseenthum, S.; Katisart, T.; Luang-In, V. Cold Plasma Treatment on Mustard Green Seeds and its Effect on Growth, Isothiocyanates, Antioxidant Activity and Anticancer Activity of Microgreens. *Int. J. Agric. Biol.* **2021**, *25*, 667–676. [CrossRef]
43. Luang-In, V.; Saengha, W.; Karirat, T.; Buranrat, B.; Matra, K.; Deeseenthum, S.; Katisart, T. Effect of cold plasma and elicitors on bioactive contents, antioxidant activity and cytotoxicity of Thai rat-tailed radish microgreens. *J. Sci. Food Agric.* **2020**, *101*, 1685–1698. [CrossRef] [PubMed]
44. Wittstock, U.; Meier, K.; Dörr, F.; Ravindran, B.M. NSP-Dependent Simple Nitrile Formation Dominates upon Breakdown of Major Aliphatic Glucosinolates in Roots, Seeds, and Seedlings of *Arabidopsis thaliana* Columbia-0. *Front. Plant Sci.* **2016**, *7*, 1821. [CrossRef] [PubMed]
45. Ting, H.-M.; Cheah, B.H.; Chen, Y.-C.; Yeh, P.-M.; Cheng, C.-P.; Yeo, F.K.S.; Vie, A.K.; Rohloff, J.; Winge, P.; Bones, A.M.; et al. The Role of a Glucosinolate-Derived Nitrile in Plant Immune Responses. *Front. Plant Sci.* **2020**, *11*, 257. [CrossRef] [PubMed]
46. Eckardt, N.A. Some Like It with Nitriles: A nitrile-specifying protein linked to herbivore feeding behavior in *Arabidopsis*. *Plant Cell* **2001**, *13*, 2565–2568. [CrossRef]
47. Halkier, B.A.; Gershenzon, J. Biology and biochemistry of glucosinolates. *Annu. Rev. Plant Biol.* **2006**, *57*, 303–333. [CrossRef]
48. Burow, M.; Losansky, A.; Muller, R.; Plock, A.; Kliebenstein, D.; Wittstock, U. The Genetic Basis of Constitutive and Herbivore-Induced ESP-Independent Nitrile Formation in *Arabidopsis*. *Plant Physiol.* **2008**, *149*, 561–574. [CrossRef]
49. Judickaitė, A.; Lyushkevich, V.; Filatova, I.; Mildažienė, V.; Žūkienė, R. The Potential of Cold Plasma and Electromagnetic Field as Stimulators of Natural Sweeteners Biosynthesis in *Stevia rebaudiana* Bertoni. *Plants* **2022**, *11*, 611. [CrossRef]
50. Jasim, S.F.; Al-zubaidi, L.A.; Abdulbaqi, N.J. The effect of cold plasma on gene expression of major genes in the biosynthesis of phenylpropanoids and essential oil contents in *Ocimum basilicum* L. *Ann. Rom. Soc. Cell Biol.* **2021**, *25*, 14996–15010.
51. Marzban, M.; Farahani, F.; Atyabi, S.M.; Noormohammadi, Z. Induced genetic and chemical changes in medicinally important plant *Catharanthus roseus* (L.) G. Don: Cold plasma and phytohormones. *Mol. Biol. Rep.* **2021**, *49*, 31–38. [CrossRef]
52. Holubová, L'.; Švubová, R.; Slováková, L'.; Bokor, B.; Kročková, V.C.; Renčko, J.; Uhrin, F.; Medvecká, V.; Zahoranová, A.; Gálová, E. Cold Atmospheric Pressure Plasma Treatment of Maize Grains—Induction of Growth, Enzyme Activities and Heat Shock Proteins. *Int. J. Mol. Sci.* **2021**, *22*, 8509. [CrossRef]
53. Pérez-Pizá, M.C.; Ibañez, V.N.; Varela, A.; Cejas, E.; Ferreyra, M.; Chamorro-Garcés, J.C.; Zilli, C.; Vallecorsa, P.; Fina, B.; Prevosto, L.; et al. Non-Thermal Plasmas Affect Plant Growth and DNA Methylation Patterns in Glycine max. *J. Plant Growth Regul.* **2021**, *41*, 2732–2742. [CrossRef]

life

MDPI

Article

Abiotic and Herbivory Combined Stress in Tomato: Additive, Synergic and Antagonistic Effects and Within-Plant Phenotypic Plasticity

Rosa Vescio [1], Roberta Caridi [1], Francesca Laudani [1], Vincenzo Palmeri [1], Lucia Zappalà [2], Maurizio Badiani [1] and Agostino Sorgonà [1,*]

1 Department Agraria, University Mediterranea of Reggio Calabria, Località Feo di Vito, 89122 Reggio Calabria, Italy
2 Department of Agriculture, Food and Environment, University of Catania, Via Santa Sofia 100, 95123 Catania, Italy
* Correspondence: asorgona@unirc.it; Tel.: +39-096-5169-4373

Citation: Vescio, R.; Caridi, R.; Laudani, F.; Palmeri, V.; Zappalà, L.; Badiani, M.; Sorgonà, A. Abiotic and Herbivory Combined Stress in Tomato: Additive, Synergic and Antagonistic Effects and Within-Plant Phenotypic Plasticity. *Life* **2022**, *12*, 1804. https://doi.org/10.3390/life12111804

Academic Editors: Hakim Manghwar and Wajid Zaman

Received: 8 October 2022
Accepted: 4 November 2022
Published: 7 November 2022

Publisher's Note: MDPI stays neutral with regard to jurisdictional claims in published maps and institutional affiliations.

Copyright: © 2022 by the authors. Licensee MDPI, Basel, Switzerland. This article is an open access article distributed under the terms and conditions of the Creative Commons Attribution (CC BY) license (https://creativecommons.org/licenses/by/4.0/).

Abstract: Background: Drought, N deficiency and herbivory are considered the most important stressors caused by climate change in the agro- and eco-systems and varied in space and time shaping highly dynamic and heterogeneous stressful environments. This study aims to evaluate the tomato morpho-physiological and metabolic responses to combined abiotic and herbivory at different within-plant spatial levels and temporal scales. Methods: Leaf-level morphological, gas exchange traits and volatile organic compounds (VOCs) profiles were measured in tomato plants exposed to N deficiency and drought, *Tuta absoluta* larvae and their combination. Additive, synergistic or antagonistic effects of the single stress when combined were also evaluated. Morpho-physiological traits and VOCs profile were also measured on leaves located at three different positions along the shoot axes. Results: The combination of the abiotic and biotic stress has been more harmful than single stress with antagonistic and synergistic but non-additive effects for the morpho-physiological and VOCs tomato responses, respectively. Combined stress also determined a high within-plant phenotypic plasticity of the morpho-physiological responses. Conclusions: These results suggested that the combined stress in tomato determined a "new stress state" and a higher within-plant phenotypic plasticity which could permit an efficient use of the growth and defense resources in the heterogeneous and multiple stressful environmental conditions.

Keywords: within-plant phenotypic plasticity; combined stresses; additive; antagonistic and synergic effects; VOCs

1. Introduction

Owing to sessile nature, plants are continually exposed to abiotic (mainly drought, heat and salinity) and biotic (pathogens and herbivory) stresses whose intensity and frequency are expected to be increased by climate change. The effects of these stresses and how the plants respond to these stressful factors, taken individually, have been extensively studied at both the morpho-physiological and molecular scale [1,2] and plant community level [3,4]. However, under field condition, these various biotic and abiotic factors are constantly changing during the plant life cycle and, above all, co-occur in nature [5]. Hence, the plants have to make decisions about fine-tuning their responses to allocate resources efficiently for responding to the more serious and different threats at any given point in time. Different studies have uncovered that plants evoke a "unique response" to the combination of abiotic and biotic stresses compared to the single stress [5–8] revealing that the plants' responses to combined stress pointed out "a new stress state" with mostly non-additive effects (i.e., synergistic or antagonistic). For example, insect herbivory antagonized the heat responses in tomato [9], whereas, in this same plant species, drought stress synergized the emission of specific VOCs in combination with aphid herbivory [10] as well with the larvae of green alder sawfly (*Monsoma pulveratum*) in *Alnus glutinosa* [11]. Drought and simulate herbivory combination synergized some morphological traits of *Pinus sylvestris*, while other ones were antagonized [12]. In addition to the "new stress state", the plants' responses to the stress are strictly dependent on the plant traits, genotypes, species, and type, intensity, frequency and duration of the stress suggesting that more investigation is needed for a better understanding of the abiotic and pest herbivore interaction, the least studied among stress combinations.

The plants' responses to the individual abiotic and biotic stress have been observed to show a modulation (induced and constitutive) with a strong spatio-temporal component (local or systemic, transient or permanent) that determined a high "within-plant variation". For example, the spatial scale of herbivore-induced changes can range from localized at the site of attack [13] to systemic throughout the entire plant or tissue type [14,15]. Additionally, the light and heat gradients determined different responses within the tree canopy [16] or the nutrient deficiency caused different morpho-physiological responses among the different root types [17,18]. The temporal scale of the plant responses can also vary: rapid or long term, ontogeny-modulated [19] and, in some cases, even trans-generational responses are evoked for herbivory [20] as well as for abiotic stress [21]. The multiple ecological role of the 'within-plant' variation was recently pointed out in the adaptation to individual abiotic and biotic gradients [22] as well as in the alteration of plant–antagonist interactions [23] so much so that it was proposed as "functional trait itself" whose influences on ecosystem functioning are still neglected [24]. In spite of this important role, the within-plant variation in response to the combined stress has not been investigated so far, to the best of our knowledge.

Since 2006, the tomato production of the Mediterranean region has been under attack by a newly introduced insect, *Tuta absoluta*, whose larvae feed on leaves, stems and fruits causing severe damage to the tomato with decreases in production both in the field and greenhouse [25]. Previous studies revealed that the low nitrogen levels and drought stress inputs to tomato negatively affected the biological traits of *T. absoluta* [26,27] but also demonstrated that the N deficiency and drought could be unfavorable to the tomato plants, suggesting that the trade-off between negative impact on *Tuta* pest and plant growth should be evaluated.

In this framework, experiments were set up to study the spatial and temporal expressions of the morpho-physiological and metabolic responses of the tomato plants to the single and/or combined abiotic (drought+N deficiency) and biotic stress (herbivory by *T. absoluta*). In particular, the present study investigates the following questions: (1) Are the morpho-physiological responses to individual stresses different from the combined ones in tomato plants? (2) Are additive, synergistic or antagonistic effects in the combined stress? (3) Do the tomato responses to the single and combined stress occur at between- or within-plant levels?

2. Materials and Methods

2.1. Experimental Procedure and Plant Material

The present study was carried out on hydroponically-grown tomato plants (*Solanum lycopersicum* L., cultivar nano S. Marzano) and consisted of two separate experiments, addressing different but interrelated questions.

The first experiment, denoted as the 'synergic, antagonistic and additive effects', aimed to determine the tomato responses to the abiotic and biotic stress, single or in combination, and their temporal evolution, and also to evaluate whether the responses to the combined stress were the results of additive, synergistic or antagonistic effects of the single stress. For this purpose, we evaluated the effects of the drought plus N deficiency (abiotic stress, ABIO), or herbivory by *T. absoluta* (biotic stress, BIO) and their combination (combined stress, COMB) with the time of exposure (0, 1, 3 and 8 days) on the morphological (leaf fresh and dry weight), physiological [water content, photosynthesis, stomatal conductance, transpiration rate and intrinsic water use efficiency (iWUE)] and metabolic traits (VOCs profile). The fresh and dry weight of the leaf are traits directly related to the plant status, while the leaf water content was strictly correlated with the plant drought tolerance [28] but also with the plant palatability [29]. The gas exchange traits (photosynthesis, stomatal conductance, transpiration and iWUE) are involved in the plant responses to drought and N deficiency [30,31] and further the photosynthesis is " ... a plant-driven response to the perception of stress rather than a secondary physiological response to tissue damage ... " highlighting strict interactions between photosynthesis, reactive oxygen species (ROS) and hormonal signaling pathways for the plant responses to insect herbivory [32]. Finally, the VOCs, as direct and indirect defense, are emitted by plants subject to both abiotic and biotic stress [33].

Tomato plants (*Solanum lycopersicum* L., cultivar nano S. Marzano) (provided by BAVICCHI S.p.a., Perugia, Italy) were exposed to the abiotic (nitrogen limitation and drought stress simulated by the use of polyethylene glycol (PEG)) and biotic stress (two first instar larvae placed in a leaf), or their combination and were considered as 'stress condition'. The control plants (CTR) were maintained at optimal N concentration, no drought and herbivory, considered as the 'optimal condition'. For the morpho-physiological analysis, we used a randomized block design in which the entire experiment yielded a total of 4 (treatments) $\times$ 4 (time of exposure) $\times$ 2 (block) $\times$ 2 (replications) = 64 samples. The block was introduced because we used two experiments at two different times. A completely randomized design was used for the VOCs profiling, in which the entire experiment was constituted by 4 (treatments) $\times$ 4 (time of exposure) $\times$ 3 (replicates) $\times$ 3 (measurements) = 144 samples. The replicates for the VOCs were obtained in three different experiments.

The second experiment, denoted as the "within-plant phenotypic plasticity", aimed to evaluate the within-plant variation of the tomato morpho-physiological and metabolic traits and how this within-plant phenotypic plasticity changed with the treatment conditions (optimal, abiotic, biotic and combined stress). For this aim, the foliar morpho-physiological and metabolic traits were evaluated on mature leaves located at three different positions along the shoot axes for each treatment. For each treatment, we used a completely randomized design in which the entire experiments yielded a total of 3 (leaves) $\times$ 1 (time of exposure) $\times$ 3 (replicates) = 9 samples. For the gas exchanges traits only, we took two measurements for each leaf, hence the experiments provided 3 (leaves) $\times$ 1 (time of exposure) $\times$ 2 (measurements) $\times$ 4 replicates = 24 samples.

The Figure S1 reported the experimental protocol schedule of both experiments including the plant growth, treatments and analysis.

2.2. Tomato Growth Conditions

Tomato seeds were surface sterilized for 15 min in 10% (*v*/*v*) sodium hypochloride, rinsed with tap water and then germinated in a Petri dish (diameter 90 mm) on filter paper with 0.1 mM $CaSO_4$. After 7 d of germination [7 days after germination (DAG)], six seedlings of uniform size were transferred into each of eight hydroponic units, each containing 4.5 L of the following aerated nutrient solution (http://www.haifa-group.com/files/Guides/tomato/Tomato.pdf (accessed on 8 February 2016) at 50% strength and adjusted to pH 6.0 with 0.1 M potassium hydroxide: 5 mM KNO_3, 1 mM NH_4NO_3, 1.44 mM $MgSO_4$, 3.99 mM $Ca(NO_3)_2$, 0.97 mM KH_2PO_4, 1 mM K_2SO_4, 25 μM H_3BO_3, 50 μM KCl, 2 μM $MnSO_4$, 4 μM $ZnSO_4{\cdot}7H_2O$, 0.5 μM $CuSO_4{\cdot}5H_2O$, 0.5 μM $(NH_4)Mo_7O_{24}{\cdot}4H_2O$, 20 μM EDTA iron(III) sodium salt.

The hydroponic units were placed in a growth chamber at 24 °C, 14 h photoperiod; photon flux rate of 300 μmol m^{-2} s^{-1}; 70% relative humidity (RH).

After 7 days (14 DAG), the nutrient solution was brought to 100% strength and the plants of each pot were thinned to four for the morpho-physiological analysis while they were left to six for VOCs analysis. The nutrient solution was renewed every 2 days.

2.3. Insect Rearing

The tomato leafminer *Tuta absoluta* (Lepidoptera: Gelechiidae) colony was maintained in climatic chambers (25°C, RH 70%, 16 h light). It was kept in cages (Bugdorm®—60 × 60 × 60 cm) containing tomato plants. Sugar and water were provided *ad libitum* to adults in rearing cages.

2.4. Abiotic Stress and Herbivory Treatment

At 28 DAG, six hydroponic units continued to receive the same nutrient solution as previously described in growth conditions, while in two hydroponic units 5% (*w*/*v*) polyethylene glycol 8000 (Sigma PEG8000, Sigma Aldrich, St. Louis, MO, USA) and 1 mM nitrogen were added for simulating the drought stress and nitrogen deficiency, respectively (ABIO group). The final PEG concentration was gradually achieved by the addition of 2.5% (*w*/*v*) PEG8000 every two days. The osmotic potential of the solutions, measured by an osmometer (Freezing point osmometer, Osmomat 3000, Gonotec, Berlin, Germany), was −0.55 MPa for 5% PEG and −0.05 MPa for the control solution (0% PEG). To obtain 1 mM N for the nitrogen deficiency, the NH_4NO_3 was not added and the KNO_3 and $Ca(NO_3)_2$ were reduced to 1 mM and 0.5 mM, respectively. In order to balance K and Ca, the K_2SO_4 and $CaSO_4$ were increased to 3 mM and the 3.5 mM, respectively. Preliminary experiments were run to ascertain that the selected PEG8000 and N concentrations did not prejudice plants' survival.

At 42 DAG, the hydroponic units were treated as follows for obtaining the whole set of treatments (Figure S1):

(1) two hydroponic units were renewed with the optimal nutrient solution (CTR group);
(2) two hydroponic units were renewed with the nutrient solution with N deficiency and PEG (ABIO group);
(3) two hydroponic units received the optimal nutrient solution but the plants were infested with *Tuta* larvae to induce the biotic stress (BIO group);
(4) two hydroponic units maintained the same nutrient solution with N deficiency and PEG and, in addition, the plants were infested by *Tuta* larvae (COMB group).

The plant infestation was obtained by placing two first instar larvae of *Tuta* in the 1st fully-developed leaf (with five leaflets) from the bottom and, to avoid larvae escaping, each infested leaf was then bagged with a nylon mesh of 4.7 cm diameter (Figure S2). We added herbivorous insects to plants after 14 days of abiotic stress treatment in order to simulate the effects of a pest outbreak which are predicted to become more frequent with climate change [34].

2.5. First Experimental: Synergic, Antagonistic and Additive Effects

2.5.1. Tomato Samplings and Measurements

At 0 (42 DAG), 1 (43 DAG), 3 (45 DAG) and 8 days from the treatments (50 DAG), the measurements/samplings were realized in order to simulate the early, intermediate and late responses, respectively. Gas exchange measurements were carried out on the terminal leaflet of the first fully-developed leaf (in presence of larvae, we used lateral leaflets) while the whole plants were used for the morphological analysis. Consecutively, three leaves for each treatment and time of exposure were sampled for the VOCs.

2.5.2. Gas Exchange Measurements

A calibrated portable photosynthesis system (LI-6400; LI-COR, Inc.; Lincoln, NE, USA) was used to measure the net CO_2 assimilation rate (A) (μmol (CO_2) m^{-2} s^{-1}), stomatal conductance (gs) (mol H_2O m^{-2} s^{-1}), and the transpiration rate (T) (mmol H_2O m^{-2} s^{-1}). These gas exchange parameters were measured at 500 cm^3 min^{-1} flow rate, 26 °C leaf temperature, CO_2 concentration 400 μmol (mol air)$^{-1}$ (controlled by CO_2 cylinder), and 1200 μmol m^{-2} s^{-1} of photosynthetically active radiation supplied by the LED light source in the leaf chamber. Each measurement was made with a minimum and maximum wait time of 120 and 200 s, respectively, and matching the infrared gas analyzers for 50 μmol (CO_2) mol (air)$^{-1}$ difference in the CO_2 concentration between the sample and the reference before every change of plants. The leaf-to-air vapor pressure difference (VPD) was set to 1.5 kPa, and continuously monitored around the leaf during measurements. It was maintained at a constant level by manipulating the humidity of incoming air as needed. All measurements were performed in the growth chamber.

Finally, the intrinsic water use efficiency (iWUE) was calculated as the rate of photosynthesis (A) divided by the rate of stomatal conductance to water (gs) [35].

2.5.3. Morphological Measurements

All the leaves of the plants were harvested, immediately weighed to obtain the leaf fresh weight (LFW, g) and then placed in an oven at 70 °C for 2 days to determine the leaf dry weight (LDW, g).

Finally, the leaf water content (LWC, %) was calculated as the following, as reported in Jin et al. [36]:

$$\text{Leaf Water content (\%)} = (\text{LFW} - \text{LDW})/\text{LFW} \times 100 \quad (1)$$

2.5.4. VOCs Analysis

VOCs from three leaves per treatment and time of exposure were profiled by headspace–solid phase microextraction (HS/SPME) method. One leaf was sealed in a 20 mL hermetic vial with butyl lid and allowed to incubate for 20 min at room temperature. The fiber (50/30 μm DVB/CAR/PDMS) (Supelco®, Bellefonte, PA, USA), previously conditioned according to the supplier's instructions, was inserted into the headspace of the vial containing the sample and allowed to adsorb leaf volatiles for 20 min. The volatiles were then desorbed by placing the fiber for 6 min into the injection port of the gas chromatography–mass spectrometry (GC-MS) system. All the SPME sampling and desorption conditions were identical for all the samples. Blanks were performed before first SPME extraction and randomly repeated during each series of measurements.

GC-MS analysis of VOCs were performed with a Thermo Fisher TRACE 1300 (Trace 1300, Thermo Fisher Scientific, Waltham, MA, USA) gas chromatograph equipped with a DB-5 capillary column (30 m × 0.25 mm; coating thickness = 0.25 μm, with 10 m of pre-column) coupled to a Thermo Fisher ISQ LT ion trap mass detector (ISQ LT, Thermo Fisher Scientific, Waltham, MA, USA) (emission current: 10 microamps; count threshold: 1 count; multiplier offset: 0 volts; scan time: 1.00 second; prescan ionization time: 100 microseconds; scan mass range: 30–300 *m/z*; ionization mode: EI).

GC–MS data were obtained under the following analytical conditions: carrier gas Helium (He 99.99%); flow rate 1 mL/min; spiltless. The initial oven temperature was 60 °C for 3 min, after which it was raised to 240 °C at 6 °C/min, and finally isothermal for 3 min. The injection port, transfer line, and ion source were kept at 250 °C, 250 °C, and 260 °C, respectively.

Qualitative identification of VOCs was performed using GC–MS reference libraries (NIST x.0). Linear retention indices (LRI) were determined from the retention times of a series of n-alkane mixture (C8-C20, Sigma Aldrich, Milan, Italy) analyzed under the same conditions reported above [37]. Percentages of the studied compounds were calculated from the peak areas in the total ion chromatograms. The relative abundance of each volatile with respect to the total amount of released compounds was estimated from its peak area against the total ions chromatogram, and expressed as a percentage, after subtracting possible contaminants.

2.5.5. Statistical Analysis

Morpho-Physiological Data

By SPSS Inc. V. 10.0, 2002 (SPSS Inc., Evanston, IL, USA), all the morpho-physiological parameters were analyzed by two-way ANOVA with the treatment (Tr) (CTR, ABIO, BIO and COMB), time of exposure (Ti) and block (Bl) as main factors and the TrxTi as interaction.

Then, Tukey's test was used to compare the means of all the parameters of each Tr and Ti. All data were tested for normality (Kolmogorov–Smirnoff test) and homogeneity of variance (Levene median test) and, where required, the data were transformed.

VOCs Data

The VOCs dataset was elaborated by using the R statistical software 3.5 [38].

Differences among treatments, time of exposure and TrxTi interaction were inferred through PERMANOVA multivariate analysis (999 permutations) using the package *vegan*. Pairwise comparisons were calculated using a custom script and correcting p values using the False Discovery Rate (FDR) method.

In order to identify VOCs key predictors that could constitute a molecular signature identifier among the treatments within each time of exposure, we used a preliminary unsupervised (Principal Component Analysis, PCA) and then supervised analysis (Sparse Projection to Latent Structure–Discriminant Analysis, sPLS-DA) by using the *mixOmics* package [39]. Statistical algorithms are detailed in Rohart et al. [39] and they account for multiple comparisons inherent in biomarker datasets, where multiple classification features are considered for a relatively small number of specimens ($p >> n$). In particular, the sPLS-DA procedure constructs artificial latent components of the predicted dataset (VOCs Table denoted X ($N \times P$)) and the response variable (denoted Y with categorical information of samples, e.g., CTR, ABIO, BIO and COMB). To predict the number of latent components (associated loading vectors) and the number of discriminants, for sPLS-DA, we used the *perf.plsda()* and *tune.splsda()* functions, respectively. We finetuned the model using five-fold cross-validation repeated 10 times to estimate the classification error rates employing two metrics, overall error rates and balanced error rates (BER), between the predicted latent variables with the centroid of the class labels (categories considered in this study) and specifying the *max.dist* (which gave the minimal classification rate in this study).

Calculation of Additive, Synergistic or Antagonistic Effects in Combined Stress

To determine whether stress combination yielded additive, synergistic or antagonistic effects respect to each of the single stress alone, we used the method of Bansal et al. [12]. To such aim, we compared the observed effects (Ob) with expected additive effects (Ex) for the plants exposed to the abiotic stress and herbivory combination (COMB) at 3 and 8 days of treatments, only. The Ob effect sizes were calculated as the absolute value of:

$$Ob = (ob - \overline{x}CTR)/\overline{x}CTR \quad (2)$$

where Ob is the value of each measured trait in each plants and treatment and $\overline{x}$CTR is the mean value of the same trait measured in CTR plants.

For each of the traits considered, the Ex additive effect sizes for the COMB treatment were defined in two steps by first determining and then summing the independent effects (In) of each treatment. The In effect sizes were calculated as the absolute value of:

$$Ind = (\overline{x}stress - \overline{x}CTR)/\overline{x}CTR \quad (3)$$

where $\overline{x}$stress is the mean values of a given trait in the presence of a single stress, and $\overline{x}$CTR is the corresponding mean value in CTR plants. Then, the Ex additive effect size for the COMB treatment was calculated by using a multiplicative risk model [40], that is the sum of the two In effects minus their product. Finally, the Ex additive values for COMB plants were compared to the actual Ob additive effects. In particular, we calculated a mean difference (±95% confidence interval) between the effect sizes of Ob and Ex was for COMB plants. When Ob-Ex > 0 and the lower 95% confidence limit was greater than zero, then the impact from the combination of both stressors was classified as synergistic. Antagonistic effects were defined when the Ob-Ex < 0 and the upper 95% confidence limit was less than zero. Finally, we classified additive effects when the 95% confidence interval crossed the zero line.

2.6. Second Experiment: Within-Plant Phenotypic Plasticity

2.6.1. Tomato Samplings and Measurements

The tomato samplings and measurements were carried out at 8 days from the treatments (50 DAG) in the leaves located at three different positions along the shoot axes: basal (B), intermediate (I) and apical leaf (A) belonging to the first, second and third node, respectively. Preliminary experiments using phloem dying as reported in Orians et al. [41] resulted that the apical leaf, but not the intermediate one, is linked to the basal one via vasculature connections (Figure S3). On such basis, we also considered the basal, intermediate and apical leaves as the local (L), no-orthostichous (nO) and orthostichous leaf (O), respectively. The basal/local leaf was used for placing the first instar larvae of *T. absoluta* for the experimental infestation.

Measurements for the gas exchanges traits were carried out on two opposite leaflets of the basal/local (B/L), intermediate/noOrthostic (I/noO) and apical/orthostic leaf (A/O) and the same leaves were subsequently collected for the morphological analysis.

All the morpho-physiological analyses were carried out as in the first experiment.

2.6.2. Statistics

Within-Plant Variance of the Morpho-Physiological Traits

The within-plant variance of the morpho-physiological traits was evaluated as in Zywiec et al. [42].

In order to estimate the partitioning of total variation of the morpho-physiological traits among- and within-treatments, we conducted a linear mixed models with treatments and plant nested within-treatments as random effects using the whole-plant data. The variance partitions among- and within-treatments and tests on the statistical significance of variance components were conducted using restricted maximum likelihood (REML).

In order to verify the effects of each treatments on morpho-physiological traits of different leaves within the plants, we analyzed the within-plant variation by applying a hierarchical partition to divide total variance into two levels of variation: among plants and among the leaves in the same plants (leaf nested within plant). All levels were considered as random effects, as required for variance partitioning. Analyses were conducted with the mixed procedure of SPSS. The replicate obtained for each leaflets sample allowed us to estimate measurement error and thus assess the variance component and statistical significance (Wald Z and p values) of this component between- and within-individual plants.

Morpho-Physiological Data

By SPSS Inc. V. 10.0, 2002 (SPSS Inc., Evanston, IL, USA), all the morpho-physiological parameters were analyzed by one way ANOVA with Tukey's test as post-hoc test ($p < 0.05$).

3. Results and Discussion

3.1. Are the Morpho-Physiological Responses to Individual Stresses Different from the Combined Ones? Are Additive, Synergistic or Antagonistic Effects in the Combined Stress?

The morpho-physiological results clearly indicated an opposite pattern in the response of the tomato plants to the single stresses with the ABIO treatment showing a more negative impact than the BIO one with respect to the CTR plants (Figures 1 and 2).

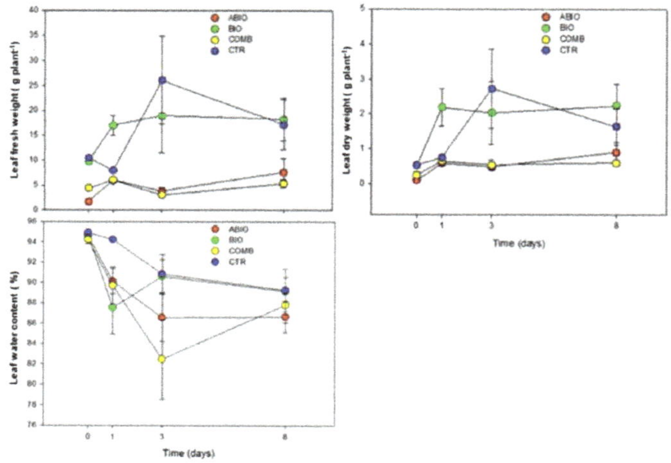

Figure 1. Morphological traits. Leaf fresh (g), dry weight (g) and leaf water content (%) of tomato plants treated with different stress: drought stress plus N deficiency (ABIO), infestation by the insect *Tuta absoluta* (BIO) and their combination (COMB) for different times of exposure (0, 1, 3 and 8 days). Unstressed control plants (CTR). Each value and its error bar indicate the mean and the standard error of the mean, respectively (N = 4).

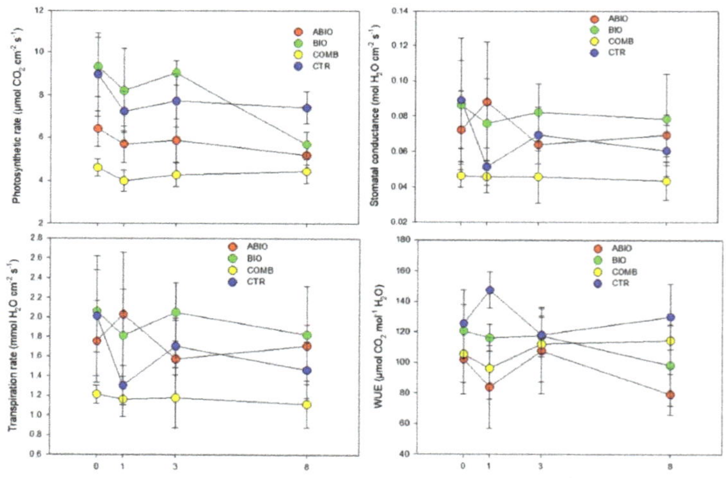

Figure 2. Gas exchange parameters. Photosynthetic rate (μmol CO_2 m^{-2} s^{-1}), stomatal conductance (mol H_2O m^{-2} s^{-1}), transpiration rate (mmol H_2O m^{-2} s^{-1}) and intrinsic water use efficiency (μmol CO_2 mol^{-1} H_2O) of the leaves of tomato plants treated with different stress for different time of exposure (0, 1, 3 and 8 days). Acronyms as in Figure 1. Each value and its error bar indicate the mean and the standard error of the mean, respectively (N = 4).

In particular, the LFW, LDW, LWC, A, and iWUE were significantly reduced in the ABIO plants with respect to the control, whereas no significant differences were observed in the presence of herbivory except than for leaf water content and iWUE (Figures 1 and 2; Tables 1 and 2).

Table 1. Results of two-way ANOVA [Treatment (Tr), time (Ti), block (Bl), TrxTi interaction (TrxTi)] on the morphological traits. Abbreviations as in Figure 1. Statistics: F- and *p*-values. Within each morphological traits and time of exposure, the different letters indicated statistical differences among the means of the treatments ($p < 0.05$, test of Tukey).

Parameters	Statistics	Time (Ti)				
		Treatments (Tr)	t0	t1	t3	t8
Leaf fresh weight	Tr 10.92 ***	ABIO	a	a	b	b
	Ti 1.40 ^{NS}	BIO	a	a	a	a
	Bl 6.08 *	COMB	a	a	b	b
	TrxTi 1.06 ^{NS}	CTR	a	a	a	a
Leaf dry weight	Tr 8.87 ***	ABIO	a	a	b	b
	Ti 2.91 *	BIO	a	a	a	a
	BI 17.17 ***	COMB	a	a	b	b
	TrxTi 1.26 ^{NS}	CTR	a	a	a	ab
Leaf water content	Tr 3.16 *	ABIO	a	ab	ab	a
	Ti 12.01 ***	BIO	a	b	a	a
	BI 52.15 ***	COMB	a	b	b	a
	TrxTi 1.50 ^{NS}	CTR	a	a	a	a

* $0.05 < p < 0.01$; *** $p < 0.001$; NS not significant.

Table 2. Results of two-way ANOVA [Treatment (Tr), time (Ti), block (Bl) TrxTi interaction (TrxTi)] on the gas exchanges traits. Abbreviations as in Figure 1. Statistics: F- and *p*-values. Within each gas exchanges traits and time of exposure, the different letters indicated statistical differences among the mean of the treatments ($p < 0.05$, test of Tukey).

Parameters	Statistics	Time (Ti)				
		Treatments (Tr)	t0	t1	t3	t8
Photosynthetic rate	Tr 17.60 ***	ABIO	a	b	bc	ab
	Ti 2.73 ^{NS}	BIO	a	a	a	ab
	BI 20.74 ***	COMB	a	b	c	b
	TrxTi 0.76 ^{NS}	CTR	a	ab	ab	a
Stomatal conductance	Tr 5.38 **	ABIO	a	a	ab	a
	Ti 0.60 ^{NS}	BIO	a	a	a	a
	BI 61.82 ***	COMB	a	b	b	a
	TrxTi 0.57 ^{NS}	CTR	a	ab	ab	a
Transpiration rate	Tr 6.94 **	ABIO	a	a	ab	ab
	Ti 0.79 ^{NS}	BIO	a	a	a	a
	BI 55.73 ***	COMB	a	a	b	b
	TrxTi 1.13 ^{NS}	CTR	a	a	ab	ab
iWUE	Tr 6.23 **	ABIO	a	b	a	b
	Ti 0.47 ^{NS}	BIO	a	b	a	b
	BI 136.54 ***	COMB	a	b	a	ab
	TrxTi 1.52 ^{NS}	CTR	a	a	a	a

** $0.01 < p < 0.001$; *** $p < 0.001$; NS not significant.

It is known that the drought stress, either alone [43] or in combination with N deficiency [44], reduced the A, g_s and LWC with negative consequence for the leaf growth of tomato plants which arise from established molecular mechanisms [45]. As observed, the BIO treatment did not produce modification of the morpho-physiological traits in comparison to the control (Figures 1 and 2; Tables 1 and 2) and this no response to the herbivory falls in the highly variable effects observed in different plant–insect combinations. For example, the leaf dry to fresh mass ratio was not changed by *Monsoma pulveratum* feeding on *Alnus glutinosa* [11] but a weak negative effect in the soybean-natural herbivory interaction was observed [46]. Further, the net photosynthesis was found to be either sharply reduced [47], or even increased [48] or even not modified [49]. In the present study, the infestation of *Tuta absoluta* could probably have caused 'indirect effects' in leaf tomato such as increase of the photosynthesis and water losses by transpiration rate, resulting in reduced iWUE and leaf water content as also observed in soybean-Japanese beetles and corn earworm caterpillars interactions [49]. However, in a previous study concerning the *Tuta*–tomato interactions, a reduction in leaflet growth was pointed out [50].

Although the plant responses to the individual abiotic stress and herbivory infestation are well understood, the information concerning the effects of stress combination is, in most of the cases, scanty or even absent. In the present study, the combination among the N deficiency and drought, on one side, and herbivory by *Tuta*, on the other side, caused the highest reduction of the tomato morpho-physiological traits respect to the control plants (Figures 1 and 2; Tables 1 and 2). This overstate effect of the combined stress could be due to the interactive responses determined by cross-talk in hormonal signaling and by the transcriptional modulation of defense-related genes. Indeed, in the interaction between *Solanum dulcamara* and the herbivory by specialist *Leptinotarsa decemlineata*, the antagonism of the specific herbivory-induced salycilic acid on the jasmonic acid (JA) was found to prevail over the synergism of the specific drought-induced abscissic acid (ABA) with consequent reduction of the defense responses observed at transcriptional levels (increase in the cell wall components and secondary metabolism) [51]. Moreover, in tomato plants subjected to both drought and herbivory by *Spodoptera exigua*, an adaptive response was observed by a transcriptional activation of the genes related to the photosynthetic machinery and chlorophyll biosynthesis causing, as a consequence, a reduction of the secondary metabolite production [51].

The temporal evolution of the plant responses to environmental stresses is fundamental for the success of the plant adaptation, although such aspect has been comparatively less studied. In the present work, differently to the single stress, the COMB treatment reduced the LFW and LDW at 3 days from the stress treatments, while the LWC, A, g_s and iWUE respond faster, being already evident at 1 day of treatment (Figures 1 and 2; Tables 1 and 2). It is likely that the combined stress in tomato plants rapidly activated the stomatal closure, to reduce the water losses, causing in turn a reduction of the photosynthetic process accompanied by a decrease in the synthesis of defense-related metabolites and all this subsequently translated into a lower leaf growth. However, this morpho-physiological pattern, although faster, could have been the final result of the signaling and molecular network which is instead activated in very rapid responses (within seconds and minutes) as observed in different abiotic- and biotic-stressed plants [21].

Figures 1 and 2 and Tables 1 and 2 also show that abiotic stress and herbivory by *Tuta* more negatively affected the physiological traits than the morphological ones. The plant physiological plasticity is more related to an enhanced ability to exploit the transient environmental resources, such as water and nutrient patches, or to produce the defense responses (secondary metabolites) to the herbivory attack at low cost by short-term adjustments [52–54]. Conversely, the plant morphological plasticity is more resource-intensive and hence more functional to long-term plant adaptation [52,54].

The VOCs emission is an important plant defense process in response to the herbivory attack [55], as well as abiotic stress [56]. HS/SPME GC-MS analysis revealed forty-five volatile compounds emitted by the tomato leaves exposed to the single (abiotic and biotic) and combined stress (Table S1). In particular, the volatile profile was mostly characterized by mono- (24% of the total) and sesquiterpenes (44%), although hydrocarbons (11%), ester (9%), alcohol (5%), ether (5%) and aldehyde (2%) were also present (Table S1). Volatile terpenoid metabolites have been recognized as having a range of specific roles in plant/environment and plant/plant interactions [57] and, in particular, in the direct and indirect defense of tomato plants against the herbivory [58]. Table S2 reported the volatiles from each treatment and time of exposure on the basis of % area of each peak over the total area of the chromatogram. In order to test the influence of the treatments and time of exposure on the volatile profiling, we run a multivariate approach that included, first, the Permanova, and then the PCA and sPLSDA that allowed to visualize the differences among the groups and to select informative and relevant volatiles. Permanova analysis indicated that the treatments, the time of exposure and their interaction determined significant differences in the volatilome of tomato plants (Table S3). Pairwise comparison among the treatments within each time of exposure revealed that the volatile profile of COMB plants was different from control ones at 3 and 8 days of exposure, while in the BIO plants such differences emerged only at 8 days of exposure ($p_{adjusted} < 0.05$; Table S4). Since differences among the treatments became evident after the two aforementioned times of exposure, we only used the volatiles dataset from the four treatments at 3 and 8 days of exposure to run the PCA. Figure S4 showed the results of PCA where it is apparent that this multivariate test was not able to separate the treatment groups owing to the high variability among samples. Therefore, data were further analyzed using sPLS-DA at each time of exposure (3 and 8 days). At 3 days of exposure, the performance step of the sPLSA-DA for the selection of the number of components suggested that three components were sufficient to sharply reduce the balanced error rate around 0.23 (Figure 3A). Further, the final model obtained by tuning process pointed out that Component 1, 2 and 3 comprised 25, 19 and 31 volatiles, respectively (Figure 3B), but with a scarce discrimination among the treatments, as highlighted by the sample plots on the first three components (Figure 3C,D).

At 8 days of exposure, three components were selected with a balanced error rate around 0.28 and with a molecular signature composed of 16, 16 and 31 VOCs selected on the first three components, respectively (Figure 4A,B). The sample plots on the three components put into evidence a discrimination among the treatments with 60% of total explained variability split up by 32%, 15% and 13% for the first, second and third components, respectively (Figure 4C,D).

In particular, plotting the first two components showed that the BIO treatment was sharply separated from the control and COMB ones by the second component (Figure 4C). Conversely, the first and third components scarcely discriminated among the treatments (Figure 4C,D). All 16 VOCs selected on the second component had a negative weight in the linear combination, and a subset of them, namely (E)-2 hexen-1 ol, (2E)-2-hexenyl propionate, (+)-4-carene, α-copaene, hexyl propionate, β-phellandrene, α-pinene, terpinolene, β-pinene, E β-caryophyllene, were found to be emitted at comparatively higher rates in BIO plants (Figure 4E). However, a higher emission was observed for dodecane, 2,6,11-trimethyl in CTR plants, the α-humulene and α-terpinene in ABIO ones, and γ-elemene and δ-cadinene in COMB ones (Figure 4E). The alcohol (E)-2 hexen-1-ol [59], and the two esters (2E)-2-hexenyl propionate [60] and hexyl propionate [61] are green leaf volatiles which are released in response to different stress conditions to aid in plant defense against herbivory and bacterial and fungal pathogens [62]. The (+)-4-carene, α-copaene, β-phellandrene, α-pinene, terpinolene, β-pinene and E β-caryophyllene are terpenes, a large family of organic compounds mostly involved in the plant defense. For example, the (+)-4-carene, α-copaene and β-phellandrene are the most abundant VOCs emitted by *Solanum* spp. in the presence of *Bactericera cockerelli* herbivory [63], and terpinolene and E β-caryophyllene were mainly produced by tomato leaves infested with *Trialeurodes vaporario-rum* [64]. Further, the α-humulene was found to be responsible of tomato repellence against *Bemisia tabaci* [65]. The dodecane, 2,6,11-trimethyl, which, on a comparative basis, was found to be highly emitted by the CTR plants of the present study (Figure 4E), could be regarded as a VOCs marker of healthy plant status, since this same evidence was obtained by Giunti et al. in olive plants [66]. Interestingly, the α-terpinene and δ-terpinene were found to be mainly discriminant of ABIO treatment (Figure 4E), thus confirming the results obtained in tomato plants (cv. Gan Liang Mao Fen 802 F1) fertilized with low levels of N [65] and in drought-stressed *Thymus vulgaris* plants [67]. Finally, the present study revealed the VOCs emitted by multi-stressed tomato plants. In particular, the γ-muurolene, δ-elemene, δ-cadinene, β-elemene, α-gauiene, z-β-caryophyllene, aromadendrene, γ-elemene, 1,3,7 Nonatriene 4,8 dimethyl (3E)-, methyl salicylate, β-cadinene and myrcene were the compounds constituting the VOCs blend emitted by tomato plants when exposed to combined N and drought stress with infestation by *Tuta absoluta* (Figure 4E,F). It should be noted that the γ-elemene and δ-cadinene were present in both sPLD-DA components while the other volatiles were only observed in the Component 1 that is the lesser discriminant (Figure 4E,F). In *Gossypium arboretum*, the δ-cadinene is a precursor of the cyclic secondary sesquiterpene aldehydes, including gossypol, that are insecticides [68]. Unlike δ-cadinene, the emission of γ-elemene was not modified in the tomato leaves exposed to pest attack [64]. Besides the δ-cadinene and γ-elemene, the other volatiles belonging to the component 1 were also of interest in plant responses to the abiotic and biotic stress. Methyl salicylate was observed to increase in double drought-stressed and aphid-infested tomato plants [10] but also in the drought-herbivory combination together with 1,3,7-nonatriene, 4,8-dimethyl-, (3E)-, with which methyl salicylate forms a couple of stress-specific VOCs [11]. The other volatiles released by the COMB plants of the present study were also present in the volatilome of different plant species exposed to the combination of two or more stresses [11,69,70].

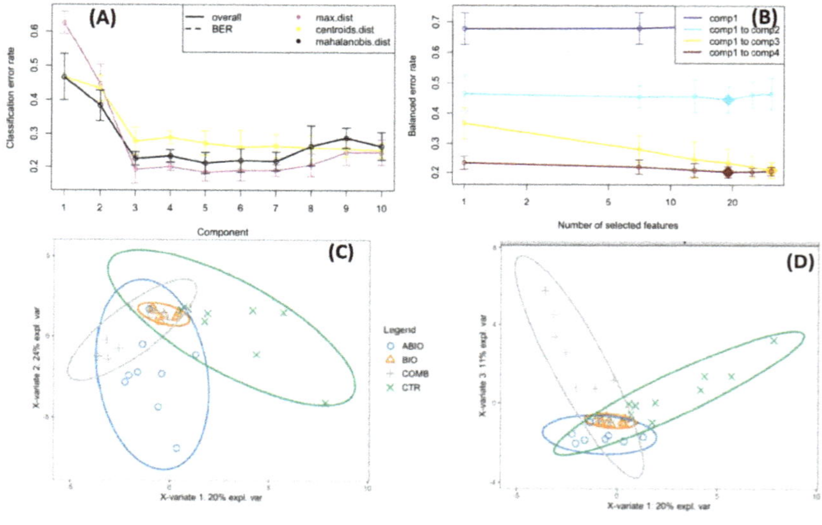

Figure 3. Sparse Projection to Latent Structure–Discriminant Analysis (sPLS-DA) of volatile organic compounds emitted from the leaves of tomato plants exposed to different stress (ABIO, BIO, COMB, see acronym sin Figure 1), or not stressed (CTR) for 3 days of exposure. (**A**) Choosing the number of components in sPLS-DA by performance test. Mean classification by overall and balanced error rate (5 cross-validation averaged 50 times) for each sPLS-DA component. (**B**) Choosing the number of volatiles for each sPLS-DA components by tuning test. Estimated classification balanced error rates for volatile dataset (5 cross-validation averaged 50 times) with respect to the number of selected volatiles for the sparse exploratory approaches. (**C**,**D**) sPLS-DA sample plot for the different components using 95% confidence ellipses. (**C**) Component 1 vs. Component 2, (**D**) Component 1 vs. Component 3.

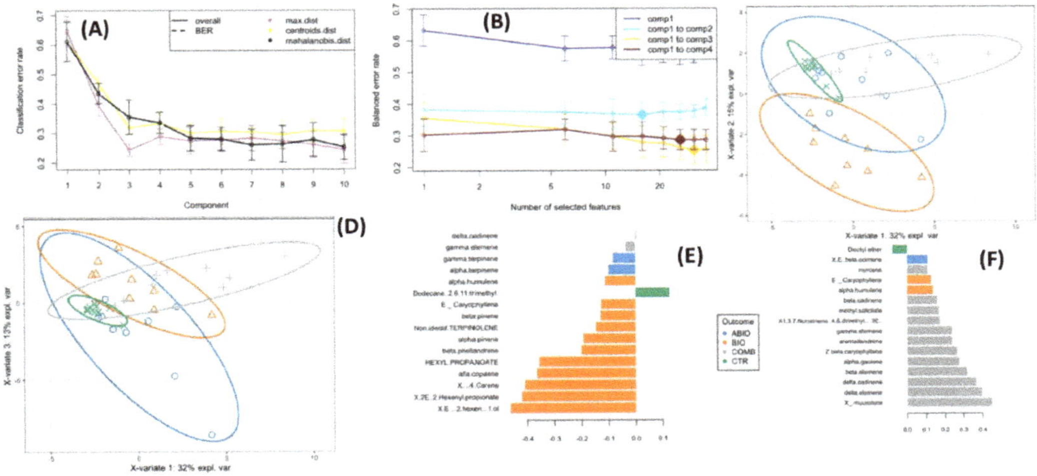

Figure 4. Sparse Projection to Latent Structure–Discriminant Analysis (sPLS-DA) of volatile organic compounds emitted from the leaves of tomato plants exposed to different stress (ABIO, BIO, COMB, see acronym sin Figure 1), or not stressed (CTR) for 3 days of exposure. (**A**) Choosing the number of components in sPLS-DA by performance test. Mean classification by overall and balanced error rate (5 cross-validation averaged 50 times) for each sPLS-DA component. (**B**) Choosing the number of volatiles for each sPLS-DA components by tuning test. Estimated classification balanced error rates for volatile dataset (5 cross-validation averaged 50 times) with respect to the number of selected volatiles for the sparse exploratory approaches. (**C**,**D**) sPLS-DA sample plot for the different components using 95% confidence ellipses. (**C**) Component 1 vs. Component 2, (**D**) Component 1 vs. Component 3. Contribution plots by loading weights of the volatiles selected for the Component 2 (**E**) and Component 1 (**F**) of the sPLS-DA. The color indicated the treatments for which the selected volatile has a maximal mean loading weight value.

3.2. Are Additive, Synergistic or Antagonistic Effects in the Combined Stress?

In analyzing the tomato morpho-physiological and metabolic responses to the combined stress, it is of interest to understand whether stress combination caused additive (i.e., equal to the sum of the single-stress effects), synergistic (i.e., higher than expected) or antagonistic effects (i.e., lower than expected) with respect to those caused by each single stress taken alone. This, in turn, would allow hypotheses to rise about the signaling pathways and molecular mechanisms underlying the plant strategy in the presence of simultaneous stress. In this respect, the additive, synergistic and antagonistic effects of abiotic (drought and N deficiency) and biotic stress in tomato plants were evaluated by the Bansal et al. method [12]. The reported results show that, in general, the physiological (photosynthesis, stomatal conductance and transpiration) and the metabolic (VOCs) traits pointed out more synergistic effects than morphological ones especially at an early time of exposure, i.e., at 1 and 3 days (Figure 5).

In particular, a synergic effect (Figure 5) was observed for the reduction of the physiological traits (photosynthesis, stomatal conductance and transpiration) (Figure 2) and for the increase in the VOCs emission (Table S2). The closure of the stomata is the first plant response to the water scarcity and it is mediated by ABA that orchestrates a network of stress-responsive metabolites and gene expression [71]. The ABA signaling pathways also interact with that of the JA one, the phytohormone that activates the signaling cascades for regulating downstream transcriptional responses to the herbivory [72,73] and, furthermore, it was observed that MeJA signaling is overlapped with ABA signaling in guard cells [74]. In the present work, such interaction between the ABA and JA signaling pathways could have caused the synergistic negative effect on the physiological traits observed in COMB plants (Figure 5). However, besides the hormonal interactions, unique and novel molecular mechanisms were also found during the stress combination in several studies. For example, the transcriptome analysis revealed that a unique set of transcripts was altered in response to the combination of drought and nematode infection [75], drought, heat stress and virus [76], infection by *Botrytis cinerea*, herbivory by chewing larvae and drought stress [77]. In the present study, the observed synergic effect of stress combination on the decrease of both photosynthetic and transpiration rates could have been determined by the stomatal closure in addition to the herbivory-induced resource reallocation to chemical defense [78] that determined more intense dark respiration [79].

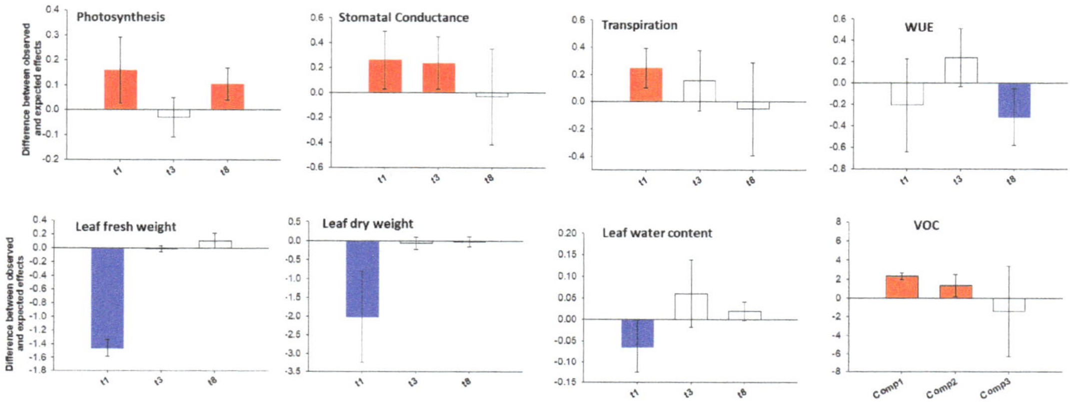

Figure 5. The combined impacts from abiotic (drought and N deficiency) and biotic stress (Infestation of *Tuta absoluta*) on morpho-physiological traits of tomato plants at 1, 3 and 8 days of treatment and VOCs emission at 8 days of treatment. The combined impact of single stressors was estimated as synergist (red color), additive (white color) or antagonistic (blue color) (greater than, equal to or less than expected effects, respectively, based on single stressor effect sizes). The vertical and error bars represent, respectively, the mean and the 95% confidence interval of the overall effect size difference between the observed and expected additive effects from combined abiotic and biotic stress on morpho-physiological and metabolic traits of tomato plants. The zero line represents the expected additive effects from combined stressors. When the means (and their 95% confidence limits) were higher than or less than the zero line, they were considered synergistic or antagonistic, respectively.

A synergic effect (Figure 5) was also involved in the increase of the metabolic traits such as the VOCs emission in COMB-treated tomato plants (Table S2). This synergic effect could be due to diverse reasons. First, the improvement of the formation reactive oxygen species (ROS) by drought stress [80] and nutrient deficiency [81] that could sensitize the VOCs response. Indeed, it is known that the VOCs are emitted by early signaling events involving the ROS during the herbivory [82]. Secondly, the abiotic stress and herbivory by *Tuta* could have increased the biosynthesis of VOCs both via hormone cross-talking and higher resource reallocation to chemical defense. For example, the ABA and JA, the main phytohormones respectively involved in the plant response to the drought and herbivory, interact among each other via molecular cross-talk [72–74] and the reallocation of plant resources to defense by modification of the gene expression profiles after herbivory was also observed [83]. Third, the increased VOCs biosynthesis in presence of both stresses could have caused their accumulation inside the leaf, leading to the formation of a steep partial pressure gradient between the atmosphere and substomatal cavities along which the VOCs could be highly emitted.

Unlike the physiologic and metabolic traits, the results obtained for the morphological traits suggested an early antagonistic effect (at 1 day of treatment) in COMB plants, evolving thereafter towards additivity (Figure 5). This result could be explained by the fact that

COMB plants prioritized their responses towards herbivory. Indeed, the COMB plants aimed to reduce their leaf water content more in the presence of *Tuta* rather than under ABIO ones compared to the CTR through a sharp increase of the leaf dry weight (Figure 1). Why? The water and the dry weight are strictly and negatively linked to the plant palatability towards the pest [84]; hence, in the presence of combined stress, tomato plants might have redirected the allocation of their resources towards the formation of carbon-based secondary compounds such as lignin and fiber contents which contribute to leaf toughness and reduce palatability [85].

3.3. Do the Tomato Responses to the Single and Combined Stress Occur at Between- or Within-Plant Levels?

Recent studies pointed out the importance of the within-individual variation of the plant responses to the abiotic and biotic stress rather than between-individual for the ecology at individual, population, and community levels [86,87]. For example, it was pointed out that a higher within-plant variation of the morpho-physiological responses permitted an improvement of the exploitation of the heterogeneously distributed resources such as light, CO_2, nutrient [24,88], an optimization of the cost-expensive defenses against herbivory and pathogens [89], and an alteration of plant–antagonist interactions [23]. In this respect, first, we assessed whether the among-treatments variance of the morpho-physiological traits and VOCs profiles of the tomato plants is higher than within-treatment ones, and then, we evaluated the between- and within-plant variance for each treatment. To do this, we used the morpho-physiological traits and VOCs observed at 8 days, that is the time at which wider and higher modifications of these traits became evident (Tables 1 and 2; Figures 1, 2 and 4). The contribution of the among- and within-treatment levels to the total variance in mean of the morpho-physiological traits and VOCs responses to the treatments considered was estimated by linear mixed models and statistically tested by restricted maximum likelihood (REML) (Figure 6).

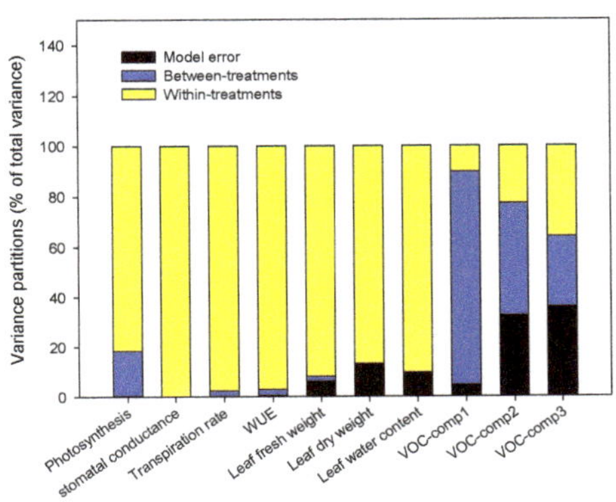

Figure 6. Dissection of total variance components of the morpho-physiological traits and VOCs responses at 8 days of treatments (control, abiotic stress, biotic stress, combined stress). Considering all treatments pooled, the contributions of treatment (blue color) and within-treatment (yellow color) level to the total variance in mean of the morpho-physiological traits and VOCs responses in the four treatments considered were estimated by linear mixed models.

Figure 6 indicated that most variance occurred at the within-treatments level, especially for the morpho-physiological traits but not for the VOCs. Hence, unlike the morpho-physiological traits, the emission of the VOCs was more dependent on the stress treatments rather than the individual plants. For such reason, only the morpho-physiological traits were used in the analysis of within-plant variation that was conducted applying a hierarchical partition to divide total variance into two levels of variation: among plants and among the leaves within the same plant. Such analysis was performed for each single treatment in order to verify its effects on within-plant phenotypic plasticity. The variance partitions varied substantially among the different treatments and traits considered (Figure 7).

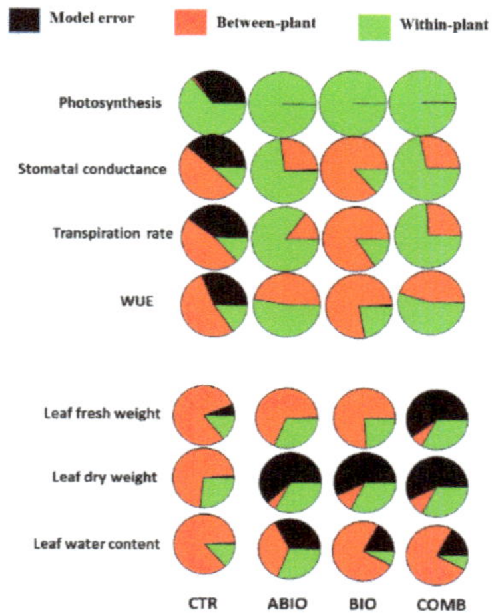

Figure 7. Nested within-treatment variance partitions (% of the total) in the morpho-physiological responses of tomato plants to different abiotic and biotic stress. The between-plant variance comprises plant within treatment (red color) and the within-plant variance involved the leaves within plant within treatment (green color). The black color indicated the model error.

In general, the physiological traits pointed out a higher within-plant variance (average 54%) while the morphological ones showed more between-plant variance (average 50%) (Figure 7). Why did the individual tomato plants modify more their leaf physiological traits than morphological ones within their shoot? This was likely owing to the physiological traits being comparatively less expensive, in terms of metabolic resources, and their modification could have been comparatively faster in response to the abiotic and biotic stress; for example, rapid local and systemic responses through specific signaling pathways have been observed in the presence of light stress [90], or herbivory [91], or during acquisition of heat tolerance [16]. Among the treatments, the stresses induced a higher within-plant variance than growing under optimal conditions: ABIO (average 58%), COMB (average 53%), BIO (average 31%) and CTR (average 22%) (Figure 7). The within-plant variation under stressful conditions could allow a better ability to exploit the stress-induced transient changes in environmental and soil resources [53,92] and the optimization of the defenses against herbivory [93].

By considering that each treatment pointed out an important within-plant variation for the morpho-physiological traits, we asked if a well-defined spatial pattern of these responses among the leaves of the tomato shoot could be revealed. In this respect, by one-way ANOVA, we compared the morpho-physiological responses to each treatment on three mature leaves located at three different positions (basal (B), intermediate (I) and apical leaf (A) placed at first, second and third node, respectively) along the shoot axes on the morpho-physiological traits for each treatments. Further, the B, I and A leaf can be also considered as local (L), no-orthostichous (nO) and orthostichous leaf (O), respectively, because the apical leaf, but not the intermediate, is linked to the basal one by vasculature connection (Figure S3). Hence, such leaf selection allowed to evaluate which among the vascular (L vs. O vs. nO) or architectural patterns (B vs. I vs. A) caused the within-plant phenotypic variability. Figure S5 depicted the vascular or architectural pattern or no pattern of the tomato responses to each treatments. The Figures 8–14 showed the results of gas exchanges (A, gs, T, and iWUE) and morphological traits (LFW, LDW, LWC) for each treatment.

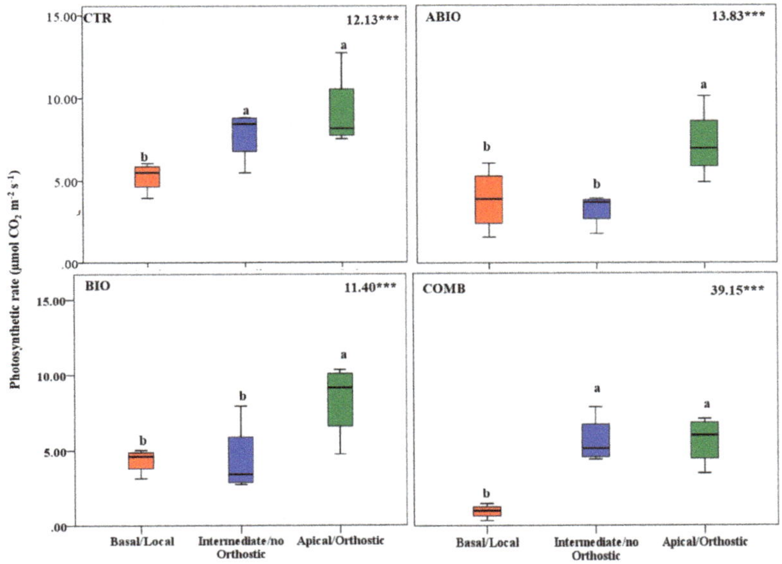

Figure 8. Photosynthetic rate of three different leaves of tomato plants exposed for 8 days at diverse stresses (abiotic (ABIO); biotic (BIO); combined (COMB)). No stresses (CTR). The box plot indicated the minimum, first quartile, median, third quartile, and maximum value. Different letters indicated significant difference among the mean groups (N = 8; $p < 0.05$ test of Tukey). The values within each panel indicated the F statistic with the p values (*** $p < 0.001$) derived from one-way ANOVA.

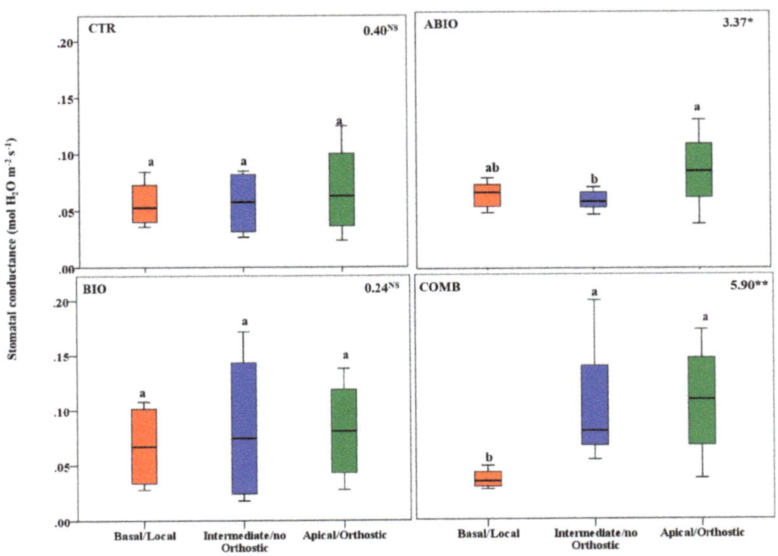

Figure 9. Stomatal conductance of three different leaves of tomato plants exposed for 8 days at diverse stresses (abiotic (ABIO); biotic (BIO); combined (COMB)). No stresses (CTR). The box plot indicated the minimum, first quartile, median, third quartile, and maximum value. Different letters indicated significant difference among the mean groups (N = 8; $p < 0.05$ test of Tukey). The values within each panel indicated the F statistic with the p values (* $0.05 < p < 0.01$; ** $0.01 < p < 0.001$; NS not significant) derived from one-way ANOVA.

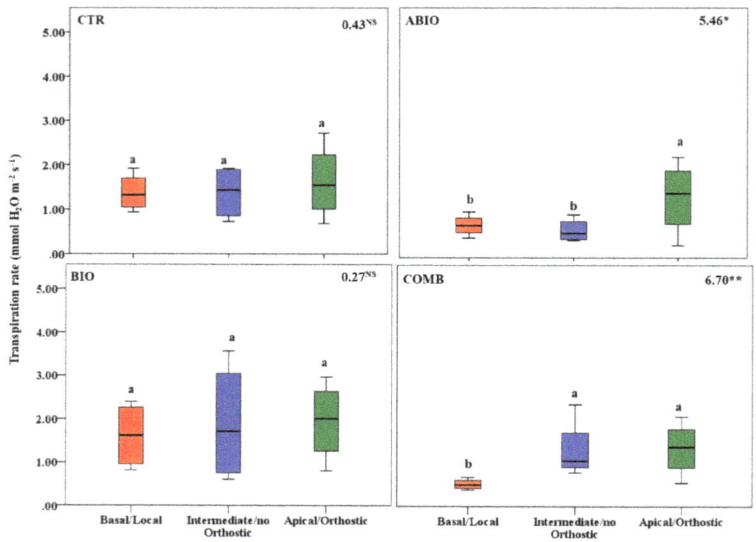

Figure 10. Transpiration rate of three different leaves of tomato plants exposed for 8 days at diverse stresses (abiotic (ABIO); biotic (BIO); combined (COMB)). No stresses (CTR). The box plot indicated the minimum, first quartile, median, third quartile, and maximum value. Different letters indicated significant difference among the mean groups (N = 8; $p < 0.05$ test of Tukey). The values within each panel indicated the F statistic with the p values (* $0.05 < p < 0.01$; ** $0.01 < p < 0.001$; NS not significant) derived from one-way ANOVA.

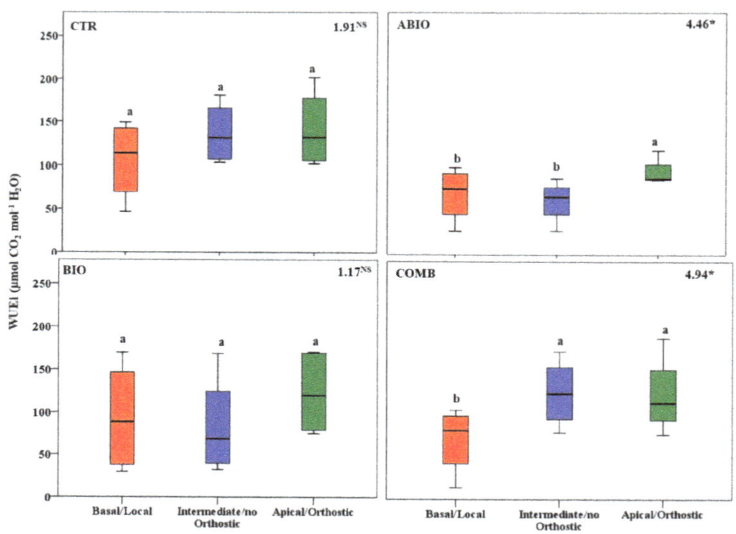

Figure 11. iWUE of three different leaves of tomato plants exposed for 8 days at diverse stresses (abiotic (ABIO); biotic (BIO); combined (COMB)). No stresses (CTR). The box plot indicated the minimum, first quartile, median, third quartile, and maximum value. Different letters indicated significant difference among the mean groups (N = 8; $p < 0.05$ test of Tukey). The values within each panel indicated the F statistic with the p values (* $0.05 < p < 0.01$; NS not significant) derived from one-way ANOVA.

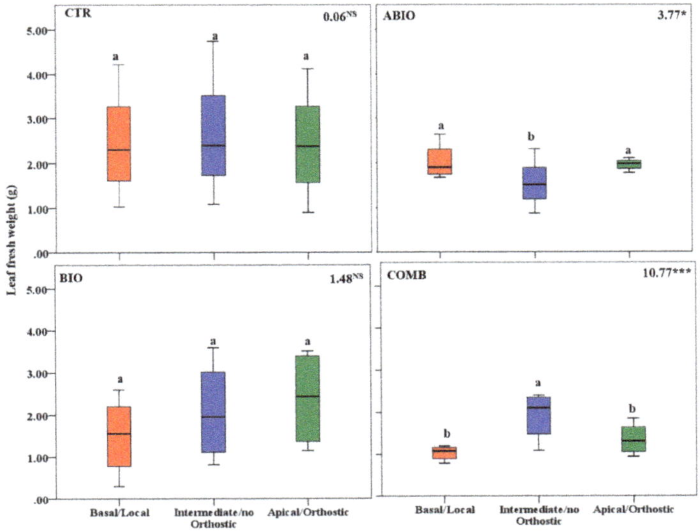

Figure 12. Leaf fresh weight of three different leaves of tomato plants exposed for 8 days at diverse stresses (abiotic (ABIO); biotic (BIO); combined (COMB)). No stresses (CTR). The box plot indicated the minimum, first quartile, median, third quartile, and maximum value. Different letters indicated significant difference among the mean groups (N = 8; $p < 0.05$ test of Tukey). The values within each panel indicated the F statistic with the p values (* $0.05 < p < 0.01$; *** $p < 0.001$; NS not significant) derived from one-way ANOVA.

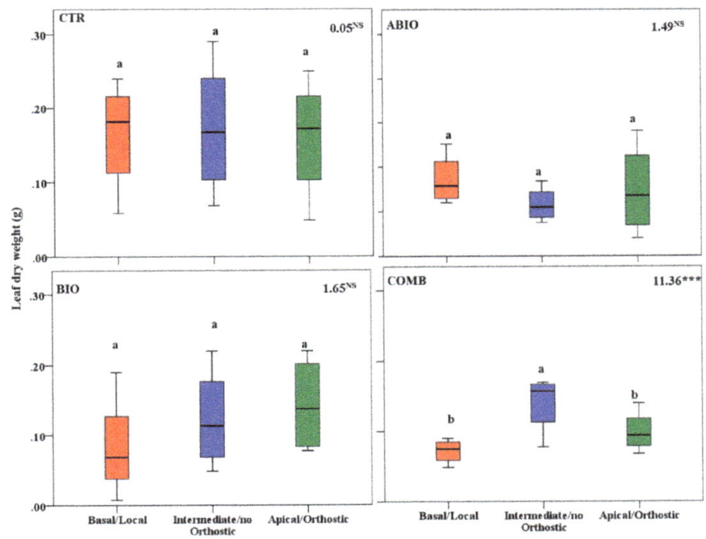

Figure 13. Leaf dry weight of three different leaves of tomato plants exposed for 8 days at diverse stresses (abiotic (ABIO); biotic (BIO); combined (COMB)). No stresses (CTR). The box plot indicated the minimum, first quartile, median, third quartile, and maximum value. Different letters indicated significant difference among the mean groups (N = 8; $p < 0.05$ test of Tukey). The values within each panel indicated the F statistic with the p values (*** $p < 0.001$; NS not significant) derived from one-way ANOVA.

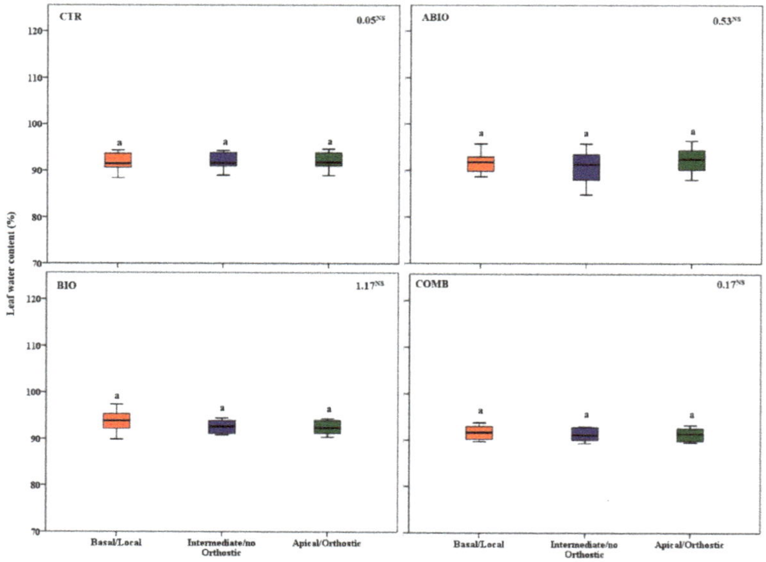

Figure 14. Leaf water content of three different leaves of tomato plants exposed for 8 days at diverse stresses (abiotic (ABIO); biotic (BIO); combined (COMB)). No stresses (CTR). The box plot indicated the minimum, first quartile, median, third quartile, and maximum value. Different letters indicated significant difference among the mean groups (N = 8; $p < 0.05$ test of Tukey). The values within each panel indicated the F statistic with the p values (NS not significant) derived from one-way ANOVA.

An overall result is that, unlike the CTR and BIO, the ABIO and the COMB treatments significantly modified all the morpho-physiological traits among the three tomato leaves, except the LWC and A (Figures 8–14). Further, by comparing the spatial patterns of plant responses (Figure S5), the physiological responses (A, g_s, T, and iWUE) to the ABIO and COMB treatments suggested an architectural pattern while the morphological ones pointed out a vascular pattern (Figures 8–14). The abiotic stressors are known to strongly influence the plant photosynthetic traits in relation to the leaf position/age, reflecting an architecture pattern, in order to preserve the highly valuable tissues, such as the young leaf [94]. The vascular pattern of the LFW and LDW in response to the COMB treatment could have been due to the ABA-JA cross-talking signaling pathways [72–74] which could have occurred between the two vascularly connected leaves (local and orthostic). In this respect, we could hypothesize that the *Tuta* larvae placed on the local leaf of tomato plants could have triggered, by vascular connection, these hormonal signaling pathways which could have redirected the photosynthetic resources towards the defense compounds rather than leaf growth ones. This kind of hormone cross-talk signaling interaction among the different vascularly connected leaves within the plant has already been observed. For example, abiotic stresses antagonized the immune responses by ABA-SA hormonal interaction in older leaves of Arabidopsis, but such effect was suppressed in the younger leaf through a signaling component of the SA pathway [95].

4. Conclusions and Future Directions

The present study has been addressed to answer questions related to the tomato responses in the presence of combined abiotic stress (drought and N deficiency) and herbivore infestation, based on the prediction that crop plants will have to face an increased incidence of both detrimental factors in a changing climate. A first result of the present study was that, respect to each single stress taken alone, the combination of drought, N deficiency and *Tuta* infestation caused a stronger negative impact on the tomato morpho-physiological traits and induced a specific VOCs blend. It is likely that hormone cross-talking regulating the signaling and metabolic systems of the plant responses could be assumed as an explanation. Interestingly, and unlike each single stress, the VOCs blend induced by stress combination contained, among others, the homoterpene 4,8-dimethyl-1,3,7-nonatriene, known to be a rare and fundamental plant alarm volatile, as well as methyl salicylate, a well-known herbivore-induced plant volatile which attracts natural enemies and affects herbivore behavior.

A second result and main outcome of the present study was the relatively rapid responses of the tomato plants to the COMB treatment and the synergistic effects for the physiological and VOCs responses, but antagonistic for the morphological ones. In this respect, no-additive effects of the single stress in tomato response to the combined stress were put into evidence. This remarkably suggests that a "new stress state" characterized by specific signaling pathways and gene expression, and probably orchestrated by hormonal interactions, could be evoked in tomato plants stressed by the combination of drought, N deficiency and *Tuta* infestation.

Finally, except for the VOCs emission, the stressful conditions induced a higher within-plant variance in tomato with the abiotic and combined stress being the most influential. The increase of the stress-induced variability of the morpho-physiological responses within the tomato plants is of interest, because it supports the view that a high within-plant variance could be of help for the defense against herbivore infestation and for maximizing the exploitation efficiency of scarce soil resources.

In deployment of the stress-adaptation strategies by the modification of the morpho-physiological traits, the plants use a signaling network consisting of several interacting pathways such as, for example, the production and detoxification of reactive oxygen species (ROS) and calcium-, phytohormone-, and MAPK-signaling pathway regulated at the multi-genic level. These metabolic pathways in plants subjected to the combination of different abiotic and biotic stress resulted in "shared" or "unique" responses with respect to those pointed out in presence of the single stress. Future research direction should be based on the investigation of these complex molecular networks that produce a "new stress state" tailored for stress combinations. By omic technologies, it is possible to identify specific genes involved in such shared and unique responses under combined stresses, which is an important step toward developing potential stress tolerance traits useful for providing multiple stress resistance to the crops.

Supplementary Materials: The following supporting information can be downloaded at: https://www.mdpi.com/article/10.3390/life12111804/s1, Figure S1: Protocol schedule including tomato growth and treatments and plant sampling events, Figure S2: Trap for *Tuta absoluta*'s larva, Figure S3: Evaluation of the degree of vascular connectivity among leaves located in different position along the shoot. Figure S4: Principal component analysis applied to volatiles emission. Figure S5: Tomato responses to the experimental conditions resembling the vascular pattern or architectural pattern or no pattern (blue line). Table S1: Volatile organic compounds identified using gas chromatography/mass spectrometry (GC/MS), Table S2: Volatile organic compounds (Area %) from the leaves of tomato plants, Table S3: Permanova results of the VOCs emission, Table S4: PERMANOVA pairwise comparison between treatments.

Author Contributions: Investigation and formal analysis, R.V.; writing and editing, R.C. and M.B.; methodology and review, F.L., V.P. and L.Z.; conceptualization, review and supervision, A.S. All authors have read and agreed to the published version of the manuscript.

Funding: This research has received partial funding from the European Commission's 7th Framework Programme under the grant agreement no. 618127, ERA-NET action ARIMNET2 2015 call, project "Sustainable Tomato Production"—(STomP).

Institutional Review Board Statement: Not applicable.

Informed Consent Statement: Not applicable.

Data Availability Statement: The authors declare that the data supporting the findings of this study are available within the article and Supplemental Materials, as well as from the corresponding author upon reasonable request.

Acknowledgments: We thank the PhD course "Scienze Agrarie, Alimentari e Forestali" of the University "Mediterranea" of Reggio Calabria and the Department AGRARIA at Università Mediterranea of Reggio Calabria for supporting RV.

Conflicts of Interest: The authors declare no conflict of interest.

References

1. Heil, M.; Bostock, R.M. Induced systemic resistance (ISR) against pathogens in the context of induced plant defences. *Ann. Bot.* **2002**, *89*, 503–512. [CrossRef] [PubMed]
2. He, M.; He, C.Q.; Ding, N.Z. Abiotic stresses: General defenses of land plants and chances for engineering multistress tolerance. *Front. Plant Sci.* **2018**, *9*, 1771. [CrossRef]
3. Maron, J.L.; Cron, E. Herbivory: Effects on plant abundance, distribution and population growth. *Proc. R. Soc. B* **2006**, *273*, 2575–2584. [CrossRef]
4. Mordecai, E.A. Pathogen impacts on plant communities: Unifying theory, concepts, and empirical work. *Ecol. Monogr.* **2011**, *81*, 429–441. [CrossRef]
5. Pandey, P.; Irulappan, V.; Bagavathiannan, M.V.; Senthil-Kumar, M. Impact of combined abiotic and biotic stresses on plant growth and avenues for crop improvement by exploiting physio-morphological traits. *Front. Plant Sci.* **2017**, *8*, 537. [CrossRef]
6. Vescio, R.; Abenavoli, M.R.; Sorgonà, A. Single and combined abiotic stress in maize root morphology. *Plants* **2021**, *10*, 5. [CrossRef]
7. Vescio, R.; Malacrinò, A.; Bennett, A.E.; Sorgonà, A. Single and combined abiotic stressors affect maize rhizosphere bacterial microbiota. *Rhizosphere* **2021**, *17*, 100318. [CrossRef]
8. Tiziani, R.; Miras-Moreno, B.; Malacrinò, A.; Vescio, R.; Lucini, L.; Mimmo, T.; Cesco, S.; Sorgonà, A. Drought, heat, and their combination impact the root exudation patterns and rhizosphere microbiome in maize roots. *Environ. Exp. Bot.* **2022**, *203*, 105071. [CrossRef]
9. Havko, N.E.; Das, M.R.; McClain, A.M.; Kapali, G.; Sharkey, T.D.; Howe, G.A. Insect herbivory antagonizes leaf cooling responses to elevated temperature in tomato. *Proc. Nat. Acad. Sci. USA* **2020**, *117*, 2211–2217. [CrossRef]
10. Catola, S.; Centritto, M.; Cascone, P.; Ranieri, A.; Loreto, F.; Calamai, L.; Balestrini, R.; Guerrieri, E. Effects of single or combined water deficit and aphid attack on tomato volatile organic compound (VOC) emission and plant-plant communication. *Environ. Exp. Bot.* **2018**, *153*, 54–62. [CrossRef]
11. Copolovici, L.; Kännaste, A.; Remmel, T.; Niinemets, Ü. Volatile organic compound emissions from *Alnus glutinosa* under interacting drought and herbivory stresses. *Environ. Exp. Bot.* **2014**, *100*, 55–63. [CrossRef] [PubMed]
12. Bansal, S.; Hallsby, G.; Löfvenius, M.O.; Nilsson, M.C. Synergistic, additive and antagonistic impacts of drought and herbivory on *Pinus sylvestris*: Leaf, tissue and whole-plant responses and recovery. *Tree Physiol.* **2013**, *33*, 451–463. [CrossRef] [PubMed]
13. Mason, C.J.; Villari, C.; Keefover-Ring, K.; Jagemann, S.; Zhu, J.; Bonello, P.; Raffa, K.F. Spatial and temporal components of induced plant responses in the context of herbivore life history and impact on host. *Funct. Ecol.* **2017**, *31*, 2034–2050. [CrossRef]
14. Orians, C.M. Herbivores, vascular pathways, and systemic induction: Facts and artifacts. *J. Chem. Ecol.* **2005**, *31*, 2231–2242. [CrossRef] [PubMed]
15. Park, S.W.; Kaimoyo, E.; Kumar, D.; Mosher, S.; Klessig, D.F. Methyl salicylate is a critical mobile signal for plant systemic acquired resistance. *Science* **2007**, *318*, 113–116. [CrossRef] [PubMed]
16. Slot, M.; Krause, G.H.; Krause, B.; Hernández, G.G.; Winter, K. Photosynthetic heat tolerance of shade and sun leaves of three tropical tree species. *Photosynth. Res.* **2019**, *141*, 119–130. [CrossRef]
17. Rubio, G.; Sorgonà, A.; Lynch, J.P. Spatial mapping of phosphorus influx in bean root systems using digital autoradiography. *J. Exp. Bot.* **2004**, *55*, 2269–2280. [CrossRef]
18. Sorgonà, A.; Abenavoli, M.R.; Gringeri, P.G.; Cacco, G. Comparing morphological plasticity of root orders in slow- and fast-growing citrus rootstocks supplied with different nitrate levels. *Ann. Bot.* **2007**, *100*, 1287–1296. [CrossRef]
19. Boege and Marquis Facing herbivory as you grow up: The ontogeny of resistance in plants. *Trends Ecol. Evol.* **2005**, *20*, 441–444. [CrossRef]
20. Holeski, L.M. Within and between generation phenotypic plasticity in trichome density of *Mimulus guttatus*. *J. Evol. Biol.* **2007**, *20*, 2092–2100. [CrossRef]
21. Kollist, H.; Zandalinas, S.I.; Sengupta, S.; Nuhkat, M.; Kangasjärvi, J.; Mittler, R. Rapid responses to abiotic stress: Priming the landscape for the signal transduction network. *Trends Plant Sci.* **2019**, *24*, 25–37. [CrossRef] [PubMed]
22. Hidalgo, J.; Rubio de Casas, R.; Muñoz, M.A. Environmental unpredictability and inbreeding depression select for mixed dispersal syndromes. *BMC Evol. Biol.* **2016**, *16*, 71. [CrossRef] [PubMed]
23. Wetzel, W.C.; Meek, M.H. Physical defenses and herbivory vary more within plants than among plants in the tropical understory shrub *Piper polytrichum*. *Botany* **2019**, *97*, 113–121. [CrossRef]

24. Herrera, C.M.; Medrano, M.; Bazaga, P. Continuous within plant variation as a source of intraspecific functional diversity: Patterns, magnitude, and genetic correlates of leaf variability in *Helleborus foetidus* (Ranunculaceae). *Am. J. Bot.* **2015**, *102*, 225–232. [CrossRef] [PubMed]
25. Desneux, N.; Luna, M.G.; Guillemaud, T.; Urbaneja, A. The invasive South American tomato pinworm, *Tuta absoluta*, continues to spread in Afro-Eurasia and beyond—The new threat to tomato world production. *J. Pest Sci.* **2011**, *84*, 403–408. [CrossRef]
26. Han, P.; Lavoir, A.V.; Le Bot, J.; Amiens-Desneux, E.; Desneux, N. Nitrogen and water availability to tomato plants triggers bottom-up effects on the leafminer *Tuta absoluta*. *Sci. Rep.* **2014**, *4*, 4455. [CrossRef]
27. Larbat, R.; Adamowicz, S.; Robin, C.; Han, P.; Desneux, N.; Le Bot, J. Interrelated responses of tomato plants and the leaf miner *Tuta absoluta* to nitrogen supply. *Plant Biol.* **2016**, *18*, 495–504. [CrossRef]
28. Ahmed, I.M.; Dai, H.; Zheng, W.; Cao, F.; Zhang, G.; Sun, D.; Wu, F. Genotypic differences in physiological characteristics in the tolerance to drought and salinity combined stress between Tibetan wild and cultivated barley. *Plant Physiol. Biochem.* **2013**, *63*, 49–60. [CrossRef]
29. Schädler, M.; Jung, G.; Auge, H.; Brandl, R. Palatability, decomposition and insect herbivory: Patterns in a successional old-field plant community. *Oikos* **2003**, *103*, 121–132. [CrossRef]
30. Flexas, J.; Carriquí, M.; Nadal, M. Gas exchange and hydraulics during drought in crops: Who drives whom? *J. Exp. Bot.* **2018**, *69*, 3791–3795. [CrossRef]
31. Schwarz, D.; Kläring, H.P.; van Iersel, M.W.; Ingram, K.T. Growth and Photosynthetic response of tomato to nutrient solution concentration at two light levels. *J. Am. Soc. Hort. Sci.* **2002**, *127*, 984–990. [CrossRef]
32. Kerchev, P.I.; Fenton, B.; Foyer, C.H.; Hancock, R.D. Plant responses to insect herbivory: Interactions between photosynthesis, reactive oxygen species and hormonal signalling pathways. *Plant Cell Environ.* **2012**, *35*, 441–453. [CrossRef] [PubMed]
33. Holopainen, J.K.; Gershenzon, J. Multiple stress factors and the emission of plant VOCs. *Trends Plant Sci.* **2010**, *15*, 176–184. [CrossRef] [PubMed]
34. IPCC. *Climate Change and Land: An IPCC Special Report on Climate Change, Desertification, Land Degradation, Sustainable Land Management, Food Security, and Greenhouse Gas Fluxes in Terrestrial Ecosystems*; IPCC: Geneva, Switzerland, 2019.
35. Seibt, U.; Rajabi, A.; Griffiths, H.; Berry, J.A. Carbon isotopes and water use efficiency: Sense and sensitivity. *Oecologia* **2008**, *155*, 441–454. [CrossRef]
36. Jin, X.; Shi, C.; Yu, C.Y.; Yamada, T.; Sacks, E.J. Determination of leaf water content by visible and near-infrared spectrometry and multivariate calibration in *Miscanthus*. *Front. Plant Sci.* **2017**, *8*, 721. [CrossRef] [PubMed]
37. Van den Dool, H.; Kratz, P.D. A generalization of the retention index system including linear temperature programmed gas-liquid partition chromatography. *J. Chromatogr.* **1963**, *11*, 463–471. [CrossRef]
38. R Core Team. *R: A Language and Environment for Statistical Computing*; R Foundation for Statistical Computing: Vienna, Austria, 2020. Available online: https://www.R-project.org/ (accessed on 3 June 2020).
39. Rohart, F.; Gautier, B.; Singh, A.; Lê Cao, K.A. mixOmics: An R package for 'omics feature selection and multiple data integration. *PLoS Comput. Biol.* **2017**, *13*, e1005752. [CrossRef]
40. Darling, E.S.; McClanahan, T.R.; Côté, I.M. Combined effects of two stressors on Kenyan coral reefs are additive or antagonistic, not synergistic. *Conserv. Lett.* **2010**, *3*, 122–130. [CrossRef]
41. Orians, C.M.; Pomerleau, J.; Ricco, R. Vascular architecture generates fine scale variation in the systemic induction of proteinase inhibitors in tomato. *J. Chem. Ecol.* **2000**, *26*, 471–485. [CrossRef]
42. Zywiec, M.; Delibes, M.; Fedriani, J.M. Microgeographical, interindividual, and intra-individual variation in the flower characters of Iberian pear *Pyrus bourgaeana* (Rosaceae). *Oecologia* **2012**, *169*, 713–722. [CrossRef]
43. Liang, G.; Liu, J.; Zhang, J.; Guo, J. Effects of drought stress on photosynthetic and physiological parameters of tomato. *J. Am. Soc. Hort. Sci.* **2020**, *145*, 12–17. [CrossRef]
44. García, A.L.; Marcelis, L.; Garcia-Sanchez, F.; Nicolas, N.; Martínez, V. Moderate water stress affects tomato leaf water relations in dependence on the nitrogen supply. *Biol. Plant.* **2007**, *51*, 707–712. [CrossRef]
45. Tamburino, R.; Vitale, M.; Ruggiero, A.; Sassi, M.; Sannino, L.; Arena, S.; Grillo, S. Chloroplast proteome response to drought stress and recovery in tomato (*Solanum lycopersicum* L.). *BMC Plant Biol.* **2017**, *17*, 40. [CrossRef] [PubMed]
46. Grinnan, R.; Carter, T.E.; Johnson, M.T.J. The effects of drought and herbivory on plant–herbivore interactions across 16 soybean genotypes in a field experiment. *Ecol. Entomol.* **2013**, *38*, 290–302. [CrossRef]
47. Zangerl, A.R.; Hamilton, J.G.; Miller, T.J.; Crofts, A.R.; Oxborough, K.; Berenbaum, M.R.; DeLucia, E.H. Impact of folivory on photosynthesis is greater than the sum of its holes. *Proc. Nat. Acad. Sci. USA* **2002**, *99*, 1088–1091. [CrossRef]
48. Thomson, V.P.; Cunningham, S.A.; Ball, M.C.; Nicotra, A.B. Compensation for herbivory by Cucumis sativus through increased photosynthetic capacity and efficiency. *Oecologia* **2003**, *134*, 167–175. [CrossRef]
49. Aldea, M.; Hamilton, J.G.; Resti, J.P.; Zangerl, A.R.; Berenbaum, M.R.; Delucia, E.H. Indirect effects of insect herbivory on leafgas exchange in soybean. *Plant Cell Environ.* **2005**, *28*, 402–411. [CrossRef]
50. Coqueret, V.; Le Bot, J.; Larbat, R.; Desneux, N.; Robin, C.; Adamowicz, S. Nitrogen nutrition of tomato plant alters leafminer dietary intake dynamics. *J. Insect Physiol.* **2017**, *99*, 130–138. [CrossRef]
51. Nguyen, D.; Poeschl, Y.; Lortzing, T.; Hoogveld, R.; Gogol-Döring, A.; Cristescu, S.M.; Steppuhn, A.; Mariani, C.; Rieu, I.; van Dam, N.M. Interactive responses of *Solanum dulcamara* to drought and insect feeding are herbivore species-specific. *Int. J. Mol. Sci.* **2018**, *19*, 3845. [CrossRef]

52. Mou, P.; Jones, R.H.; Tan, Z.; Bao, Z.; Chen, H. Morphological and physiological plasticity of plant roots when nutrients are both spatially and temporally heterogeneous. *Plant Soil* **2013**, *364*, 373–384. [CrossRef]
53. Puglielli, G.; Catoni, R.; Spoletini, A.; Varone, L.; Gratani, L. Short-term physiological plasticity: Trade-off between drought and recovery responses in three Mediterranean *Cistus* species. *Ecol. Evol.* **2017**, *7*, 10880–10889. [CrossRef]
54. Wagner, M.R.; Mitchell-Olds, T. Plasticity of plant defense and its evolutionary implications in wild populations of *Boechera stricta*. *Evolution* **2018**, *72*, 1034–1049. [CrossRef] [PubMed]
55. Foti, V.; Araniti, F.; Manti, F.; Alicandri, E.; Giuffrè, A.M.; Bonsignore, C.P.; Castiglione, E.; Sorgonà, A.; Covino, S.; Paolacci, A.R.; et al. Profiling volatile terpenoids from Calabrian Pine stands infested by the Pine processionary Moth. *Plants* **2020**, *9*, 1362. [CrossRef] [PubMed]
56. Loreto, F.; Schnitzler, P. Abiotic stresses and induced BVOCs. *Trends Plant Sci.* **2010**, *15*, 154–166. [CrossRef] [PubMed]
57. Tholl, D. Biosynthesis and biological functions of terpenoids in plants. In *Biotechnology of Isoprenoids*; Springer: Cham, Switzerland, 2015; Volume 148, pp. 63–106.
58. Silva, D.B.; Weldegergis, B.T.; Van Loon, J.J.; Bueno, V.H. Qualitative and quantitative differences in herbivore-induced plant volatile blends from tomato plants infested by either *Tuta absoluta* or *Bemisia tabaci*. *J. Chem. Ecol.* **2017**, *43*, 53–65. [CrossRef] [PubMed]
59. Deglow, E.; Borden, J. Green leaf volatiles disrupt and enhance response to aggregation pheromones by the ambrosia beetle, *Gnathotrichus sulcatus* (Coleoptera: Scolytidae). *Can. J. For. Res.* **2011**, *28*, 1697–1705. [CrossRef]
60. Ruther, J. Retention index database for identification of general green leaf volatiles in plants by coupled capillary gas chromatography−mass spectrometry. *J. Chromatogr.* **2000**, *890*, 313–319. [CrossRef]
61. Rodriguez-Saona, C.; Parra, L.; Quiroz, A.; Isaacs, R. Variation in highbush blueberry floral volatile profiles as a function of pollination status, cultivar, time of day and flower part: Implications for flower visitation by bees. *Ann. Bot.* **2011**, *107*, 1377–1390. [CrossRef]
62. Muhammad, H.; Zamri, Z.; Ismanizan, I. Green leaf volatiles: Biosynthesis, biological functions and their applications in biotechnology. *Plant Biotechnol. J.* **2015**, *13*, 727–739.
63. Mayo-Hernández, J.; Ramírez-Chávez, E.; Molina-Torres, J.; Guillén-Cisneros, M.L.; Rodríguez-Herrera, R.; Hernández-Castillo, F.; Flores-Olivas, A.; Valenzuela Soto, J.H. Effects of *Bactericera cockerelli* herbivory on volatile emissions of three varieties of *Solanum lycopersicum*. *Plants* **2019**, *8*, 509. [CrossRef]
64. López, Y.I.Á.; Martínez-Gallardo, N.A.; Ramírez-Romero, R.; López, M.G.; Sánchez-Hernández, C.; Délano-Frier, J.P. Cross-kingdom effects of plant plant signaling via volatile organic compounds emitted by tomato (*Solanum lycopersicum*) plants infested by the greenhouse whitefly (*Trialeurodes vaporariorum*). *J. Chem. Ecol.* **2012**, *38*, 1376–1386. [CrossRef]
65. Islam, M.N.; Hasanuzzaman, A.T.M.; Zhang, Z.F.; Zhang, Y.; Liu, T.X. High level of nitrogen makes tomato plants releasing less volatiles and attracting more *Bemisia tabaci* (Hemiptera: Aleyrodidae). *Front. Plant Sci.* **2017**, *8*, 466. [CrossRef] [PubMed]
66. Giunti, G.; Benelli, G.; Conte, G.; Mele, M.; Caruso, G.; Gucci, R.; Flamini, G.; Canale, A. VOCs-mediated location of olive fly larvae by the braconid parasitoid Psyttalia concolor: A multivariate comparison among VOC bouquets from three olive cultivars. *BioMed Res. Int.* **2016**, *2016*, 7827615. [CrossRef] [PubMed]
67. Mahdavi, A.; Moradi, P.; Mastinu, A. Variation in terpene profiles of *Thymus vulgaris* in water deficit stress response. *Molecules* **2020**, *25*, 1091. [CrossRef] [PubMed]
68. Tan, X.P.; Liang, W.Q.; Liu, C.J.; Luo, P.; Heinstein, P.; Chen, X.Y. Expression pattern of (+)-delta-cadinene synthase genes and biosynthesis of sesquiterpene aldehydes in plants of *Gossypium arboreum* L. *Planta* **2000**, *210*, 644–651. [CrossRef]
69. Errard, A.; Ulrichs, C.; Kühne, S.; Mewis, I.; Drungowski, M.; Schreiner, M.; Baldermann, S. Single- versus multiple-pest infestation affects differently the biochemistry of tomato (*Solanum lycopersicum* 'Ailsa Craig'). *J. Agric. Food Chem.* **2015**, *63*, 10103–10111. [CrossRef] [PubMed]
70. Weldegergis, B.T.; Zhu, F.; Poelman, E.H.; Dicke, M. Drought stress affects plant metabolites and herbivore preference but not host location by its parasitoids. *Oecologia* **2015**, *177*, 701–713. [CrossRef]
71. Takahashi, F.; Kuromori, T.; Urano, K.; Yamaguchi-Shinozaki, K.; Shinozaki, K. Drought stress responses and resistance in plants: From cellular responses to long-distance intercellular communication. *Front. Plant Sci.* **2020**, *11*, 556972. [CrossRef]
72. Nguyen, D.; Rieu, I.; Mariani, C.; van Dam, N.M. How plants handle multiple stresses: Hormonal interactions underlying responses to abiotic stress and insect herbivory. *Plant Mol. Biol.* **2016**, *91*, 727–740. [CrossRef] [PubMed]
73. Yang, J.; Duan, G.; Li, C.; Liu, L.; Han, G.; Zhang, Y.; Wang, C. The crosstalks between jasmonic acid and other plant hormone signaling highlight the involvement of Jasmonic Acid as a core component inplant response to biotic and abiotic stresses. *Front. Plant Sci.* **2019**, *10*, 1349. [CrossRef]
74. Hossain, M.A.; Munemasa, S.; Uraji, M.; Nakamura, Y.; Mori Izumi, C.; Murata, Y. Involvement of endogenous abscisic acid in Methyl Jasmonate-induced stomatal closure in Arabidopsis. *Plant Physiol.* **2011**, *156*, 430–438. [CrossRef]
75. Atkinson, N.J.; Lilley, C.J.; Urwin, P.E. Identification of genes involved in the response of Arabidopsis to simultaneous biotic and abiotic stresses. *Plant Physiol.* **2013**, *162*, 2028–2041. [CrossRef] [PubMed]
76. Prasch, C.M.; Sonnewald, U. Simultaneous application of heat, drought, and virus to Arabidopsis plants reveals significant shifts in signaling networks. *Plant Physiol.* **2013**, *162*, 1849–1866. [CrossRef] [PubMed]

77. Coolen, S.; Proietti, S.; Hickman, R.; Davila Olivas, N.H.; Huang, P.P.; Van Verk, M.C.; Van Pelt, J.A.; Wittenberg, A.H.; De Vos, M.; Prins, M.; et al. Transcriptome dynamics of Arabidopsis during sequential biotic and abiotic stresses. *Plant J.* **2016**, *86*, 249–267. [CrossRef]
78. Nabity, P.D.; Zavala, J.A.; DeLucia, E.H. Indirect suppression of photosynthesis on individual leaves by arthropod herbivory. *Ann. Bot.* **2009**, *103*, 655–663. [CrossRef] [PubMed]
79. Schmidt, L.; Schurr, U.; Röse, U.S.R. Local and systemic effects of two herbivores with different feeding mechanisms on primary metabolism of cotton leaves. *Plant Cell Environ.* **2009**, *32*, 893–903. [CrossRef] [PubMed]
80. Chaves, M.M.; Flexas, J.; Pinheiro, C. Photosynthesis under drought and salt stress: Regulation mechanisms from whole plant to cell. *Ann. Bot.* **2009**, *103*, 551–560. [CrossRef] [PubMed]
81. Shin, R.; Berg, R.H.; Schachtman, D.P. Reactive oxygen species and root hairs in Arabidopsis root response to nitrogen, phosphorus and potassium deficiency. *Plant Cell Physiol.* **2005**, *46*, 1350–1357. [CrossRef]
82. Arimura, G.I.; Ozawa, R.; Maffei, M.E. Recent advances in plant early signaling in response to herbivory. *Int. J. Mol. Sci.* **2011**, *12*, 3723–3739. [CrossRef]
83. Thompson, G.A.; Goggin, F.L. Transcriptomics and functional genomics of plant defence induction by phloem-feeding insects. *J. Exp. Bot.* **2006**, *57*, 755–766. [CrossRef]
84. Elger, A.; Willby, N.J. Leaf dry matter content as an integrative expression of plant palatability: The case of freshwater macrophytes. *Funct. Ecol.* **2003**, *17*, 58–65. [CrossRef]
85. Grime, J.P.; Cornelissen, J.H.C.; Thompson, K.; Hodgson, J.G. Evidence of a causal connection between antiherbivore defence and the decomposition rate of leaves. *Oikos* **1996**, *77*, 489–494. [CrossRef]
86. Herrera, C.M. *Multiplicity in Unity: Plant Subindividual Variation and Interactions with Animals*; University of Chicago Press: Chicago, IL, USA, 2009.
87. Herrera, C.M. The ecology of subindividual variability in plants: Patterns, processes, and prospects. *Web Ecol.* **2017**, *17*, 51–64. [CrossRef]
88. Osada, N.; Yasumura, Y.; Ishida, A. Leaf nitrogen distribution in relation to crown architecture in the tall canopy species, Fagus crenata. *Oecologia* **2014**, *175*, 1093–1106. [CrossRef]
89. Meldau, S.; Erb, M.; Baldwin, I.T. Defence on demand: Mechanisms behind optimal defence patterns. *Ann. Bot.* **2012**, *110*, 1503–1514. [CrossRef] [PubMed]
90. Devireddy, A.R.; Zandalinas, S.I.; Gómez-Cadenas, A.; Blumwald, E.; Mittler, R. Coordinating the overall stomatal response of plants: Rapid leaf-to-leaf communication during light stress. *Sci. Signal.* **2018**, *11*, 518. [CrossRef]
91. Girón-Calva, P.S.; Li, T.; Koski, T.M.; Klemola, T.; Laaksonen, T.; Huttunen, L.; Blande, J.D. A role for volatiles in intra- and inter-plant interactions in birch. *J. Chem. Ecol.* **2014**, *40*, 1203–1211. [CrossRef]
92. Meier, I.C.; Leuschner, C. Genotypic variation and phenotypic plasticity in the drought response of fine roots of European beech. *Tree Physiol.* **2008**, *28*, 297–309. [CrossRef]
93. Rodriguez-Saona, C.R.; Rodriguez-Saona, L.E.; Frost, C.J. Herbivore induced volatiles in the perennial shrub, Vaccinium corymbosum, and their role in inter-branch signaling. *J. Chem. Ecol.* **2009**, *35*, 163–175. [CrossRef] [PubMed]
94. Schurr, U.; Walter, A.; Rascher, U. Functional dynamics of plant growth and photosynthesis–from steady-state to dynamics–from homogeneity to heterogeneity. *Plant Cell Environ.* **2006**, *29*, 340–352. [CrossRef]
95. Berens, M.L.; Wolinska, K.W.; Spaepen, S.; Ziegler, J.; Nobori, T.; Nair, A.; Krüler, V.; Winkelmüller, T.M.; Wang, Y.; Mine, A.; et al. Balancing trade-offs between biotic and abiotic stress responses through leaf age-dependent variation in stress hormone cross-talk. *Proc. Nat. Acad. Sci. USA* **2019**, *116*, 2364–2373. [CrossRef] [PubMed]

life

MDPI

Article

Application of Potassium after Waterlogging Improves Quality and Productivity of Soybean Seeds

Muhammad Abdullah Al Mamun [1], Julekha [1], Umakanta Sarker [2,*], Muhammad Abdul Mannan [1], Mohammad Mizanur Rahman [3], Md. Abdul Karim [1], Sezai Ercisli [4], Romina Alina Marc [5] and Kirill S. Golokhvast [6,*]

1 Department of Agronomy, Faculty of Agriculture, Bangabandhu Sheikh Mujibur Rahman Agricultural University, Gazipur 1706, Bangladesh
2 Department of Genetics and Plant Breeding, Faculty of Agriculture, Bangabandhu Sheikh Mujibur Rahman Agricultural University, Gazipur 1706, Bangladesh
3 Department of Soil Science, Faculty of Agriculture, Bangabandhu Sheikh Mujibur Rahman Agricultural University, Gazipur 1706, Bangladesh
4 Department of Horticulture, Faculty of Agriculture, Ataturk University, 25240 Erzurum, Turkey
5 Food Engineering Department, Faculty of Food Science and Technology, University of Agricultural Sciences and Veterinary Medicine, 400372 Cluj-Napoca, Romania
6 Siberian Federal Scientific Center of Agrobiotechnology RAS, 2b Centralnaya street, Krasnoobsk 630501, Russia
* Correspondence: umakanta@bsmrau.edu.bd (U.S.); golokhvast@sfsca.ru (K.S.G.)

Abstract: Potassium (K) improves the stress tolerance of crop plants, which varies on the timing of K application and crop varieties. Soybean is a promising crop that can easily fit with the cropping pattern during kharif I season, when water logging occurs due to sudden rain. Therefore, an experiment was conducted to determine the effect of K management on the productivity and seed quality of soybean under normal and waterlogged conditions. The treatments comprised three factors, namely soybean genotypes (BU Soybean-1 and BU Soybean-2), waterlogging (WL) (control and WL for 4 days at the flowering stage (FS)), and K application (full dose as basal and 50% as basal +50% as top dress after termination of the flooding). The trial was laid out in a randomized complete block design with three replications. Findings revealed that BU Soybean-1 produced a higher number of pods and seeds pod^{-1} under control conditions with basal application of K. On the other hand, BU Soybean-2 produced taller plants and heavier grain, improving grain and straw yield under WL conditions when K was top dressed. The varieties absorbed a higher amount of nitrogen, phosphorus, and potassium under control conditions compared to WL when K was top dressed. Similarly, the seed protein content of both varieties was higher in the control condition with a top dressing of K. However, a higher percentage of seed germination was obtained from BU Soybean-2 in the control condition with a top dressing of K. Further, more electrical conductivity and more mean germination time were recorded in the case of BU Soybean-2 under WL with the basal application of K. Split application of 50% of recommended K fertilizer after the recession of flood water could be suggested for improved grain yield in flood-affected soybean growing areas.

Keywords: potassium; soybean; water logging; yield

Citation: Mamun, M.A.A.; Julekha; Sarker, U.; Mannan, M.A.; Rahman, M.M.; Karim, M.A.; Ercisli, S.; Marc, R.A.; Golokhvast, K.S. Application of Potassium after Waterlogging Improves Quality and Productivity of Soybean Seeds. *Life* **2022**, *12*, 1816. https://doi.org/10.3390/life12111816

Academic Editors: Kousuke Hanada, Hakim Manghwar and Wajid Zaman

Received: 21 October 2022
Accepted: 3 November 2022
Published: 7 November 2022

Publisher's Note: MDPI stays neutral with regard to jurisdictional claims in published maps and institutional affiliations.

Copyright: © 2022 by the authors. Licensee MDPI, Basel, Switzerland. This article is an open access article distributed under the terms and conditions of the Creative Commons Attribution (CC BY) license (https://creativecommons.org/licenses/by/4.0/).

1. Introduction

Soybean (*Glycine max* L.) has been cultivated since the early 1970s in Bangladesh, when the Mennonite Central Committee worked in the district of greater Noakhali. Recently, the cultivation of soybean has been extended dramatically from only 5000 ha in 2005 [1] to 57,670.85 ha in 2020–2021 in Bangladesh [2]. In Bangladesh, consciousness about the high protein and nutrient content of soybean is increasing day by day [3,4]. Plants are natural sources of biochemicals with numerous phenolics, antioxidants, vitamins, flavonoids, minerals, numerous pigments, dietary fiber, protein, and carbohydrates [5–10] for human health benefits. One hundred grams of dry seed of soybean contains 30–50% protein, 277 mg calcium, 15.7 mg iron, 280 mg magnesium, 704 mg phosphorus, 1797 mg potassium and 375 μg folic acid. Diversified adaptation strategies and nutritional value make soybean more popular to growers worldwide. However, many soybean-growing countries have encouraged farmers to produce more food through supporting soybean cultivation due to its multiple use and positive impact on soil [11]. Moreover, if we compare the soybean yield, we find that average yield is 1.8 t ha^{-1} in Bangladesh, while the global average is 2.76 t ha^{-1} [12,13].

In Bangladesh, the optimum sowing time for soybean is mid-December to mid-January, and the crop is harvested during April [14]. Planting soybean in January suffers from waterlogging (WL) at the pod formation stage in March due to a change in rainfall pattern. For example, heavy torrential rains in April 2017 and super cyclone Amphan in 2020 damaged soybean crops at the late pod development stage (physiological maturity) in the greater Noakhali and Bhola areas. In 2016, there was a cyclone called NADA in early November that caused heavy rains and delayed soybean sowing, which was supposed to be occur directly after harvesting Aman rice.

Soybean prefers adequate soil oxygen for maximum productivity, but WL reduces the amount of oxygen available to the plant. The WL condition is critical water stress, which affects the adaptation of soybean and reduces grain yield because it induces a significant detrimental effect on morphological and biochemical attributes of soybean. Different plant processes such as photosynthesis, accumulation of dry matter, plant growth, and formation of flowers and pods are marked as disturbed under the WL conditions [15–17]. Any abiotic stress such as WL/drought/salinity reduces the production of crops [18], by creating oxidative damage [19,20] by reactive oxygen species (ROS) [21], which eventually generate change, i.e., membrane, DNA, and protein damage, nutrient imbalance [22,23], and diminution in photosynthetic rates and changes color pigments [24–26]. To mitigate ROS, the plant has enhanced both enzymatic and non-enzymatic antioxidants, such as tocopherols betalain, ascorbic acids, carotenoids, betacyanin, betaxanthin, chlorophyll *a*, chlorophyll *b*, beta-carotene, phenolic and flavonoids [27–32], which detoxify the ROS. Under the WL condition, the plant suffers from a deficiency of oxygen, carbon dioxide, and light. Nitrogen (N) fixation is greatly affected by WL since the nodules of soybean fix N in the soil [33,34]. Under the WL condition, plants show different symptoms such as yellowing of leaves due to chlorosis, cell damage due to necrosis, and defoliation. However, the height of the plant is drastically reduced and N fixation is retarded under excess soil moisture conditions. As a result, 20–39% yield of soybean is reduced when soybean plants are exposed to flood at the R5 stage [35]. Soil WL caused an 18% yield loss in soybean around the world when flooded during the late vegetative stage [36]. Similarly, the grain yield of soybean decreased at a higher rate when it was exposed to WL in the R4 stage as compared to the R1 stage [37]. In the case of WL at the V2 and V3 growth stage, yield loss was 20% as reported by Sullivan et al. [38]. The yield loss in soybean was due to severe disease incidence, hypoxia, stunting of shoot height, reduced root and nodule formation under WL conditions. For better crop establishment, judicial application of input is essential [39]. Potassium (K) is one of the most important macronutrients for crop growth and development. This nutrient element has a great role in cereal and legume crops. It is used to uptake water and maintain cell turgidity. The formation and translocation of starch are also regulated by K within the plant. Translocation of nutrient and protein synthesis are also influenced by K. It helps the soybean plants to cope with different stresses, diseases, pests, and balanced uptake of other nutrients. It also helps in enzyme activation during nodulation [40] and has a prominent functionality in N and P uptake. Potassium enhances the photosynthetic rate and carbohydrate production, translocation, and metabolism. Therefore, it ultimately improves grain quality and yield. The root activity of plants is reduced under excess soil moisture conditions due to the low amount of K [41].

However, the demand for K in the plant is closely related to internal metabolic paths or the rate of phosphate recycling [42].

The K element has a viable function in the development, growth, and production processes of the plant [43], as it imparts its role in the morphologic and physiological characteristics of living plant cells [44], rather than only being part of the plant structure [45] Thus, proper time of K application can protect soybean crops from nutrient deficiency and can help recover from lodging and flood damage. Under moisture stress, K improves the stress tolerance [46] and increases the dry matter [47] and yield. Therefore, it is imperative to ensure the proper time of K application for sustainable agricultural production in K-deficient soil. Thus, the present experiment was undertaken to determine the effect of K management on the yield and seed quality of soybean under excess soil moisture conditions.

2. Materials and Methods

2.1. Location

A field experiment was conducted at the agronomy research field of Bangabandhu Sheikh Mujibur Rahman Agricultural University Gazipur, Bangladesh, during the rabi season 2020. The experimental site was located between 24°09′ N and 90°26′ E under the sub-tropical climatic zone with an elevation of 8.4 m from sea level. The textural class was silty clay, containing 40% clay, 45% silt, and 15% sand, having a pH of 6.1, soil organic matter 1.20%, total N 0.11%, available P 7.21 ppm, exchangeable K 0.19 meq/100 g soil, and available S of 11 ppm. The climate is sub-tropical in nature, characterized by moderately low temperatures associated with scanty rainfall during winter. The air temperature is low in the early crop growth stage and increased gradually from January to June. Total monthly precipitation is minimum up to April but dramatically increases from May (Table 1). Prior to experimentation, a soil sample was collected from the field to determine the physicochemical properties of the soil. The soil belongs to the Salna series under the Shallow Red Brown Terrace of the Argo-ecological zone (AEZ) Madhupur Tract (AEZ-28).

Table 1. Temperature, precipitation, and relative humidity of the research site during the experiment period.

Months and Metrological Events	January	February	March	April	May	June
Average temperature (°C)	17.8	20.0	25.2	28.3	29.0	30.2
Maximum temperature (°C)	28.8	26.5	32.3	34.7	33.4	33.4
Minimum temperature (°C)	12.8	13.5	18.5	21.8	24.8	26.7
Relative humidity (%)	87	73	80	82	84	84.5
Total precipitation (mm)	31.8	2.3	16	40.2	290.5	416.3

2.2. Land Preparation and Layout

The land was prepared very well by deep and cross plowing with tractor-drawn disc-plow and rotavator followed by laddering. All uprooted weeds and stubbles were incorporated into the soil. After one week, the plots were prepared as per design. Before layout preparation, the land was fertilized with urea, triple superphosphate, muriate of potash, gypsum, and zinc sulfate at 60, 170, 100, 100, and 10 kg ha^{-1}, respectively. The fertilizers were uniformly incorporated into the plot before sowing seeds. However, muriate of potash was applied as per treatment. The unit plot size was 3 m × 4 m. Ridges were made around each plot to restrict the lateral movement of water. The blocks and unit plots were separated by 1.0 m and 1.5 m spacing, respectively.

2.3. Experimental Treatments and Design

The treatments comprised three factors, Factor A (soybean varieties): (i) BU Soybean-1 and (ii) BU Soybean-2, Factor B (WL): (i) control and (ii) WL for 4 days at the flowering stage), and Factor C (K application): (i) full dose as basal and (ii) 50% as basal +50% as top dress after the termination of the WL. The experiment was laid out in a randomized complete block design with three replications.

2.4. Seed Sowing and Crop Culture

Seeds of soybean varieties under this experiment were collected from the Department of Agronomy, Bangabandhu Sheikh Mujibur Rahman Agricultural University, Gazipur Bangladesh. The germination of the collected seed was 95%, which was confirmed by a germination test before sowing in the main field. Sowing was performed manually in lines maintaining the spacing of 30 cm from line to line and 5 cm from plant to plant. Immediately after sowing seeds, the plots were lightly irrigated to ensure uniform emergence. The seedlings emerged within five days after sowing (DAS). Thinning was performed during the appearance of the first trifoliate leaf stage, keeping one uniform and healthy seedling after every 5 cm distance in each row. Weeding and mulching were performed to keep the crop free from weeds. A sufficient amount of water was applied in each plot by supplemental irrigation twice per week up to the flowering stage of the crop.

2.5. Imposition of WL Stress

WL plots were surrounded by 30 cm deep polythene anchored into the ground and extending 30 cm above the ground to hold water. The WL treatment was imposed at the flowering stage (60 DAS). The WL stress was induced by flooding the plots completely up to 5 cm above the ground level. The treatments were continued up to 4 days of WL. Afterward, water was drained out from the treated plots. In the control (non-stress) treatment, water was applied twice per week.

2.6. Harvesting and Sampling Crops

Harvesting was performed at physiological maturity of the crop (turned brownish and became hard). A total of five plants were considered as a sample of those respective varieties for recording yield contributing characters. At each sampling, five plants were randomly selected from a single row. To avoid the border effect, the first and last rows of the plot were discarded during sampling. The sample plants were collected randomly. For taking yield data, a 1.8 m^2 area was harvested, and seeds were threshed and dried. The grain weight was taken and adjusted at 14% moisture content. The plant stems were dried and straw was recorded.

2.7. Quantification of Yield Data

Plant height was measured by a meter scale of 100 cm. The plants were cut from the ground level. The height of five plants was measured from the base to the tip of the main shoot, and the height of the five plants was averaged. All pods from the five sample plants were counted, and the average value was taken. The pod having at least one seed was counted as a filled pod and the pod having no seed was counted as an unfilled pod. Pod length was measured on a small scale of 30 cm. After collecting all the pods, ten pods were selected randomly, the length was measured, and the mean value was recorded. After separating the seeds from the pods, they were counted by hand. Then, the average value was recorded. Weight of 100 seeds was recorded for each variety treatment-wise. Total seeds from a 1.8 m^2 area were weighed by an electrical balance. The weight of seeds was converted to t ha^{-1}. The moisture content of seeds was measured using a digital moisture meter and adjusted to 14% moisture using formula (1).

$$\text{Adjusted weight} = \frac{W \times (100 - M_1)}{(100 - M_2)} \times 100 \tag{1}$$

where W is the fresh weight, and M_1 and M_2 are the fresh and adjusted moisture percent of the grain, respectively.

For taking straw yield, a 1.8 m^2 area was harvested, and seeds were separated. The plant stems were dried, and straw was recorded and converted to t ha^{-1}. Harvest index (HI) was determined using the following Formula (2):

$$HI = \frac{\text{Grain yield}}{\text{Grain yield} + \text{Straw yield}} \quad (2)$$

2.8. Determination of Seed Quality Data

Germination of seed is the most important criterion of seed quality. One hundred seeds harvested from different treatments were used and replicated three times. Seeds were placed in a 9 cm petri dish containing filter paper soaked with distilled water. The petri-dishes were placed in an incubator at 30 °C until the completion of germination. Seedlings were counted every day, and a seed was considered to be germinated as the seed coat ruptured and the radical came out 2 mm in length. The final germination count was made according to ISTA [48]. Germination percentages were calculated by using the following Formula (3):

$$\text{Germination (\%)} = \frac{\text{No. of seeds germinated}}{\text{No. of seeds incubated for germination}} \times 100 \quad (3)$$

The simplest method is to make preliminary germination counts at a standard time before germination is completed. The seed sample that produces the largest number of germinated seeds at the preliminary count will produce the fastest growing seedlings and the fastest stand establishment. The speed of germination of the seed sample was monitored by counting the germinated seedling at an interval of 24 h and counting until germination was completed. An index of the speed of germination was then calculated by adding the quotients of the daily counts divided by the number of days of germination. Thereafter, a germination index (GI) was computed by using the following Formula (4) to know the seed vigor [49].

$$GI = \frac{n}{d} \quad (4)$$

where n = number of seedlings emerging on the day 'd', d = day after planting

Seed vigor index (SVI) was calculated by using the following Formula (5):

$$SVI = \frac{\text{Seedling length (cm) Germination (\%)}}{100} \quad (5)$$

Mean germination time (MGT) was calculated by the formula (6):

$$MGT = \frac{n1 \times d1 + n2 \times d2 + n3 \times d3 + - - - - - - - - -}{\text{Total number of days}} \quad (6)$$

where n = number of germinated seeds, d = number of days

2.9. Determination of Seed Coat Leakage in Seeds

The electrical conductivity (EC) of the soybean seeds was tested using the standard procedure to determine the quality of the seeds. The seeds were weighed on an analytical balance, immersed in 75 mL of deionized water in plastic cups, and kept in a germination chamber at 25 °C for 24 h. After the seed-soaking period, the electrical conductivity of the soaking solutions was determined in a conductivity meter. The results obtained were divided by the mass of fifty seeds and expressed in µS cm^{-1} g^{-1} of seeds [50].

2.10. Determination of Nutrient Composition in Seeds

The soybean seeds were dried at 70 °C for 72 h and ground by a Wiley Mill. The ground sample was digested in concentrated H_2PO_4, and the total N concentration was determined by the micro Kjeldahl method [51]. The concentration of P and K was analyzed by digesting a 0.2 g ground sample with 6 mL of 5:2 HNO_3:$HClO_4$ [51]. Total nutrient uptake was determined by the following formula (7):

$$\text{Nutrient in grain}\left(\text{kg ha}^{-1}\right) = \frac{\text{Nutrient in grain (\%) Grain yield}\left(\text{kg ha}^{-1}\right)}{100} \tag{7}$$

The amount of protein present in seed samples was calculated from the N concentration of the seeds following formula (8).

$$\text{Protein (\%)} = \text{N (\%)} \times 5.71 \tag{8}$$

2.11. Statistical Analysis

Replication-wise means data were obtained by averaging the sample mean [52]. All data were expressed as mean ± standard deviation of the triplicate measurements [53]. The collected data of different parameters were compiled and subjected to analysis of variance by using CropStat 7.2 statistical package program. The treatment means were compared using the DMRT at a 5% level of significance [54].

3. Results

3.1. Plant Height and Pod Production

There was no significant difference between control and WL treatment regarding plant height of soybean when K was applied either basally or top dressed (Table 2). However, BU Soybean-2 produced a taller plant under WL condition (48.86 cm) and BU Soybean-1 gave the shorter plant under control condition (24.28 cm) when K was top dressed (48.86 cm). In the case of pod production, BU Soybean-1 produced a numerically higher number of pods under control (18.86 and 18.53 pods plant^{-1} with basal and top dressing of K application, respectively) than WL condition (Table 2). The longest pod (4.31 cm) was measured in BU Soybean-2 under control when K was top dressed.

Table 2. Effect of variety, fertilizer, and WL on plant height and pod production of soybean.

Soybean Varieties	Growing Condition	Plant Height (cm)		Pods Plant^{-1}		Pod Length (cm)	
		Basal K	Top Dress K	Basal K	Top Dress K	Basal K	Top Dress K
BU Soybean-1	Control	25.20 c	24.28 c	18.86	18.53	3.98 ab	3.56 c
	WL	24.78 c	24.48 c	14.40	14.73	3.76 b	3.63 c
BU Soybean-2	Control	42.83 b	41.20 b	14.46	15.66	4.31 a	3.78 b
	WL	48.14 a	48.86 a	14.93	16.20	4.16 a	4.18 a

WL, water logging, WL, water logging, Figures with similar letters in a column did not vary significantly.

3.2. Production Seeds Plant^{-1} and 100-Seed Weight

Although seeds pod^{-1} did not vary significantly due to interaction of variety, K and WL, BU Soybean-2 produced a higher number of seeds pod^{-1} when K was applied basally under control (2.80 pod^{-1}) followed by WL (2.60 pod^{-1}) condition (Table 3). In the case of BU Soybean-2, the seeds plant^{-1} was 42.26 and 40.50 with the basal application of K in control and WL condition, respectively. The 100-seed weight of both genotypes was higher under the control condition compared to WL. However, the 100-seed weight of BU Soybean-2 was higher (22.04 g) when K was top dressed followed by basal application (19.73 g) in control.

3.3. Grain and Straw Yield of Soybean

The interaction of variety, K, and WL exhibited a significant effect on the grain and straw yield of soybean (Figure 1). BU Soybean-2 produced the highest grain yield compared to BU Soybean-1 in all growing conditions and modes of K application. This variety produced 2.63 and 2.84 t ha^{-1} grain in control and 1.65 and 1.71 t ha^{-1} under WL condition with basal and top dressing of K, respectively. BU Soybean-1 produced 1.33 and 1.36 t ha^{-1}

under control and 1.24 and 1.23 t ha^{-1} under WL conditions when K was applied basally and top dressing, respectively (Figure 1). BU Soybean-2 produced a higher amount of straw yield compared to BU Soybean-1, and the straw yield of both genotypes was higher under control than WL condition (Figure 1). The straw yield of BU Soybean-2 was the highest (3.28 t ha^{-1}) under control conditions when K was top dressed followed by basal application (3.15 t ha^{-1}). Similar to grain yield, BU Soybean-1 gave higher straw yield under control than WL condition. BU Soybean-1 produced 1.66 and 1.63 t ha^{-1} under control and 1.66 and 1.45 t ha^{-1} under WL conditions when K was applied basally and top dressing, respectively (Figure 1). The HI of BU Soybean-2 was higher (0.46) under control conditions when K was top dressed. The lower (0.42) was obtained from WL control when K was applied basally in BU Soybean-1.

Table 3. Effect of variety, fertilizer, and WL on seed production and 100-seed weight of soybean.

Soybean Varieties	WL	Seeds Pod^{-1}		Seeds Plant^{-1}		100-Seed Weight (g)	
		Basal K	Top Dress K	Basal K	Top Dress K	Basal K	Top Dress K
BU Soybean-1	Control	2.66	2.45	50.54 a	45.48 ab	11.81 c	11.10 c
	WL	2.33	2.00	32.98 c	30.40 c	8.88 d	9.75 d
BU Soybean-2	Control	2.80	2.13	42.26 ab	39.08 b	19.73 a	22.04 a
	WL	2.60	2.60	40.50 ab	33.09 c	13.81 c	16.44 b

WL, water logging, WL, water logging, Figures with similar letters in a column did not vary significantly.

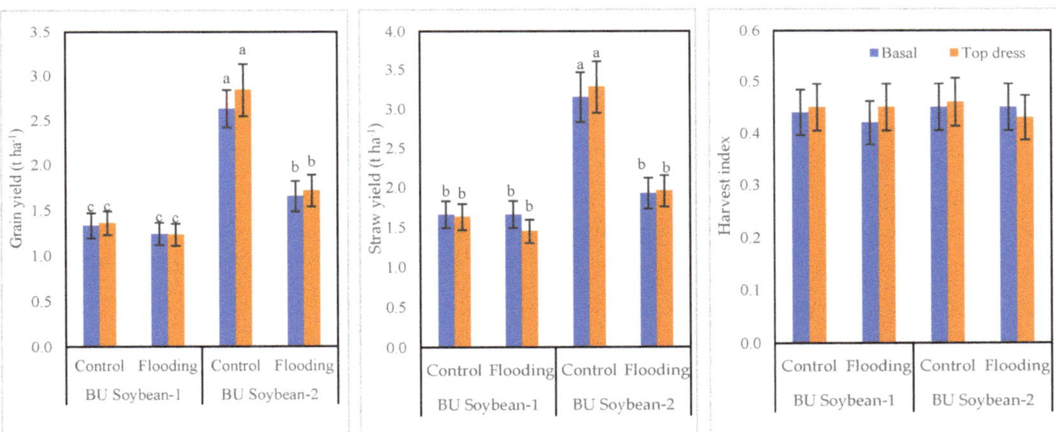

Figure 1. Interaction effect of variety, K and WL on grain and straw yield of soybean. Bar graphs indicate mean value ± standard error. Bars with similar letters did not differ significantly at $p < 0.05$ level.

3.4. Nutrient Accumulation in Soybean Seed

The interaction of variety, K, and WL exhibited a significant effect on the N, P, and K content of soybean grains. Both varieties absorbed a higher amount of N under control conditions compared to WL (Figure 2). Between two soybean varieties, BU Soybean-2 accumulated a higher amount of N in grain. BU Soybean-2 absorbed the highest amount of N (188.04 kg ha^{-1}) when K was top dressed followed by basal application under control conditions (170.93 kg ha^{-1}). This variety accumulated 90.73 and 80.88 kg N ha^{-1} in grain under WL conditions when K was applied basally and top dressing, respectively. The N content of the grains of BU Soybean-1 was accumulated at 67.73 and 72.30 kg ha^{-1} in control and 60.35 and 59.09 kg N ha^{-1} in WL condition with basal and top dressing of K fertilizer, respectively.

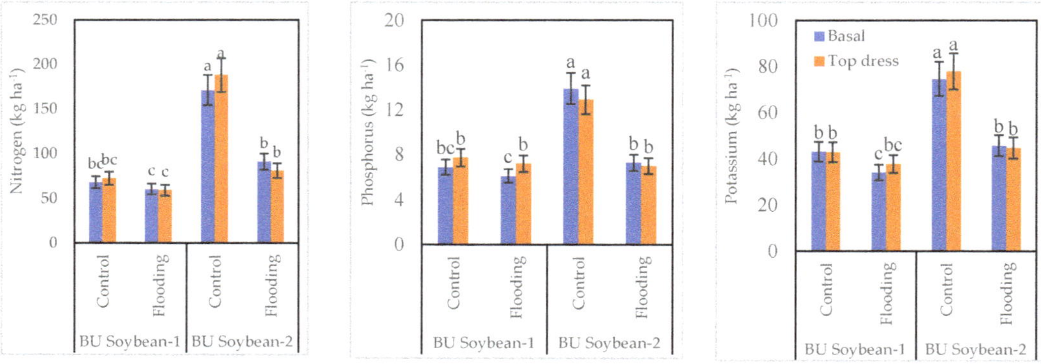

Figure 2. Interaction effect of variety, K and WL on nutrient contents in soybean seeds, Bar graphs indicate mean value ± standard error. Bars with similar letters did not differ significantly at $p < 0.05$ level.

Similar to N accumulation, a higher amount of P was taken up by BU Soybean-2 under control than WL condition (Figure 2). The P absorption in the grain of BU Soybean-2 was the highest (13.89 kg ha^{-1}) under control conditions when K was applied basally followed by top dressing (12.90 kg ha^{-1}). Again, BU Soybean-2 accumulated 7.28 and 7.00 kg P ha^{-1} in basal and top dressing of K, respectively, under WL conditions. In the case of BU Soybean-1, a higher amount of P was taken up under top dressed treatment in both growing conditions. BU Soybean-1 absorbed 7.73 and 7.20 kg P ha^{-1} under top dressing and 6.88 and 6.10 kg P ha^{-1} under basal application of K fertilizer in control and WL conditions, respectively (Figure 2).

Both varieties absorbed a higher amount of K under control conditions compared to WL. Between two soybean varieties, BU Soybean-2 accumulated a higher amount of K in grain. BU Soybean-2 absorbed the highest amount of K (77.89 kg ha^{-1}) when K was top dressed followed by basal application under control conditions (74.52 kg ha^{-1}). This variety accumulated 45.65 and 44.69 kg K ha^{-1} in grain under WL conditions when K was applied basally and top dressing, respectively. The K content of the grains of BU Soybean-1 was 42.96 and 42.77 kg ha^{-1} in control and 34.07 and 37.75 kg K ha^{-1} in WL condition with basal and top dressing of K fertilizer, respectively.

3.5. Protein Content and EC of Soybean Seed

The interaction of soybean, K, and WL exhibited a significant effect on the protein percentage of soybean (Table 4). Generally, BU Soybean-2 contained a higher amount of protein as compared to BU Soybean-1. Similarly, both soybean varieties produced higher amounts of protein under control conditions with top dress K application. The lowest amount of protein (28.48%) was found in BU Soybean-1 under WL condition when K was top dressed. BU Soybean-2 gave the highest amount of EC (129 µS cm^{-1} g^{-1}) under WL condition when K was applied basally. The second highest EC (125 µS cm^{-1} g^{-1}) was also obtained from BU Soybean-2. The lowest value of EC (82 µS cm^{-1} g^{-1}) was found in BU Soybean-1 under control when K was top dressed (Table 4). Between two varieties, BU Soybean-2 produced the heaviest seed. Under control conditions, both varieties produced the highest amount of seed weight compared to WL. BU Soybean-2 produced the heaviest seed (220.40 mg seed^{-1}) under control conditions when K was top dressed and the lowest (138.16 mg seed^{-1}) in WL when K was applied basally (Table 4). In the case of BU Soybean-1, the heaviest (118.13 mg seed^{-1}) seed was observed under control conditions, and the lighter one (88.86 mg seed^{-1}) in WL condition when K was applied basally.

Table 4. Effect of variety, fertilizer, and WL on protein, EC, and weight of soybean seed.

Soybean Varieties	Flooding	Protein (%)		EC (μS cm^{-1} g^{-1})		Seed Weight (mg $seed^{-1}$)	
		Basal K	Top Dress K	Basal K	Top Dress K	Basal K	Top Dress K
BU Soybean-1	Control	30.19 b	31.48 b	108 a	82 c	118.13 b	111.06 b
	WL	28.85 c	28.48 c	89 c	92 b	88.86 c	97.50 c
BU Soybean-2	Control	38.60 a	39.35 a	106 b	92 b	197.33 a	220.40 a
	WL	32.61 b	31.83 b	129 a	125 a	138.16 ab	164.46 ab

WL, water logging, WL, water logging, Figures with similar letters in a column did not vary significantly.

3.6. Germination and Seed Vigor Index of Soybean

The interaction of variety, K, and WL exhibited a significant effect on the germination index of soybean (Figure 3). Between the two varieties, BU Soybean-2 gave the highest germination index compared to BU Soybean-1. The germination index of BU Soybean-2 was higher (34.14) under the WL condition, while it was lower (29.56) in control when K was applied basally. In the case of BU Soybean-1, the germination index was 7.63 and 16.35 in the basal application and 16.35 and 12.50 in top dressed treatment under control and WL, respectively (Figure 3). In BU Soybean-2, the highest percentage of germination (80.0%) was found under the flooded condition when K was top dressed and a lower percentage (62.0%) in control when K was applied basally. In the case of BU Soybean-1, a higher germination percentage (34.00%) of germination was found in WL, and the lower one (14.66%) was observed in control when K was applied basally. BU Soybean-2 gave higher seed vigor index compared to BU Soybean-1. BU Soybean-2 gave the highest seed vigor index (8.45) under WL condition when K was top dressed, while the lowest one was observed in BU Soybean-1 in control when K was applied basally.

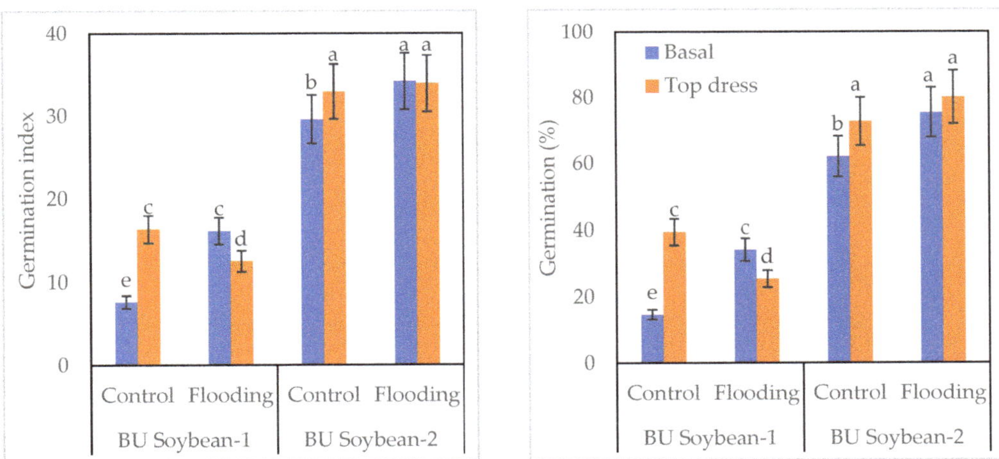

Figure 3. Effect of variety, K and WL on germination of soybean seeds. Bar graphs indicate mean value ± standard error. Bars with similar letters did not differ significantly at $p < 0.05$ level.

Between the two varieties, MGT was higher in BU Soybean-2 compared to BU Soybean-1. The MGT of BU Soybean-2 was 48.86 days under WL condition when K was top dressed followed by basal application. Similarly, MGT was 40.26 days in top dress treatment and 34.56 days for basal application of K in the control plot. However, MGT for BU Soybean-1 was 24.66 days when K was top dressed, while it was 8.60 days in basal K application plot under control conditions. Moreover, this variety needed 20.53 and 16.50 days for MGT in basal and top dress plots, respectively, under flooded conditions (Figure 4).

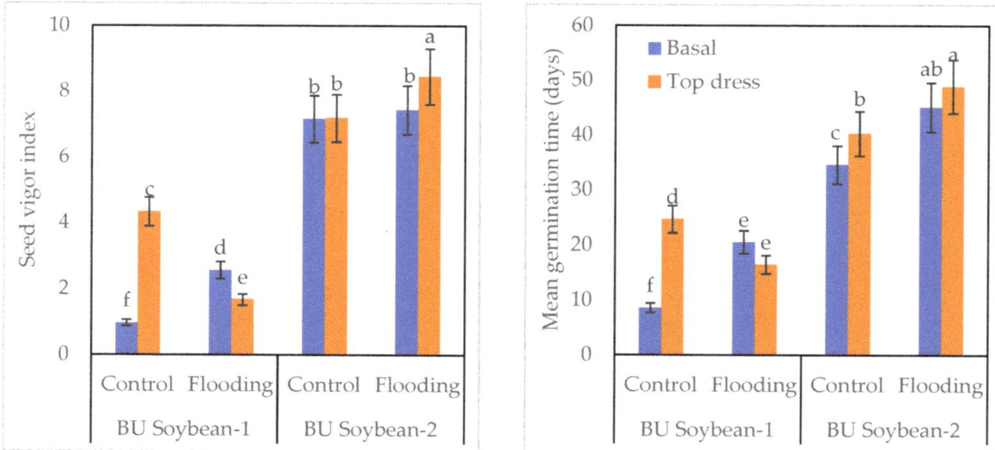

Figure 4. Effect of variety, fertilizer, and WL on vigor index and mean germination time of soybean seeds. Bar graphs indicate mean value ± standard error. Bars with similar letters did not differ significantly at $p < 0.05$ level.

4. Discussion

Nutrient element K has a viable function in the development, growth, and production process of plants [43,55]. In the present experiment, the plant height of BU Soybean-1 decreased from 25.20 cm in control to 24.78 cm in WL condition when K was applied basally. However, taller plants were measured in both testing soybean materials when K was top dressed. Moreover, BU Soybean-2 produced taller plants than BU Soybean-1 under both growing conditions and modes of K applications. The genetic difference was responsible for the different plant heights of the two varieties. The WL-induced decrease in plant height was noted in soybean [56]. Jin-Woong et al. [57] reported that the reduction in plant height under WL conditions was probably due to oxygen deficiency, anaerobic conditions, less root activity, and inhibition of synthesis and transport of photosynthetic assimilates.

Production of pods plant^{-1} is an important yield-contributing characteristic. In this study, the number of pods plant^{-1} decreased due to the imposition of WL treatment. However, BU Soybean-1 produced a higher number of pods than BU Soybean-2 under control conditions. The WL induced several physiological disturbances in growth and pod formation [17]. Jin-Woong et al. [57] and Sathi et al. [58] found that the number of pod plant^{-1} was sharply reduced due to the imposition of WL. In this study, the application of K after the recession of flood water provokes the production of more pods plant^{-1} (Table 2). The supplementation of K increased photosynthetic capacity and Chl content reported, resulting in taller plants and maximum pods plant^{-1} [59,60]. On the other hand, the longer-sized pod was produced by BU Soybean-2 more than BU Soybean-1 under both growing conditions and mode of K application. This indicated that BU Soybean-2 can produce pods of longer length with a fewer number of pods plant^{-1}. However, the lower number of pods plant^{-1} under WL conditions resulted in a lower yield [61–63]. The better field performance under WL conditions in terms of pod production with the split application of K after the recession of flood was also supported by previous findings [64,65].

The production of seed plant^{-1} or pod^{-1} and individual seed weight is directly related to grain yield. BU soybean-1 produced a higher number of seeds plant^{-1} under control compared to WL. However, both soybean varieties performed better for seed production under control with basal K application. Many experiments have explored the influence of the basal application of K on the yield and quality of wheat [66,67]. Comparatively, smaller seeds were found under WL conditions in both varieties. Ara et al. [37] and Sathi et al. [58]

found that when the plants were subjected to WL stress, 100-seed weight decreased in comparison to the control condition. However, BU Soybean-2 produced a lower number of seeds plant^{-1} but gave bigger-sized seeds. Seed weight is a genetic characteristic, and BU Soybean-2 is a bold grain soybean variety. However, split application of K fertilizer increased the 100-grain weight of soybean varieties under WL conditions (Table 3).

This indicated that the split application of K improved the production of seed through the use of another nutrient element by soybean plants. Ahmed et al. [68] reported that the test weight of maize and soybean increased by 8 and 4%, respectively, due to a higher amount of K application. The application of K at a higher rate increased photosynthesis and accumulation of a greater amount of photosynthate to grain [65–69], as split application of K after recession of flood water favors roots to absorb more minerals from the soil. On the other hand, K also helps to increase photosynthesis and production of more photo-assimilates that are ultimately stored in the seed. Thus, the 100-seed weight of BU Soybean-2 increased when K was applied after removal of flood.

The yield of soybean reduced significantly under WL conditions. The reduction of yield-contributing characteristics under WL (Tables 2 and 3) resulted in lower yield (Figure 1). Islam et al. [70] also found that the number of pods plant^{-1}, seed weight, and seed yield in mungbean were significantly affected due to soil WL stress. Amin et al. [39] and Vineela [63] recorded a significant decrease in seed yield in mungbean due to WL. In this experiment, BU Soybean-2 produced 3.6% more yield under WL conditions when K was applied after the recession of flood water as compared to the basal application of K fertilizer (Figure 1). This indicated that K fertilizer can reduce the detrimental effect of flooding. Vyas et al. [60] reported that K application significantly improved the seed yield of soybean. Uddin et al. [71] found that test weight and seed increased by K application. The greater yield and high-quality grains obtained due to K application might be due to increased photosynthesis, greater carbohydrate translocation toward the sink, and metabolism [44,72,73].

Similar to yield, the straw yield of BU Soybean-2 also increased by split application of K fertilizer under control as well as WL condition (Figure 1). This variety produced 1.92 and 1.95 t ha^{-1} straw in basal and top dressing of K, respectively, under WL conditions. A similar finding was also reported by Farhad et al. [74]. However, the detrimental effect of WL on the grain yield of soybean was also found by Beutler et al. [75] and Koger et al. [76]. Although there was no significant change in HI found in this study, Youn et al. [77] reported that HI increased only in WL soybeans. On the contrary, a reduction of HI due to WL was reported in mungbean [63,78]. According to Nguyen et al. [79], WL stress during the vegetative stage of soybean growth causes a reduction in grain yield of approximately 17–40%, and that the reproductive stage led to a 40–57% yield reduction. A strong positive relationship between K fertilizer input and grain yield has been shown [80].

Nutrient accumulation in seeds was more when K was applied after the recession of flood water (Figure 2). Ahmed et al. [68] found that all K applications improved the total N accumulation in plants. This helped to release ammonium ions from the soil and made N more available to plants. Increasing the K application increased the N, P, and K content of plants [81]. Board et al. [82] reported that the effects of WL on P were minor. It is now well established that metabolic energy is required for the active transport of N, P, and K through the root system [83,84]. Under hypoxic conditions, the stored metabolic energy of root cells is appreciably reduced, thereby suppressing the active transport of these nutrients [85,86].

Both testing varieties absorbed a higher amount of N, P, and K under control conditions compared to flooding. WL inhibits the uptake of most essential nutrients in the soil and thus leads to deficiencies in N, P, K, Mg, and Ca [87]. The N contents decreased markedly in different parts of the cotton plant under waterlogged conditions and exogenously applied K showed considerable improvement in N contents in all plant parts [88]. The application of K under WL conditions improved the accumulation of other plant nutrients such as K^+, Ca^{2+}, N, Mn^{2+}, and Fe^{2+} [59].

BU Soybean-2 accumulated the highest amount of protein under control conditions with a top dressing of K fertilizer. The second highest protein content was also obtained from BU Soybean-2 under control conditions with basal K application (Table 4). Vyas et al. [60] found that protein content was significantly higher with a split application of 37.5 kg as basal + 37.5 kg K_2O ha^{-1} at the flowering stage of the crop. Alam et al. [65] reported that the grain protein content of wheat was significantly influenced due to the application of different levels of K. During the seed quality test, BU Soybean-2 gave a higher EC value compared to BU Soybean-1. The higher EC value indicated the higher injury of the seeds during flooding. Wuebker et al. [89] observed that seeds absorbed water (imbibition) and reduced germination under WL conditions. Several reports showed a negative correlation between germination percentage and WL stress [90,91]. The germination test showed that exogenous application of K favored seed germination (Figures 3 and 4) when the field was waterlogged for up to 4 days. Seed lots with a greater germination index is considered to be more vigorous [52]. The survival rate and germination percent were quickly lost due to the WL condition where the amount of oxygen was very low [92]. Potassium is a good catalyst for seed germination and emergence. Potassium nitrate, potassium chloride, and dipotassium hydrogen phosphate are the common K salts used in seed priming [83].

Heavy rainfall, high water tables and poor drainage create water logging in areas of the world [93]. Low-oxygen in soil under waterlogged conditions limits yield of soybean [94]. Waterlogging impacts around 10–12% of agricultural soils [95], and about 6 million tons of grain per year are lost due to this stress with economic losses of approximately US $ 1.5 billion annually [96].

In Bangladesh, the cropping intensity is high during the rabi (November to March) season, which is the best time for soybean cultivation. Different natural calamities start during the month of March, when the soybean crops attain pod formation to the maturity stage. In 2017, heavy rainfall occurred during the month of April, and the torrential rains damaged the soybean at the pod development stage in the Noakhali and Bhola districts of Bangladesh. Therefore, the application of K after flooding will reduce damage to soybean and increase production, improve farmer income and ensure national food security in Bangladesh. However, the intensity of flooding damage caused during the vegetative stage and its impact on seed quality should be addressed in future research.

5. Conclusions

Waterlogging showed a detrimental effect on pods and seeds $plant^{-1}$, pod length, 100-seed weight, grain and straw yield, nutrient, and protein accumulation in soybean grain. On the other hand, flood-affected seeds had higher germination percent, seed vigor index, and electrical conductivity, and needed more mean germination time for both soybean varieties. Basal application of potassium fertilizer improved height of plant, pods, and seeds $plant^{-1}$, gave higher electrical conductivity and needed more mean germination time of both soybean varieties. On the contrary, top dressing (50% as basal + 50% as top dress after the termination of flooding) increased 100-seed weight, grain and straw yield, nutrient, protein accumulation in grain, germination percent, and seed vigor index of soybean under both control and flooding. Therefore, it might be concluded that exogenous application of K fertilizer after the recession of flood water could be recommended for higher grain yield in flood-affected soybean growing areas.

Author Contributions: Conceptualization, M.A.A.M. and M.A.K.; methodology, M.A.A.M. and U.S.; software, M.A.A.M.; validation, J., U.S., S.E., R.A.M., K.S.G. and M.A.A.M.; formal analysis, M.A.A.M. and U.S.; investigation, J.; M.A.A.M. and M.A.K.; resources, M.A.A.M.; data curation, U.S.; writing—original draft preparation, M.A.A.M. and U.S.; writing—review and editing, M.A.A.M., M.M.R., M.A.K., U.S., S.E., R.A.M. and K.S.G.; visualization, M.A.A.M. and M.A.M.; supervision, M.A.K. All authors have read and agreed to the published version of the manuscript.

Funding: This work was supported by the Decree No. 220 by the Government of the Russian Federation (Mega-grant No 220-2961-3099).

Institutional Review Board Statement: Not applicable.

Informed Consent Statement: Not applicable.

Data Availability Statement: Data recorded in the current study are available in all tables and figures of the manuscript.

Acknowledgments: The authors would like to extend their sincere appreciation to the the Decree No. 220 by the Government of the Russian Federation (Mega-grant No 220-2961-3099).

Conflicts of Interest: The authors declare no conflict of interest.

References

1. Satter, M.A.; Rahman, M.M.; Rashid, M.H.; Ali, M.S.; Alam, M.S. *Krishi Projukti Hatboi (Handbook on Agro-Technology)*; Bangladesh Agricultural Research Institute: Gazipur, Bangladesh, 2005.
2. BBS (Bangladesh Bureau of Statistics). *Yearbook of Agricultural Statistics-2021*; Bangladesh Bureau of Statistics: Dhaka, Bangladesh, 2021.
3. Mannan, M.A.; Karim, M.A.; Haque, M.M.; Khaliq, Q.A.; Higuchi, H.; Nawata, E. Response of soybean to salinity: I. Genotypic variations in salt tolerance at the vegetative stage. *Trop. Agric. Dev.* **2012**, *56*, 117–122.
4. Hossain, M.S.; Mamun, M.A.A.; Khaliq, Q.A.; Higuchi, H.; Nawata, E.; Karim, M.A. Waterlogging induced changes in morpho-physiology of soybean. *Res. Trop. Agric.* **2019**, *1*, 53–54.
5. Sarker, U.; Oba, S. Polyphenol and Flavonoid Profiles and Radical Scavenging Activity in Selected Leafy Vegetable *Amaranthus gangeticus*. *BMC Plant Biol.* **2020**, *20*, 499. [CrossRef] [PubMed]
6. Sarker, U.; Oba, S. Nutritional and Bioactive Constituents and Scavenging Capacity of Radicals in *Amaranthus hypochondriacus*. *Sci. Rep.* **2020**, *10*, 19962. [CrossRef] [PubMed]
7. Sarker, U.; Oba, S.; Ercisli, S.; Assouguem, A.; Alotaibi, A.; Ullah, R. Bioactive Phytochemicals and Quenching Activity of Radicals in Selected Drought-Resistant *Amaranthus tricolor* Vegetable Amaranth. *Antioxidants* **2022**, *11*, 578. [CrossRef]
8. Sarker, U.; Oba, S. Antioxidant Constituents of Three Selected Red and Green Color Amaranthus Leafy Vegetable. *Sci. Rep.* **2019**, *9*, 18233. [CrossRef]
9. Sarker, U.; Oba, S.; Daramy, M.A. Nutrients, Minerals, Antioxidant Pigments and Phytochemicals, and Antioxidant Capacity of the Leaves of Stem Amaranth. *Sci. Rep.* **2020**, *10*, 3892. [CrossRef]
10. Sarker, U.; Oba, S. Nutrients, Minerals, Pigments, Phytochemical, and Radical Scavenging Activity in *Amaranthus blitum* Leafy Vegetable. *Sci. Rep.* **2020**, *10*, 3868. [CrossRef]
11. Siamabele, B. The significance of soybean production in the face of changing climates in Africa. *Cogent Food Agric.* **2021**, *7*, 1933745. [CrossRef]
12. BBS (Bangladesh Bureau of Statistics). *Statistical Pocket Book of Bangladesh*; Bangladesh Bureau of Statistics, Ministry of Planning, Government of People's Republic of Bangladesh: Dhaka, Bangladesh, 2020; p. 117.
13. Terzic, D.; Popovic, V.; Tatic, M.; Vasileva, V.; Dekic, V.; Ugrenovic, V.; Popovic, S.; Avdic, P. Soybean area, yield and production in world. XXII Eco-Conference, 2018. *Ecol. Mov. Novi Sad* **2018**, *8*, 135–145.
14. Miah, A.A.; Karim, M.A.; Mamun, M.A.A.; Khan, M.A.R.; Akter, N.; Haque, M.M. Planting time effect on phenology and yield of early maturing dwarf soybean genotypes. *Bangladesh J. Ecol.* **2020**, *2*, 19–24.
15. Pociecha, E.; Kościelniak, J.; Filek, W. Effects of root flooding and stage of development on the growth and photosynthesis of field bean (*Vicia faba* L.). *Acta Physiol. Plant.* **2008**, *30*, 529–535. [CrossRef]
16. Celik, G.; Turhan, E. Genotypic variation in growth and physiological responses of common bean (*Phaseolus vulgaris* L.) seedlings to flooding. *Afr. J. Biotechnol.* **2011**, *10*, 7372–7380.
17. Hasanuzzaman, M.; Nahar, K.; Rahman, A.; Mahmud, J.A.; Hossain, M.S.; Fujita, M. Soybean production and environmental stresses. In *Environmental Stresses in Soybean Production*; Academic Press: Cambridge, MA, USA, 2016; pp. 61–102.
18. Sarker, U.; Oba, S. Salinity Stress Enhances Color Parameters, Bioactive Leaf Pigments, Vitamins, Polyphenols, Flavonoids and Antioxidant Activity in Selected *Amaranthus* Leafy Vegetables. *J. Sci. Food Agric.* **2019**, *99*, 2275–2284. [CrossRef]
19. Sarker, U.; Oba, S. Catalase, Superoxide Dismutase and Ascorbate-Glutathione Cycle Enzymes Confer Drought Tolerance of *A. tricolor*. *Sci. Rep.* **2018**, *8*, 16496. [CrossRef]
20. Sarker, U.; Oba, S. The Response of Salinity Stress-Induced *A. tricolor* to Growth, Anatomy, Physiology, Non-Enzymatic and Enzymatic Antioxidants. *Front. Plant Sci.* **2020**, *11*, 559876. [CrossRef]
21. Sarker, U.; Oba, S. Drought Stress Effects on Growth, ROS Markers, Compatible Solutes, Phenolics, Flavonoids, and Antioxidant Activity in *Amaranthus tricolor*. *Appl. Biochem. Biotechnol.* **2018**, *186*, 999–1016. [CrossRef]
22. Sarker, U.; Oba, S. Drought Stress Enhances Nutritional and Bioactive Compounds, Phenolic Acids and Antioxidant Capacity of *Amaranthus* Leafy Vegetable. *BMC Plant Biol.* **2018**, *18*, 258. [CrossRef]
23. Sarker, U.; Islam, M.T.; Oba, S. Salinity Stress Accelerates Nutrients, Dietary Fiber, Minerals, Phytochemicals and Antioxidant Activity in *Amaranthus tricolor* Leaves. *PLoS ONE* **2018**, *13*, 0206388. [CrossRef]
24. Sarker, U.; Oba, S. Response of Nutrients, Minerals, Antioxidant Leaf Pigments, Vitamins, Polyphenol, Flavonoid and Antioxidant Activity in Selected Vegetable Amaranth under Four Soil Water Content. *Food Chem.* **2018**, *252*, 72–83. [CrossRef]

25. Sarker, U.; Oba, S. Augmentation of Leaf Color Parameters, Pigments, Vitamins, Phenolic Acids, Flavonoids and Antioxidant Activity in Selected *Amaranthus tricolor* under Salinity Stress. *Sci. Rep.* **2018**, *8*, 12349. [CrossRef] [PubMed]
26. Hossain, M.N.; Sarker, U.; Raihan, M.S.; Al-Huqail, A.A.; Siddiqui, M.H.; Oba, S. Influence of Salinity Stress on Color Parameters, Leaf Pigmentation, Polyphenol and Flavonoid Contents, and Antioxidant Activity of *Amaranthus lividus* Leafy Vegetables. *Molecules* **2022**, *27*, 1821. [CrossRef] [PubMed]
27. Sarker, U.; Lin, Y.P.; Oba, S.; Yoshioka, Y.; Ken, H. Prospects and potentials of underutilized leafy Amaranths as vegetable use for health promotion. *Plant Physiol. Biochem.* **2022**, *182*, 104–123. [CrossRef]
28. Sarker, U.; Rabbani, M.G.; Oba, S.; Eldehna, W.M.; Al-Rashood, S.T.; Mostafa, N.M.; Eldahshan, O.A. Phytonutrients, Colorant Pigments, Phytochemicals, and Antioxidant Potential of Orphan Leafy *Amaranthus* Species. *Molecules* **2022**, *27*, 2899. [CrossRef] [PubMed]
29. Sarker, U.; Oba, S.; Alsanie, W.F.; Gaber, A. Characterization of Phytochemicals, Nutrients, and Antiradical Potential in Slim Amaranth. *Antioxidants* **2022**, *11*, 1089. [CrossRef]
30. Sarker, U.; Iqbal, M.A.; Hossain, M.N.; Oba, S.; Ercisli, S.; Muresan, C.C.; Marc, R.A. Colorant Pigments, Nutrients, Bioactive Components, and Antiradical Potential of Danta Leaves (*Amaranthus lividus*). *Antioxidants* **2022**, *11*, 1206. [CrossRef]
31. Hassan, J.; Rajib, M.M.R.; Sarkar, U.; Akter, M.; Khan, M.; Khandaker, S.; Khalid, F.; Rahman, G.; Ercisli, S.; Muresan, C.; et al. Optimizing textile dyeing wastewater for tomato irrigation through physiochemical, plant nutrient uses and pollution load index of irrigated soil. *Sci. Rep.* **2022**, *12*, 10088. [CrossRef]
32. Sarker, U.; Azam, M.G.; Talukder, M.Z.A. Genetic Variation in Mineral Profiles, Yield Contributing Agronomic Traits, and Foliage Yield of Stem Amaranth. *Genetika* **2022**, *54*, 91–108. [CrossRef]
33. Jung, G.; Matsunami, T.; Oki, Y.; Kokubun, M. Effects of waterlogging on nitrogen fixation and photosynthesis in supernodulating soybean cultivar Kanto 100. *Plant Prod. Sci.* **2008**, *11*, 291–297. [CrossRef]
34. Matsunami, T.; Jung, G.H.; Oki, Y.; Kokubun, M. Effect of waterlogging during vegetative stage on growth and yield in supernodulating soybean cultivar sakukei 4. *Plant Prod. Sci.* **2007**, *10*, 112–121. [CrossRef]
35. Rhine, M.D.; Stevens, G.; Shannon, G.; Wrather, A.; Sleper, D. Yield and nutritional responses to waterlogging of soybean cultivars. *Irrig. Sci.* **2010**, *28*, 135–142. [CrossRef]
36. Komatsu, S.; Makino, T.; Yasue, H. Proteomic and biochemical analyses of the cotyledon and root of flooding-stressed soybean plants. *PLoS ONE* **2013**, *8*, e65301. [CrossRef] [PubMed]
37. Ara, R.; Mannan, M.A.; Khaliq, Q.A.; Miah, M.U. Waterlogging tolerance of soybean. *Bangladesh Agron. J.* **2016**, *18*, 105–109. [CrossRef]
38. Sullivan, M.; Van Toai, T.; Fausey, N.; Beuerlein, J.; Parkinson, R.; Soboyejo, A. Evaluating on-farm flooding impacts on soybean. *Crop Sci.* **2001**, *41*, 93–100. [CrossRef]
39. Amin, M.R.; Karim, M.A.; Khaliq, Q.A.; Islam, M.R.; Aktar, S. The influence of waterlogging period on yield and yield components of mungbean (*Vigna radiata* L. Wilczek). *Agriculturists* **2017**, *15*, 88–100. [CrossRef]
40. Divito, G.A.; Sadras, V.O. How do phosphorus, potassium and sulphur affect plant growth and biological nitrogen fixation in crop and pasture legumes? A meta-analysis. *Field Crop. Res.* **2014**, *156*, 161–171. [CrossRef]
41. Amin, R. Nutrient Management and Yield Performance of Mungbean Genotypes under Soil Flooding Conditions. Ph.D. Thesis, Bangabandhu Sheikh Mujibur Rahman Agricultural University, Gazipur, Bangladesh, 2014.
42. Chen, Y.L.; Dunbabin, V.M.; Diggle, A.J.; Siddique, K.H.; Rengel, Z. Phosphorus starvation boosts carboxylate secretion in P-deficient genotypes of *Lupinus angustifolius* with contrasting root structure. *Crop Past. Sci.* **2013**, *64*, 588–599. [CrossRef]
43. Minjian, C.; Haiqiu, Y.; Hongkui, Y.; Chunji, J. Difference in tolerance to potassium deficiency between two maize inbred lines. *Plant Prod. Sci.* **2007**, *10*, 42–46. [CrossRef]
44. Pettigrew, W.T. Potassium influences on yield and quality production for maize, wheat, soybean and cotton. *Physiol. Plant.* **2008**, *133*, 670–681. [CrossRef]
45. Wang, M.; Zheng, Q.; Shen, Q.; Guo, S. The critical role of potassium in plant stress response. *Int. J. Mol. Sci.* **2013**, *14*, 7370–7390. [CrossRef]
46. Wei, J.; Li, C.; Li, Y.; Jiang, G.; Cheng, G.; Zheng, Y. Effects of external potassium (K) supply on drought tolerances of two contrasting winter wheat cultivars. *PLoS ONE* **2013**, *8*, e69737. [CrossRef]
47. Amanullah, A.; Iqbal, A.; Iqbal, M. Impact of potassium rates and their application time on dry matter partitioning, biomass and harvest index of maize (*Zea mays*) with and without cattle dung application. *Emir. J. Food Agric.* **2015**, *27*, 447–453. [CrossRef]
48. ISTA (International Seed Testing Association). Available online: https://www.seedtest.org/en/home.html (accessed on 31 January 2019).
49. Agrawal, R.L. *Seed Technology*; Oxford & IBH Publishing Co., Pvt., Ltd.: New Delhi, India, 2005; pp. 565–590.
50. Marcos Filho, J.; Vieira, R.D. Seed vigor tests: Procedures-conductivity tests. In *Seed Vigor Tests Handbook*; Association of Official Seed Analysts: Ithaca, NY, USA, 2009; pp. 186–200.
51. Yoshida, S.; Forno, D.A.; Cock, J.H.; Gomez, K.A. *Laboratory Manual for Physiological Studies of Rice*, 3rd ed.; International Rice Research Institute: Manila, Philippines, 1976.
52. Kulsum, U.; Sarker, U.; Rasul, M.G. Genetic variability, heritability and interrelationship in salt-tolerant lines of T. Aman rice. *Genetika* **2022**, *54*, 761–776. [CrossRef]

53. Hasan, M.J.; Kulsum, M.U.; Sarker, U.; Matin, M.Q.I.; Shahin, N.H.; Kabir, M.S.; Ercisli, S.; Marc, R.A. Assessment of GGE, AMMI, Regression, and Its Deviation Model to Identify Stable Rice Hybrids in Bangladesh. *Plants* **2022**, *11*, 2336. [CrossRef]
54. Gomez, K.A.; Gomez, A.A. *Statistical Procedure for Agricultural Research*, 2nd ed.; John Willey and Sons: Singapore, 1984; pp. 28–192.
55. Marschner, H. *Marschner's Mineral Nutrition of Higher Plants*; Academic Press: Cambridge, MA, USA, 2011.
56. Kim, Y.; Seo, C.W.; Khan, A.L.; Mun, B.G.; Shahzad, R.; Ko, J.W.; Lee, I.J. Ethylene mitigates waterlogging stress by regulating glutathione biosynthesis-related transcripts in soybeans. *BioRxiv* **2018**, 252312. [CrossRef]
57. Jin-Woong, C.H.O.; Ji, H.C.; Yamakawa, T. Comparison of photosynthetic response of two soybean cultivars to soil flooding. *J. Facul. Agric. Kyushu Univ.* **2006**, *51*, 227–232.
58. Sathi, K.S.; Masud, A.A.C.; Falguni, M.R.; Ahmed, N.; Rahman, K.; Hasanuzzaman, M. Screening of Soybean Genotypes for Waterlogging Stress Tolerance and Understanding the Physiological Mechanisms. *Adv. Agric.* **2022**, *2022*, 5544665. [CrossRef]
59. Ashraf, M.A.; Ahmad, M.S.A.; Ashraf, M.; Al-Qurainy, F.; Ashraf, M.Y. Alleviation of waterlogging stress in upland cotton (*Gossypium hirsutum* L.) by exogenous application of potassium in soil and as a foliar spray. *Crop Past. Sci.* **2011**, *62*, 25–38. [CrossRef]
60. Vyas, A.K.; Billore, S.D.; Ramesh, A.; Joshi, O.P.; Gupta, G.K.; Sharma, A.N.; Imas, P. Role of potassium in balanced fertilization of soybean-wheat cropping system. In Proceedings of the Regional Seminar on Recent Advances in Potassium Nutrition Management for Soybean Based Cropping Systems, Indore, India, 28–29 September 2007; National Research Centre for Soybean: Indore, India, 2008; pp. 28–29.
61. Solaiman, Z.; Colmer, T.D.; Loss, S.P.; Thomson, B.D.; Siddique, K.H.M. Growth responses of cool-season grain legumes to transient waterlogging. *Aust. J. Agric. Res.* **2007**, *58*, 406–412. [CrossRef]
62. Kuswantoro, H. Response of soybean genotypes to waterlogging. *J. Agron. Indones.* **2011**, *39*, 19–23.
63. Vineela, V. Effect of Waterlogging on Growth and Yield of Green Gram (*Vigna radiata* L.). Ph.D. Thesis, Acharya N.G. Ranga Agricultural University, Rajendra Nagarp, Hyderabad, India, 2013.
64. Mengel, K.; Kirkby, E.A.; Kosegarten, H.; Appel, T. Nitrogen. In *Principles of Plant Nutrition*; Springer: Dordrecht, The Netherlands, 2001; pp. 397–434.
65. Alam, M.R.; Ali, M.A.; Molla, M.S.H.; Momin, M.A.; Mannan, M.A. Evaluation of different levels of potassium on the yield and protein content of wheat in the high Ganges River floodplain soil. *Bangladesh J. Agric. Res.* **2009**, *34*, 97–104. [CrossRef]
66. Khan, A.A.; Inamullah; Jan, M.T. Impact of various nitrogen and potassium levels and application methods on grain yield and yield attributes of wheat. *Sarhad J. Agric.* **2014**, *30*, 35–46.
67. Khan, A.; Wang, L.; Ali, S.; Tung, S.A.; Hafeez, A.; Yang, G. Optimal planting density and sowing date can improve cotton yield by maintaining reproductive organ biomass and enhancing potassium uptake. *Field Crop. Res.* **2017**, *214*, 164–174. [CrossRef]
68. Ahmed, A.; Aftab, S.; Hussain, S.; Nazir Cheema, H.; Liu, W.; Yang, F.; Yang, W. Nutrient accumulation and distribution assessment in response to potassium application under maize–soybean intercropping system. *Agronomy* **2020**, *10*, 725. [CrossRef]
69. Ahmad, M.; Riaz, A.; Ishaque, M.; Malik, A.U. Response of maize hybrids to varying potassium application in Pakistan. *Pak. J. Agric. Sci.* **2009**, *46*, 179–184.
70. Islam, M.R.; Akter, N.; Parvej, S.S.; Haque, K.S. Growth and yield response of mungbean (*Vigna radiata* L. Wilczek) genotypes to wet puddling, flooding and saturated soil culture. *J. Plant Sci.* **2014**, *2*, 311–316.
71. Uddin, S.; Sarkar, M.; Rahman, M. Effect of nitrogen and potassium on yield of dry direct-seeded rice cv. NERICA 1 in aus season. *Int. J. Agron. Plant Prod.* **2013**, *4*, 69–75.
72. Zörb, C.; Senbayram, M.; Peiter, E. Potassium in agriculture–status and perspectives. *J. Plant Physiol.* **2014**, *171*, 656–669. [CrossRef]
73. Lu, Z.; Lu, J.; Pan, Y.; Lu, P.; Li, X.; Cong, R.; Ren, T. Anatomical variation of mesophyll conductance under potassium deficiency has a vital role in determining leaf photosynthesis. *Plant Cell Environ.* **2016**, *39*, 2428–2439. [CrossRef]
74. Farhad, I.S.M.; Islam, M.N.; Hoque, S.; Bhuiyan, M.S.I. Role of potassium and sulphur on the growth, yield and oil content of soybean (*Glycine max* L.). *Acad. J. Plant Sci.* **2010**, *3*, 99–103.
75. Beutler, A.N.; Giacomeli, R.; Alberto, C.M.; Silva, V.N.; da Silva Neto, G.F.; Machado, G.A.; Santos, A.T.L. Soil hydric excess and soybean yield and development in Brazil. *Aust. J. Crop Sci.* **2014**, *8*, 1461–1466.
76. Koger, C.H.; Zablotowicz, R.M.; Weaver, M.A.; Tucker-Patterson, M.R.; Krutz, J.L.; Walker, T.W.; Street, J.E. Effect of winter flooding on weeds, soybean yield, straw degradation, and soil chemical and biochemical characteristics. *Am. J. Plant Sci.* **2013**, *4*, 10–18. [CrossRef]
77. Youn, J.T.; Van, K.J.; Lee, J.E.; Kim, W.H.; Yun, H.T.; Kwon, Y.U.; Lee, S.H. Waterlogging Effects on Nitrogen Accumulation and N_2 Fixation of Super nodulating Soybean Mutants. *J. Crop Sci. Biotechnol.* **2008**, *11*, 111–118.
78. Ullah, M.J. Effect of waterlogging on growth and yield of mungbean cv. kanti (*Vigna radiata*). *Legume Res. Int. J.* **2006**, *29*, 196–200.
79. Nguyen, V.T.; Vuong, T.D.; VanToai, T.; Lee, J.D.; Wu, X.; Mian, M.R.; Nguyen, H.T. Mapping of quantitative trait loci associated with resistance to *Phytophthora sojae* and flooding tolerance in soybean. *Crop Sci.* **2012**, *52*, 2481–2493. [CrossRef]
80. Dong, H.; Kong, X.; Li, W.; Tang, W.; Zhang, D. Effects of plant density and nitrogen and potassium fertilization on cotton yield and uptake of major nutrients in two fields with varying fertility. *Field Crop. Res.* **2010**, *119*, 106–113. [CrossRef]
81. Ali, S.; Hafeez, A.; Ma, X.; Tung, S.A.; Chattha, M.S.; Shah, A.N.; Yang, G. Equal potassium-nitrogen ratio regulated the nitrogen metabolism and yield of high-density late-planted cotton (*Gossypium hirsutum* L.) in Yangtze River valley of China. *Ind. Crop. Prod.* **2019**, *129*, 231–241. [CrossRef]

82. Board, J.E. Waterlogging effects on plant nutrient concentrations in soybean. *J. Plant Nutr.* **2008**, *31*, 828–838. [CrossRef]
83. Mohammadi, G.R. The effect of seed priming on plant traits of late-spring seeded soybean (*Glycine max* L.). *Am. Eurasian J. Agric. Environ. Sci.* **2009**, *5*, 322–326.
84. Taiz, L.; Zeiger, E. Responses and adaptations to abiotic stress. In *Plant Physiology*, 5th ed.; Sinauer Associates, Inc.: Sunderland, MA, USA, 2010; pp. 755–778.
85. Vodnik, D.; Strajnar, P.; Jemc, S.; Maček, I. Respiratory potential of maize (*Zea mays* L.) roots exposed to hypoxia. *Environ. Exp. Bot.* **2009**, *65*, 107–110. [CrossRef]
86. Colmer, T.D.; Greenway, H. Ion transport in seminal and adventitious roots of cereals during O_2 deficiency. *J. Exp. Bot.* **2011**, *62*, 39–57. [CrossRef] [PubMed]
87. Smethurst, C.F.; Garnett, T.; Shabala, S. Nutritional and chlorophyll fluorescence responses of lucerne (*Medicago sativa*) to waterlogging and subsequent recovery. *Plant Soil* **2005**, *270*, 31–45. [CrossRef]
88. Elzenga, J.T.M.; van Veen, H. Waterlogging and plant nutrient uptake. In *Waterlogging Signaling and Tolerance in Plants*; Springer: Berlin/Heidelberg, Germany, 2010; pp. 23–35.
89. Wuebker, E.F.; Mullen, R.E.; Koehler, K. Flooding and temperature effects on soybean germination. *Crop Sci.* **2001**, *41*, 1857–1861. [CrossRef]
90. Yaklich, R.W.; Abdul-Baki, A.A. Variability in Metabolism of Individual Axes of Soybean Seeds and Its Relationship to Vigor 1. *Crop Sci.* **1975**, *15*, 424–426. [CrossRef]
91. Maryam, A.; Nasreen, S. A review: Water logging effects on morphological, anatomical, physiological and biochemical attributes of food and cash crops. *Int. J. Water Resour. Environ. Sci.* **2012**, *1*, 113–120.
92. Parolin, P. Seed germination and early establishment of 12 tree species from nutrient-rich and nutrient-poor Central Amazonian floodplains. *Aquat. Bot.* **2001**, *70*, 89–103. [CrossRef]
93. Jitsuyama, Y. Hypoxia-responsive root hydraulic conductivity infuences soybean cultivar-specifc waterlogging tolerance. *Am. J. Plant Sci.* **2017**, *8*, 770. [CrossRef]
94. Collaku, A.; Harrison, S.A. Losses in wheat due to waterlogging. *Crop Sci.* **2002**, *42*, 444–450. [CrossRef]
95. Kaur, G.; Singh, G.; Motavalli, P.P.; Nelson, K.A.; Orlowski, J.M.; Golden, B.R. Impacts and management strategies for crop production in Waterlogged/Flooded soils: A review. *Agron. J.* **2020**, *112*, 1475–1501. [CrossRef]
96. Wu, C.; Mozzoni, L.A.; Moseley, D.; Hummer, W.; Ye, H.; Chen, P.; Nguyen, H. Genome-wide association mapping of fooding tolerance in soybean. *Mol. Breed.* **2020**, *40*, 4. [CrossRef]

life

MDPI

Article

In Vitro Multiplication and Cryopreservation of *Penthorum chinense* Shoot Tips

Rabbi A. K. M. Zilani [1,2], Hyoeun Lee [1], Elena Popova [3] and Haenghoon Kim [1,*]

1 Department of Agricultural Life Science, Sunchon National University, Suncheon 57922, Korea
2 Agricultural Training & Management Development Institute, Kaliakoir 1750, Bangladesh
3 K.A. Timiryazev Institute of Plant Physiology of Russian Academy of Sciences, Moscow 127276, Russia
* Correspondence: cryohkim@scnu.ac.kr; Tel.: +82-061-7505185; Fax: +82-061-7503210

Abstract: This study provides alternative approaches toward ex situ conservation by means of in vitro seed germination and the multiplication of *Penthorum chinense* Pursh using nodal explants. An overlay of a liquid medium on top of a gelled medium significantly increased the growth of shoots and roots, while the presence of activated charcoal or growth regulators (benzyl adenine and α-naphthaleneacetic acid) decreased the growth. Shoot tips of in vitro plantlets were cryopreserved using a droplet-vitrification method. The standard procedure included preculture with 10% sucrose for 31 h and with 17.5% sucrose for 17 h, osmoprotection with loading solution C4-35% (17.5% glycerol + 17.5% sucrose, w/v) for 20 min, cryoprotection with alternative plant vitrification solution (PVS) A3-70% (29.2% glycerol + 11.7% DMSO + 11.7% EG + 17.4% sucrose, w/v) at 0 °C for 30 min, cooling the samples in liquid nitrogen using aluminum foil strips and rewarming by plunging into pre-heated (40 °C) unloading solution (35% sucrose) for 40 min. A three-step regrowth procedure starting with ammonium-free medium followed by ammonium-containing medium with and without growth regulators was essential for the regeneration of cryopreserved shoot tips. The species was found to be very sensitive to the chemical cytotoxicity of permeating cryoprotectants during cryoprotection and to ammonium-induced oxidant stress during initial regrowth steps. Improvement of donor plant vigor by using apical sections and liquid overlay on top of the solid medium for propagation, improved shoot tip tolerance to osmotic stress and increased post-cryopreservation regeneration up to 64% were observed following PVS B5-85% (42.5% glycerol + 42.5% sucrose) treatment for 60 min. The systematic approach used in this study enables fast optimization of the in vitro growth and cryopreservation procedure for a new stress-sensitive wild plant species.

Keywords: alternative plant vitrification solution; ammonium-free medium; cytotoxicity; droplet-vitrification; endangered species; liquid overlay; regrowth medium

Citation: Zilani, R.A.K.M.; Lee, H.; Popova, E.; Kim, H. In Vitro Multiplication and Cryopreservation of *Penthorum chinense* Shoot Tips. *Life* **2022**, *12*, 1759. https://doi.org/10.3390/life12111759

Academic Editors: Hakim Manghwar and Wajid Zaman

Received: 9 October 2022
Accepted: 27 October 2022
Published: 1 November 2022

Publisher's Note: MDPI stays neutral with regard to jurisdictional claims in published maps and institutional affiliations.

Copyright: © 2022 by the authors. Licensee MDPI, Basel, Switzerland. This article is an open access article distributed under the terms and conditions of the Creative Commons Attribution (CC BY) license (https://creativecommons.org/licenses/by/4.0/).

1. Introduction

Penthorum chinense Pursh is a perennial herb that occurs in swamps and propagates by means of underground rhizomes and, to minor extent, through seeds [1]. The species belongs to Penthoraceae family, and the genus *Penthorum* has only two accepted species [2]. The plant has potential as a traditional herbal medicine with antioxidant properties and anti-cancer, anti-blebbing and hepatoprotective actions, as well as a food supplement [3–6].

Decreasing populations, habitat loss and climate change are major threats to this species [7]. Moreover, its high medicinal and ornamental value make this species worthy of integrated conservation interventions [7]. In addition to in situ conservation efforts, in vitro seed germination and in vitro culture of this valuable wild species can be considered as complementary strategies [8]. Pence et al. [9] also emphasized the importance of in vitro propagation methods in providing materials for the reintroduction and restauration of populations, and research in wild species that produce limited amounts of seeds.

Cryopreservation is a long-term method of storing living biological materials in liquid nitrogen (LN, −196 °C), after which the conserved samples may be recovered to produce new plants [10]. Cryopreservation combined with in vitro technologies offers a solid basis for developing effective conservation and restoration strategies for endangered species producing seeds that are non-orthodox, dormant or insufficient in quantity [8,9,11].

In a modern cryopreservation method called droplet-vitrification, samples are subjected to a series of pre-LN treatments with progressively increasing concentrations of cryoprotectants (CPAs). After rewarming (post-LN), this process is repeated "in reverse" (from a higher to lower concentration of CPAs) when samples are bathed in an unloading solution and then transferred to a regrowth medium [12]. To simplify the procedure and avoid laborious condition screening at both the pre-LN and post-LN stages, we developed a systematic approach that covers a range of plant species and starts with a limited number of standard procedures using alternative vitrification solutions (VSs) to adopt the cryopreservation protocol to new taxons [12]. This approach was successfully tested with a number of endangered plants, e.g., *Kalopanax septemlobus* (Thunb.) Koidz. [13], *Betula lenta* [14], *Castilleja levisecta* Greenm. [15], *Aster altaicus* var. *uchiyamae* [16] and *Lupinus rivularis* Douglas ex Lindl. [17] with over 60% of average regrowth achieved for all species.

In order to develop a droplet-vitrification protocol for a given plant material, it is necessary to evaluate the sensitivity of samples to cytotoxicity induced by cryoprotection with highly concentrated VSs, which is the main limiting factor for high post-cryopreservation regrowth [12]. Hence, optimization of cryoprotection conditions is crucial in cryopreservation studies. Therefore, the majority of the cryopreservation studies for the new species are focused on the optimization of preculture and cryoprotectant treatments [10]. Regretfully, very few studies have explored the effect of post-cryopreservation (regrowth) conditions (light, medium composition and growth regulators); however, at least for some species, these factors may have a major impact on the regeneration of healthy plantlets from cryopreserved materials [16,18,19]. Thus, in this study we also investigated the effect of sequential regrowth steps and ammonium-free regrowth medium to improve the regeneration of healthy plantlets from cryopreserved shoot tips.

To the best of our knowledge, this is the first report on developing a complementary conservation approach for *Penthorum chinense* through in vitro seed germination, in vitro propagation, and cryopreservation of shoot tips of in vitro propagated plantlets.

2. Materials and Methods

2.1. Plant Material

2.1.1. In Vitro Seed Germination and Establishment of In Vitro Culture

A small quantity (about 0.1 g) of mature seeds of *Penthorum chinense* Pursh were received from the National Institute of Biological Resources, Incheon, Republic of Korea. For the in vitro germination, the seeds were sterilized with 1% (*v*/*v*) NaOCl for 7 min and then washed with sterile distilled water for 10 min. Then, the seeds were inoculated on Murashige and Skoog (MS) medium [20] with 30 g L^{-1} sucrose, 3.5 g L^{-1} gelrite and 1 g L^{-1} activated charcoal, pH 5.8 (hereafter referred to as 'MSF') in 300 mL Gaooze^{TM} culture vessels (Korea Scientific Technology Industry, Suwon, Korea). Seeds were germinated in a culture room at 25 °C under a 16/8 h light/dark photoperiod and 40 μE m^{-2} s^{-1} light intensity (one fluorescent lamp, 40 W).

The germinated plantlets were cultured for six weeks on the same medium at 25 °C under a 16 h photoperiod and 60 μE m^{-2} s^{-1} light intensity (two lamps) and further propagated using nodal sections. To facilitate the growth of healthy plants, 15 mL of liquid MSF medium was added on top of the solid medium at day 10 of every subculture. Developed plants were further propagated using nodal segments (6/vessel) with 7-week subculture intervals.

2.1.2. In Vitro Propagation Using Nodal Segments

To establish the in vitro propagation system, six combinations of in vitro culture conditions were examined, including MS medium strength (full, 1/2), activated charcoal (AC, 1.0 g L^{-1}), overlay of liquid medium on top of the gelled solid medium at day 10 (Liquid), growth regulators (benzyl adenine (BA) 0.7 mg L^{-1} + α-naphthaleneacetic acid (NAA) 1.0 mg L^{-1}), light intensity 1 or 2 lamps (40 and 60 μE m^{-2} s^{-1}, respectively). The medium was supplemented with 30 g L^{-1} sucrose, and pH of the medium was adjusted to 5.8 prior to autoclaving. Nodal cuttings (5–6 mm in length) were inoculated into Gaooze™ culture vessels (seven cuttings per vessel). Three replicates were used for each treatment and all the experiments were repeated thrice. The height (cm) and dry weight (g) of shoots and roots were individually measured after six weeks. Dry weight was measured after drying plant samples to a constant weight at 7 h at 75 °C in an oven.

2.2. *Droplet-Vitrification Cryopreservation Procedure*

2.2.1. Standard Droplet-Vitrification Procedure

Shoot tips (1.3 mm in length, 1–2 lateral leaves) were extracted from 3–4-day-old node cuttings. Except where otherwise stated, explants were precultured in a liquid MS medium with 10% sucrose (S-10%) and 17.5% sucrose (S-17.5%) for 31 h and 17 h, respectively, osmoprotected with C4-35% solution (17.5% glycerol + 17.5% sucrose) at 25 °C for 20 min, and then cryoprotected with vitrification solution A3-70% (29.2% glycerol + 11.7% dimethyl sulfoxide + 11.7% ethylene glycol + 17.4% sucrose) on ice for 30 min.

Shoot tips were then placed in 5 μL droplets of A3-70% on aluminum foil strips (7 mm × 20 mm), before being plunged directly into LN for a minimum of 1 h. For rewarming and unloading, foil strips containing the shoot tips were transferred to 20 mL pre-heated (40 °C) 35% sucrose (S-35%) solution and kept for 40 min, with sucrose solution been replaced by the new solution after the first 15 min. Cryoprotected control (LNC) shoot tips were treated with the same procedure except without cooling in LN, and unloaded in S-35% solution before being transferred to recovery medium. The explants retrieved from S-35% were blotted dry and transferred to recovery medium 1 (ammonium nitrate (NH_4NO_3)-free MS medium + 1 g L^{-1} casein hydrolysate + 1 mg L^{-1} gibberellic acid (GA_3) + 0.5 mg L^{-1} BA with 30 g L^{-1} sucrose and 3.5 g L^{-1} gelrite, hereafter referred to as "RM1") in SPL culture vessels (90 mm × 40 mm, SPL Life Sciences, Gyeonggi-do, Korea). Shoot tips were kept at 25 °C in the dark for recovery. After 5 days, explants were moved to recovery medium 2 (normal (NH_4NO_3-containing) MS medium + 1 g L^{-1} casein hydrolysate + 1 mg L^{-1} GA_3 + 0.5 mg L^{-1} BA with 30 g L^{-1} sucrose and 3.5 g L^{-1} gelrite, RM2) and cultured for 3 weeks under the same conditions described above for establishment. Developed shoots were further transferred to growth hormone-free MS medium (MSF) for further regeneration.

2.2.2. Experimental Design in Droplet-Vitrification Procedure

A set of 16 treatments was designed to optimize the droplet-vitrification protocol based on analysis of the sensitivity of the material to various treatments (standard procedure, 11 pre-LN and 4 post-LN variants). The variants tested are listed in Table 1. Other conditions remained the same as in the standard procedure (indicated as "standard" in Table 1). The composition of the cryoprotectant (CPA) solutions is given in Table 2. For the treatment of container modification, shoot tips were cryopreserved in 2 mL cryovials with 0.5 mL VS A3-70%, kept in LN for 1 h and rewarmed in a pre-heated (40 °C) water bath; shoot tips were withdrawn from the vials, unloaded in S-35% solution and recovered as described above in the standard droplet-vitrification procedure.

2.2.3. Further Optimization of Droplet-Vitrification Procedure

To improve the LN regeneration, subcultured plantlets were revitalized by using of apical section, instead of nodal section, and applying of liquid overlay (MSF medium) on top of gelled medium after 2 weeks of inoculation. After two cycles of 5–6 weeks subcultures, shoot tips were extracted from 3–4-day-old node cuttings and subjected to cryopreservation using the standard droplet-vitrification procedure. Based upon the results in first round experiments, both four-component VS of "A" series (PVS2, AS-70%, A3-80%) were cryoprotected on ice, and a PVS3 dilution (B5-85%) was treated at room temperature. The composition of each VS was listed in Table 2.

Table 1. Experimental design applied in cryopreservation of *Penthorum chinense* shoot tips using a systematic approach in the droplet-vitrification procedure (standard procedure and 15 additional treatments).

Protocol Step	Treatments	Treatment Code
Preculture	No preculture	No-PC
	10% sucrose 31 h → 25% sucrose 17 h	S-10% → S-25%
	10% sucrose 31 h → 17.5% sucrose 17 h	S-10% → S-17.5%, standard
	10% sucrose 48 h	S-10%
Osmoprotection and container	No osmoprotection	No-OP
	C4-35% 20 min, Aluminum foil strips	OP/foil, standard
	C4-35% 20 min, Cryovial (2 mL)	Vial
Cryoprotection (Vitrification solution, VS)	A1-73.7% (PVS2) ice, 30 min	A1-73.7% (PVS2, 30 min)
	A3-90% ice, 30 min	A3-90% (30 min)
	A3-80% ice, 30 min	A3-80% (30 min)
	A3-70% ice, 30 min	A3-70% (30 min), standard
	A3-70% ice, 60 min	A3-70% (60 min)
	B1-100 (PVS3) 25 °C, 30 min	B1-100% (PVS3, 30 min)
	B5-80% 25 °C, 30 min	B5-80% (30 min)
Regrowth	RM1, 5d, dark → RM2, 3w2d, 1 L → MSF,2w, 2 L	RM1-RM2-MSF, standard
	RM2, 5d, dark → RM2, 3w2d, 1 L → MSF, 2w, 2 L	RM2-RM2-MSF
	RM1, 5d, dark → RM2, 1w2d, 1 L → RM2, 1w, 2 L → MSF, 1w, 2 L	RM1-RM2-RM2-MSF
	RM1, 5d, dark → RM2, 3w2d, 1 L → MSF, 2w, 2 L → MSF, 2w, 2 L	RM1-RM2-MSF-MSF
	RM2, 5d, dark → RM2, 3w2d, 1 L → RM2, 2w, 2 L	RM2~

RM1—MS medium without NH_4NO_3 + casein hydrolysate 1 g L^{-1} + GA_3 1 mg L^{-1} + BA 0.5 mg L^{-1}; RM2—MS medium + casein hydrolysate 1 g L^{-1} + GA_3 1 mg L^{-1} + BA 0.5 mg L^{-1}; MSF—MS medium without growth regulators; RM2~—no medium change; d, days; w, weeks; 1L and 2L, light provided by 1 and 2 fluorescent lamps (40 and 60 μE m^{-2} s^{-1}, respectively); "Standard" indicates treatments composing the standard procedure where other stages are the same as in the standard protocol.

2.2.4. Assessment of Shoot Tip Recovery and Statistical Analysis

Survival was evaluated two weeks following cryopreservation by counting the number of shoot tips showing elongation, or formation of new leaves. Regeneration (shoot development) was determined after 8 weeks, when the shoot tips had developed into normal plantlets (≥8 mm) with fully developed leaves and roots.

Ten to thirteen shoot tips were used per experimental conditions and the experiments were replicated 3–4 times. Data were analyzed by analysis of variance (ANOVA) and Duncan's multiple range test ($p < 0.05$) following arcsine transformation using SAS 9.1 software (SAS, Raleigh, NC, USA). Results are presented as average values with their standard deviation.

Table 2. Composition of cryoprotectant solutions used for preculture, osmoprotection, cryoprotection and unloading.

Protocol Step	Solutions	Composition (%, w/v)	Total Concentration (%, w/v)
Preculture and unloading	S-10%	S* 10.0	10.0
	S-35%	S 35.0	35.0
Osmoprotection	C4-35%	G 17.5 + S 17.5	35.0

Table 2. *Cont.*

Protocol Step	Solutions	Composition (%, w/v)	Total Concentration (%, w/v)
Cryoprotection (vitrification solution)	A1-73.7% (PVS2)	G 30.0 + DMSO 15.0 + EG 15.0 + S 13.7	73.7
	A3-90%	G 37.5 + DMSO 15.0 + EG 15.0 + S 22.5	90.0
	A3-80%	G 33.3 + DMSO 13.3 + EG 13.3 + S 20.1	80.0
	A3-70%	G 29.2 + DMSO 11.7 + EG 11.7 + S 17.4	70.0
	B1-100% (PVS3)	G 50.0 + S 50.0	100.0
	B5-80%	G 40.0 + S 40.0	80.0
	B5-85%	G 42.5 + S 42.5	85.0

* S, sucrose; G, glycerol; DMSO, dimethyl sulfoxide; EG, ethylene glycol. All solutions were made on the basis of MS medium; pH was adjusted to 5.8 before filter-sterilization.

3. Results

3.1. Establishment of In Vitro Culture and Propagation System

To establish the in vitro propagation system, nodal segments were cultured in six condition variants (standard + 5 alternative conditions; Table 1) for 6 weeks. Significant differences in growing pattern were found among the treatments after 4–5 weeks. Half-strength MS medium (treatment 1 in Table 3) resulted in similar or slightly higher length of shoots and roots and dry weight of shoots and roots compared to the full-strength MS medium (treatment 2), though the differences were insignificant (Table 3). An overlay of liquid MSF medium on top of gelled medium (treatment 3) produced significantly greater length of shoots and roots compared to gelled medium only, while the dry weight of shoots and roots was not significantly different. Removing of the activated charcoal (treatment 4) had a positive effect on the length of shoots (21.6 vs. 11.2 cm) and roots (11.6 vs. 4.1 cm), while dry weight of shoots were not significantly affected. Addition of growth regulators (treatment 5) showed a notable negative effect on the in vitro growth of shoots and roots. Light intensity (one vs. two lamps) was not a significant factor for in vitro growth (treatment 2 vs. treatment 6).

Table 3. Height and dry weight of in vitro grown *Penthorum chinense* plantlets depending on growth medium and culture conditions.

No.	Culture Medium	AC	Illumination	Height (cm)		Dry Weight (g)	
				Shoots	Roots	Shoots	Roots
1	**1/2MSF**	+	2 L	$13.2 \pm 4.8^{b,c}$	6.5 ± 3.4^{b}	$0.057 \pm 0.020^{a,b}$	$0.006 \pm 0.006^{b,c}$
2	**MSF**	+	2 L	$11.2 \pm 3.2^{c,d}$	4.1 ± 2.5^{c}	$0.051 \pm 0.010^{a,b}$	$0.005 \pm 0.004^{b,c}$
3	**MSF + Liquid**	+	2 L	16.2 ± 3.7^{b}	8.3 ± 1.8^{b}	0.066 ± 0.024^{a}	0.006 ± 0.005^{b}
4	**MSF**	−	2 L	21.6 ± 2.3^{a}	11.6 ± 1.9^{a}	$0.056 \pm 0.012^{a,b}$	0.012 ± 0.005^{a}
5	**MS + BA0.7 + NAA1.0**	+	2 L	7.9 ± 1.7^{d}	1.1 ± 0.4^{d}	0.036 ± 0.007^{b}	0.001 ± 0.001^{c}
6	**MSF**	+	1 L	$11.8 \pm 3.1^{c,d}$	5.1 ± 2.7^{c}	0.042 ± 0.007^{b}	$0.002 \pm 0.002^{b,c}$
	Pr < p			$p < 0.0001$	$p < 0.0001$	$p < 0.0052$	$p < 0.001$

MSF—standard MS medium with 30 g L^{-1} sucrose and 3.5 g L^{-1} gelrite; 1/2MSF—same medium with half-strength mineral salts; MSF + Liquid—overlay of liquid medium on top of the gelled solid medium added at day 10; MS + BA0.7 + NAA1.0—MS medium with 30 g L^{-1} sucrose, 3.5 g L^{-1} gelrite, 0.7 mg L^{-1} benzyl adenine (BA) and 1.0 mg L^{-1} α-naphthaleneacetic acid (NAA); AC—activated charcoal (1.0 g L^{-1}); 1 L and 2 L—illumination of 40 and 60 μE m^{-2} s^{-1}, provided by 1 or 2 fluorescent lamps, respectively; Means with the same letters (a,b,c,d) in each column are not significantly different by Least Significant Difference Test (LSDT, $p < 0.05$).

In conclusion, conditions combining the most favorable factors: half-strength MS medium without AC with an overlay of liquid medium on top of solid medium with two lamps (60 μE m^{-2} s^{-1}) was recommended for the in vitro culture of *P. chinense* plantlets.

3.2. Droplet-Vitrification Procedure for Shoot Tip Cryopreservation

3.2.1. Effect of Preculture

Among the preculture treatments tested, preculture with moderate sucrose concentrations (S-10% and S-17.5%) did not significantly improve the survival and regeneration of both cryoprotected (LNC) and cryopreserved (LN) shoot tips, compared to treatment without preculture (No-PC) (Table 4). Preculture with higher sucrose concentration (S-25%) produced even lower survival and regeneration of both the cryoprotected control (LNC, 35.1% survival, 18.7% regeneration) and cryopreserved (LN, 29.3% survival, 8.2% regeneration) shoot tips. This study suggests that *P. chinense* shoot tips are sensitive to osmotic stress and, moreover, preculture with moderate concentration of sucrose (S-10 and S-17.5%) was not effective for the acquisition of osmotic adaptation to further cryoprotectant treatments.

Table 4. Effect of preculture treatments on survival and regeneration of cryoprotected control (LNC) and cryopreserved (LN) *Penthorum chinense* shoot tips.

Preculture Treatments	LNC		LN	
	Survival	Regeneration	Survival	Regeneration
No-PC	71.1 ± 9.6 ^{a,b}	36.4 ± 15.2 ^{a}	73.3 ± 9.6 ^{a}	33.6 ± 13.8 ^{a}
S-10% → S-25%	35.1 ± 9.4 ^{c}	18.7 ± 7.4 ^{a}	29.3 ± 8.6 ^{b}	8.2 ± 4.1 ^{b}
S-10% → S-17.5%, standard	81.8 ± 7.3 ^{a}	35.0 ± 8.7 ^{a}	67.6 ± 14.4 ^{a}	35.0 ± 11.2 ^{a}
S-10%	67.3 ± 7.4 ^{b}	35.0 ± 5.0 ^{a}	61.3 ± 11.3 ^{a}	30.0 ± 7.1 ^{a}
$Pr < p$	$p < 0.0001$	ns	$p < 0.0052$	$p < 0.05$

No-PC, no preculture; S-10% → S-25%, 10% sucrose (S-10%) and 25% sucrose (S-25%) for 31 h and 17 h, respectively; S-10% → S-17.5%, 10% sucrose (S-10%) and 17.5% sucrose (S-17.5%) for 31 h and 17 h; S-10%, 10% sucrose for 48 h. After preculture, shoot tips were osmoprotected with C4-35% for 20 min, and cryoprotected with VS A3-70% on ice for 30 min before storage in LN; Means with the same letters (a,b,c) in each column are not significantly different by Least Significant Difference Test ($p < 0.05$); ns, non-significant.

3.2.2. Effect of Osmoprotection and Container in Cooling/Warming

Even after the best preculture treatment based on Table 4, shoot tips were sensitive to osmotic stress induced by direct exposure to VS A3-70% without osmoprotection (Table 5, No-OP) with 52.2% survival and 13.3% regeneration before cryopreservation (LNC). After liquid nitrogen exposure (LN), survival and regeneration level were similar to those of cryoprotected control (LNC), indicating that the low shoot tip viability was caused by osmotic stress induced by direct exposure to VS, rather than freezing injury. Application of osmoprotection solution (OP) for 20 min significantly ($p < 0.05$) increased survival and regeneration of LNC shoot tips.

Cooling and rewarming using 2 mL cryovials (Vial) produced 45.2% lower survival and 22.5% lower regeneration compared to the aluminum foil strips (Table 5), which indicated that shoot tips were insufficiently cryoprotected via standard treatment with VS A3-70% before being immerged into LN.

3.2.3. Effect of Cryoprotection Treatment (Vitrification Solution)

When cryoprotection was performed with PVS2 and its variants, four-component vitrification solutions of "A" series, on ice for 30 min, although there was no significance, a relatively lower rate of survival of both LNC and LN shoot tips was observed with original PVS2 (A1-73.7%) and the alternative A3-90%, possibly due to their cytotoxicity (Table 6). Dilution of VS A3-90% to 70% (A3-70%) produced the highest regeneration of both LNC (48.3%) and LN (49.8%) shoot tips. Longer cryoprotection (60 min) with A3-70% also resulted in lowest regeneration of both LNC (31.1%) and LN (26.2%), reflecting the cytotoxicity. Original PVS3 (B1-100%) was also toxic to shoot tips (Table 6), and thus its dilution to 80% (B5-80%) improved the regeneration of LNC and LN shoot tips by 16.6% and 8.4%, respectively.

Table 5. Effect of osmoprotection (OP) and cooling/warming container on survival and regeneration of cryoprotected control (LNC) and cryopreserved (LN) *Penthorum chinense* shoot tips.

Osmoprotection and Container	LNC		LN	
	Survival	Regeneration	Survival	Regeneration
No-OP	52.2 ± 10.5 [b]	13.3 ± 9.4 [b]	46.9 ± 9.6 [a]	16.7 ± 12.5 [a]
OP/foil, standard	81.8 ± 7.3 [a]	35.0 ± 8.7 [a]	67.6 ± 14.4 [a]	35.0 ± 11.2 [a]
Vial	-	-	22.4 ± 13.2 [b]	12.5 ± 13.0 [a]
$Pr < p$	$p < 0.05$	$p < 0.05$	$p < 0.05$	ns

No-OP, no osmoprotection; OP/foil, standard, osmoprotection with C4-35% for 20 min, and using the aluminum foil strips for cooling and rewarming; Vial, osmoprotection with C4-35% for 20 min, and using the 2 mL cryovial; Means with the same letters (a,b) in each column are not significantly different by Least Significant Difference Test ($p < 0.05$); ns, non-significant.

Table 6. Effect of cryoprotection treatments (vitrification solution) on the survival and regeneration of cryoprotected control (LNC) and cryopreserved (LN) *Penthorum chinense* shoot tips.

Cryoprotection Treatments	LNC		LN	
	Survival	Regeneration	Survival	Regeneration
A1-73.7% (PVS2), 30 min	64.2 ± 9.8 [a]	35.8 ± 15.1 [a]	68.9 ± 7.3 [a]	41.1 ± 9.9 [a]
A3-90%, 30 min	75.8 ± 7.1 [a,b]	44.2 ± 8.9 [a]	73.9 ± 5.7 [a]	39.3 ± 6.4 [a]
A3-80%, 30 min	70.0 ± 8.2 [a,b]	44.2 ± 12.7 [a]	64.6 ± 9.3 [a]	36.8 ± 11.6 [a]
A3-70%, 30 min, standard	88.3 ± 11.2 [a]	48.3 ± 25.2 [a]	71.2 ± 8.8 [a]	49.8 ± 14.6 [a]
A3-70%, 60 min	60.0 ± 11.5 [b]	31.1 ± 11.1 [a]	59.0 ± 10.7 [a]	26.2 ± 6.2 [a]
B1-100 (PVS3), 30 min	68.9 ± 14.6 [a,b]	29.2 ± 11.8 [a]	75.2 ± 7.8 [a]	32.5 ± 11.8 [a]
B5-80%, 30 min	78.3 ± 6.9 [a,b]	45.8 ± 13.8 [a]	70.2 ± 9.3 [a]	40.9 ± 12.8 [a]
$Pr < p$	$p < 0.05$	ns	ns	ns

See Table 2 for vitrification solution composition. Means with the same letters (a,b) in each column are not significantly different by Least Significant Difference Test ($p < 0.05$); ns, non-significant.

Although there was no significance, PVS2 and its variants produced slightly higher LN regeneration compared to PVS3 and its variant (36.8–49.8% vs. 32.5–40.9%). Among the VSs tested, the highest survival and regeneration of LNC and LN shoot tips were recorded with a dilution of A3-90% to 70% (A3-70%) for 30 min (Table 6), indicating that *P. chinense* shoot tips are very sensitive to both osmotic stress and chemical toxicity induced by highly concentrated VSs during cryoprotection.

3.2.4. Effect of Step-Wise Regrowth on Different Media

Overall, the regrowth conditions tested moderately affected the survival of both LNC and LN shoot tips, and significantly ($p < 0.05$) affected the regeneration of both LNC and LN treatments. Initial regrowth on ammonium-free medium for 5 days (RM1) followed by standard medium supplemented with growth regulators (RM2) and, finally, hormone-free medium (MSF) (Table 7, RM1-RM2-MSF, standard) showed relatively high regeneration (43.6% LNC and 35% LN). A more frequent transferring to a new medium of RM2 or MSF (RM1-RM2-RM2-MSF, RM1-RM2-MSF-MSF) did not significantly improve the regeneration. Initial regrowth on ammonium-containing medium with growth regulators without transferring to a new medium (RM2~) was harmful for the regeneration of both LNC and LN (0.0% and 3.3%, respectively). Similarly, initial regrowth with ammonium-containing medium with growth regulators followed by the same transfer steps as in the standard treatment (RM2-RM2-MSF) produced poor regeneration of LNC 10.0% and LN 5.2%. These results suggest that the initial regrowth with ammonium-containing regrowth medium was always harmful for the regeneration of LNC and LN shoot tips. Moreover, sequential transfer to fresh regrowth medium, per se, was not beneficial for improving the regeneration of both LNC and LN shoot tips. Hence, initial (5 days) regrowth on ammonium-free medium followed by step-wise transfer to ammonium containing medium and then medium without growth regulators was beneficial for regeneration of normal plants after cryoprotection and cryopreservation.

Table 7. Effect of regrowth treatments on survival and regeneration of cryoprotected control (LNC) and cryopreserved (LN) *Penthorum chinense* shoot tips.

Regrowth Medium	LNC		LN	
	Survival	Regeneration	Survival	Regeneration
RM1-RM2-MSF, standard	100.0 ± 0.0 ^{a}	43.6 ± 8.8 ^{a}	73.9 ± 2.6 ^{a}	35.0 ± 4.8 ^{a}
RM2-RM2-MSF	70.7 ± 0.4 ^{b}	10.0 ± 5.8 ^{b}	56.9 ± 5.7 ^{b}	5.2 ± 3.4 ^{b}
RM1-RM2-RM2-MSF	96.7 ± 3.3 ^{a}	49.0 ± 5.0 ^{a}	78.8 ± 0.7 ^{a}	35.6 ± 5.1 ^{a}
RM1-RM2-MSF-MSF	97.5 ± 2.5 ^{a}	45.0 ± 4.1 ^{a}	83.9 ± 3.7 ^{a}	39.4 ± 3.6 ^{a}
RM2~ (no medium change)	70.7 ± 0.4 ^{b}	0.0 ± 0.0 ^{b}	57.8 ± 7.2 ^{b}	3.3 ± 3.3 ^{b}
Pr < p	*p < 0.05*	*p < 0.05*	*p < 0.05*	*p < 0.05*

RM1—MS medium without NH_4NO_3 + casein hydrolysate 1 g L^{-1} + GA_3 1 mg L^{-1} + BA 0.5 mg L^{-1} + AC 1 g L^{-1}; RM2—MS medium + casein hydrolysate 1 g L^{-1} + GA_3 1 mg L^{-1} + BA 0.5 mg L^{-1}; MSF—MS medium without growth regulators. Means with the same letters (a,b) are not significantly different by Least Significant Difference Test ($p < 0.05$).

3.3. Further Optimization of Droplet-Vitrification Procedure

As an attempt to increase regeneration after cryopreservation, we modified the subculture of donor plants by inoculating the apical section, instead of the nodal section, and applying liquid overlay on top of the gelled medium. Donor plants subcultured in this manner were vitalized, and their subculture duration could be reduced from 7 to 5–6 weeks (Figure 1B,C). Shoot tips excised from these revitalized node cuttings were more tolerate to osmotic stress and chemical toxicity of CPAs and thus the duration of osmoprotection treatment with C4-35% could be increased from 20 min to 40 min. At the same time, the concentration of the VS and cryoprotection duration were also increased. With these modified conditions, the highest regeneration after cryopreservation (64.2%) was produced using cryoprotection with VS B5-85% for 60 min. The second-best treatment (53.6% regeneration) applied VS A3-80% for 60 min (Figure 2). The regeneration of shoot tips cryopreserved after cryoprotection with A3-70% for 30 min (30%) or A3-80% for 30 min (31.9%) was slightly lower than the values for the previous experiments, i.e., 49.8% or 36.8%%, respectively. This result reflects that vigorously grown donor plants allowed a higher concentration and longer duration of osmoprotection and cryoprotection treatments. The cryopreserved shoot tips following the cryoprotection with B5-85% for 60 min were regrown to become normal plantlets with a step-wise regrowth medium of RM1-RM2-MSF (Figure 1D).

Figure 1. (**A**), *Penthorum chinense* plant in wild habitat; (**B**), in vitro-cultured plantlets under the subculture condition of MSF for six weeks; (**C**), in vitro-grown plantlet before node cutting; (**D**), regrowth of cryopreserved shoot apices following cryoprotection with B5-85% RT for 60 min and RM1-RM2-MSF regrowth using SPL cultures (90 mm × 40 mm) for six weeks.

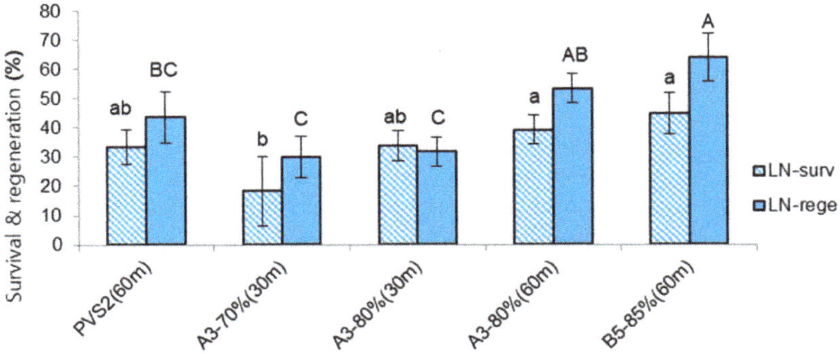

Figure 2. Effect of cryoprotection (vitrification solution composition and conditions of treatment) on survival (surv) and regeneration (rege) of cryopreserved (LN) *Penthorum chinense* shoot tips. See Table 2 for composition of vitrification solutions; m—minutes. Different letters (a,b for LN-surv; A,AB,BC,C for LN-rege) on each graph differ by Least Significant Difference Test ($p < 0.05$).

4. Discussion

4.1. In Vitro Culture and Propagation System

The in vitro germination of *P. chinense* seeds and the multiplication of plantlets using node sections as well as cryopreservation were investigated in this study as an alternative ex situ conservation approach for this species.

Among various in vitro culture media and conditions tested, an overlay of liquid medium on top of gelled medium most effectively promoted the growth of shoots and roots, while adding activated charcoal or growth regulators was not beneficial. Pullman and Skryabina [21] reported that an overlay of liquid medium on gelled medium 14 days after subculture improved embryogenic tissue initiation in conifers, possibly due to allowing nutrients replenishment, adjustment of pH, hormones, etc. Liquid overlay also stimulated the multiplication of ginger plantlets [22]. In the combination of solid and liquid media, direct contact with the liquid medium may promote the growth, particularly when the root system and its function are not sufficiently developed. When node cuttings were cultured for the in vitro tuberization of *Solanum tuberosum* (L.) cultivars, a liquid overlay on top of the solid medium produced significantly greater length and dry weight of shoots and roots compared to the other treatments, i.e., solid (gelled), static liquid, and wall-supported liquid media (unpublished data). Gelling agents may affect the physiochemical characteristics of the culture medium due to differences in the diffusion rate of nutrients, elemental and organic impurities and gel hardness [23]. As a gelling agent, gelrite was used in this study since it supported faster growth of shoots compared to agar, perhaps due to impurity of agar which contains agropectin [24]. Agar gelation also slowed the absorption of 2,4-dichlorophenoxyacetic acid (2,4-D) and abscisic acid (ABA) compared to gelrite [21].

Activated charcoal is often used in tissue culture to improve tissue growth and development via, among other factors, the absorption of inhibitory compounds in the culture medium [25]. In several orchid, yam and *Lycium* species, the addition of AC enhanced multiplication by increasing plant height, rooting and protocorm development from seeds [26–28], but inhibited plant regeneration in *Athyrium niponicum* var. pictum [29]. Addition of AC in regrowth medium had a beneficial effect on the recovery of cryopreserved *Lavandula* cells [30] and Norway spruce somatic embryos [31]. In the present study, the adding of AC was designed as the standard condition, since AC was beneficial for the formation of normal plantlet during the in vitro germination and in vitro culture establishment stages in preliminary experiment. However, this study revealed that the AC had a negative effect on the length and dry weight of *P. chinense* plantlets after 6-week subculture, possibly through the absorption of needed substances (vitamins, minerals, etc.) or some other, still unknown, reasons. In conclusion, a combination of half-strength MS medium without AC with an overlay of liquid medium with two lamps (60 μE m^{-2} s^{-1}) was recommended for the in vitro multiplication of *P. chinense* plantlets.

4.2. Development of Droplet-Vitrification Protocol

For the acquisition of osmotic tolerance, preculture treatment to a final concentration of 0.3 M (10% w/v) or 0.5 M (17.5% w/v) sucrose has usually been applied in vitrification procedures [32]. Preculture with high concentration of sucrose affected cell metabolism, causing alterations in gene expression [33], the accumulation of specific amino acids and soluble sugars [34–36], and changes in protein and fatty acid composition [33,37].

In this study, the regeneration of cryopreserved *P. chinense* shoot tips was not significantly improved by stepwise preculture with 10% sucrose → 17.5% sucrose. A higher concentration of sucrose in preculture medium (10% sucrose → 25% sucrose) was even detrimental for regrowth (Table 4). This indicates that the species is sensitive to osmotic stress, even imposed by sucrose treatment. Likewise, shoot tips of another endangered wild species habituated in wetland, *Pogostemon yatabeanus*, were very sensitive to osmotic stress, and 10% sucrose was chosen as the best preculture condition [38]. Mallón et al. [39] noted that preculture with 0.25–0.3 M sucrose was effective for regrowth, while preculture with a higher sucrose concentration resulted in reduced regrowth of cryoprotected shoot tips in critically endangered species of the Asteraceae family.

Osmoprotection (or loading) treatment (incubating the explants with moderately concentrated CPAs before vitrification solution treatment) increased the osmotic tolerance of the cells and minimize osmotic damage caused by the VS [40–42]. Though a mixture of 2 M glycerol (18.0%) plus 0.4 M sucrose (13.7%) has been most frequently used as an osmoprotectant solution [43], an alternative formulation, C-35%, produced higher LN regeneration in osmotically sensitive materials, such as chrysanthemum shoot tips [44]. Osmoprotection was particularly important for hairy roots of *Rubia akane*, which have been shown to be highly susceptible to the cytotoxicity of VSs [45]. In the present study, osmoprotection treatment with C4-35% for 20 min increased regeneration of cryoprotected (21.7%) and cryopreserved (18.3%) shoot tips (Table 5), but there was no significant difference in regeneration. In the preliminary experiment, a longer loading treatment (40 min) was detrimental. Alternatively, a sequential osmoprotection from lower to higher concentration or longer exposure at 0 °C instead of 25 °C might be beneficial for a sufficient osmoprotection of sensitive material [46].

Cytotoxicity of the highly concentrated VSs is a limiting factor for the successful regeneration of differentiated propagules such as shoot tips; therefore, identifying the nature of cytotoxicity, whether biochemical and/or osmotic [47], is necessary for establishing the reliable solution-based vitrification protocols for a given plant material. Although PVS2 [32,48] is the most commonly used VS, PVS3 is recommended for larger explants [32]. Additionally, their alternative variants were also successfully tested [49]. Dilution of a PVS2 variant, A3-90%, to 80% (A3-80%) or 70% (A3-70%) has been successfully applied for sensitive materials, such as tiny shoot tips of *Pogostemon yatabeanus* (Makino) Press [35], *Aster altaicus* [16], or root cultures of *Rubia akane* [50]. In the present study, *P. chinense* shoots were very sensitive to both osmotic stress and chemical toxicity induced by original PVS3, PVS2 or its variant A3-90%, and thus cryoprotection with the diluted VS B5-80% or A3-70% for 30 min produced higher survival and regeneration of both LNC and LN shoot tips. Based on the classification proposed in our previous research [12], *P. chinense* shoots are very sensitive to osmotic stress and chemical cytotoxicity.

Compared to 'classical' vitrification using cryovials, droplet-vitrification using aluminum foil strips resulted in higher survival and regeneration of *P. chinense* shoots. With insufficient cryoprotection using VSs of a lower concentration and a relatively shorter duration, higher cooling and rewarming rates provided by aluminum foil strips and a preheated (40 °C) unloading solution (droplet-vitrification) were superior to using cryovials (vitrification) and ensured the inhibition of crystallization and recrystallization events [51].

To obtain the high survival and fast regrowth of cryopreserved shoot tips, congenial regrowth conditions are essential; in particular, the initial recovery media and culture conditions are critical in many species [16,18,30]. Recovery could eventually occur in initial darkness, since dark incubation contributed to the avoidance of photo-oxidation, which could be harmful to the tissue [41]. Though plant growth regulators have been widely applied for the recovery of cryopreserved shoot tips [52,53], sequential transferring to a hormone-free medium was beneficial for the normal regeneration of both cryoprotected and cryopreserved shoot tips of *Aster altaicus* [16]. The viability of cryopreserved rice cells exponentially depended on the concentration of ammonium ions on the liquid regrowth medium during the initial seven days after slow freezing [54]. Substitution of ammonium nitrate (NH_4NO_3) by potassium nitrate (KNO_3) improved post-cryopreservation recovery from 22.5% to 53% in *Betula pendula* shoot tips [30]. In the PVS2-based vitrification method, recovery of LN sweet-potato shoot tips was significantly improved by initial regrowth on ammonium-free regrowth medium (32% vs. 93%) [55].

The present study highlights the critical effect of ammonium-free medium at the initial regrowth stages on the successful recovery of cryoprotected and cryopreserved shoot tips in three-steps regrowth procedure. Initial regrowth on the ammonium-containing medium (treatments RM2-RM2-MSF and RM2~ in Table 7) produced very low regeneration of both cryoprotected (0–10%) and cryopreserved (3.3–5.2%) shoot tips. Though the mechanism of the critical effect of ammonium during the first recovery steps is not yet clear, it is hypothesized that any type of injury during cryopreservation, i.e., osmotic stress, chemical toxicity, crystallization, etc. may be associated with oxidative stress. In this case, the initial regrown on ammonium-containing medium triggers ammonium-induced oxidative stress, which amplifies the stress already caused by cryopreservation and causes a failure in recovery of both the LNC and LN shoot tips of sensitive species like *P. chinense* [55]. This study also indicates that ammonium-induced oxidative stress occurred mainly at the stage of cryoprotection with VSs (LNC) before cooling and warming in LN.

Further optimization of the developed cryopreservation procedure was performed through vitalization of donor plants including using apical sections instead of nodal cuttings for propagation and a liquid medium overlay on top of the solid medium. Shoot tips excised from these donor plants were able to withstand longer osmoprotection treatment (40 min instead of 20 min) and longer exposure to VSs (up to 60 min), which ensured better cryoprotection and resulted in higher (64.2%) regrowth after cryopreservation. The physiological condition of donor plant material usually plays a vital role in the success of cryopreservation. With the endangered plant *Castilleja levisecta*, for example, shoot tips excised from 6- or 12-day-old in vitro plants showed better response to cryopreservation than shoot tips from older plant material [15]. In *Chrysanthemum morifolium* cv. Peak, the optimal age of donor plants for cryopreservation was 4–5.5 weeks for apical shoot tips and 7 weeks for axillary shoot tips [56].

5. Conclusions

In this study, we established, for the first time, the procedure for the in vitro seed germination and multiplication of *Penthorum chinense* using node cuttings and developed cryopreservation of the shoot tips as complimentary options for ex situ conservation of this valuable medicinal species. The results highlight that *P. chinense* shoots were very sensitive to osmotic stress and chemical toxicity induced by highly concentrated vitrification solutions. The highest survival and regeneration of both cryoprotected and cryopreserved shoot tips were obtained following 40 min osmoprotection and treatment with alternative vitrification solution B5-85% for 60 min. Regeneration of 64.2% after cryopreservation can be considered relatively high for the wetland species.

A three-step regrowth starting with ammonium-free medium during the initial five days was critically important for the regeneration of healthy plantlets both from cryoprotected and cryopreserved shoot tips. The results open the door for the more effective conservation of this species and highlight the importance of alternative vitrification solutions, the vigor of donor plants and regrowth on ammonium-free medium for species habituating in the wetlands and thus sensitive to severe osmotic and oxidative stressed provoked by cryopreservation. Considering the complexity of diverse factors during the course of in vitro propagation, cryopreservation and regrowth, machine learning can be applied in modeling, predicting, and optimizing the procedure for a new species [57].

Author Contributions: Conceptualization, H.K.; investigation, R.A.K.M.Z. and H.L.; data curation, writing, R.A.K.M.Z., E.P., H.K.; funding acquisition, H.K. All authors have read and agreed to the published version of the manuscript.

Funding: This work was supported by Research Promotion Program of Sunchon National University (SCNU).

Data Availability Statement: Not applicable.

Conflicts of Interest: The authors declare no conflict of interest.

References

1. The Plant List: A Working List of All Plant Species. Available online: http://www.theplantlist.org/ (accessed on 9 September 2022).
2. Mahesh, T.; Menon, V.P. Quercetin alleviates oxidative stress in streptozotocin-induced diabetic rats. *Phytother. Res.* **2004**, *18*, 123–127. [CrossRef] [PubMed]
3. Moon, Y.J.; Wang, X.; Morris, M.E. Dietary flavonoids: Effects on xenobiotic and carcinogen metabolism. *Toxicol. Vitro* **2006**, *20*, 187–210. [CrossRef]
4. Kang, D.; Lundström, A.; Steiner, H. Trichoplusia ni attacin A, a differentially displayed insect gene coding for an antibacterial protein. *Gene* **1996**, *174*, 245–249. [CrossRef]
5. Sigurdsson, S.; Nordmark, G.; Göring, H.H.; Lindroos, K.; Wiman, A.C.; Sturfelt, G.; Jönsen, A.; Rantapää-Dahlqvist, S.; Möller, B.; Kere, J.; et al. Polymorphisms in the tyrosine kinase 2 and interferon regulatory factor 5 genes are associated with systemic lupus erythematosus. *Am. J. Hum. Genet.* **2005**, *76*, 528–537. [CrossRef] [PubMed]
6. *Dictionary of Traditional Chinese Materia Medica (Chinese-English)*, 2nd ed.; Human Science & Technology Press: Changsha, China, 1993.
7. Shin, H.J. Ecological study of habitat restoration and environmental characteristics of *Pethorum. chinense*. Ph.D. Thesis, Changwon National University, Changwon, Korea, 2011.
8. Northcutt, C.; Davies, D.; Gagliardo, R.; Bucalo, K.; Determann, R.O.; Cruse-Sanders, J.M.; Pullman, G.S. Germination in vitro, micropropagation, and cryogenic storage for three rare pitcher plants: *Sarracenia. oreophila* (Kearney) Wherry (federally endangered), *S. leucophylla* Raf., and *S. purpurea* spp. *venosa* (Raf.) Wherry. *HortScience* **2012**, *47*, 74–80. [CrossRef]
9. Pence, V.C.; Ballesteros, D.; Walters, C.; Reed, B.M.; Philpott, M.; Dixon, K.W.; Pritchard, H.W.; Culley, T.M.; Vanhove, A.C. Cryobiotechnologies: Tools for expanding long-term ex situ conservation to all plant species. *Biol. Conserv.* **2020**, *250*, 108736. [CrossRef]
10. Kaviani, B.; Kulus, D. Cryopreservation of endangered ornamental plants and fruit crops from tropical and subtropical regions. *Biology* **2022**, *11*, 847. [CrossRef]
11. Tanaka, D.; Niino, T.; Uemura, M. Cryopreservation of plant genetic Resources. *Adv. Exp. Med. Biol.* **2018**, *1081*, 355–369. [CrossRef]
12. Kim, H.H.; Lee, S.C. Personalisation of droplet-vitrification protocols for plant cells: A systematic approach to optimising chemical and osmotic effects. *CryoLetters* **2012**, *33*, 271–279.
13. Shin, D.J.; Kong, H.; Popova, E.V.; Moon, H.K.; Park, S.Y.; Park, S.U.; Lee, S.C.; Kim, H.H. Cryopreservation of *Kalopanax. septemlobus* embryogenic callus using vitrification and droplet-vitrification. *CryoLetters* **2012**, *33*, 402–410.
14. Rathwell, R.; Popova, E.; Shukla, M.R.; Saxena, P.K. Development of cryopreservation methods for cherry birch (*Betula. lenta* L.), an endangered tree species in Canada. *Can. J. For. Res.* **2016**, *46*, 1284–1292. [CrossRef]
15. Salama, A.; Popova, E.; Jones, M.P.; Shukla, M.R.; Fisk, N.S.; Saxena, P.K. Cryopreservation of the critically endangered golden paintbrush (*Castilleja. levisecta* Greenm.): From nature to cryobank to nature. *In Vitro Cell. Dev. Biol.-Plant* **2018**, *54*, 69–78. [CrossRef]
16. Choi, C.H.; Popova, E.; Lee, H.; Park, S.U.; Ku, J.; Kang, J.H.; Kim, H.H. Cryopreservation of endangered wild species, *Aster. altaicus* var. uchiyamae Kitam, using droplet-vitrification procedure. *CryoLetters* **2019**, *40*, 113–122. [PubMed]
17. Popova, E.V.; Shukla, M.R.; McIntosh, T.; Saxena, P.K. In vitro and cryobiotechnology approaches to safeguard *Lupinus rivularis* Douglas ex Lindl., an endangered plant in Canada. *Agronomy* **2021**, *11*, 37. [CrossRef]
18. Turner, S.R.; Touchell, D.H.; Senaratna, T.; Bunn, E.; Tan, B.; Dixon, K.W. Effects of plant growth regulators on survival and recovery growth following cryopreservation. *CryoLetters* **2001**, *22*, 163–174.
19. Bettoni, J.C.; Kretzschmar, A.A.; Bonnart, R.; Shepherd, A.; Volk, G.M. Cryopreservation of 12 *Vitis* species using apical shoot tips derived from plants grown in vitro. *HortScience* **2019**, *54*, 976–981. Available online: https://journals.ashs.org/hortsci/view/journals/hortsci/54/6/article-p976.xml (accessed on 21 October 2022). [CrossRef]

20. Murashige, T.; Skoog, F. A revised medium for rapid growth and bio assays with tobacco tissue cultures. *Physiol Plant* **1962**, *15*, 473–497. [CrossRef]
21. Pullman, G.S.; Skryabina, A. Liquid medium and liquid overlays improve embryogenic tissue initiation in conifers. *Plant Cell Rep.* **2007**, *26*, 873–887. [CrossRef]
22. Hussien, F.A.; Osman, M.A.; Idris, T.I. The influence of liquid media support, gelling agents and liquid overlays on performance of in vitro cultures of ginger (*Zingiber. officinale*). *Intl. J. Sci. Res. Pub.* **2014**, *4*, 2250–3153.
23. Jain, A.; Poling, M.D.; Smith, A.P.; Nagarajan, V.K.; Lahner, B.; Meagher, R.B.; Raghothama, K.G. Variations in the composition of gelling agents affect morphophysiological and molecular responses to deficiencies of phosphate and other nutrients. *Plant Physiol.* **2009**, *150*, 1033–1049. [CrossRef]
24. Das, N.; Tripathi, N.; Basu, S.; Bose, C.; Maitra, S.; Khurana, S. Progress in the development of gelling agents for improved culturability of microorganisms. *Front. Microbiol.* **2015**, *6*, 698. [CrossRef]
25. Thomas, T.D. The role of activated charcoal in plant tissue culture. *Biotechnol. Adv.* **2008**, *26*, 618–631. [CrossRef] [PubMed]
26. Pacek-Bieniek, A.; Dyduch-Siemińska, M.; Rudaś, M. Influence of activated charcoal on seed germination and seedling development by asymbiotic method in *Zygostates. grandiflora* (Lindl.) Mansf. (Orchidaceae). *Folia Hortic.* **2010**, *22*, 45–50. [CrossRef]
27. Polzin, F.; Sylvestre, I.; Déchamp, E.; Ilbert, P.; Etienne, H.; Engelmann, F. Effect of activated charcoal on multiplication of African yam (*Dioscorea. cayenensis.-rotundata*) nodal segments using a temporary immersion bioreactor (RITA®). *In Vitro Cell. Dev. Biol.-Plant* **2014**, *50*, 210–216.
28. Kim, J.K.; Park, S.U. Enhancement of in vitro rooting through growth media, gelling agents and activated charcoal in *Lycium. Chinense*. *Online J. Biol. Sci.* **2017**, *17*, 151–156. [CrossRef]
29. Shin, S.L.; Lee, C.H. Effect of medium components and culture methods on shoot regeneration from *Athyrium niponicum*. *Kor. J. Plant Res.* **2011**, *24*, 113–120. [CrossRef]
30. Kuriyama, A.; Kuriyama, K.; Kuriyama, F.; Kuriyama, M. Sensitivity of cryopreserved *Lavandula vera* cells to ammonium ion. *J. Plant Physiol.* **1996**, *148*, 693–695. [CrossRef]
31. Pullman, G.S.; Gupta, P.K.; Timmis, R.; Carpenter, C.; Kreitinger, M.; Welty, E. Improved Norway spruce somatic embryo development through the use of abscisic acid combined with activated carbon. *Plant Cell Rep.* **2005**, *24*, 271–279. [CrossRef]
32. Sakai, A.; Engelmann, F. Vitrification, encapsulation-vitrification and droplet-vitrification: A review. *CryoLetters* **2007**, *28*, 151–172.
33. Carpentier, S.C.; Witters, E.; Laukens, K.; van Onckelen, H.; Swennen, R.; Panis, B. Banana (*Musa* spp.) as a model to study the meristem proteome: Acclimation to osmotic stress. *Proteomics* **2007**, *7*, 92–105. [CrossRef]
34. Hitmi, A.; Barthomeuf, C.; Sallanon, H. Cryopreservation of *Chrysanthemum cinerariaefolium* shoot tips. Effects of pretreatment conditions and retention of biosynthetic capacity. *CryoLetters* **1999**, *20*, 109–120.
35. Suzuki, M.; Ishikawa, M.; Okuda, H.; Noda, K.; Kishimoto, T.; Nakamura, T.; Ogiwara, I.; Shimura, I.; Akihama, T. Physiological changes in gentian axillary buds during two-step preculturing with sucrose that conferred high levels of tolerance to desiccation and cryopreservation. *Ann. Bot.* **2006**, *97*, 1073–1081. [CrossRef] [PubMed]
36. Zhu, G.Y.; Geuns, J.M.; Dussert, S.; Swennen, R.; Panis, B. Change in sugar, sterol and fatty acid composition in banana meristems caused by sucrose-induced acclimation and its effects on cryopreservation. *Physiol Plant* **2006**, *128*, 80–94. [CrossRef]
37. Jitsuyama, Y.; Suzuki, T.; Harada, T.; Fujikawa, S. Sucrose incubation increases freezing tolerance of asparagus (*Asparagus officinalis* L.) embryogenic cell suspensions. *CryoLetters* **2002**, *23*, 103–112.
38. Lee, H.E.; Popova, E.; Park, H.N.; Park, S.U.; Kim, H.H. Optimization of a cryopreservation method for the endangered Korean species *Pogostemon yatabeanus* using a systematic approach: The key role of ammonium and growth regulators. *Plants* **2021**, *10*, 2018. [CrossRef] [PubMed]
39. Mallón, R.; Bunn, E.; Turner, S.R.; González, M.L. Cryopreservation of *Centaurea. ultreiae* (Compositae) a critically endangered species from Galicia (Spain). *CryoLetters* **2008**, *29*, 363–370. [PubMed]
40. Rall, W.F.; Fahy, G.M. Ice-free cryopreservation of mouse embryos at −196 °C by vitrification. *Nature* **1985**, *313*, 573–575. [CrossRef]
41. Benson, E.E.; Reed, B.M.; Brennan, R.M.; Clacher, K.A.; Ross, D.A. Use of thermal analysis in the evaluation of cryopreservation protocols for *Ribes nigrum* L. germplasm. *CryoLetters* **1996**, *17*, 347–362.
42. Volk, G.M.; Maness, N.; Rotindo, K. Cryopreservation of garlic (*Allium sativum* L.) using plant vitrification solution 2. *CryoLetters* **2004**, *25*, 219–226.
43. Matsumoto, T.; Sakai, A.; Takahashi, C.; Yamada, K. Cryopreservation of in vitro-grown apical meristems of wasabi (*Wasabia japonica*) by encapsulation-vitrification method. *CryoLetters* **1995**, *16*, 189–206.
44. Kim, H.H.; Lee, Y.G.; Park, S.U.; Lee, S.C.; Baek, H.J.; Cho, E.G.; Engelmann, F. Development of alternative loading solutions in droplet-vitrification procedures. *CryoLetters* **2009**, *30*, 291–299. [PubMed]
45. Lambert, E.; Goossens, A.; Panis, B.; van Labeke, M.C.; Geelen, D. Cryopreservation of hairy root cultures of *Maesa lanceolata* and *Medicago truncatula*. *Plant Cell Tissue Organ Cult.* **2009**, *96*, 289–296. [CrossRef]
46. Park, S.U.; Kong, H.; Shin, D.J.; Bae, C.H.; Lee, S.C.; Bae, C.H.; Rha, E.S.; Kim, H.H. Development of vitrification protocol in *Rubia akane* (Nakai) hairy roots using a systematic approach. *CryoLetters* **2014**, *35*, 138–144. [PubMed]
47. Fahy, G.M.; MacFarlane, D.R.; Angell, C.A.; Meryman, H.T. Vitrification as an approach to cryopreservation. *Cryobiology* **1984**, *21*, 407–426. [CrossRef]
48. Sakai, A.; Kobayashi, S.; Oiyama, I. Cryopreservation of nucellar cells of navel orange (*Citrus. sinensis.* Osb. var. *brasiliensis.* Tanaka) by vitrification. *Plant Cell Rep.* **1990**, *9*, 30–33.

49. Kim, H.H.; Lee, Y.G.; Shin, D.J.; Ko, H.C.; Gwag, J.G.; Cho, E.G.; Engelmann, F. Development of alternative plant vitrification solutions in droplet-vitrification procedures. *CryoLetters* **2009**, *30*, 320–334. [CrossRef]
50. Kim, H.H.; Popova, E.V.; Shin, D.J.; Bae, C.H.; Baek, H.J.; Park, S.U.; Engelmann, F. Development of a droplet-vitrification protocol for cryopreservation of *Rubia. akane* (Nakai) hairy roots using a systematic approach. *CryoLetters* **2012**, *33*, 506–517.
51. Kim, H.H.; Lee, J.K.; Yoon, J.W.; Ji, J.J.; Nam, S.S.; Hwang, H.S.; Cho, E.G.; Engelmann, F. Cryopreservation of garlic bulbil primordia by the droplet-vitrification procedure. *CryoLetters* **2006**, *27*, 143–153.
52. Touchell, D.; Turner, S.R.; Senaratna, T.; Bunn, E.; Dixon, K.W. Cryopreservation of Australian Species—The Role of Plant Growth Regulators. In *Cryopreservation of Plant Germplasm II Biotechnology in Agriculture and Forestry*; Springer: Berlin/Heidelberg, Germany, 2002; pp. 373–390.
53. Mukherjee, P.; Mandal, B.B.; Bhat, K.V.; Biswas, A.K. Cryopreservation of Asian *Dioscorea. bulbifera* L. and *D. alata* L. by vitrification: Importance of plant growth regulators. *CryoLetters* **2009**, *30*, 100–111.
54. Kuriyama, A.; Watanabe, K.; Ueno, S.; Mitsuda, H. Inhibitory effect of ammonium ion on recovery of cryopreserved rice cells. *Plant Sci.* **1989**, *64*, 231–235. [CrossRef]
55. Pennycooke, J.C.; Towill, L.E. Medium alterations improve regrowth of sweet potato (*Ipomoea. batatas* [L.] lam.) shoot tips cryopreserved by vitrification and encapsulation-dehydration. *CryoLetters* **2001**, *1*, 381–389.
56. Lee, Y.G.; Popova, E.; Cui, H.Y.; Kim, H.H.; Park, S.U.; Bae, C.H.; Lee, S.C.; Engelman, F. Improved cryopreservation of chrysanthemum (*Chrysanthemum. morifolium*) using droplet-vitrification. *CryoLetters* **2011**, *32*, 487–497. [PubMed]
57. Hesami, M.; Naderi, R.; Tohidfar, M. Introducing a hybrid artificial intelligence method for high-throughput modeling and optimizing plant tissue culture processes: The establishment of a new embryogenesis medium for chrysanthemum, as a case study. *Appl. Microbiol. Biotechnol.* **2020**, *104*, 10249–10263. [CrossRef] [PubMed]

life

MDPI

Article

Dispersal Kernel Type Highly Influences Projected Relationships for Plant Disease Epidemic Severity When Outbreak and At-Risk Populations Differ in Susceptibility

Paul M. Severns

Department of Plant Pathology, University of Georgia, Athens, GA 30602, USA; paul.severns@uga.edu

Abstract: In silico study of biologically invading organisms provide a means to evaluate the complex and potentially cryptic factors that can influence invasion success in scenarios where empirical studies would be difficult, if not impossible, to conduct. I used a disease event simulation program to evaluate whether the two most frequently used types of plant pathogen dispersal kernels for epidemiological projections would provide complementary or divergent projections of epidemic severity when the hosts in a disease outbreak differed from the hosts in the at-risk population in the degree of susceptibility. Exponential dispersal kernel simulations of wheat stripe rust (*Puccinia striiformis* var *trittici*) predicted a relatively strong and dominant influence of the at-risk population on the end epidemic severity regardless of outbreak disease levels. Simulations using a modified power law dispersal kernel gave projections that varied depending on the amount of disease in the outbreak and some interactions were counter-intuitive and opposite of the exponential dispersal kernel projections. Although relatively straightforward, the disease spread simulations in the present study strongly suggest that a more biologically accurate dispersal kernel generates complexity that would not be revealed by an exponential dispersal gradient and that selecting a less accurate dispersal kernel may obscure important interactions during biological invasions.

Keywords: disease gradient; disease outbreak; *Puccinia*; wheat stripe rust; plant epidemic; dispersal ecology

Citation: Severns, P.M. Dispersal Kernel Type Highly Influences Projected Relationships for Plant Disease Epidemic Severity When Outbreak and At-Risk Populations Differ in Susceptibility. *Life* **2022**, *12*, 1727. https://doi.org/10.3390/life12111727

Academic Editors: Hakim Manghwar and Wajid Zaman

Received: 10 October 2022
Accepted: 26 October 2022
Published: 28 October 2022

Publisher's Note: MDPI stays neutral with regard to jurisdictional claims in published maps and institutional affiliations.

Copyright: © 2022 by the author. Licensee MDPI, Basel, Switzerland. This article is an open access article distributed under the terms and conditions of the Creative Commons Attribution (CC BY) license (https://creativecommons.org/licenses/by/4.0/).

1. Introduction

Complex systems are difficult to study empirically, but its components can be understood or at least statistically described in a way that the information can be used to create models to project responses under scenarios that may be impossible to create experimentally [1–3]. For invasion biology, models are an important tool for projecting the spatio-temporal patterns of a biological invasion, and they can also facilitate investigations into difficult to study factors and how they may suppress or encourage organism invasion. The insights gained from carefully constructed models containing well-established ties to biologically realistic mechanisms can be crucial for implementing mitigation strategies to control the invading organism [4,5]. While in silico studies are an obvious departure from on-the-ground empirical study and require simplifying assumptions, they are an important method to understand and project the impacts and spatio-temporal patterns associated with biological invasions. A wait and see strategy for the empirical study of a biological invasion is simply not pro-active enough given the world-wide loss of biodiversity, human life, and ecosystem changes that are now the text-book outcomes of uncontrolled biological invasions.

To create a potentially useful statistical model of organism invasion, the stages of the invading organism in terms of colonization, reproduction, and dispersal need to be integrated and preferably run in a spatially explicit, virtual landscape [6]. Demographic rates (e.g., vital rates) and colonization probability can be measured through observation and/or estimated through direct empirical study, manipulative experimentation, and/or in combination with in silico methods that use sensitivity analyses and pattern-oriented modeling [2,6]. One of the more difficult, but critically important, aspects of biological invasion models to measure and/or estimate is the dispersal kernel. The dispersal kernel is a mathematical function that is used to statistically describe how an organism disperses through a landscape over time. Dispersal kernels are notoriously difficult to measure and to accurately parameterize for organisms that are prone to rare, long-distance dispersal events. The challenge to represent the rarer successful long-distance dispersal events are that the successful events are sparse and embedded within a large expanse of absences and this is type of data is information poor compared to the area near an invasion source which contains a relatively large number of successful dispersal events over shorter distances. Because these long-distance events are rare, they can be easily underestimated by a dispersal kernel but be biologically meaningful for the patterns of invasion spread [7].

Disease epidemics are considered a form of biological invasion [8] and they share similar factors that influence epidemic severity as well as control philosophies [9–13]. For plant pathogens, there are two primary types of dispersal kernels that have been used to project the spread of aerially vectored plant diseases (primarily fungal diseases). One family of dispersal kernels are those with functions that are exponentially bound (e.g., exponential, double exponential) and the other family consists of those functions that are not bound by an exponential (e.g., modified power law, modified Pareto distribution) [7,14]. Exponential functions have longer distributional tails (more kurtotic) than a normal distribution and describe the decrease of inoculum/disease dispersed from a source over a greater expanse with rapidly decreasing disease levels as the distance from the source increases. Exponential family dispersal kernels eventually terminate when either the fitted function to empirically collected data crosses the x-axis or the probability of occurrence reaches zero (for probability density functions). The non-exponentially bound dispersal kernels are comparatively more leptokurtic (fatter, thicker, or heavier tails) than the exponentially bounded functions, with the distribution's tails extending for a much greater distance at very low predicted probabilities. In comparison, these more kurtotic functions expand the small probability of long-distance dispersal events over a much greater distance than exponential kernels.

It is a mathematically demonstrated outcome that if the amount of host is approximately continuous and homogenously distributed, and in a sufficiently sized area, that exponentially bound functions will produce disease invasion fronts that move through the host population with a constant rate following a short period of acceleration [14,15]. These exponentially bound dispersal kernel functions simplify to a diffusion rate, a constant rate of disease spread over space, and this property facilitates straightforward predictive diffusion-based epidemiological projections. However, there is also empirical evidence that wind vectored plant diseases are inadequately described by an exponential function (the function's tails are significantly truncated compared to the actual observed dispersal gradient) and that non-exponentially bound functions (dispersal kernels with much longer distribution tails) are biologically more appropriate [7,16–18]. In contrast to the exponential family of dispersal kernels, the long-tailed, non-exponentially bound dispersal kernels produce disease invasions with fronts that appear to always increase in velocity over space until host and/or space become limiting, and therefore cannot be represented by a rate constant, even as a simplification [14,16,18]. Provided the same raw data which were modeled under the same environmental (and host) conditions, these two dispersal kernel types not only generate different rates of organism spread but they also predict markedly different patterns of disease abundance with respect to its source [14–17].

The issue of disease susceptibility, especially as it pertains to understanding and projecting the spread of disease, is an important topic given that vaccinations are expected to generate specific outcomes in the at-risk population and disease resistance bred into plants should suppress disease. However, this issue is not straightforward to study empirically, as between field borders can differ in cultivar composition, fields may be intercropped, cultivar mixtures can be planted, and even alternating rows of different cultivars and fungicide treatments (a cost saving technique that lowers fungicide application rates) are not uncommon grower practices. For such scenarios, it reasonable to ask whether there is a suppressive or facilitative influence (and whether this impact may be predictable) on subsequent epidemic severity when disease disperses from the outbreak into an at-risk population where host resistance is either greater or lower than that of the outbreak host population. For the purposes of this manuscript, I consider the outbreak to be the area (and its host plants) that the initial disease generation occupies and the at-risk population to be all hosts outside of the outbreak. It is possible that the answer to this question could be purely demographic in nature - simply that the reduction or increase in relative reproductive rates is the primary determinant of later disease severity in an at-risk population. However, the shape and degree of dispersal kernel kurtosis can generate a strong impact on the subsequent patterns of spread and the spatial patterns of disease intensification from an outbreak as it spreads into the at-risk population [14,16,19–21].

I used a series of in silico experiments to understand whether the dispersal kernel type, exponentially bound or non-exponentially bound, substantively influences the patterns of disease projections when disease transitions between outbreak and at-risk host populations that differ in disease susceptibility. I focused on wheat stripe rust, an economically important, world-wide, disease of wheat caused by the fungus *Puccinia striiformis* var *tritici* (hereafter *Pst*), a well-studied and relatively well characterized plant pathosystem from an epidemiological perspective. In particular, I was interested in evaluating whether one or both dispersal kernel types (exponentially bound or non-exponentially bound) could yield relationships that are consistently predictable over a range of disease outbreak levels and whether those projections are similar enough to suggest that a simplified approximation could be made about the potential interactions. For example, it is possible that the overall difference in susceptibility between the outbreak and at-risk populations proportionally increases or decreases the amount of disease in the at-risk population according to a predictable linear relationship.

2. Materials and Methods

2.1. Wheat Stripe Rust

Wheat stripe rust (WSR) is caused by the fungus *Puccinia striiformis* var *tritici* (*Pst*) and it is an obligate parasite of its host plant (obligate plant pathogens require relatively healthy and vigorous hosts for disease to occur). WSR can be encountered wherever wheat is grown [22–25] and its alternative host plants appear to be *Berberis* spp. [26,27]. However, it is unlikely that *Berberis* spp. are necessary for WSR epidemics as *Pst* spores can overwinter in the soil and thatch when conditions are mostly above freezing [23,28]. *Pst* produces spores (~10 to 20 microns which appear to be somewhat environmentally resilient to temperature and some UV light exposure [29]) that are borne on uredinia in small aggregates referred to as pustules. Groups of pustules form lesions, which are presented linearly on the upper and lower leaf surfaces, and elongate over time parallel to wheat leaf veins, yielding the "striped" appearance of WSR. Spores are produced in large amounts, several hundred or more uridineospores/day per square millimeter of lesion [25], and R_0 (the basic reproductive number, the mean number of daughter infections arising from a single infection) can be very high (ranging from 35 to ~800) depending on host availability and pre-existing disease levels [30,31]. Disease occurs as long as the wheat plant can physiologically support either new infections or the expansion of existing lesions. As WSR outbreaks intensify, the disease grows at an exponential rate [10], but successful dispersal events can occur over large distances even from relatively small outbreaks [32] and rarer long-distance events are known at continental scales [28,33].

Although wheat stripe rust can be theoretically well-managed through the appropriate timing of fungicide applications [34], WSR epidemics can cause massive damage on susceptible wheat cultivars [23,24,28,35]. Unfortunately, there is also recent evidence that some *Pst* lineages have evolved fungicide resistant mutations [36], which has caused considerable problems for the management other wheat fungal diseases on wheat such as Septoria leaf blotch (*Zymoseptoria trittici*) [37–40], eyespot (*Oculimacula* spp.) [41], and wheat blast (*Magnaporthe oryzae*) [42]. With the increasing incidence of fungicide resistant wheat plant diseases across the world, including *Pst*, control will probably be accomplished through the breeding of durable disease resistance [43]. This means that understanding how disease susceptibility may alter epidemic behavior is an important aspect to understand going forward.

2.2. Wheat Stripe Rust Disease Spread Simulations

I used an updated and highly modified version of the plant disease simulation program EPIMUL [44] to run the in silico projected disease spread experiments. EPIMUL is a spatially explicit, compartmental disease event simulator, which is parameterized to represent real space in a wheat field. Each compartment was filled with virtual host plants that were similar in density to production fields and previously published empirical studies of WSR spread. Plants within the compartment were assigned properties (e.g., density, disease carrying capacity, latent and infectious periods, disease reproduction rates, infection probability, outbreak or at-risk population) and effective disease spores were distributed across this landscape according to a specified dispersal kernel. The epidemiological variables used in the present model originated from published intensive field studies performed in western and central Oregon, USA (see below for simulation and parameter details). For this study, I used deterministic simulations as I was interested in the mean differences between scenarios rather than focusing on the variation within a single scenario and how that variation overlaps with a slightly different set of parameter values. In EPIMUL, stochasticity is built into the dispersal gradient as a Poisson resampling of the original dispersal gradient [10]. In previous simulations, the mean disease levels over space and time in each compartment from 100 stochastic simulations was nearly equivalent to one deterministic run in EPIMUL [10], so while there was information in variability to be gained from stochastic simulations this approach was not necessary given the goals of the present study.

Compartment parameters for the simulations were consistent with previous WSR simulations [17,45] and updated with more accurate parameter values when supported by newer published data. The simulation field size was 800 × 800 compartments, with each compartment having dimensions of 1.52 m × 1.52 m (the width of a wheat planter) and each compartment had a carrying capacity of 200,000 infection sites, which is the average number of sites estimated from a standard wheat planting density over the life of the average wheat plant [17]. I used a latent and infectious period of 12 days, which is common for WSR outbreaks in the late spring and early summer when conditions are optimal for the disease in central Oregon. R_0, the basic reproduction number [46], which is the mean number of daughter infections arising from a single mother infection, was set at 70 for the completely susceptible genotype and reduced proportionally with a decrease in susceptibility (an increase in disease resistance). This method is described below in a separate paragraph. The fully susceptible host R_0 = 70 is consistent with previous experiments featuring fully susceptible and partially susceptible wheat genotypes [45,47] and studies of WSR development [30] over a range of environmental conditions that were comparable to central Oregon.

I used two different dispersal kernels to simulate WSR disease spread in the exact same virtual field arrangement to understand the influence of each dispersal kernel type on epidemic projections. The first gradient was the modified power law dispersal kernel reported by Farber et al. [32]. This is the most accurately and precisely described dispersal gradient for WSR available in the published literature. For the modified power law, the dispersal kernel was described by the formula $y = a\,(x + c)^{-b}$ where "a" was a value that adjusts the amount of disease produced at the source; b modified the steepness of the dispersal kernel; and the c value allowed for the power law dispersal kernel to have a non-zero value when x = 0 and also modified the kernel shape. For the modified power law simulations, the values of each variable were: a = 425, b = 2.28, c = 0.23. The exponential function was calculated from the original data used by Farber et al. [32] (which was originally and appropriately best-fit to the power law kernel above) and an exponential model was forced on the Farber et al. [32] raw data with the method traditionally used by plant pathologists to fit disease gradients to an exponential kernel [48]. The exponential dispersal kernel was described by the following formula: y = a exp (−bx); for this study a = 19.2, b = 0.1903. WSR infections were dispersed equally (radially) from the source using the downwind dispersal gradient reported by Farber et al. [32].

I also evaluated the potential influence of the amount of disease in the outbreak on the projections. Disease levels in the outbreak can have a strong and dominant impact on the severity of the subsequent WSR epidemic in the at-risk population, in field experiments [45,47] and in simulations [10]. It is possible there were dispersal kernel × outbreak disease level interactions that influence epidemic severity when host populations differ in disease susceptibility. The outbreak levels of disease in my simulations were set at 0.05%, 1.0%, and 5.0% of the total sites available (disease carrying capacity), and these values span the range of biologically reasonable outbreak levels (0.05% and 1.0%) and exceptionally high outbreak levels (5%).

I set up two virtual landscapes that were used with both dispersal kernels and each disease outbreak level to project the interactions of epidemic severity given the differences in host disease susceptibility in a standardized landscape. Both fields contained an outbreak (focus) that was one compartment (1.52 m × 1.52 m) in the center of an 800 × 800 compartment landscape. All compartments other than the outbreak represent the at-risk host population. In one scenario, the outbreak compartment was always 100% susceptible but the at-risk population host susceptibility varied in increments of 10% (from 100% to 10%). A susceptibility of 0 would not generate disease in the model as these hosts are completely resistant and useless in the present study. In the second scenario, the at-risk population was always 100% susceptible but the outbreak varied in susceptibility by increments of 10% (Figure 1). To compare the relative effect of the transition from populations of host that differed in susceptibilities, an internal control, I simulated disease spread in monocultures for the same increments of susceptibility (e.g., 10% focus to 10% at-risk, 50% focus to 50% at-risk, 100% focus to 100% at-risk).

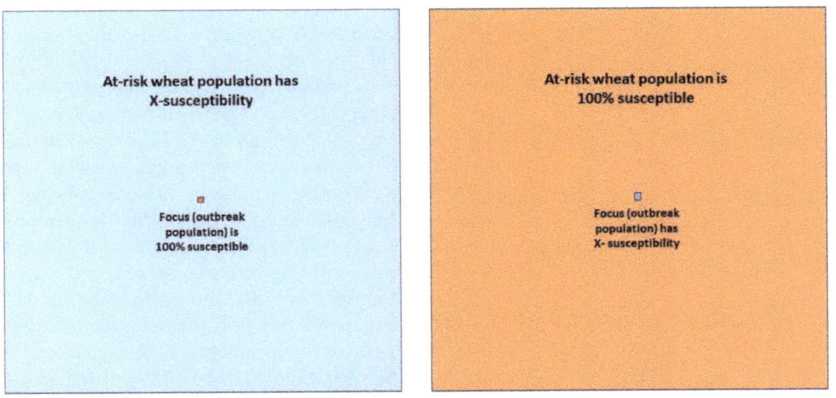

Figure 1. Schematic representation of two different landscape simulation scenarios where the outbreak (focus) and the at-risk population differed in the degree of quantitative resistance (susceptibility) by increments of 10% through the proportional reduction of R_0 (see methods below). The left field depicts the scenario where the outbreak (focus) is comprised of a 100% WSR susceptible genotype and the at-risk population (the remainder of the field) decreases in the degree of susceptibility by increments of 10%. The right field depicts the opposite scenario where the focus is comprised of host plants that are variably susceptible and the at-risk population is 100% susceptible. Note that the focus and the at-risk field is not to the scale of the simulations. In the simulations the focus is considerably smaller relative to the at-risk field size.

To model the differences in host susceptibility within the two landscape scenarios, I proportionally decreased R_0 in 10% increments and assigned the desired levels of susceptibility to the outbreak and at-risk compartments (e.g., 100% susceptible hosts had an $R_0 = 70$, 10% susceptible hosts have an $R_0 = 7$). I held the infection probability the same for the dispersed effective spores which, in combination with a decreased R_0, reduced their capacity for disease production if infected. Although, this approach is overly simplistic, as biologically resistance can arise from different mechanisms (e.g., reduced infection probability, reduced virulence, smaller lesions, lower sporulation rates), proportionally reducing the R_0 is a straightforward method to represent hypothetical quantitative resistance from any mechanism and evaluate the resultant patterns of epidemic progression. To index the relative amount of disease that accumulated in the at-risk population from the outbreak after five disease generations (60 days), I calculated the area under the disease gradient (AUDG) for a 1×301 compartment area extending from the outbreak in a straight line (Figure 1). I subtracted the amount of disease in the outbreak compartment to arrive at an end epidemic AUDG value for the at-risk population. Calculating the amount of disease along a transect in the simulations mimics empirical studies of plant disease spread that sample disease at points along a straight line from the source [16,17,45,47,48]. I plotted the AUDG values for all different combinations of outbreak disease levels, monocultures, and disease accumulated in the at-risk populations and grouped the simulations by the two different landscape scenarios for comparison.

3. Results

There were general patterns of disease increase that were consistent regardless of the dispersal model. Overall, the AUDG values (an index of relative epidemic severity) was predictably greater when the amount of disease in the outbreak was greater (Figure 2). Additionally, when the outbreak and at-risk populations were both at 100% susceptibility, the projected amount of disease was the greatest observed, and when either the outbreak or at-risk population was comprised of host plants with 10% susceptibility, epidemic severity was the lowest observed (Figure 2). However, projections from the two dispersal kernel types yielded differently shaped responses in the amount of disease accumulated after 5 generations.

The exponential dispersal kernel projected relatively consistent epidemic responses when disease developed from an outbreak and intensified over time in at-risk population which differed from the outbreak in the degree of disease susceptibility (Figure 2D–F). When the at-risk population was 100% susceptible, the at-risk population susceptibility exerted a dominant influence on the amount of disease that accumulated over time in the at-risk population, regardless of host susceptibility in the outbreak (Figure 2D–F orange lines). When the outbreak was 100% susceptible, the at-risk population degree of susceptibility also strongly influenced the amount of disease that accumulated in the at-risk population (Figure 2D–F blue lines). For both landscape scenarios, the projected relationships were approximately linear at the lower outbreak disease levels (0.05% and 1%), suggesting that host susceptibility of the at-risk population drives epidemic severity in a potentially straightforward and predictable manner. Only at the greatest outbreak disease level (5%), did the projected relationships become more curvilinear (Figure 2F), but the influence of the at-risk host population susceptibility on the end epidemic severity was consistent with lower outbreak disease levels.

There was no consistent pattern of end epidemic severity when disease was dispersed with the modified power law (Figure 2A–C). For each outbreak disease level, the influence of either the outbreak or the at-risk population's level of susceptibility appeared to generate different projections of end epidemic severity (Figure 2A–C). In all circumstances, including the monocultures, the projected epidemic severity relationships did not appear to behave in any obvious generalizable manner and different outbreak disease levels projected different relationships. For example, at the lowest outbreak disease level, 0.05%, the projected epidemic severities were strongly curvilinear while at the greatest outbreak disease level

(5% disease) the projected relationships were more linear than curvilinear compared with lower outbreak disease levels. Unlike the exponential dispersal kernel simulations, the outbreak could either strongly influence the epidemic outcome (Figure 2A), be roughly equivalent in its influence to that of the at-risk population (Figure 2B), or be slightly suppressed by the level of susceptibility of the at-risk population (Figure 2C).

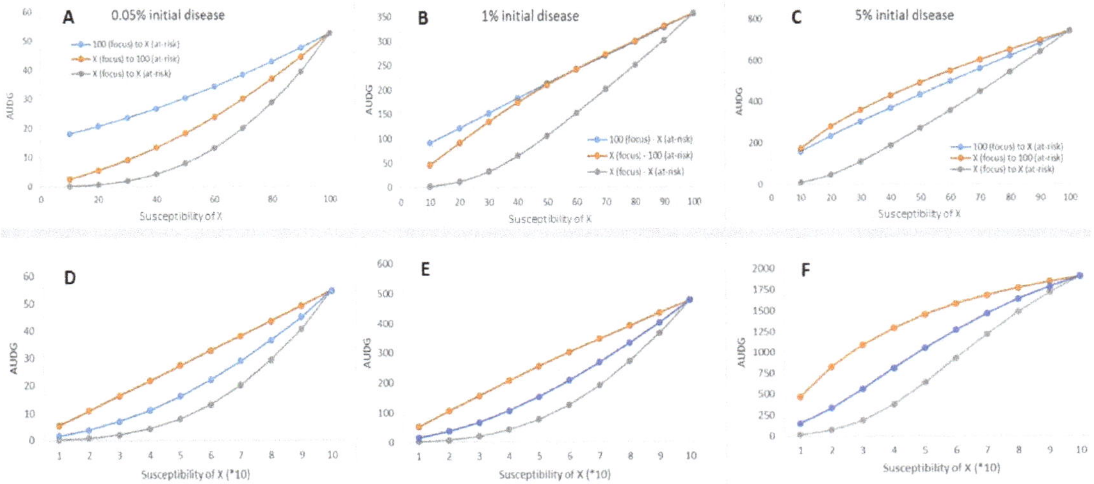

Figure 2. (**A**) (0.05% outbreak disease levels), (**B**) (1% outbreak disease levels), (**C**) (5% outbreak disease levels) are the trends projected from AUDG (area under the disease gradient) values generated from the two field scenario simulations where the outbreak and at-risk populations varied in their relative degree of susceptibility (quantitative resistance to WSR) using the modified power law dispersal kernel. (**D**) (0.05% outbreak disease levels), (**E**) (1% outbreak disease levels), (**F**) (5% outbreak disease levels) are the same projections and disease summary statistics that were generated by the simulations using the exponential dispersal kernel (note that the susceptibility is the same, but it is presented differently on the x-axis, with (*10) to draw attention that these figures were generated from the exponential dispersal kernel). Gray dots and trend lines represent simulation results from monocultures (e.g., 10% focus to 10% at-risk, to 100% focus to 100% at-risk), the blue dots and trend lines represent the scenario where the focus is 100% susceptible but the at-risk population has variable susceptibility, and the orange dots and trend line represents the scenario where the focus varies in susceptibility but the at-risk population is 100% susceptible.

4. Discussion

WSR disease spread simulations, which were calibrated against well-characterized demographic, epidemiological and dispersal parameters, yielded conflicting projections of how disease susceptibility may alter epidemic severity when the outbreak and at-risk host populations differ in their degree of resistance. These differences emerged as an interaction between the dispersal kernel type and the amount of disease that founded the outbreak. Host arrangement, including the virtual field size, compartment number and its dimensions, the location of the outbreak within the virtual field, the area from which disease was estimated, the latent and infectious periods, host density, generation time, infection probability and R_0 (for 100% susceptible genotypes) were standardized throughout the simulations. Disease was also dispersed in a radially symmetric manner from the focus (there was no asymmetric anisotropy; e.g., upwind, downwind, changing wind directions and magnitude) to keep the scenarios as straightforward as possible for comparison. Only the dispersal kernel, the amount of disease in the focus at the outbreak onset, and the susceptibility of host plants in specific compartments (through the proportional reduction of R_0) were modified. Despite this degree of standardization and constant conditions that are obvious departures from a "real life" WSR outbreak, simulations suggested that rule of thumb guidelines for predicting where, when and how much disease may be generated may be possible for organisms with exponential dispersal kernels but unrealistic for organisms characterized by long-distance dispersal. The penalty for over-simplification (a truncated dispersal gradient) was a suite of facile but potentially seriously misleading epidemic projections. These projections were attractive for suggesting a potential predictable pattern of disease spread, whereas attempting to reflect a more biological realistic scenario (through a well-fit dispersal kernel) gave a less intuitive assemblage of epidemic projections. There appeared to be an important tradeoff threshold between convenient interpretation and attempting to reasonably represent the biological reality of long-distance dispersal due specifically to the dispersal kernel.

For the sake of disease management and projecting the impacts of having a mosaic landscape of hosts that differ in disease resistance, it would be convenient if WSR was dispersed according to an exponential function. If WSR dispersal was realistically approximated by an exponential type kernel, understanding and projecting WSR impacts would be relatively tractable, as the constant rate of disease spread after an initial and short period of increasing velocity [15] could be approximated by a diffusion rate [49]. Diffusion rate projections are often applied in invasive species models [7,50,51] as they are in plant epidemiological models [48,49,52]. Furthermore, the exponential dispersal kernel simulations consistently projected a dominant influence of the at-risk population disease resistance properties (susceptibility) on the end epidemic severity (Figure 2). Conceptually, disease resistance properties and host density within the at-risk population are the fundamental underpinnings and assumptions of effective modern disease management tactics, such as quarantine zones, vaccinations, and ring culls [1,53–55]. Although, the prioritization of the at-risk population to control disease outbreaks is intuitive, as on-the-ground approaches often prioritize protecting and modifying the at-risk population to contain and dampen the impacts of any disease outbreak, it may be only a partial solution [55].

In contrast to the relatively consistent and potentially straightforward projections of the exponential dispersal kernel simulations, the modified power law dispersal kernel projections were markedly variable and not intuitively predictable over the range of conditions evaluated. Modified power law kernel simulations suggested that the outbreak may generate a strong and dominant influence on the resulting epidemic severity when compared to the at-risk population, especially at low outbreak disease levels. These disease projection results are counter to most disease mitigation approaches which are directed towards treating and prioritizing the at-risk population (e.g., quarantines, vaccinations, ring culling). There is theoretical [56], empirical [10,45,47] and in silico support [45] for a dominant influence of the outbreak on the epidemic severity in the at-risk population, but the mechanisms governing this phenomenon are not yet well-understood. However, at higher outbreak disease levels, >1% of the total possible infections at the outbreak onset, the power law dispersal kernel projections suggested that the outbreak and at-risk population susceptibility properties may exert a roughly equal contribution to the end epidemic severity and at higher outbreak disease levels the at-risk population was projected to have a greater influence than the outbreak (Figure 2A vs. Figure 2B,C). These results suggest that the contributions of the outbreak and the at-risk population may be highly context dependent and challenging to predict if non-exponentially bound, heavy-tailed dispersal kernels are used to more realistically account for the potential of long-distance dispersal. This suggests that with increasing biological accuracy in the dispersal kernel, epidemic projections will likely become both more complex, context dependent, and unfortunately maybe necessarily nuanced.

The landscape scenarios I used in this study were straightforward (Figure 1) and relatively simple compared with the host spatial complexity of cultivar mixtures, intercropping, and a patchwork landscape mosaic of variably sized agricultural fields featuring different wheat cultivars interspersed with fields/habitats without WSR host plants. Regardless, when applying a modified power law dispersal kernel, which empirical data strongly supports over an exponential dispersal kernel [17,18,32], it is clear that the most biologically accurate of the two dispersal kernels yields unintuitive WSR disease projections even though there was no anisotropic disease dispersal (a well-known feature of wind dispersed pathogens [49,57]), homogenous host distribution, and host plants with invariant physiological states (both of which are not biologically true). As attractive, convenient, and readily interpretable as the disease projections from the exponential model appear to be, such biologically inaccurate models have the distinct potential to lead epidemiological understanding and the resulting control management practices down a deceptive path.

Funding: P.M. Severns was supported by the Advancing plant epidemiology for the growers of Georgia and beyond [project accession no. 1023738] from the USDA National Institute of Food and Agriculture.

Institutional Review Board Statement: Not applicable.

Informed Consent Statement: Not applicable.

Data Availability Statement: Data from the simulations are available upon request.

Acknowledgments: I thank Katheryn Sackett for providing the parameter values for the exponentially fit equation and three anonymous reviewers who helped to improve this manuscript.

Conflicts of Interest: The author declares no conflict of interest. The funders had no role in the design of the study; in the collection, analyses, or interpretation of data; in the writing of the manuscript, or in the decision to publish the results.

References

1. Keeling, M.J.; Woolhouse, M.E.J.; May, R.M.; Davies, G.; Grenfell, B.T. Modelling vaccination strategies against foot-and-mouth disease. *Nature* **2003**, *421*, 136–142. [CrossRef] [PubMed]
2. Grimm, V.; Revilla, E.; Berger, U.; Jeltsch, F.; Mooij, W.M.; Railsback, S.F.; Thulke, H.H.; Weiner, J.; Wiegand, T.; DeAngelis, D.L. Pattern-oriented modeling of agent-based complex systems: Lessons from ecology. *Science* **2005**, *310*, 987–991. [CrossRef]
3. Bzdok, D.; Altman, N.; Krzywinski, M. Statistics versus machine learning. *Nat. Methods* **2018**, *15*, 233–234. [CrossRef] [PubMed]
4. Sardain, A.; Sardain, E.; Leung, B. Global forecasts of shipping traffic and biological invasions to 2050. *Nat. Sustain.* **2019**, *2*, 274–282. [CrossRef]
5. Manzoor, S.A.; Griffiths, G.; Lukac, M. Land use and climate change interaction triggers contrasting trajectories of biological invasion. *Ecol. Indic.* **2021**, *120*, 106936. [CrossRef]
6. Pili, A.N.; Tingley, R.; Chapple, D.G.; Schumaker, N.H. VirToad: Simulating the spatiotemporal population dynamics and management of a global invader. *Landsc. Ecol.* **2022**, *37*, 2273–2292. [CrossRef]
7. Clobert, J.; Baguette, M.; Benton, T.G.; Bullock, J.M. *Dispersal Ecology and Evolution*; Oxford University Press: Oxford, UK, 2012.
8. Simberloff, D. The role of propagule pressure in biological invasions. *Annu. Rev. Ecol. Evol. Syst.* **2009**, *40*, 81–102. [CrossRef]
9. Holle, B.V.; Simberloff, D. Ecological resistance to biological invasion overwhelmed by propagule pressure. *Ecology* **2005**, *86*, 3212–3218. [CrossRef]
10. Severns, P.M.; Sackett, K.; Mundt, C.C. Outbreak propagule pressure influences the landscape spread of a wind-dispersed, epidemic-causing, plant pathogen. *Landsc. Ecol.* **2015**, *30*, 2111–2119. [CrossRef]
11. Howard, P.L. Human adaptation to invasive species: A conceptual framework based on a case study metasynthesis. *Ambio* **2019**, *48*, 1401–1430. [CrossRef]
12. Espinoza, B.; Castillo-Chavez, C.; Perrings, C. Mobility restrictions for the control of epidemics: When do they work? *PLoS ONE* **2020**, *15*, e0235731. [CrossRef]
13. Lucardi, R.D.; Bellis, E.S.; Cunard, C.E.; Gravesande, J.K.; Hughes, S.C.; Whitehurst, L.E.; Worthy, S.J.; Burgess, K.S.; Marsico, T.D. Seeds attached to refrigerated shipping containers represent a substantial risk of nonnative plant species introduction and establishment. *Sci. Rep.* **2020**, *10*, 15017. [CrossRef] [PubMed]
14. Kot, M.; Lewis, M.A.; van den Driessche, P. Dispersal data and the spread of invading organisms. *Ecology* **1996**, *77*, 2027–2042. [CrossRef]
15. Frantzen, J.; Van den Bosch, F. Spread of organisms: Can travelling and dispersive waves be distinguished? *Basic Appl. Ecol.* **2000**, *1*, 83–92. [CrossRef]
16. Ferrandino, F.J. Dispersive epidemic waves: I. Focus expansion within a linear planting. *Phytopathology* **1993**, *83*, 795–802. [CrossRef]
17. Sackett, K.E.; Mundt, C.C. Primary disease gradients of wheat stripe rust in large field plots. *Phytopathology* **2005**, *95*, 983–991. [CrossRef]

18. Mundt, C.C.; Sackett, K.E.; Wallace, L.D.; Cowger, C.; Dudley, J.P. Long-distance dispersal and accelerating waves of disease: Empirical relationships. *Am. Nat.* **2009**, *173*, 456–466. [CrossRef] [PubMed]
19. Lindström, T.; Håkansson, N.; Wennergren, U. The shape of the spatial kernel and its implications for biological invasions in patchy environments. *Proc. R. Soc. B Biol. Sci.* **2011**, *278*, 1564–1571. [CrossRef] [PubMed]
20. Byrne, A.W.; Quinn, J.L.; O'Keeffe, J.J.; Green, S.; Paddy Sleeman, D.; Wayne Martin, S.; Davenport, J. Large-scale movements in European badgers: Has the tail of the movement kernel been underestimated? *J. Anim. Ecol.* **2014**, *83*, 991–1001. [CrossRef]
21. Sullivan, L.L.; Michalska-Smith, M.J.; Sperry, K.P.; Moeller, D.A.; Shaw, A.K. Consequences of ignoring dispersal variation in network models for landscape connectivity. *Conserv. Biol.* **2021**, *35*, 944–954. [CrossRef]
22. Wan, A.; Zhao, Z.; Chen, X.; He, Z.; Jin, S. Wheat stripe rust epidemic and virulence of *Puccinia striiformis* f. sp. *tritici* in China in 2002. *Plant Dis.* **2004**, *88*, 896–904. [CrossRef] [PubMed]
23. Chen, X.M. Epidemiology and control of stripe rust on wheat. *Can. J. Plant Pathol.* **2005**, *27*, 314–337. [CrossRef]
24. Wellings, C.R. *Puccinia stiiformis* in Australia: A review of the incursion, evolution and adaptation of stripe rust in the period 1979–2006. *Aust. J. Agric. Res.* **2007**, *58*, 567–575. [CrossRef]
25. Milus, E.A.; Kristensen, K.; Hovmøller, M.S. Evidence for increased aggressiveness in a recent widespread strain of *Puccinia striiformis* f. sp. *tritici* causing stripe rust of wheat. *Phytopathology* **2009**, *99*, 89–94. [CrossRef] [PubMed]
26. Wang, M.N.; Chen, X.M. First report of Oregon grape (*Mahonia aquifolium*) as an alternate host for the wheat stripe rust pathogen (*Puccinia striiformis* f. sp. *tritici*) under artificial inoculation. *Plant Dis.* **2013**, *97*, 839. [CrossRef]
27. Rodriguez-Algaba, J.; Walter, S.; Sørensen, C.K.; Hovmøller, M.S.; Justesen, A.F. Sexual structures and recombination of the wheat rust fungus *Puccinia striiformis* on *Berberis vulgaris*. *Fungal Genet. Biol.* **2014**, *70*, 77–78. [CrossRef]
28. Zeng, S.M.; Luo, Y. Long-distance spread and interregional epidemics of wheat stripe rust in China. *Plant Dis.* **2006**, *90*, 980–988. [CrossRef]
29. Maddison, A.C.; Manners, J.G. Lethal effects of artificial ultraviolet radiation on cereal rust uredospores. *Trans. Br. Mycol. Soc.* **1973**, *60*, 471–494. [CrossRef]
30. Papastamati, K.; van den Bosch, F. The sensitivity of epidemic growth rate to weather variables, with an application to yellow rust on wheat. *Phytopathology* **2006**, *97*, 202–210. [CrossRef]
31. Farber, D.H.; Medlock, J.; Mundt, C.C. Local dispersal of *Puccinia striiformis* f. sp. *tritici* from isolated source lesions. *Plant Pathol.* **2017**, *66*, 28–37. [CrossRef]
32. Farber, D.H.; De Leenheer, P.; Mundt, C.C. Dispersal kernels may be scalable: Implications from a plant pathogen. *J. Biogeogr.* **2019**, *46*, 2042–2055. [CrossRef] [PubMed]
33. Brown, J.K.; Hovmøller, M.S. Aerial dispersal of pathogens on the global and continental scales and its impact on plant disease. *Science* **2002**, *297*, 537–541. [CrossRef] [PubMed]
34. Carmona, M.; Sautua, F.; Pérez-Hérnandez, O.; Reis, M. Role of fungicide applications on the integrated management of wheat stripe rust. *Front. Plant Sci.* **2020**, *11*, 733. [CrossRef]
35. Figueroa, M.; Hammond-Kosack, K.E.; Solomon, P.S. A review of wheat diseases-a field perspective. *Mol. Plant Pathol.* **2018**, *19*, 1523–1536. [CrossRef]
36. Cook, N.M.; Chang, S.; Woodman, T.L.; Warren, R.; Oliver, R.P.; Saunders, D.G. High frequency of fungicide resistance-associated mutations in the wheat yellow rust pathogen *Puccinia striiformis* f. sp. *tritici*. *Pest Manag. Sci.* **2021**, *77*, 3358–3371. [CrossRef]
37. Hawkins, N.J.; Fraaije, B.A. Fitness penalties in the evolution of fungicide resistance. *Annu. Rev. Phytopathol.* **2018**, *56*, 339–360. [CrossRef]
38. Sykes, E.M.; Sackett, K.E.; Severns, P.M.; Mundt, C.C. Sensitivity variation and cross-resistance of *Zymoseptoria tritici* to azole fungicides in North America. *Eur. J. Plant Pathol.* **2018**, *151*, 269–274. [CrossRef]
39. McDonald, M.C.; Renkin, M.; Spackman, M.; Orchard, B.; Croll, D.; Solomon, P.S.; Milgate, A. Rapid parallel evolution of azole fungicide resistance in Australian populations of the wheat pathogen *Zymoseptoria tritici*. *Appl. Environ. Microbiol.* **2019**, *85*, e01908-18. [CrossRef]
40. Garnault, M.; Duplaix, C.; Leroux, P.; Couleaud, G.; David, O.; Walker, A.S.; Carpentier, F. Large-scale study validates that regional fungicide applications are major determinants of resistance evolution in the wheat pathogen *Zymoseptoria tritici* in France. *New Phytol.* **2021**, *229*, 3508–3521. [CrossRef]
41. Leroux, P.; Gredt, M.; Remuson, F.; Micoud, A.; Walker, A.S. Fungicide resistance status in French populations of the wheat eyespot fungi *Oculimacula acuformis* and *Oculimacula yallundae*. *Pest Manag. Sci.* **2013**, *69*, 15–26. [CrossRef]
42. Castroagudín, V.L.; Ceresini, P.C.; de Oliveira, S.C.; Reges, J.T.; Maciel, J.L.; Bonato, A.L.; McDonald, B.A. Resistance to QoI fungicides is widespread in Brazilian populations of the wheat blast pathogen *Magnaporthe oryzae*. *Phytopathology* **2015**, *105*, 284–294. [CrossRef] [PubMed]
43. Mundt, C.C. Pyramiding for resistance durability: Theory and practice. *Phytopatology* **2018**, *108*, 792–802. [CrossRef] [PubMed]
44. Kampmeijer, P.; Zadocks, J.C. *EPIMUL, a Simulator of Foci and Epidemics in Mixtures of Resistant and Susceptible Plants, Mosaics and Multilines*; Centre for Agricultural Publishing and Documentation: Wageningen, The Netherlands, 1977.
45. Severns, P.M.; Mundt, C.C. Delays in epidemic outbreak control cost disproportionately large treatment footprints to offset. *Pathogens* **2022**, *11*, 393. [CrossRef] [PubMed]
46. Diekmann, O.; Heesterbeek, J.A.P. *Mathematical Epidemiology of Infectious Diseases: Model Building, Analysis and Interpretation*; John Wiley & Sons: Hoboken, NJ, USA, 2000.

47. Severns, P.M.; Estep, L.K.; Sackett, K.E.; Mundt, C.C. Degree of host susceptibility in the initial disease outbreak influences subsequent epidemic spread. *J. Appl. Ecol.* **2014**, *51*, 1622–1630. [CrossRef]
48. Madden, L.V.; Hughes, G.; van den Bosch, F. *The Study of Plant Disease Epidemics*; American Phytopathological Society: St. Paul, MN, USA, 2007.
49. Aylor, D.E. *Aerial Dispersal of Pollen and Spores*; APS Press: St. Paul, MN, USA, 2017; 418p.
50. Neubert, M.G.; Parker, I.M. Projecting rates of spread for invasive species. *Risk Anal.* **2004**, *24*, 817–831. [CrossRef]
51. Srivastava, V.; Lafond, V.; Griess, V.C. Species distribution models (SDM): Applications, benefits and challenges in invasive species management. *CABI Rev.* **2019**, *24*, 7. [CrossRef]
52. Atallah, S.S.; Gómez, M.I.; Conrad, J.M.; Nyrop, J.P. A plant-level, spatial, bioeconomic model of plant disease diffusion and control: Grapevine leafroll disease. *Am. J. Agric. Econ.* **2015**, *97*, 199–218. [CrossRef]
53. Anderson, R.M.; May, R.M. The invasion, persistence and spread of infectious diseases within animal and plant communities. *Phil. Trans. R. Soc. Lond. B Biol. Sci.* **1986**, *314*, 533–570. [CrossRef]
54. Shearer, F.M.; Moyes, C.L.; Pigott, D.M.; Brady, O.J.; Marinho, F.; Deshpande, A.; Longbottom, J.; Browne, A.J.; Kraemer, M.U.; O'Reilly, K.M.; et al. Global yellow fever vaccination coverage from 1970 to 2016: An adjusted retrospective analysis. *Lancet Infect. Dis.* **2017**, *17*, 1209–1217. [CrossRef]
55. Ogden, N.H.; Wilson, J.R.; Richardson, D.M.; Hui, C.; Davies, S.J.; Kumschick, S.; Le Roux, J.J.; Measey, J.; Saul, W.C.; Pulliam, J.R. Emerging infectious diseases and biological invasions: A call for a One Health collaboration in science and management. *R. Soc. Open Sci.* **2019**, *6*, 181577. [CrossRef]
56. Zadoks, J.C.; van den Bosch, F. On the spread of plant disease: A theory on foci. *Annu. Rev. Phytopathol.* **1994**, *32*, 503–521. [CrossRef] [PubMed]
57. Isard, S.A.; Chamecki, M. A physically based theoretical model of spore deposition for predicting spread of plant diseases. *Phytopathology* **2016**, *106*, 244–253. [CrossRef] [PubMed]

life

MDPI

Article

Effects of *Rhizobium leguminosarum* Thy2 on the Growth and Tolerance to Cadmium Stress of Wheat Plants

Dilara Maslennikova [1,2,*], Karina Nasyrova [2], Olga Chubukova [1,2], Ekaterina Akimova [1], Andrey Baymiev [1], Darya Blagova [1], Almaz Ibragimov [1] and Oksana Lastochkina [1]

1 Ufa Federal Research Center, Institute of Biochemistry and Genetics, Russian Academy of Sciences, 450054 Ufa, Russia
2 Department of Molecular Technologis, Ufa State Petroleum Technical University, 450000 Ufa, Russia
* Correspondence: dishaoil@mail.ru; Tel.: +8-347-2356088

Abstract: Cadmium (Cd) stress is an obstacle for crop production, quality crops, and sustainable agriculture. An important role is played by the application of eco-friendly approaches to improve plant growth and stress tolerance. In the current study, a pre-sowing seed treatment with *Rhizobium leguminosarum* strains, isolated from the leguminous plants *Phaseolus vulgaris* (strain Pvu5), *Vicia sylvatica* (strain VSy12), *Trifolium hybridium* (strain Thy2), and *T. pratense* (strain TPr4), demonstrated different effects on wheat (*Triticum aestivum* L.) plant growth under normal conditions. Among all tested strains, Thy2 significantly increased seed germination, seedling length, fresh and dry biomass, and leaf chlorophyll (Chl) content. Further analysis showed that Thy2 was capable of producing indole-3-acetic acid and siderophores and fixing nitrogen. Under Cd stress, Thy2 reduced the negative effect of Cd on wheat growth and photosynthesis and had a protective effect on the antioxidant system. This was expressed in the additional accumulation of glutathione and proline and the activation of glutathione reductase. In addition, Thy2 led to a significant reduction in oxidative stress, which was evidenced by the data on the stabilization of the ascorbate content and the activity of ascorbate peroxidase. In addition, Thy2 markedly reduced Cd-induced membrane lipid peroxidation and electrolyte leakage in the plants. Thus, the findings demonstrated the ability of the *R. leguminosarum* strain Thy2, isolated from *T. hybridium* nodules, to exert a growth-promoting and anti-stress effect on wheat plants. These results suggest that the Thy2 strain may enhance wheat plant growth by mitigating Cd stress, particularly through improving photosynthesis and antioxidant capacity and reducing the severity of oxidative damage. This may provide a basic and biological approach to use the Thy2 strain as a promising, eco-friendly candidate to combat Cd stress in wheat production.

Keywords: *Rhizobium leguminosarum*; PGPR; *Triticum aestivum* L.; cadmium stress; tolerance; ascorbate; glutathione; malondialdehyde; chlorophylls

Citation: Maslennikova, D.; Nasyrova, K.; Chubukova, O.; Akimova, E.; Baymiev, A.; Blagova, D.; Ibragimov, A.; Lastochkina, O. Effects of *Rhizobium leguminosarum* Thy2 on the Growth and Tolerance to Cadmium Stress of Wheat Plants. *Life* **2022**, *12*, 1675. https://doi.org/10.3390/life12101675

Academic Editors: Hakim Manghwar and Wajid Zaman

Received: 7 October 2022
Accepted: 18 October 2022
Published: 21 October 2022

Publisher's Note: MDPI stays neutral with regard to jurisdictional claims in published maps and institutional affiliations.

Copyright: © 2022 by the authors. Licensee MDPI, Basel, Switzerland. This article is an open access article distributed under the terms and conditions of the Creative Commons Attribution (CC BY) license (https://creativecommons.org/licenses/by/4.0/).

1. Introduction

Increasing agricultural crop productivity, especially cereals such as wheat, is relevant and extremely important [1–4]. One of the guiding principles of modern agricultural production is the introduction of environmentally sound and harmless approaches that are safe for human health [1,5–7]. Of particular importance and interest for solving this problem is the use of microbiological approaches that are based on the use of the potential of plants and microorganisms and the biological mechanisms of interaction between the components of plant–microbial systems [1,8,9]. The use of biologicals based on plant growth-promoting rhizobacterial (PGPR) strains of the genus of *Rizobium leguminosarum*, demonstrating beneficial properties of plants, is of interest. They can be successful symbionts for various plant species [10–15]. The presence of the bacteria *R. leguminosarum* can regulate in plants such processes as the increased associative fixation of molecular nitrogen [1,15,16], additional production of physiologically active compounds, including phytohormones [8,16], and improvement of the water regime of plants [16]. In addition, rhizobacteria can participate in the dissolution of hard-to-reach phosphorus compounds [17], produce antibiotic compounds that protect roots from bacterial and fungal infections [9], and modulate stress reactions in plants, thereby increasing their resistance to adverse external factors and other influences [16,18].

The complex and positive effects of the influence of PGPR on plants is widely used in plant growing practices as a seed inoculation (bacterization) technique [4]. It should be noted that the pre-sowing inoculation of seeds with microbial preparations that accelerate the growth and development of plants is an important element of crop cultivation technology aimed at obtaining friendly seedlings of cereal crops and, as a result, increasing yields [4]. It does not require complex equipment, and it provides a stable positive effect. It can be noted that the effect of microbiopreparations on plants is being studied quite actively [4,19]. However, insufficient attention has been paid to the study of the response of cereal crops to bacterization by growth-stimulating microorganisms. The application of PGPR is a sustainable approach to improving the physiological processes of crops and overcoming abiotic stresses. The effectiveness of the use of a particular bacterial preparation is often related to a particular crop, and this does not guarantee a positive effect on another crop or variety [10–14,16]. The maximum effect from the use of associative strains can be obtained only based on a careful selection of strains that have a greater positive effect on a particular culture. While it is important to evaluate both the growth-stimulating and protective effects of the strain, photosynthesis, components of the system's regulating redox metabolism, and osmoprotection play an important role in the growth and development of plants under normal growing conditions [14,20]. Under stress conditions, timely changes in the operation of these systems determine the level of adaptation of a plant to stress, and the survival rate of that plant.

In connection with the growth of technogenic production, an acute problem is the pollution of soils with heavy metal ions, in particular, cadmium (Cd) ions [6,7]. Mining, smelting, and synthetic fertilizers are the main sources of Cd toxicity. Cd inhibits plant growth. Moreover, the accumulation of Cd in wheat grains poses a health risk to humans [7]. Cd ions negatively affect the seed germination, growth, grain quality, and productivity of wheat plants. In addition, Cd ions inhibit photosynthetic processes and cause the development of oxidative stress in cells [7,21]. The literature describes many different mitigation strategies to combat Cd toxicity in wheat [14,16,20]. Due to Cd ions entering wheat plants from through soil, a number of these strategies aim to reduce Cd content and its availability to wheat roots. Inexpensive and accessible strategies are often used to achieve this, such as the application of fertilizers or biochar to the soil [3,7]. In addition, the pre-sowing treatment of seeds with various phytohormones, nanoparticles, lasers, and microwaves is used [3,4,7]. In order to increase the resistance of plants to Cd, plants are treated with microbial growth regulators [1]. The use of PGPR effectively reduces the level of the negative effects of Cd on plants [10–15]. It was found that strains of the genus of *Rhizobium leguminosarum*, i.e., the rhizospheric bacterial strains Pvu5 (isolated from *Phaseolus vulgaris*), VSy12 (*Vicia sylvatica*), Thy2 (*Trifolium hybridium*), and TPr4 (*T. pratense*), stimulated the growth of roots of cucumber seedlings [22]. At the same time, the *R. leguminosarum* strain VSy12 also stimulated the growth of the roots of tomato and amaranth seedlings, while the *R. leguminosarum* strain Pvu5 only stimulated the growth of amaranth, and none of the strains exhibited growth-promoting activity on carrot roots. These results indicate that the studied strains have growth-stimulating activities for dicot plants. These strains were obtained by the researchers in [22] and there is no information about their effect on monocot plants such as wheat. This is very surprising as wheat is an important crop all over the world. The current research aims to investigate the effects of the *R. leguminosarum* strains Pvu5, VSy12, Thy2, and TPr4 on wheat growth and their ability to alleviate Cd stress.

2. Materials and Methods

2.1. Bacterial Strains

The *Rhizobium leguminosarum* strains Pvu5, VSy12, Thy2, and TPr4 were obtained from the Collection of Symbiotic Microorganisms "Symbiont" of the Institute of Biochemistry and Genetics of the Ufa Federal Research Center of the Russian Academy of Sciences (Ufa, Russia). The root nodule rhizobia strains were earlier isolated from wild-growing leguminous plants of the Southern Urals: the *R. leguminosarum* strain Pvu5—from *Phaseolus vulgaris;* the *R. leguminosarum* strain VSy12—from *Vicia sylvatica;* the *R. leguminosarum* strain Thy2—from *Trifolium hybridium;* and the *R. leguminosarum* strain TPr4—from *T. pratense.*

The isolation of pure cultures of the bacteria from nodules of the leguminous plants was carried out by the standard method [23], with some modifications [24]. DNA from the bacteria was isolated by cell lysis in a 1% Triton X100 (Serva, Heidelberg, Germany) and 1% Chelex 100 (Bio-Rad, Hercules, CA, USA) suspension [25]. Nucleotide sequences were determined on an Applied Biosystems 3500 automatic sequencer (Applied Biosystems, Inc., Sequencer Applied Biosystems, Inc., Waltham, MA, USA) using Big Dye Terminator v. 3.1.

Fragments of the 16S rRNA gene, approximately 1500 bp, were amplified using the universal primers fD1 5′-cccgggatccaagcttaaggaggtgatccagcc-3′ and rD1 5′-ccgaattcgtcgacaacagagttgatcctggctcag-3′. Phylogenetic analysis of the studied strains was performed using the MegaBlast program based on the data from the multiple alignments of the sequenced fragments of the 16S rRNA gene [26].

2.2. Inoculum Preparation and Seed Treatment

Cells of the *R. leguminosarum* strains Pvu5, VSy12, Thy2, and TPr4 were cultured for 2 days in glass flasks in liquid yeast mannitol YM medium (wt. % in an aqueous solution: mannitol, 1; yeast extract, 0.04; sodium chloride (NaCl), 0.01; and magnesium sulfate ($MgSO_4$), 0.01) at 28 °C and 140 rpm to a concentration of 10^8 CFU mL^{-1}. To obtain an inoculum at a concentration of 10^5 CFU mL^{-1}, the stock 10^8 CFU mL^{-1} was diluted with sterile water and monitored at 600 nm (SmartSpecTM Plus spectrophotometer, Bio-Rad, Oceanside, CA, USA).

2.3. Analysis of Plant Growth-Promoting (PGP) Characteristics of the Tested Bacterial Strains

A determination of indole-3-acetic acid (IAA) production was made using the Salkowski reagent [27]. The purified and freshly grown cultures on the average slopes of Luria-Bertani (LB) were transferred into tubes containing 5 mL of LB broth supplemented by 1 mg mL^{-1} of L-tryptophan (L-TRP). It was incubated at 28 ± 1 °C for 2 days. Then, the culture was centrifuged for 5 min at 1300 rpm. The Salkowski reagent (2% ferric chloride ($FeCl_3$) at 0.5 M in 35% perchloric acid ($HClO_4$)) was added to the supernatant at a rate of 1:2. After 20–25 min, when the color of the supernatant containing the IAA turned red, the color absorption was measured with a spectrophotometer (BioSpec_Mini, Shimadzu, Japan) at 535 nm. The uninoculated broth served as a control. The typical curve was prepared with 1, 2, 3, 4, 5, 10, 15, and 20 mg L^{-1} IAA.

The production of siderophores was detected by the universal method described in [28] using blue agar plates containing Chromium Azurol S (CAS) dye. To prepare 100 mL of the medium, 6.5 g of liquid CAS was added to 5 mL of water and mixed with 1 mL of a solution containing 1 mM of $FeCl_3$ and 10 mM of hydrochloric acid (HCl). Then, 4 mL of a solution containing 7.3 mg of hexadecyltrimethylammonium bromide (HDTMA) was added to the CAS solution. The mixture was autoclaved and then added to 100 mL of LB agar medium with a pH = 6.8. Bacteria were grown on Petri dishes with "blue agar" for 7 days. The formation of a yellow halo around the blue colony was an indication of siderophore production.

The phosphate solubilization activity was tested in a Pikovskaya medium (PVK). This medium consisted of (g L^{-1}): (calcium orthophosphate ($Ca_3(P0_4)_2$)—5, glucose—20, NaCl—0.2, magnesium sulfate ($MgS0_4$)-0.1, manganese sulfate ($MnS0_4$)—traces, ferrous sulfate ($FeS0_4$)-traces, and agar-agar-20) [29]. The bacteria were cultivated in this environment for seven days.

Nitrogen fixation was tested using Ashby's nitrogen-free medium [30]. The bacteria were grown at 28 °C for 3 days.

2.4. *Plant Materials and Growth Conditions*

Wheat seeds (*Triticum aestivum* L., cv. Salavat Yulaev) were provided by the Chishminsky Breeding Station UFRC RAS (Bashkortostan, Russia). The seeds were sterilized in 96% ethanol for 1 min, then washed with sterile water for 2–3 min (until the smell of alcohol disappeared). Thereafter, the seeds were immersed into solutions of the strains Pvu5, VSy12, TPr4, and Thy2 (10^5 CFU mL^{-1}) or water (control) for 30 min. The seeds (100 pieces) were grown for three days on filter paper moistened with water under illumination (200 μmoL m^{-2} s^{-1}) at 22–23 °C with a long-day photoperiod (16 h light/8 h dark) to select a strain that had a significant growth-stimulating effect.

To assess the protective effect of the bacterial strain, the control and inoculated seeds were grown in Petri dishes (on filter paper moistened with water (Control) and Cd—1 mM $(CH_3COO)_2Cd$) [5] under a 16/8 h light/dark photoperiod regime (200 mmoL m^{-1} s^{-1}) at 22–24 °C for three days. After germination, the 3-day old seedlings were transplanted into beakers with 10% Hoagland–Arnon solution and grown for 7 days. In stressful variants of the experiment, the plants were constantly grown on a 1 mM Cd solution prepared on the 10% Hoglanad–Arnon. Plant samples of the 7-day old seedlings (roots, shoots, or whole seedlings) were taken to assess their physiological-biochemical attributes. To determine the content of the photosynthetic pigments, chlorophyll (Chl) a and Chl b, the plants' leaves were taken. Whole plants were used to determine the content of glutathione (GSH), ascorbic acid (AsA), and proline and the activity of glutathione reductase (GR) and ascorbate peroxidase (APX), as well as the level of lipid peroxidation (i.e., malondialdehyde (MDA) content) and electrolyte leakage (EL).

2.5. *Assessment of Growth Parameters*

The germination rate was determined by how many seeds had germinated after seven days [31]. The fresh weight (FW) of the roots and shoots were recorded on the same day. For dry weight (DW) measurements, the samples were placed in a 70 °C furnace. After 48 h, measurements (DW) were carried out [32]. Data regarding shoot and root length were measured with the help of a meter rod [31]. Each variant included 30 seedlings in three biological replicates.

2.6. *Measurement of Photosynthetic Pigments*

The fresh harvested plant leaves (0.05 g) were homogenized in 90% ethanol (10 mL) with the addition of calcium carbonate ($CaCO_3$) and then filtered. The optical density of the filtered extracts was measured using a SmartSpecTM Plus spectrophotometer (Bio-Rad, Oceanside, CA, USA) at 663 nm (Chl a) and 646 nm (Chl b). The Chl a and Chl b contents were expressed as mg g^{-1} FW [33].

2.7. *Measurement of Non-Enzymatic Antioxidants*

2.7.1. Glutathione (GSH) and Oxidized Glutathione (GSSG) Contents

The GSH and GSSG contents from one plant mount were determined using the spectrofluoromeric method. This method is based on obtaining an o-phthalic aldehyde fluorescent product (Sigma, Australia) in accordance with the pH of the medium. Whole plant samples (approximately 0.5 g) were homogenized in 4 mL of a mixture consisting of 0.1 M potassium phosphate buffer (pH 8.0) and 25% metaphosphoric acid (HPO_3) solution in a ratio of 3.75:1 (by volume) as recommended in [34]. The homogenate was centrifuged for 10 min at 8000× *g*, and then the supernatant was centrifuged again for 5 min at 13,000× *g*. The quantitative assessment of the GSH and GSSG in the obtained supernatant was carried out by applying the reagents specified in [35]. To assess the GSH and GSSG contents, the kinetics of the fluorescence intensity of the formed complexes were recorded at pH 8.0 and pH 12.0, respectively, using an EnSpire Model 2300 Multilabel Microplate Reader (PerkinElmer, Boston, MA, USA) at 420 nm (excitation wavelength 350 nm). The levels of glutathione forms were expressed in μmoL mg^{-1} protein.

2.7.2. Ascorbic Acid (AsA) Content

The amount of AsA was determined by the titration method [35]. The wheat samples (10 g) were crushed in a porcelain mortar, extracted with 10 mL of distilled water, agitated, and filtered through a paper filter. An amount of 20 mL of filtrate was taken into a conical flask and 1 mL of 2% hydrochloric acid (HCl), 0.5 mL of 1% potassium iodide (KI), and 2 mL of 0.5% starch were added, and then stirred. The resultant mixture was titrated with 0.001 moL L^{-1} potassium iodate (KIO_3) until it reached a stable blue color staining. The concentration of AsA was reported as mg% FW [35].

2.8. Measurement of the Enzymatic Antioxidants

2.8.1. Glutathione Reductase (GR) Activity

GR (EC 1.6.4.2) activity was assessed by the ability of the enzyme to catalyze the reduction of GSSG using NADPH as a reductant and measuring the formation of NADP+, as expressed in units of nmoL min^{-1} mg^{-1} protein; in the calculation, the extinction coefficient for nicotinamide adenine dinucleotide phosphate (NADPH) was equal to 6.22 mM^{-1} 1 cm^{-1} [36]. Seedlings (0.25 g) were ground in liquid nitrogen by adding 0.75 mL of 50 mM Tris-HCl buffer (pH 7.6). The homogenate was centrifuged at 22,000× *g* at 4 °C for 30 min and the supernatant was used for the enzyme assay. The reaction mixture contained 50 mM Tris-HCl buffer (pH 7.6) (1.91 mL), 0.15 mM NADPH (0.02 mL), and enzyme extract (0.05 mL). Before measurement, 1 mM GSSG (0.02 mL) was added to the experimental cuvette. The reaction was monitored by a decreased in absorbance of NADPH at 340 nm.

2.8.2. Ascorbate Peroxidase (APX) Activity

The activity of ascorbate peroxidase (APX, EC 1.11.1.11) was estimated by monitoring the oxidation of the ascorbate at 290 nm. [37]. The reaction mixture (2.93 mL) consisted of 50 mM phosphate buffer (pH 7.0), 17 mM (0.03 mL) ascorbic acid ($C_6H_8O_6$), 0.03 mL ethylenediamine tetra acetic acid (EDTA), and 0.01 mL extract. The response started with the addition of 0.03 mL of 0.06% H_2 O_2 and was determined in the first 100 s. The data received were expressed as μmoL ascorbate oxidized mg^{-1} protein min^{-1}.

The activity of all the studied antioxidant enzymes was presented, taking into account the protein content of the sample. The total soluble protein was estimated according to the Bradford method [38]. Bovine serum albumin (BSA) served as the standard. For spectrophotometric analyses, a UNICO 2800 spectrophotometer (United Products @ Instruments, Middlesex, NJ, USA) was used.

2.9. Malondialdehyde (MDA) Content

To determine the MDA concentration in the wheat seedlings, samples of 0.5 g were ground in distilled water and then homogenized in 20% trichloroacetic acid ($C_2HCl_3O_2$). The homogeneous samples were centrifuged (10,000× *g*, 10 min), and the supernatant was mixed with 0.5% thiobarbituric acid ($C_4H_4N_2O_2S$) prepared in 20% $C_2HCl_3O_2$ and kept in a boiling water bath (100 °C for 30 min) and then quickly cooled. The absorbance was spectrophotometrically (SmartSpecTM Plus, Bio-Rad, Oceanside, CA, USA) measured at 532 nm and 600 nm. The MDA content was calculated using an extinction coefficient of 155 mM^{-1} cm^{-1} and expressed as nmoL g^{-1} FW.

2.10. Measurement of Electrolyte Leakage (EL)

The state of permeability of the cell membranes was evaluated by following the EL from plant tissues registered with the use of an OK 102/1 conductivity meter HI8733 (Hanna Instruments, Inc., Sarmeola di Rubano, Padova, Italy) by measuring the ohmic resistance of the water extracts on a constant current [35]. The plant samples (1 g) were washed with running water and cut into equally sized fragments; thereafter, they were washed with running water for 3 min, rinsed with distilled water, slightly dried, supplied with 20 mL distilled water, and incubated for 1 h at 25 °C. Then, the samples were filtered and the electroconductivity of the obtained solution was measured and expressed as μSi g^{-1} FW.

2.11. Proline Content

The level of proline was evaluated according to [39], with modifications [40]. Fresh plants (0.5 g) and boiling water (2.0 mL) were added to the test tubes. The test pieces were placed in a water bath and boiled over a period of 30 min. After, the tubes were removed and cooled. The extract (1 mL) was mixed with an equal volume of acid ninhydrin ($C_9H_6O_4$) solution and glacial acetic acid (CH_3COOH). The samples were then incubated at 100 °C for 1 h and cooled in an ice container. The optical densities of the solutions were measured at 520 nm (SmartSpecTM Plus spectrophotometer, Bio-Rad, Hercules, CA, USA). The proline contents (μmoL g^{-1} FW) were calculated using a calibration curve.

2.12. Statistical Analysis

All microbiological, molecular, biochemical, and physiological experiments were performed in three or more bioassays and three or four analytical tests. The arithmetic average values and confidence intervals calculated from the standard error are shown in the table and graphs (± SEM). Statistically significant differences between the mean values were evaluated using analysis of variance (ANOVA), followed by the Tukey test ($p < 0.05$).

3. Results

3.1. Effect of Rhizobacteria Strains on the Growth Parameters and Leaf Chlorophyll Content in Wheat Plants under Normal Conditions

Analysis was carried out on *R. leguminosarum* (strains Pvu5, VSy12, Thy2, and TPr4)-treated wheat seeds 7 days post-germination under normal conditions. It was revealed that the strain Thy2 increased the vigor of seed germination by 4% (Table 1), the length of seedlings by 120% (Figure 1A), and the FW and DW by 115%. Seed pre-treatment with TPr4 increased seed germination, the length of seedlings, and biomass, though not significantly. At the same time, the strain VSy12 had an inhibitory effect on these parameters. Taking together, these results demonstrated that wheat seeds differently responded to treatment with tested strains, showing a positive response to the Thy2 and TPr4 strains and a negative response to the VSy12 strain. For the Pvu5 strain, the studied parameters were at the control level (Table 1 and Figure 1).

Table 1. Effect of the strains of *R. leguminosarum* on wheat seed germination. The presented data are the averages of three repetitions ($n = 100$).

	R. leguminosarum **Strains**				
	Control	**Pvu5**	**VSy12**	**Thy2**	**TPr4**
Seed germination, %	96 ± 2	94 ± 1	80 ± 2	99 ± 1	97 ± 1

The appearance of the plants presented in Figure 2 demonstrates that the Thy2-pretreated plants looked larger compared to plants of the control and the other bacterial treatments. Plants grown from seeds pretreated with the strain TPr4 were significantly shorter than the Thy2-treated plants, but taller than the Pvu5- and VSy12-treated ones. The plant growth of the Pvu5-pretreated plants was at the level of the control, while the plants pretreated with the Vsy12 strain appeared to grow much slower than control (Figure 2).

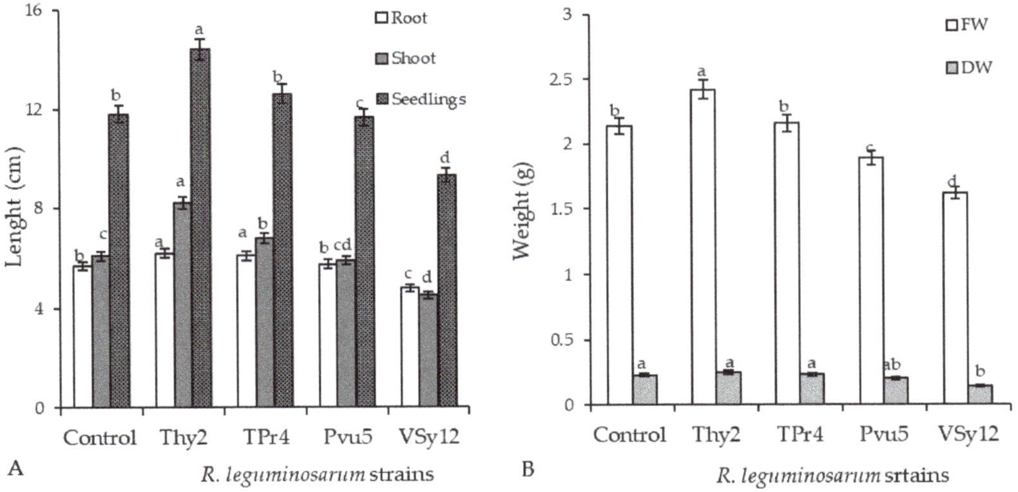

Figure 1. Effect of the strains of *R. leguminosarum* on the length of 7-day-old wheat seedlings: (**A**) fresh weight FW and (**B**) dry weight DW under normal growth conditions. The bars indicate the mean values of three replicates ± SEM. Different letters indicate a significant difference between the means at the level of $p < 0.05$.

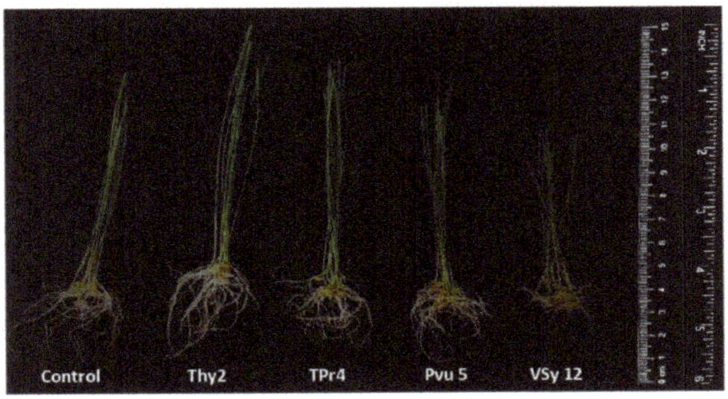

Figure 2. Appearance of the *T. aestivum* L. 7-day-old wheat seedlings. The seeds were inoculated with different strains of *R. leguminosarum*.

It was revealed that in the wheat leaves pretreated with Thy2, the content of Chl a increased by 160% and Chl b by 140% above the control values (Figure 3). Seed treatment with the strain TPr4 increased the content of Chl a and Chl b, but significantly less than the strain Thy2. The content of chlorophyll in wheat leaves pretreated with Pvu5 was at the level of the control values. The strain VSy12 reduced the content of Chl a by 70% and Chl b by 83% relative to the control level (Figure 3).

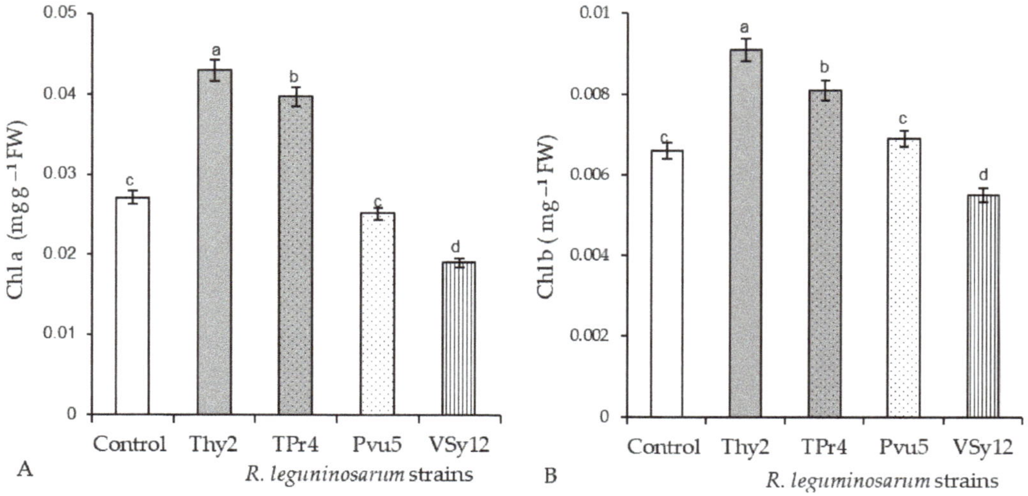

Figure 3. Effects of the strains of *R. leguminosarum* on the contents of Chl a (**A**) and Chl b (**B**) in the leaves of the 7-day-old wheat seedlings. The bars indicate the mean values of three replicates ± SEM. Different letters indicate a significant difference between the means at the level of $p < 0.05$.

3.2. *Thy2 Strain Exerts the Main PGP Traits*

The results showed that the Thy2 strain had the ability to produce siderophores (Table 2, Figure 4A) and IAA and was capable of nitrogen fixation (Figure 4B). The ability to solubilize phosphate was not detected in the tested strain (Table 2).

Table 2. Plant growth promotion activity of the Thy2 strain.

PGP Traits	Thy2
IAA (mg L^{-1})	10 ± 0.5
P solubization (mg L^{-1})	−
Siderophore production	+
Nitrogen fixation	+

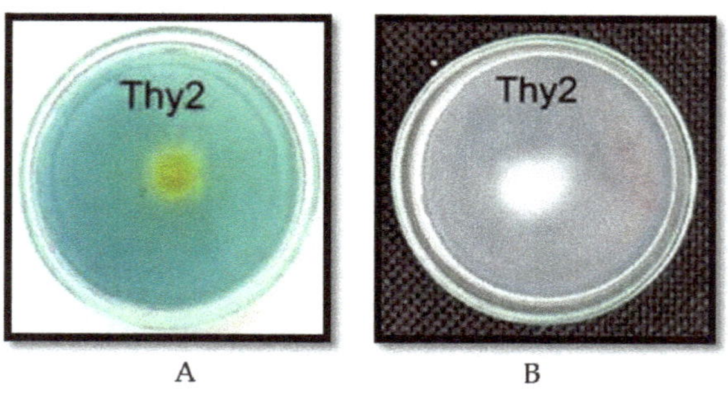

Figure 4. Siderophore production (**A**) and nitrogen fixation (**B**) activity of the Thy2 strain.

3.3. Evaluation of the Protective Effect of the Thy2 Strain on the Growth and Physio-Biochemical Parameters of Wheat Plants

3.3.1. Effect of Thy2 Strain on Wheat Growth under Cd Stress

Cd stress leads to a decrease in seed germination of 60% (Table 3). Two-fold reductions in the length and biomass (FW and DW) of the stressed wheat plants were revealed (Figures 5 and 6). Wheat grown from the seeds inoculated with the Thy2 strain showed improved seed germination that was 58% higher than the uninoculated ones under the same stress conditions (Table 3). In addition, the bacterial-inoculated and Cd-stressed plants were longer and heavier than the uninoculated and the same stressed variants of the experiment by more than 1.5 times (Figures 5 and 6).

Table 3. Effect of Cd stress on the seed germination (seven days after sowing) percentage of the uninoculated (control) and Thy2 strain-inoculated wheat plants. The presented data are the averages of three repetitions.

Variant	Seed Germination, %
Control	95 ± 1
1 mM Cd	56 ± 2
Thy2	99 ± 1
Thy2 + Cd	89 ± 3

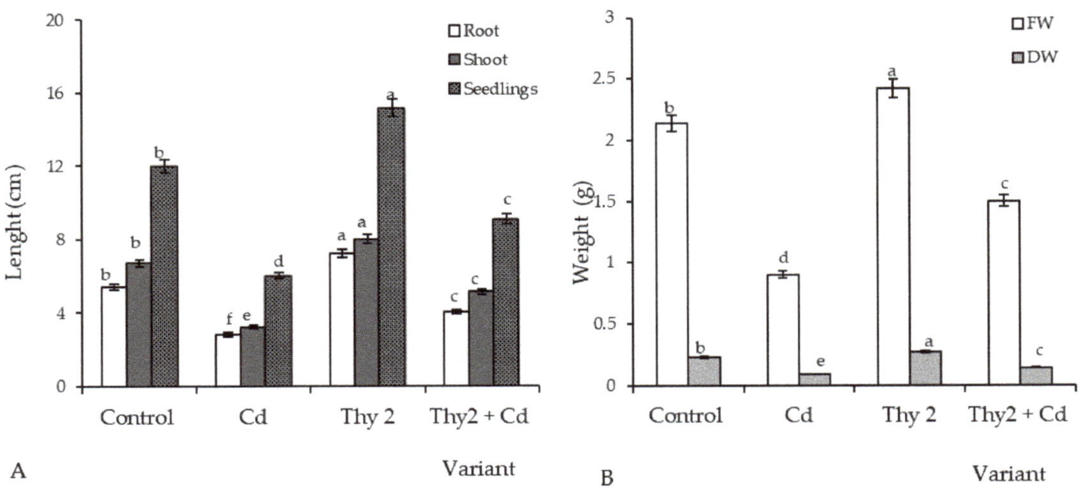

Figure 5. Influence of the presence of cadmium (Cd) on the length (**A**) and weight (**B**) of the uninoculated and inoculated (with *R. leguminosarum* Thy2) 7-day-old wheat plants. The bars indicate the mean values of three replicates ± SEM. Different letters indicate a significant difference between the means at the level of $p < 0.05$.

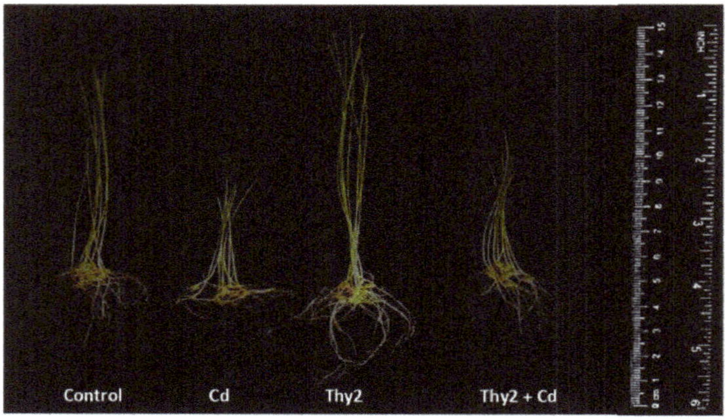

Figure 6. Appearance of the pretreated (with the *R. leguminosarum* strain Thy2) and untreated 7-day-old wheat plants grown under normal and Cd stress conditions.

3.3.2. Effect of the Thy2 Strain on the Content of Chlorophyll in the Leaves of Wheat Plants under Cd Stress

The incubation of plants in the Cd solution led to an almost two-fold drop in the content of Chl a (Figure 7A) and Chl b (Figure 7B) in the leaves. In the plants pretreated with the Thy2 strain, the content of chlorophyll was 62–67% higher than that in the stressed plants (Figure 7). The appearance of the plants treated with Thy2, as shown in Figure 6, demonstrated that the pretreated plants appeared physiologically to be more safely stressed.

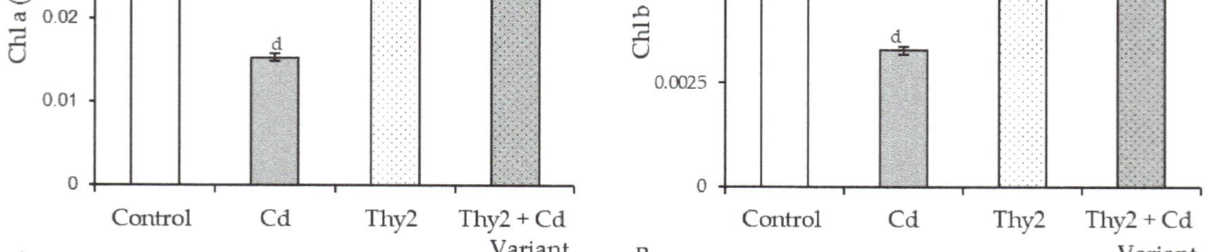

Figure 7. Effect of Thy2 on the content of Chl a (**A**) and Chl b (**B**) in the leaves of the 7-day-old wheat seedlings under Cd stress. The bars indicate the mean values of three replicates ± SEM. Different letters indicate a significant difference between the means at the level of $p < 0.05$.

3.3.3. Effect of the Thy2 Strain on the Content of Non-Enzymatic Antioxidants in the Wheat Plants under Cd Stress

Exposure to Cd stress led to a two-fold fall in GSH and the same level of GSSG accumulation (Figure 8A). Cd stress led to a two-fold drop in ascorbic acid (AsA) content (Figure 7B). Seed treatment with the Thy2 strain resulted in a slight accumulation of GSH and AsA under normal growth conditions. At the same time, the content of GSSG remained at the level of the control values. Seed pretreatment with *R. leguminosarum* Thy2 led to an additional GSH accumulation (by 15% relative to the control) and a decrease in the level of GSSG (by 60%) (Figure 8A). The level of AsA in these plants was two times higher than that in the stressed plants, but it did not reach the control value (Figure 8B).

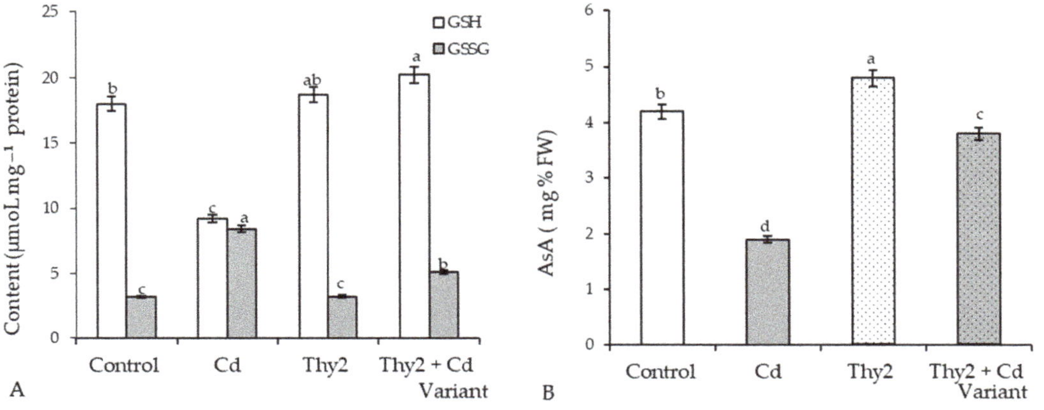

Figure 8. Effect of Thy2 on the content of GSH, GSSG (**A**), and AsA (**B**) in the 7-day-old wheat seedlings under normal and Cd stress conditions. The bars indicate the mean values of three replicates ± SEM. Different letters indicate a significant difference between the means at the level of $p < 0.05$.

3.3.4. Effect of Thy2 on the Activity of Enzymatic Antioxidants in the Wheat Plants under Cd Stress

Cd stress led to an increase of 1.8 and 3 times the activity of GR and APX in wheat plants, respectively. Pretreatment with the Thy2 strain did not affect the activities of enzymes, as their levels remained at the control values (Figure 9). Under stress conditions, the plants pretreated with Thy2 were characterized by additional GR activation (130%) of the stress level (Figure 9A), while the APX activity in these plants was 76% lower than its level in the stressed control (untreated) plants (Figure 9B).

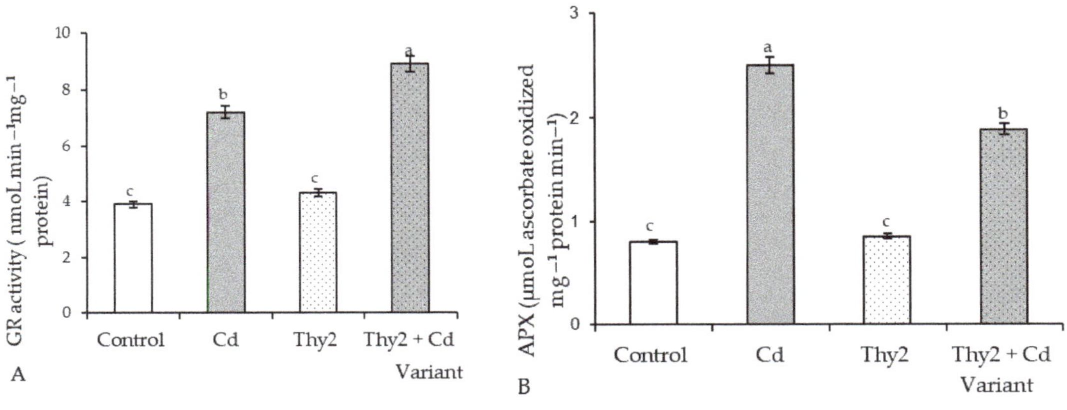

Figure 9. Effect of Thy2 on the activity of GR (**A**) and APX (**B**) in the 7-day-old wheat seedlings under normal and Cd stress conditions. The bars indicate the mean values of three replicates ± SEM. Different letters indicate a significant difference between the means at the level of $p < 0.05$.

3.3.5. Effect of Thy2 on Proline in the Wheat Plants under Cd Stress

It was revealed that Cd stress resulted in a three-fold proline rise in wheat seedlings (Figure 10). However, pretreatment with the Thy2 strain reduced the concentration of stress-induced proline accumulation by 68% (Figure 10). Under normal growth conditions, pre-treatment with the Thy2 strain left the proline at the control level.

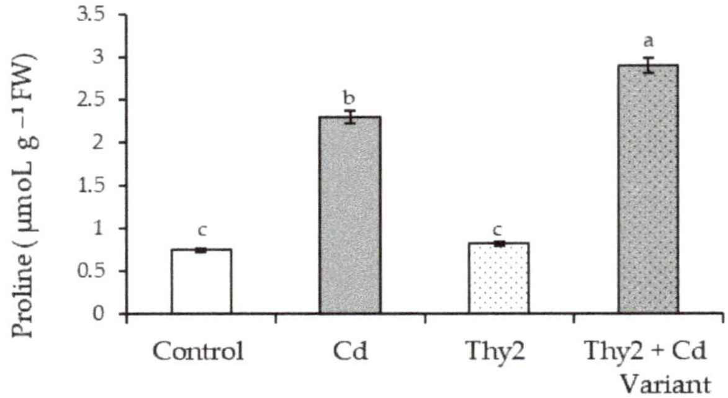

Figure 10. Effect of Thy2 on the proline content of the 7-day-old wheat seedlings under normal and Cd stress conditions. The bars indicate the mean values of three replicates ± SEM. Different letters indicate a significant difference between the means at the level of $p < 0.05$.

3.3.6. Effect of Thy2 on the Membrane Stability in the Wheat Plants under Cd Stress

The results showed that Cd stress led to a more than two-fold increase in membrane lipid peroxidation (as judged by MDA concentration) and leakage of electrolytes (EL) from plant tissues (Table 4). However, pretreatment with the Thy2 strain significantly mitigated the stress-caused damages. Thus, the content of MDA and the leakage of electrolytes was higher (relative to the control) by 1.5 times and by 1.6 times, respectively. At the same time, treatment with the Thy2 strain itself did not affect the state of the membrane structures.

Table 4. Malondialdehyde (MDA) content and electrolyte leakage (EL) in the untreated (control) and Thy2-pretreated 7-day-old wheat seedlings under Cd stress.

Variants	MDA (nmoL g^{-1} FW)	EL (μS^{-1} FW)
Control	45 ± 4.2	50 ± 5.2
Cd	101 ± 8.9	120 ± 9.9
Thy2	47 ± 4.9	52 ± 5.3
Thy + Cd	68 ± 4.9	80 ± 7.5

4. Discussion

An important indicator for assessing the prospects for the use of bacteria is the assessment of their influence on seed germination and growth parameters. During the work, it was found that the Thy2 strain stimulated the germination of seeds (Table 1) and the growth of roots and shoots (Figure 1A). It should be noted that other strains had an ambiguous effect on the growth rates of wheat. This corresponds to the literature data indicating the absence of a universal stimulating effect of bacteria on the germination and further development of plants [41]. In addition, the Thy2 strain caused a significant accumulation of both the fresh and dry biomass of wheat plants (Figure 1B). It is well known that dry biomass is a product of the photosynthetic activity of plants [2]. The Thy2 strain increased the content of chlorophylls a and b, which indicated an increased level of photo- synthetic activity in the leaves of these plants (Figure 3). Thus, based on the data obtained, it can be concluded that the Thy2 strain stimulates the growth of wheat plants, which was confirmed by the visual assessment of the tested plants (Figures 1 and 2). Moreover, the Thy2 strain showed PGP characteristic such as IAA synthesis, siderophores production, and nitrogen fixation activity. The production of IAA is an essential tool for PGPR to stimulate and facilitate plant growth [42]. IAA synthesizing rhizobacteria stimulates the growth of the root system and the number of lateral roots, which leads to a better acquisition of nutrients and an increase in the development and productivity of plants [43]. Park et al. 2021 [44] obtained a bacterial strain from food waste that synthesized IAA (16.6 mg L^{-1}) and demonstrated a growth-promoting effect on apple mint and chrysanthemums. Our results showed that the Thy2 strain synthesized 10 mg L^{-1} and also stimulated wheat growth (Table 2). Moreover, the Thy2 strain showed the ability to produce siderophores, which is another one of the most important signs of PGPR traits that are involved in plant growth stimulation [16,18]. Siderophores are molecules of low molecular weight (<1000 Da) with a great specificity and affinity for chelate or the link Fe3+. They play an important role in stimulating plant growth, enhancing sustainability and protecting against pathogens. [45]. Since the Thy2 strain was found to grow on a nitrogen-free medium, it has the potential for nitrogen fixation. Nitrogen is a vital element for plant growth; it is required for the synthesis of macromolecules such as amino acids, nucleic acids, and chlorophyll. According to the literature data, nitrogen fixation is one of the most important functions of *Rizobium leguminosarum* L. [1]. The obtained data proved that the Thy2 strain has PGP traits (Table 2).

In addition, the results showed that Thy2 had a protective effect on the growth of wheat seedlings exposed to Cd stress (Table 3). Inoculation with Thy2 increased germination percentage, the size of seedlings (roots, shoots), and their fresh and dry weights under Cd conditions when compared to the control (Figures 5 and 6). Further, the Thy2 strain significantly reduced the Cd-caused degradation of chlorophylls a and b (Figure 7).

The application of PGPR increased the wheat growth and photosynthesis and decreased Cd uptake both in shoots and roots [1]. These studies showed that microbes can be used for the reduction of Cd toxicity in plants. Rhizobia, along with other amendments such as biochar, have enhanced the growth and yield of crops under either normal or stressful conditions [46,47].

Thereby, Thy2 has the ability to decrease the adverse effects of Cd stress in wheat plants and increase the level of photosynthetic activity. This is also evidenced by the appearance of the plants. Many studies have shown that PGPR treatment has a positive effect on plants, even in the presence of heavy metals (HMs) in the medium. For example, in the alfalfa plant *Medicago sativa* L. treated with PGPR and grown in the presence of Cu, Pb, and Zn, shoot length increased by 22–77% and shoot biomass increased up to 220% compared with untreated plants [14,20]. Treatment of the *Atriplex halimus* and *Arthrocnemum macrostachyum* plants growing in the presence of HMs led to an improvement in their morphometric parameters compared to untreated controls. This may be due to the fact that microorganisms produce siderophores, improve plant nutrition by nitrogen fixation, and boost plant growth by secreting auxins [10,12,13]. Thus, the resistance of plants to the toxic effect of Cd may be due to more efficient root growth because of the positive effects of the substances released by microorganisms.

It is well known that the antioxidant system plays a fundamental role in maintaining the redox homeostasis of plants under stress [48]. As expected, the presence of Cd in the growth medium led to the development of oxidative stress [7,48], which was accompanied by the depletion of glutathione (GSH) and ascorbate (AsA) pools, as well as the stress-induced activation of GR and APX. The overaccumulation of ROS led to the excessive synthesis of MDA and an increase in the permeability of the membrane structures. An excess of MDA, the end product of lipid peroxidation, and depletion of GSH, which is a fundamental molecule regulating mitosis [3,35], led to the inhibition of plant growth under stress. Seed treatment with the Thy2 bacteria contributed to the additional accumulation GSH (Figure 8A) and the activation of GR (Figure 9A), the key enzyme for maintaining GSH pools under stress [35]. In addition, the stabilization of AsA-APX components were observed in these plants (Figures 8B and 9B), which indicates that these plants survive stress more easily. The contents of AsA and GSH can characterize the resistance and adaptive capacity of plants in response to any type of stress [48]. Thus, pretreatment with the Thy2 strain increased the adaptation of the wheat plants to the effects of the Cd ions. This was also confirmed by data on the stabilization of the state of the membrane structures of these plants.

Another indicator of the physiological state of plants is the level of accumulation of proline, which performs the function of being an osmoprotectant and an antioxidant [48,49]. Exogenous proline treatment enhanced *Brassica juncea*'s tolerance to Cd via a decrease of Cd accumulation and the reestablishment of redox homeostasis [50]. Proline is an important metabolite for plant adaptation, protection, and tolerance to Cd stress. The accumulation of proline in plants is recognized as a strategy for counteracting Cd stress by adjusting osmotic potential, stabilizing membrane structures, and reducing oxidative stress [7]. It can be assumed that a significant contribution to the realization of the resistance of wheat plants when treated with the Thy2 strain under conditions of Cd stress is the additional accumulation of proline (Figure 10). This effect of the bacteria provides a decrease in the damaging effect of Cd ions to wheat (Table 4).

5. Conclusions

The present study showed that seed inoculation with *R. leguminosarum* Thy2, isolated from *T. hybridium* nodules, had a growth-promoting and protective effect on wheat plants under Cd stress. Particularly, the Thy2 strain exhibits PGP properties, producing IAA, siderophores, and fixing N, and its application in a liquid formulation enhanced wheat growth and biomass compared to the control under Cd stress. Moreover, the Thy2 strain contributed to the additional accumulation of proline and glutathione and the activation of glutathione reductase in wheat plants. These led to a decrease in oxidative stress and stabilized membrane structures, measured in terms of the MDA and electrolyte leakage. Thus, *R. leguminosarum* Thy2, isolated from *T. hybridium* nodules, revealed several growth-promoting traits and induced resistance in wheat plants under Cd stress through improved photosynthesis and antioxidant capacity, and it reduced the severity of oxidative damages. Evaluation of the current study suggests that the Thy2 strain can potentially be utilized as a promising alternative and an environmentally friendly approach to facilitating wheat growth and tolerance under Cd stress. With that, further field experiments are required to evaluate its full potential for mitigating heavy metals-caused stress in plants.

Author Contributions: Conceptualization: D.M.; supervision: D.M. and O.L.; methodology: D.M., D.B., E.A., K.N. and O.C.; software: A.I., D.B., and A.B.; validation: D.M., D.B. and O.L.; formal analysis: K.N., A.I. and D.B.; investigation: D.M., D.B., O.L. and O.C.; visualization: D.M., K.N., A.I., A.B. and D.B.; resources: O.C., A.B. and E.A.; data curation: D.M. and O.L. writing—original draft: D.M., D.B., O.C. and O.L.; review and editing: D.M., O.L., D.B. and O.C. All authors have read and agreed to the published version of the manuscript.

Funding: This research received no external funding.

Institutional Review Board Statement: Not applicable.

Informed Consent Statement: Not applicable.

Data Availability Statement: Not applicable.

Acknowledgments: The research was carried out within the framework of the state assignment of Russia (registration number AAAA-A21-121011990120-7).

Conflicts of Interest: The authors declare no conflict of interest.

References

1. Hassan, W.; Bashir, S.; Ali, F.; Ijaz, M.; Hussain, M.; David, J. Role of ACC-deaminase and/or nitrogen fixing rhizo- bacteria in growth promotion of wheat (*Triticum aestivum* L.) under cadmium pollution. *Environ. Earth Sci.* **2016**, *75*, 267. [CrossRef]
2. El-Hendawy, S.; Al-Suhaibani, N.; Alotaibi, M.; Hassan, W.; Elsayed, S.; Tahir, M.U.; Mohamed, A.I.; Schmidhalter, U. Estimating growth and photosynthetic properties of wheat grown in simulated saline field conditions using hyperspectral reflectance sensing and multivariate analysis. *Sci. Rep.* **2019**, *9*, 16473. [CrossRef] [PubMed]

3. Zhou, M.; Li, Z. Recent advances in minimizing cadmium accumulation in wheat. *Toxics* **2022**, *10*, 187. [CrossRef]
4. Cardarelli, M.; Woo, S.L.; Rouphael, Y.; Colla, G. Seed treatments with microorganisms can have a siostimulant effect by Influencing germination and seedling growth of crops. *Plants* **2022**, *11*, 259. [CrossRef]
5. Shakirova, F.M.; Allagulova, C.R.; Maslennikova, D.R.; Klyuchnikova, E.O.; Avalbaev, A.M.; Bezrukova, M.V. Salicylic acid-induced protection against cadmium toxicity in wheat plants. *Environ. Exp. Bot.* **2016**, *122*, 19–28. [CrossRef]
6. Haider, F.U.; Liqun, C.; Coulter, J.A.; Sardar, A.C.; Wu, J.; Zhang, R.; Wenjun, M.; Farooq, M. Cadmium toxicity in plants: Impacts and remediation strategies. *Ecotoxicol. Environ. Saf.* **2021**, *211*, 111887. [CrossRef] [PubMed]
7. Zulfiqar, U.; Ayub, A.; Hussain, S.; Waraich, E.A.; El-Esawi, M.A.; Ishfaq, M.; Ahmad, M.; Ali, N.; Maqsood, M.F. Cadmium toxicity in plants: Recent progress on morpho-physiological effects and remediation strategies. *J. Soil Sci. Plant Nutr.* **2022**, *22*, 212–269. [CrossRef]
8. Kudoyarova, G.; Arkhipova, T.; Korshunova, T.; Bakaeva, M.; Loginov, O.; Dodd, I.C. Phytohormone mediation of interactions between plants and non-symbiotic growth promoting bacteria under edaphic stresses. *Front. Plant Sci.* **2019**, *10*, 1368. [CrossRef]
9. Mumtaz, M.Z.; Ahmad, M.; Mehmood, K.; Sheikh, A.S.; Malik, A.; Hussain, A.; Nadeem, S.M.; Zahir, A.Z. Role of plant growth-promoting rhizobacteria in combating abiotic and biotic stresses in plants. In *Microbial Biotechnology for Sustainable Agriculture*; Springer: Singapore, 2022; pp. 43–105.
10. Sharma, R.K.; Archana, G. Cadmium minimization in food crops by cadmium resistant plant growth promoting rhizobacteria. *Appl. Soil Ecol.* **2016**, *107*, 66–78. [CrossRef]
11. Paredes-Páliz, K.; Rodríguez-Vázquez, R.; Duarte, B.; Caviedes, M.A.; Mateos-Naranjo, E.; Redondo-Gómez, S.; Caçador, M.I.; Rodríguez-Llorente, I.D.; Pajuelo, E. Investigating the mechanisms underlying phytoprotection by plant growth-promoting rhizobacteria in *Spartina densiflora* undermetal stress. *Plant Biol.* **2018**, *20*, 497–506. [CrossRef]
12. Mesnoua, M.; Mateos-Naranjo, E.; Pérez-Romero, J.A.; Barcia-Piedras, J.M.; Lotmani, B.; Redondo-Gómez, S. Combined effect of Cr-toxicity and temperature rise on physiological and biochemical responses of *Atriplex halimus* L. *Plant Physiol. Biochem.* **2018**, *132*, 675–682. [CrossRef]
13. Navarro-Torre, S.; Rodríguez-Llorente, I.D.; Meddich, A.; Redondo-Gómez, S.; Pajuelo, E. Safe cultivation of Medicago sativa in metal-polluted soils from semi-arid regions assisted by heat- and metallo-resistant PGPR. *Microorganisms* **2019**, *7*, 212. [CrossRef]
14. Konkolewska, A.; Piechalak, A.; Ciszewska, L.; Antos-Krzemińska, N.; Skrzypczak, T.; Hanć, A.; Sitko, K.; Małkowski, E.; Barałkiewicz, D.; Małecka, A. Combined use of companion planting and PGPR for the assisted phytoextraction of trace metals (Zn, Pb, Cd). *Environ. Sci. Pollut. Res. Int.* **2020**, *12*, 13809–13825. [CrossRef] [PubMed]
15. Ali, B.; Wang, X.; Saleem, M.H.; Sumaira; Hafeez, A.; Afridi, M.S.; Khan, S.; Zaib-Un-Nisa; Ullah, I.; Amaral Júnior, A.T.d.; et al. PGPR-mediated salt tolerance in maize by modulating plant physiology, antioxidant defense, compatible solutes accumulation and bio-surfactant producing genes. *Plants* **2022**, *11*, 345. [CrossRef] [PubMed]
16. Khan, N.F.; Rasool, A.; Mansoor, S.; Saleem, S.; Baba, T.R.; Haq, S.M.; Rehman, S.A.; Adetunji, C.O.; Popesc, S.M. Potential applications of Rhizobacteria as eco-friendly biological control, plant growth promotion and soil metal bioremediation. In *Sustainable Crop Production Recent Advances*; Meena, V., Choudhary, M., Meena, S.K., Yadav, R.P., Eds.; IntechOpen Limited: London, UK, 2022; pp. 104–170. [CrossRef]
17. Kalayu, G. Phosphate solubilizing microorganisms: Promising approach as biofertilizers. *Int. J. Agron.* **2019**, *2019*, 4917256. [CrossRef]
18. Meena, M.; Swapnil, P.; Divyanshu, K.; Kumar, S.; Tripathi, Y.N.; Zehra, A.; Marwal, A.; Upadhyay, R.S. PGPR-mediated induction of systemic resistance and physiochemical alterations in plants against the pathogens: Current perspective. *J. Basic Microbiol.* **2020**, *60*, 828–861. [CrossRef] [PubMed]
19. Vishwakarma, K.; Kumar, V.; Tripathi, D.K.; Sharma, S. Characterization of rhizobacterial isolates from *Brassica juncea* for multitrait plant growth promotion and their viability studies on carriers. *Environ. Sustain.* **2018**, *1*, 253–265. [CrossRef]
20. Rizwan, M.; Ali, S.; Abbas, T.; Zia-ur-Rehman, M.; Hannan, F.; Keller, C.; Al-Wabel, M.I.; Ok, Y.S. Cadmium minimization in wheat: A critical review. *Ecotoxicol. Environ. Saf.* **2016**, *130*, 43–53. [CrossRef]
21. Abedi, T.; Mojiri, A. Cadmium uptake by wheat (*Triticum aestivum* L.): An overview. *Plants* **2020**, *9*, 500. [CrossRef]
22. Vershinina, Z.R.; Khakimova, L.R.; Lavina, A.M.; Karimova, L.R.; Baimiev, A.K.; Serbaeva, E.R.; Safronova, V.I.; Shaposhnikov, A.I. Effect of constitutive expression of the Rapal gene on formation of bacterial biofilms and growth—Stimulating activity of rhizobia. *Microbiology* **2019**, *1*, 54–62. [CrossRef]
23. Vincent, J.M. *A Manual for the Practical Study of Root Nodule Bacteria*; Blackwell Science: Oxford, UK, 1970; 164p.
24. Baymiev, A.K.; Ptitsyn, K.G.; Baimiev, A.K. Influence of the introduction of Caragana Arborescens on the composition of its root-nodule bacteria. *Microbiology* **2010**, *79*, 115–120. [CrossRef]
25. Akimova, E.S.; Gumenko, R.S.; Vershinina, Z.R.; Baymiev, A.K. Genetic markers for search of rhizobia based on symbiotic genes. *Microbiology* **2017**, *86*, 640–646. [CrossRef]
26. Baymiev, A.K.; Akimova, E.S.; Gumenko, R.S.; Vladimirova, A.A.; Muldashev, A.A.; Chemeris, A.V.; Baymiev, A.K. Genetic diversity andphylogeny of root nodule bacteria isolated from nodules of plants of the *Lupinaster* genus inhabiting the southern Urals. *Russ. J. Genet.* **2019**, *55*, 45–51. [CrossRef]
27. Malik, D.K.; Sindhu, S.S. Production of indole acetic acid by *Pseudomonas* sp.: Effect of coinoculation with *Mesorhizobium* sp. Cicer on nodulation and plant growth of chickpea (*Cicer arietinum*). *Physiol. Mol. Biol. Plants* **2011**, *17*, 25–32. [CrossRef] [PubMed]

28. Schwyn, B.; Neilands, J.B. Universal chemical assay for the detection and determination of siderophores. *Anal. Biochem.* **1987**, *160*, 47–56. [CrossRef]
29. Pikovskaya, R.I. Mobilization of phosphorus in soil in connection with vital activity of some microbial species. *Mikrobiologiya* **1948**, *17*, 362–370.
30. Netrusov, A.I.; Egorova, M.A.; Zakharchuk, L.M. *A Practical Course in Microbiology*; Academia Publishing: Moscow, Russia, 2005.
31. Ahmad, I.; Akhtar, M.J.; Zahir, Z.A.; Jamil, A. Efect of cadmium on seed germination and seedling growth of four wheat (*Triticum aestivum* L.) cultivars. *Pak. J. Bot.* **2012**, *44*, 1569–1574.
32. Lastochkina, O.; Pusenkova, L.; Yuldashev, R.; Babaev, M.; Garipova, S.; Blagova, D.; Khairullin, R.; Aliniaeifard, S. Effects of *Bacillus subtilis* on some physiological and biochemical parameters of *Triticum aestivum* L. (wheat) under salinity. *Plant Physiol. Biochem.* **2017**, *121*, 80–88. [CrossRef]
33. Jeffrey, S.; Humphrey, G. New spectrophotometric equations for determining chlorophylls a, b, c1 and c2 in higher plants, algae and natural phytoplankton. *Biochem. Physiol. Pflanz.* **1975**, *167*, 191–194. [CrossRef]
34. Hissin, P.J.; Hilf, R.A. A Fluorometric method for determination of oxidize and reduced glutathione in tissues. *Anal. Biochem.* **1976**, *74*, 214–226. [CrossRef]
35. Maslennikova, D.; Lastochkina, O. Contribution of ascorbate and glutathione in endobacteria *Bacillus subtilis*-mediated drought tolerance in two *Triticum aestivum* L. genotypes contrasting in drought sensitivity. *Plants* **2021**, *10*, 2557. [CrossRef]
36. Rao, M.V.; Paliyath, G.; Ormrod, D.P. Ultraviolet-B- and ozone-induced biochemical changes in antioxidant enzymes of *Arabidopsis thaliana*. *Plant Physiol.* **1996**, *110*, 125–136. [CrossRef] [PubMed]
37. Verma, S.; Dubey, R.S. Lead toxicity induces lipid peroxidation and alters the activities of antioxidant enzymes in growing rice plants. *Plant Sci.* **2003**, *164*, 645–655. [CrossRef]
38. Bradford, M.M. A rapid and sensitive methods for quantitation of microgram quantities of protein utilizing the princi- ple of protein dye binding. *Anal. Biochem.* **1976**, *74*, 248–254. [CrossRef]
39. Bates, L.S.; Waldern, R.P.; Teare, D. Rapid determination of free proline for water-stress studies. *Plant Soil* **1973**, *39*, 205–207. [CrossRef]
40. Kalinkina, L.G. Proline accumulation in cells of marine and freshwater chlorella in depending on the concentration of NaCl in the medium and the growth rate of algae. *Plant Physiol.* **1985**, *32*, 42–52.
41. Bakaeva, M.D.; Kuzina, E.V.; Rafikova, G.F.; Chetverikova, D.V.; Stolyarova, E.A.; Mukhamatd'yarova, S.R.; Kudoyarova, G.R. Influence of bacteria-destructors of oil hydrocarbons on the germination and growth of plants. *Ekobiotekh* **2019**, *2*, 175–183.
42. Lebrazi, S.; Fadil, M.; Chraibi, M.; Fikri-Benbrahim, K. Screening and optimization of indole-3-acetic acid production by Rhizobium sp. strain using response surface methodology. *J. Genet. Eng. Biotechnol.* **2020**, *18*, 21. [CrossRef]
43. Egamberdieva, D. Indole-acetic acid production by root associated bacteria and its rile in plant growth and development. In *Auxins: Structure, Biosynthesis and Functions*; Nova Publishers: Hauppauge, NY, USA, 2012.
44. Park, S.; Kim, A.-L.; Hong, Y.-K.; Shin, J.-H.; Joo, S.-H. A highly efficient auxin-producing bacterial strain and its effect on plant growth. *J. Genet. Eng. Biotechnol.* **2021**, *19*, 179. [CrossRef]
45. Singh, P.; Chauhan, P.K.; Upadhyay, S.K.; Singh, R.K.; Dwivedi, P.; Wang, J.; Jain, D.; Jiang, M. Mechanistic insights and potential use of siderophores producing microbes in rhizosphere for mitigation of stress in plants grown in degraded land. *Front. Microbiol.* **2022**, *13*, 898979. [CrossRef]
46. Ahmad, M.T.; Asghar, H.N.; Saleem, M.; Khan, M.Y.; Zahir, Z.A. Synergistic effect of rhizobia and biochar on growth and physiology of maize. *Agron. J.* **2015**, *107*, 2327–2334. [CrossRef]
47. Akhtar, S.S.; Andersen, M.N.; Naveed, M.; Zahir, Z.A.; Liu, F. Interactive effect of biochar and plant growth-promoting bacterial endophytes on ameliorating salinity stress in maize. *Fund. Plant Biol.* **2015**, *42*, 770–781. [CrossRef]
48. Zhu, T.; Li, L.; Duan, Q.; Liu, X.; Chen, M. Progress in our understanding of plant responses to the stress of heavy metal cadmium. *Plant Signal. Behav.* **2020**, *16*, 1836884. [CrossRef] [PubMed]
49. Lastochkina, O.; Garshina, D.; Ivanov, S.; Yuldashev, R.; Khafizova, R.; Allagulova, C.; Fedorova, K.; Avalbaev, A.; Maslennikova, D.; Bosacchi, M. Seed priming with endophytic *Bacillus subtilis* modulates physiological responses of two different *Triticum aestivum L.* cultivars under drought stress. *Plants* **2020**, *9*, 1810. [CrossRef] [PubMed]
50. Wang, Y.; Tan, P.; Chang, L.; Yue, Z.; Zeng, C.; Li, M.; Liu, Z.; Dong, X.; Yan, M. Exogenous proline mitigates toxic effects of cadmium via the decrease of cadmium accumulation and reestablishment of redox homeostasis in *Brassica juncea*. *BMC Plant Biol.* **2022**, *22*, 182. [CrossRef] [PubMed]

life

MDPI

Article

PKS5 Confers Cold Tolerance by Controlling Stomatal Movement and Regulating Cold-Responsive Genes in Arabidopsis

Chengyan Sun †, Lin Zhu †, Linlin Cao †, Huimin Qi, Huijuan Liu, Fengyun Zhao and Xiuli Han *

School of Life Sciences and Medicine, Shandong University of Technology, Zibo 255049, China
* Correspondence: hanhxl@sdut.edu.cn; Tel.: +86-533-2781-329; Fax: +86-533-3188-608
† These authors contributed equally to this work.

Abstract: Cold stress limits plant growth and development; however, the precise mechanisms underpinning plant acclimation to cold stress remain largely unknown. In this study, the Ser/Thr protein kinase SOS2-LIKE PROTEIN KINASE5 (PKS5) was shown to play a positive role in plant responses to cold stress. A PKS5 loss-of-function mutant (*pks5-1*) exhibited elevated sensitivity to cold stress, as well as a lower survival rate and increased ion leakage. Conversely, PKS5 gain-of-function mutants (*pks5-3*, *pks5-4*) were more tolerant to cold stress and exhibited higher survival rates and decreased ion leakage. Stomatal aperture analysis revealed that stomatal closure was slower during the first 25 min after cold exposure in *pks5-1* compared to wild-type, whereas *pks5-3* and *pks5-4* displayed accelerated stomatal closure over the same time period. Further stomatal aperture analysis under an abscisic acid (ABA) treatment showed slower closure in *pks5-1* and more rapid closure in *pks5-3* and *pks5-4*. Finally, expression levels of cold-responsive genes were regulated by PKS5 under cold stress conditions, while cold stress and ABA treatment can regulate *PKS5* expression. Taken together, these results suggest that PKS5 plays a positive role in short-term plant acclimation to cold stress by regulating stomatal aperture, possibly via ABA pathways, and in long-term acclimation by regulating cold-responsive genes.

Keywords: cold stress; PKS5; stomatal aperture

Citation: Sun, C.; Zhu, L.; Cao, L.; Qi, H.; Liu, H.; Zhao, F.; Han, X. PKS5 Confers Cold Tolerance by Controlling Stomatal Movement and Regulating Cold-Responsive Genes in Arabidopsis. *Life* **2022**, *12*, 1633. https://doi.org/10.3390/life12101633

Academic Editors: Hakim Manghwar, Wajid Zaman and Balazs Barna

Received: 15 September 2022
Accepted: 16 October 2022
Published: 19 October 2022

Publisher's Note: MDPI stays neutral with regard to jurisdictional claims in published maps and institutional affiliations.

Copyright: © 2022 by the authors. Licensee MDPI, Basel, Switzerland. This article is an open access article distributed under the terms and conditions of the Creative Commons Attribution (CC BY) license (https://creativecommons.org/licenses/by/4.0/).

1. Introduction

Cold stress is a major environmental factor restricting plant growth and development [1]. Cold stress decreases plant growth and yields by damaging cell structures and by inhibiting cell activities, for example, by damage to cell membranes and proteins by ice crystals, and by cold inhibition of photosynthesis [2]. Plants have evolved complex mechanisms to adapt to cold stress and improve their tolerance to freezing. The ICE1-CBF pathway plays a key role in cold stress responses in diverse plant species [1]. Upon exposure to cold stress, transcription factor ICE1 stimulates expression of *CBF* genes within 3 h. Within 24 h, CBFs activate the expression of cold-regulated (*COR*) genes that facilitate cold tolerance in plants [3,4]. Stomatal apertures are also thought to be involved in cold stress responses in plants [5,6]. Cold stress damage can be partially alleviated by H_2S regulation of stomatal movement in concert with MPK4 [5,7]. At low soil temperatures (≤2 °C), photosynthesis rates and stomatal conductance were significantly reduced in high-elevation grasslands, indicating a role for stomatal movement in cold stress response [8]. Similarly, stomatal conductance decreased in Scots pine seedlings within 45 min of a cold stress treatment, whereas conductance started to increase with the extension of cold treatment time [6].

Abscisic acid (ABA) is an important plant hormone that regulates growth, development, and stress responses [9,10]. Under normal physiological conditions, ABA signaling is restricted by the inhibition of SnRK activity by clade A PP2Cs. Upon exposure to stress, ABA accumulates quickly and is recognized by its intracellular receptors (PYLs), then PYLs and PP2Cs form complexes to further release SnRK2 activity [11]. Protein kinases play roles in ABA-regulated stress responses. They include SnRK2.6, which is involved in regulating stomatal closure under osmotic and drought stresses [9], and Raf-like protein kinases, which regulate SnRK2 activity under osmotic stress [12]. ABA can regulate stomatal movements, and thereby photosynthetic rates, under various stress conditions, such as drought, salt, and cold exposure [13,14]. Although the mechanisms underlying ABA-regulated stomatal movements under osmotic and drought stresses have been well characterized [9], the regulatory mechanisms controlling stomatal movements under cold stress remain largely unknown.

SOS2-like protein kinase5 (PKS5) is a Ser/Thr protein kinase that plays an important role in the regulation of plant physiological activities. In the absence of salt stress, PKS5 negatively regulates plasma membrane (PM) H^+-ATPase activity by preventing 14-3-3 protein binding to AHA2 [15]. PKS5 also inhibits SOS2 kinase activity by the promotion of 14-3-3 protein binding to SOS2 [16]. However, under salt stress, PKS5 interacts with J3 protein to release the activity of PM H^+-ATPase, and PKS5 inhibition of SOS2 activity is released by the interaction between 14-3-3 and PKS5 [16,17]. PKS5 is also involved in ABA signal transduction via phosphorylation of abscisic acid-insensitive5 (ABI5), which regulates seed germination [18]. The stress-responsive roles of PKS5 in ABA signal transduction and PM H^+-ATPase modulation suggest a role for PKS5 in stomatal movement regulation under stress conditions. PKS5 is thought to participate in stomatal movement regulation via formation of CBL5–PKS5 complexes and by stimulation of slow anion channel-associated1 (SLAC1) anion channel activity in stomatal guard cells [19]. Although changes in stomatal movement have been observed during cold stress [5], the underlying regulatory mechanisms remain unknown.

Whether PKS5 plays a role in plant cold stress response has not been reported. In this study, Arabidopsis PKS5 was found to act as a positive regulator during the response to cold stress. Seedlings of PKS5 loss-of-function mutants exhibited a cold-sensitive phenotype and slower stomatal closure under cold stress than wild-type plants. These results, together with the involvement of PKS5 in ABA signaling in previous studies and the similar pattern of stomatal movement between cold stress and ABA treatment in this study, suggest that PKS5 mediates plant cold responses, at least partially, via the regulation of ABA signal transduction.

2. Materials and Methods

2.1. Plant Materials and Growth Conditions

The following Arabidopsis strains were used in this study: *pks5-1* (SALK_108074) mutant and its wild-type Col-0, tilling mutants of *pks5-3* and *pks5-4* and their wild-type Col *erecta105* (BigM) [16].

Plants were grown on MS medium at 22 °C in a controlled environment growth chamber (YKNJ, Hefei youke, China) under a 16-h-light/8-h-dark photoperiod with the light intensity of 144 $\mu mol \cdot m^{-2} \cdot s^{-1}$. The growth chamber was equipped with LED light sources composed of red, blue, and far-red lights, with peak blue light at 460 nm and peak red light at 665 nm.

2.2. Freezing Tolerance Assay

The freezing tolerance assay was performed as described previously [3]. Arabidopsis seedlings were grown on MS medium containing 0.4% phytagel for two weeks at 22 °C. For the non-acclimated freezing tolerance treatment, plants were subjected to freezing at −5.5 °C for 5 h, then transferred to 4 °C under dark conditions for 12 h, and then shifted to normal growth condition of 22 °C with 16-h-light/8-h-dark photoperiod for the recovery of growth. For the cold-acclimated freezing tolerance treatment, plants were first pretreated at 4 °C for 3 d, then subjected to freezing at −9.5 °C for 6 h, then transferred to 4 °C under dark conditions for 12 h, and then shifted to normal growth conditions of 22 °C with 16-h-light/8-h-dark photoperiod for recovery of growth. Representative images were taken using a Nikon D5000 camera during the recovery growth at 22 °C under a 16-h-light/8-h-dark photoperiod.

For survival rate analysis, freezing-treated seedlings were recovered under normal growth condition of 22 °C with 16-h-light/8-h-dark photoperiod. The seedlings that could still grow new leaves were recorded as survivors, and the survival rates were calculated by the ratio of survived seedlings to total seedlings.

2.3. Ion Leakage Assay

The ion leakage assay was performed as previously described [3]. The freezing-treated seedlings were placed in a 15 mL tube containing 10 mL deionized water, whose electrical conductivity was detected and recorded as S0. The tube containing seedlings was vacuumed for about 5 min until the seedlings were totally immersed in the water, and then the tube was incubated for another 15 min on a shaking table at room temperature to obtain the electrical conductivity of S1. The tube was further boiled at 100 °C for 10 min, and the electrical conductivity was detected, which was recorded as S2. Finally, ion leakage was calculated by the formula: $(S1 - S0)/(S2 - S0) \times 100$.

2.4. Stomatal Aperture Assay

The stomatal aperture assay was performed as previously described with minor changes [20]. Briefly, leaves of four-week-old seedlings were used for the assay. The abaxial epidermis was obtained by placing the abaxial surface of the leaf on a tape and removing mesophyll cells and adaxial epidermis quickly with another tape. Epidermal strips were then floated onto the opening buffer of 30 mM KCl, 10 mM MES-KOH, pH 6.15 for 1 h at 22 °C. For the stomatal aperture assay cold stressed leaves, strips were placed in the same opening buffer that had been pre-cooled at 4 °C for at least 1 h, and then investigated at the indicated times. For the stomatal aperture assay after ABA treatment, strips were moved from the opening buffer to the opening buffer with 100 μmol ABA, and then investigated at the indicated times. To maintain the accuracy of the experiments, the images of stoma at the indicated times were taken quickly, and then another ABA-treated strips at the indicated times were quickly put on the microscope and images were taken. Representative images were taken using a Nikon D5000 camera coupled to a Nikon Eclipse 55i microscope (magnification 20×). The stomatal aperture was measured using ImageJ.

2.5. qRT-PCR Analysis

Total RNAs were extracted from 14-d-old seedlings grown on MS medium using TRIzol reagent (Invitrogen). The extracted RNA was treated with RNase-free DNase I (Takara, Kusatsu, Japan) and reverse transcriptase (Promega, Madison, WI, USA) to remove genomic DNA and perform reverse transcription according to the manufacturer's protocols. cDNA was then used for qPCR analysis. qPCR was performed on a 7500 real time PCR system (Life Technologies, USA) using the SYBR Premix Ex Taq Kit (Takara) according to the manufacturer's protocol. Actin was used as an internal control and the relative expression levels of the detected genes were calculated as described previously [3]. Primers used in this study are listed in Table 1.

Table 1. Primers for RT-PCR in this study.

Primer	Sequence (5′-3′)
PKS5-F	GAAGGTGCTAAAGTTGATGTATGGTCT
PKS5-R	CGTCATCGTGGAACTTGATCTGTTT
CBF1-F	GCATGTCTCAACTTCGCTGA
CBF1-R	ATCGTCTCCTCCATGTCCAG
CBF2-F	TGACGTGTCCTTATGGAGCTA
CBF2-R	CTGCACTCAAAAACATTTGCA
CBF3-F	GATGACGACGTATCGTTATGGA
CBF3-R	TACACTCGTTTCTCAGTTTTACAAAC

Table 1. *Cont.*

Primer	Sequence (5′-3′)
COR15A-F	GCTTCAGATTTCGTGACGGATAAAAC
COR15A-R	GCAAAACATTAAAGAATGTGACGGTG
KIN1-F	ACCAACAAGAATGCCTTCCA
KIN1-R	CCGCATCCGATACACTCTTT
RD29A-F	GCCGAGAAACTTCAGATTGG
RD29A-R	CCATTCCTCCTCCTCCTTTC
ACTIN2/8-F	GGTAACATTGTGCTCAGTGGTGG
ACTIN2/8-R	AACGACCTTAATCTTCATGCTGC

3. Results

3.1. PKS5 Is Essential for Plant Freezing Tolerance

PKS5 has multiple regulatory roles in plant physiological processes. To identify whether PKS5 was involved in plant cold stress responses, a PKS5 loss-of-function mutant, *pks5-1*, was assessed in a freezing tolerance assay. As shown in Figure 1A, compared with wild-type Col-0 plants, *pks5-1* mutants exposed to a freezing treatment displayed elevated leaf withering under both non-acclimated and cold-acclimated conditions, indicating that PKS5 was essential for tolerance to cold stress.

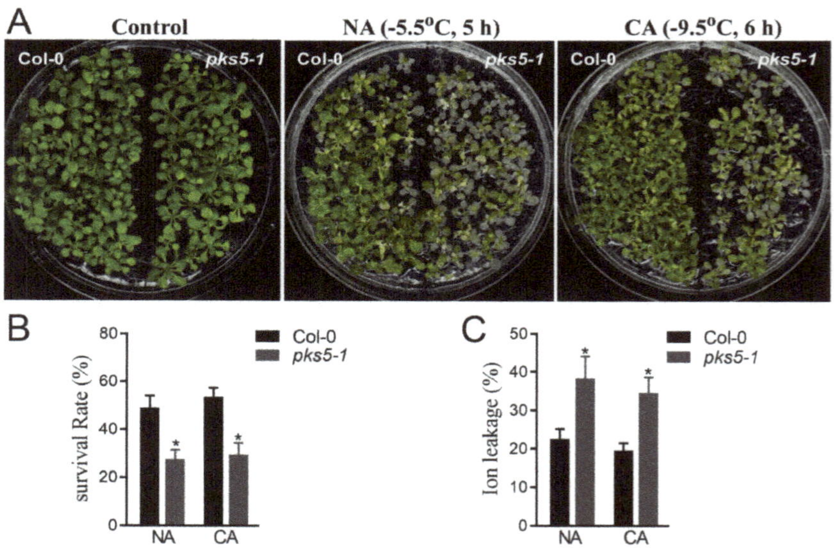

Figure 1. A deficiency of PKS5 impairs plant cold stress response. (**A**) Freezing phenotypes *pks5-1* under non−acclimated (NA) and cold−acclimated (CA) conditions. The wild−type Col−0 and *pks5-1* were grown on MS medium at 22 °C for 2 weeks before being subjected to the freezing treatment. For the NA treatment, seedlings were treated at −5.5 °C for 5 h; for the CA treatment, seedlings were pretreated at 4 °C for 3 d and then treated at −9.5 °C for 6 h. The freezing−treated seedlings were then transferred to 4 °C under a dark condition for 12 h, and then shifted to a normal growth condition at 22 °C for recovery of growth. Representative images were taken during the recovery growth at 22 °C. (**B**) Survival rates of *pks5-1* under NA and CA conditions. (**C**) Ion leakages of *pks5-1* under NA and CA conditions. Student's *t*−test was used to analyze the statistical significance; each bar is the mean ± SD of three biological replications. Significant differences ($p \leq 0.05$) in (**B**,**C**) are indicated by asterisks.

Freezing treatment impairs cell structure and activity during cold exposure, and plants exposed to freezing exhibit decreased survival rates even after resumption of growth at normal temperatures. Compared with Col-0 seedlings, *pks5-1* mutant seedlings displayed a decreased survival rate after freezing under both non-acclimated and cold-acclimated conditions (Figure 1B).

Previous research reported that membrane damage caused by cold stress resulted in ion flow out of cells [3]. Ion leakage analysis of freeze-treated seedlings showed that *pks5-1* seedlings had a higher ion leakage rate than Col-0 plants under both non-acclimated and cold-acclimated conditions (Figure 1C). The cold-sensitive phenotype, decreased survival rate, and increased ion leakage exhibited by the *pks5-1* mutant under cold stress are indicative of an essential role for PKS5 in cold tolerance responses, with PKS5 playing a positive role in the response to cold stress.

3.2. Increases in PKS5 Activity Enhance Plant Freezing Tolerance

The regulatory role of PKS5 in the response to cold stress was further assessed using two previously developed PKS5 gain-of-function mutants, *pks5-3* and *pks5-4*, which have elevated PKS5 activity levels [17]. Seedlings of *pks5-3*, *pks5-4*, and the corresponding wild-type (BigM), were exposed to freezing treatment. Compared with BigM, *pks5-3* and *pks5-4* displayed a freezing-tolerant phenotype under both non-acclimated and cold-acclimated conditions (Figure 2A). This result indicates that increasing PKS5 activity can improve plant freezing tolerance.

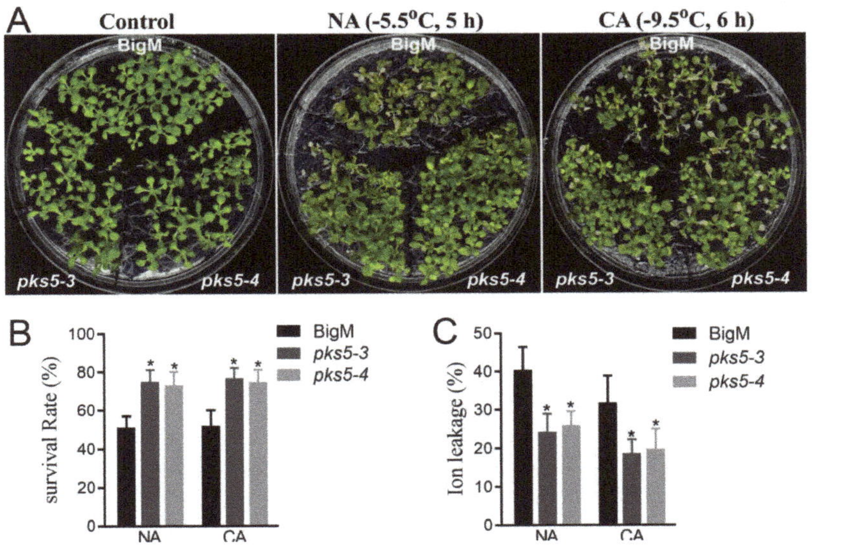

Figure 2. The increase of PKS5 activity improves plant cold stress response. (**A**) Freezing phenotypes *pks5-3* and *pks5-4* under NA and CA conditions. The wild−type BigM, *pks5-3* and *pks5-4* were grown on MS medium at 22 °C for 2 weeks before being subjected to freezing treatment. For the NA treatment, seedlings were treated at −5.5 °C for 5 h; for the CA treatment, seedlings were pretreated at 4 °C for 3 d and then treated at −9.5 °C for 6 h. The freezing−treated seedlings were then transferred to 4 °C under a dark condition for 12 h, and then shifted to a normal growth condition at 22 °C for recovery of growth. Representative images were taken during the recovery growth at 22 °C. (**B**) Survival rates of *pks5-3* and *pks5-4* under NA and CA conditions. (**C**) Ion leakage of *pks5-3* and *pks5-4* under NA and CA conditions. Student's t−test was used to analyze the statistical significance; each bar is the mean ± SD of three biological replications. Significant differences ($p \leq 0.05$) in (**B**,**C**) are indicated by asterisks.

Analysis of seedling survival after freezing treatment showed that *pks5-3* and *pks5-4* displayed higher survival rates than BigM under both non-acclimated and cold-acclimated conditions (Figure 2B). The ion leakage assay of the freeze-treated seedlings also showed that, compared with BigM, *pks5-3* and *pks5-4* displayed lower ion leakage after freezing treatment (Figure 2C).

These results indicate that increasing PKS5 activity enhances plant freezing tolerance and suggest that PKS5 positively regulates freezing tolerance in Arabidopsis.

3.3. PKS5 Regulates Stomatal Movements under Cold Stress

Photosynthesis and stomatal movement are regulated during plant responses to cold stress [8]. To investigate whether PKS5 can regulate plant cold stress responses via the regulation of stomatal movement, stomatal aperture in response to cold stress was assessed in loss-of-function *pks5-1* seedlings. Stomata in both Col-0 and loss-of-function *pks5-1* seedlings were closed in 25 min after initiation of cold exposure; however, *pks5-1* stomata closed more slowly than those in Col-0, with a significant difference in stomatal closure observed between Col-0 and *pks5-1* at 10 min after cold initiation (Figure 3A–C). This result indicates that loss of PKS5 function impairs the regulation of stomatal movement during exposure to cold stress.

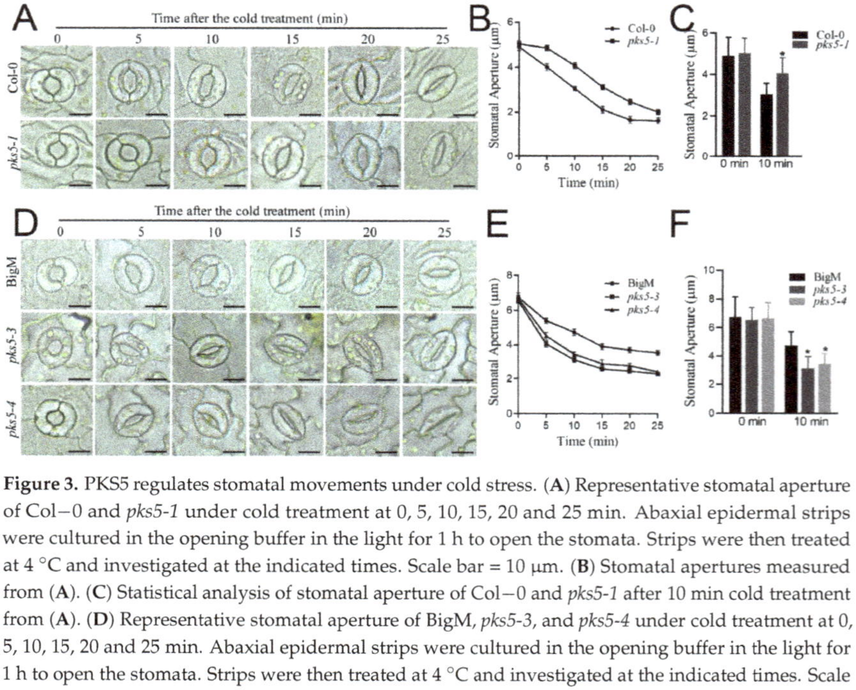

Figure 3. PKS5 regulates stomatal movements under cold stress. (**A**) Representative stomatal aperture of Col−0 and *pks5-1* under cold treatment at 0, 5, 10, 15, 20 and 25 min. Abaxial epidermal strips were cultured in the opening buffer in the light for 1 h to open the stomata. Strips were then treated at 4 °C and investigated at the indicated times. Scale bar = 10 μm. (**B**) Stomatal apertures measured from (**A**). (**C**) Statistical analysis of stomatal aperture of Col−0 and *pks5-1* after 10 min cold treatment from (**A**). (**D**) Representative stomatal aperture of BigM, *pks5-3*, and *pks5-4* under cold treatment at 0, 5, 10, 15, 20 and 25 min. Abaxial epidermal strips were cultured in the opening buffer in the light for 1 h to open the stomata. Strips were then treated at 4 °C and investigated at the indicated times. Scale bar = 10 μm; (**E**) Stomatal apertures measured from (**D**). (**F**) Statistical analysis of stomatal aperture of BigM, *pks5-3*, and *pks5-4* after 10 min cold treatment from D. Student's t−test was used to analyze the statistical significance; each bar is the mean ± SD of three biological replications ($n > 50$). Significant differences ($p \leq 0.05$) in (**C**,**F**) are indicated by asterisks.

The role of PKS5 in stomatal movement was assessed further using *pks5-3* and *pks5-4* gain-of-function mutants. Stomata in *pks5-3*, *pks5-4*, and BigM were closed in 25 min after cold treatment initiation (Figure 3D). However, closure of stomata occurred more quickly in *pks5-3* and *pks5-4* than in BigM, and a significant difference in stomatal closure was observed between BigM and *pks5-3* and *pks5-4* at 10 min after cold initiation (Figure 3E,F). This suggests that increasing PKS5 activity increases the rate of stomatal closure under cold stress.

Together, these observations indicate that PKS5 plays a regulatory role in stomatal movements during plant cold stress responses.

3.4. PKS5 Mediates ABA-Regulated Stomatal Movements

ABA levels were previously shown to increase upon exposure to cold stress, and ABA is known to play a role in the regulation of stomatal movement [13,21,22]. To investigate whether the PKS5-regulated stomatal movement observed after cold exposure was related to ABA signaling, stomatal apertures were examined after ABA treatment in the PKS5 loss-of-function mutant, *pks5-1*, and gain-of-function mutants *pks5-3* and *pks5-4*. ABA treatment induced stomatal closure in both Col-0 and *pks5-1*, but *pks5-1* stomata closed more slowly than those in Col-0 (Figure 4A,B). A significant difference in stomatal closure was observed between Col-0 and *pks5-1* at 10 min after treatment (Figure 4C). ABA treatment also induced stomatal closure in *pks5-3*, *pks5-4*, and BigM, and *pks5-3* and *pks5-4* stomata closed more quickly than those in BigM (Figure 4D,E). A significant difference in stomatal closure was observed between BigM and the *pks5-3* and *pks5-4* mutants at 10 min after treatment (Figure 4F). These results suggest that the regulatory role of PKS5 in stomatal movement is related to ABA signaling, and that cold-induced ABA accumulation may contribute to the regulation of stomatal movement by PKS5 under cold stress.

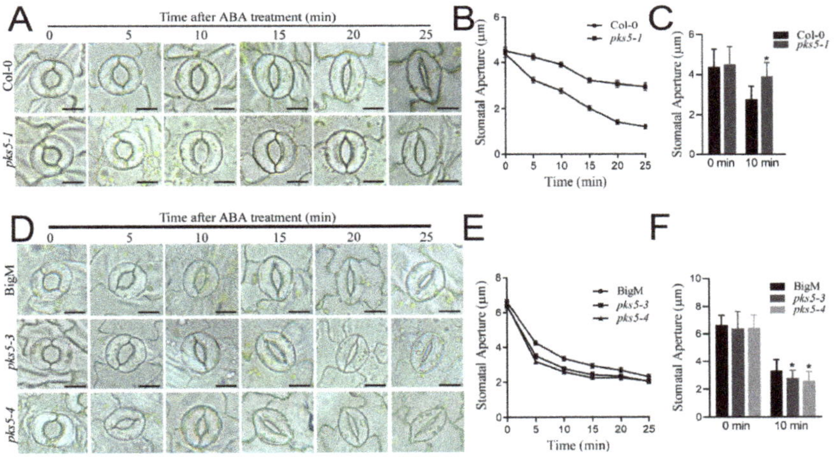

Figure 4. PKS5 mediates ABA−regulated stomatal movements. (**A**) Representative stomatal aperture of Col−0 and *pks5-1* under ABA treatment at 0, 5, 10, 15, 20 and 25 min. Abaxial epidermal strips were cultured in the opening buffer in the light for 1 h to open the stomata. Strips were then treated with ABA−containing buffer and investigated at the indicated times. Scale bar = 10 μm. (**B**) Stomatal apertures measured from (**A**). (**C**) Statistical analysis of stomatal aperture of Col−0 and *pks5-1* after 10 min ABA treatment from (**A**). (**D**) Representative stomatal aperture of BigM, *pks5-3*, and *pks5-4* under ABA treatment at 0, 5, 10, 15, 20 and 25 min. Abaxial epidermal strips were cultured in the opening buffer in the light for 1 h to open the stomata. Strips were then treated with ABA−containing buffer and investigated at the indicated times. Scale bar = 10 μm. (**E**) Stomatal apertures measured from (**D**). (**F**) Statistical analysis of stomatal aperture of BigM, *pks5-3*, and *pks5-4* after 10 min ABA treatment from (**D**). Student's t−test was used to analyze the statistical significance; each bar is the mean ± SD of three biological replications ($n > 50$). Significant differences ($p \leq 0.05$) in (**C**,**F**) are indicated by asterisks.

3.5. Cold-Responsive Genes Regulated by PKS5 under Cold Stress

Cold-responsive genes, such as *CBF* and *COR*, are upregulated upon exposure to cold stress in plants [1]. To explore whether *CBF* and *COR* genes were involved in the PKS5-regulated cold stress response, expression levels of cold-responsive genes, including

CBF1, *CBF2*, *CBF3*, *COR15A*, *KIN1*, and *RD29A*, were examined in wild-type, *pks5-1*, *pks5-3*, and *pks5-4* plants. At 22 °C, gene expression was comparable between *pks5-1* and Col-0, and between *pks5-3*, *pks5-4*, and BigM (Figure 5A–D). However, cold-induced expression of *CBF* genes was lower in *pks5-1*, and higher in *pks5-3* and *pks5-4*, compared to their respective wild-types (Figure 5A,C). Moreover, three CBF-regulated genes, *COR15A*, *KIN1*, and *RD29A*, also exhibited lower expression in *pks5-1*, and higher expression in *pks5-3* and *pks5-4*, compared to their respective wild-types (Figure 5B,D). These results suggest that PKS5 has a positive regulatory role in acclimation of plants to cold stress, and that this regulation is mediated, at least partially, via regulation of expression of *CBF* and CBF-regulated genes.

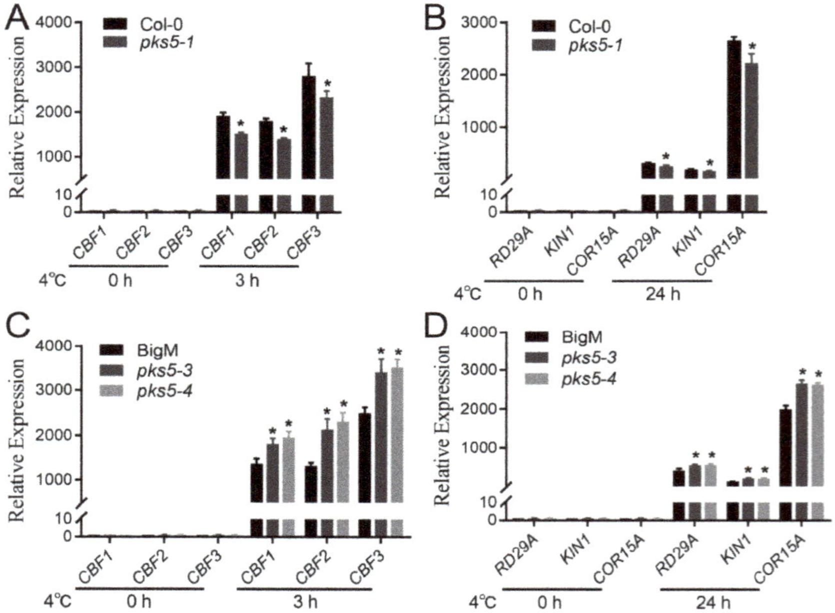

Figure 5. PKS5 regulates cold−responsive genes under cold stress. (**A**) Expression level analysis of *CBF* genes in *pks5-1*. (**B**) Expression level analysis of CBF target genes in *pks5-1*. (**C**) Expression level analysis of *CBF* genes in *pks5-3* and *pks5-4*. (**D**) Expression level analysis of CBF target genes in *pks5-3* and *pks5-4*. Seedlings grown for 2 weeks on MS medium at 22 °C were treated at 4 °C for 3 h (**A**,**C**) to analyze *CBF* gene expression and were treated at 4 °C for 24 h to analyze CBF target gene expression. Gene expression levels in untreated wild−type seedlings (Col−0 in A and B, BigM in (**C**,**D**)) were set to 1. Student's t−test was used to analyze the statistical significance; each bar is the mean ± SD of three biological replications. Significant differences ($p \leq 0.05$) in (**A**–**D**) are indicated by asterisks.

3.6. Cold Stress and ABA Treatment Regulate PKS5 Expression

To further explore whether the expression of *PKS5* was induced by cold stress, expression levels of *PKS5* were analyzed under freezing treatment. The results showed that a freezing treatment of −5.5 °C induced elevated expression of *PKS5*, especially during the transition period at 4 °C overnight (Figure 6A). In this study and previous studies, cold acclimation at 4 °C for 3 d could significantly improve plant freezing tolerance [3]. Consistent with the improved freezing phenotype, *PKS5* expression level could also be induced during the cold acclimation process (Figure 6B). Freezing treatment on cold-acclimated seedlings also showed that *PKS5* expression could be induced by the freezing treatment, especially during the transition period at 4°C overnight (Figure 6C). However, the freezing-treated, cold-acclimated seedlings showed a substantially greater increase of *PKS5* expression level (about 8-fold) compared with that in non-acclimated seedlings (about 4-fold) during the transition period at 4 °C overnight (Figure 6A,C). These results indicate that the *PKS5* gene can be induced by cold stress.

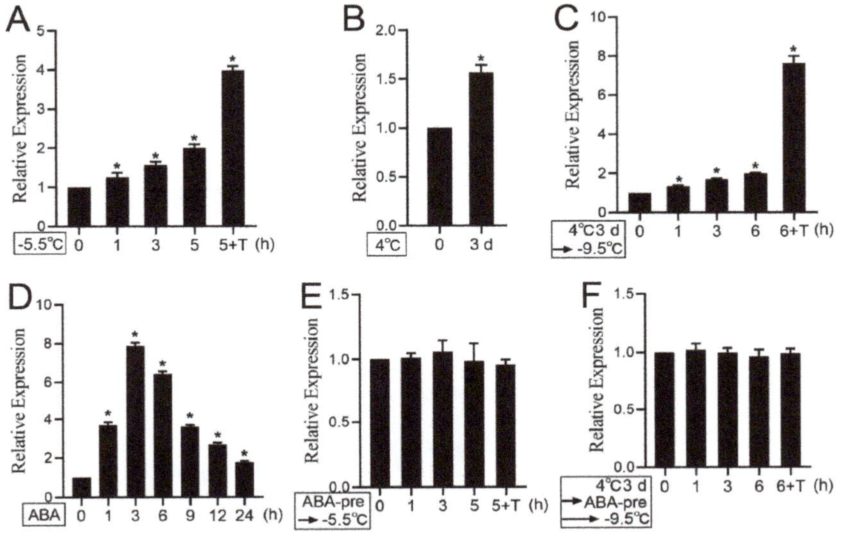

Figure 6. Expression level analysis of *PKS5* under cold stress and ABA treatment. (**A**) Expression level analysis of *PKS5* under cold stress in non−acclimated seedlings. (**B**) Expression level analysis of *PKS5* after cold acclimation. (**C**) Expression level analysis of *PKS5* under cold stress in cold−acclimated seedlings. (**D**) The expression level analysis of *PKS5* under ABA treatment. (**E**) ABA effect on *PKS5* expression under freezing stress in non−acclimated seedlings. (**F**) ABA effect on *PKS5* expression under freezing stress in cold−acclimated seedlings. The two−week old Col−0 seedlings grown on MS medium at 22 °C were subjected to various treatments. For freezing treatment in non−acclimated seedlings, seedlings were frozen at −5.5 °C and collected for *PKS5* expression level analysis. For the freezing treatment in cold−acclimated seedlings, seedlings underwent a cold acclimation process of 4 °C for 3 d, and were then subjected to freezing at −9.5 °C. For the ABA treatment, seedlings were first sprayed with ddH_2O containing 10 μM ABA, and then the seedlings were collected for *PKS5* expression. For the ABA−pretreatment, seedlings were pretreated with 10 μM ABA for 3 h, and then non−acclimated seedlings and cold−acclimated seedlings were subjected to freezing. T: transition period at 4 °C under dark for 12 h. Actin was used as an internal control. Gene expression levels in wild−type Col−0 seedlings at 0 h were set to 1. Student's *t*−test was used to analyze the statistical significance; each bar is the mean ± SD of three biological replications. Significant differences ($p \leq 0.05$) in (**A–D**) are indicated by asterisks.

To further investigate ABA effect on *PKS5* expression, *PKS5* expression was first examined by ABA treatment under normal temperature. The result showed that *PKS5* expression could be induced by ABA treatment, with the highest induced level at 3 h (Figure 6D). To investigate the effect of cold-induced ABA accumulation on PKS5 expression, seedlings of non-acclimated and cold-acclimated were both pretreated with 10 μM ABA for 3 h, and then treated with freezing stress; however, no increase in *PKS5* expression was observed in both non-acclimated seedlings and cold-acclimated seedlings (Figure 6E,F). These results suggest that cold-induced PKS5 expression might be regulated by ABA accumulated under cold stress.

4. Discussion

Exposure of plants to cold stress stimulates a cascade of short-term and long-term physiological activities that mitigate cellular damage. Regulation of gene expression in response to cold exposure is well understood, and includes *CBF* genes, which are upregulated within 1–3 h of exposure, and *COR* genes, which are upregulated later, 12–24 h after cold exposure [3]. Previous research showed that cold stress induced stomatal movement. Stomatal conductance in Scots pine seedlings first decreased within 45 min of the freezing period of freeze-thaw treatment, then subsequently increased with the extension of freezing treatment time [6]. In this study, the PKS5 loss-of-function mutant *pks5-1*, which exhibited a cold-sensitive phenotype, exhibited slower stomatal closure during the first 25 min after cold exposure than wild-type seedlings, and also exhibited lower expression of *CBF* and *COR* genes 3 h and 24 h after cold exposure, respectively. Conversely, PKS5 gain-of-function mutants *pks5-3* and *pks5-4*, which were cold-tolerant, exhibited faster stomatal closure during the first 25 min after cold exposure than wild-type, and also exhibited higher expression of *CBF* and *COR* genes 3 h and 24 h after cold exposure, respectively. Stomatal aperture regulation was examined at 10 min intervals from 25 min to 3 h after cold exposure, but no clear regulatory pattern was apparent (data not shown). These observations in Arabidopsis are similar to previous research results examining stomatal closure in Scots pine seedlings, which also responded to cold stress by regulating stomatal closure shortly after cold exposure [6]. This study confirms the involvement of PKS5 in stomatal aperture regulation under cold stress in Arabidopsis; however, the detailed regulatory mechanisms require further elucidation.

Cold stress can be divided into chilling stress (0–15 °C) and freezing stress (<0 °C) [1]. Since freezing stress leads to ice formation, and is more harmful to the plant, plants have evolved cold acclimation to response to freezing stress. *CBFs* expression is rapidly upregulated under chilling stress at 4 °C, which further activates downstream *COR* genes, leading to an increase of freezing tolerance via the biosynthesis of osmolytes, such as soluble sugars [1,21]. It is reported that 14-3-3 proteins negatively regulate freezing tolerance [23]. Under cold stress, plasma membrane kinase CRPK1 phosphorylates 14-3-3 proteins, which are then translocated from cytosol into nucleus to facilitate the degradation of CBF proteins [23]. It is interesting that PKS5 can interact with 14-3-3 proteins under salt stress, and 14-3-3 proteins can bind to salt-induced Ca^{2+} and repress PKS5 activity. Although the CBF cold signaling pathway has been extensively studied, the upstream components regulated by calcium signaling remain largely unknown. Whether PKS5-regulated cold stress response involves calcium signaling needs further study.

PKS5 was previously shown to have a role in ABA signal transduction, and ABA treatment in this study resulted in similar stomatal closure patterns to those seen after cold treatment. However, the downstream or upstream elements involved in PKS5-regualted stomatal movement, and the mechanism by which PKS5 mediates ABA-regulated stomatal movements, require further investigation. Under normal cultivation conditions, PKS5 can interact with and phosphorylate SOS2 at Ser-294, inhibiting SOS2 activity [16]. PKS5 can also phosphorylate PM H^+-ATPase at Ser-931 to inhibit PM H^+-ATPase activity [15]. Conversely, upon exposure to salt stress, PKS5 can interact with J3 and 14-3-3 to activate PM H^+-ATPase and SOS2 [16,17]. During ABA-regulated seed germination, PKS5 can interact with ABI5 and phosphorylate ABI5 at Ser-42 to play a positive role in ABA signaling [18]. PKS5 also participates in plant defense responses via phosphorylation of NPR1 (Nonexpressor of Pathogenesis-Related gene 1) at the C-terminal region [24]. The guard cell anion channel SLAC1 has a fundamental role in the regulation of stomatal aperture control, and PKS5 can form CBL5–CIPK11 complexes to stimulate SLAC1 activity [19]. These PKS5-interacting proteins may participate in PKS5-regulated stomatal aperture closure under cold stress conditions.

Cold stress induces a rapid increase of cytosolic calcium (Ca^{2+}) and the accumulation of ABA. Whether 14-3-3 proteins can bind to cold-induced Ca^{2+} and regulate PKS5 activity may help decode Ca^{2+} signaling in the cold stress response. SLAC1 is required for plant guard cell S-type anion channel function in stomatal signaling [25], and PKS5 can regulate

SLAC1 anion channel activity through formation of the CBL5-PKS5 complex [19]. Whether SLAC1 plays a role in the cold stress response, and whether PKS5 involves this process, are still unknown and need further study. PM H^+-ATPase in plant is an essential enzyme with multiple physiological functions, and is highly regulated, mainly by phosphorylation [26]. PM H^+-ATPase regulates stomatal movements under various conditions, such as jasmonate-regulated stomatal closure [27], light-induced stomatal opening [28], and salt stress response [17]. Although PM H^+-ATPase has been reported to be involved in cold stress responses [29], there is still no direct evidence about how it regulates cold stress responses. PKS5 might be a link between PM H^+-ATPase and cold stress response. Whether PKS5 plays a role mediating Ca^{2+} signaling in the cold stress response, and which transcriptional factors (TFs) mediate its regulation of *CBFs* and *CORs* genes, needs further study. The model depicted in Figure 7, as a combination of this study and previous studies, may help explain the mechanism of plant responses to cold stress.

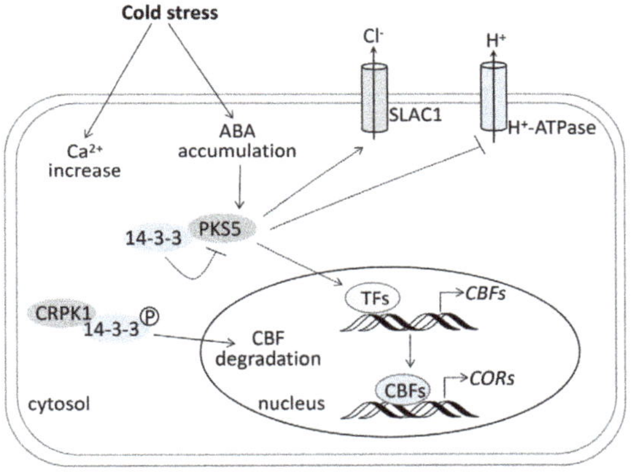

Figure 7. Model for PKS5 in regulating freezing tolerance in Arabidopsis. Cold stress induces rapid increase of cytosolic calcium (Ca^{2+}) and the accumulation of ABA. Cold−induced ABA accumulation, on the one hand, induces *PKS5* expression under cold stress and, on the other hand, regulates PKS5 role in the regulation of stomatal movement under cold stress. PKS5 may stimulate SLAC1 anion channel activity through the formation of CBL5−PKS5 complex, or directly inhibit PM H^+−ATPase activity by the phosphorylation of AHA2 at Ser−931 to regulate stomatal movement under cold stress in short−term plant acclimation to cold stress. In long−term plant acclimation to cold stress, PKS5 can indirectly upregulate the expression of *CBFs* and *CORs* genes through unknown transcriptional factors (TFs) to increase the osmolyte synthesis to improve plant freezing tolerance. The 14−3−3 proteins play negative roles in plant response to cold stress, which, on the one hand, inhibit PKS5 activity through their interaction and, on the other hand, can be phosphorylated by CRPK1 and translocated into nucleus, leading to CBF degradation. Broken arrows indicate activation unconfirmed to occur under cold stress.

Author Contributions: X.H. and F.Z. designed the experiments; C.S., L.Z., L.C., H.Q. and H.L. performed the experiments; C.S., L.Z. and L.C. analyzed the data and wrote the manuscript. All authors have read and agreed to the published version of the manuscript.

Funding: This work was supported by the National Natural Science Foundation of China (32000219) and the National Natural Science Foundation of Shandong province (ZR2020MC021).

Institutional Review Board Statement: Not applicable.

Informed Consent Statement: Not applicable.

Data Availability Statement: Not applicable.

Conflicts of Interest: The authors declare no conflict of interest.

References

1. Shi, Y.; Ding, Y.; Yang, S. Molecular regulation of CBF signaling in cold acclimation. *Trends Plant Sci.* **2018**, *23*, 623–637. [CrossRef] [PubMed]
2. Wang, X.; Song, Q.; Liu, Y.; Brestic, M.; Yang, X. The network centered on ICEs play roles in plant cold tolerance, growth and development. *Planta* **2022**, *255*, 81. [CrossRef]
3. Li, H.; Ding, Y.; Shi, Y.; Zhang, X.; Zhang, S.; Gong, Z.; Yang, S. MPK3- and MPK6-mediated ICE1 phosphorylation negatively regulates ICE1 stability and freezing tolerance in Arabidopsis. *Dev. Cell* **2017**, *43*, 630–642.e634. [CrossRef] [PubMed]
4. Zhang, H.; Zhu, J.; Gong, Z.; Zhu, J.K. Abiotic stress responses in plants. *Nat. Rev. Genet.* **2022**, *23*, 104–119. [CrossRef] [PubMed]
5. Du, X.; Jin, Z.; Liu, D.; Yang, G.; Pei, Y. Hydrogen sulfide alleviates the cold stress through MPK4 in Arabidopsis thaliana. *Plant Physiol. Biochem.* **2017**, *120*, 112–119. [CrossRef]
6. Lindfors, L.; Hölttä, T.; Lintunen, A.; Porcar-Castell, A.; Nikinmaa, E.; Juurola, E. Dynamics of leaf gas exchange, chlorophyll fluorescence and stem diameter changes during freezing and thawing of Scots pine seedlings. *Tree Physiol.* **2015**, *35*, 1314–1324. [CrossRef]
7. Du, X.; Jin, Z.; Liu, Z.; Liu, D.; Zhang, L.; Ma, X.; Yang, G.; Liu, S.; Guo, Y.; Pei, Y. H_2S persulfidated and increased kinase activity of MPK4 to response cold stress in Arabidopsis. *Front. Mol. Biosci.* **2021**, *8*, 635470. [CrossRef] [PubMed]
8. Gobel, L.; Coners, H.; Hertel, D.; Willinghofer, S.; Leuschner, C. The role of low soil temperature for photosynthesis and stomatal conductance of three graminoids from different elevations. *Front. Plant Sci.* **2019**, *10*, 330. [CrossRef]
9. Chen, K.; Li, G.J.; Bressan, R.A.; Song, C.P.; Zhu, J.K.; Zhao, Y. Abscisic acid dynamics, signaling, and functions in plants. *J. Integr. Plant Biol.* **2020**, *62*, 25–54. [CrossRef]
10. Verma, V.; Ravindran, P.; Kumar, P.P. Plant hormone-mediated regulation of stress responses. *BMC Plant Biol.* **2016**, *16*, 86. [CrossRef]
11. Wang, Y.G.; Fu, F.L.; Yu, H.Q.; Hu, T.; Zhang, Y.Y.; Tao, Y.; Zhu, J.K.; Zhao, Y.; Li, W.C. Interaction network of core ABA signaling components in maize. *Plant Mol. Biol.* **2018**, *96*, 245–263. [CrossRef] [PubMed]
12. Fabregas, N.; Yoshida, T.; Fernie, A.R. Role of Raf-like kinases in SnRK2 activation and osmotic stress response in plants. *Nat. Commun.* **2020**, *11*, 6184. [CrossRef] [PubMed]
13. Agurla, S.; Gahir, S.; Munemasa, S.; Murata, Y.; Raghavendra, A.S. Mechanism of stomatal closure in plants exposed to drought and cold Stress. *Adv. Exp. Med. Biol.* **2018**, *1081*, 215–232. [PubMed]
14. Waadt, R.; Seller, C.A.; Hsu, P.K.; Takahashi, Y.; Munemasa, S.; Schroeder, J.I. Plant hormone regulation of abiotic stress responses. *Nat. Rev. Mol. Cell Biol.* **2022**, *23*, 680–694. [CrossRef]
15. Fuglsang, A.T.; Guo, Y.; Cuin, T.A.; Qiu, Q.; Song, C.; Kristiansen, K.A.; Bych, K.; Schulz, A.; Shabala, S.; Schumaker, K.S.; et al. Arabidopsis protein kinase PKS5 inhibits the plasma membrane H^+-ATPase by preventing interaction with 14-3-3 protein. *Plant Cell* **2007**, *19*, 1617–1634. [CrossRef]
16. Yang, Z.; Wang, C.; Xue, Y.; Liu, X.; Chen, S.; Song, C.; Yang, Y.; Guo, Y. Calcium-activated 14-3-3 proteins as a molecular switch in salt stress tolerance. *Nat. Commun.* **2019**, *10*, 1199. [CrossRef]
17. Yang, Y.; Qin, Y.; Xie, C.; Zhao, F.; Zhao, J.; Liu, D.; Chen, S.; Fuglsang, A.T.; Palmgren, M.G.; Schumaker, K.S.; et al. The Arabidopsis chaperone J3 regulates the plasma membrane H^+-ATPase through interaction with the PKS5 kinase. *Plant Cell* **2010**, *22*, 1313–1332. [CrossRef]
18. Zhou, X.; Hao, H.; Zhang, Y.; Bai, Y.; Zhu, W.; Qin, Y.; Yuan, F.; Zhao, F.; Wang, M.; Hu, J.; et al. SOS2-LIKE PROTEIN KINASE5, an SNF1-RELATED PROTEIN KINASE3-Type protein kinase, is important for abscisic acid responses in Arabidopsis through phosphorylation of ABSCISIC ACID-INSENSITIVE5. *Plant Physiol.* **2015**, *168*, 659–676. [CrossRef]
19. Saito, S.; Hamamoto, S.; Moriya, K.; Matsuura, A.; Sato, Y.; Muto, J.; Noguchi, H.; Yamauchi, S.; Tozawa, Y.; Ueda, M.; et al. N-myristoylation and S-acylation are common modifications of Ca^{2+}-regulated Arabidopsis kinases and are required for activation of the SLAC1 anion channel. *New Phytol.* **2018**, *218*, 1504–1521. [CrossRef]
20. Merlot, S.; Leonhardt, N.; Fenzi, F.; Valon, C.; Costa, M.; Piette, L.; Vavasseur, A.; Genty, B.; Boivin, K.; Muller, A.; et al. Constitutive activation of a plasma membrane H^+-ATPase prevents abscisic acid-mediated stomatal closure. *EMBO J.* **2007**, *26*, 3216–3226. [CrossRef]
21. Ding, Y.; Li, H.; Zhang, X.; Xie, Q.; Gong, Z.; Yang, S. OST1 kinase modulates freezing tolerance by enhancing ICE1 stability in Arabidopsis. *Dev. Cell* **2015**, *32*, 278–289. [CrossRef] [PubMed]
22. Wei, H.; Jing, Y.; Zhang, L.; Kong, D. Phytohormones and their crosstalk in regulating stomatal development and patterning. *J. Exp. Bot.* **2021**, *72*, 2356–2370. [CrossRef] [PubMed]
23. Liu, Z.; Jia, Y.; Ding, Y.; Shi, Y.; Li, Z.; Guo, Y.; Gong, Z.; Yang, S. Plasma Membrane CRPK1-Mediated Phosphorylation of 14-3-3 Proteins Induces Their Nuclear Import to Fine-Tune CBF Signaling during Cold Response. *Mol. Cell* **2017**, *66*, 117–128.e115. [CrossRef] [PubMed]
24. Xie, C.; Zhou, X.; Deng, X.; Guo, Y. PKS5, a SNF1-related kinase, interacts with and phosphorylates NPR1, and modulates expression of WRKY38 and WRKY62. *J. Genet. Genom. = Yi Chuan Xue Bao* **2010**, *37*, 359–369. [CrossRef]
25. Vahisalu, T.; Kollist, H.; Wang, Y.-F.; Nishimura, N.; Chan, W.-Y.; Valerio, G.; Lamminmäki, A.; Brosché, M.; Moldau, H.; Desikan, R.; et al. SLAC1 is required for plant guard cell S-type anion channel function in stomatal signalling. *Nature* **2008**, *452*, 487–491. [CrossRef]

26. Falhof, J.; Pedersen, J.T.; Fuglsang, A.T.; Palmgren, M. Plasma Membrane H(+)-ATPase Regulation in the Center of Plant Physiology. *Mol. Plant* **2016**, *9*, 323–337. [CrossRef]
27. Yan, S.; McLamore, E.S.; Dong, S.; Gao, H.; Taguchi, M.; Wang, N.; Zhang, T.; Su, X.; Shen, Y. The role of plasma membrane H(+) -ATPase in jasmonate-induced ion fluxes and stomatal closure in Arabidopsis thaliana. *Plant J. Cell Mol. Biol.* **2015**, *83*, 638–649. [CrossRef]
28. Wang, Y.; Noguchi, K.; Ono, N.; Inoue, S.; Terashima, I.; Kinoshita, T. Overexpression of plasma membrane H+-ATPase in guard cells promotes light-induced stomatal opening and enhances plant growth. *Proc. Natl. Acad. Sci. USA* **2014**, *111*, 533–538. [CrossRef]
29. Kim, H.S.; Oh, J.M.; Luan, S.; Carlson, J.E.; Ahn, S.J. Cold stress causes rapid but differential changes in properties of plasma membrane H+-ATPase of camelina and rapeseed. *J. Plant Physiol.* **2013**, *170*, 828–837. [CrossRef]

life

MDPI

Review

Molecular and Physiological Mechanisms to Mitigate Abiotic Stress Conditions in Plants

Baljeet Singh Saharan [1,*,†], Basanti Brar [1,†], Joginder Singh Duhan [2], Ravinder Kumar [2], Sumnil Marwaha [3], Vishnu D. Rajput [4,*] and Tatiana Minkina [4]

1 Department of Microbiology, CCS Haryana Agricultural University, Hisar 125004, India
2 Department of Biotechnology, Ch. Devi Lal University, Sirsa 125055, India
3 ICAR-National Research Centre on Camel, Bikaner 334001, India
4 Academy of Biology and Biotechnology, Southern Federal University, 344090 Rostov-on-Don, Russia
* Correspondence: baljeetsaharan@hau.ac.in (B.S.S.); rajput.vishnu@gmail.com (V.D.R.); Tel.: +91-9416-545-332 (B.S.S.); +7-918-589-00-93 (V.D.R.)
† These authors contributed equally to this work and both are considered as first author.

Abstract: Agriculture production faces many abiotic stresses, mainly drought, salinity, low and high temperature. These abiotic stresses inhibit plants' genetic potential, which is the cause of huge reduction in crop productivity, decrease potent yields for important crop plants by more than 50% and imbalance agriculture's sustainability. They lead to changes in the physio-morphological, molecular, and biochemical nature of the plants and change plants' regular metabolism, which makes them a leading cause of losses in crop productivity. These changes in plant systems also help to mitigate abiotic stress conditions. To initiate the signal during stress conditions, sensor molecules of the plant perceive the stress signal from the outside and commence a signaling cascade to send a message and stimulate nuclear transcription factors to provoke specific gene expression. To mitigate the abiotic stress, plants contain several methods of avoidance, adaption, and acclimation. In addition to these, to manage stress conditions, plants possess several tolerance mechanisms which involve ion transporters, osmoprotectants, proteins, and other factors associated with transcriptional control, and signaling cascades are stimulated to offset abiotic stress-associated biochemical and molecular changes. Plant growth and survival depends on the ability to respond to the stress stimulus, produce the signal, and start suitable biochemical and physiological changes. Various important factors, such as the biochemical, physiological, and molecular mechanisms of plants, including the use of microbiomes and nanotechnology to combat abiotic stresses, are highlighted in this article.

Keywords: microbiota; natural farming; physical factors; physiological changes; signal transduction and stressed conditions

Citation: Saharan, B.S.; Brar, B.; Duhan, J.S.; Kumar, R.; Marwaha, S.; Rajput, V.D.; Minkina, T. Molecular and Physiological Mechanisms to Mitigate Abiotic Stress Conditions in Plants. *Life* **2022**, *12*, 1634. https://doi.org/10.3390/life12101634

Academic Editors: Hakim Manghwar and Wajid Zaman

Received: 29 September 2022
Accepted: 15 October 2022
Published: 19 October 2022

Publisher's Note: MDPI stays neutral with regard to jurisdictional claims in published maps and institutional affiliations.

Copyright: © 2022 by the authors. Licensee MDPI, Basel, Switzerland. This article is an open access article distributed under the terms and conditions of the Creative Commons Attribution (CC BY) license (https://creativecommons.org/licenses/by/4.0/).

1. Introduction

Stress is defined as a stimulus that inhibits the growth of plants and their metabolism and development at the time of both abiotic and biotic stress [1]. Abiotic stresses, such as higher or insufficient water supply, low and high temperature, heavy metals, ultraviolet radiation and salinity, are damaging to plant development and growth, and cause considerable losses in agricultural productivity worldwide [2,3]. When the stress threshold is exceeded, the plant is stressed, followed by activation of physiological, biochemical, morphological, and molecular-level mechanisms. The activation of these mechanisms can show the development of a fresh physiological state and the restoration of homeostasis in plants [4]. Due to abiotic stress, it has been estimated that crop production yield decreases by up to 70% in several commercially significant crops and they execute at just 30% of their genetic makeup in terms of yield [5].

Plants survive in environments that are repeatedly changing and frequently not suitable for plant growth as well as development. These harsh situations for the growth and development of plants arise mainly due to abiotic stress [6]. Moreover, abiotic stresses are anticipated to intensify and occur more frequently in the near future due to climate change, which may cause severe salinization of more than 50% of soil of the arable domain by 2050 [7,8]. Therefore, productive agricultural land and crop yields may gradually decline due to increasing temperature and recurrent flooding over several decades, particularly in the mid-latitudes [7,8].

Along with these factors, anthropogenic activities may increase the amounts of contaminants in the water, soil and air, with which plants must contend. It has been estimated that abiotic stress factors have a greater than 90% impact on plants and crop growth during their growing season in rural areas [9]. On the other hand, the rapid growth of the population has increased the demand for food and other necessary resources. Therefore, to develop stress-resistant cultivars that can endure abiotic stress and feed the expanding population, knowing plants' stress responses is crucial. When plants faces variety of stresses, activates the stress signal and respond accordingly [9]. Plants with abiotic stress have primary signals for ion toxicity detection, low proline and chlorophyll content, low CO_2 assimilation, and osmotic effects, etc., in the cells (Figure 1). Secondary consequences of these abiotic stresses are complicated and comprise oxidative stresses that harm various cellular components such as nucleic acids, proteins present in membranes and lipids, and metabolite malfunction. Thus, different abiotic stresses have distinctive and overlapping signals [10].

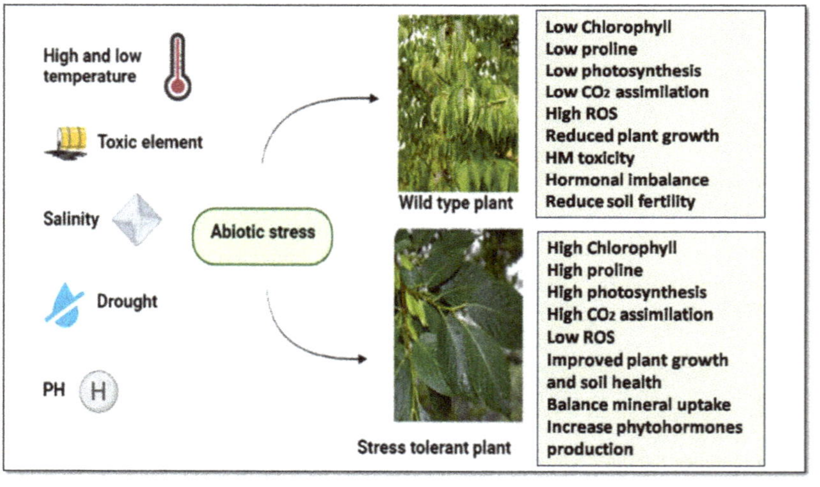

Figure 1. Plant behaviors during abiotic stress conditions.

Drought and salt stress affect water potential homeostasis and distribution of ions at cellular as well as molecular levels. Alterations in water and ion homeostasis can cause growth inhibition, molecular damage, and even death [10]. Primary stress signals trigger some cellular responses; however, the rest are triggered by secondary stress signals. One of the vital features of these signals is the hyperosmotic signal, which increases phytochrome and abscisic acid in plants and provides a protective role during different abiotic stresses such as drought and salt stress [10]. Plants facing cold or chilling stress first indicate a change in cell membrane structure that affects the plant development, then disrupts protein or protein complex stability and lowers the ROS scavenging enzyme activity. These mechanisms cause photo-inhibition, decreased photosynthesis, and considerable membrane damage [10–12]. In addition, stress triggers protein synthesis and gene expression, as it triggers RNA secondary structure formations [13]. All these components of plant activity are critical for stress tolerance for minimizing the internal damage in the new stress environment so that homeostatic conditions must be reestablished and growth must be restored, though at a slower rate [5].

Considerable achievements have been made towards the knowledge of plant cellular and molecular mechanisms under different stress conditions [14]. Recognition of a stress environment causes changes in the expression of genes, resulting in changes in the composition of the plant proteomes, transcriptomes, and metabolomes. Plants' response to different abiotic stress is not a simple pathway; however, it is an intricate integrated circuit comprising several pathways and precise tissue and cellular compartments and their interactions with additional cofactors, as well as signaling molecules for management of a particular response to a current stimulus. Thiourea (TU), a synthetic plant growth regulator with a composition of 36% nitrogen and 42% sulfur, has received much interest for its participation in plant stress tolerance. Thiourea helps to modulate some of the pathways associated with plants' resistance to abiotic stress. Understanding the processes during TU-induced tolerance may help improve crop production under stress situations [15].

E3-ubiquitin ligases control abiotic stress responses either positively or negatively. Additionally, the target protein and the outcome of the changes in UPS-mediated breakdown, activity regulation, or translocation depend on the involvement of ubiquitin ligases-E3 enzymes in plants' abiotic stress behavior. Consequently, recognizing and depicting ubiquitin ligases aims is vital during any stress response investigation [16]. The latest developments in understanding the molecular mechanisms underlying plant behaviors to abiotic stress highlight the complexity of these mechanisms, which involve a variety of processes including sensing, signal activation, transcription with transcript processing, and translation and posttranslational changes for combating the abiotic stress situations in plants (Table 1). The improved knowledge and use of various approaches, including genetic, chemical, and microbial techniques, enhance crop production and agricultural sustainability [17].

Table 1. Different studies concerning abiotic stress tolerance in plants.

Abiotic Stress Type	Mechanism and Key Parameters Studied	Plants/Crops	References
Cold stress	The CBF (C repeat binding factor) transcriptional cascade, CBF expression and CBF-independent regulons mediate the transcriptional regulation and pre-mRNA processing, export, and degradation involved in post-regulatory mechanisms	Arabidopsis	[18]
Low-temperature stress	Alter hormonal expression		[19]
Heat and drought stress	Enhancing the accumulation of carbohydrates		[20]
Heat stress	Autophagy plays a vital role in cellular homeostasis, metabolism, and other processes		[21]
Heat stress	ceRNA networks are mediated by the differentially expressed circRNA, by the influence of various important genes, and participate in response to hydrogen peroxide, heat stress, and phytochrome signaling pathway		[22]
Water stress	Lipid peroxidation decreases with scavenging reactive oxygen species and higher excitation energy dissolution due to photochemical quenching with reduced excitation pressure		[23]
Drought stress	The physiological activities and antioxidant protective systems modulate CarMT gene overexpression		[24]
Drought stress	H_2S endogenous production rate increases and a noteworthy transcriptional reorganization of pertinent miRNAs		[25]
Drought stress	A transmembrane potassium ions efflux as well as calcium and chloride ions influxes are induced due to endogenous hydrogen sulfides		[26]

Table 1. *Cont.*

Abiotic Stress Type	Mechanism and Key Parameters Studied	Plants/Crops	References
Cold and drought stress	Dehydrins concentrated in roots and stems	Blueberry	[27]
Heat stress	Lower accumulation of H_2O_2 and damage to cells	Strawberry	[28,29]
Salinity stress	Plant response is positively regulated due to OsH1RP1-ring finger protein 1	Maize	[30]
Cold stress	Changes in DNA methylation	Rice	[31]
Heat stress	Lipid peroxidation as well as antioxidant enzymes in roots and leaves		[32]
Drought stress	E3-ubiquitin breakdown		[16,33]
Heat stress	Candidate genes as well as quantitative trait loci		[34]
Cold stress	Linear electron transport chain is downregulated and PSII is repressed, as represented by the lowering in the PSII photochemistry efficiency along with electron transport efficiency	Hibiscus	[35]
Water deficit and heat stress	Contribution of ferredoxin-mediated cyclic pathway and chlororespiration		[36]
Salinity stress	Simultaneous expression of variable expressed genes		[37]
Cold stress	Enhances epidermal cell density, stomatal density and index, width of xylem vessel and phloem tissue and sclerenchyma	Candyleaf	[38]
Salt stress	Accumulation of biomass, ions concentration in tissue and steviolglycosides		[39]
Drought stress	The use of steviol glycosides enhances the harvest index		[40]
Salinity and drought stress	Sodium chloride serves as an activator, and mannitol works for the downregulation of genes involved in the steviol glycosides synthesis pathways that alter the steviol glycosides production		[41]
Cold stress	Photosynthetic electron transport chain protection by the sub-cellular antioxidant system	Wheat	[42]
Heat and drought stress	Signaling of phytohormone and epigenetic control		[43,44]
Salinity stress	Maintenance of osmoprotectants, photosynthetic activity and sodium/potassium ions ratio		[45,46]
Cold stress	WRKY gene expression	Grapevine	[47]
Heat stress	HSPs genes, along with antioxidant enzyme expression	Tomato	[48]
Salinity stress	Reduced the accumulation of different ions such as sodium, magnesium, and zinc in leaves and roots	Commonbean	[49]
Drought stress	Enhances plant metabolism with water relation parameters, antioxidant enzyme water relation parameters, activities of antioxidant enzymes and yield per plant increases	Lemon grass	[50,51]
Salinity stress	Antioxidants in leaves and lipid peroxidation	Tomato	[52]
Salinity stress	Biomass production as well as stomatal conductance		[53]
Drought stress	Sugar and amino acid content accumulation	Alfalfa	[54]

2. Abiotic Stresses and Crop Plants

Generally, various stresses act simultaneously, such as combined water, heat, salt, heavy metals, and other light stresses. As a result, these changes interfere with the regular plant metabolism function and the source–sink interaction, which lowers plant growth, metabolism, and production [55]. Moreover, these stresses alter the expression sequence of several plant genes and have huge effects on crop production worldwide, reducing the usual yields of important crops such as wheat and rice [16,56].

2.1. Salt Stress

Soil salinity is a huge risk for agriculture in areas where water shortage and poor drainage systems of irrigated farms lower the productivity of crops significantly [57]. According to the Food and Agricultural Organization (FAO 2016) [58], more than 6% of total global land and 19.5% of total irrigated land is already affected by salt conditions. Both human and natural factors can cause soil salinity. Of a total 932.2 Mha salt-affected soils globally, 76.6 Mha soil salinization has been caused by human beings [59]. The salt-affected lands having higher amounts of either exchangeable sodium or soluble salts, both perhaps due to insufficient leaching of cations that forms the base. The chief soluble salts that act as anions are sulfate ($SO_2{}^{-4}SO_4{}^{-}$), carbonate ($CO_2{}^{-3}CO_3{}^{-}$), chloride (Cl^{-}) and nitrate ($NO^{-3}NO_3{}^{-}$) salts and the cations are potassium, calcium, magnesium, and sodium. Many metals, such as selenium, lithium, boron, strontium, silica, fluorine, rubidium, manganese, aluminum, and molybdenum are present in hyper-saline soil water, some of which can be harmful to human, animal, and plant health [6,57].

Salt accumulation slows down the growth of plants and lowers plants' absorption capacity for nutrients and water, a consequence of osmotic stress. The level of resistance to salt stress changes from species to species. Cereal crops such as barley can resist up to 250 mM NaCl; moderately salt-resistant crops are bread wheat, maize, sorghum, while wheat is less resistant to salinity [60]. Plant development is reduced by subsequent salt exposure during two phases: ionic toxicity and osmotic stress [61].

2.2. Drought Stress

Rainfall distribution is unequal due to climate change, which is responsible for the major stress reported: drought, which is the most prevalent abiotic stress globally, reducing grain output drastically. It has a devastating impact on the ability to fulfill the food requirements of the increasing worldwide population. Drought stress is linked to a lack of water and cellular dehydration. A decrease in water potential, stomatal closure, and turgor pressure results in poor plant growth and development [62]. Low-water stress affects biochemical and physiological functions, including ion acquisition and chlorophyll synthesis, photosynthesis, respiration, and carbohydrate and nutrient metabolism, resulting in reduced plant growth [63]. Plant adaptation to low-water stress is a process that involves physiological and biochemical alterations in the plant system. Under drought conditions, plants limit their shoot development and metabolic demands. In maize, yield reduction is observed up to 40%, and in wheat, 21%, with around 40% water shortage or reduction [64]. A yield decline ranging from 34 to 68% was reported in cowpea during drought stress [65].

2.3. Cold Stress

This stress has been identified as the main abiotic stress, which reduces agricultural crop productivity by decreasing crop quality and postharvest life. Cold stress consists of chilling from 0–15 °C and freezing at 0 °C, negatively affecting plant growth and agriculture production [66,67]. During the comparison of both freezing and chilling stress, it has been found that freezing stress is far more harmful to plants. Typically, freezing's harmful effects start with the formation of a nucleation of ice within the cells, then progressively grows and forms ice crystals, causing water leakage and cell dehydration [68,69]. Several major crops are still unable to cope with cold acclimation. For example, tomato (*Solanum lycopersicum*), maize (*Zea mays*), rice (*Oryza sativa*), cotton (*Gossypium hirusutum*), and soybean (*Glycine max*) cannot withstand lower temperatures and have the capacity to grow and survive only in tropical and subtropical areas [70]. Hence, cold stress has a negative impact on plant growth along with plant metabolism and development, resulting in the reduction of crop yields globally [69,71].

2.4. Heat Stress

During heat stress, plants are highly sensitive; at extremely high temperatures, plants die. Generally, plants perform better at their optimum temperature; below and above the optimum temperature plant growth and development are severely affected [72]. Most biochemical and enzymatic reactions double at temperatures from 20 °C to 30 °C and change with every 10 °C. Temperatures above and below the optimum range reduce the reaction rate because enzymes are denatured and inactivated progressively. One or two degrees of temperature change have a huge impact on plant growth and development, particularly reproduction, causing a negative impact on early stages of male gametophyte formation in various crops, including wheat, rice, sorghum, barley, maize, and chickpea [5,73]. Heat stress causes male sterility and spikelet production abnormalities in rice and wheat [5]. In wheat and rice, cold and heat stress results in tapetum breakdown, and changes in the callose walls of microspores, exine formation and metabolism of carbohydrate, ultimately ensuing in male sterility [5,74]. However, temperature stress shows no adverse effects on female gametophyte development [75].

2.5. Heavy Metals Stress

Metal poisoning is one of the main environmental risks that impairs plants' ability to operate and engage in normal metabolic activities. Heavy metals (HMs) are a group of non-biodegradable, persistent inorganic substances having an atomic weight of more than 20 and a density of more than 5 g cm^{-3}, which affect and pollute the food chains, irrigation, soils, aquifers, and nearby atmosphere before having mutagenic, cytotoxic, and genotoxic effects on human, plant, and animal health [76,77].

Toxic metals are accumulated in the agricultural soils due to excessive use of chemical fertilizers along with increasing industrialization, showing harmful consequences to the soil–plant interaction system [78]. These metals concentrate and enter the plant system at a slow rate via water and air and enter the food chains over a certain time [79]. This poses a considerable threat to the natural food web and biogeochemical cycle [80]. The unprecedented in vivo heavy metals accumulation and bioaccumulation in the environment presents a dilemma for all plants and organisms. Toxic concentrations of HM can interact with various important cellular molecules, including nucleoproteins and DNA, causing excessive production of ROS [6,81]. This will result in serious plant changes, e.g., proteolysis, shoot chlorosis, and lipid peroxidation [80,82]. Under abiotic stress, it was thought that using osmolytes, nanoparticles, mineral nutrients, hydrogels, antioxidants, protectants, potassium, and plant growth hormones such as uniconazole and salicylic acid would boost plant production [83,84]. Additionally, plants may adapt to the negative impacts of droughts by applying plant hormones such as brassinolide (BR), gibberellic acid (GA), auxins, ABA, cytokinins, JA and ethylene, which govern several beneficial reactions in plants [83,84].

3. Sensing and Responding Mechanisms of Plants during Abiotic Stress Conditions

Biological molecules that act as sensors detect undesirable environmental changes and elicit quick stimuli to abiotic stress by signal molecules that activate the system. Abiotic stress causes additional Ca^{2+} to enter the cytoplasm from apoplastic sources. Ca^{2+} entrance passages are one kind of sensor for detecting stress signals [85–88]. The Ca^{+}, nitric oxide, and reactive oxygen species (ROS) also work as messenger substances that activate plant responses during cold stress. Reactive oxygen species such as hydrogen peroxide, hydroxyl radicals, and superoxides are formed in response to various kinds of stress [89]. Receptor-like kinases have an intracellular and extracellular domain at which ligand binding occurs and protein-with-protein binding will occur. When a sensor protein attaches to the extracellular domain, the histidine residues found in the intracellular domain self-phosphorylate and are activated. The activated ligand or sensor proteins may then induce signal-specific cellular response via a MAPKs cascade. Intracellular signaling, i.e., protein dephosphorylation and phosphorylation, regulate a broad range of cellular functions: enzyme activation, macromolecule assemble, protein localization, and degradation [90,91]. When plants detect abiotic stress, signaling cascades are activated, followed by activation of kinase cascades, the assembly of ROS and plant hormones accumulation, leading to the induction of precise set of genes responsible for combating the plant abiotic stress, as shown in Figure 1.

The metabolism of cytokinins, ABA, and ethylene are all impacted by stress, and these molecules then interact with certain kinases to control various biological functions, from controlling shoot development under stress to stomata movement [92]. Abiotic stress in plants, e.g., low water conditions (drought stress), low temperature, and excessive salinity stimulate the expression of the huge range of genes present in plants. With gene expression, different proteins are formed in different plant parts that prevent damage to the cell and activate the large number of genes essential for several abiotic resistance processes in plants. Various kinds of proteins are produced, such as chaperones and late embryogenesis abundant proteins (LEA proteins), which are primarily involved in the development of tolerance. At the same time, stress-associated genes are all concerned with generating the stress response [93]. Plant genes responsible for stress are regulated at three stages, i.e., transcriptional, posttranscriptional, and posttranslational.

3.1. Gene Regulation at the Transcriptional Level

Transcriptional regulation, consisting of chromatin modification and remodeling enhancers and promoters, has regular binding sites positioned downstream and upstream of the coding area known as cis-regulatory and trans-regulatory elements, typically transcription factors. Various abiotic stresses cause alterations in the methylation pattern of histone proteins and DNA, which repress or promote gene transcription. Promoters are unique sequences that have a regulatory role, binding RNA polymerase and other transcription factors to initiate transcription [94]. C-repeat binding factors (CBF), dehydration responsive element binding (DREB), MYB, zinc finger families, and leucine zipper (bZIP) are examples of the regulatory elements concerned in plant defense systems along with genes responsible for stress during binding of the responsive gene promoter's cis-element [95]. Drought tolerance was improved by *Oryza sativa* WRKY 11 (trans-regulatory element) overexpression under heat shock protein 101 (HSP 101) promoter control [96,97]. A significant discovery is new cis-acting elements, C-repeat and DRE, which respond to low temperature, drought (low water stress), and excess salt stress [98]. C-repeat binding factor proteins have been isolated progressively since their discovery by identification of DNA-associated proteins which attach the DRE and CRT motifs [99]. C-repeat binding factors 1–3 are cold-induced CBF genes found in *Arabidopsis* and are located on chromosome IV in tandem; CBF1-3 include Apetala2 or ethylene responsive type transcription factors, which directly bind with the CRT/DRE-conserved motifs present in the promoters of CBF regulons, also known as COR genes, and trigger gene expression during a low-temperature environment [99,100]. Transgenic *Arabidopsis* with CBF1 overexpression has more COR expression and is more resistant to freezing [101]. Orthologous expression of CBFs has been reported in various plants types such as tomato, wheat, rice, maize, and barley, along with heterologous expression of CBFs in *Arabidopsis*, which also improves their freezing tolerance mechanism [102], and suggests that it shows tolerance to cold only in those plants with CBF genes from tomato only. Hence, it is reported that CBF1-3 play an essential role in modulating cold tolerance and it is not more conserved in every species, but also it is species-specific [102].

3.2. Gene Regulation at Posttranscriptional Level

Posttranscriptional gene regulation refers to the regulation that happens between the stages of pre-mRNA and mRNA translation. It involves four levels: (A) pre-messenger RNA processing via capping as well as splicing along with polyadenylation; (B) nucleoplasmic trafficking of mRNA; (C) mRNA turnover and stability; (D) translocation of mRNA [103]. Alternative splicing is another strategy that plays an essential role in gene regulation during heat and cold stress, e.g., a gene stabilized 1 (STAT1) encoding a nuclear pre-mRNA provides cold resistance in *A. thaliana* [18,104]. Cold responsive (COR) gene regulation is essential for posttranscriptional regulation in plants. A dead box gene expression regulator (RCF1) plays an important role in the correct pre-mRNA splicing of various COR genes during low-temperature stress [105]. It has been investigated that alternate splicing pathways alter gene expression to cope with temperature stress [106]. Small RNA segments consisting of 20 to 25 nucleotides are formed from non-coding dsRNA precursors via dicer-like (DCL) RNases, generating several posttranscriptional gene silencing processes. One of these processes, mediated by 21 nucleotide microRNAs, cleaves mRNAs or blocks their translation [107].

3.3. Gene Regulation at the Posttranslational Level

Protein phosphorylation, ubiquitination, and sumoylation are the posttranscriptional activities that play an essential role in modifying plant behavior to many abiotic stresses. SNF1-related protein kinase (SnRKs) and MAPKs known as mitogen-activated protein kinases are key players in various signal transduction cascades initiated by osmotic stress and dehydration via phosphorylation of particular residues [108]. They include XERICO and SnRK2 gene codes for an H2-type zinc finger, and E3-ubiquitin ligases, which are associated in ABA-dependent response during water stress, such as stomata closure [109].

Posttranslational histone modifications and DNA methylation are linked to gene expression changes in response to chilling or cold stress. Histone protein acetylation and deacetylation are due to histone acetyl transferase (HAT) and histone deacetylases (HDAs) being involved in the plant during cold stress [110]. HAD 6 of *Arabidopsis* is overexpressed due to low temperature and positively enhances the freezing resistance [111]. During low-temperature stress, HDAs come into sight directly with maize DREB1 gene activation and histone hyper-acetylation. It has been reported that histone acetylation of ZmDREB1Aas well as ZmCOR413 in maize and OsDREB1 gene in rice histone is activated by lower temperature [112,113]. In the case of *Arabidopsis*, the RNA-mediated methylation 4 (RMP4) protein was found to play an important role in RNA-directed DNA methylation by combining with RNA polymerase Pol II and Pol V [114]. During cold stress, gene RDM4 is essential for Pol II possession at CBF2 and CBF3 gene promoters to fight abiotic stress in plants [115].

4. Crucial Signal Transduction Mechanism for Abiotic Stress

Abiotic stresses, which induce signal transduction pathways, such as drought, cold, light, heat and salt, are classified into three types. The first one is MAPK modules, which are used in osmotic/oxidative stress signaling to create antioxidant substances, ROS scavenging enzymes, and osmolytes; the second is calcium-dependent signaling, which promotes the triggers of LEA-type genes; followed by calcium-dependent SOS (salt overlay sensitive), which maintains ion homeostasis, as shown in Figure 2.

4.1. Oxidative or Osmotic Stress Signaling in Plants

All stresses involving salt, heat, drought, oxidative, and cold stress cause the formation of ROS species and cause significant damage to plants [116]. A greater level of ROS functions as a signal and one of the preventive plant responses is the production of ROS scavengers. Osmotic stress triggers various protein kinases, such as MAPKs, in restoring osmotic homeostasis. As a result, osmotic stress activates sensor or receptor proteins such as G protein-coupled receptor proteins, tyrosine and histidine kinases, which activate the MAPK network and signaling cascade, and are associated with the production of more osmolytes, required during osmotic stress. The primary function of osmolytes in cell turgor pressure maintenance is to act as a driving force for water uptake. Well-suited solutes such as proline, glycine betaine, mannitol, and trehalose will serve as ROS and chemical chaperones by stabilizing membrane proteins [117].

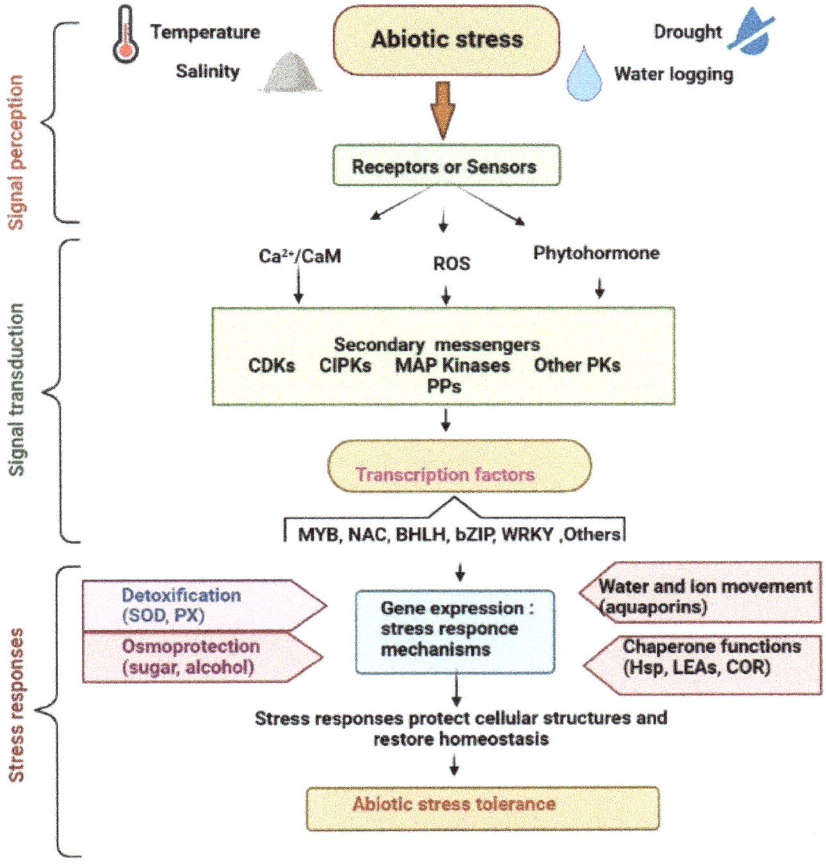

Figure 2. Signaling mechanisms for abiotic stress tolerance in plants.

The MAP kinase network modulates and transmits the signals from the cell surface to the nucleus. Three kinases are triggered consecutively by the upstream kinase in the core MAPK cascades. The MAP kinase phosphorylates on threonine and serine residues after the activation of MAPKKK, and once activated the MAPK either transfers to the nucleus directly and triggers the transcription factor, stimulating other signal substances to regulate gene expression to cope with stress [118].

4.2. Ca^{2+}-Dependent Activation of LEA Genes

Abiotic stress increases with calcium ions' entry into the cytoplasm. Calcium ion entry channels serve as sensors for abiotic stress detection. Calcium ions trigger CDPKs (calcium-dependent protein kinases) and are represented by multigene families; their expression stages are modulated both geographically and temporarily during complete development. The CDPK pathway plays an important role in the production of large numbers of anti-desiccation proteins via the activation of LEA-type genes, indicating the damage and repair processes, which are distinct from the pathways that regulate osmolyte synthesis [119,120]. Seeds are naturally desiccated during maturation to minimize desiccation shock at the time of germination by accumulating high levels of LEA proteins [121]. Water deficiency, low temperature, excessive osmolarity, and low-temperature stress cause crop plants to accumulate LEA protein. These proteins are utilized to protect proteins from denaturation or renaturation, maintain membrane integrity and protein structure, and sequester ions in affected tissues. Many scientific publications state that chaperones and LEA proteins protect macromolecules against dehydration, such as lipids, enzymes, and mRNA [122,123]. LEA proteins specialize in membrane desiccation defense, while antioxidant enzymes and compounds have a role in desiccation tolerance. Both LEA proteins and osmolytes work in association with membrane structure and protein stabilization by conferring favored hydration during moderate desiccation conditions and changing water levels to protect against abiotic stress in plants [124].

4.3. Calcium Ion-Dependent SOS Signaling

Plants response to high salt concentration, change the ion transporter channel, helps in the regulation of ion homeostasis during salinity. Higher intracellular or extracellular sodium ions function during the SOS pathway, primarily increasing a cytoplasmic calcium ions signal, which affects the expression and activity of different ion transporters, including K^+, sodium ions, and H^+. A shift in turgor is used as the input for osmotic stress signaling. Salt stress signaling pathways including osmotic and ionic homeostasis signaling routes, detoxification pathways, and growth maintenance pathways are reported in *Arabidopsis* [125,126]. During abiotic stress signaling, evidence shows that CDPKs and SOS3 of calcium ion sensors have a main role in coupling an inorganic signal with a specified protein phosphorylation pathway and seem important for plant salinity resistance [127].

5. Functions of the Microbiome in Abiotic Stress Management

Plants are not isolated entities and are not able to live alone. Instead, they coexist with various microbes, including bacteria, fungi, protists, viruses, and other microorganisms [128]. These microorganisms coexist in various plant tissues, forming the plant's microbiome, which lives in three different places: the phyllosphere, endosphere, and rhizosphere [127]. It is becoming increasingly proven that mycorrhizal fungi and useful soil bacteria, including PGPB and PGPR, have a significant role in sustainable farming and agriculture by stimulating plant growth and increasing plant tolerance to abiotic stresses [127]. Many microbes are essential in plant development, metabolism, and growth under abiotic stress situations (Table 2; Figure 3).

5.1. Arbuscular Mycorrhizal Fungi (AMF)

The development and health of plants may be impacted by AMF, which are symbiotic soil-borne fungi [128]. Agroecosystem services related to AMF still have significant knowledge gaps that need to be filled to be optimized, even though they are currently thought to have significant potential stability and sustainability in the agriculture system, and huge progress has been reported in the understanding of the arbuscular mycorrhizal symbiosis.

It has been found that AMF plays an essential role during abiotic stress, when various stresses are combined [129]. Recently, the role of AMF for protection aligned with abiotic stress in tomatoes has been reported [130]. In conjunction with halophytes and glycophytes, AMF can protect against salt stress [131]. A researcher [132] demonstrated a contrary higher dependence of glycophytes versus AMF compared to halophytes during salt stress. The advantages of AMF under salt stress were the subject of meta-analyses of various genes required for stress tolerance [130,133] (Auge et al. 2014, Chandrasekaran et al. 2021). The osmolyte, carbohydrate, and antioxidant systems can all be improved by AMF [134]. A high K^+/Na^+ level is maintained when they are in symbiosis with plants, preventing the absorption or transfer of harmful Na^+ [135].

AMF can also affect how efficiently carbon is used by maintaining larger grain yields, faster rates of stomatal conductance, net photosynthesis, and poorer internal water use efficiency during salinity stress [136]. AMF-associated mechanisms have recently been discussed in [137], ranging from the uptake of nutrients to increased water use efficiency, from osmoprotectant and ionic homeostasis to enhanced photosynthetic efficiency, from cell structure protection to strengthening functions along with triggering antioxidant metabolism, till phytochrome profile modulation. AM-colonized plants show a modification in plant metabolism, specifically, an enhancement in proline amount with greater H_2O_2 and isoprene emission in contrast to not-inoculated plants [135]. In the context of water stress conditions, AMF also has shown a beneficial outcome in tomato plants; however, the effects varied depending on the features and fungi species taken into account [138,139]. When three AMFs from various genera were examined to see how they affected tomato resistance to salt or drought stress, it became clear that all studied AMFs shared certain responses; however, others were unique to individual isolates [140].

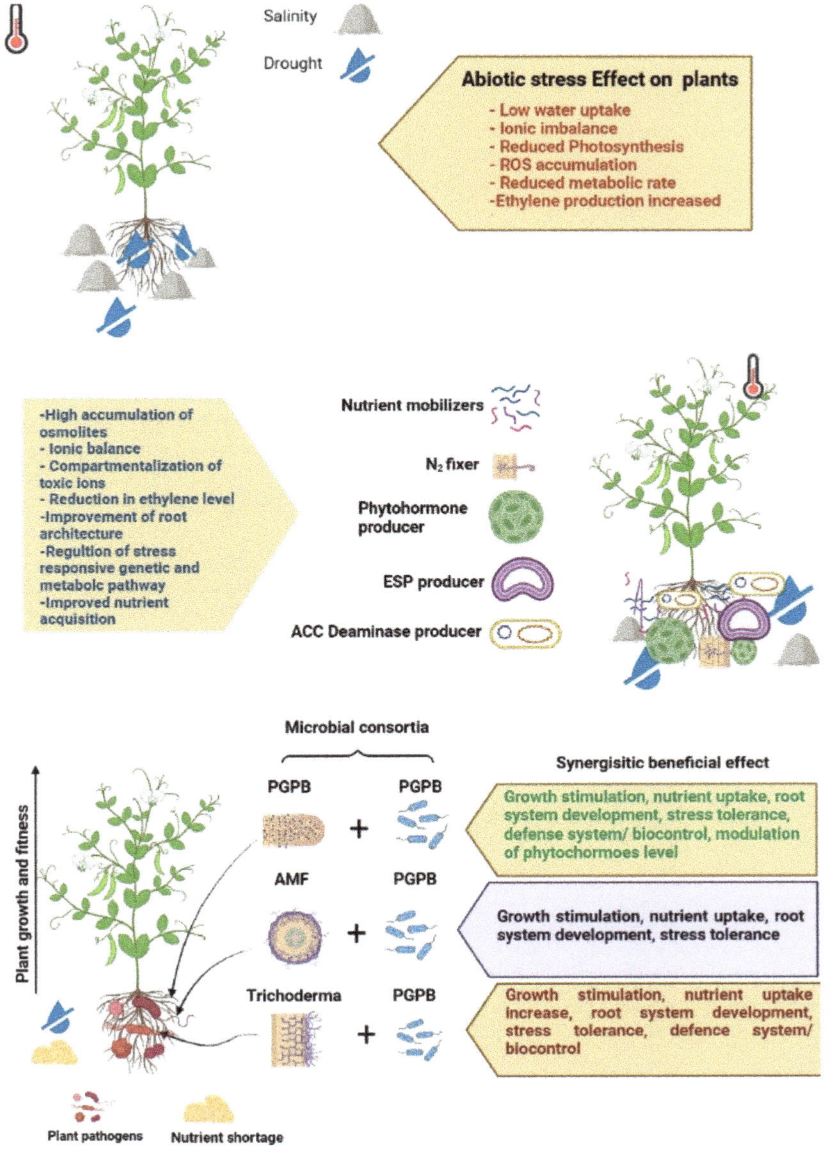

Figure 3. Functions of different microbes in combating abiotic stress conditions.

Table 2. Microbiome-mediated abiotic stress resistance and mechanism in plants.

Type of Microbes	Abiotic Stress Type	Plant and Tolerance Mechanism	References
Pseudomonas putida P45	Drought	Sunflower (*Helianthus annuus*) showed EPS production and enhanced soil aggregation	[141]
Pseudomonas	Drought	Pea (*Pisum sativum*) plants reduced the production of ethylene	[142]
Azospirillium sp.	Drought	Wheat (*Triticum aestivum*) has better water relations	[143]
AM fungi	Drought and salinity	Sorghum (*Sorghum bicolor*) showed better water relations	[144]
Scytonema	Coastal salinity	Rice (*Oryza sativa*) with extracellular products and gibberellic acid	[145]
Burkholderia phytofirmans	Cold	Grapevine (*Vitis vinifera*) with ACC-deaminase synthesis	[146]
Burkholderia sp. and *Methylobacterium oryzae*	Cd and Ni toxicity	Tomato (*Solanum lycopersicum*) with lower uptake along with translocation of heavy metals	[147]
Pseudomonas fluorescences	Salinity	Groundnut (*Arachis hypogaea*) with ACC-deaminase synthesis	[148]
Rhizobium tropici; P. polymyxa	Drought	Common bean (*Phaseolus vulgaris*) with change in hormonal composition and stomatal conductance	[149]
Glomus intraradices and *Pseudomonas mendocina*	Drought	Lettuce (*Lactuca sativa*) has antioxidant status improved	[150]
Pseudomonas strain AMK-P6	Heat	Sorghum (*Sorghum bicolor*) with better biochemical status due to activation of heat shock proteins	[151]
Glomus sp. and *Bacillus megaterium*	Drought	*Trifolium* (*Trifolium repens*) with proline and IAA production	[152]
Paraphaeosphaeria quadriseptata	Drought	*Arabidopsis* (*Arabidopsis thaliana*) has HSP-heat shock protein induction	[153]
Bacillus subtilis	Salinity	*Arabidopsis (Arabidopsis thaliana)* has reduced root Na^+ import by reduced transcriptional expression of AtHTK1 (a high-affinity KC transporter) genes	[153]
Pseudomonas putida	Salinity	Cotton (*Gossypium hirusutum)* stopped the salinity-associated accumulation of ABA in seedlings	[154]
Glomus etunicatum and *Glomus clarum*	Salinity	Wheat (*Triticum aestivum),* Chilli (*Capsicum annum*) and mung bean (*Vigna radiata)* have increased KC concentration in root and reduced NaC in shoots and root	[155]
PGPRs	Heat	Clover (*Trifolium repens*) plants with greater nitrogen fixation	[156]
Bacillus licheformis	Drought	*Capsicum annum* with expression and activation of stress-related proteins and genes	[157]
Bacillus thuringiensis	Drought	Wheat (*Triticum aestivum)* showed the organic compound production	[158]
Pantoea dispersa and *Azospirillium brailense*	Salinity	*Capsicum annuum* has an increase in photosynthesis rates well as stomatal conductance	[159]
Burkholderia phytofirmans and *Enterobacter* sp.	Drought	Maize (*Zea mays*) showed an increased rate of shoot and root biomass	[160]
Pseudomonas koreensis strain	Salinity	Soyabean (*Glycine max*) has increase KC level and decreased Na^+ level	[161]
Enterobacter intermedius	Zn toxicity	White mustard (*Sinapis alba*) with ACC deaminase, IAA IAA, hydrocyanic acid, and solubilization of phosphate	[162]

Table 2. *Cont.*

Type of Microbes	Abiotic Stress Type	Plant and Tolerance Mechanism	References
Serratia sp. and *Bacillus cereus*	Drought	Cucumber *(Cucumis sativa)* showed the production of genes responsible for the synthesis of proline, an antioxidant enzyme, and monodehydroascorbate	[163] Wang et al.; 2012
Photobacterium spp.	Mercury toxicity	Common reed (*Phragmites australis*) showed activity of IAA and mercury reductase	[164]
Rhizobium leguminosarum and *Pseudominas brassicacerum*	Zinc toxicity	Mustard (*Brassica juncea*) with metal chelating molecules	[165]
Rhizobium	Salinity	Asian rice (*Oryza sativa*) with RAB 18 salt stress-associated gene expression	[166]
PB 50 strain of *B. megaterium*	Drought	Rice (*Oryza sativa*) showed better plant growth under osmotic stress, plants protected via stomatal closure with enhanced soluble sugar, carotenoid content and protein content	[167]
Bacillus albus and *Bacillus cereus*	Drought	Maize (*Zea mays*) seeds have a higher germination rate and increased seedling length with reduced toxic effects	[168]
Gluconacetobacter diazotrophics (Pal5)	Drought	Rice (*Oryza sativa* L.) shows gor, P5CR, BADH and cat genes expression with increase glycine betaine and proline content	[169]
Penicillium sp. and *Calcoaceticus*	Drought	Foxtail millet (*Setaria italica*) with increased glycine betaine and proline content, sugars and chlorophyll a and b with the decrease in lipid oxidation	[170]
Streptomyces pactum and *actinomyces*	Drought	Wheat (*T. aestivum*) reduces stress via an enhancement in sugar levels and antioxidant enzymes	[171]
B. amyloliquefaciens; Pseudomonas putida	Drought	Chick pea (*Cicer arietinum*) with better photosynthesis, chlorophyll content, biomass and osmolyte content	[172]
PGPR consortium	Salinit	Common bean (*Phaseolus vulgaris*) with available iron content in the soil increased	[173,174]
B. gladioli; P. aeruginosa	Cd toxicity	Tomato (*Solanum lycopersicum*) with higher expression of metal transporter genes	[175]
Mesorhizobium; Rhizobium	Salinity	Chickpea (*Cicer arietinum*) with enhanced nitrogen fixation	[176]
Bacillus megaterium	Osmotic	Maize (*Zea mays*) with higher expression of 2 ZmPIP isoforms in roots	[177]

Additionally, in drought circumstances, AMF has valuable effects on maize, promoting plant development and photosynthesis by considerably increasing mineral absorption, assimilation and chlorophyll content, compatible solute concentration, and triggering the antioxidant defensive system [129]. The structures and functions of the photosystem PSII and PSI are less likely to be harmed by water stress due to AMF, as has also been shown [178–180]. The latest research discovered that the stimulation of fungal genes' encoding for SOD scavenging enzymes and non-enzymatic defenses such as glutaredoxin, metallothioein, etc., had a protective function against ROS burst [181]. The contrary impacts of AMF on aquaporin expression and plant hormone levels rely on the type of fungus, aquaporin, and applied stress [182]. Since dryness and salt cause generalized stressors, it stands to reason that AMF, which enables plants to survive under high-salinity environments, makes plants resistant to drought [182]. Transcriptomics has helped understand how several crop species, including rice [182,183], tomatoes [184]), grapevines [185] (Balestrini et al. 2017), as well as wheat [186], regulate expression of fungal as well as plant genes in AM interactions.

Transcriptomic analysis has recently been used in tomato roots colonized by AM to confirm the contribution of the AM symbiosis to tomato plant response, nematode function, and water stress, and revealed unique information regarding the response to AM symbiosis. A function for arbuscular mycorrhizal colonization in initiating defensive behavior against root-knot nematodes was also suggested by alterations in the tomato gene expression associated with nematode attack during AM symbiosis [187].

5.2. Actinomycetes

Several studies have shown *Actinomycetes* to thrive in various stressful environments, including drought, high temperatures, and high salt. They significantly reduce the harmful effects of abiotic stresses while encouraging plant development [188]. Streptomyces play a synergistic role in wheat crops under severe abiotic conditions and promote plant development and their capacity to withstand such conditions [189]. Furthermore, because of their unusual shape, they mix with the soil particles in the rhizosphere and create a solid link with the plants. This approach assists the plant in strained soil and enables more effective utilization of water and nutrients in the rhizosphere. These bacteria increase their capacity to withstand abiotic stresses by an array of mechanisms, comprising changes in root and cell wall morphology, 1-ACC deaminase activity, and the ability to protect from oxidative damage with the production of proline and glycine betaine, which aid in osmoregulation processes [190]. It has been found that wheat seeds soaked with streptomyces inoculum showed greatly enhanced shoot length, root depth, and high root and shoot biomass, and appreciably boosted the germination of roots under high-saline conditions [191]. Tomato plants inoculated with *Streptomyces* sp. PGPA39, during salt stress, showed a substantial reduction in leaf proline concentration and enhancement in plant biomass compared to the non-treated plants [192].

Treating maize plants with Actinomycetes during low-water conditions produced plants with higher growth, survival rate, shoot and root dry weight, and chlorophyll content than untreated plants [193]. Furthermore, Hasegawa and colleagues [194] demonstrated that endophytic *Actinomycetes* II improves drought resistance in *Kalmia latifolia* L. (mountain laurel) by causing callose buildup and lignification in the cell wall. According to next-generation sequencing methods used for microbial communities' identification, it has been investigated that Actinobacteria have regularly been discovered as one of the bacteria most found in soils [195]. However, although actinomycetes have played an essential role in reducing abiotic plant stress, slight knowledge regarding the dynamics of their interactions with plants limits the potential for using these microorganisms in agriculture.

5.3. PGPR/Plant Growth-Promoting Bacteria

The plant rhizosphere is a highly delicate and dynamic ecosystem, with different housing types of microorganisms that play various roles in plant development and survival [196]. Numerous distinctive and illustrative PGPR bacteria, comprising the *Bacillus*, *Pseudomonas*, *Enterobacter*, *Agrobacterium*, *Klebsiella*, *Azotobacter*, *Ervinia*, *Bradyrhizobium*, *Burkholderia*, and *Serratia*, have been isolated and described [197,198]. Conversely, several variables, including plant growth stage, species, cultivation method, soil ecology, and testing settings may affect how effective PGPR treatment is at enhancing plant growth. Rhizobacteria that promote plant development can lessen the drought stress effects and increase yield in treated plants [199,199]. It has been reported that *Rhizobacterium* induces drought endurance and resilience, known as RIDER, which results in biochemical alterations, and allows PGPR to decrease the drought stress effects [200,201]. Phytohormones development, bacterial exopolysaccharides formation, cyclic metabolic pathway conventions, and optimization of the antioxidant defense system are associated with the deposition of various carbon-based substances, including amino acids; polyamines, sugars, and heat shock protein production are examples of RIDER mechanisms [202,203].

Recently, it has been investigated that PGPB, along with other plant microbiomes, acts as a biological technique for decreasing plant salt stress [173]. *Rhizobacteria* that encourage plant development can successfully lessen the effects of drought on *Zea mays* and wheat grass. Two bacterial strains, namely *Enterobacter* sp. 16i and *Bacillus* sp. *12D6*, were employed to counteract drought stress in plants and it was reported that *Bacillus* sp. *12D6* performed better under drought stress [204]. *Azospirillum* (GQ255950)-treated maize plants perform better and have more root and shoot fresh weight biomass along with proline, soluble sugars, amino acids, and osmotic levels compared to non-treated maize plants [205]. A significant biotic stress element that reduces agricultural output is drought. In prior research, *Pseudomonas libanensis* EU-LWNA-33 was used to treat wheat plants, and measurements of growth parameters showed that the root length and biomass were enhanced after treatment. Additionally, biochemical analysis revealed that at 75% stress, levels of proline increased up to two times while levels of glycine betaine were enhanced 1.2-fold [206].

Plant survival under various types of stress is challenging and depends specifically on the plant's root microbiome. It has been demonstrated that bacteria isolated from harsh environmental circumstances contain traits that make them resistant to salt stress. *P. fluorescens* was isolated from Saharan rhizosphere soil and showed a PGPB property in maize under salt stress. Numerous microorganisms living on plants with high salt content exhibit adaptations to salinity stress and do well in these environments [207]. One study was conducted in a greenhouse where the soil was treated with three isolated bacteria from rice fields. The soil showed how drought stress might change microbial interactions, and soil microbial interactions with plants were changed with water limitation, e.g., *Actinobacteria* and *Chloroflexi* were abundant in the changing patterns, while *Acidobacteria* and *Deltaproteobacteria* were lost. Plant survival under drought circumstances results from compartment-specific reorganization [208].

Several methods of PGPB might respond to drought stress effects and promote phytohormones as well as solute production, chlorophyll synthesis, and activation of 1-aminocyclopropane-1-carboxylate deaminase, exo-polysaccharides, and mineral solubilization [209]. *Bacillus amyloliquefaciens* GB03 and *Pseudomonas flurescencs* WCS417R- treated plants showed that activating the antioxidant defense system considerably decreased the effects of drought stress [210].

Rhizobacteria that encourage plant development, including *Enterobacter*, *Bacillus*, *Pseudomonas* and *Moraxella*, have been identified. Under drought stress, wheat plants treated with *Bacillus* species showed enhanced auxin production up to 25.9 g mL^{-1} along with an increase in field capability of up to 10% as well as increased crop output of 34% [211]. In Uttar Pradesh, India, PGPR strains including *Pseudomonas putida* and *B. amyloliquefaciens* have been identified in alkaline soils, and the combined impact was studied in chickpea plants, which produced plants with boosted biomass, photosynthesis, osmolyte, and chlorophyll content, and reduced abiotic stress, as well as being eco-friendly to the environment [172]. Wheat seedlings treated with *Azospirillum brasilense* Sp245 had enhanced water control, greater mineral content, higher amounts of K, Mg, and Ca, and a 12.4% overall improvement in grain yield when compared to untreated plants [143].

Foxtail millet crops were used in an experiment, treated with *Penicillium* and *Acinetobacter*, which revealed increased physiological growth and a reduction in the harmful effects of drought stress [170]. During this study, buildup of proline, osmolytes, and glycine betaine with elevated levels of chlorophyll content was reported that helped reduce drought stress [170].

6. Nanoparticles' Application in Combating Abiotic Stress in Plants

Numerous plant responses are triggered by abiotic stress, ranging from growth and morphological changes to crop output and yield [212]. Nanotechnology is one of the newest and most promising methods for treating abiotic stress conditions in plants, such as HMs stress, heat stress, drought stress, and salinity stress. It is also an environmentally friendly technique. With the increasing cellular antioxidants, nutrient uptake, photosynthetic efficiency, and molecular as well as biochemical pathways, NPs dramatically increase the ability of plants to withstand abiotic stress [213]. Modernizing the agricultural sector with prospective applications for improving plant growth as well as development under stress conditions, nanotechnology has recently made strides [213].

Several studies have been reported on the possible use of NPs in the remediation of HM-contaminated soil [214–216]. A study of FeO NPs found they can relieve drought stress [217,218] and reduce Cd toxicity in wheat plants by enhancing biomass, chlorophyll levels, and antioxidant enzymes [216]. Si NPs have been reported to reduce HM-induced phytotoxicity in wheat as well as rice and pea [219,220]. To lessen the detrimental effects of HMs on plant growth and development, new nanoremediation techniques must be developed.

It has been found that Si NP treatment enhances plants' ability to withstand drought stress [221]. By modifying the morpho-physiological characteristics of barley plants, Si NPs showed promising drought stress recovery [222]. It has been revealed that in saline- and water-deficient circumstances, Si NPs improved cucumber growth and yield [223]. Wheat plants under drought stress exhibited higher relative water content, CAT, SOD activity, yield, and biomass after application of chitosan [224]. Silver nanoparticles were used to lessen the harmful effects of drought in lentil (*Lens culinaris* Medic.) plants [225]. Abscisic acid administration aided by Si NPs was described as an efficient management technique to increase drought resistance in *Arabidopsis thaliana* [226].

Wheat plants' growth, chlorophyll concentrations, and antioxidant enzymes were improved by treatment with FeO NPs, which reduced the effects of salt stress [216]. It has been found that manganese NP seed-priming controls salt stress by altering the molecular reactions in pepper (*C. annuum* L.) plants [227]. To better understand how NPs work to increase plants' resistance to salinity and other abiotic stress in plants, more molecular and physiological research is required. Numerous studies have demonstrated the possible use of NPs to increase the ability of plants to withstand heat stress [228,229]. Recently it has been reported that the use of organically produced Se NPs (100 g/mL) enhanced wheat development by enhancing plant tolerance to heat stress [230]. Researchers have discovered that silver nanoparticles considerably improved the morphological characteristics of wheat plants under heat stress conditions [231]. Overall, using metallic NPs as nanofertilizers can help plants be more resilient to heat stress, which is important for sustainable agriculture.

7. Conclusions

Evidence from various studies regarding molecular, biochemical, and physio-morphological characteristics, plant responses to diverse types of abiotic stresses, and the microbe interaction mitigation methods in plants have been examined. These studies have improved our knowledge of the processes behind gene cascades, microbial interactions, and metabolic pathways, along with augmentation of different proteins, enzymes, metabolites, and the down- and upregulation of numerous genes. This paper provides dynamic information about plants' response to different types of abiotic stress. This paper gives innovative ideas for improving the current methods to mitigate abiotic stress by employing various gene regulation-based and microbial-mediated plant interactions, facilitating improved germination, increased ability to withstand and mitigate adverse environment conditions, and better yield in plants. By supplying vital nutrients, nanotechnology can be a cost effective and promising way to increase plants' ability to withstand abiotic stress. However, because of their widespread use, there are some possible worries regarding their adverse impacts on the environment.

Plants have evolved built-in adaptation mechanisms to deal with various complex abiotic challenges. With the aid of science and technological advancements, it is now feasible to comprehend gene function, establish gene-manipulation strategies, and generate plant characteristics to combat abiotic stress. Signaling pathways must be seen as intricate networks. Hence, abiotic stress signal-transduction cascades are better understood by molecular investigations of the signaling molecules. Abiotic stresses have a detrimental effect on plants that is exacerbated by ongoing climate change, reducing agricultural production. Abiotic stress tolerance is a multigenic response that involves stress-responsive gene expression, signal transduction, and sensing. As a result of these coordinated efforts, agricultural production, productivity, and food security will improve.

8. Future Perspectives

Genetic modification and recombinant techniques should be integrated with traditional and marker-assisted breeding practices in the future to produce plants with modified characteristics that cope with adverse conditions. In addition to this, nanotechnology will be the most potent tool for combating the abiotic stresses in plants. Moreover, finding new adaptable germplasm is crucial for directing breeding programs to identify plants for a changing climate and future research should focus on developing NPs that are inexpensive, nontoxic, self-degradable, and eco-friendly using green methods. By addressing the effects of climate change, particularly stress caused by drought, heat and cold, these combined initiatives will significantly advance the cause of food security through increased agricultural output and productivity.

Author Contributions: Conceptualization, B.S.S. and B.B.; validation, J.S.D., V.D.R., R.K. and S.M.; formal analysis, B.S.S.; data curation, B.B. and S.M.; writing—original draft preparation, B.B., V.D.R., J.S.D. and T.M.; writing—review and editing, J.S.D., R.K.; visualization, S.M.; supervision, B.S.S.; project administration B.S.S. and B.B. equally contributed to MS and are considered both first authors. All authors have read and agreed to the published version of the manuscript.

Funding: This research received no external funding.

Institutional Review Board Statement: Not applicable.

Informed Consent Statement: Not applicable.

Data Availability Statement: Not applicable.

Acknowledgments: Authors are highly grateful to the Department of Microbiology, CCS HAU, Hisar, for providing infrastructural support for preparing the manuscript. V.D.R. and T.M. acknowledge support from the laboratory «Soil Health» of the Southern Federal University with the financial support of the Ministry of Science and Higher Education of the Russian Federation, agreement No. 075-15-2022-1122.

Conflicts of Interest: The authors declare no conflict of interest.

References

1. Umar, O.B.; Ranti, L.A.; Abdulbaki, A.S.; Bola, A.L.; Abdulhamid, A.K.; Biola, M.R.; Victor, K.O. Stresses in Plants: Biotic and Abiotic. In *Current Trends in Wheat Research*; Ansari, M., Ed.; IntechOpen: London, UK, 2021.
2. Wania, S.H.; Kumar, V.; Shriram, V.; Sah, S.K. Phytohormones and their metabolic engineering for abiotic stress tolerance in crop plants. *Crop J.* **2016**, *4*, 162–176. [CrossRef]
3. Mei, H.; Cheng-Qiang, H.; Nai-Zheng, D. Abiotic Stresses: General defenses of land plants and chances for engineering multistress tolerance. *Front. Plant Sci.* **2018**, *7*, 1771. [CrossRef]
4. Jogaiah, S.; Govind, S.R.; Tran, L.S.P. Systems biology-based approaches toward understanding drought tolerance in food crops. *Crit. Rev. Biotechnol.* **2013**, *33*, 23–29. [CrossRef] [PubMed]
5. Fahad, S.; Bajwa, A.A.; Nazir, U.; Anjum, S.A.; Farooq, A.; Zohaib, A.; Sadia, S.; Nasim, W.; Adkins, S.; Saud, S.; et al. Crop production under drought and heat stress: Plant responses and management options. *Front. Plant Sci.* **2017**, *8*, 1147. [CrossRef]
6. Sihag, S.; Brar, B.; Joshi, U.N. Salicylic acid induces amelioration of chromium toxicity and affects antioxidant enzyme activity in *Sorghum bicolor* L. *Int. J. Phytoremediation* **2019**, *21*, 293–304. [CrossRef]
7. Fedoron, N.V.; Battisti, D.S.; Beachy, R.N.; Cooper, P.J.; Fischhon, D.A.; Hodges, C.N.; Knauf, V.C.; Lobell, D.; Mazur, B.J.; Molden, D.; et al. Radically rethinking agriculture for the 21st century. *Science* **2010**, *327*, 833–834. [CrossRef]
8. Kundzewicz, Z.; Ulbrich, U.; Brucher, T.; Graczyk, D.; Kruger, A.; Leckebusch, G.C. Summer floods in Central Europe—Climate change track? *Nat. Hazards* **2005**, *36*, 165–189. [CrossRef]
9. Cramer, G.R.; Urano, K.; Delrot, S.; Pezzotti, M.; Shinozaki, K. Effects of abiotic stress on plants: A systems biology perspective. *BMC Plant Biol.* **2011**, *11*, 163–176. [CrossRef]
10. Mathivanan, S. Abiotic Stress-Induced Molecular and Physiological Changes and Adaptive Mechanisms in Plants. In *Abiotic Stress in Plants*; Fahad, S., Saud, S., Chen, Y., Wu, C., Wang, D., Eds.; IntechOpen: London UK, 2021. [CrossRef]

11. Ruelland, E.; Vaultier, M.N.; Zachowski, A.; Hurry, V. Cold signalling and cold acclimation in plants. *Adv. Bot. Res.* **2009**, *49*, 35–150.
12. Siddiqui, K.S.; Cavicchioli, R. Cold-adapted enzymes. *Annu. Rev. Biochem.* **2006**, *75*, 403–433. [CrossRef]
13. Rajkowitsch, L.; Chen, D.; Stampfl, S.; Semrad, K.; Waldsich, C.; Mayer, O.; Jantsch, M.F.; Konrat, R.; Bläsi, U.; Schroeder, R. RNA chaperones, RNA annealers and RNA helicases. *RNA Biol.* **2007**, *4*, 118–130. [CrossRef]
14. Hadiarto, T.; Tran, L.S.P. Progress studies of drought-responsive genes in rice. *Plant Cell Rep.* **2011**, *30*, 297–310. [CrossRef]
15. Waqas, M.A.; Kaya, C.; Riaz, A.; Farooq, M.; Nawaz, I.; Wilkes, A.; Li, Y. Potential mechanisms of abiotic stress tolerance in crop plants induced by thiourea. *Front. Plant Sci.* **2019**, *10*, 1336. [CrossRef]
16. Melo, F.V.; Oliveira, M.M.; Saibo, N.J.; Lourenço, T.F. Modulation of abiotic stress responses in rice by E3-ubiquitin ligases: A promising way to develop stress-tolerant crops. *Front. Plant Sci.* **2021**, *12*, 368. [CrossRef]
17. Zhang, H.; Zhu, J.; Gong, Z.; Zhu, J.K. Abiotic stress responses in plants. *Nat. Rev. Genet.* **2022**, *23*, 104–119. [CrossRef]
18. Hwarari, D.; Guan, Y.; Ahmad, B.; Movahedi, A.; Min, T.; Hao, Z.; Lu, Y.; Chen, J.; Yang, L. ICE-CBF-COR signaling cascade and its regulation in plants responding to cold stress. *Int. J. Mol. Sci.* **2022**, *23*, 1549. [CrossRef]
19. Eremina, M.; Rozhon, W.; Poppenberger, B. Hormonal control of cold stress responses in plants. *Cell. Mol. Life Sci.* **2016**, *73*, 797–810. [CrossRef]
20. Rizhsky, L.; Liang, H.; Shuman, J.; Shulaev, V.; Davletova, S.; Mittler, R. When Defense Pathways Collide. The Response of *Arabidopsis* to a Combination of Drought and Heat Stress. *Plant Physiol.* **2004**, *134*, 1683–1696. [CrossRef]
21. Zhang, Y.; Min, H.; Shi, C.; Xia, G.; Lai, Z. Transcriptome analysis of the role of autophagy in plant response to heat stress. *PLoS ONE* **2021**, *16*, e0247783. [CrossRef]
22. Pan, T.; Sun, X.; Liu, Y.; Li, H.; Deng, G.; Lin, H.; Wang, S. Heat stress alters genome-wide profiles of circular RNAs in *Arabidopsis*. *Plant Mol. Biol.* **2018**, *96*, 217–229. [CrossRef]
23. Sperdouli, I.; Moustakas, M. Spatio-temporal heterogeneity in *Arabidopsis thaliana* leaves under drought stress. *Plant Biol.* **2012**, *14*, 118–128. [CrossRef]
24. Dubey, A.K.; Kumar, N.; Kumar, A.; Ansari, M.A.; Ranjan, R.; Gautam, A.; Sahu, N.; Pandey, V.; Behera, S.K.; Mallick, S.; et al. Over-expression of CarMT gene modulates the physiological performance and antioxidant defense system to provide tolerance against drought stress in *Arabidopsis thaliana* L. *Ecotoxicol. Environ. Saf.* **2019**, *171*, 54–65. [CrossRef]
25. Shen, J.; Xing, T.; Yuan, H.; Liu, Z.; Jin, Z.; Zhang, L.; Pei, Y. Hydrogen sulfide improves drought tolerance in *Arabidopsis thaliana* by microRNA expressions. *PLoS ONE* **2013**, *8*, e77047. [CrossRef]
26. Jin, Z.; Wang, Z.; Ma, Q.; Sun, L.; Zhang, L.; Liu, Z.; Liu, D.; Hao, X.; Pei, Y. Hydrogen sulfide mediates ion fluxes inducing stomatal closure in response to drought stress in *Arabidopsis thaliana*. *Plant Soil* **2017**, *419*, 141–152. [CrossRef]
27. Panta, G.; Rieger, M.; Rowland, L. Effect of cold and drought stress on blueberry dehydrin accumulation. *J. Hortic. Sci. Biotechnol.* **2001**, *76*, 549–556.
28. Ergin, S.; Kesici, M.; Gulen, H. Changes in H_2O_2 and peroxidase activities in strawberry plants under heat stress. *Harran TarımveGıdaBilimleri Derg.* **2012**, *16*, 25–35.
29. Gulen, H.; Eris, A. Effect of heat stress on peroxidase activity and total protein content in strawberry plants. *Plant Sci.* **2004**, *166*, 739–744. [CrossRef]
30. Kim, J.H.; Lim, S.D.; Jang, C.S. *Oryza sativa* heat-induced RING finger protein 1 (OsHIRP1) positively regulates plant response to heat stress. *Plant Mol. Biol.* **2019**, *99*, 545–559. [CrossRef]
31. Pan, Y.; Wang, W.; Zhao, X.; Zhu, L.; Fu, B.; Li, Z. DNA methylation alterations of rice in response to cold stress. *Plant Omics J.* **2011**, *4*, 364–369.
32. De Azevedo Neto, A.D.; Prisco, J.T.; Eneas-Filho, J.; de Abreu, C.E.B.; Gomes-Filho, E. Effect of salt stress on antioxidative enzymes and lipid peroxidation in leaves and roots of salt-tolerant and salt-sensitive maize genotypes. *Environ. Exp. Bot.* **2006**, *56*, 87–94. [CrossRef]
33. Fang, Y.; Xiong, L. General mechanisms of drought response and their application in drought resistance improvement in plants. *Cell. Mol. Life Sci.* **2015**, *72*, 673–689. [CrossRef] [PubMed]
34. Kilasi, N.L.; Singh, J.; Vallejos, C.E.; Ye, C.; Jagadish, S.K.; Kusolwa, P.; Rathinasabapathi, B. Heat stress tolerance in rice (*Oryza sativa* L.): Identification of quantitative trait loci and candidate genes for seedling growth under heat stress. *Front. Plant Sci.* **2018**, *9*, 1578. [CrossRef] [PubMed]
35. Paredes, M.; Quiles, M.J. The effects of cold stress on photosynthesis in *Hibiscus* plants. *PLoS ONE* **2015**, *10*, e0137472. [CrossRef] [PubMed]
36. Munoz, R.; Quiles, M.J. Water deficit and heat affect the tolerance to high illumination in hibiscus plants. *Int. J. Mol. Sci. Mar.* **2013**, *7*, 5432–5444. [CrossRef]
37. Ni, L.; Wang, Z.; Liu, X.; Wu, S.; Hua, J.; Yin, Y.; Li, H.; Gu, C. Transcriptome Analysis of Salt Stress in Hibiscus hamabo Sieb. et Zucc Based on Pacbio Full-Length Transcriptome Sequencing. *Int. J. Mol. Sci.* **2022**, *23*, 138. [CrossRef]
38. Hajihashemi, S.; Noedoost, F.; Geuns, J.; Djalovic, I.; Siddique, K.H. Effect of cold stress on photosynthetic traits, carbohydrates, morphology, and anatomy in nine cultivars of *Stevia rebaudiana*. *Front. Plant Sci.* **2018**, *9*, 1430. [CrossRef]
39. Debnath, M.; Ashwath, N.; Midmore, D.J. Physiological and morphological responses to abiotic stresses in two cultivars of *Stevia rebaudiana* (Bert.) Bertoni. *S. Afr. J. Bot.* **2019**, *123*, 124–132. [CrossRef]

40. Karimi, M.; Ahmadi, A.; Hashemi, J.; Abbasi, A.; Tavarini, S.; Pompeiano, A.; Angelini, L.G. The positive role of steviol glycosides in stevia (*Stevia rebaudianaBertoni*) under drought stress condition. *Plant Biosyst.* **2016**, *150*, 1323–1331. [CrossRef]
41. Pandey, M.; Chikara, S.K. Effect of Salinity and Drought Stress on Growth Parameters, Glycoside Content and Expression Level of Vital Genes in Steviol Glycosides Biosynthesis Pathway of *Stevia rebaudiana* (Bertoni). *Int. J. Genet.* **2015**, *7*, 153–160.
42. Li, X.; Cai, J.; Liu, F.; Dai, T.; Cao, W.; Jiang, D. Cold priming drives the sub-cellular antioxidant systems to protect photosynthetic electron transport against subsequent low temperature stress in winter wheat. *Plant Physiol. Biochem.* **2014**, *82*, 34–43. [CrossRef]
43. Ni, Z.; Li, H.; Zhao, Y.; Peng, H.; Hu, Z.; Xin, M.; Sun, Q. Genetic improvement of heat tolerance in wheat: Recent progress in understanding the underlying molecular mechanisms. *Crop J.* **2018**, *6*, 32–41. [CrossRef]
44. Schmidt, J.; Tricker, P.J.; Eckermann, P.; Kalambettu, P.; Garcia, M.; Fleury, D. Novel alleles for combined drought and heat stress tolerance in wheat. *Front. Plant Sci.* **2020**, *10*, 1800. [CrossRef]
45. Hussain, N.; Ghaffar, A.; Zafar, Z.U.; Javed, M.; Shah, K.H.; Noreen, S.; Manzoor, H.; Iqbal, M.; Hassan, I.F.Z.; Bano, H.; et al. Identification of novel source of salt tolerance in local bread wheat germplasm using morpho-physiological and biochemical attributes. *Sci. Rep.* **2021**, *11*, 10854. [CrossRef]
46. Omrani, S.; Arzani, A.; Esmaeilzadeh Moghaddam, M.; Mahlooji, M. Genetic analysis of salinity tolerance in wheat (*Triticum aestivum* L.). *PLoS ONE* **2022**, *17*, e0265520. [CrossRef]
47. Wang, L.; Zhu, W.; Fang, L.; Sun, X.; Su, L.; Liang, Z.; Wang, N.; Londo, J.P.; Li, S.; Xin, H. Genome-wide identification of WRKY family genes and their response to cold stress in *Vitis Vinifera*. *BMC Plant Biol.* **2014**, *14*, 103. [CrossRef]
48. Yu, W.; Wang, L.; Zhao, R.; Sheng, J.; Zhang, S.; Li, R.; Shen, L. Knockout of SlMAPK3 enhances tolerance to heat stress involving ROS homeostasis in tomato plants. *BMC Plant Biol.* **2019**, *19*, 1–13. [CrossRef]
49. Niron, H.; Barlas, N.; Salih, B.; Türet, M. Comparative Transcriptome, Metabolome, and Ionome Analysis of Two Contrasting Common Bean Genotypes in Saline Conditions. *Front. Plant Sci.* **2020**, *11*, 2007. [CrossRef]
50. Stoeva, N.; Kaymakanova, M. Effect of salt stress on the growth and photosynthesis rate of bean plants (*Phaseolus vulgaris* L.). *J. Cent. Eur. Agric.* **2008**, *9*, 385–391.
51. Singh, M.; Khan, M.M.A.; Uddin, M.; Naeem, M.; Qureshi, M.I. Proliferating effect of radiolytically depolymerized carrageenan on physiological attributes, plant water relation parameters, essential oil production and active constituents of *Cymbopogon flexuosus* Steud. under drought stress. *PLoS ONE* **2017**, *12*, e0180129. [CrossRef]
52. Shalata, A.; Tal, M. The effect of salt stress on lipid peroxidation and antioxidants in the leaf of the cultivated tomato and its wild salt-tolerant relative *Lycopersicon pennellii*. *Physiol. Plant.* **1998**, *104*, 169–174. [CrossRef]
53. Maggio, A.; Raimondi, G.; Martino, A.; De Pascale, S. Salt stress response in tomato beyond the salinity tolerance threshold. *Environ. Exp. Bot.* **2007**, *59*, 276–282. [CrossRef]
54. Schubert, S.; Serraj, R.; Plies-Balzer, E.; Mengel, K. Effect of drought stress on growth, sugar concentrations and amino acid accumulation in N2-fixing alfalfa (Medicago sativa). *J. Plant Physiol.* **1995**, *146*, 541–546. [CrossRef]
55. Khan, F.; Hakeem, K.R. Cell signaling during drought and salt stress. In *Plant Signaling: Understanding the Molecular Crosstalk*; Springer: New Delhi, India, 2014; pp. 227–239.
56. Abhinandan, K.; Skori, L.; Stanic, M.; Hickerson, N.M.; Jamshed, M.; Samuel, M.A. Abiotic stress signaling in wheat—An inclusive overview of hormonal interactions during abiotic stress responses in wheat. *Front. Plant Sci.* **2018**, *9*, 734. [CrossRef]
57. Kumar, P.; Sharma, P.K. Soil salinity and food security in India. *Front. Sustain. Food Syst.* **2020**, *4*, 533781. [CrossRef]
58. FAO. *The State of Food and Agriculture—Climate Change, Agriculture and Food Security*; Food and Agricultural Organization of the United Nations: Rome, Italy, 2016.
59. Shahid, S.A.; Zaman, M.; Heng, L. Soil salinity: Historical perspectives and a world overview of the problem. In *Guideline for Salinity Assessment, Mitigation and Adaptation Using Nuclear and Related Techniques*; Springer: Cham, Switzerland, 2018; pp. 43–53. [CrossRef]
60. Maas, E.V.; Honman, G.J. Crop salt tolerance—Current assessment. *J. Irrig. Drain. Div.* **1977**, *103*, 115–134. [CrossRef]
61. Yildiz, C.; Acikgoz, H. A kernel extreme learning machine-based neural network to forecast very short-term power output of an on-grid photovoltaic power plant. *Energy Sources Part A Recovery Util. Environ. Eff.* **2021**, *43*, 395–412. [CrossRef]
62. Farooq, M.; Wahid, A.; Kobayashi, N.; Fujita, D.; Basra, S.M.A. Plant drought stress: Effects, mechanisms and management. *Agron. Sustain. Dev.* **2009**, *29*, 185–212. [CrossRef]
63. Li, X.N.; Pu, H.C.; Liu, F.L.; Zhou, Q.; Cai, J.; Dai, T.B.; Cao, W.; Jiang, D. Winter wheat photosynthesis and grain yield responses to spring freeze. *Agron. J.* **2015**, *107*, 1002–1010. [CrossRef]
64. Daryanto, S.; Wang, L.; Jacinthe, P.A. Global synthesis of drought enacts on maize and wheat production. *PLoS ONE* **2016**, *11*, e0156362. [CrossRef]
65. Farooq, M.; Gogoi, N.; Barthakur, S.; Baroowa, B.; Bharadwaj, N.; Alghamdi, S.S. Drought stress in grain legumes during reproduction and grain filling. *J. Agron. Crop Sci.* **2017**, *203*, 81–102. [CrossRef]
66. Guo, X.Y.; Liu, D.F.; Chong, K. Cold signaling in plants: Insights into mechanisms and regulation. *J. Integr. Plant Biol.* **2018**, *60*, 45–756. [CrossRef] [PubMed]
67. Liu, H.; Yu, C.; Li, H.; Ouyang, B.; Wang, T.; Zhang, J.; Wang, X.; Ye, Z. Overexpression of ShDHN, a dehydrin gene from *Solanum habrochaites* enhances tolerance to multiple abiotic stresses in tomato. *Plant Sci.* **2015**, *231*, 198–211. [CrossRef] [PubMed]
68. Dowgert, M.F.; Steponkus, P.L. Behavior of the plasma membrane of isolated protoplasts during a freeze-thaw cycle. *Plant Physiol.* **1984**, *75*, 1139–1151. [CrossRef] [PubMed]

69. Pearce, R.S. Plant freezing and damage. *Ann. Bot.* **2001**, *87*, 417–424. [CrossRef]
70. Chinnusamy, V.; Zhu, J.; Zhu, J.K. Cold stress regulation of gene expression in plants. *Trends Plant Sci.* **2007**, *12*, 444–451. [CrossRef]
71. Zhang, Q.; Chen, Q.; Wang, S.; Hong, Y.; Wang, Z. Rice and cold stress: Methods for its evaluation and summary of cold tolerance-related quantitative trait loci. *Rice* **2014**, *7*, 1–12. [CrossRef]
72. Liliane, T.N.; Charles, M.S. Factors affecting yield of crops. In *Agronomy-Climate Change Food Security*; Amanullah, Ed.; Intech Open: London, UK, 2020. [CrossRef]
73. Liu, X.; Zhou, Y.; Xiao, J.; Bao, F. Effects of chilling on the structure, function and development of chloroplasts. *Front. Plant Sci.* **2018**, *22*, 1715. [CrossRef]
74. Mamun, E.A.; Alfred, S.; Cantrill, L.C.; Overall, R.L.; Sutton, B.G. Effects of chilling on male gametophyte development in rice. *Cell Biol. Int.* **2006**, *30*(7), 583–591. [CrossRef]
75. Fahad, S.; Chen, Y.; Saud, S.; Wang, K.; Xiong, D.; Chen, C.; Wu, C.; Shah, F.; Nie, L.; Huang, J. Ultraviolet radiation effect on photosynthetic pigments, biochemical attributes, antioxidant enzyme activity and hormonal contents of wheat. *J. Food Agric. Environ.* **2013**, *11*, 1635–1641.
76. Ali, H.; Khan, E.; Sajad, M.A. Phytoremediation of heavy metals-concepts and applications. *Chemosphere* **2013**, *91*, 869–881. [CrossRef]
77. Rascio, N.; Navari-Izzo, F. Heavy metal hyperaccumulating plants: How and why do they do it? And what makes them so interesting? *Plant Sci.* **2011**, *18*, 169–181. [CrossRef]
78. Nievola, C.C.; Carvalho, C.P.; Carvalho, V.; Rodrigues, E. Rapid response of plants to temperature changes. *Temperature* **2017**, *4*, 371–405. [CrossRef]
79. Rauwane, M.; Ntushelo, K. Understanding biotic stress and hormone signalling in cassava (*Manhot esculenta*): Potential for using hyphenated analytical techniques. *Appl. Sci.* **2020**, *20*, 8152. [CrossRef]
80. Haider, F.U.; Liqun, C.; Coulter, J.A.; Cheema, S.A.; Wu, J.; Zhang, R.; Farooq, M. Cadmium toxicity in plants: Impacts and remediation strategies. *Ecotoxicol. Environ. Saf.* **2021**, *211*, 111887. [CrossRef]
81. Emamverdian, A.; Ding, Y.; Mokhberdoran, F.; Xie, Y. Heavy metal stress and some mechanisms of plant defense response. *Sci. World J.* **2015**, *2015*, 756120. [CrossRef]
82. Mani, A.; Sankaranarayanan, K. Heavy Metal and Mineral Element-Induced Abiotic Stress in Rice Plant. In *Rice Crop—Current Developments*; Shah, F., Khan, Z.H., Iqbal, A., Eds.; IntechOpen: London, UK, 2018. [CrossRef]
83. Lamaoui, M.; Jemo, M.; Datla, R.; Bekkaoui, F. Heat and drought stresses in crops and approaches for their mitigation. *Front. Chem.* **2018**, *6*, 26. [CrossRef]
84. Hewedy, O.A.; Abdel Lateif, K.S.; Seleiman, M.F.; Shami, A.; Albarakaty, F.M.; M El-Meihy, R. Phylogenetic diversity of *Trichoderma* strains and their antagonistic potential against soil-borne pathogens under stress conditions. *Biology* **2020**, *9*, 189. [CrossRef]
85. Zhu, J.K. Abiotic Stress Signaling and Responses in Plants. *Cell* **2016**, *167*, 313–324. [CrossRef]
86. Plieth, C.; Hansen, U.P.; Knight, H.; Knight, M.R. Temperature sensing by plants: The primary characteristics of signal perception and calcium response. *Plant J.* **1999**, *18*, 491–497. [CrossRef]
87. Knight, H.; Trewavas, A.J.; Knight, M.R. Cold calcium signaling in *Arabidopsis* involves two cellular pools and a change in calcium signature after acclimation. *Plant Cell* **1996**, *8*, 489–503.
88. Knight, M.R.; Knight, H. Low-temperature perception leading to gene expression and cold tolerance in higher plants. *New Phytologist.* **2012**, *195*, 737–751. [CrossRef]
89. Tyystjarvi, E. Photoinhibition of photosystem II. *Int. Rev. Cell Mol. Biol.* **2013**, *300*, 243–303.
90. Xiong, L.; Zhu, J.K. Abiotic stress signal transduction in plants: Molecular and genetic perspectives. *Physiol. Plant.* **2001**, *112*, 152–166. [CrossRef]
91. Sha Valli Khan, P.S.; Nagamallaiah, G.V.; Dhanunjay Rao, M.; Sergeant, K.; Hausman, J.F. Abioticstress tolerance in plants insights from proteomics. In *Emerging Technologies and Management of Crop Stress Tolerance*; Ahmad, P., Ed.; Elsevier Academic Press: London, UK, 2014; Volume 2, pp. 23–68.
92. Skalak, J.; Nicolas, K.L.; Vankova, R.; Hejatko, J. Signal integration in plant abiotic stress responses via multistep phosphorelay signaling. *Front. Plant Sci.* **2021**, *12*, 644823. [CrossRef]
93. Magwanga, R.O.; Lu, P.; Kirungu, J.N.; Lu, H.; Wang, X.; Cai, X.; Zhou, Z.; Zhang, Z.; Salih, H.; Wang, K.; et al. Characterization of the late embryogenesis abundant (LEA) proteins family and their role in drought stress tolerance in upland cotton. *BMC Genet.* **2018**, *15*, 6. [CrossRef]
94. Juven-Gershon, T.; Hsu, J.Y.; Theisen, J.W.; Kadonaga, J.T. The RNA polymerase II core promoter—The gateway to transcription. *Curr. Opin. Cell Biol.* **2008**, *20*, 253–259. [CrossRef]
95. Hu, H.; Dai, M.; Yao, J.; Xiao, B.; Li, X.; Zhang, Q.; Xiong, L. Overexpressing a NAM, ATAF, and CUC (NAC) transcription factor enhances drought resistance and salt tolerance in rice. *Proc. Natl. Acad. Sci. USA* **2006**, *103*, 12987–12992. [CrossRef]
96. Phukan, U.J.; Jeena, G.S.; Shukla, R.K. WRKY transcription factors: Molecular regulation and stress responses in plants. *Front. Plant Sci.* **2016**, *3*, 760. [CrossRef]
97. Wu, X.; Shiroto, Y.; Kishitani, S.; Ito, Y.; Toriyama, K. Enhanced heat and drought tolerance in transgenic rice seedlings overexpressing OsWRKY11 under the control of HSP101 promoter. *Plant Cell Rep.* **2009**, *28*, 21–30. [CrossRef]

98. Singh, D.; Laxmi, A. Transcriptional regulation of drought response: A tortuous network of transcriptional factors. *Front. Plant Sci.* **2015**, *6*, 895. [CrossRef]
99. Liu, Q.; Kasuga, M.; Sakuma, Y.; Abe, H.; Miura, S.; Yamaguchi-Shinozaki, K.; Shinozaki, K. Two transcription factors, DREB1 and DREB2, with an EREBP/AP2 DNA binding domain separate two cellular signal transduction pathways in drought- and low-temperature-responsive gene expression, respectively. *Arabidopsis. Plant Cell* **1998**, *10*, 1391–1406. [CrossRef] [PubMed]
100. Medina, J.; Bargues, M.; Terol, J.; Perez-Alonso, M.; Salinas, J. The *Arabidopsis* CBF gene family is composed of three genes encoding AP2 domain-containing proteins whose expression is regulated by low temperature but not by abscisic acid or dehydration. *Plant Physiol.* **1999**, *119*, 463–470. [CrossRef]
101. Jaglo-Ottosen, K.R.; Gilmour, S.J.; Zarka, D.G.; Schabenberger, O.; Thomashow, M.F. *Arabidopsis* CBF1 overexpression induces COR genes and enhances freezing tolerance. *Science* **1998**, *280*, 104–106. [CrossRef] [PubMed]
102. Shi, Y.; Ding, Y.; Yang, S. Molecular regulation of CBF signaling in cold acclimation. *Trends Plant Sci.* **2018**, *23*, 623–637. [CrossRef] [PubMed]
103. Floris, M.; Mahgoub, H.; Lanet, E.; Robaglia, C.; Menand, B. Posttranscriptional regulation of gene expression in plants during abiotic stress. *Int. J. Mol. Sci.* **2009**, *10*, 3168–3185. [CrossRef] [PubMed]
104. Lee, B.H.; Kapoor, A.; Zhu, J.; Zhu, J.K. Stabilized1, a stress-upregulated nuclear protein, is required for pre-mRNA splicing, mRNA turnover, and stress tolerance in *Arabidopsis*. *Plant Cell* **2006**, *18*, 1736–1749. [PubMed]
105. Guan, Q.; Wu, J.; Zhang, Y.; Jiang, C.; Liu, R.; Chai, C.; Zhu, J.A. Dead box RNA helicase is critical for pre-mRNA splicing, cold-responsive gene regulation, and cold tolerance in *Arabidopsis*. *Plant Cell* **2013**, *25*, 342–356. [CrossRef] [PubMed]
106. Calixto, C.P.G.; Guo, W.B.; James, A.B.; Tzioutziou, N.A.; Entizne, J.C.; Panter, P.E.; Knight, H.; Nimmo, H.G.; Zhang, R.; Brown, J.W. Rapid and dynamic alternative splicing impacts the *Arabidopsis* cold response transcriptome. *Plant Cell* **2018**, *30*, 1424–1444.
107. Sunkar, R.; Kapoor, A.; Zhu, J.K. Posttranscriptional induction of two Cu/Zn superoxide dismutase genes in *Arabidopsis* is mediated by downregulation of miR398 and important for oxidative stress tolerance. *Plant Cell* **2006**, *18*, 2051–2065.
108. Zhu, J.K. Salt and drought stress signal transduction in plants. *Annu. Rev. Plant Biol.* **2002**, *53*, 247–273.
109. Yoshida, R.; Umezawa, T.; Mizoguchi, T.; Takahashi, S.; Takahashi, F.; Shinozaki, K. The regulatory domain of SRK2E/OST1/SnRK2.6 interacts with ABI1 and integrates abscisic acid (ABA) and osmotic stress signals controlling stomatal closure in *Arabidopsis*. *J. Biologic. Chem.* **2006**, *281*, 5310–5318. [CrossRef]
110. Kim, J.M.; Sasaki, T.; Ueda, M.; Sako, K.; Seki, M. Chromatin changes in response to drought, salinity, heat, and cold stresses in plants. *Front. Plant Sci.* **2015**, *6*, 114. [CrossRef]
111. To, T.K.; Nakaminami, K.; Kim, J.M.; Morosawa, T.; Ishida, J.; Tanaka, M.; Yokoyama, S.; Shinozaki, K.; Seki, M. Arabidopsis HDA6 is required for freezing tolerance. *Biochem. Biophys. Res. Commun.* **2011**, *406*, 414–419. [CrossRef]
112. Roy, D.; Paul, A.; Roy, A.; Ghosh, R.; Ganguly, P.; Chaudhuri, S. Deferential acetylation of histone H3 at the regulatory region of OsDREB1b promoter facilitates chromatin remodelling and transcription activation during cold stress. *PLoS ONE* **2004**, *9*, e100343.
113. Hu, Y.; Zhang, L.; Zhao, L.; Li, J.; He, S.B.; Zhou, K.; Yang, F.; Huang, M.; Jiang, L.; Li, L. Trichostatin A selectively suppresses the cold-induced transcription of the ZmDREB1 gene in maize. *PLoS ONE* **2011**, *6*, e22132. [CrossRef]
114. He, X.J.; Hsu, Y.F.; Zhu, S.H.; Liu, H.L.; Pontes, O.; Zhu, J.H.; Cui, X.; Wang, C.S.; Zhu, J.K. A conserved transcriptional regulator is required for RNA-directed DNA methylation and plant development. *Genes Dev.* **2009**, *23*, 2717–2722. [CrossRef]
115. Chan, Z.L.; Wang, Y.P.; Cao, M.J.; Gong, Y.H.; Mu, Z.X.; Wang, H.Q.; Hu, Y.; Deng, X.; He, X.J.; Zhu, J.K. RDM4 modulates cold stress resistance in *Arabidopsis* partially through the CBF-mediated pathway. *New Phytol.* **2016**, *209*, 1527–1539. [CrossRef]
116. Sunkar, R.; Bartels, D.; Kirch, H.H. Overexpression of a stress-inducible dehydrogenase gene from *Arabidopsis thaliana* in transgenic plants improves stress tolerance. *Plant J.* **2003**, *35*, 452–464. [CrossRef]
117. Wang, W.; Vinocur, B.; Altman, A. Plant responses to drought, salinity and extreme temperatures: Towards genetic engineering for stress tolerance. *Planta* **2003**, *218*, 1–14. [CrossRef]
118. Xiong, L.; Schumaker, K.S.; Zhu, J.K. Cell signaling during cold, drought, and salt stress. *Plant Cell* **2002**, *14*, 165–183. [CrossRef]
119. Serrano, R.; Gaxiola, R.; Rios, G.; Forment, J.; Vicente, O.; Ros, R. Salt stress proteins identified by a functional approach in yeast. *Monatshefte Chem./Chem. Mon.* **2003**, *134*, 1445–1464. [CrossRef]
120. Fahad, S.; Nie, L.; Chen, Y.; Wu, C.; Xiong, D.; Saud, S.; Hongyan, L.; Cui, K.; Huang, J. Crop plant hormones and environmental stress. *Sustain. Agric. Rev.* **2015**, *15*, 371–400.
121. Garay-Arroyo, A.; Colmenero-Flores, J.M.; Garciarrubio, A.; Covarrubias, A.A. Highly hydrophilic proteins in prokaryotes and eukaryotes are common during conditions of water deficit. *J. Biol. Chem.* **2000**, *275*, 5668–5674. [CrossRef]
122. Fahad, S.; Bano, A. Effect of salicylic acid on physiological and biochemical characterization of maize grown in the saline area. *Pak. J. Bot.* **2012**, *44*, 1433–1438.
123. Eimer, M. Transgenic drought- and salt-tolerant plants. *Genet. Eng. Newsl.* **2004**, *15*, 1–14.
124. Serrano, R.; Montesinos, C. Molecular bases of desiccation tolerance in plant cells and potential applications in food dehydration. *Food Sci. Technol. Int.* **2003**, *9*, 157–161. [CrossRef]
125. Quintero, F.J.; Ohta, M.; Shi, H.; Zhu, J.K.; Pardo, J.M. Reconstitution in yeast of the Arabidopsis SOS signaling pathway for Na^+ homeostasis. *Proc. Natl. Acad. Sci. USA* **2002**, *25*, 9061–9066. [CrossRef]
126. Seifikalhor, M.; Aliniaeifard, S.; Shomali, A.; Azad, N.; Hassani, B.; Lastochkina, O.; Li, T. Calcium signaling and salt tolerance are diversely entwined in plants. *Plant Signal Behav.* **2019**, *14*, 1665455. [CrossRef]

127. Sandrini, M.; Nerva, L.; Sillo, F.; Balestrini, R.; Chitarra, W.; Zampieri, E. Abiotic stress and belowground microbiome: The potential of omics approaches. *Int. J. Mol. Sci.* **2022**, *23*, 1091. [CrossRef]
128. Antoine, S.; Heriche, M.; Boussageon, R.; Noceto, P.A.; van Tuinen, D.; Wipf, D.; Courty, P.E. A historical perspective on mycorrhizal mutualism emphasizing arbuscular mycorrhizas and their emerging challenges. *Mycorrhiza* **2021**, *31*, 637–653.
129. Begum, N.; Ahanger, M.A.; Su, Y.; Lei, Y.; Mustafa, N.S.A.; Ahmad, P.; Zhang, L. Improved drought tolerance by AMF inoculation in maize (*Zea mays*) involves physiological and biochemical implications. *Plants* **2019**, *8*, 579. [CrossRef] [PubMed]
130. Chandrasekaran, M.; Boopathi, T.; Manivannan, P. Comprehensive assessment of ameliorative effects of AMF in alleviating abiotic stress in tomato plants. *J. Fungi* **2021**, *7*, 303. [CrossRef] [PubMed]
131. Kosova, K.; Vitamvas, P.; Urban, M.O.; Prasil, I.T. Plant proteome responses to salinity Stress-Comparison of glycophytes and halophytes. *Funct. Plant Biol.* **2013**, *40*, 775–786. [CrossRef] [PubMed]
132. Pan, J.; Peng, F.; Tedeschi, A.; Xue, X.; Wang, T.; Liao, J.; Zhang, W.; Huang, C. Do halophytes and glycophytes differ in their interactions with arbuscular mycorrhizal fungi under salt stress? A meta-analysis. *Bot. Stud.* **2020**, *61*, 13. [CrossRef]
133. Auge, R.M.; Toler, H.D.; Saxton, A.M. Arbuscular mycorrhizal symbiosis and osmotic adjustment in response to NaCl stress: A meta-analysis. *Front. Plant Sci.* **2014**, *5*, 562.
134. Evelin, H.; Giri, B.; Kapoor, R. Ultrastructural evidence for AMF mediated salt stress mitigation in *Trigonella foenum-graecum*. *Mycorrhiza* **2013**, *23*, 71–86. [CrossRef]
135. Evelin, H.; Kapoor, R.; Giri, B. Arbuscular mycorrhizal fungi in alleviation of salt stress: A review. *Ann. Bot.* **2009**, *104*, 1263–1280. [CrossRef]
136. Eroglu, C.G.; Cabral, C.; Ravnskov, S.; Bak Topbjerg, H.; Wollenweber, B. Arbuscular mycorrhiza influences carbon-use efficiency and grain yield of wheat grown under pre- and post-anthesis salinity stress. *Plant Biol.* **2020**, *22*, 863–871. [CrossRef]
137. Sharma, K.; Gupta, S.; Thokchom, S.D.; Jangir, P.; Kapoor, R. Arbuscular mycorrhiza-mediated regulation of polyamines and aquaporins during abiotic stress: Deep insights on the recondite players. *Front. Plant Sci.* **2021**, *12*, 1072. [CrossRef]
138. Volpe, V.; Chitarra, W.; Cascone, P.; Volpe, M.G.; Bartolini, P.; Moneti, G.; Pieraccini, G.; Di Serio, C.; Maserti, B.; Guerrieri, E.; et al. The association with two different arbuscular mycorrhizal fungi differently affects water stress tolerance in tomato. *Front. Plant Sci.* **2018**, *9*, 1480. [CrossRef]
139. Chitarra, W.; Pagliarani, C.; Maserti, B.; Lumini, E.; Siciliano, I.; Cascone, P.; Schubert, A.; Gambino, G.; Balestrini, R.; Guerrieri, E. Insights on the impact of arbuscular mycorrhizal symbiosis on tomato tolerance to water stress. *Plant Physiol.* **2016**, *171*, 1009–1023. [CrossRef]
140. Rivero, J.; Alvarez, D.; Flors, V.; Azcon-Aguilar, C.; Pozo, M.J. Root metabolic plasticity underlies functional diversity in mycorrhiza-enhanced stress tolerance in tomato. *New Phytol.* **2018**, *220*, 1322–1336. [CrossRef]
141. Sandhya, V.; Ali, S.K.Z.; Minakshi, G.; Reddy, G.; Venkateswarlu, B. Alleviation of drought stress effects in sunflower seedlings by the exopolysaccharides producing *Pseudomonas putida* strain GAP-P45. *Biol. Fertil. Soils* **2009**, *46*, 17–26. [CrossRef]
142. Arshad, M.; Sharoona, B.; Mahmood, T. Inoculation with *Pseudomonas* spp. containing ACC deaminase partially eliminate the effects of drought stress on growth, yield and ripening of pea (*Pisum sativum* L.). *Pedosphere* **2008**, *18*, 611–620. [CrossRef]
143. Creus, C.M.; Sueldo, R.J.; Barassi, C.A. Water relations and yield in *Azospirillum*-inoculated wheat exposed to drought in the field. *Can. J. Bot.* **2004**, *82*, 273–281. [CrossRef]
144. Cho, E.K.; Hong, C.B. Over-expression of tobacco NtHSP70-1 contributes to drought-stress tolerance in plants. *Plant Cell Rep.* **2006**, *25*, 349–358. [CrossRef]
145. Rodriguez, A.A.; Stella, A.M.; Storni, M.M.; Zulpa, G.; Zaccaro, M.C. Effect of cyanobacterial extracellular products and gibberellic acid on salinity tolerance in *Oryza sativa* L. *Saline Syst.* **2006**, *2*, 7. [CrossRef]
146. Barka, E.A.; Nowak, J.; Clement, C. Enhancement of chilling resistance of inoculated grapevine plantlets with a plant growth-promoting rhizobacterium *Burkholderiaphytofirmans* strain PsJN. *Appl. Environ. Microbiol.* **2006**, *72*, 7246–7252. [CrossRef]
147. Madhaiyan, M.; Poonguzhali, S.; Sa, T. Metal tolerating methylotrophic bacteria reduces nickel and cadmium toxicity and promotes plant growth of tomato (*Lycopersicon esculentum* L.). *Chemosphere* **2007**, *69*, 220–228. [CrossRef]
148. SaravanaMartins, S.J.; Rocha, G.A.; De Melo, H.C.; De Castro Georg, R.; Ulhoa, C.J.; De Campos Dianese, É.; Oshiquiri, L.H.; da Cunha, M.G.; da Rocha, M.R.; de Araújo, L.G.; et al. Plant-associated bacteria mitigate drought stress in soybean. *Environ. Sci. Pollut. Res.* **2018**, *25*, 13676–13686. [CrossRef]
149. Figueiredo, M.V.B.; Burity, H.A.; Martinez, C.R.; Chanway, C.P. Alleviation of drought stress in common bean (*Phaseolus vulgaris* L.) by coinoculation with *Paenibacilluspolymyxa* and *Rhizobium tropici*. *Appl. Soil Ecol.* **2008**, *40*, 182–188. [CrossRef]
150. Kohler, J.; Hernández, J.A.; Caravaca, F.; Roldán, A. Plant growth promoting rhizobacteria and arbuscular mycorrhizal fungi modify alleviation biochemical mechanisms in water stressed plants. *Funct. Plant. Biol.* **2008**, *35*, 141–151. [CrossRef]
151. Ali, S.Z.; Sandhya, V.; Grover, M.; Kishore, N.; Rao, L.V.; Venkateswarlu, B. Pseudomonas sp. strain AKM-P6 enhances tolerance of sorghum seedlings to elevated temperatures. *Biol. Fert. Soils* **2009**, *46*, 45–55. [CrossRef]
152. Marulanda, A.; Porcel, R.; Barea, J.M.; Azcon, R. Drought tolerance and antioxidant activities in lavender plants colonized by native drought tolerant or drought sensitive *Glomus* species. *Microb. Ecol.* **2007**, *54*, 543–552. [CrossRef]
153. McLellan, C.A.; Turbyville, T.J.; Wijeratne, K.; Kerschen, A.; Vierling, E.; Queitsch, C.; Whiteshell, L.; Gunatilaka, A.A.L. A rhizosphere fungus enhances *Arabidopsis* thermotolerance through production of an HSP90 inhibitor. *Plant Physiol.* **2007**, *145*, 174–182. [CrossRef]

154. Zhang, H.; Kim, M.S.; Sun, Y.; Dowd, S.E.; Shi, H.; Paré, P.W. Soil bacteria confer plant salt tolerance by tissue-specific regulation of the sodium transporter HKT1. *Mol. Plant Microbe Interact.* **2008**, *21*, 737–744. [CrossRef]
155. Yao, L.X.; Wu, Z.S.; Zheng, Y.Y.; Kaleem, I.; Li, C. Growth promotion and protection against salt stress by *Pseudomonas putida* Rs-198 on cotton. *Eur. J. Soil Biol.* **2010**, *46*, 49–54. [CrossRef]
156. Daei, G.; Ardekani, M.R.; Rejali, F.; Teimuri, S.; Miransari, M. Alleviation of salinity stress on wheat yield, yield components, and nutrient uptake using arbuscular mycorrhizal fungi under field conditions. *J. Plant Physiol.* **2009**, *66*, 617–625. [CrossRef]
157. Lazzarotto, P.; Calanca, P.; Semenov, M.; Fuhrer, J. Transient responses to increasing CO_2 and climate change in an unfertilized grass clover sward. *Clim. Res.* **2010**, *41*, 221–232. [CrossRef]
158. Lim, J.H.; Kim, S.D. Induction of drought stress resistance by multifunctional PGPR Bacillus licheniformis K11 in pepper. *Plant Pathol. J.* **2013**, *29*, 201–208. [CrossRef]
159. Timmusk, S.; El-Daim, I.A.A.; Copolovici, L.; Tanilas, T.; Kannaste, A.; Behers, L.; Nevo, E.; Seisenbaeva, G.; Stenström, E.; Niinemets, Ü. Drought-tolerance of wheat improved by rhizosphere bacteria from harsh environments: Enhanced biomass production and reduced emissions of stress volatiles. *PLoS ONE* **2014**, *9*, e96086. [CrossRef] [PubMed]
160. del Amor, F.; Cuadra-Crespo, P. Plant growth-promoting bacteria as a tool to improve salinity tolerance in sweet pepper. *Funct. Plant Biol.* **2012**, *39*, 82–90. [CrossRef] [PubMed]
161. Naveed, M.; Mitter, B.; Reichenauer, T.G.; Wieczorek, K.; Sessitsch, A. Increased drought stress resilience of maize through endophytic colonization by *Burkholderiaphytofirmans* PsJN and *Enterobacter* sp FD17. *Environ. Exp. Bot.* **2014**, *97*, 30–39. [CrossRef]
162. Kasotia, A.; Varma, A.; Choudhary, D.K. *Pseudomonas*-mediated mitigation of salt stress and growth promotion in *Glycine max*. *Agric. Res.* **2015**, *4*, 31–41. [CrossRef]
163. Plociniczak, T.; Sinkkonen, A.; Romantschuk, M.; Piotrowska-seget, Z. Characterization of *Enterobacter intermedius* MH8b and its use for the enhancement of heavy metal uptake by *Sinapsis alba* L. *Appl. Soil Ecol.* **2013**, *63*, 1–7. [CrossRef]
164. Wang, C.; Yang, W.; Wang, C.; Gu, C.; Niu, D.; Liu, H.X.; Wang, Y.P.; Guo, J.H. Induction of drought tolerance in cucumber plants by a consortium of three plant growth promoting rhizobacterium strains. *PLoS ONE* **2012**, *7*, e52565. [CrossRef]
165. Mathew, D.C.; Ho, Y.N.; Gicana, R.G.; Mathew, G.M.; Chien, M.C.; Huang, C.C. A rhizosphere-associated symbiont, *Photobacterium* spp. strain MELD1, and its targeted synergistic activity for phytoprotection against mercury. *PLoS ONE* **2015**, *10*, e0121178. [CrossRef]
166. Adediran, G.A.; Ngwenya, B.T.; Mosselmans, J.F.W.; Heal, K.V. Bacteria–zinc co-localization implicates enhanced synthesis of cysteine-rich peptides in zinc detoxification when *Brassica juncea* is inoculated with accumulation in N2-fixing alfalfa (*Medicago sativa*). *J. Plant. Physiol.* **2016**, *146*, 541–546.
167. Jha, Y.; Sablok, G.; Subbarao, N.; Sudhakar, R.; Fazil, M.H.U.T.; Subramanian, R.B.; Squartini, A.; Kumar, S. Bacterial-induced expression of RAB18 protein in *Orzya sativa* salinity stress and insights into molecular interaction with GTP ligand. *J. Mol. Recognit.* **2014**, *27*, 521–527. [CrossRef]
168. Arun, K.D.; Sabarinathan, K.G.; Gomathy, M.; Kannan, R.; Balachandar, D. Mitigation of drought stress in rice crop with plant growth-promoting abiotic stress-tolerant rice phyllosphere bacteria. *J. Basic Microbiol.* **2020**, *60*, 768–786. [CrossRef]
169. Ashry, N.M.; Alaidaroos, B.A.; Mohamed, S.A.; Badr, O.A.; El-Saadony, M.T.; Esmael, A. Utilization of drought-tolerant bacterial strains isolated from harsh soils as a plant growth-promoting rhizobacteria (PGPR). *Saudi J. Biol. Sci.* **2022**, *29*, 1760–1769. [CrossRef]
170. Filgueiras, L.; Silva, R.; Almeida, I.; Vidal, M.; Baldani, J.I.; Meneses, C.H.S.G. *Gluconacetobacterdiazotrophicus* mitigates drought stress in *Oryza sativa* L. *Plant Soil* **2020**, *451*, 57–73. [CrossRef]
171. Kour, D.; Rana, K.L.; Yadav, A.N.; Sheikh, I.; Kumar, V.; Dhaliwal, H.S.; Saxena, A.K. Amelioration of drought stress in Foxtail millet (*Setariaitalica* L.) by P-solubilizing drought-tolerant microbes with multifarious plant growth promoting attributes. *Environ. Sustain.* **2020**, *3*, 23–34. [CrossRef]
172. Li, H.; Guo, Q.; Jing, Y.; Liu, Z.; Zheng, Z.; Sun, Y.; Xue, Q.; Lai, H. Application of *Streptomyces pactum* Act12 enhances drought resistance in wheat. *J. Plant Growth Regul.* **2020**, *39*, 122–132. [CrossRef]
173. Kumar, M.; Mishra, S.; Dixit, V.; Kumar, M.; Agarwal, L.; Chauhan, P.S.; Nautiyal, C.S. Synergistic effect of *Pseudomonas putida* and *Bacillus amyloliquefaciens* ameliorates drought stress in chickpea (*Cicer arietinum* L.). *Plant Signal. Behav.* **2016**, *11*, e1071004. [CrossRef]
174. Kumar, A.; Singh, S.; Gaurav, A.K.; Srivastava, S.; Verma, J.P. Plant growth-promoting bacteria: Biological tools for the mitigation of salinity stress in plants. *Front. Microbiol.* **2020**, *11*, 1216. [CrossRef]
175. Ferreira, M.J.; Silva, H.; Cunha, A. Siderophore-producing rhizobacteria as a promising tool for empowering plants to cope with iron limitation in saline soils: A review. *Pedosphere* **2019**, *29*, 409–420. [CrossRef]
176. Khanna, K.; Jamwal, V.L.; Gandhi, S.G.; Ohri, P.; Bhardwaj, R. Metal resistant PGPR lowered Cd uptake and expression of metal transporter genes with improved growth and photosynthetic pigments in *Lycopersicon esculentum* under metal toxicity. *Sci. Rep.* **2019**, *9*, 5855. [CrossRef]
177. Chaudhary, D.; Sindhu, S.S. Amelioration of salt stress in chickpea (*Cicer arietinum* L.) by coinculation of ACC deaminase-containing rhizospheric bacteria with *Mesorhizobium* strains. *Legume Res.* **2017**, *40*, 80–86. [CrossRef]
178. Marulanda, A.; Azcón, R.; Chaumont, F.; Ruiz-Lozano, J.M.; Aroca, R. Regulation of plasma membrane aquaporins by inoculation with a *Bacillus megaterium* strain in maize (*Zea mays* L.) plants under unstressed and salt-stressed conditions. *Planta* **2010**, *232*, 533–543. [CrossRef]

179. Balestrini, R.; Brunetti, C.; Chitarra, W.; Nerva, L. Photosynthetic traits and nitrogen uptake in crops: Which is the role of arbuscular mycorrhizal fungi? *Plants* **2020**, *9*, 1105. [CrossRef]
180. Mathur, S.; Tomar, R.S.; Jajoo, A. Arbuscular mycorrhizal fungi (AMF) protects photosynthetic apparatus of wheat under drought stress. *Photosynth. Res.* **2019**, *139*, 227–238. [CrossRef]
181. Zou, Y.N.; Wu, Q.S.; Kuca, K. Unravelling the role of arbuscular mycorrhizal fungi in mitigating the oxidative burst of plants under drought stress. *Plant Biol.* **2021**, *23*, 50–57. [CrossRef]
182. Qin, Y.; Druzhinina, I.S.; Pan, X.; Yuan, Z. Microbially mediated plant salt tolerance and microbiome-based solutions for saline agriculture. *Biotechnol. Adv.* **2016**, *34*, 1245–1259. [CrossRef]
183. Fiorilli, V.; Vallino, M.; Biselli, C.; Faccio, A.; Bagnaresi, P.; Bonfante, P. Host and non-host roots in rice: Cellular and molecular approaches reveal differential responses to arbuscular mycorrhizal fungi. *Front. Plant Sci.* **2015**, *6*, 636. [CrossRef]
184. Fiorilli, V.; Catoni, M.; Miozzi, L.; Novero, M.; Accotto, G.P.; Lanfranco, L. Global and cell-type gene expression profiles in tomato plants colonized by an arbuscular mycorrhizal fungus. *New Phytol.* **2009**, *184*, 975–987. [CrossRef]
185. Balestrini, R.; Salvioli, A.; Dal Molin, A.; Novero, M.; Gabelli, G.; Papparelli, E.; Marroni, F.; Bonfante, P. Impact of an arbuscular mycorrhizal fungus versus a mixed microbial inoculum on the transcriptome reprogramming of grapevine roots. *Mycorrhiza* **2017**, *7*, 417–430. [CrossRef]
186. Fiorilli, V.; Vannini, C.; Ortolani, F.; Garcia-Seco, D.; Chiapello, M.; Novero, M.; Domingo, G.; Terzi, V.; Morcia, C.; Bagnaresi, P.; et al. Omics approaches revealed how arbuscular mycorrhizal symbiosis enhances yield and resistance to leaf pathogen in wheat. *Sci. Rep.* **2018**, *8*, 9625. [CrossRef]
187. Balestrini, R.; Rosso, L.C.; Veronico, P.; Melillo, M.T.; De Luca, F.; Fanelli, E.; Colagiero, M.; Salvioli, A.; Ciancio, A.; Pentimone, I. Transcriptomic responses to water deficit and nematode infection in mycorrhizal tomato roots. *Front. Microbiol.* **2019**, *10*, 1807. [CrossRef]
188. Grover, M.; Bodhankar, S.; Maheswari, M.; Srinivasarao, C. 6 Actinomycetes as mitigators of climate change and abiotic stress. In *Plant Growth Promoting Actinobacteria*; Subramaniam, G., Arumugam, S., Rajendran, V., Eds.; Springer: Singapore, 2016; pp. 203–212.
189. Yandigeri, M.S.; Meena, K.K.; Singh, D.; Malviya, N.; Singh, D.P.; Solanki, M.K.; Yadav, A.K.; Arora, D.K. Drought-tolerant endophytic Actinobacteria promote growth of wheat (*Triticum aestivum*) under water stress conditions. *Plant Growth Regul.* **2012**, *68*, 411–420. [CrossRef]
190. Singh, R.P.; Pandey, D.M.; Jha, P.N.; Ma, Y. ACC deaminase producing rhizobacterium *Enterobacter cloacae* ZNP-4 enhance abiotic stress tolerance in wheat plant. *PLoS ONE* **2022**, *17*, e0267127. [CrossRef] [PubMed]
191. Aly, M.M.; El-Sabbagh, S.M.; El-Shouny, W.A.; Ebrahim, M.K.H. Physiological response of Zea mays to NaCl stress with respect to *Azotobacter chroococcum* and *Streptomyces niveus*. *Pak. J. Biol. Sci.* **2003**, *6*, 2073–2080. [CrossRef]
192. Palaniyandi, S.A.; Damodharan, K.; Yang, S.H.; Suh, J.W. Streptomyces sp. Strain PGPA39 alleviates salt stress and promotes growth of 'Micro Tom' tomato plants. *J. Appl. Microbiol.* **2014**, *117*, 766–773. [CrossRef] [PubMed]
193. Ghavami, N.; Alikhani, H.A.; Pourbabaei, A.A.; Besharati, H. Effects of two new siderophore-producing rhizobacteria on growth and iron content of maize and canola plants. *J. Plant Nutr.* **2017**, *40*, 736–746. [CrossRef]
194. Hasegawa, S.; Meguro, A.; Toyoda, K.; Nishimura, T.; Kunoh, H. Drought tolerance of tissue-cultured seedlings of mountain laurel (*Kalmia latifolia* L.) induced by an endophytic actinomycete II. Acceleration of callose accumulation and lignification. *Actinomycetologica* **2005**, *19*, 13–17. [CrossRef]
195. Johnston-Monje, D.; Lundberg, D.S.; Lazarovits, G.; Reis, V.M.; Raizada, M.N. Bacterial populations in juvenile maize rhizospheres originate from both seed and soil. *Plant Soil* **2016**, *405*, 337–355. [CrossRef]
196. Genitsaris, S.; Stefanidou, N.; Leontidou, K.; Matsi, T.; Karamanoli, K.; Mellidou, I. Bacterial communities in the rhizosphere and phyllosphere of halophytes and drought-tolerant plants in mediterranean ecosystems. *Microorganisms* **2020**, *8*, 1708. [CrossRef]
197. Ullah, M.A.; Mahmood, I.A.; Ali, A.; Nawaz, Q.; Sultan, T.; Zaman, B.U. Effect of inoculation methods of biozote-max (plant growth promoting rhizobacteria-pgpr) on growth and yield of rice under naturally salt-affected soil. *Res. Plant Biol.* **2017**, *7*, 24–26. [CrossRef]
198. Grover, M.; Bodhankar, S.; Sharma, A.; Sharma, P.; Singh, J.; Nain, L. PGPR mediated alterations in root traits: Way toward sustainable crop production. *Front. Sustain. Food Syst.* **2021**, *4*, 618230. [CrossRef]
199. Vanamala, P.; Sultana, U.; Sindhura, P.; Gul, M.Z. Plant Growth-Promoting Rhizobacteria (PGPR): A Unique Strategy for Sustainable Agriculture. In *Handbook of Research on Microbial Remediation and Microbial Biotechnology for Sustainable Soil*; IGI Global: Hershey, PA, USA, 2021; pp. 332–357.
200. Ansari, F.A.; Ahmad, I. Alleviating drought stress of crops through PGPR: Mechanism and application. In *Microbial Interventions in Agriculture and Environment*; Singh, D.P., Gupta, V.K., Prabha, R., Eds.; Springer: Singapore, 2019; pp. 341–358. [CrossRef]
201. Khan, N.; Bano, A.; Shahid, M.A.; Nasim, W.; Babar, A. Interaction between PGPR and PGR for water conservation and plant growth attributes under drought condition. *Biologia* **2018**, *73*, 1083–1098. [CrossRef]
202. Sinha, D.; Mukherjee, S.; Mahapatra, D. Multifaceted potential of Plant Growth Promoting Rhizobacteria (PGPR): An overview. In *Handbook of Research on Microbial Remediation and Microbial Biotechnology for Sustainable Soil*; Ahmad Malik, J., Ed.; IGI Global: Hershey, PA, USA, 2021; pp. 205–268. [CrossRef]
203. Monteiro, G.; Nogueira, G.; Neto, C.; Nascimento, V.; Freitas, J. Promotion of Nitrogen Assimilation by Plant Growth-Promoting Rhizobacteria. In *Nitrogen in Agriculture-Physiological, Agricultural and Ecological Aspects*; Intech Open: London, UK, 2021.

204. Jochum, M.D.; Mcwilliams, K.L.; Borrego, E.J.; Kolomiets, M.V.; Niu, G.; Pierson, E.A.; Jo, Y.K. Bioprospecting plant growth-promoting rhizobacteria that mitigate drought stress in grasses. *Front. Microbiol.* **2019**, *10*, 2106. [CrossRef]
205. Bano, Q.; Ilyas, N.; Bano, A.; Zafar, N.; Akram, A.; Hassan, F. Effect of *Azospirillum* inoculation on maize (*Zea mays* L.) under drought stress. *Pak. J. Bot.* **2013**, *45*, 13–20.
206. Kour, D.; Rana, K.L.; Sheikh, I.; Kumar, V.; Yadav, A.N.; Dhaliwal, H.S.; Saxena, A.K. Alleviation of drought stress and plant growth promotion by *Pseudomonas libanensis* EU-LWNA-33, a drought-adaptive phosphorus-solubilizing bacterium. *Proc. Natl. Acad. Sci. India B Biol. Sci.* **2020**, *90*, 785–795. [CrossRef]
207. Zerrouk, I.Z.; Benchabane, M.; Khelifi, L.; Yokawa, K.; Ludwig-Muller, J.; Baluska, F. *Pseudomonas* strain isolated from date-palm rhizospheres improves root growth and promotes root formation in maize exposed to salt and aluminum stress. *J. Plant Physiol.* **2016**, *191*, 111–119. [CrossRef]
208. Santos-Medellin, C.; Edwards, J.; Liechty, Z.; Nguyen, B.; Sundaresan, V. Drought stress results in a compartment-specific restructuring of the rice root-associated microbiomes. *MBio* **2017**, *8*, e00764-17. [CrossRef]
209. Priyanka, J.P.; Goral, R.T.; Rupal, K.S.; Saraf, M. Rhizospheric microflora: A natural alleviator of drought stress in agricultural crops. In *Plant Growth Promoting Rhizobacteria for Sustainable Stress Management*; Sayyed, R.Z., Ed.; Springer: Singapore, 2019; pp. 103–115. [CrossRef]
210. Chiappero, J.; Del Rosario Cappellari, L.; Alderete, L.G.S.; Palermo, T.B.; Banchio, E. Plant growth promoting rhizobacteria improve the antioxidant status in *Mentha piperita* grown under drought stress leading to an enhancement of plant growth and total phenolic content. *Industr. Crops Prod.* **2019**, *139*, 111553. [CrossRef]
211. Raheem, A.; Shaposhnikov, A.; Belimov, A.A.; Dodd, I.C.; Ali, B. Auxin production by rhizobacteria was associated with improved yield of wheat (*Triticum aestivum* L.) under drought stress. *Arch. Agron. Soil Sci.* **2018**, *64*, 574–587. [CrossRef]
212. Hanif, S.; Saleem, M.F.; Sarwar, M.; Irshad, M.; Shakoor, A.; Wahid, M.A.; Khan, H.Z. Biochemically triggered heat and drought stress tolerance in rice by proline application. *J Plant Growth Regul.* **2021**, *40*, 305–312. [CrossRef]
213. Rana, R.A.; Siddiqui, M.N.; Skalicky, M.; Brestic, M.; Hossain, A.; Kayesh, E.; Popov, M.; Hejnak, V.; Gupta, D.R.; Mahmud, N.U.; et al. Prospects of nanotechnology in improving the productivity and quality of horticultural crops. *Horticulturae* **2021**, *7*, 332. [CrossRef]
214. Tripathi, D.K.; Singh, V.P.; Prasad, S.M.; Chauhan, D.K.; Dubey, N.K. Silicon nanoparticles (SiNp) alleviate chromium (VI) phytotoxicity in *Pisum sativum* (L.) seedlings. *Plant Physiol. Biochem.* **2015**, *96*, 189–198. [CrossRef]
215. Hussain, A.; Ali, S.; Rizwan, M.; Zia ur Rehman, M.; Javed, M.R.; Imran, M.; Chatha, S.A.S.; Nazir, R. Zinc oxide nanoparticles alter the wheat physiological response and reduce the cadmium uptake by plants. *Environ. Pollut.* **2018**, *242*, 1518–1526. [CrossRef]
216. Manzoor, N.; Ahmed, T.; Noman, M.; Shahid, M.; Nazir, M.M.; Ali, L.; Alnusaire, T.S.; Li, B.; Schulin, R.; Wang, G. Iron oxide nanoparticles ameliorated the cadmium and salinity stresses in wheat plants, facilitating photosynthetic pigments and restricting cadmium uptake. *Sci. Total Environ.* **2021**, *769*, 145221. [CrossRef]
217. Hussain, A.; Ali, S.; Rizwan, M.; Rehman, M.Z.; ur Qayyum, M.F.; Wang, H.; Rinklebe, J. Responses of wheat (Triticum aestivum) plants grown in a Cd contaminated soil to the application of iron oxide nanoparticles. *Ecotoxicol. Environ. Saf.* **2019**, *173*, 156–164. [CrossRef]
218. Adrees, M.; Khan, Z.S.; Ali, S.; Hafeez, M.; Khalid, S.; ur Rehman, M.Z.; Hussain, A.; Hussain, K.; Chatha, S.A.S.; Rizwan, M. Simultaneous mitigation of cadmium and drought stress in wheat by soil application of iron nanoparticles. *Chemosphere* **2020**, *238*, 124681. [CrossRef]
219. Gao, M.; Zhou, J.; Liu, H.; Zhang, W.; Hu, Y.; Liang, J.; Zhou, J. Foliar spraying with silicon and selenium reduces cadmium uptake and mitigates cadmium toxicity in rice. *Sci. Total Environ.* **2018**, *631–632*, 1100–1108. [CrossRef]
220. Rehman, M.Z.; Rizwan, M.; Hussain, A.; Saqib, M.; Ali, S.; Sohail, M.I.; Shafiq, M.; Hafeez, F. Alleviation of cadmium (cd) toxicity and minimizing its uptake in wheat (Triticum aestivum) by using organic carbon sources in cd-spiked soil. *Environ. Pollut.* **2018**, *241*, 557–565. [CrossRef]
221. Ashkavand, P.; Tabari, M.; Zarafshar, M.; Tomaskova, I.; Struve, D. Effect of SiO_2 nanoparticles on drought resistance in hawthorn seedlings. *For. Res. Pap.* **2018**, *76*, 350–359. [CrossRef]
222. Ghorbanpour, M.; Mohammadi, H.; Kariman, K. Nanosilicon-based recovery of barley (Hordeum vulgare) plants subjected to drought stress. *Environ. Sci. Nano* **2020**, *7*, 443–461. [CrossRef]
223. Alsaeedi, A.; El-Ramady, H.; Alshaal, T.; El-Garawany, M.; Elhawat, N.; Al-Otaibi, A. Silica nanoparticles boost growth and productivity of cucumber under water deficit and salinity stresses by balancing nutrients uptake. *Plant Physiol. Biochem.* **2019**, *139*, 1–10. [CrossRef]
224. Behboudi, F.; Tahmasebi-Sarvestani, Z.; Kassaee, M.Z.; Modarres-Sanavy, S.A.M.; Sorooshzadeh, A.; Mokhtassi-Bidgoli, A. Evaluation of chitosan nanoparticles effects with two application methods on wheat under drought stress. *J. Plant Nutr.* **2019**, *42*, 1439–1451. [CrossRef]
225. Das, A.; Das, B. Nanotechnology a potential tool to mitigate abiotic stress in crop plants. In *Abiotic and Biotic Stress in Plants*; De Oliveira, A., Ed.; IntechOpen: London, UK, 2019.
226. Sun, D.; Hussain, H.I.; Yi, Z.; Rookes, J.E.; Kong, L.; Cahill, D.M. Delivery of abscisic acid to plants using glutathione responsive mesoporous silica nanoparticles. *J. Nanosci. Nanotechnol.* **2017**, *18*, 1615–1625. [CrossRef] [PubMed]

227. Ye, Y.; Cota-Ruiz, K.; Hernández-Viezcas, J.A.; Valdés, C.; Medina-Velo, I.A.; Turley, R.S.; Peralta-Videa, J.R.; Gardea-Torresdey, J.L. Manganese nanoparticles control salinity-modulated molecular responses in Capsicum annuum L. through priming: A sustainable approach for agriculture. *ACS Sustain. Chem. Eng.* **2020**, *8*, 1427–1436. [CrossRef]
228. Wu, J.; Wang, T. Synergistic effect of zinc oxide nanoparticles and heat stress on the alleviation of transcriptional gene silencing in Arabidopsis thaliana. *Bull. Environ. Contam. Toxicol.* **2020**, *104*, 49–56. [CrossRef] [PubMed]
229. Thakur, S.; Asthir, B.; Kaur, G.; Kalia, A.; Sharma, A. Zinc oxide and titanium dioxide nanoparticles influence heat stress tolerance mediated by antioxidant defense system in wheat. *Cereal Res. Commun.* **2021**, *49*, 1–12. [CrossRef]
230. El-Saadony, M.T.; Saad, A.M.; Najjar, A.A.; Alzahrani, S.O.; Alkhatib, F.M.; Shafi, M.E.; Selem, E.; Desoky, E.S.M.; Fouda, S.E.; El-Tahan, A.M.; et al. The use of biological selenium nanoparticles to suppress Triticum aestivum L. crown and root rot diseases induced by Fusarium species and improve yield under drought and heat stress. *Saudi J. Biol. Sci.* **2021**, *28*, 4461–4471. [CrossRef]
231. Iqbal, M.; Raja, N.I.; Mashwani ZU, R.; Hussain, M.; Ejaz, M.; Yasmeen, F. Effect of silver nanoparticles on growth of wheat under heat stress. *Iran. J. Sci. Technol. Trans. A Sci.* **2019**, *43*, 387–395. [CrossRef]

life

MDPI

Review

Organic Amendments for Mitigation of Salinity Stress in Plants: A Review

Md. Najmol Hoque [1,†], Shahin Imran [2,†], Afsana Hannan [3], Newton Chandra Paul [2], Md. Asif Mahamud [4], Jotirmoy Chakrobortty [5], Prosenjit Sarker [6], Israt Jahan Irin [2], Marian Brestic [7,8,*] and Mohammad Saidur Rhaman [9,*]

1 Department of Biochemistry and Molecular Biology, Khulna Agricultural University, Khulna 9100, Bangladesh
2 Department of Agronomy, Khulna Agricultural University, Khulna 9100, Bangladesh
3 Department of Genetics and Plant Breeding, Bangladesh Agricultural University, Mymensingh 2202, Bangladesh
4 Department of Agricultural Chemistry, Khulna Agricultural University, Khulna 9100, Bangladesh
5 Department of Soil Science, Khulna Agricultural University, Khulna 9100, Bangladesh
6 Department of Crop Botany, Khulna Agricultural University, Khulna 9100, Bangladesh
7 Department of Botany and Plant Physiology, Czech University of Life Sciences, Kamycka 129, 16500 Prague, Czech Republic
8 Institute of Plant and Environmental Studies, Slovak University of Agriculture, A. Hlinku 2, 94976 Nitra, Slovakia
9 Department of Seed Science and Technology, Bangladesh Agricultural University, Mymensingh 2202, Bangladesh
* Correspondence: marian.brestic@uniag.sk (M.B.); saidursst@bau.edu.bd (M.S.R.)
† These authors contributed equally to this work.

Abstract: Natural and/or human-caused salinization of soils has become a growing problem in the world, and salinization endangers agro-ecosystems by causing salt stress in most cultivated plants, which has a direct effect on food quality and quantity. Several techniques, as well as numerous strategies, have been developed in recent years to help plants cope with the negative consequences of salt stress and mitigate the impacts of salt stress on agricultural plants. Some of them are not environmentally friendly. In this regard, it is crucial to develop long-term solutions that boost saline soil productivity while also protecting the ecosystem. Organic amendments, such as vermicompost (VC), vermiwash (VW), biochar (BC), bio-fertilizer (BF), and plant growth promoting rhizobacteria (PGPR) are gaining attention in research. The organic amendment reduces salt stress and improves crops growth, development and yield. The literature shows that organic amendment enhances salinity tolerance and improves the growth and yield of plants by modifying ionic homeostasis, photosynthetic apparatus, antioxidant machineries, and reducing oxidative damages. However, the positive regulatory role of organic amendments in plants and their stress mitigation mechanisms is not reviewed adequately. Therefore, the present review discusses the recent reports of organic amendments in plants under salt stress and how stress is mitigated by organic amendments. The current assessment also analyzes the limitations of applying organic amendments and their future potential.

Keywords: bio-fertilizer; ionic homeostasis; organic amendments; salinity; vermicompost

Citation: Hoque, M.N.; Imran, S.; Hannan, A.; Paul, N.C.; Mahamud, M.A.; Chakrobortty, J.; Sarker, P.; Irin, I.J.; Brestic, M.; Rhaman, M.S. Organic Amendments for Mitigation of Salinity Stress in Plants: A Review. *Life* **2022**, *12*, 1632. https://doi.org/10.3390/life12101632

Academic Editors: Hakim Manghwar and Wajid Zaman

Received: 11 September 2022
Accepted: 14 October 2022
Published: 18 October 2022

Publisher's Note: MDPI stays neutral with regard to jurisdictional claims in published maps and institutional affiliations.

Copyright: © 2022 by the authors. Licensee MDPI, Basel, Switzerland. This article is an open access article distributed under the terms and conditions of the Creative Commons Attribution (CC BY) license (https://creativecommons.org/licenses/by/4.0/).

1. Introduction

Soil salinity is a key abiotic stress that interferes with crop growth, development, and yield through altering morphological, physio-biochemical, and molecular processes [1–6]. Every year, 1–2% of cultivable land is reduced due to soil salinity and worldwide, about 800 million hectares (23%) of total arable lands are affected by soil salinity [7,8]. It is predicted that salinity will affect 50% of the world's arable land by 2050 [9]. It has been reported that the rise in groundwater levels with high salt content, inefficient drainage and irrigation systems, and the overuse of fertilizers are responsible for soil salinity [10]. Plants use a number of methods to counteract salt stress in order to survive in an ever changing environment. Metabolic adjustments, increasing Na^+ efflux or Na^+ compartmentalization to vacuoles, scavenging of free radicals, the safeguarding of cellular machinery, ionic homeostasis maintenance, certain proteins expression and the up-regulation of their genes and so on are the plant adaptation mechanisms to salinity stress [11–14]. Additionally, it is widely recognized that using microRNAs (miRNAs) is a significant tactic that can affect post-transcriptional gene regulation under a variety of environmental conditions, including salt. Salt stress interaction is strongly controlled by post-translational gene regulations because various gene transcripts are differentially regulated by miRNAs during salt stress [15]. Furthermore, microRNAs serve important roles in embryogenesis, morphogenesis, life cycle stage transformation, flower formation, increases fruit ripening, boosts anthocyanin production, vegetative and reproductive stage transitions, tillering and branching, and enhances salinity stress tolerance in plants [15–18].

To reduce excess soil salinity, plant scientists are employing techniques such as subsoiling, mixing sand, seed bed preparation, and salt scraping, as well as modern agronomic practices, hydrophilic polymer, gypsum, sulfur acids, green manuring, humic substance, farm yard manures, irrigation system, and salt-tolerant crops [19–22]. Recently, different organic amendments such as the application of vermi-compost (VC), vermi-wash (VW), biochar (BC), plant growth promoting rhizobacteria (PGPR), and bio-fertilizers (BF) are being used widely to ameliorate the negative consequences of soil salinity [5,6,23–26]. For instance, VC enhances morphological traits, chlorophyll content, antioxidant enzyme activities, and improves salinity tolerance of maize plants [27]. Several studies showed that BF and BC enhance plant growth progressions under salinity stress by improving antioxidant enzyme activities, and reduces oxidative damage in different plants [5,28,29]. In addition, the inoculation of PGPR under salt stress accelerates microbial population and gene expression in the rhizosphere, boosts biomass production and enhances the salt tolerance of different plants [6,30,31]. The organic amendments mitigate salt stress via a wide range of mechanisms, including the regulation of ionic homeostasis, antioxidant enzyme activities, and the reduction of oxidative damage. Several studies described that PGPR and BC relieved the negative effects of salinity by increasing the photosynthetic rate, antioxidant enzyme functions, secondary metabolites accumulation, and decreasing ROS in plants [6,32–34]. Organic amendments such as VC and VW include a variety of plant growth-regulating components such as micro and macro elements, vitamins, enzymes, and hormones that have been shown to reduce the harmful effects of salts on plants [25]. Furthermore, several studies have stated that VC and VW have been shown to reduce soil salinity through the enhancement of antioxidant enzymes and to lessen electrolyte leakage and oxidative stress [35,36]. Similar to other organic supplements, BF has been shown to attain a better environment through fixing atmospheric nitrogen, phosphate and potassium solubilization or mineralization, releasing o plant growth regulating materials, producing antibiotics, and degrading organic matter in the soil, all of which contribute to increased plant salinity stress tolerance [37–39].

The world's population grows significantly every day. To feed the increasing millions, researchers are attempting to develop modern, effective agronomic and eco-friendly organic ways to reduce salinity stress and boost crop yields. Furthermore, investigating environmentally safe and sustainable methods to lessen the negative consequences of salinity is necessary due to the ongoing environmental degradation. The literature suggests that organic approaches can alter biochemical and molecular systems to enable plants to withstand salinity stress, and these strategies are proving to be quite effective. As a result, this report highlighted the recent findings about organic amendments like VC, VW, BC, BF, and PGPR used for salinity stress mitigation. Correspondingly, keeping in view the role of organic amendments during saline conditions, this review explores the potentiality of the modern and widely used organic amendments for the alleviation of salt stress and their regulatory mechanisms. Despite the fact that other reviews of organic amendments for reducing salt stress have been published independently, this study offers a thorough analysis of all commonly used organic amendments for reducing salt stress in a single frame.

2. Organic Amendments for Salinity Stress Mitigation

2.1. Vermicompost and Vermiwash

The VC is an organic fertilizer that is prepared through the conversion of organic wastes by worms [40], and is rich in different enzymes including humic and fulvic acids [41]. It contains a number of plant growth regulating substances (micro and macro elements, vitamins, enzymes, and hormones) and has anti-stress effects [42,43]. Earthworms in VC have beneficial effects on soil qualities such as physical, chemical, and biological properties [44], as well as increasing plant development and production by making nutrients available to the plant [45].

The VW and vermicompost leachate (VCL) are two important derivatives prepared from vermicompost. The VW is a clear, translucent, pale-yellow fluid obtained by passing water through a column of the vermi-worms' excreta, which contains mucus secretions as well as micronutrients from decomposed organic sources [46]. Khan et al. [47] reported that VW has been utilized as an organic fertilizer for plants, and is a rich source of amino acids, vitamins, N, P, Mg, Zn, Fe, Cu, auxins and cytokinins. The VCL is a liquid that is collected when water drains over decomposing solids [48]. This liquid may drain out from a traditional compost bin or a worm bin. Leachate is used as soil drench after dilution. VCL appears to be an effective and environmental friendly VC derivative for lowering salt's harmful impact on bean seedlings [49]. It was demonstrated that vermicompost promoted seed germination, root and shoot growth, proline accumulation, and oxidative stress management (Figure 1, [50–52]). In addition, VCL alleviates salt stress by enhancing photosynthetic efficiency, promoting antioxidant enzyme activity, and reducing electrolyte leakage [42]. Among the various organic amendments practices, VC and VW are low-cost techniques to reduce the detrimental impact of salts on plants (Table 1, [53]). The VC has been shown to reduce salt toxicity and enhance the emergence rate and seedling growth of different plants [54,55]. The VC enhances soil organic matter in salt soils by decreasing electrical conductivity (EC), bulk density and increasing field capacity, saturated hydraulic conductivity, and cation exchange capacity [56].

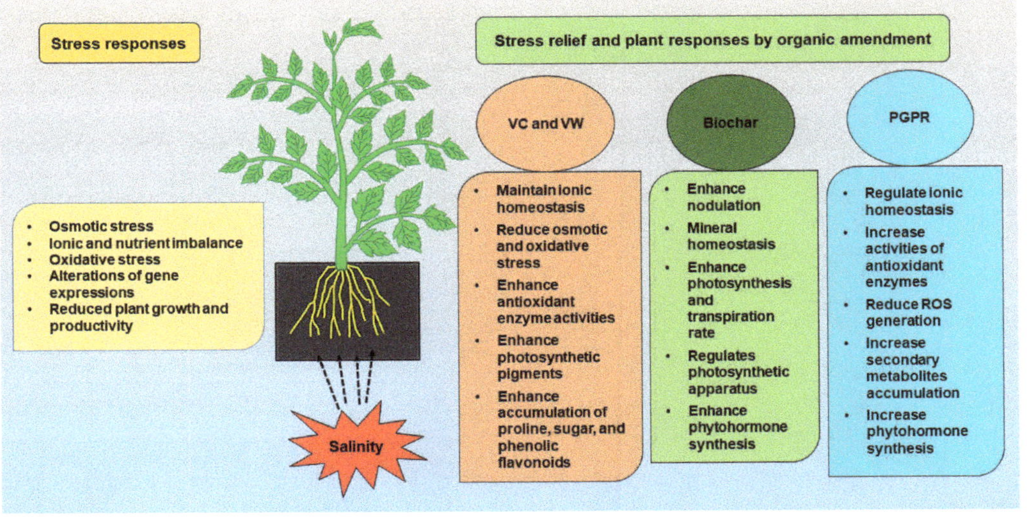

Figure 1. Stress relief and possible plant responses by VC, VW, BC, and PGPR.

A number of studies show that VC enhanced salinity tolerance and improved the morphological characteristics such as stem and root length, fresh and dry weight of root-shoot, vigor index, leaf area, and dry weight per plant [54,57–59]. It has been reported that VC boosted Na^+ exclusion and K^+ accumulation, alleviated stomatal restriction, raised leaf pigment concentrations, improved root activity, decreased oxidative damage in Fountain Grass, and improved salt tolerance [60]. In addition, VC application in maize and tomato plants under salt stress improved Chl a, Chl b, Total Chl, carotenoids, CAT, POD, and SOD and lowered H_2O_2 and MDA [27,61]. The addition of VC and VW to potato improved growth metrics, plant height, stem diameter, and tuber features, reducing the impact of salt stress [35]. Liu et al. [62] found that in coastal saline soil, VC application in maize increased nutrient availability and soil macro-aggregates by up to 91.02 percent. The soil amendment of VC increased exchangeable K^+, Ca^{2+}, and Mg^{2+}, plant height, total dry matter content, and decreased exchangeable Na^+ in the saline soil [63–65]. In addition, VC application reduced salt-induced injuries of plants grown in saline soil by increasing relative water content, stomatal conductance, chlorophyll-a, superoxide dismutase (SOD), ascorbate peroxidase (APX), and catalase (CAT) activities and decreased electrolyte leakage and malondialdehyde (MDA) levels [55,66–68].

Table 1. Application of vermi-compost (VC) and its derivatives for minimizing soil salinity.

Plant Species	Stress Level	Treatment and Application Methods	Effects of Amendments	References
Tomato (*Solanum lycopersicum* L.)	NaCl @ 150 mM	VC @ 6 mL/L	Improved foliar growth, increased water content of the leaves, reduced osmotic potential at the root level and Na content of the leaves; promoted the accumulation of proline and total sugars.	[69]
	NaCl @ 125 mM	VCL @ 18 mLL^{-1}	Improved plant growth and lowered Na^+ deposition in salt-stressed plants; delayed young leaf senescence.	[70]
	NaCl @ 0, 50 and 150 mM	VC @ 10, and 20%	Increased shoot length, stem diameter, leaves number, root length, shoot and root fresh, dry weight, Chl a, Chl b and carotenoid; increased Cat; decreased MDA; improved salinity tolerance.	[61]
Potato (*Solanum tuberosum* L.)	NaCl @ 15, 20, and 25 mM	VC @ 300, 580, and 860 g plant^{-1}; VW @ 5-, 10-, and 15-mL plant^{-1}	The addition of VC and VW increased the height of the plant and the diameter of the stem. VC reduced salinity effects on the plant.	[35]
	2.85 dSm^{-1}	Exogenous VC, proline and glycine betaine	Increased growth, yield, bio-constituents and antioxidant enzymatic activity. Improved salt tolerance of potatoes	[71]
Maize (*Zea mays* L.)	NaCl @ 0, 50, 100, 150 and 200 mM	VC with bacteria having ACC deaminase activity	Improved seed germination and the growth of seedlings; increased proline, chlorophyll content and alleviated the salt stress.	[72]
	Coastal saline soil	VC with humic acid fertilizer	Increased soil macro-aggregates, improved soil nutrient availability and maize nutrient uptake.	[62]
	NaCl @ 6, and 12 dS m^{-1}	VC @ 5, and 10%	Increased root, shoot fresh and dry weight; increased Chl a, Chl b, total Chl, carotenoids; increased CAT, SOD, POD activities; decreased H_2O_2, MDA content; increased salinity tolerance.	[27]

Table 1. *Cont.*

Plant Species	Stress Level	Treatment and Application Methods	Effects of Amendments	References
Moldavian dragonhead (*Dracocephalum moldavica* L.)	NaCl @ 0,50, 100 and 150 mM	VC @ 0, 5, 10 and 15% (*v*/*v*)	Increased plant biomass, chlorophyll content and proline accumulation. Reduced the effects of high sodium chloride concentrations.	[59]
Lemon verbena (*Lippia citriodora*)	NaCl @70 mM	VC @ 0%, 10% and 30% of pot volume	Alleviated salt stress by improving the growth and phenolic compounds of the plants.	[73]
Basil (*Ocimum basilicum* L.)	NaCl @ 0, 50 and 100 mM	Humates VC @ 0 and 1/60 *v*/*v*	Enhanced shoots and roots length, fresh and dry biomass of root, stem, leaf and leaf area. Reduced salinity.	[74]
Smoke tree (*Cotinus coggygria* Scop.)	NaCl @ 1, 4 and 7 $dS.m^{-1}$	VC @ 80% *v/v* soil + 20% *v*/*v*	Increased fresh and dry weight of shoots, increased leaf area; Reduced sodium and chloride of leaf and increased potassium. Increased salt tolerance of plant.	[58]
Fenugreek (*Trigonellafoenum-graecum* L.)	NaCl @ 0, 100 and 200 mM	VC @ 0, 5 and 10 weight%	Increased number of seed per pod, number of pods, number of sub branch and plant height. Reduced salinity effects.	[57]
Wheat (*Triticum durum* Desf. cv. Yelken)	High salt stress	VC and fish flour (1:1)	Enhanced growth, seed vigor and total phenolic-flavonoids, chlorophyl-carotenoids contents, and increased phenylalanine ammonialyase (PAL), peroxidase (POD) activities. VC decreased salinity effects.	[75]
	Coastal salinity	Soil amendment of VC	Increased soil macro-aggregates; Improved shoot biomass, grain yield, soil physical, chemical and biological properties. Ameliorated salt-induced stress.	[76]
	Saline soil	VC @ 10.0-ton ha^{-1}; Biochar @ 10-ton ha^{-1}	Improved relative water content, total chlorophyll, stomatal conductance, leaf K^+ concentration; Reduced oxidative stress, leaf Na^+ concentration, and proline content; improved yield related traits, productivity, soil water level and chemical properties. Eliminated the detrimental effects of soil salinity.	[68]
Rice (*Oryza sativa*)	Soil salinity	VC and rice husk ash @ 1000 kg per Rai for both	Increased exchangeable K^+, Ca^{2+} and Mg^{2+} in soil; reduced electrical conductivity and risen tillers per clump; improved the physiological and biochemical responses. Increased the rice growth in salt affected area.	[44]

Table 1. *Cont.*

Plant Species	Stress Level	Treatment and Application Methods	Effects of Amendments	References
Lettuce (*Lactuca sativa*)	NaCl @ 0, 4 and 8 dS m^{-1}	VC @ 0, 2.5 and 5% (*w/w*)	Enhanced soil organic matter, available P, total N, available K and the cation exchange capacity of the soils; Increased field capacity, available water capacity, saturated hydraulic conductivity, total porosity, and aggregate stability; Decreased EC values and the bulk density of the soils.	[56]
	NaCl @ 8.32 dS/m	VC 50% and pulverized eggshell 12.5%	Decreased soil salinity for about 77%; fasten the seed germination and seedling growth.	[77]
	NaCl @ 4, 8 dSm^{-1}	VC 5% (*w/w*)	Increased P, K, Mg, Fe, Mn and Zn concentrations; decreased Na contents. Reduced toxic effects of salinity on the plant.	[65]
	NaCl @ 4 and 8 dSm^{-1}	VC @ 0, 2.5 and 5% (*w/w*)	Increased relative water content, stomatal conductance, chlorophyll *a* content; decreased electrolyte leakage, malondialdehyde (MDA) contents; increased superoxide dismutase (SOD) and catalase (CAT) activities. Alleviated the salt stress.	[67]
Pot marigold (*Calendula officinalis* L.)	NaCl @ 0, 50, 100, 150 and 200 mM	VC @ 0%, 5%, 10%, 15% and 20%	Increased the activity of the antioxidant system; increased proline and chlorophyll content. Reduced salinity impacts and boost-up yield.	[78]
Noni (*Morinda citrifolia* L.)	Salinity stress @ 0.5, 1.5, 3.0 and 4.5 dS m^{-1}	Substrates with humus; 33.33 and 66.66% of humus	Decreased the intensification of electrical conductivity of irrigation water; mitigated the negative effects of salts on plants.	[79]
Bean (*Phaseolus vulgaris* L.)	NaCl @ 20, 40, 60 and 80 mmol l^{-1}	VC: Sand = 0:100; 10:90; 25:75; 50:50 and 75:25	Increased photosynthetic rate and potassium (K^+) and calcium (Ca^{2+}) concentration in leaf and root; improved the growth of bean plants. Alleviated negative effects of salinity.	[49]
Pomegranate (*Punica granatum* L.)	NaCl @ 0, 30, and 60 mM	Vermicompost leachate (VCL) foliar spray	Leaf area, photosynthetic efficiency, and shoot and root fresh and dry weight significantly increased; improved the activity of antioxidant enzymes; reduced oxidative stress and electrolyte leakage. VCL alleviated the damage caused by salt stress	[42]
Tall fescue turfgrass (*Festuca arundinacea* cv Queen)	NaCl @ 0, 3, 6 and 12 dS/m	VC @ 0, 100, 200 and 300 g	Activities of CAT and APX were increased; leaf area, shoot length and dry shoot weight were highest. Reduced the effects of high concentrations of sodium chloride in saline soils.	[66]
Onion (*Allium cepa* L. cv. Metan)	NaCl @ 50 and 100 mM	Seed Priming with VC	Higher germination, seedling growth, CAT, SOD and APX activities were found in VC treated seeds. VC mitigated salinity effects	[55]
Bell pepper (*Capsicum annuum* L.)	NaCl @ 160 mM	Addition of 7 mL/L VCL	Increased sugar concentration in roots and proline content in leaves; increased leaf fresh weight. VCL enhanced the property of salt-stress resistance in bell peppers.	[52]

Table 1. *Cont.*

Plant Species	Stress Level	Treatment and Application Methods	Effects of Amendments	References
Medicago (*Medicago rigidula* L.)	NaCl @ 0, 50 and 100 mM	VC @ 0, 10, 20 and 30%.	Increased plant survival capacity, chlorophyll contents, shoot dry weight; maximize leaf area values.	[80]
Sunflower (*Helianthus annuus* L.)	EC: 0.5, 4.8 and 8.6 dS/m	VC @ 1 kg/pot	Increased plant growth, yield, nitrate and protein content; decreased sodium (Na^+), chloride (Cl^-), ammonium; Increased N-assimilation.	[64]
Borage (*Borago officinalis*)	NaCl @ 0, 4, 8 and 12 dSm^{-1}	VC @ 0, 6, 12 and 18% (*w*/*w*) of soil	Increased chlorophyll b, carotenoids and MDA contents and reduced the negative effects of salinity.	[81]
Milk thistle (*Silybum marianum* L.)	NaCl @ 0, −2, −4, −6, and −8 bar	Superabsorbent polymers with VC coats	Increased seedlings emergence rate, plant vigor index, shoot dry weight, leaf area, specific leaf area, relative water content, and total chlorophyll.	[54]
Sugarcane commercial variety of 'Bululawang (BL)'	NaCl @ 4.12 dS/m	VC @ 0, 10, 20 t/ha) and nitrogen fertilizer @ 50, 75 and 100 kg N/ha	Increased N, K uptakes and the growth of sugarcane and alleviated salinity effects.	[82]
Rapeseed (*Brassica napus* L.)	NaCl @ 100 mM	VCL (1:10, *v*/*v*)	VCL was shown to improve seed germination and management of oxidative stress.	[51]
Fountain Grass (*Pennisetum alopecuroides*)	NaCl @ 5.0 g per kg soil	VC	Enhanced Na^+ exclusion and K^+ accumulation, relieved stomatal limitation, increased leaf pigment contents, enhanced electron transport efficiency and net photosynthesis, improved root activity, and minimized the oxidative damage.	[60]

2.2. Biochar

BC is a carbon-rich organic substance with a porous structure, a wide surface area, and a high ion exchange ability that improves the physical qualities of agricultural soil [83,84]. A number of studies found that BC application improves the different physio-biochemical processes such as photosynthesis, hormonal and enzymatic activity in plants and decreases the harmful effects of salt stress on plants (Figure 1 and Table 2, [5,22,83–85]).

Morphological attributes such as seedling emergence, plant height, shoot biomass, root biomass, leaf area, dry matter, and yield of plants under salinity stress have been shown to be improved by BC incorporation [32,86,87]. Moreover, BC application boosted photosynthetic rate, stomatal conductance, and transcription rate under salinity stress conditions in wheat [84], sorghum [87], quinoa [83], and eggplant [32]. On the other hand, the availability and uptake of different nutrients such as N, P, K in maize [88] and P, K, Fe, Mn, Zn, and Cu in tomato [89] improved by the utilization of BC as amendment to saline soil.

Furthermore, in saline conditions, BC traps excess Na^+ in soil, releasing mineral nutrients and decreasing osmotic stress [86]. Studies showed that the use of BC lowered the concentration of Na^+ and decreased the Na^+/K^+ ratio in a variety of plants, assisting in the reduction of the negative effects of salt on plants [84,90]. Moreover, under salinity stress, BC application improves osmotic balance by increasing water holding capacity and CO_2 assimilation, which ultimately results in a better photosynthetic rate, stomatal conductance, and transcription rate [32,83,86]. It has been reported that the leaf photosynthesis and net assimilation rate of the rice population was greatly aided by biochar's potential positive effects on chlorophyll content, leaf N content, leaf area index, photosynthetic potential, stomatal conductance, and transpiration rates [91,92]. Additionally, the biochar treatment greatly enhanced the salt tolerance of cabbage seedlings and dramatically raised LRWC, Chl a, Chl b, and total Chl while reducing sucrose, proline, H_2O_2, and MDA [29]. In addition, BC application reduced the effects of salt by lowering the levels of phytohormones such ABA, ACC, and JA, as well as increasing the amount of IAA in beans [93]. Similarly, Nikpour-Rashidabad et al. [94] reported that BC improved the vascular cylinder, parenchyma, IAA/ABA and IAA/ACC ratios to ameliorate the effects of salinity. Furthermore, under salt stress, the treatment with BC enhanced nodulation, nitrogen content, rubisco activity, GDH, GS, GOGAT, and NR activities in various parts of the soybean plant [85]. Given the findings of the preceding investigations, BC appears to be a promising strategy for reducing salt stress and increasing plant growth and biomass in a variety of plants.

Table 2. Application of biochar (BC) for minimizing soil salinity.

Plant Species	Stress Level	Treatment and Application Methods	Effects of Amendments	References
Wheat (*Triticum aestivum* L.)	Saline water irrigation @ 10 dSm^{-1}	BC @ 10, 20, 30 t/ha	Increased and relative water content, photosynthesis. Decreased Na^+/K^+, and leaf senescence.	[84]
	NaCl @ 3000 ppm	Soybean straw BC	Increased plant growth, grain yield and biomass production; increased leaf chlorophyll content, water use efficiency, PSII efficiency, and net photosynthesis rate; decreased electrolyte leakage, H_2O_2, MDA; increased CAT, APX, SOD, GR activities; improved salinity tolerance.	[95]
Quinoa (*Chenopodium quinoa* L.)	Saline water irrigation @ 400 mM	BC@ 5% (*w/w*)	Increased photosynthesis, stomatal conductance, WUE and K^+ content. Decreased ABA and Na^+ content.	[83]
Eggplant (*Solanum melongena* L.)	Saline water irrigation @ 2 and 4 dSm^{-1}	Hardwood BC @ 5%, Softwood BC @ 5%	Increased biomass, photosynthesis and stomatal conductance. Decreased leaf temperature and electrolyte leakage in leaf tissue.	[32]
Maize (*Zea mays* L.)	Saline soil	Wheat straw BC @ 12 t/ha	Increased LAI, Chlorophyll content, K, P and N content. Reduced MDA, soluble sugar, ascorbic acid and proline content.	[88]
	Saline soil	BC @ 5% (*w/w*)	Increased photosynthesis and stomatal conductance, K^+ content and K^+/Na^+. Decreased ABA and Na^+ content.	[96]
Soybean (*Glycine max* L.)	NaCl @ 5 and 10 dSm^{-1}	BC @ 50 and 100 g kg^{-1} soil	Improved nodulation, chlorophyll content, N content, rubisco activity, GDH, GS, GOGAT, and NR activities.	[85]
Bean (*Phaseolus vulgaris* L.)	NaCl @ 6 and 12 dSm^{-1}	BC @ 10% and 20% *w/w*	Decreased Na^+ concentration, PAO activity, polyamines, ABA, ACC and JA; enhanced IAA content.	[93]
Mungbean (*Vigna radiata* L.)	NaCl @ 5 and 10 dS m^{-1}	BC @ 50 and 100 g kg^{-1}	Increased and relative water content, IAA content, vascular cylinder, cortical parenchyma areas, IAA/ABA and IAA/ACC ratios; decreased ABA and ACC.	[94]
Sorghum (*Sorghum bicolor* L.)	NaCl @ 0.26, 5.8 and 12.6 dSm^{-1}	Soil mixer @ 2.5%, 5% and 10% (*w/w*) of total mass	Increased photosynthesis, stomatal conductance, transpiration rate CAT, POD, and SOD activity.	[87]
	NaCl @ 0.8, 4.1, and 7.7 dS m^{-1}	BC @ 0, 2.5, 5, and 10% (*w/w*)	Increased saddling emergence percentage, dry matter accumulation, and relative water content. Mitigated salinity stress.	[86]

Table 2. *Cont.*

Plant Species	Stress Level	Treatment and Application Methods	Effects of Amendments	References
Potato (*Solanum tuberosum* L.)	NaCl @ 25 and 50 mM	BC @ 5% *w*/*w* of total mass	Increased photosynthesis, stomatal conductance, leaf water potential, K^+ content; decreased Na^+, Na^+/K^+ ratio and ABA concentration.	[97]
Rice (*Oryza sativa*)	Saline soil	BC @ 0%, 1.5%, 3.0% and 4.5% *w*/*w*	Increased biomass, grain yield; decreased in leaf Na^+ concentration and Na^+/K^+ ratio; increased in leaf K^+ concentration; decreased ABA, MDA content; increased leaf photosynthesis rates (Pn), transpiration rates (Tr), stomatal conductance (Gs); improved salinity tolerance.	[91]
	NaCl @ 3 g per kg soil	BC application	Decreased the value of EC, soluble Na^+ and Cl^- contents; increased CEC, SOM, HA, total nitrogen, and total phosphorus contents in the soil; increased soil microbial community.	[92]
Cabbage (*Brassica oleracea*)	NaCl @ 0 and 150 mM	BC @ 0%, 2.5%, and 5%	Increased stem diameter, leaf area, shoot fresh weight, root fresh weight, shoot dry weight, and root dry weight; decreased malondialdehyde (MDA), hydrogen peroxide (H_2O_2), proline, and sucrose content; reduced Cl and Na concentration, and reactive oxygen species (ROS) production; increased CAT and SOD activities.	[29]
Borage (*Borago officinalis*)	NaCl @ 1250, 2500, 5000, and 7500 mg per kg soil	BC @ 5%	Decreased leaf water potential (Yw), osmotic potential (Ys), water saturation deficit (WSD); increased relative water content (RWC), water content (WC), and water retention capacity (WTC); increased K^+, and K^+/Na^+ ratio; decreased MDA, H_2O_2; increased POD, SOD activities; improved salinity tolerance.	[98]

2.3. Bio-Fertilizer

BFs are one kind of fertilizer that contains living cells from various microorganisms and can transform via biological mechanisms; nutrients are converted from the inaccessible to the accessible form [99,100]. Recently, many studies have described the potential of BF in salt tolerance enhancement (Table 3). The application of BF in wheat seedlings lessened the negative effects of salinity by increasing chlorophyll content and decreasing proline content, and improved plant growth and yield [28,37]. Under salt stress, amaranth enhanced plant height and biomass production [101]. It has also been reported that the application of BF to lavender enhanced its capacity to withstand salt stress by increasing morphological attributes and RWC, Chl a, Chl b, and total Chl content as well as its essential oil output [102]. Similarly, BF application on wheat (*Triticum aestivum* L.), okra (*Abelmoschus esculentus* L.), yellow passion fruit (*Passiflora edulis*), cowpea (*Vigna unguiculata* L.), corn (*Zea mays* L.), and olive plants (*Olea europaea* L.) enhanced growth and yield metrics, micro and macronutrient content, and relieved salinity-related detrimental effects [28,37,38,103–106]. Souza et al. [104] showed that in yellow passion fruit, BF application reduced the salt stress and enhanced the absolute growth rate, side branches, and yield. In addition, BF application to olive and papaya plants increased growth and plant biomass, improved osmotic adjustments between root and soil, increased microbial activity in the rhizosphere zone, and reduced the toxic effects of salts [106,107].

BF application increased antioxidant activity through the up-regulating of POX, SOD, and CAT, and reduced MDA and H_2O_2 production in lettuce (*Lactuca sativa* L.), safflower (*Carthamus tinctorius* L.) and cowpea (*Vigna unguiculata* L.) [105,108,109]. Al-Taey and Majid [108] found that the functions of POD, CAT, SOD, and MDA were increased as a result of the increased salinity stress in lettuce (*Lactuca sativa* L.). It has been reported that BF ameliorates the effects of salt stress via the production of phytohormones (IAA, CK, and ABA) and secondary metabolites (proline) in plants [109–111].

Overall, the discussion concluded that BFs increased plant growth and production while also inducing salt tolerance by enhancing antioxidant enzyme activities, secondary metabolite accumulation and phytohormone synthesis.

Table 3. Bio-fertilizer used for mitigating soil salinity.

Plant Species	Stress Level	Treatment and Application Methods	Effects of Amendments	References
Wheat (*Triticum aestivum* L.)	NaCl @ 0, 3000, 6000, 9000 ppm	Cerealien, Phosphorien and Cerealien + Phosphorien in addition mix-up with wheat grains.	Increased growth, dry matter accumulation, and yields. Decreased proline content. Improved salinity tolerance.	[37]
	NaCl @ 0, 2.76, 5.53, and 8.3 dSm^{-1}	Four (04) biofertilizer treatments were applied: not at all biofertilizer; seed injection by *Azotobacter chroococcum* Beijerinck strain 5; *Pseudomonas putida* (Trevisan) Migula strain 186; joint inoculation of *Azotobacter* + *Pseudomonas*	Increased chlorophyll index, relative water content, and grain yield. Concentrated dry matter, stem reserve mobilizations to grain yield and decreased proline content.	[28]
Lettuce (*Lactuca sativa* L.)	Irrigated with saline water @ 1.2 dSm^{-1}	Biofertilizer @ 5 kg/ha	Increased POD, CAT, MDA, SOD activities. Decreased disruption of endohormones, osmotic stress and mitigates salinity stress.	[108]
Geranium plant (*Pelargonium graveolens* L.)	Irrigated with saline water $NaCl_1$: $NaCl_2$ (1:1)	(Half dose of compost + Bio) & (full dose of peanut compost + Bio) added to the pot.	Increased oil percentage but N, P, K contents remained unchanged. Improved yield and mitigated salinity stress.	[112]
Okra (*Abelmoschus esculentus* L.)	Irrigated with saline water with 3 levels 0.47, 2, & 4 dSm^{-1}	Biofertilizers + Ascorbic acid @ 100 & 200 mgL^{-1} was applied.	Increased chlorophyll content, growth and yield but deceased ascorbic acid and proline content in okra plants.	[38]
Barley (*Hordeum vulgare*) & Broad beans (*Vicia faba*)	Irrigated with saline water @ 0, −1, −3, −5 Mpa	Seeds were presoaked with biofertilizer (2 mL of nanomaterial + 10 mL cyanobacterial (algal culture) + 10 mL rhizobacterial strain + 10 mL MeSA) for one day and 12 h and then added to the saline soil.	Increased bioavailability of nutrients, production of growth hormones and bio-stimulants. Decreased Na^+, Cl^-, and proline concentrations ultimately reduced salinity.	[113]
Yellow passion fruit (*Passiflora edulis*)	Irrigated with saline water (EC 0.35 & 4 dSm^{-1})	Soil applied biofertilizer @ 0 and 50%	Increased absolute growth rate, period for pruning the side branches, and yield, and decreased the adverse effect of salinity.	[104]
Soybean (*Glycine max* L.)	Saline water @ 3.13, 6.25, 9.38 dSm^{-1}	Seeds were inoculated with bio-fertilizers and applied on the field.	Increased ascorbic acid, total indoles, a- amylase activity and polyphenol oxidase, decreased total soluble phenols, total soluble sugars and free proline. Decreased the salinity effects.	[114]

Table 3. *Cont.*

Plant Species	Stress Level	Treatment and Application Methods	Effects of Amendments	References
Safflower (*Carthamus tinctorius* L.)	NaCl @ 250 mM	Coated seeds with biofertilizers & sugars were applied to the pot.	Increased antioxidant enzymes (SOD, CAT, POD, and APX), decreased proline and malondialdehyde (MDA). Improved salinity tolerance	[109]
Peanut (*Arachis hypogaea* L.)	Irrigated with saline water @ 0.5, 1.5, 2.5, 3.5, 4.5 and 5.5 dSm^{-1}	Biofertilizer @ 15, 30 and 45	In the peanut, it promoted higher vegetative growth and improved photosynthesis rate. Decreased soil salinity and improved yield.	[115]
Cowpea (*Vigna unguiculata* L.)	NaCl @ 25, 50, 100, 200, and 300 mM	Biofertilizers mixed with sand @ 0.8 g/Kg	Increased growth parameters, total pigments, protein, proline contents and activities of SOD and CAT. Reduced H_2O_2 production and alleviated salinity stress.	[105]
Pitombeira seedlings (*Talisia esculenta*)	NaCl @ 0.8, 2, 4, 6, 8 dSm^{-1}	Biofertilizer @ 10% of the total volume	Increased plant height, stem diameter, number of leaves, leaf area, total leaf area, Dickson quality index, dry mass of root and stem. Mitigated the harmful effects of salinity.	[116]
Cotton (*Gossypium hirsutum* L.)	NaCl @ 15 dSm^{-1}	Seeds were coated with biofertilizers.	Increased shoot growth, root growth and yield. Decreased leaf gas exchange characteristics.	[117]
Corn (*Zea mays* L.)	Irrigated with saline water @ 0.47, 2.50, and 3.90 dSm^{-1}	Biofertilizer "Halix" was applied as an inoculum to corn seeds before cultivation.	Increased the concentrations of macro and micronutrients, total chlorophyll, and ascorbic acid in maize plants, as well as mitigated the negative effects of salinity on corn.	[104]
Olive (*Olea europaea* L.)	Irrigated with saline water @ 2000, 3000 and 4000 ppm	Biofertilization treatments control, *Azotobacter* chroococcum, Mycorrhizae (*Glomus macrocarbium*) and mix of *Azotobacter chroococcum* + Mycorrhizae	Enhanced growth and plant biomass, improved microbial activity in the rhizosphere zone. Decreased intensity of salt toxic effects.	[106]
Papaya (*Carica papaya* L.)	Irrigated with saline water @ 0.5, 1, 2, 3 and 4 dSm^{-1}	Biofertilizer applied @ 10% of the substrate volume.	Enhanced growth and plant biomass, provided greater osmotic adjustments between root and soil solution, increased absorption efficiency of water and essential nutrients stimulating plants to grow. Decreased intensity of salt toxic effects on growth.	[107]
Amaranth (*Amaranthus tricolor* L.)	NaCl @ 0, 2500, 5000, 7500, and 10,000 ppm	*Bacillus* sp., *Lactobacillus* sp., *Saccharomyces* sp., *Streptomyces* sp., *Azospirillum* sp., *Pseudomonas* sp., *Azotobacter* sp., *Rhizobium* sp.	Increased plant height, number of leaves, and stem metaxylem diameter.	[101]
Lavender (*Lavandula angustifolia*)	NaCl @ 0, 50, and 100 mM	Azotobacter, Azospirillum, and a combination of Azotobacter and Azospirillum	Increased plant height, stem length, root length, fresh weight, dry weight, relative water content, chlorophyll a, chlorophyll b, total chlorophyll, and essential oil yield; improved salinity tolerance.	[102]

2.4. PGPR

Plant-growth-promoting rhizobacteria (PGPR) are microorganisms that colonize plant roots and are used as chemical alternatives in agricultural fields for crop production and protection [6,118]. PGPR, which are resistant to salinity, help the plants to endure salty conditions. These plant-associated rhizobacteria can synthesize a variety of substances, including extracellular polymeric substance, 1-aminocyclopropane-1-carboxylate deaminase, phytohormones, antioxidants and volatile chemical compounds [6,30]. Gao et al. [119] reported that rhizosphere bacteria reduce salt stress while promoting plant development by supplying nitrogen, phosphate, potassium, auxin, cytokinin, and abscisic acid to plants. During several field tests, crops grown under saline soil conditions responded favorably to the utilization of PGPR in terms of growth and yield (Table 4). Kumawat et al. [120] in his study revealed that PGPR increased seed germination, height of the plant, biomass, and chlorophyll contents under salt stress that ameliorate the negative effects of soil salinity. Water potential and stomatal opening is a crucial plant physiological activity for their survival which even salinity-stressed condition were found to be modified by PGPR to compensate salt stress [121,122]. For example, *Enterobacter cloacae, Pseudomonas fluorescence, Bacillus pumilus, and Exiguobacterium aurantiacum* were found to greatly alleviate the toxic effect of salt stress in *Triticum aestivum* plants [122,123]. Moreover, Ali et al. [124] reported that under salt stress circumstances, *Enterobacter cloacae* PM23 boosted maize growth, biomass, photosynthetic pigment contents, carotenoids, and relative water content compared to control treatment. Similar effects were observed when co-inoculation of *Rhizobium* sp. and *Enterococcus mundtii* in *Vigna radiata* were carried out and obtained the grain production was improved under saline stress by regulating ion homeostasis [120]. Additionally, when infected with *B. megaterium*, *Solanum lycopersicum* and *Arabidopsis thaliana* both grew roots, shoots, and more leaves under salt stress [125,126]. Furthermore, *S. marcescens* inoculation enhanced *Triticum aestivum* shoot length, fresh weight, and chlorophyll (Chl) content [127]. Under saline stress conditions, the *Enterobacter cloacae* in *Brassica napus* enhanced seedling development [128]. Inoculating *Triticum aestivum* with *Pseudomonas fluorescens* led to similar outcomes, as did inoculating *Oryza sativa* with *Alcaligenes faecalis*, *B. pumilus*, and *Ochrobactrum* sp. [129]. In addition, the application of some PGPR has been shown to improve nodule formation and fix nitrogen in plants under salt stress [130]. For example, *Rhizobium* sp. and *Bradyrhizobium japonicum*'s co-inoculation improved root nodule formation in *Glycine max* compared to control conditions, resulting in increased stress tolerance, plant growth, and higher yield [130]. Likewise, *Bacillus aryabhattai* and *Azotobacter vinelandii* inoculation enhanced root nodule numbers and N-contents in *Trifolium repens* compared with the non-inoculated plants [131]. However, PGPR not only increased nodule numbers but also increased plant dry weight, shoot dry weight, the extent of nitrogen yield and protein content in some applications [132].

Many studies described that PGPR can alleviate the salt-induced growth inhibition of plants by positively regulating ion homeostasis and antioxidant enzyme activity, improving photosynthetic attributes, secondary metabolite accumulation, and oxidative stress reduction (Figure 1 and Table 4, [6,118,133,134]). For instance, the use of PGPR reduced the negative effects of salinity in pea (*Pisum sativum*) by enhancing the plants' proline and soluble sugar contents while lowering sodium (Na^+) contents, which in turn reduced the amount of electrolyte leakage and H_2O_2 content [135,136]. In addition, the harmful effects of salinity are reduced by PGPR via declining lipid peroxidation and ROS in wheat plants [137]. Singh et al. [125] and Kumawat et al. [120] reported that PGPR alters the selectivity of Na^+, K^+, and Ca^{2+} under salt stress and thus maintains ionic balance due to ion homeostasis. Moreover, inoculating *Pseudomonas* sp. or *Glutamicibacter* sp. with the halophte *Suaeda fruticosa* led to noticeably greater shoot dry weight and decreased buildup of Na^+ and Cl^- in shoots of salt-treated plants [138]. Similarly, the *Piriformospora indica* inoculation in *Zea mays* decreased K^+ flow from roots while increasing K^+ concentration in shoots under saline condition; this effect may be linked to a high-affinity K^+ transporter where PGPR produced a proton-driven force through H^+-ATPase [139]. Moreover, *Azotobacter* isolates under salinity stress had greater K^+/Na^+ proportions in shoots and decreased Na^+ and Cl^- amounts in maize leaves [140]. Additionally, some PGPR lowered Cl_2 and NO_2 concentrations and increased the K^+/Na^+ ratio, which contributed to enhanced stomatal conductance, and maintained hormonal balance and photosynthesis under salt stress [6,141]. It has also been reported that PGPR incorporation enhances the synthesis of the phytohormone that improves salt stress tolerance [6]. Such as, Some PGPR i.e., *Thalassobacillus denorans*, *Oceanobacillus kapialis*, *Pseudomonas* strains, *Bacillus tequilensis* and *Bacillus aryabhattai* synthesized more auxin and ABA, accumulated osmolytes in cell cytoplasm that sustain their cell turgor to make ensure plant growth under osmotic stress in *Oryza sativa* to endure high saline conditions [142]. In addition to the mechanisms mentioned above, in order to survive under salt stress, PGPR may alter salt tolerant gene expressions. The expression of TaABARE, TaOPR1, TaMYB, TaWRKY, TaST, SOS1, SOS4, TaNHX1, TaHAK, and TaHKT1 genes were up-regulated in PGPR inoculated plants leading to the expression of stress related genes [143,144]. According to the findings, salinity tolerance genes ZmNHX1, ZmNHX2, ZmN HX3, ZmWRKY58, and ZmDREB2A were up-regulated, as well as the antioxidants ZmGR1 (Zea mays glutathione reductase) and ZmAPX1's (Zea mays ascorbate peroxidase) transcript levels [145,146]. Moreover, salinity tolerance was increased when PGPR enhanced antioxidant enzymes' gene expression such as CAT, POD, APX, MnSOD, GR and GPX in inoculated plants [143]. Furthermore, according to Ali et al. [124], the inoculation of maize with *Enterobacter cloacae* PM23 increased APX, SOD, POD, total soluble sugars, and proteins while decreasing flavonoids and phenolic contents under salt stress. Additionally, in *Suaeda fruticosa* under high salinity, *Glutamicibacter* sp. inoculation dramatically decreased MDA levels while enhancing the activities of SOD, CAT, APX, and GR. Habib et al. [147] reported that salinity circumstances in the okra plant led to greater synthesis of APX and CAT by *B. megaterium* and *Enterobacter* sp. It has also been found that the treatment of *Arabidopsis* seedlings with *Enterobacter* sp. increased APX function and boosted salt tolerance [148]. Thus, from the reports of the studies it is apparent that the exogenous application PGPR could bring positive growth and yield results within the plants under saline condition and can be considered as a promising modern agronomic tactic to develop the plants survival under a saline environment. In future, dealing with extensive molecular research may reveal the efficacy of PGPR isolates and mechanisms to improve its stress responsive capability within the short duration in a wide area for sustainable agricultural production.

Table 4. Effects of PGPR on plant growth enhancement and salinity stress mitigation.

Plant Species	PGPR Inoculation	Salinity Stress	Effects of Inoculation	References
Wheat (*Triticum aestivum*)	*Pseudomonas fluorescence*, *Bacillus pumilus*, and *Exiguobacterium aurantiacum*	10% NaCl solution	Maximum root growth and dry biomass was observed; higher in proline and total soluble proteins contents; antioxidant activity improved; improved water and osmotic potential.	[122]
	Enterobacter cloacae	10% and 15% NaCl solution	Decreased the accumulation of Na^+ and increased K^+ uptake in shoots and roots; higher K^+/Na^+ ratios; improved antioxidant activity.	[123]
	Bacillus subtilis and *Arthrobacter* sp.	2–6 dSm^{-1}	Improved antioxidant activity; increased in dry biomass, total soluble sugars and proline content.	[135]
	Dietzia natronolimnaea	100 and 150 mM NaCl	Modulated the expression of stress responsive genes; improved ion transporters TaNHX1, TaHAK, and TaHKT1; improved the activities of antioxidant enzymes.	[143]

Table 4. *Cont.*

Plant Species	PGPR Inoculation	Salinity Stress	Effects of Inoculation	References
Maize (*Zea mays*)	*Serratia marcescens*	150–200 mM NaCl	Higher osmo-protectants and growth parameters; higher K^+/Na^+ ratios; increased SOD, APX, and CAT activity.	[127]
	Kocuria rhizophila	100 and 200 mM NaCl	Improved IAA and the ABA activity; upregulation of salt tolerant genes ZmNHX1, ZmNHX2, ZmNHX3, ZmWRKY58 and ZmDREB2A; higher K^+/Na^+ ratios; improved the growth parameters; higher chlorophyll, proline, and total soluble sugar content.	[146]
	Azotobacter chroococcum	0, 2.93 and 5.85 g NaCl/kg soil	Increased in biomass and stomatal conductance; higher $K^{+/}Na^+$ ratios; improved antioxidant enzyme activity.	[140]
	Enterobacter cloacae	0, 300, 600, and 900 mM NaCl	Enhanced plant growth, biomass, and photosynthetic pigments under salinity stress; enhanced radical scavenging capacity, RWC, soluble sugars, proteins, secondary metabolite content.	[124]
	Piriformospora indica	500 µM KCl and 100 µM $CaCl_2$	Higher biomass and stomatal conductance; lower K^+ efflux from roots and higher potassium content in shoots.	[139]
Soybean (*Glycine max*)	*Rhizobium* sp. *Bradyrhizobium japonicum* and *Hydrogenophaga* sp.	100, 250, and 500 mM NaCl solution	Higher shoot biomass at the vegetative stage, reproductive stages; improved seed weight and shoot K^+/Na^+ ratio.	[130]
	Methylobacterium aminovorans and *Methylobacterium rhodinum; Bradyrhizobium japonicum* and *Bacillus megaterium*	0.170 dSm^{-1}	Increased nodule numbers and dry weight of nodules; significantly increased in N, P and K; higher number of pods, seed index and seed yield.	[132]
Rice (*Oryza sativa*)	*Pseudomonas pseudoalcaligenes and Bacillus pumilus*	5, 10, 15, 20, and 25 g NaCl L^{-1}	Reduced lipid peroxidation and superoxide dismutase activity; reduced plant cell membrane index cell caspase-like protease activity, and programmed cell death.	[137]
	Bacillus amyloliquefaciens	120 and 250 mM NaCl	Higher synthesis of amino acids; improved endogenous SA and ABA; improved plant physiology.	[142]
Mung bean (*Vigna radiate*)	*Rhizobium* sp. and *Enterococcus mundtii*	10% NaCl solution	Higher seed germination and seedling growth and biomass; enhanced chlorophyll content and macro-micronutrient uptake; improved soil physical, chemical and biological parameters.	[120]
Barley (*Hordeum vulgare*)	Bacillus megatherium, Pseudomonas fluorescens, Bacillus circulans, Paenibacillus polymyxa, Azotobacter chroococcum, *Azospirillum* sp. Paenibacillus polymyxa2, Azospirillum brasilense, *Hyderella* sp.	250, 500 or 1000 mM NaCl	Alleviated the deleterious effect of salinity; higher dry masses and relative water content.	[133]

Table 4. *Cont.*

Plant Species	PGPR Inoculation	Salinity Stress	Effects of Inoculation	References
Pea (*Pisum sativum*)	*Acinetobacter bereziniae*, *Enterobacter ludwigii*, and *Alcaligenes faecalis*	75 mM, 100 mM and 150 mM NaCl	Improved the growth parameters; higher chlorophyll, proline, and total soluble sugar content; improved electrolyte leakage; improved the activities of antioxidant enzymes.	[136]
Pepper (*Capsicum annuum*)	*Azospirillum brasilense* and *Pantoea dispersa*	40, 80 and 120 mM NaCl	Higher K^+ /Na^+ ratio; improved leaf photosynthesis and stomatal conductance.	[141]
Burclover (*Medicago* sp.)	*Bacillus megaterium*, *E. medicae*, *Ensifer Medicae* and *B. megaterium*	0–2000 mM NaCl	Improved IAA and the ACC deaminase activity; higher chlorophyll, proline, and total soluble sugar content.	[125]
Okra (*Abelmoschus esculentus*)	*Bacillus megaterium* and *Enterobacter* sp.	75 mM NaCl	Enhanced ROS-scavenging enzyme activity; increased antioxidant enzyme SOD, APX, and CAT; upregulation of ROS pathway genes CAT, APX, GR, and DHAR.	[147]
Suaeda fruticosa	*Glutamicibacter* sp. and *Pseudomonas* sp.	600 mM NaCl	Increased shoot K^+ and Ca^{2+} content; lowered shoot MDA concentration and less accumulation of Na^+ and Cl^- in shoots.	[138]
Rapeseed (*Brassica napus*)	*Enterobacter cloacae*	50 and 100 mM NaCl	Promoted seed germination and seedling growth; improved chlorophyll, water potential and other physiological activity.	[128]
Avena sativa, *Medicago sativa*, *and Cucumis sativus*	*Advenella incenata*, *Providencia re-Ttgeri*, *Acinetobacter calcoaceticus*, and *Serratia plymuthica*	Salinity stress	Enhanced ROS-scavenging enzyme activity; increased SOD, APX, and CAT activity; enhanced plant growth, and photosynthetic pigments. enhanced RWC and proteins, content	[145]
Tomato (*Solanum lycopersicum*)	*Bacillus megaterium*	200 mM NaCl	Improved the growth parameters and biomass; higher chlorophyll, proline, and total soluble sugar content.	[126]

3. Limitation of Organic Amendments and Future Perspectives

The organic amendments have particular physiochemical features and their application in soil has a great influence on soil properties as well as plant growth and development. It has been apparent from the above discussion that external organic activities operate as a powerful growth regulator, enhancing plant growth performance in salt-stressed environments. Although organic farming approaches are particularly beneficial for agricultural production in salt, they do have certain drawbacks.

- Its preparations, which are organically altered as natural weathering processes, need more labor, time, space, and raw resources.
- More experienced and skilled people, as well as scientific understanding, are required to maintain environmental conditions such as temperature, moisture, and respiration.
- Some organic methods, such as vermicomposting, biochar, and bio-fertilizer, emit a foul stench and attract flies. On the worm-feeding materials, harmful molds and bacteria are frequently produced in some cases.

Regardless of the fact that organic amendments take longer to prepare and require additional forms of management. In order to increase plant nutrition in salt-stressed agroecosystems, numerous proactive and preventive strategies have been used over time, with well-defined adverse effects. Numerous methods, such as organic amendments, have shown to be highly efficient in easing different agricultural restrictions, such as salt stress. Multidisciplinary approaches and solutions, driven not only by plant and agri-environmental scientists but also by experts from other fields (remote sensing, artificial intelligence, machine learning, big data analyses, etc.), can produce very helpful tools for detecting, guarding against, and controlling salinization, thereby minimizing the harm brought on by salt stress. However, many reports have been published about the role of organic amendments in the alterations of physio-biochemical reactions to plants under salt stress. To counteract salt stress, however, additional approaches and strategies based on breeding assisted by genetic markers, genome editing, and advanced biotechnological procedures can be applied in addition to all standard treatments. In addition, there is currently a paucity of knowledge regarding how secondary metabolites, distinct stress-responsive genes, and the primary metabolic pathways that govern due to salinity. More studies are required to better understand the morphological, physio-biochemical, transcriptomics, and proteomics of organic amendment application in a saline environment to improve crop productivity.

4. Conclusions

Abiotic stressors are significant barriers that reduce agricultural yields around the world. One of the most damaging environmental variables limiting agricultural productivity is salinity. Salinity-induced oxidative stress and Na+ ion absorption lead to cellular damage including ionic instability, which inhibits growth and has detrimental effects on the morphological and biochemical characteristics of plants. It is absolutely necessary to look for environmentally sound and long-term solutions to reduce the negative effects of salt on plants. However, these negative impacts of salinity were lessened by the applications of VC, VW, BC, BF, and PGPR. It is clear from the discussion that VC, VW, BC, BF, and PGPR promote plant growth and increase salt tolerance by maintaining ionic homeostasis, enhancing antioxidant enzyme activities, lowering osmotic and oxidative stress, and regulating gene expression, all of which lead to improved plant growth and productivity. Although several studies on the regulatory functions of VC, VW, BC, BF, and PGPR in various crops under salt stress have been carried out, there is still much that needs to be investigated at the molecular, biochemical, and physiological level.

Author Contributions: Conceptualization, M.N.H., S.I. and M.S.R.; writing—original draft preparation, M.N.H., S.I., A.H., N.C.P., M.A.M., J.C., P.S., I.J.I. and M.S.R.; writing—review and editing, S.I., M.B. and M.S.R.; visualization, S.I., M.B. and M.S.R. All authors have read and agreed to the published version of the manuscript.

Funding: The APC was supported by the project VEGA (number: 1/0664/22).

Institutional Review Board Statement: Not applicable.

Informed Consent Statement: Not applicable.

Data Availability Statement: Not applicable.

Conflicts of Interest: The authors declare that they have no conflict of interest.

References

1. Zörb, C.; Geilfus, C.M.; Dietz, K.J. Salinity and crop yield. *Plant Biol.* **2019**, *21*, 31–38. [CrossRef] [PubMed]
2. Rhaman, M.S.; Imran, S.; Rauf, F.; Khatun, M.; Baskin, C.C.; Murata, Y.; Hasanuzzaman, M. Seed priming with phytohormones: An effective approach for the mitigation of abiotic stress. *Plants* **2020**, *10*, 37. [CrossRef] [PubMed]
3. Kawakami, Y.; Imran, S.; Katsuhara, M.; Tada, Y. Na^+ transporter SvHKT1; 1 from a halophytic turf grass is specifically upregulated by high Na^+ concentration and regulates shoot Na^+ concentration. *Int. J. Mol. Sci.* **2020**, *21*, 6100. [CrossRef] [PubMed]
4. Malhi, G.S.; Kaur, M.; Kaushik, P.; Alyemeni, M.N.; Alsahli, A.A.; Ahmad, P. Arbuscular mycorrhiza in combating abiotic stresses in vegetables: An eco-friendly approach. *Saudi J. Biol. Sci.* **2021**, *28*, 1465–1476. [CrossRef]
5. Imran, S.; Sarker, P.; Hoque, M.N.; Paul, N.C.; Mahamud, M.A.; Chakrobortty, J.; Tahjib-Ul-Arif, M.; Latef, A.A.; Hasanuzzaman, M.; Rhaman, M.S. Biochar actions for the mitigation of plant abiotic stress. *Crop Pasture Sci.* **2022**. [CrossRef]
6. Hoque, M.N.; Hannan, A.; Imran, S.; Paul, N.C.; Mondal, M.; Sadhin, M.; Rahman, M.; Bristi, J.M.; Dola, F.S.; Hanif, M.; et al. Plant Growth-Promoting Rhizobacteria-Mediated Adaptive Responses of Plants Under Salinity Stress. *J. Plant Growth Regul.* **2022**, *28*, 1–20. [CrossRef]

7. Wallender, W.W.; Tanji, K.K. Nature and extent of agricultural salinity and sodicity. In *Agricultural Salinity and Management*, 2nd ed.; American Society of Civil Engineers: Reston, VA, USA, 2012; pp. 1–25. [CrossRef]
8. Alqahtani, M.; Roy, S.J.; Tester, M. Increasing Salinity Tolerance of Crops. In *Encyclopedia of Sustainability Science and Technology*; Springer: New York, NY, USA, 2019. [CrossRef]
9. Mustafa, G.; Akhtar, M.S.; Abdullah, R. Global concern for salinity on various agro-ecosystems. In *Salt Stress, Microbes, and Plant Interactions: Causes and Solution*; Springer: Singapore, 2019; pp. 1–19. [CrossRef]
10. Shahid, S.A.; Zaman, M.; Heng, L. Introduction to soil salinity, sodicity and diagnostics techniques. In *Guideline for Salinity Assessment, Mitigation and Adaptation Using Nuclear and Related Techniques*; Springer: Cham, Switzerland, 2018; pp. 1–42. [CrossRef]
11. Munns, R.; Tester, M. Mechanisms of salinity tolerance. *Ann. Rev. Plant Biol.* **2008**, *59*, 651–681. [CrossRef] [PubMed]
12. Muchate, N.S.; Nikalje, G.C.; Rajurkar, N.S.; Suprasanna, P.; Nikam, T.D. Plant salt stress: Adaptive responses, tolerance mechanism and bioengineering for salt tolerance. *Bot. Rev.* **2016**, *82*, 371–406. [CrossRef]
13. Imran, S.; Tsuchiya, Y.; Tran, S.T.; Katsuhara, M. Identification and Characterization of Rice OsHKT1; 3 Variants. *Plants* **2021**, *10*, 2006. [CrossRef]
14. Alkharabsheh, H.M.; Seleiman, M.F.; Hewedy, O.A.; Battaglia, M.L.; Jalal, R.S.; Alhammad, B.A.; Schillaci, C.; Ali, N.; Al-Doss, A. Field crop responses and management strategies to mitigate soil salinity in modern agriculture: A review. *Agronomy* **2021**, *11*, 2299. [CrossRef]
15. Islam, W.; Waheed, A.; Naveed, H.; Zeng, F. MicroRNAs mediated plant responses to salt stress. *Cells* **2022**, *11*, 2806. [CrossRef] [PubMed]
16. Cao, C.; Long, R.; Zhang, T.; Kang, J.; Wang, Z.; Wang, P.; Sun, H.; Yu, J.; Yang, Q. Genome-wide identification of microRNAs in response to salt/alkali stress in medicagotruncatula through high-throughput sequencing. *Int. J. Mol. Sci.* **2018**, *19*, 4076. [CrossRef] [PubMed]
17. Kuang, L.; Shen, Q.; Wu, L.; Yu, J.; Fu, L.; Wu, D.; Zhang, G. Identification of microRNAs responding to salt stress in barley by high-throughput sequencing and degradome analysis. *Environ. Exp. Bot.* **2019**, *160*, 59–70. [CrossRef]
18. Ma, Y.; Xue, H.; Zhang, F.; Jiang, Q.; Yang, S.; Yue, P.; Wang, F.; Zhang, Y.; Li, L.; He, P.; et al. The miR156/SPL module regulates apple salt stress tolerance by activating MdWRKY100 expression. *Plant Biotechnol. J.* **2021**, *19*, 311–323. [CrossRef]
19. Shahid, S.A.; Zaman, M.; Heng, L. Salinity and sodicity adaptation and mitigation options. In *Guideline for Salinity Assessment, Mitigation and Adaptation Using Nuclear and Related Techniques*; Springer: Cham, Switzerland, 2018; pp. 55–89. [CrossRef]
20. Shilev, S. Plant-growth-promoting bacteria mitigating soil salinity stress in plants. *Appl. Sci.* **2020**, *10*, 7326. [CrossRef]
21. Meena, K.K.; Bitla, U.M.; Sorty, A.M.; Singh, D.P.; Gupta, V.K.; Wakchaure, G.C.; Kumar, S. Mitigation of salinity stress in wheat seedlings due to the application of phytohormone-rich culture filtrate extract of methylotrophic actinobacterium *Nocardioides* sp. NIMMe6. *Front. Microbiol.* **2020**, *11*, 2091. [CrossRef] [PubMed]
22. Bhowmik, U.; Kibria, M.G.; Rhaman, M.S.; Murata, Y.; Hoque, M.A. Screening of rice genotypes for salt tolerance by physiological and biochemical characters. *Plant Sci. Today* **2021**, *8*, 467–472. [CrossRef]
23. Kanwal, S.; Ilyas, N.; Shabir, S.; Saeed, M.; Gul, R.; Zahoor, M.; Batool, N.; Mazhar, R. Application of biochar in mitigation of negative effects of salinity stress in wheat (*Triticum aestivum* L.). *J. Plant Nutr.* **2018**, *41*, 526–538. [CrossRef]
24. Rasheed, F.; Anjum, N.A.; Masood, A.; Sofo, A.; Khan, N.A. The key roles of salicylic acid and sulfur in plant salinity stress tolerance. *J. Plant Growth Regul.* **2020**, *30*, 1–4. [CrossRef]
25. Hannan, A.; Hoque, M.N.; Hassan, L.; Robin, A.H. Adaptive Mechanisms of Root System of Rice for Withstanding Osmotic Stress. In *Recent Advances Rice Research*; IntechOpen: London, UK, 2020. [CrossRef]
26. Ali, M.; Kamran, M.; Abbasi, G.H.; Saleem, M.H.; Ahmad, S.; Parveen, A.; Malik, Z.; Afzal, S.; Ahmar, S.; Dawar, K.M.; et al. Melatonin-induced salinity tolerance by ameliorating osmotic and oxidative stress in the seedlings of two tomato (*Solanum lycopersicum* L.) cultivars. *J. Plant Growth Regul.* **2021**, *40*, 2236–2248. [CrossRef]
27. Alamer, K.H.; Perveen, S.; Khaliq, A.; Zia Ul Haq, M.; Ibrahim, M.U.; Ijaz, B. Mitigation of salinity stress in maize seedlings by the application of vermicompost and sorghum water extracts. *Plants* **2022**, *11*, 2548. [CrossRef] [PubMed]
28. Khalilzadeh, R.; Seyed, S.R.; Jalilian, J. Growth, physiological status, and yield of salt stressed wheat (*Triticum aestivum* L.) plants affected by biofertilizer and cycocel applications. *Arid. Land Res. Manag.* **2017**, *32*, 71–90. [CrossRef]
29. Ekinci, M.; Turan, M.; Yildirim, E. Biochar mitigates salt stress by regulating nutrient uptake and antioxidant activity, alleviating the oxidative stress and abscisic acid content in cabbage seedlings. *Turk. J. Agric. For.* **2022**, *46*, 28–37. [CrossRef]
30. Kumar, A.; Singh, S.; Gaurav, A.K.; Srivastava, S.; Verma, J.P. Plant growth-promoting bacteria: Biological tools for the mitigation of salinity stress in plants. *Front. Microbiol.* **2020**, *11*, 1216. [CrossRef] [PubMed]
31. Chauhan, P.S.; Lata, C.; Tiwari, S.; Chauhan, A.S.; Mishra, S.K.; Agrawal, L.; Chakrabarty, D.; Nautiyal, C.S. Transcriptional alterations reveal *Bacillus amyloliquefaciens*-rice cooperation under salt stress. *Sci. Rep.* **2019**, *9*, 11912. [CrossRef]
32. Parkash, V.; Singh, S. Potential of biochar application to mitigate salinity stress in eggplant. *Hortc. Sci.* **2020**, *55*, 1946–1955. [CrossRef]
33. Kerbab, S.; Silini, A.; Chenari, B.A.; Cherif-Silini, H.; Eshelli, M.; El Houda, R.N.; Belbahri, L. Mitigation of NaCl stress in wheat by rhizosphere engineering using salt habitat adapted PGPR halotolerant bacteria. *Appl. Sci.* **2021**, *11*, 1034. [CrossRef]
34. Ha-Tran, D.M.; Nguyen, T.T.; Hung, S.H.; Huang, E.; Huang, C.C. Roles of plant growth-promoting rhizobacteria (PGPR) in stimulating salinity stress defense in plants: A review. *Int. J. Mol. Sci.* **2021**, *22*, 3154. [CrossRef] [PubMed]

35. Pérez-Gómez, J.D.; Abud-Archila, M.; Villalobos-Maldonado, J.J.; Enciso-Saenz, S.; Hernández de, L.H.; Ruiz-Valdiviezo, V.M.; Gutiérrez-Miceli, F.A. Vermicompost and vermiwash minimized the influence of salinity stress on growth parameters in potato plants. *Compost Sci. Util.* **2017**, *25*, 282–287. [CrossRef]
36. Ruiz-Lau, N.; Oliva-Llaven, M.A.; Montes-Molina, J.A.; Gutiérrez-Miceli, F.A. Mitigation of Salinity Stress by Using the Vermicompost and Vermiwash. In *Ecological and Practical Applications for Sustainable Agriculture*; Springer: Singapore, 2020; pp. 345–356. [CrossRef]
37. Mahmoud, A.A.; Mohamed, H.F. Impact of biofertilizers application on improving wheat (*Triticum aestivum* L.) resistance to salinity. *Res. J. Agric. Biol. Sci.* **2008**, *4*, 520–528.
38. Mahdy, A.M.; Nieven, O.F. Interactive effects between biofertilizer and antioxidant on salinity mitigation and nutrition and yield of okra plants (*Abelmoschus esculentus* L.). *J. Soil Sci. Agric. Eng.* **2012**, *3*, 189–205. [CrossRef]
39. Sinha, R.K.; Valani, D.; Chauhan, K.; Agarwal, S. Embarking on a second green revolution for sustainable agriculture by vermiculture biotechnology using earthworms: Reviving the dreams of Sir Charles Darwin. *Int. J. Agric. Health Saf.* **2014**, *1*, 50–64.
40. Kiyasudeen, K.; Ibrahim, M.H.; Quaik, S.; Ismail, S.A. Vermicompost, its applications and derivatives. In *Prospects of Organic Waste Management and the Significance of Earthworms*; Springer: Cham, Switzerland, 2016; pp. 201–230.
41. Arancon, N.Q.; Lee, S.; Edwards, C.A.; Atiyeh, R. Effects of humic acids derived from cattle, food and paper-waste vermicomposts on growth of greenhouse plants: The 7th international symposium on earthworm ecology · Cardiff · Wales · 2002. *Pedobiologia* **2003**, *47*, 741–744. [CrossRef]
42. Bidabadi, S.S.; Dehghanipoodeh, S.; Wright, G.C. Vermicompost leachate reduces some negative effects of salt stress in pomegranate. *Int. J. Recycl. Org. Waste Agric.* **2017**, *6*, 255–263. [CrossRef]
43. Koozehgar, K.M.; Ardakani, M.R. Effects of vermicomposting and compost tea on nitrogen, phosphorus, and potassium yield and uptake of *Mentha aquatic* L. inoculated with mycorrhizal fungi Glomus moseae. *Iran. J. Plant Physiol.* **2017**, *11*, 10–19.
44. Pengkam, C.; Iwai, C.B.; Kume, T. Effects of Vermicompost and Rice Husk Ash on the Change of Soil Chemical Properties and the Growth of Rice in Salt Affected Area. *Int. J. Environ. Rural. Dev.* **2019**, *10*, 129–132. [CrossRef]
45. Demir, Z.; Tursun, N.; Işık, D. Role of different cover crops on DTPA-extractable micronutrients in an apricot orchard. *Turk. J. Agric. Food Sci. Technol.* **2019**, *7*, 698–706. [CrossRef]
46. Ansari, A.A.; Ismail, S.A. Role of earthworms in vermitechnology. *J. Agric. Technol.* **2012**, *8*, 403–415.
47. Khan, M.H.; Meghvansi, M.K.; Gupta, R.; Veer, V.; Singh, L.; Kalita, M.C. Foliar spray with vermiwash modifies the arbuscular mycorrhizal dependency and nutrient stoichiometry of Bhut jolokia (*Capsicum assamicum*). *PLoS ONE* **2014**, *9*, e92318. [CrossRef]
48. Nath, G.; Singh, K. Effect of vermiwash of different vermicomposts on the kharif crops. *J. Cent. Eur. Agric.* **2012**, *13*, 379–402. Available online: https://hrcak.srce.hr/83274 (accessed on 10 September 2022). [CrossRef]
49. Beykkhormizi, A.; Abrishamchi, P.; Ganjeali, A.; Parsa, M. Effect of vermicompost on some morphological, physiological and biochemical traits of bean (*Phaseolus vulgaris* L.) under salinity stress. *J. Plant Nutr.* **2016**, *39*, 883–893. [CrossRef]
50. Chinsamy, M.; Kulkarni, M.G.; Van Staden, J. Garden-waste-vermicompost leachate alleviates salinity stress in tomato seedlings by mobilizing salt tolerance mechanisms. *Plant Growth Regul.* **2013**, *71*, 41–47. [CrossRef]
51. Benazzouk, S.; Djazouli, Z.E.; Lutts, S. Vermicompost leachate as a promising agent for priming and rejuvenation of salt-treated germinating seeds in *Brassica napus*. *Commun. Soil Sci. Plant Anal.* **2019**, *50*, 1344–1357. [CrossRef]
52. Ahmadi, N.; Akbari, E. The preventive impact of vermicompost on bell pepper (*Capsicum annuum* L.) salinity resistance: An evaluation. *Afr. J. Agric. Res.* **2021**, *17*, 46–56. [CrossRef]
53. Bassaco, A.C.; Antoniolli, Z.I.; Júnior, B.D.; Eckhardt, D.P.; Montagner, D.F.; Bassaco, G.P. Chemistry characterization from animal origin residues and Eisenia andrei behaviour. *Ciênc. Natura* **2015**, *37*, 45–51. [CrossRef]
54. Ebrahimi, M.H.; Taghvaei, M.; Sadeghi, H.; Zarei, M. Effect of organic coats with superabsorbent polymers on improving the germination and early vigor Milk thistle (*Silybum marianum* L.) seeds under salinity stress. *Desert* **2019**, *24*, 207–215. [CrossRef]
55. Muhie, S.H.; Yildirim, E.; Memis, N.; Demir, I. Vermicompost priming stimulated germination and seedling emergence of onion seeds against abiotic stresses. *Seed Sci. Technol.* **2020**, *48*, 153–157. [CrossRef]
56. Demir, Z. Alleviation of adverse effects of sodium on soil physicochemical properties by application of vermicompost. *Compost Sci. Util.* **2020**, *28*, 100–116. [CrossRef]
57. Barahouee, M.; Sabbagh, E. Influence of vermicompost and salt stress on some characteristics of fenugreek (*Trigonellafoenum-graecum* L.). *Int. J. Agric. Biosci.* **2017**, *6*, 60–63.
58. Banadkooki, A.M.; Ardakani, M.D.; Shirmardi, M.; Momenpour, A. Effects of cow manure and vermicompost on growth characteristics of smoke tree (*Cotinus coggygria* Scop.) under salt stress under greenhouse. *Iranian J. Poplar Res.* **2019**, *26*, 483–495. [CrossRef]
59. Gohari, G.; Mohammadi, A.; Duathi, K.H. Effect of vermicompost on some growth and biochemical characteristic of *Dracocephalum moldavica* L. under water salinity stress. *J. Agric. Sci Sustain. Prod.* **2019**, *29*, 151–168.
60. Song, X.; Li, H.; Song, J.; Chen, W.; Shi, L. Biochar/vermicompost promotes Hybrid *Pennisetum* plant growth and soil enzyme activity in saline soils. *Plant Physiol. Biochem.* **2022**, *183*, 96–110. [CrossRef]
61. Bziouech, S.A.; Dhen, N.; Helaoui, S.; Ammar, I.B.; Dridi, B.A.M. Effect of vermicompost soil additive on growth performance, physiological and biochemical responses of tomato plants (*Solanum lycopersicum* L. var. Firenze) to salt stress. *Emir. J. Food Agric.* **2022**, *34*, 316–328. [CrossRef]

62. Liu, M.; Wang, C.; Wang, F.; Xie, Y. Maize (*Zea mays*) growth and nutrient uptake following integrated improvement of vermicompost and humic acid fertilizer on coastal saline soil. *Appl. Soil Ecol.* **2019**, *142*, 147–154. [CrossRef]
63. Oo, A.N.; Iwai, C.B.; Saenjan, P. Soil properties and maize growth in saline and nonsaline soils using cassava-industrial waste compost and vermicompost with or without earthworms. *Land Degrad. Dev.* **2015**, *26*, 300–310. [CrossRef]
64. Jabeen, N.; Ahmad, R. Growth response and nitrogen metabolism of sunflower (*Helianthus annuus* L.) to vermicompost and biogas slurry under salinity stress. *J. Plant Nutr.* **2017**, *40*, 104–114. [CrossRef]
65. Demir, Z.; Kiran, S. Effect of vermicompost on macro and micro nutrients of lettuce (*Lactuca sativa* var. Crispa) under salt stress conditions. *Kahramanmaraş. Sütçü. İmam. Üniversitesi. Tarım. Doğa. Dergisi.* **2020**, *23*, 33–43. [CrossRef]
66. Adamipour, N.; Heiderianpour, M.B.; Zarei, M. Application of vermicompost for reducing the destructive effects of salinity stress on tall fescue turfgrass (*Festuca arundinacea* Schreb. 'Queen'). *J. Soil Plant Interact. Isfahan Uni. Technol.* **2016**, *7*, 35–47. [CrossRef]
67. Kiran, S. Alleviation of adverse effects of salt stress on lettuce (*Lactuca sativa* var. crispa) by application of vermicompost. *Acta Sci. Pol. Hortorum Cultus* **2019**, *5*, 153–160. [CrossRef]
68. Hafez, E.M.; Omara, A.E.; Alhumaydhi, F.A.; El-Esawi, M.A. Minimizing hazard impacts of soil salinity and water stress on wheat plants by soil application of vermicompost and biochar. *Physiol. Plant* **2020**, *172*, 587–602. [CrossRef]
69. Benazzouk, S.; Lutts, S.; Djazouli, Z.E. Alleviation of salinity stress by Vermicompost extract in *Solanum lycopersicum* L. by mobilizing salt tolerance mechanisms. *AgroBiologia* **2018**, *8*, 1136–1144.
70. Benazzouk, S.; Dobrev, P.I.; Djazouli, Z.E.; Motyka, V.; Lutts, S. Positive impact of vermicompost leachate on salt stress resistance in tomato (*Solanum lycopersicum* L.) at the seedling stage: A phytohormonal approach. *Plant Soil.* **2020**, *446*, 145–162. [CrossRef]
71. Ezzat, A.S.; Badway, A.S.; Abdelkader, A.E. Sequenced vermicompost, glycine betaine, proline treatments elevate salinity tolerance in potatoes. *Middle East J. Agric. Res.* **2019**, *8*, 126–138.
72. Zhou, J.; Ahmed, N.; Cheng, Y.; Qin, C.; Chen, P.; Zhang, C.; Zhang, L. Effect of inoculation of strains with acc deaminase isolated from vermicompost on seed germination and some physiological attributes in maize (*Zea mays* L.) exposed to salt stress. *Pak. J. Bot.* **2019**, *51*, 1169–1177. [CrossRef]
73. Mohsenzadeh, S.; Zamanpour, S.H. Evaluation of Municipal Solid Waste Compost and Agricultural Waste Vermicompost by Growth of *Lippia citriodora* Under Salinity Stress. *J. Environ. Sci. Stud.* **2019**, *4*, 2135–2143.
74. Reyes-Pérez, J.J.; Murillo-Amador, B.; Nieto-Garibay, A.; Troyo-Diéguez, E.; Rueda-Puente, E.O.; Hernández-Montiel, L.G.; Preciado, R.P.; Beltrán, M.A.; Rodríguez, F.F.; López Bustamante, R.J. Use of humates of vermicompost to reduce the effect of salinity on growth and development of basil (*Ocimum basilicum* L.). *Rev. Mex. Cienc. Agrí.* **2016**, *7*, 1375–1387.
75. Yücel, N.C.; Chİtİlova, M. Improving wheat performance by fish flour and vermicompost priming against salt stress. *Int. J. Agric. Biol.* **2017**, *19*, 1483–1488.
76. Liu, M.; Wang, C.; Wang, F.; Xie, Y. Vermicompost and humic fertilizer improve coastal saline soil by regulating soil aggregates and the bacterial community. *Arch. Agron. Soil Sci.* **2019**, *65*, 281–293. [CrossRef]
77. Zurbano, L.Y. Response of lettuce (*Lactuca sativa*) on saline soil amended with vermicompost and pulverized eggshell. *Indian J. Sci. Technol.* **2018**, *11*, 1–8. [CrossRef]
78. Adamipour, N.; Khosh-Khui, M.; Salehi, H.; Rho, H. Effect of vermicompost on morphological and physiological performances of pot marigold (*Calendula officinalis* L.) under salinity conditions. *Adv. Hortic. Sci.* **2019**, *33*, 345–358. [CrossRef]
79. Santos, D.G.; Diniz, B.L.; Diniz, M.A.; Silva, J.H.; Oliveira, W.N.; Ferreira, R.M. Growth and chlorophyll in noni seedlings irrigated with saline water in substrate with vermicompost. *Rev. Bras. Eng. Agricola Ambient.* **2019**, *23*, 586–590. [CrossRef]
80. Akhzari, D.; Pessarakli, M.; Khedmati, M. Effects of vermicompost and salinity stress on growth and physiological traits of *Medicago rigidula* L . *J. Plant Nutr.* **2016**, *39*, 2106–2114. [CrossRef]
81. Sorkhi, F. Effect of vermicompost fertilizer on antioxidant enzymes and chlorophyll contents in *Borago officinalis* under salinity stress. *Iran. J. Plant Physiol.* **2021**, *11*, 3589–3598. [CrossRef]
82. Djajadi, D.; Syaputra, R.; Hidayati, S.N.; Khairiyah, Y. Effect of vermicompost and nitrogen on N, K, Na uptakes and growth of sugarcane in saline soil. *Agrivita. J. Agric. Sci.* **2020**, *42*, 110–119. [CrossRef]
83. Yang, A.; Akhtar, S.S.; Li, L.; Fu, Q.; Li, Q.; Naeem, M.A.; He, X.; Zhang, Z.; Jacobsen, S.E. Biochar mitigates combined effects of drought and salinity stress in quinoa. *J. Agron.* **2020**, *10*, 912. [CrossRef]
84. Huang, M.; Zhang, Z.; Zhai, Y.; Lu, P.; Zhu, C. Effect of straw biochar on soil properties and wheat production under saline water irrigation. *Agronomy* **2019**, *9*, 457. [CrossRef]
85. Farhangi-Abriz, S.; Torabian, S. Biochar improved nodulation and nitrogen metabolism of soybean under salt stress. *Symbiosis* **2018**, *74*, 215–223. [CrossRef]
86. Ibrahim, M.E.H.; Ali, A.Y.A.; Elsiddig, A.M.I.; Zhou, G.; Nimir, N.E.A.; Agbna, G.H.; Zhu, G. Mitigation effect of biochar on sorghum seedling growth under salinity stress. *Pak. J. Bot.* **2021**, *53*, 387–392. [CrossRef]
87. Ibrahim, M.E.H.; Ali, A.Y.A.; Zhou, G.; Elsiddig, A.M.I.; Zhu, G.; Nimir, N.E.A.; Ahmad, I. Biochar application affects forage sorghum under salinity stress. *Chil. J. Agric. Res.* **2020**, *80*, 317–325. [CrossRef]
88. Lashari, M.S.; Ye, Y.; Ji, H.; Li, L.; Kibue, G.W.; Lu, H.; Zheng, J.; Pan, G. Biochar–manure compost in conjunction with pyroligneous solution alleviated salt stress and improved leaf bioactivity of maize in a saline soil from central China: A 2-year field experiment. *J. Sci. Food Agric.* **2015**, *95*, 1321–1327. [CrossRef]

89. Usman, A.R.; Al-Wabel, M.I.; Abdulaziz, A.H.; Mahmoud, W.A.; EL-Naggar, A.H.; Ahmad, M.; Abdulelah, A.F.; Abdulrasoul, A.O. Conocarpus biochar induces changes in soil nutrient availability and tomato growth under saline irrigation. *Pedosphere* **2016**, *26*, 27–38. [CrossRef]
90. Ali, S.; Rizwan, M.; Qayyum, M.F.; Ok, Y.S.; Ibrahim, M.; Riaz, M.; Arif, M.S.; Hafeez, F.; Al-Wabel, M.I.; Shahzad, A.N. Biochar soil amendment on alleviation of drought and salt stress in plants: A critical review. *Environ. Sci. Pollut. Res.* **2017**, *24*, 12700–12712. [CrossRef] [PubMed]
91. Jin, F.; Piao, J.; Che, W.; Li, X.; Zhang, C.; Wang, Q.; Hua, S. Peanut shell biochar increases rice yield in highly saline-alkali paddy fields by regulating of leaf ionic concentration and improving leaf photosynthesis rate. *Plant Soil* **2022**, preprint. [CrossRef]
92. Huang, J.; Zhu, C.; Kong, Y.; Cao, X.; Zhu, L.; Zhang, Y.; Ning, Y.; Tian, W.; Zhang, H.; Yu, Y.; et al. Biochar Application Alleviated Rice Salt Stress via Modifying Soil Properties and Regulating Soil Bacterial Abundance and Community Structure. *Agronomy* **2022**, *12*, 409. [CrossRef]
93. Farhangi-Abriz, S.; Torabian, S. Biochar increased plant growth-promoting hormones and helped to alleviates salt stress in common bean seedlings. *J. Plant Growth Regul.* **2018**, *37*, 591–601. [CrossRef]
94. Nikpour-Rashidabad, N.; Tavasolee, A.; Torabian, S.; Farhangi-Abriz, S. The effect of biochar on the physiological, morphological and anatomical characteristics of mung bean roots after exposure to salt stress. *Arch. Biol. Sci.* **2019**, *71*, 321–327. [CrossRef]
95. Soliman, M.H.; Alnusairi, G.S.; Khan, A.A.; Alnusaire, T.S.; Fakhr, M.A.; Abdulmajeed, A.M.; Aldesuquy, H.S.; Yahya, M.; Najeeb, U. Biochar and selenium nanoparticles induce water transporter genes for sustaining carbon assimilation and grain production in salt-stressed wheat. *J. Plant Growth Regul.* **2022**, 1–22. [CrossRef]
96. Akhtar, S.S.; Andersen, M.N.; Naveed, M.; Zahir, Z.A.; Liu, F. Interactive effect of biochar and plant growth-promoting bacterial endophytes on ameliorating salinity stress in maize. *Funct. Plant. Biol.* **2015**, *42*, 770–781. [CrossRef]
97. Akhtar, S.S.; Andersen, M.N.; Liu, F. Biochar mitigates salinity stress in potato. *J. Agron. Crop Sci.* **2015**, *201*, 368–378. [CrossRef]
98. Farouk, S.; AL-Huqail, A.A. Sustainable biochar and/or melatonin improve salinity tolerance in borage plants by modulating osmotic adjustment, antioxidants, and ion homeostasis. *Plants* **2022**, *11*, 765. [CrossRef] [PubMed]
99. Hegde, D.M.; Dwivedi, B.S.; Sudhakara Babu, S.N. Biofertilizers for cereal production in India: A review. *Indian J. Agric. Sci.* **1999**, *69*, 73–83.
100. Vessey, J.K. Plant growth promoting rhizobacteria as biofertilizers. *Plant Soil.* **2003**, *255*, 571–586. [CrossRef]
101. Riesty, O.S.; Siswanti, D.U. Effect of biofertilizer on growth and metaxylem diameter of *Amaranthus tricolor* L. in salinity stress condition. *Biogenes. J. Ilm. Biol.* **2021**, *9*, 178–188. [CrossRef]
102. Khatami, S.A.; Kasraie, P.; Oveysi, M.; Moghadam, H.R.T.; Ghooshchi, F. Mitigating the adverse effects of salinity stress on lavender using biodynamic preparations and bio-fertilizers. *Ind. Crops Prod.* **2022**, *183*, 114985. [CrossRef]
103. Mahdy, A.M.; Fathi, N.O.; Kandil, M.M.; Elnamas, A.E. Synergistic effects of biofertilizers and antioxidants on growth and nutrients content of corn under salinity and water-deficit stresses. *Alex. Sci. Exch. J.* **2012**, *33*, 292–304. [CrossRef]
104. Souza, J.T.; Cavalcante, L.F.; Nunes, J.C.; Bezerra, F.T.; da Silva, N.J.A.; Silva, A.R.; Oresca, D.; Cavalcante, A.G. Effect of saline water, bovine biofertilizer and potassium on yellow passion fruit growth after planting and on soil salinity. *Afr. J. Agric. Res.* **2016**, *11*, 2994–3003. [CrossRef]
105. AlAbdallah, N.M.; Basalah, M.O.; Roushdy, S.S. The promotive effect of algal biofertilizers on growth and some metabolic activities of (*Vigna unguiculata* L.) under salt stress conditions. *Egypt. J. Exp. Biol.* **2017**, *13*, 187–195. [CrossRef]
106. El-Shazly, M.; Ghieth, W.M. Effect of some biofertilizers and humic acid application on olive seedlings growth under irrigation with saline water. *Alex. Sci. Exch. J.* **2019**, *40*, 263–279. [CrossRef]
107. de Lima-Neto, A.J.; Cavalcante, L.F.; Mesquita, F.D.; Souto, A.G.; dos Santos, G.P.; dos Santos, J.Z.; de Mesquita, E.F. Papaya seedlings irrigation with saline water in soil with bovine fertilizer. *Chil. J. Agric. Res.* **2015**, *76*, 236–242. [CrossRef]
108. Al-Taey, D.K.; Majid, Z.Z. Study effect of kinetin, bio-fertilizers and organic matter application in lettuce under salt stress. *J. Glob. Pharma. Technol.* **2018**, *10*, 148–164.
109. Yasmin, H.; Mazher, J.; Azmat, A.; Nosheen, A.; Naz, R.; Hassan, M.N.; Noureldeen, A.; Parvaiz, A.P. Combined application of zinc oxide nanoparticles and biofertilizer to induce salt resistance in safflower by regulating ion homeostasis and antioxidant defence responses. *Ecotoxicol. Environ. Saf.* **2021**, *218*, 112262. [CrossRef] [PubMed]
110. Albdaiwi, R.N.; Khyami-Horani, H.; Ayad, J.Y. Plant growth-promoting Rhizobacteria: An emerging method for the enhancement of wheat tolerance against salinity stress. *Jordan J. Biol. Sci.* **2019**, *12*, 525–534.
111. Goswami, M.; Deka, S. Isolation of a novel rhizobacteria having multiple plant growth promoting traits and antifungal activity against certain phytopathogens. *Microbiol. Res.* **2020**, *240*, 126516. [CrossRef] [PubMed]
112. Leithy, S.; Gaballah, M.S.; Gomaa, A.M. Associative impact of bio- and organic fertilizers on geranium plants grown under saline conditions. *Int. J. Acad. Res.* **2009**, *1*, 17–23.
113. El Semary, N.A.; Alouane, M.H.; Nasr, O.; Aldayel, M.F.; Alhaweti, F.H.; Ahmed, F. Salinity Stress Mitigation using encapsulated biofertilizers for sustainable agriculture. *Sustainability* **2020**, *12*, 9218. [CrossRef]
114. Abdelhamid, M.; Gaballah, M.S.; Rady, M.; Gomaa, A. Biofertilizer and ascorbic acid alleviated the detrimental effects of soil salinity on growth and yield of soybean. In Proceedings of the Second Science with Africa Conference, Addis Ababa, Ethiopia, 22–25 June 2010; pp. 73–81.

115. Oliveira, D.S.; Dias, T.J.; Edvania Pereira de Oliveira, E.P.; Santos, H.C.; Magalhaes, W.B.; Matos, B.F.; Sousa, L.M.C.; Filho, J.S.M. Growth and physiology of peanut (*Arachis hypogaea* L.) irrigated with saline water and biofertilizer application times. *Afr. J. Agric. Res.* **2016**, *11*, 4517–4524. [CrossRef]
116. Véras, M.L.M.; da Silva, A.R.; de Sousa, A.L.; de Melo, F.J.S.; da Silva, I.T.H.; Dias, T.J. Growth and dry matter of pitombeira seedlings under salinity levels and application of biofertilizer. *Com. Sci.* **2017**, *8*, 486–492. [CrossRef]
117. Amjad, M.; Akhtar, J.; Rashid, M.S. Evaluating the effectiveness of biofertilizer on salt tolerance of cotton (*Gossypium hirsutum* L.). *Arch. Agron. Soil Sci.* **2014**, *61*, 1165–1177. [CrossRef]
118. Prasad, M.; Srinivasan, R.; Chaudhary, M.; Choudhary, M.; Jat, L.K. Plant growth promoting rhizobacteria (PGPR) for sustainable agriculture: Perspectives and challenges. In *PGPR Amelioration in Sustainable Agriculture*; Woodhead Publishing: Sawston, UK, 2019; pp. 129–157. [CrossRef]
119. Gao, Y.; Zou, H.; Wang, B.; Yuan, F. Progress and applications of plant growth-promoting bacteria in salt tolerance of crops. *Int. J. Mol. Sci.* **2022**, *23*, 7036. [CrossRef]
120. Kumawat, K.C.; Sharma, P.; Nagpal, S.; Gupta, R.K.; Sirari, A.; Nair, R.M.; Bindumadhava, H.; Singh, S. Dual microbial inoculation, a game changer?—Bacterial biostimulants with multifunctional growth promoting traits to mitigate salinity stress in Spring Mungbean. *Front. Microbiol.* **2021**, *11*, 3491. [CrossRef]
121. Simonin, K.A.; Burns, E.; Choat, B.; Barbour, M.M.; Dawson, T.E.; Franks, P.J. Increasing leaf hydraulic conductance with transpiration rate minimizes the water potential drawdown from stem to leaf. *J. Exp. Bot.* **2015**, *66*, 1303–1315. [CrossRef]
122. Nawaz, A.; Shahbaz, M.; Imran, A.; Marghoob, M.U.; Imtiaz, M.; Mubeen, F. Potential of salt tolerant PGPR in growth and yield augmentation of wheat (*Triticum aestivum* L.) under saline conditions. *Front. Microbiol.* **2020**, *11*, 2019. [CrossRef]
123. Singh, R.P.; Jha, P.; Jha, P.N. Bio-inoculation of plant growth-promoting rhizobacterium Enterobacter cloacae ZNP-3 increased resistance against salt and temperature stresses in wheat plant (*Triticum aestivum* L.). *J. Plant Growth Regul.* **2017**, *36*, 783–798. [CrossRef]
124. Ali, B.; Wang, X.; Saleem, M.H.; Sumaira; Hafeez, A.; Afridi, M.S.; Khan, S.; Zaib-Un-Nisa; Ullah, I.; AmaralJúnior, A.T.D.; et al. PGPR-Mediated salt tolerance in maize by modulating plant physiology, antioxidant defense, compatible solutes accumulation and bio-surfactant producing genes. *Plants* **2022**, *11*, 345. [CrossRef] [PubMed]
125. Chinnaswamy, A.; Coba de la Peña, T.; Stoll, A.; de la Peña Rojo, D.; Bravo, J.; Rincón, A.; Lucas, M.M.; Pueyo, J.J. A nodule endophytic *Bacillus megaterium* strain isolated from *Medicago polymorpha* enhances growth, promotes nodulation by *Ensifer medicae* and alleviates salt stress in alfalfa plants. *Ann. Appl. Biol.* **2018**, *172*, 295–308. [CrossRef]
126. Nascimento, F.X.; Hernández, A.G.; Glick, B.R.; Rossi, M.J. Plant growth-promoting activities and genomic analysis of the stress-resistant *Bacillus megaterium* STB1, a bacterium of agricultural and biotechnological interest. *Biotechnol. Rep.* **2020**, *25*, e00406. [CrossRef]
127. Singh, R.P.; Jha, P.N. The multifarious PGPR *Serratia marcescens* CDP-13 augments induced systemic resistance and enhanced salinity tolerance of wheat (*Triticum aestivum* L.). *PLoS ONE* **2016**, *11*, e0155026. [CrossRef] [PubMed]
128. Li, H.; Lei, P.; Pang, X.; Li, S.; Xu, H.; Xu, Z.; Feng, X. Enhanced tolerance to salt stress in canola (*Brassica napus* L.) seedlings inoculated with the halotolerant *Enterobacter cloacae* HSNJ4. *Appl. Soil Ecol.* **2017**, *119*, 26–34. [CrossRef]
129. Bal, H.B.; Nayak, L.; Das, S.; Adhya, T.K. Isolation of ACC deaminase producing PGPR from rice rhizosphere and evaluating their plant growth promoting activity under salt stress. *Plant Soil* **2013**, *366*, 93–105. [CrossRef]
130. Ilangumaran, G.; Schwinghamer, T.D.; Smith, D.L. Rhizobacteria from root nodules of an indigenous legume enhance salinity stress tolerance in soybean. *Front. Sustain. Food. Syst.* **2021**, *4*, 617978. [CrossRef]
131. Matse, D.T.; Huang, C.H.; Huang, Y.M.; Yen, M.Y. Effects of coinoculation of Rhizobium with plant growth promoting rhizobacteria on the nitrogen fixation and nutrient uptake of *Trifolium repens* in low phosphorus soil. *J. Plant Nutr.* **2020**, *43*, 739–752. [CrossRef]
132. Omara, A.E.; Hauka, F.; Afify, A.; Nour, E.M.; Kassem, M. The role of some PGPR strains to biocontrol *Rhizoctonia solani* in soybean and enhancement the growth dynamics and seed yield. *Env. Biodivers. Soil Secur.* **2017**, *1*, 47–59. [CrossRef]
133. Kasim, W.A.; Gaafar, R.M.; Abou-Ali, R.M.; Omar, M.N.; Hewait, H.M. Effect of biofilm forming plant growth promoting rhizobacteria on salinity tolerance in barley. *Ann. Agric. Sci.* **2016**, *6*, 217–227. [CrossRef]
134. Bharti, N.; Barnawal, D. Amelioration of salinity stress by PGPR: ACC deaminase and ROS scavenging enzymes activity. In *PGPR Amelioration in Sustainable Agriculture*; Woodhead Publishing: Sawston, UK, 2019; pp. 85–106. [CrossRef]
135. Upadhyay, S.K.; Singh, J.S.; Saxena, A.K.; Singh, D.P. Impact of PGPR inoculation on growth and antioxidant status of wheat under saline conditions. *Plant Biol.* **2012**, *14*, 605–611. [CrossRef] [PubMed]
136. Sapre, S.; Gontia-Mishra, I.; Tiwari, S. Plant growth-promoting rhizobacteria ameliorates salinity stress in pea (*Pisum sativum*). *J. Plant Growth Regul.* **2022**, *41*, 647–656. [CrossRef]
137. Jha, Y.; Subramanian, R.B. PGPR regulate caspase-like activity, programmed cell death, and antioxidant enzyme activity in paddy under salinity. *Physiol. Mol. Biol. Plants* **2014**, *20*, 201–207. [CrossRef]
138. Hidri, R.; Mahmoud, O.M.-B.; Zorrig, W.; Mahmoudi, H.; Smaoui, A.; Abdelly, C.; Azcon, R.; Debez, A. Plant growth-promoting rhizobacteria alleviate high salinity impact on the halophyte *Suaeda fruticosa* by modulating antioxidant defense and soil biological activity. *Front. Plant Sci.* **2022**, *13*, 821475. [CrossRef] [PubMed]
139. Yun, P.; Xu, L.; Wang, S.S.; Shabala, L.; Shabala, S.; Zhang, W.Y. *Piriformospora indica* improves salinity stress tolerance in *Zea mays* L. plants by regulating Na^+ and K^+ loading in root and allocating K^+ in shoot. *Plant Growth Regul.* **2018**, *86*, 323–331. [CrossRef]

140. Rojas-Tapias, D.; Moreno-Galván, A.; Pardo-Díaz, S.; Obando, M.; Rivera, D.; Bonilla, R. Effect of inoculation with plant growth-promoting bacteria (PGPB) on amelioration of saline stress in maize (*Zea mays*). *Appl. Soil Ecol.* **2012**, *61*, 264–272. [CrossRef]
141. del Amor, F.M.; Cuadra-Crespo, P. Plant growth-promoting bacteria as a tool to improve salinity tolerance in sweet pepper. *Funct. Plant Biol.* **2011**, *39*, 82–90. [CrossRef]
142. Shahzad, R.; Khan, A.L.; Bilal, S.; Waqas, M.; Kang, S.M.; Lee, I.J. Inoculation of abscisic acid-producing endophytic bacteria enhances salinity stress tolerance in *Oryza sativa*. *Environ. Exp. Bot.* **2017**, *136*, 68–77. [CrossRef]
143. Bharti, N.; Pandey, S.S.; Barnawal, D.; Patel, V.K.; Kalra, A. Plant growth promoting rhizobacteria *Dietzia natronolimnaea* modulates the expression of stress responsive genes providing protection of wheat from salinity stress. *Sci. Rep.* **2016**, *6*, 34768. [CrossRef]
144. Kabiraj, A.; Majhi, K.; Halder, U.; Let, M.; Bandopadhyay, R. Role of Plant Growth-Promoting Rhizobacteria (PGPR) for crop stress management. In *Sustainable Agriculture in the Era of Climate Change*; Springer: Cham, Switzerland, 2020; pp. 367–389. [CrossRef]
145. Li, H.; Qiu, Y.; Yao, T.; Ma, Y.; Zhang, H.; Yang, X. Effects of PGPR microbial inoculants on the growth and soil properties of *Avena sativa*, *Medicago sativa*, and *Cucumis sativus* seedlings. *Soil Tillage Res.* **2020**, *199*, 104577. [CrossRef]
146. Li, X.; Sun, P.; Zhang, Y.; Jin, C.; Guan, C. A novel PGPR strain *Kocuria rhizophila* Y1 enhances salt stress tolerance in maize by regulating phytohormone levels, nutrient acquisition, redox potential, ion homeostasis, photosynthetic capacity and stress-responsive genes expression. *Environ. Exp. Bot.* **2020**, *174*, 104023. [CrossRef]
147. Habib, S.H.; Kausar, H.; Saud, H.M. Plant growth-promoting rhizobacteria enhance salinity stress tolerance in okra through ROS-scavenging enzymes. *Biomed Res. Int.* **2016**, *2016*, 6284547. [CrossRef] [PubMed]
148. Kim, K.; Jang, Y.J.; Lee, S.M.; Oh, B.T.; Chae, J.C.; Lee, K.J. Alleviation of salt stress by Enterobacter sp. EJ01 in tomato and Arabidopsis is accompanied by up-regulation of conserved salinity responsive factors in plants. *Mol. Cells* **2014**, *37*, 109. [CrossRef] [PubMed]

life

MDPI

Communication

Identification of Two GDSL-Type Esterase/Lipase Genes Related to Tissue-Specific Lipolysis in *Dendrobium catenatum* by Multi-Omics Analysis

Xinqiao Zhan [1,2,*], Yichun Qian [2] and Bizeng Mao [2,3,4]

1 Institute of Biopharmaceuticals, Taizhou University, Taizhou 318000, China
2 Institute of Biotechnology, Zhejiang University, Hangzhou 310000, China
3 Ministry of Agriculture Key Lab of Molecular Biology of Crop Pathogens and Insects, Hangzhou 310000, China
4 Key Laboratory of Biology of Crop Pathogens and Insects of Zhejiang Province, Hangzhou 310000, China
* Correspondence: joezhan@tzc.edu.cn

Abstract: *Dendrobium catenatum* is an important herb and widely cultivated in China. GDSL-Type Esterase/Lipase proteins (GELPs) are widely distributed in plants and play crucial roles in stress responses, plant growth, and development. However, no identification or functional analysis of GELPs was reported in *D. catenatum*. This study identifies 52 GELPs in *D. catenatum* genome, which is classified into four groups by phylogenetic analysis. Four conservative blocks (Ser-Gly-Asn-His) are found in most GELP domains. Transcriptome analysis reveals the expression profiles of *GELPs* in different organs and flowering phases. Co-expression analysis of the transcriptome and lipidome identifies a GELP gene, *Dca016600*, that positively correlates with 23 lipids. The purified Dca016600 protein shows the optimum pH is active from 8.0 to 8.5, and the optimum temperature is active from 30 °C to 40 °C. The kinetic study provides V_{max} (233.43 μmol·min^{-1}·mg^{-1}) and K_m (1.49 mM) for substrate *p*-nitrophenyl palmitate (*p*-NPP). Integrated analysis of the transcriptome and proteome identifies a GELP gene, *Dca005399*, which is specially induced by freezing. Interestingly, *Dca005399* shows high expression in symbiotic germination seeds and sepals. This study provides new insights into the function of *D. catenatum* GELPs in plant development and stress tolerance.

Keywords: *Dendrobium catenatum*; lipase; multi-omics; expression pattern; gene family

Citation: Zhan, X.; Qian, Y.; Mao, B. Identification of Two GDSL-Type Esterase/Lipase Genes Related to Tissue-Specific Lipolysis in *Dendrobium catenatum* by Multi-Omics Analysis. *Life* **2022**, *12*, 1563. https://doi.org/10.3390/life12101563

Academic Editors: Wajid Zaman and Hakim Manghwar

Received: 24 September 2022
Accepted: 7 October 2022
Published: 9 October 2022

Publisher's Note: MDPI stays neutral with regard to jurisdictional claims in published maps and institutional affiliations.

Copyright: © 2022 by the authors. Licensee MDPI, Basel, Switzerland. This article is an open access article distributed under the terms and conditions of the Creative Commons Attribution (CC BY) license (https://creativecommons.org/licenses/by/4.0/).

1. Introduction

GDSL-Type Esterase/Lipase proteins (GELPs) are a variety of hydrolytic enzymes with broad substrate specificity and regiospecificity, with thioesterase, protease, arylesterase, and lysophospholipase activity [1]. The conserved GDSL motif contains four invariant important catalytic residues Ser, Gly, Asn, and His, which are also named SGNH hydrolases [2]. GELPs contain many members and widely exist in plants, such as more than 100 members in rice [3], 105 members in *Arabidopsis* [4], and 194 members in soybean [2]. GELPs have been suggested to play crucial roles in plant development and metabolism. In rice, most GELPs are highly expressed in germinating seeds and are responsible for lipid homeostasis [5]. GELPs also modulate phytohormone signaling in plant growth. A GDSL lipase gene (*LIP1*) in *Arabidopsis* is induced by GA and repressed by DELLA proteins, which mediates the enhanced germination potential [6]. GELPs are involved in auxin-induced processes of suberin polymerization and degradation in root development [7]. *MHZ11* encodes a GDSL-family lipase with acyl-hydrolyzing activity and is induced by ethylene. MHZ11 also acts with the ethylene receptor ETHYLENE RESPONSE SENSOR2 (OsERS2) and impairs CONSTITUTIVE TRIPLE RESPONSE2 (OsCTR2) phosphorylation for triggering ethylene signaling in rice [8]. The *DAD1* (*defective in anther dehiscence1*) gene belongs to the GELP family and encodes a particular phospholipase A1 (PLA1) that participates in jasmonic acid (JA) biosynthesis and linolenic acid metabolism. The defects of *DAD1* lead to anthers dehiscence, pollen maturation, and flower opening [9]. Moreover, GELPs are closely associated with stomata development and are involved in plant response to abiotic stress. A total of 19 putative GELPs control stomatal dynamics, development, and plant water composition in *Arabidopsis* [10]. GELP is required for wax biosynthesis of stomatal cuticular and affected plant drought tolerance in *Arabidopsis* [11]. Soybean GELP28 can enhance the drought and salt tolerance of plants [2], but its biological function is still unknown in vivo.

Dendrobium is a large subfamily of orchids including *D. chrysotoxum*, *D. huoshanense*, *D. catenatum*, etc. [12,13]. *D. catenatum* is an important herb in southeast China, which has valuable medicinal components, such as polysaccharides, alkaloids, terpenoids, and flavonoids [14]. Recently, a total of 74 terpene metabolites are identified in *D. catenatum* and a high content of amyrenones is first found in the root [15]. Amyrenones have anti-hyperglycemic, lipid-lowering, and anti-obesity effects *in vivo* [16]. Thus, *D. catenatum* growth and development deserve attention. Our previous transcriptomic analysis hints that photosynthesis and membrane lipids are affected during freezing treatment (FT) and post-recovery freezing (FR) [17]. Proteome and lipidome analyses were further performed to investigate the lipid turnover during freezing and thawing. GELP family members play important roles in plant growth and lipid metabolic regulation [1,18]. Thus, we want to create a supporting basis for the functional prediction of the GELPs family in *D. catenatum* and identify the key candidate GELPs genes for further detailed functional study.

2. Results

2.1. Comprehensive Identification of GDSL Esterase/Lipase Protein (GELP) Family

Based on the phylogenetic analysis of 52 GELPs from *D. catenatum* genome, the GELP family was divided into four groups (Figure 1a). A total of 20 conserved motifs were predicted with MEME and displayed in Figure S1. Motifs 1, 2, 3, and 4, respectively, represented the conserved blocks I, II, III, and V of the GELP family, which are present in almost all proteins (Figure 1b). To investigate the expression profiles of the *GELP* family in *D. catenatum* growth, two transcriptome data sets were selected for analysis [19,20]. The organ-specific expression patterns indicated that half of *GELPs* were primarily expressed in flowers (Figure 1c). Furthermore, expression levels of *GELPs* were detected during different flowering phases. About half of all GELPs were highly expressed in S1, and the rest were highly expressed in S2 and S3, respectively (Figure 1d). GELPs biological activity was tightly correlated with lipid metabolism. Seven lipid categories were identified from the lipidome (Table S1). Clustering analyses of the lipidome of four tissues revealed that most of the lipids were highly accumulated in leaf tissue. Three sphingolipids and one glycerolipid were highly accumulated in the root. Fatty acid (FA), ceramides (Cer), and hexosyl sphingosine (Hex1SPH) were highly accumulated in flowers (Figure 1e). These results hinted that organ-specific expression patterns of GELPs were associated with lipid tissue distribution.

2.2. Identification and Characterization of Dca016600

To explore the key *GELPs* involved in lipid metabolism, the co-expression analysis revealed that 38 lipids and 35 *GELPs* were highly correlated (Pearson correlation coefficient > 0.9 or <−0.9), which generated 72 positively correlated pairs and 11 negatively correlated pairs (Figure 2a). Among them, *Dca016600* had 23 positively correlated pairs with lipids (Figure 2a). A total of 504 GELPs from six species of plants were used to construct a phylogenetic tree (Figure S2). This result indicated that Dca016600 was close with *O. sativa* and *P. equestris* homologs (Figure 2b). However, these homologous proteins had no reported function. The recombinant Dca016600-His protein, which had a molecular weight of 33.28 kDa (signal peptide was cut off), was expressed in *E. coli* and was purified for lipase activity assay (Figure S3). Dca016600 was active from pH 7.0 to pH 9.0, with the optimum pH at 8.0 and 8.5 (Figure 2c). The optimum temperature of purified Dca016600 was 30 °C, but it was still active at 40 °C (Figure 2d). Under optimum reaction conditions, 30 °C and pH 8.0, V_{max} and K_m of purified Dca016600 were detected and calculated as 233.43 $\mu mol \cdot min^{-1} \cdot mg^{-1}$ and 1.49 mM by Lineweaver-Burk plot, respectively (Figures 2e and S4). These results suggest that Dca016600 may be responsible for intracellular lipid catabolism.

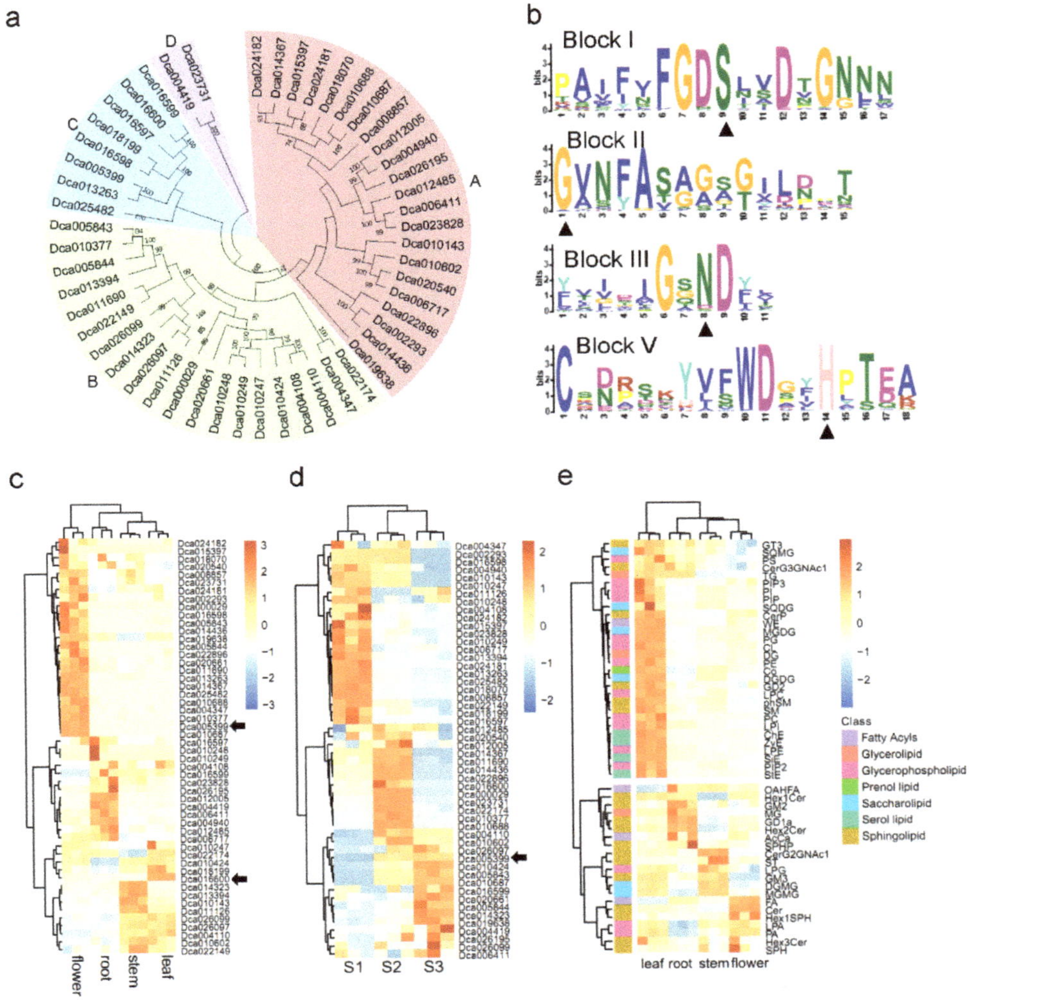

Figure 1. Comprehensive identification of GELP family. (**a**) Phylogenetic analysis of GELPs in *D. catenatum*. A total of 52 GELPs were used to construct the unrooted maximum-likelihood phylogenies. (**b**) Domain organization of GELPs. Conservative blocks (Block I, II, III, and V) of GELPs were shown from up and down. Conservative amino acid residues Ser-Gly-Asn-His in blocks are marked by black triangles. (**c**) Expression patterns of GELP genes in different tissues. (**d**) Expression patterns of GELP genes in three flowering phases. S1, the flower buds were green in the early developmental stage; S2, the flowers had purple pigmentation in the columns and the lips; S3, the sepals and petals had turned yellow and red. (**e**) Lipidome analysis of four tissues in *D. catenatum*. Raw data is shown in Table S1. Lipid abbreviations is listed in Table S2. Color scales represented the values of $\log_2$ in gene expression levels or lipid content.

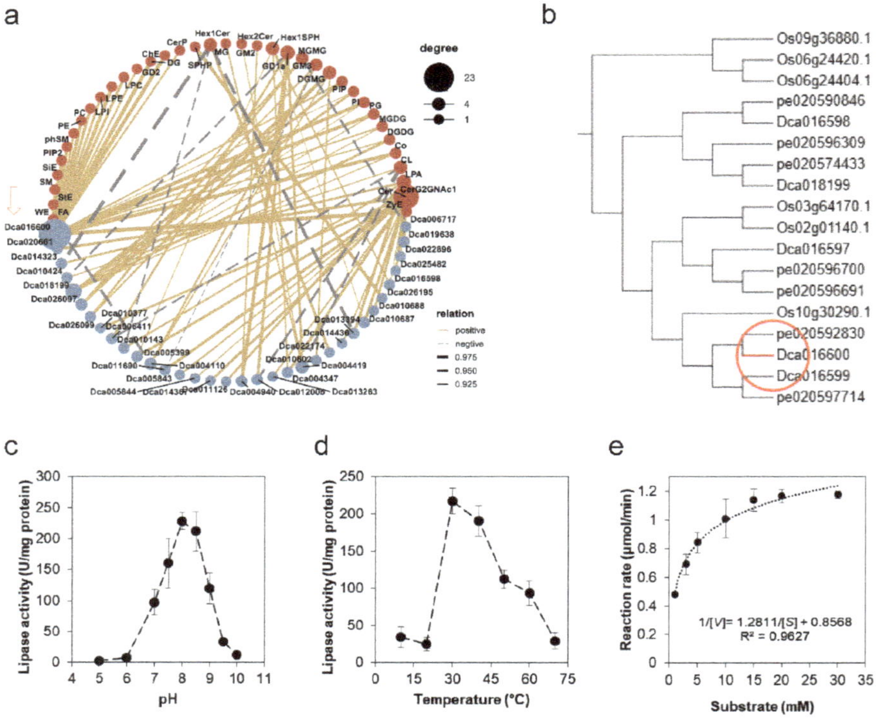

Figure 2. Identification and characterization of Dca016600. (**a**) Correlation analysis of lipidome and GELP expression levels. The dotted and solid lines, respectively, represent positive and negative correlations. The thickness of the line is determined by a Pearson correlation coefficient >0.9 or <−0.9, respectively. The dot sizes and colors represent the correlated number of lipids and genes. (**b**) Phylogenetic analysis of Dca016600 was intercepted to Supplemental Figure S2. Effects of pH (**c**), temperature (**d**), and the effect of substrate concentrations (**e**) for Dca016600 activity. Data represent the mean ± SD of three independent experiments. Km and Vmax values are determined using Lineweaver-Burk plot in Figure S4.

2.3. Identification and Expression Analysis of Dca005399

D. catenatum was greatly affected by cold damage during winter. Our previous study had been performed using metabolome and transcriptome to reveal the response of *D. catenatum* during freezing (FT) and post-freezing recovery (FR) [17]. Furthermore, only one lipase, Dca005399, was significantly changed during FT and FR at the transcriptional and translation levels (Figure 3a,b). The protein and mRNA expression levels of Dca005399 were significantly decreased in FT vs. CK, while significantly increased in FR vs. CK. Phylogenetic analysis showed that Dca005399 was close with AT3G16370 of *Arabidopsis* (Figure 3c). *AT3G16370* (*GGL19*) was preferentially expressed in leaf guard cells, filaments, and sepals [10]. Our data also found that *Dca005399* was highly expressed in flowers, especially in sepals (Figures 1c and 3d). Interestingly, *Dca005399* showed evident expression in seed germination, and especially high expression in symbiotic germination seed (Figure 3d). These results suggested that Dca005399 had important roles in cold stress, plant development and growth.

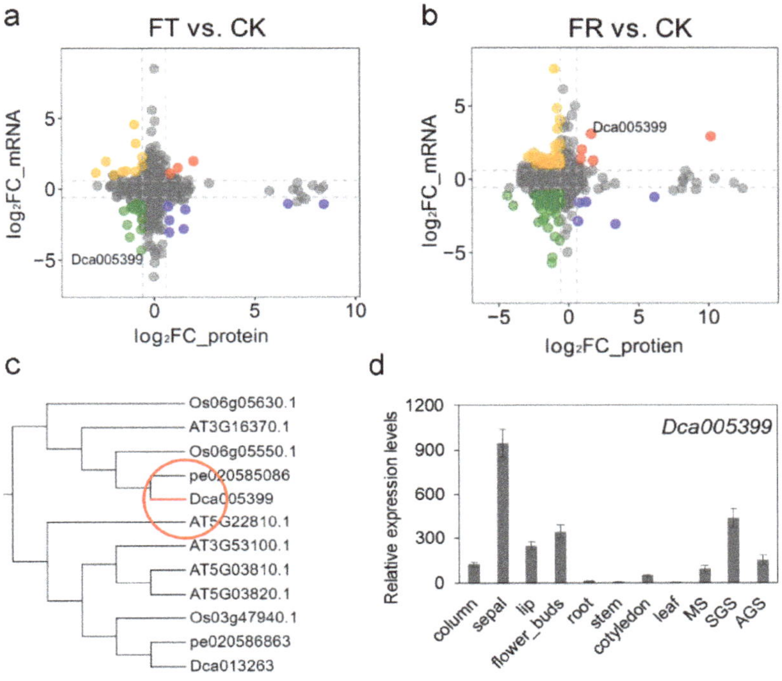

Figure 3. Identification and expression analysis of Dca005399. (**a**) The relationship between changes in protein and mRNA abundances in FT vs. CK. (**b**) The relationship between changes in protein and mRNA abundances in FR vs. CK. The colored points indicate significant upregulation or downregulation of protein and mRNA levels. CK, control condition; FT, freezing treatment; FR, post-freezing recovery. (**c**) Phylogenetic analysis of Dca005399 was intercepted to Supplemental Figure S2. (**d**) Tissues expression analysis of *Dca005399*. Data represent the mean ± SD of three independent experiments. MS, mature seed; SGS, symbiotic germination seed; AGS, asymbiotic germination seed.

3. Discussion

GELPs have broad substrate specificity and maintain a high number of family members. More than 100 members have been identified in different plant species [2,4,8]. In *Arabidopsis*, the phylogenetic analysis reveals that 105 GELPs are divided into four classes and half of them are expressed in special tissues. For example, the flower-specific expressed genes, *AtGELP42* and *AtGELP83*, improve pollen hydration on the stigma in the early pollination stage [4,21]. A total of 194 GELP genes are identified in the soybean genome and most of them show very low or no transcriptional abundance in plant growth and different tissues. Among them, the overexpression of *GmGELP28* enhances the drought and salt tolerance in plants [2]. However, the number of GELPs in *D. catenatum* and its close specie *P. equestris* are less than half of GELPs in rice, *Arabidopsis*, and *S. moellendorffi* (Figure S2). The endosperm accumulates different types of storage compounds to support the seedling during early post-germinative growth [22]. The hydrolysis of stored lipids in the endosperm by lipase plays a crucial role during seed germination [5]. In all orchid species, the endosperm is absent from the seed, including *D. catenatum* [12]. The lack of endosperm in orchids may therefore be related to the reduction in the GELPs family. Another possible reason may be that *D. catenatum* reference genome is not very well assembled and leads to the deficiency of sequences annotation.

GELPs have been identified in several important economic crops, and several GELPs have been successfully cloned and characterized, primarily in *Arabidopsis*, rice, and tomato [23]. However, there have been no reports on lipases from orchids. We integrate multi-omics data sets and identify two key GELPs in *D. catenatum*. The purified Dca016600 protein shows the optimum temperature is active from 30 °C to 40 °C, and the optimum pH is active from 8.0 to 8.5 (Figure 2). *Dca016600* is primarily expressed in leaves and has a highly positive correlation with 23 lipids (Figures 1c and 2a). These results provide a valuable reference for the study of the Dca016600 function. Another lipase Dca005399 is close to GGL19 of *Arabidopsis* (Figure 3c). *GGL19* is widely expressed in various tissues of each growth phase of *Arabidopsis*, including the early seedling stage, true leaves, and reproductive stage [10]. While *Dca005399* is primarily expressed in mature flowers (Figure 1c,d). In detail, we find that *Dca005399* is highly expressed in sepals (Figure 3d). Consistent with the flowers of other orchids, *D. catenatum* has several distinguishing features in its floral morphology. The columns are derived from the fusion of stamens and pistils. The three petal-like sepals are light green during the early developmental stages of flowering and turn yellow during the full-bloom stage [19]. Thus, *Dca005399* may also be involved in regulating floral organ development or fragrance composition, possibly with redundancy. Moreover, *Dca005399* is involved in seed germination and shows an especially high expression in symbiotic germination seeds (Figure 3d). GELPs participating in the hydrolysis of stored lipids in the initial stage of seed germination have been reported [5]. It is well known that the seeds of almost all orchids rarely germinate in natural conditions. *D. catenatum* seeds depend on mycorrhizal fungi to induce their germination [12]. Thus, we speculate that *Dca005399* mediates the regulation of symbiotic germination and is induced by infection of mycorrhizal fungi.

In recent years, various studies have combined multi-omics data sets to reveal biological progress in plants. Multi-omics technologies, including genome, epigenome, transcriptome, proteome, and metabolome, provide more possibilities to study non-model species. The GELP family contains plenty of members in *D. catenatum*. By routine gene expression analysis, it is hard to find the key candidate genes during stress. With integrated transcriptome, lipidome, and proteome analysis, we screen two GELPs may involve the regulation of different tissues and environments. *Dca016600* is primarily expressed in leaves and has lipase activity in vivo. *Dca005399* is primarily expressed in flowering and is specially induced in symbiotic germination. Although orchid plants lack endosperm, we think that it is important to lipid metabolism in *D. catenatum* seed germination.

4. Materials and Methods

4.1. Plant Materials

D. catenatum (two-year-old) was grown in soil in the greenhouse of Zhejiang University (Hangzhou, China) under conditions of 25 ± 2 °C (12 h light/12 h dark), 80 μmol photons $m^{-2}s^{-1}$, and 65–75% relative humidity [14]. Surface-sterilized seeds of *D. catenatum* were germinated on 1% (w/v) Murashige and Skoog (MS) agar medium for asymbiotic germination. For symbiotic germination testing, surface-sterilized seeds were cultured with *Tulasnella* sp. In oatmeal agar medium (0.25% oatmeal and 1% agar) under 25 ± 2 °C (12 h light/12 h dark). Four tissues and seed samples of *D. catenatum* were immediately frozen in liquid nitrogen for detection.

4.2. Lipidomic Analysis

Lipids were extracted according to a previous study [24]. The lipidomics and data analyses were performed by Shanghai Applied Protein Technology Co., Ltd. (Shanghai, China) [25]. Briefly, samples were grounded into powder in liquid nitrogen and mixed into 440 μL internal standard solution. A volume of 800 μL of methyl tert-butyl ether (MTBE) was incubated with extraction for 30 min at room temperature. After centrifugation, the organic solvent layer was dried under nitrogen. The lipid extracts were re-dissolved in 200 μL 10% ACN/isopropanol and 3 μL of the solution was injected into UHPLC (Nexera LC-30A, Shimadzu, Japan) using CSH C18 column (1.7 μm, 2.1 mm × 100 mm, Waters, Milford, MA, USA). The filtrate was separated by a linear gradient of 30% to 100% ACN/isopropanol (1:9, *v*/*v*) containing 0.1% formic acid and 0.1 mM ammonium formate with a flow rate of 300 µL min^{-1}. ESI parameters of Q-Exactive Plus (Thermo Scientific, Waltham, MA, USA) are set as follows: 300 °C source temperature; 350 °C capillary temperature, 3000 V ion spray voltage, 200–1800 *m*/*z* scan range, 50% S-Lens RF level. Lipid species were identified by LipidSearch Software (Thermo Scientific, Waltham, MA, USA) based on 5 ppm mass tolerance of fragment and 5% product ion threshold.

4.3. Phylogenetic Analysis

To excavate the homologs of GELPs in the *D. catenatum*, the hidden Markov model (HMM) file of GELP (PF00657) was provided from the PFAM website (http://pfam.xfam.org/; accessed on 1 September 2022). HMMER 3.0 was used to search the GELPs genes from *D. catenatum* reference genome [12] and the cutoff value was set to 0.01. The phylogenetic tree was calculated using the Neighbor-Joining (NJ) method of MEGA X, with the following parameters: Poisson model, pairwise deletion, and 1000 bootstrap replications [26]. The iTOL webpage tool (https://itol.embl.de/; accessed on 12 September 2022) was used to draw the phylogenetic tree [27]. The MEME online program (http://meme.nbcr.net/meme/intro.html; accessed on 12 September 2022) was used to identify the conserved motifs. The SignalP web server (http://www.cbs.dtu.dk/services/SignalP/; accessed on 12 September 2022) was used to analyze signal peptides.

4.4. Dca016600 Activity Analysis

Dca016600 sequence was amplified from *D. catenatum* cDNA using the forward primer 5′- CATATGTCTGGTGGCTGTGGATTTGATCCTC-3′ paired with the reverse primer 5′- AAGCTT ATTTAGTGATGCACCATATTTCTGG-3′. The fragment was ligated into the pET28a vector by *Nde*I and *Hind*III digestion. The construct was transformed into *E. coli* BL21 for Dca016600 protein expression. The protein was purified using Ni-NTA resin (Sangon, China) according to the previous study [24]. Purified protein was used for lipase activity assays according to the methods described previously, one enzyme unit was defined as the amount of enzyme that produced 1 µmol of *p*-nitrophenyl per min [28]. Briefly, 985 µL of substrate solution containing 30 µM *p*-NPP and 50 mM Tris-HCl buffer (pH 8.0) was incubated at 30 °C for 10 min. The substrate solution then was mixed with 10 µL of 0.5 M $CaCl_2$ and 5 µL of enzyme solution (contained 1 µg protein) at 30 °C for 10 min. Reactions were stopped by the addition of 200 µL of methanol. UV-visible detection was performed at 405 nm. To assess the effect of pH on the enzyme activity, the substrate solution was chosen 50 mM different buffers (pH 5.0–6.0 citrate, pH 7.0 sodium phosphate, pH 8.0 Tris-HCl, and pH 9.0–10.0 Glycine-NaOH). To assess the kinetic curve of the enzyme activity, the substrate solution contained 1, 3, 5, 10, 15, 20 and 30 mM *p*-NPP, respectively. Reaction conditions were as above.

4.5. Real-Time Quantitative PCR

Total RNA was extracted from seed samples using the TransZol reagent (TransGen Biotech, Beijing, China). RNA solution was treated with DNaseI (NEB, Hert, UK) to clear DNA. First-strand cDNA was transcripted from the RNA template by reverse transcription using the TIANscriptRTKit according to the manufacturer's instructions (TransGen Biotech, Beijing, China). The real-time quantitative PCR processes were performed according to our previous study [29].

4.6. Data Analysis

Transcriptome data set of FT and FR were supported by a previous study [17]. Proteome and lipidome data were treated by hierarchical clustering using the R package pheatmap (v1.0.12) and by PCA using the R package FactoMineR (v2.6) according to our previous study [15]. GO enrichment analysis was used in the R package GOplot (v1.0.2) and clusterProfiler (v4.2.2). For DAPs (differential accumulation proteins) selection, protein levels of two comparisons were determined by FC (fold change) > 1.5 or FC < 0.7 and with a statistical significance (p-value < 0.05). DALs (differential accumulation lipids) of comparisons were selected by FC > 2 or FC < 0.5, with a statistical significance (p-value < 0.05).

Supplementary Materials: The following are available online at https://www.mdpi.com/article/10.3390/life12101563/s1, Figure S1: The conserved motifs analysis of GELPs. Four blocks (Block I, II, III and V) were showed in Figure 1a, Figure S2: Phylogenetic analysis of GELPs in *Arabidopsis thaliana* (115 GELPs), *Dendrobium catenatum* (52 GELPs), *Oryza sativa* (122 GELPs), *Phalaenopsis equestris* (61 GELPs), *Selaginella moellendorffi* (145 GELPs), and *Chlamydomonas reinhardtii* (9 GELPs). Total of 504 GELPs were used to construct the unrooted maximum likelihood phylogenies, Figure S3: Dca016600 was expressed in E. coli cells and was purified for the enzymatic activity assay, Figure S4: Lineweaver-Burk plot for Dca016600 activity, Table S1: Lipidomics classification in four tissues, Table S2: Lipid abbreviations list.

Author Contributions: X.Z. conceived and designed the experiment. Y.Q. collected samples. X.Z. analyzed data. X.Z. wrote the manuscript and B.M. provided the materials. All authors have read and agreed to the published version of the manuscript.

Funding: This work was financially supported by National Natural Science Foundation of China 32200304 (to X.Z.) and Key Research and Development Projects of Zhejiang Province 2018C02034 (to B.M.).

Institutional Review Board Statement: Not applicable.

Informed Consent Statement: Not applicable.

Data Availability Statement: All data generated or analyzed during this study are included in this published article and its Additional files. The datasets generated and analyzed during the current study are available from the corresponding author on reasonable request.

Acknowledgments: We thank Shanghai Applied Protein Technology Co., Ltd. provided lipidome detection.

Conflicts of Interest: The authors declare no conflict of interest.

References

1. Akoh, C.C.; Lee, G.C.; Liaw, Y.C.; Huang, T.H.; Shaw, J.F. GDSL family of serine esterases/lipases. *Prog. Lipid Res.* **2004**, *43*, 534–552. [CrossRef] [PubMed]
2. Su, H.G.; Zhang, X.H.; Wang, T.T.; Wei, W.L.; Wang, Y.X.; Chen, J.; Zhou, Y.B.; Chen, M.; Ma, Y.Z.; Xu, Z.S.; et al. Genome-Wide Identification, Evolution, and Expression of GDSL-Type Esterase/Lipase Gene Family in Soybean. *Front. Plant Sci.* **2020**, *11*, 726. [CrossRef] [PubMed]
3. Chepyshko, H.; Lai, C.P.; Huang, L.M.; Liu, J.H.; Shaw, J.F. Multifunctionality and diversity of GDSL esterase/lipase gene family in rice (Oryza sativa L. japonica) genome: New insights from bioinformatics analysis. *BMC Genom.* **2012**, *13*, 309. [CrossRef] [PubMed]
4. Lai, C.P.; Huang, L.M.; Chen, L.O.; Chan, M.T.; Shaw, J.F. Genome-wide analysis of GDSL-type esterases/lipases in Arabidopsis. *Plant Mol. Biol.* **2017**, *95*, 181–197. [CrossRef] [PubMed]
5. Dolui, A.K.; Vijayaraj, P. Functional Omics Identifies Serine Hydrolases That Mobilize Storage Lipids during Rice Seed Germination. *Plant Physiol.* **2020**, *184*, 693–708. [CrossRef] [PubMed]
6. Rombolá-Caldentey, B.; Rueda-Romero, P.; Iglesias-Fernández, R.; Carbonero, P.; Oñate-Sánchez, L. Arabidopsis DELLA and two HD-ZIP transcription factors regulate GA signaling in the epidermis through the L1 box cis-element. *Plant Cell* **2014**, *26*, 2905–2919. [CrossRef]
7. Ursache, R.; De Jesus Vieira Teixeira, C.; Dénervaud Tendon, V.; Gully, K.; De Bellis, D.; Schmid-Siegert, E.; Grube Andersen, T.; Shekhar, V.; Calderon, S.; Pradervand, S.; et al. GDSL-domain proteins have key roles in suberin polymerization and degradation. *Nat. Plants* **2021**, *7*, 353–364. [CrossRef]
8. Zhao, H.; Ma, B.; Duan, K.-X.; Li, X.-K.; Lu, X.; Yin, C.-C.; Tao, J.-J.; Wei, W.; Zhang, W.-K.; Xin, P.-Y.; et al. The GDSL Lipase MHZ11 Modulates Ethylene Signaling in Rice Roots. *Plant Cell* **2020**, *32*, 1626–1643. [CrossRef]
9. Ishiguro, S.; Kawai-Oda, A.; Ueda, J.; Nishida, I.; Okada, K. The defective in anther dehiscence1 Gene Encodes a Novel Phospholipase A1 Catalyzing the Initial Step of Jasmonic Acid Biosynthesis, Which Synchronizes Pollen Maturation, Anther Dehiscence, and Flower Opening in Arabidopsis. *Plant Cell* **2001**, *13*, 2191–2209. [CrossRef]
10. Xiao, C.; Guo, H.; Tang, J.; Li, J.; Yao, X.; Hu, H. Expression Pattern and Functional Analyses of Arabidopsis Guard Cell-Enriched GDSL Lipases. *Front. Plant Sci.* **2021**, *12*, 748543. [CrossRef]
11. Tang, J.; Yang, X.; Xiao, C.; Li, J.; Chen, Y.; Li, R.; Li, S.; Lü, S.; Hu, H. GDSL lipase occluded stomatal pore 1 is required for wax biosynthesis and stomatal cuticular ledge formation. *New Phytol.* **2020**, *228*, 1880–1896. [CrossRef]

12. Zhang, G.Q.; Liu, K.W.; Li, Z.; Lohaus, R.; Hsiao, Y.Y.; Niu, S.C.; Wang, J.Y.; Lin, Y.C.; Xu, Q.; Chen, L.J.; et al. The Apostasia genome and the evolution of orchids. *Nature* **2017**, *549*, 379–383. [CrossRef] [PubMed]
13. Zhang, Y.; Zhang, G.-Q.; Zhang, D.; Liu, X.-D.; Xu, X.-Y.; Sun, W.-H.; Yu, X.; Zhu, X.; Wang, Z.-W.; Zhao, X.; et al. Chromosome-scale assembly of the Dendrobium chrysotoxum genome enhances the understanding of orchid evolution. *Hortic. Res.* **2021**, *8*, 183. [CrossRef] [PubMed]
14. Zhan, X.; Qi, J.; Zhou, B.; Mao, B. Metabolomic and transcriptomic analyses reveal the regulation of pigmentation in the purple variety of Dendrobium officinale. *Sci. Rep.* **2020**, *10*, 17700. [CrossRef] [PubMed]
15. Zhan, X.; Qian, Y.; Mao, B. Metabolic Profiling of Terpene Diversity and the Response of Prenylsynthase-Terpene Synthase Genes during Biotic and Abiotic Stresses in Dendrobium catenatum. *Int. J. Mol. Sci.* **2022**, *23*, 6398. [CrossRef]
16. Ferreira, R.; Guilhon-Simplicio, F.; Acho, L.D.R.; Batista, N.Y.; Guedes-Junior, F.D.C.; Ferreira, M.S.L.; Barcellos, J.F.M.; Veiga-Junior, V.F.; Lima, E.S. Anti-hyperglycemic, lipid-lowering, and anti-obesity effects of the triterpenes α and β-amyrenones in vivo. *Avicenna J. Phytomedicine* **2021**, *11*, 451–463.
17. Zhan, X.; Qi, J.; Shen, Q.; He, B.; Mao, B. Regulation of phenylpropanoid metabolism during moderate freezing and post-freezing recovery in Dendrobium officinale. *J. Plant Interact.* **2022**, *17*, 290–300. [CrossRef]
18. Li-Beisson, Y.; Shorrosh, B.; Beisson, F.; Andersson, M.X.; Arondel, V.; Bates, P.D.; Baud, S.; Bird, D.; Debono, A.; Durrett, T.P.; et al. Acyl-lipid metabolism. *Arab. Book* **2013**, *11*, e0161. [CrossRef]
19. He, C.; Liu, X.; Teixeira da Silva, J.A.; Liu, N.; Zhang, M.; Duan, J. Transcriptome sequencing and metabolite profiling analyses provide comprehensive insight into molecular mechanisms of flower development in Dendrobium officinale (*Orchidaceae*). *Plant Mol. Biol.* **2020**, *104*, 529–548. [CrossRef]
20. Wang, Y.; Chen, Y.; Wei, Q.; Wan, H.; Sun, C. Phylogenetic relationships of sucrose transporters (SUTs) in plants and genome-wide characterization of SUT genes in Orchidaceae reveal roles in floral organ development. *PeerJ* **2021**, *9*, e11961. [CrossRef]
21. Updegraff, E.P.; Zhao, F.; Preuss, D. The extracellular lipase EXL4 is required for efficient hydration of Arabidopsis pollen. *Sex. Plant Reprod.* **2009**, *22*, 197–204. [CrossRef] [PubMed]
22. Miray, R.; Kazaz, S.; To, A.; Baud, S. Molecular Control of Oil Metabolism in the Endosperm of Seeds. *Int. J. Mol. Sci.* **2021**, *22*, 1621. [CrossRef] [PubMed]
23. Shen, G.; Sun, W.; Chen, Z.; Shi, L.; Hong, J.; Shi, J. Plant GDSL Esterases/Lipases: Evolutionary, Physiological and Molecular Functions in Plant Development. *Plants* **2022**, *11*, 468. [CrossRef] [PubMed]
24. Zhan, X.; Shen, Q.; Wang, X.; Hong, Y. The sulfoquinovosyltransferase-like enzyme SQD2.2 is involved in flavonoid glycosylation, regulating sugar metabolism and seed setting in rice. *Sci. Rep.* **2017**, *7*, 4685. [CrossRef]
25. Wang, C.; Tong, Y.; Wen, Y.; Cai, J.; Guo, H.; Huang, L.; Xu, M.; Feng, M.; Chen, X.; Zhang, J.; et al. Hepatocellular Carcinoma-Associated Protein TD26 Interacts and Enhances Sterol Regulatory Element-Binding Protein 1 Activity to Promote Tumor Cell Proliferation and Growth. *Hepatology* **2018**, *68*, 1833–1850. [CrossRef]
26. Kumar, S.; Stecher, G.; Li, M.; Knyaz, C.; Tamura, K. MEGA X: Molecular Evolutionary Genetics Analysis across Computing Platforms. *Mol. Biol. Evol.* **2018**, *35*, 1547–1549. [CrossRef]
27. Letunic, I.; Bork, P. Interactive tree of life (iTOL) v3: An online tool for the display and annotation of phylogenetic and other trees. *Nucleic Acids Res.* **2016**, *44*, W242–W245. [CrossRef]
28. Rashid, N.; Shimada, Y.; Ezaki, S.; Atomi, H.; Imanaka, T. Low-temperature lipase from psychrotrophic Pseudomonas sp. strain KB700A. *Appl. Environ. Microbiol.* **2001**, *67*, 4064–4069. [CrossRef]
29. Zhan, X.; Shen, Q.; Chen, J.; Yang, P.; Wang, X.; Hong, Y. Rice sulfoquinovosyltransferase SQD2.1 mediates flavonoid glycosylation and enhances tolerance to osmotic stress. *Plant Cell Environ.* **2019**, *42*, 2215–2230. [CrossRef]

life

MDPI

Review

Seaweed-Derived Phenolic Compounds in Growth Promotion and Stress Alleviation in Plants

Omolola Aina [1], Olalekan Olanrewaju Bakare [2,3], Augustine Innalegwu Daniel [1,4], Arun Gokul [5], Denzil R. Beukes [6], Adewale Oluwaseun Fadaka [1], Marshall Keyster [3] and Ashwil Klein [1,*]

1 Plant Omics Laboratory, Department of Biotechnology, University of the Western Cape, Robert Sobukwe Road, Bellville 7530, South Africa
2 Department of Biochemistry, Faculty of Basic Medical Sciences, Olabisi Onabanjo University, Sagamu 121001, Ogun State, Nigeria
3 Environmental Biotechnology Laboratory, Department of Biotechnology, University of the Western Cape, Robert Sobukwe Road, Bellville 7530, South Africa
4 Department of Biochemistry, Federal University of Technology, P.M.B 65, Minna 920101, Niger State, Nigeria
5 Department of Plant Sciences, Qwaqwa Campus, University of the Free State, Phuthaditjhaba 9866, South Africa
6 School of Pharmacy, University of the Western Cape, Bellville 7535, South Africa
* Correspondence: aklein@uwc.ac.za; Tel.: +27-21-959-3327

Simple Summary: This review paper discusses the importance of phenolic compounds isolated from seaweed in improving plant growth and controlling the negative effects of environmental and biological factors.

Abstract: Abiotic and biotic stress factors negatively influence the growth, yield, and nutritional value of economically important food and feed crops. These climate-change-induced stress factors, together with the ever-growing human population, compromise sustainable food security for all consumers across the world. Agrochemicals are widely used to increase crop yield by improving plant growth and enhancing their tolerance to stress factors; however, there has been a shift towards natural compounds in recent years due to the detrimental effect associated with these agrochemicals on crops and the ecosystem. In view of these, the use of phenolic biostimulants as opposed to artificial fertilizers has gained significant momentum in crop production. Seaweeds are marine organisms and excellent sources of natural phenolic compounds that are useful for downstream agricultural applications such as promoting plant growth and improving resilience against various stress conditions. In this review, we highlight the different phenolic compounds present in seaweed, compare their extraction methods, and describe their downstream applications in agriculture.

Keywords: abiotic stress; biotic stress; phenolic compounds; seaweed

Citation: Aina, O.; Bakare, O.O.; Daniel, A.I.; Gokul, A.; Beukes, D.R.; Fadaka, A.O.; Keyster, M.; Klein, A. Seaweed-Derived Phenolic Compounds in Growth Promotion and Stress Alleviation in Plants. *Life* **2022**, *12*, 1548. https://doi.org/10.3390/life12101548

Academic Editors: Hakim Manghwar and Wajid Zaman

Received: 8 September 2022
Accepted: 1 October 2022
Published: 6 October 2022

Publisher's Note: MDPI stays neutral with regard to jurisdictional claims in published maps and institutional affiliations.

Copyright: © 2022 by the authors. Licensee MDPI, Basel, Switzerland. This article is an open access article distributed under the terms and conditions of the Creative Commons Attribution (CC BY) license (https://creativecommons.org/licenses/by/4.0/).

1. Introduction

Seaweeds or macroalgae are photosynthetic multicellular organisms, found in the subtidal or intertidal part of the marine environment, free-floating or attached to surfaces such as rocks [1]. These attributes allow them to serve a variety of functions such as the provision of oxygen for the utilization of aerobic organisms, the baseline for the aquatic food chain, and the elimination of pollutants from water for other marine animals, and their relative abundance can be used to gauge the health of the marine habitat [2]. There are approximately 10,000 seaweeds and are classified based on their photosynthetic pigment as red seaweed (Rhodophyta), green seaweed (Chlorophyta), and brown seaweed (Ochrophyta) [3]. History records that seaweed has been used as far back as 300BC as a source of food [4], medicine [5], cosmetics [6], skin care [7], and agriculture [8]. Presently, seaweed has gained wide acceptance globally in various sectors such as the health food sector, pharmaceutical sector, cosmetic sector, and agriculture due to its antioxidant, anti-inflammatory, antitumoral, hypocholesterolemic, anticoagulant, antiviral, and antimicrobial properties [9]. Seaweed is used in agriculture as a bio-stimulant because a low concentration is required to induce a positive response in plant growth and increase plants' tolerance to various stress factors [10]. However, the impact of stress on the realization of these functions is not clear because of limited research on this aspect.

Seaweeds, especially those in the intertidal zone, are constantly being subjected to severe stress conditions such as herbivore attack, high salinity, water loss, microbial attack, and ultraviolet rays [11]. However, the negative impact of these stress conditions cannot be detected due to numerous secondary metabolites (phytochemicals) synthesized by seaweed for defense and protection [12]. Phytochemicals are compounds produced by plants essentially to offer immunity against adverse external factors but also play a vital role in plant growth and development. These phytochemicals include phenolics, alkaloids, terpenoids, saponins, glucosides, curcumins, and steroids [13]. Among the numerous phytochemicals synthesized by seaweed for survival in their habitat, phenolic compounds are the most abundant [14]. Their significant role in defense and growth regulation is due to the following properties they possess: anti-aging [15], anti-inflammation [16], antioxidant [17], antiproliferative [18], antimutagenic [19], anthelmintic [20], antigenotoxic [21], and antimicrobial effects [22]. Thus, the understanding of the stress patterns concerning the production of these phytochemicals could shed light on the mechanism of stress tolerance and the role of environmental factors in the physiology and ecology of seaweeds.

Phenolic compounds, one of the phytochemicals produced in seaweeds, are made up of an aromatic ring with one or more hydroxyl functional groups and their structure varies from simple molecules to high-molecular-weight molecules [23]. The bioactivity of phenolic compounds is determined by the position of the hydroxy group, the number of hydroxyl groups, and the number of phenyl rings in the structure [24]. Several research works have been conducted in which phenolic compounds were isolated from seaweed and they include phlorotannin [25], flavonoids [26], phenolic acids [27], bromophenol [28], and phenolic terpenoids [29]. Seaweed-derived phenolic compounds have a wide variety of applications in the health industry for the treatment of various ailments and diseases; the food industry as preservatives and food additives; the cosmetic industry as active ingredients in cosmetics; the packaging industry to inhibit the growth of microbes; the textile industry as a source of dye; and agriculture to promote plant growth and resistance to abiotic stress factors [30]. However, there is a dearth of research on the roles of these vital compounds synthesized from seaweeds to alleviate biotic and abiotic stress in crop plants to improve agricultural productivity to meet the demands of our booming population.

Stress in plants can be classified into abiotic and biotic and it causes physiological, morphological, and biochemical changes such as reduced rate of photosynthesis, altered gene expression, slow growth rate, and impairment in the electron transport chain [31]. Abiotic stresses are caused by environmental conditions such as drought, extreme temperatures, ultraviolet rays, and salinity, while biotic stress is caused by living organisms such as herbivores, fungi, insects, bacteria, and bacteria [32]. Stress is triggered by unfavorable external factors, which harm the growth, development, and metabolism of an organism. Abiotic and biotic stress are the major factors causing a drastic decline in crop productivity by approximately 50 percent and this continues to get worse yearly due to climatic changes and global warming [33,34]. As a result of this, there is a significant threat to food security and availability for the world's population, which is expected to reach ten billion by 2050. A component of the United Nations 2030 Outline for Sustainable Growth is to develop agricultural practices with sustainable food production which would meet the increasing demand for food in such a way that hunger and malnutrition are controlled without an adverse effect on the environment [35].

Agrochemicals which include fungicides, herbicides, fertilizers, and pesticides are utilized for protecting plants against abiotic and biotic stress and to also increase crop yields [36]. Although these chemicals have increased food security, making more food available for human consumption, and thus reducing hunger, they have a negative short-term and long-term effect on animals, humans, and the ecosystem [37]. For example, organophosphorus pesticides such as glyphosate have been reported to cause various diseases such as endocrine disorders, neurological problems in infants, dementia, cardiovascular diseases, and cancer in humans [36]. There have also been several documented cases of plants developing resistance to pesticides over time. To ensure that there is an adequate food supply for the increasing human population while also protecting the environment, there is a need to utilize natural compounds such as phenolic compounds which are nontoxic, biodegradable, and as effective or better than agrochemicals [38].

Exogenous application of phenolic compounds can play an important role in increasing plants' growth and mitigating the effect of abiotic and biotic stress in plants through various mechanisms such as facilitating the lignification of plant cell walls which promotes shoot length and prevents pathogens from penetrating the host plant, influencing the activity of certain enzymes such as antioxidant enzymes and the synthesis of certain compounds such as proline and phenolic compounds [39]. Singh [40] reported that there was a significant increase in the various growth parameters (shoot length, root length, total chlorophyll, and total carotenoid content) of rice seeds primed with rutin and gallic acid. Jones [41] observed that chia seedlings were better adapted to salt stress when treated with caffeic acid. In a similar study performed by Nguyen [42], it was noted that foliar application of vanillic acid and p-hydroxybenzoic acid improves the tolerance of rice seedlings to drought compared to untreated seedlings [40–42]. Here, we comparatively review different phenolic compound extraction methods from seaweeds and highlight the impact of these compounds towards improving plant growth under abiotic and biotic stress conditions. In addition, this review would also expand the knowledge base of plant biologists on the innovative use of seaweed-derived phenolic compounds to maximize crop yield towards sustainable food production for the ever-growing human population.

2. Description and Classification of Seaweed

Seaweeds, also known as macroalgae, are multicellular organisms in the marine or coastal environment that can be found attached to rocks, logs of wood, or free-floating. they perform several functions such as food, shelter, and reproductive sites for other marine animals such as sea urchins and invertebrates [3]. Their abundance, alongside other organisms, can be used as an indicator of the well-being of the marine environment. They lack true roots, stems, and leaves but have either a flexible stipe, a stronghold fast, or blades that function like a root and enable them to attach to surfaces [9]. An estimated 10,000 different types of macroalgae have been discovered. They are often categorized based on their photosynthetic pigment into three taxa (Figure 1) namely: green algae (Chlorophyta), red algae (Rhodophyta), and brown algae (Ochrophyta, class Phaeophyceae) [43].

The Chlorophyta (green algae) are widely distributed in a variety of water bodies ranging from the Arctic region, lakes, oceans, and the Antarctic. However, about 90% are reported to inhabit freshwater bodies [44]. Their size varies from microscopic ones attached to other seaweed to large macroscopic ones and they are also the foundation of the aquatic food chain. They are generally characterized as eukaryotic algae that are multicellular, oxygenic, and photosynthetic with chlorophyll (a, b) as the dominant pigment with others which are smaller such as carotenes and xanthophylls. Examples of significant green algae are *Ulva* species (sea lettuce), *Caulerpa* species, and *Chaetomorpha* species [9].

The brown seaweeds (phylum Ochrophyta, class Phaeophyceae) are the largest and most developed of the seaweeds. They are mostly found in the marine environment, especially in cold to temperate waters. Brown algae are affected the most by climatic conditions which influence their phytochemical content in different geographical zones. The presence of a pigment known as fucoxanthin contributes to its distinctive brownish color. Examples of common species of brown seaweeds are laminaria species, Ecklonia species, *Undaria* species, *Himanthalia* species, *Sargassum* species, and *Dictyota* species [45].

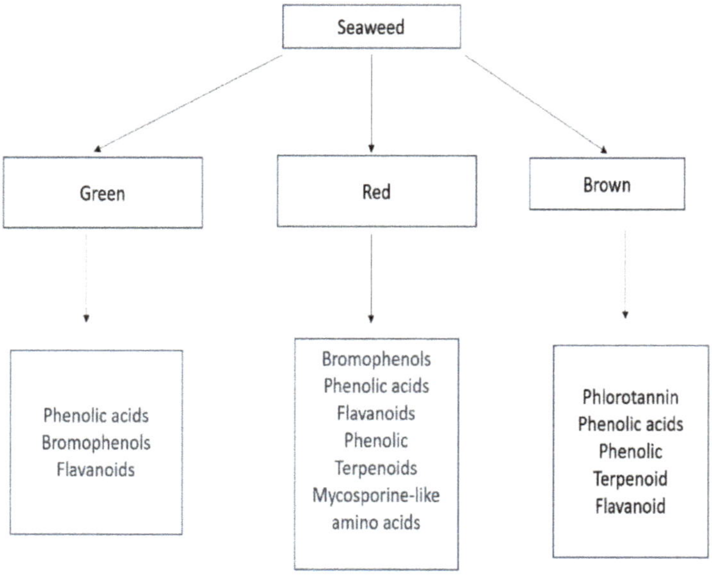

Figure 1. Classification of the main types of seaweed and their respective phenolic compounds.

The phylum Rhodophyta (red seaweed) is the most abundant seaweed which has adapted to living in almost all the water bodies from fresh water, tropical, temperate, and arctic water. However, they are dominant in tropical and temperate regions. Apart from chlorophyll a, they have an additional pigment known as phycoerythrin, which makes them tolerate low light intensities. As a result of this, they can survive in deep waters and absorb light when chlorophyll "a" is no longer active [46]. Examples of common rhodophytes are coralline red seaweed, *Porphyra*, genus *Gracilaria*, and genus *Rhodymenia* [9].

3. Phenolic Compounds as an Important Bioactive Compound in Seaweed

The broad bioactivity of seaweed has been linked to the presence of a plethora of inorganic compounds and organic compounds which include carbohydrates, protein, lipids, vitamins, hormones, betaines, and phytochemicals (terpenoids, steroids, alkaloids, and phenolic compounds) which are more abundant than the amount found in any terrestrial plants [43]. Phenolics are among the most numerous and important bioactive compounds synthesized by seaweed, particularly brown seaweed. They are produced for protection against various abiotic and biotic stress such as ultraviolet radiation, extreme temperature, salinity, pathogenic infection, and herbivory. These phenolics also contribute immensely to the growth and development of seaweeds [24].

Seaweed phenolics have a basic structure of a hydroxy group attached to an aromatic ring. They are categorized by the number of carbon atoms and benzene rings in a compound as well as their solubility. Phenolic compounds with a phenol ring, such as phenolic acids and phloroglucinol, are classified as simple phenolic compounds, whereas those with multiple phenols, such as phlorotannin, are classified as polyphenols [14,24]. The different type of phenolic compounds in seaweed and the class of seaweed where they are found are further illustrated in Figure 2.

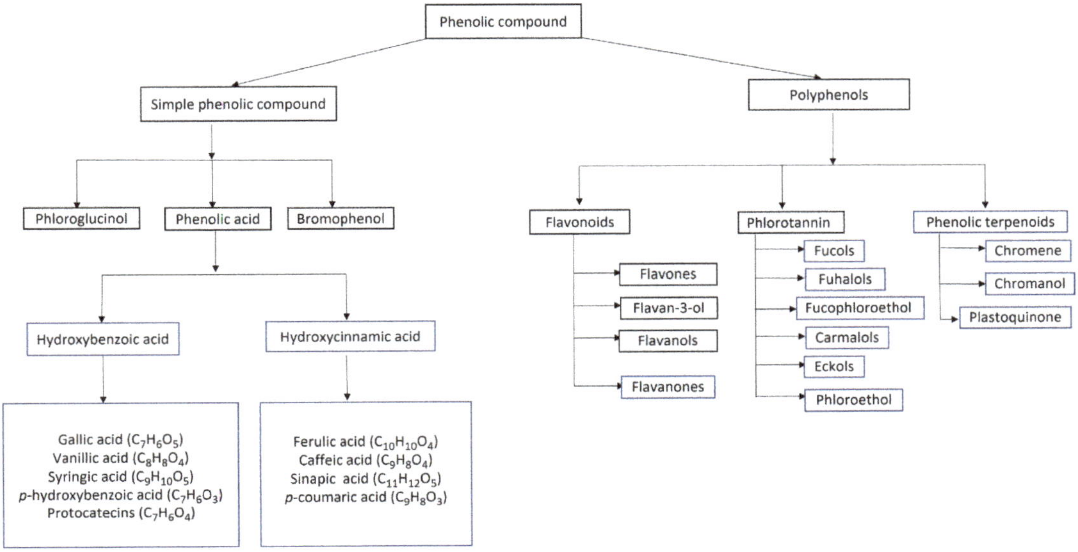

Figure 2. Classification of the main phenolic compounds in seaweed.

Extraction of Phenolic Compounds from Seaweed

Seaweeds are typically snap frozen in liquid nitrogen immediately after harvesting to halt all metabolic reactions that may result in the loss of phenolic compounds before the extraction process [47]. Following that, the seaweed samples can be extracted directly or dried before extraction. Vacuum drying, airdrying, freeze drying, or oven drying the seaweed prevents further loss of phenolic compounds and microbial growth, and ensures long-term preservation [47]. The method used to dry the seaweed has been shown to affect its phenolic content. For example, vacuum-dried *Sargassum polycystum* had the highest total phenol content compared to the other drying techniques [48].

Several methods can be used for extracting phenolic compounds from seaweed. However, due to their structural similarity, the extent of solubility, and the large molecular mass of some compounds, the chosen extraction technique should target the phenolic compound of interest. For instance, phlorotannin typically forms complexes with other metabolites within the seaweed cell walls. Therefore, an extraction method that would obtain extracts rich in phlorotannin should be adopted [45].

The common methods of extracting phenolic compounds from seaweed can be classified into traditional (conventional) and non-conventional (modern) methods. The conventional extraction techniques include Soxhlet extraction, maceration, and the percolation method while the nonconventional (novel) techniques include microwave-assisted extraction, enzyme-assisted extraction, ultrasound-assisted extraction, sub-critical water extraction, and subcritical CO_2 extraction. The advantages and disadvantages of each extraction method are summarized in Table 1.

Table 1. Comparing the advantages and limitations of the extraction procedures.

Extraction Method	Advantages	Disadvantages	References
Maceration	Simple to operate and inexpensive	It requires the use of lots of organic solvents which makes it not eco-friendly.	[49]
Soxhlet extraction method	It requires the use of a smaller volume of solvent compared to other traditional extraction methods such as maceration. The solvent can be recovered and reused.	It is time-consuming. It is not eco-friendly. It causes the degradation of thermolabile compounds. Only one sample can be processed at a time.	[49]
Microwave-assisted extraction	It involves the use of small volumes of solvent which makes it environmentally friendly and cost-effective. It is very fast, producing a high yield of the desired phenolic compound within a short time.	It operates under high temperature and microwave power which could denature heat-sensitive compounds. It requires extra separation procedures to remove solid impurities.	[50,51]
Ultrasound-assisted extraction	It also involves the use of small volumes of solvent which makes it environmentally friendly and cost-effective. It is suitable for extracting thermolabile compounds because it operates at a low temperature. The equipment used is inexpensive and easily affordable compared to other nonconventional extraction techniques. It can be scaled up for industrial applications. It is very fast, producing a high yield of the desired phenolic compound within a short period.	There may be inconsistency with the distribution of sound or mechanical waves within the medium.	[24,52]
Supercritical CO_2 extraction	It can be separated from the extract completely without leaving toxic remains. It is very fast and produces a high yield within a very short period. It is eco-friendly because no organic solvents are used. Carbon dioxide has a low critical temperature which makes it suitable for extracting thermolabile compounds. It can be used for small-scale and large-scale purposes. The resulting extract is devoid of inorganic salts and heavy metals because they cannot be extracted by carbon dioxide.	The equipment is highly sophisticated and expensive. It cannot be used to extract polar compounds due to the low polarity of carbon dioxide. However, polar solvents such as methanol are added in small quantities to supercritical CO_2 to enhance their extraction.	[14,53]
Supercritical water extraction	It is eco-friendly because it uses water as its nontoxic solvent. It is very fast producing a high yield within a short operating time. It can be used for extracting polar compounds.	It requires the use of highly sophisticated and expensive equipment. It operates under high temperature and pressure which could denature thermolabile compounds.	[49,54]
Enzyme-assisted extraction method	It can be used for small-scale and large-scale production. Toxic chemicals are not utilized during the extraction process, which makes them eco-friendly. It produces a high yield of the desired phenolic compound. It can be used in conjunction with other extraction methods to obtain a higher yield.	The enzymes used could be expensive which limits their use industrially.	[55,56]

4. Application of Phenolic Compounds in Agriculture

4.1. Role of Seaweed-Derived Phenolic Compound in Promoting Plant Growth

Phenolic compounds have been used extensively to improve plant growth and increase crop yield. Data obtained from numerous studies and the literature reveal that phenolic compounds exhibit growth-promoting properties as a result of their positive impact on various phases of plant growth and developmental processes which include seed germination, shoot length, root length, plant biomass, photosynthetic pigments, and plant metabolism [57,58]. It has also been reported that the concentration used has a significant impact. While low concentrations stimulate plant growth, high concentrations tend to inhibit it. The following mechanism of action have been proposed for their growth-promoting activity: (1) by promoting cell wall formation either as precursors for lignin or by stimulating the synthesis of lignin, (2) by regulating the synthesis and breakdown of auxin in plants, (3) by stimulating leaf expansion, (4) by promoting callus growth and increasing the growth of plant roots [59].

Seed germination is a major aspect and the main determining factor of plant growth and productivity. Germinating seeds require nutrients for the growth and synthesis of the needed cellular components. These are supplied by the enzymatic hydrolysis of the food reserves such as carbohydrates, proteins, and lipids stored within the endosperm [60]. The germination process is triggered when the seeds absorb water from the environment thereby activating the synthesis of the enzymes namely α-amylase, β-amylase, catalase, protease, and peroxidase which are required for the catabolism of the food reserves into simpler molecules that can easily be absorbed by the developing seeds [61]. Phenolic compounds are known to stimulate the activity of these enzymes, thereby enhancing the rate of seed germination. In an experiment conducted by Rengasamy [59], eckol and phloroglucinol isolated from the brown algae *Eclonia maxima* were used to treat maize seeds. It was observed that there was an increased rate of germination in the treated seeds compared to the control. This was attributed to an increase in the activity of the enzyme α-amylase in the roots of eckol and phloroglucinol-treated maize seedlings, which catalyzes the breakdown of starch to simple sugars. The sugar produced was transported to the embryo to supply the needed energy for metabolism [61].

Phenolic compounds have been reported to promote the development of adventitious root and root lengthening in plants by regulating the activity of the phytohormone indole-3-acetic acid (auxin) which is the principal hormone responsible for the process. The activity of auxin is inhibited either by conjugation or decarboxylation in a reaction catalyzed by the enzyme indole-3-acetic acid oxidase (IAA oxidase). Phenolic compounds influence the root-lengthening activity of auxin by preventing the decarboxylation reaction and by acting as a cofactor that promotes the breakdown of the enzyme IAA oxidase [62]. In a recent experiment performed by Aremu [62], two types of phlorotannins, namely eckol and phloroglucinol, were isolated from *Ecklonia maxima* and their effect on *Eucomis autumnalis* was determined. It was observed that exogenous application of the isolated polyphenols caused an increase in the auxin level, which resulted in an approximately 1.5 times increase in the root length of the treated plant.

Furthermore, phenolic compounds have been shown to promote shoot lengthening by stimulating the synthesis and deposition of lignin on cell walls and cause an increase in the activity of photosynthetic pigments (chlorophyll 'a', chlorophyll 'b', total chlorophyll, and carotenoid) in plants. In an experiment performed by Rengasamy [63] on maize, polyphenols (eckol and phloroglucinol) isolated from *Ecklonia maxima* caused an increase in shoot length, root length, and photosynthetic pigment on treated maize when compared to control plants. Briand and Salamagne [64] also evaluated the effect of phlorotannin isolated from *Fucus vesiculosus* on soybean plants grown in the field. It was noted that it promoted the growth of vegetative (aerial parts) and increased pod formation, thus causing an increase in crop yield [64]. Kulkarni [65] reported that eckol increased the total chlorophyll, carotenoid, and protein content of treated spinach plant (*Spinacea oleracea* L.) which improved the crop yield and nutritional value.

4.2. Phenolic Compounds and Abiotic Stress Intervention in Plants

Plants, as sessile organisms, are constantly exposed to abiotic and biotic stress which disrupt plant metabolism, cause stunted growth, alter plant genetic composition, and cause a reduction in crop yield globally by approximately 50 percent and 30%, respectively [66]. Abiotic stress is environmental factors such as salinity, drought, ultraviolet light, temperature, and heavy metal accumulation, while biotic stress is the consequences of damages triggered by the action of living organisms such as bacteria, fungi, insects, nematodes, and viruses on plants [67]. These stressors have been a major concern due to their detrimental effect on plant growth and development, and it is imperative to develop an effective approach, which would enable them to withstand these adverse conditions. In the subsections below, we highlight some abiotic and biotic stresses and the phenolic compounds which have been used to alleviate their impact on plants.

4.2.1. Drought

Drought is an environmental stress factor that occurs due to inadequate moisture in the soil as a result of dry weather or scarcity of surface and underground water. The effect of drought as evidenced in all the stages of plant growth and development includes a reduction in plant growth parameters (chlorophyll content, root length, shoot length, and leaf surface area), reduced germination rate, loss of cell turgor, accumulation of reactive oxygen species, and impairment of cell division which eventually reduce crop yield and availability to the population [68,69].

The use of phenolic acids to improve plants' resistance to drought stress has been reported [69,70]. According to Sun [69], treating cucumber seedlings with 50μM of cinnamic acid reduced the effect of drought stress by inducing the enzymatic (catalase, ascorbate peroxidase, superoxide dismutase, and monodehydroascorbate reductase) and nonenzymatic (proline, ascorbate, reduced glutathione, and soluble sugars) defense mechanism in plants to scavenge the generated free radicals directly and indirectly, respectively. The application of vanillic acid and *p*-hydroxybenzoic acid via foliar application increased the total chlorophyll, total carotenoid content, and total antioxidant capacity thereby reversing the damaging effect of drought in treated rice plants [70]. It is worth noting that the phenolics described above have been identified and characterized in seaweed using liquid chromatography–mass spectrometry (LC-MS) [71,72].

4.2.2. Salinity

Salinity is abiotic stress due to the buildup of salt especially sodium chloride in the soil. The common causes of salinity are human activities (bad irrigation practices, poor drainage services, and bad cultivation practices), climate change, land topography, rock weathering, and seawater deposit. Salt stress causes an accumulation of sodium and chloride ions in the leaves, as well as a decrease in the concentrations of phosphate, calcium, nitrogen, potassium, and magnesium ions. This increases the production of free radicals, disrupts ion balance, causes osmotic imbalance, and alters metabolic processes, ultimately leading to stunted growth and a massive reduction in crop yield [73].

Exogenous application of phenolic compounds has been shown to reverse the detrimental effects of salinity in plants and different types of phenolics have been quantified in seaweed. Babich [74] isolated vanillic acid and gallic acid from seaweed. Exogenous application of vanillic acid to salt-stressed tomato seedlings mitigates the adverse effect of salinity by influencing the antioxidant defense mechanism (enzymatic and nonenzymatic) which prevents lipid peroxidation and membrane damage; regulates the Na^+/K^+ balance by stimulating the absorption of potassium ion and preventing the accumulation of sodium ion; and the activity of important regulatory enzymes such as proline dehydrogenase and pyrroline-5-carboxylate synthase which increases proline synthesis and the relative water content [75].

In a study conducted by Ozfidan-Konakci [76], 0.75 and 1.5 mM of gallic acid alleviated polyethylene glycol and sodium-chloride-induced stress in three-week-old rice seedlings. It was reported that gallic acid reversed the detrimental effect of salinity by inducing the antioxidant defense mechanism (ascorbate peroxidase, catalase, superoxide dismutase, glutathione reductase, and peroxidase) which prevented the build-up of hydrogen peroxide and lipid peroxidation [76].

4.2.3. Extreme Temperature

Extreme temperature is a significant abiotic stress factor that influences all stages of plant development (germination, reproduction, growth, and yield) because all the metabolic processes within the plant cell are temperature-dependent. The impact of rapidly increasing temperatures on crop yields has become more frequent and intense, especially in Africa, which is expected to warm faster than the rest of the world, with an increase in the average temperature of 3–6 °C by the end of the century [77]. Although plants react to heat stress differently, the following response is common in all plants: production of reactive oxygen species, which reduces pollen viability which affects reproduction; inhibits photosynthesis; reduces seed germination; and reduces plant growth; and the denaturing of proteins, which affects enzymatic reactions in plants and reduction in crop yield [78].

Phenolic compounds have been shown to prevent the adverse effect of temperature extremities in plants and these compounds have been detected in seaweed.

Agregán [27] identified phenolic compounds including ferulic acid from three different brown seaweeds using liquid chromatography–diode array detection coupled to negative electrospray ionization–tandem mass spectrometry (LC-DAD–ESI-MS/MS).

According to Cheng [79], exogenous application of ferulic acid prevented the adverse effect of extreme temperature in blueberry (*Vaccinium corymbosum* L. cv. Bluecrop) seedlings. It was observed that ferulic acid treatment enhanced the transcription of genes encoding the synthesis of the antioxidant enzymes (catalase, superoxide dismutase, ascorbate peroxidase, glutathione reductase, and guaiacol peroxidase), and increased the cellular concentration of proline and soluble sugars. The increased cellular level of antioxidant enzymes prevented the accumulation of reactive oxygen species while proline and soluble sugars increased the osmotic potential and relative water content in the cell, thus preventing the detrimental effect of heat stress in the blueberry plant [79]. Exogenous application of salicylic acid reversed the effect of heat stress in ornamental pepper by activating the antioxidant defense mechanism which was evidenced by increased chlorophyll (photosynthesis rate), increased germination rate, and reduced reactive oxygen species [80].

4.2.4. Heavy Metal

The effect of heavy metal stress on agricultural land due to the continuous application of fertilizers, mining, poor irrigation practices, and industrial waste is becoming a major concern globally. Recently, there has been a build-up of heavy metals such as cadmium, zinc, mercury, arsenic, lead, copper, nickel, and aluminum in the soil. Although some of these metals are needed by plants for various biochemical processes, high concentrations have the following adverse effect on plants: alteration of cell homeostasis, accumulation of reactive oxygen species, interruption of the electron transport chain, cell membrane damage, decrease in plant growth, and reduced crop yield [81].

Phenolic compounds, known to prevent the damages caused in plants due to heavy metal stress, have been identified in different types of seaweeds. Chakraborty [82] detected salicylic acid and other phenolic compounds such as quercetin, syringic acid, and gallic acid in *Turbinaria ornate* and *Turbinaria conoides*.

Several experimental works have been performed using phenolic compounds to mitigate the effect of heavy metal stress. For example, exogenous application of salicylic acid prevented the detrimental effect of nickel stress in mustard plants by enhancing the activities of the antioxidant enzymes and the glyoxylate enzymatic system (glyoxalase I and glyoxalase II) which improved photosynthesis and plant growth [57].

4.3. Phenolic Compounds and Biotic Stress Intervention in Plants

4.3.1. Phenolic Compounds and Fungal Diseases

Fungal diseases are more prevalent than other biotic stressors and have a greater negative impact on crop production. Fungal infections have been linked to some of the world's greatest famines in history. An estimated 8,000 fungi species have been identified, accounting for more than 80% of post-harvest and pre-harvest infections [83]. Fungi infect plants through the stomata; wounds; development of special organs known as appressoria, to penetrate and attach to cuticles; and by secreting hydrolytic enzymes (cellulases, cutinases, proteases, and pectinases) which enable them to invade other parts of the plant such the epidermal cell wall and cuticle. Fungal infection causes impairment of H^+-ATPase, osmotic imbalance, decreased rate of photosynthesis, reduced crop yield, and plant death [84].

The use of phenolics to mitigate fungal-induced physiological damages in plants has been reported. Zhong [85] identified phenolics such as *p*-hydroxybenzoic acid and protocatechuic acid in red seaweed (*Dasya* sp., *Grateloupia* sp., and *Centroceras* sp), brown seaweed (*Sargassum* sp and *Ecklonia* sp.,), and green seaweed (*Ulva* sp.).

These phenolics have been used to prevent early blight disease of tomato caused by *Alternaria solani* by activating the enzymatic and nonenzymatic defense mechanism, which regulates cellular homeostasis, and antioxidant balance [86]. The aforementioned compounds also enhanced the accumulation of salicylic acid within the host cell which promoted the synthesis of pathogenesis-related proteins [86]. Table 2 highlights similar examples of the antifungal effect of exogenous application of phenolic compounds.

Table 2. Derivative phenolic compounds, uses, and their bioactivity.

Phenolic Compound	Plant Species	Type of Stress	Mechanism of Action	Reference
Salicylic acid	Safflower (*Carthamus tinctorius* L.).	Abiotic stress (drought)	Stimulated the nonenzymatic defense system. Increased synthesis of osmolytes. Increased synthesis of proline.	[87]
Vanillic acid	Blueberry (*Vaccinium corymbosum* L.)	Abiotic stress (drought)	Increased the transcription of genes encoding the synthesis of antioxidant enzymes in leaves. Increased the concentration of proline and soluble sugars. Decreased the concentration of malondialdehyde, superoxide anion, and hydrogen peroxide. Improved the relative water content.	[88]
***p*-hydroxybenzoic acid and vanillic acid**	Rice (*Oryza sativa*)	Abiotic stress (drought)	Increased the synthesis of chlorophyll "a", "b", carotenoids, and total phenolic compounds. Promoted plant growth rate. Enhanced the synthesis of phytoalexin momilactone (MA and MB) which increased tolerance to drought.	[70]
Vanillic acid	Tomato (*Solanum lycopersicum* L. cv. Pusa Ruby)	Abiotic stress (salinity)	Enhanced the glyoxalase system, thus preventing the accumulation of methylglyoxal. Activated the antioxidant defense mechanism thereby preventing lipid peroxidation and accumulation of reactive oxygen species. Increased rate of photosynthesis. Regulated the cellular Na^+/K^+ concentration. Improved the relative water content.	[75]

Table 2. *Cont.*

Phenolic Compound	Plant Species	Type of Stress	Mechanism of Action	Reference
Coumarin	Wheat (*Triticum aestivum*)	Abiotic stress (drought)	Enhanced the activity of peroxidase, thus preventing oxidative stress. Regulated the osmotic level in the cell by regulating cellular Na^{+}/K^{+} concentration. Increased synthesis of phenylalanine ammonia-lyase enzyme which increased endogenous synthesis of phenolic compound.	[89]
Ferulic acid	Blueberry seedlings (*Vaccinium corymbosum*)	Abiotic stress (Extreme temperature)	Improved plant growth. Enhanced the transcription of genes encoding for the synthesis of antioxidant enzymes (glutathione peroxidase and superoxide dismutase) which decreased lipid peroxidation and build-up of reactive oxygen species. Increased relative water content due to increased concentration of proline and soluble sugars.	[79]
Salicylic acid	*Vigna angularis*	Abiotic stress (salinity)	Increased the relative water content due to increased synthesis of glycine betaine, proline, and soluble sugar. Enhanced the enzymatic and nonenzymatic antioxidant defense mechanism. Reduction in the cellular concentration of sodium and chloride ion.	[90]
Gallic acid	Wheat (*Triticum aestivum* L.)	Abiotic stress (salinity)	Enhanced the activity of the antioxidant enzymes, thereby reducing reactive oxygen species and lipid peroxidation. Improved plant growth. Enhanced photosynthesis by increasing the chlorophyll content. Improved the relative water content.	[91]
Apigenin	Rice (*Oryza sativa L*)	Abiotic stress (salinity)	Enhanced the activity of the enzymatic (ascorbate peroxidase and catalase) and nonenzymatic defense system (endogenous flavonoids and carotenoids) thereby preventing lipid peroxidation and accumulation of reactive oxygen species. Increases the transcription of genes encoding for the synthesis of Na^{+} transporter protein, thus regulating the concentration of Na^{+}/K^{+} in the cells.	[92]
Salicylic acid	Ornamental pepper (*Capsicum annuum* L.)	Abiotic stress (extreme temperature)	Increased chlorophyll content increased the rate of photosynthesis. Activated the enzymatic and nonenzymatic defense mechanism, thus preventing the accumulation of reactive oxygen species. Prevented degradation of cellular structures by regulating osmotic balance.	[80]

Table 2. *Cont.*

Phenolic Compound	Plant Species	Type of Stress	Mechanism of Action	Reference
Salicylic acid	Mustard plant (*Brassica juncea* L.Czern. & Coss. cv. Type 59)	Abiotic stress (heavy metal)	Increased rate of photosynthesis, thus improving plant growth. Increased activity of antioxidant enzymes which prevented oxidative stress. Activated the glyoxylate system (glyoxalase I and glyoxalase II enzymes) which reduced the accumulation of toxic methylglyoxal.	[57]
Gallic acid	Sunflower (*Helianthus annuus*)	Abiotic stress (heavy metal)	Prevented absorption of cadmium ion by the root. Enhanced the activity of glutathione reductase, catalase, and ascorbate peroxidase which alleviated oxidative stress and increased plant growth.	[58]
Rutin	*Amaranthus hypochondriacus*	Abiotic stress (heavy metal)	Enhanced the synthesis of glutathione and promoted the conversion of glutathione to phytoalexins which chelate metal and prevent its accumulation within the cell. Prevents degradation of the cell membrane by inhibiting lipid peroxidation.	[93]
Gallic acid	Tea plant (*Camellia sinensis* cv. Longjing 43)	Biotic stress (*Ectropis obliqua* larvae)	Activated the phenylpropanoid and jasmonic acid pathway which stimulated the synthesis of metabolites such as epigallocatechin-3-gallate, naringenin, and astragalin that prevented the larvae from feeding on tea plants.	[94]
Salicylic acid	Green pepper (*Capsicum annuum*)	Biotic stress (antifungal)	Stimulates some immune responses in host plants such as the expression of the pathogenesis-related (PR) gene, thus inducing system resistance against the fungi. Exhibiting fungitoxic effect on the fungi and activating the synthesis of enzymes which promote the production of defense compounds.	[95]
Eckol	Cabbage (Brassica oleracea)	Biotic (insect repelling)	Increased the enzyme myrosinase which prevented cabbage aphid (*Brevicoryne brassicae*) from attacking the leaves.	[63]
Caffeic acid	Tobacco (*Nicotiana tobaccum*)	Biotic stress (antibacterial)	Increased activity of peroxidase and phenylalanine ammonia-lyase which increased the deposit of lignin in the host cell wall, thus preventing bacteria invasion. Prevented the formation of biofilm in the plant root by inhibiting the expression of *epsE and lecM genes.*	[96]
Salicylic acid	Pakchoi *(Brassicaceae)*	Biotic stress (antifungal)	Promoting the activity of antioxidant enzymes by increasing the expression of the respective gene. Increased concentration of proline and soluble protein which regulates the relative water content in the root and leaves cells.	[97]

Table 2. *Cont.*

Phenolic Compound	Plant Species	Type of Stress	Mechanism of Action	Reference
p-coumaric acid	Chinese cabbage (*Brassica rapa* var. *pekinensis*)	Biotic stress (antibacterial)	Promotes the expression of the CHS and HCT genes, thereby increasing the synthesis of endogenous phenolic compounds such as flavonoids, sinapic acid, and ferulic acid, which protects the plant from bacterial infection and promotes plant growth.	[98]

4.3.2. Phenolic Compounds and Bacteria Diseases

Plant bacterial infections are less common than fungi infections, but they are also economically significant. There are approximately 200 pathogenic bacteria species known to cause plant diseases, and they can be classified as endogenous bacteria which infect the xylem and phloem tissues, thus interrupting the transportation of water and nutrients within the plant, or exogenous bacteria, which mainly infect the intercellular spaces (apoplast) [99].

Phytopathogenic bacteria are transmitted to plants through a variety of means, including water, wind, insects, animals, and humans; however, they require openings such as stomata or wounds to penetrate the host plant. Once inside the plant, they cause diseases by synthesizing enzymes that degrade the host cell membrane and cell wall, and injecting toxins and proteins that lead to the death of the host cell [100].

Phenolic compounds can be used to reverse the detrimental effect of pathogenic bacteria in plants (Table 2) and these phenolics have been identified and quantified in seaweed [101].

According to Li [96] exogenous application of caffeic acid inhibited *Ralstonia solanacearum* infection in tobacco plants by preventing the expression of epsE and lecM genes and the formation of biofilms. In vitro, caffeic acid enhanced the activity of the enzyme phenylalanine ammonia-lyase and peroxidase, which led to the accumulation of hydroxyproline and lignin.

4.3.3. Phenolic Compounds Used to Control Viral Diseases

Viral infections pose a significant challenge in agriculture due to their ability to undergo mutations and produce new variants rapidly. Furthermore, viral infections are difficult to comprehend due to the wide range of symptoms seen in host plants. Viral infections are only transmitted via vectors humans or insects, and they penetrate the host plant through wounds [102]. Viral infections influence major biochemical and physiological processes in the host plant by altering the host genetic material. Symptoms of the viral disease include wrinkling of leaves, stunted growth, phyllody, necrotic and chlorotic lesions on leaves, wilting, and the development of irregular growth patterns known as enations (galls) [103].

Apart from salicylic acid, the use of phenolic compounds to regulate plants' resistance against viral infections remains scant. Zhang [104] showed that salicylic acid promoted plant growth and enhanced the resistance of wild soybean (*Glycine soja*) to soybean mosaic virus by stimulating the synthesis of antioxidant enzymes (catalase, peroxidase, ascorbate peroxidase, and superoxide peroxidase) and promoting the transcription of resistance-related genes (GmPR-1, GmNPR1, GmPR-10, GmEDS1, GmPR-2, and GmICS1) in the host plant. This clearly demonstrates a knowledge gap that should be addressed in future research studies.

4.3.4. Phenolic Compounds Used against Herbivore and Insect Attack

Herbivores and insects pose a significant threat to plant growth and development causing approximately a 15 percent loss in crop yield annually. They further create avenues for subsequent infection by phytopathogens which increases the severity of their attack [105].

Exogenous application of gallic acid to tea plants (*Camellia sinensis*) triggered the phenyl propanoic and jasmonic acid signaling pathway which protected the plant from herbivore attack (*Ectropis obliqua* Caterpillars) by enhancing the synthesis of three antifeeding metabolites, namely epigallocatechin-3-gallate, naringenin, and naringenin [94]. These phenolic compounds have been identified and characterized from seaweed using liquid chromatography–mass spectrometry (LC-MS) [106,107]. Exogenous application of eckol stimulates the synthesis of the enzyme myrosinase which prevents the cabbage aphid (Brevicoryne brassicae) from attacking the host leaves [63]. According to a study conducted by Jan [108], kaempferol and quercetin display pesticidal effects when applied exogenously to susceptible rice strains (TN 1 strain) by reducing the vulnerability to whitebacked planthoppers by preventing the insect from feeding on the host plant and preventing egg hatching [108].

5. Conclusions

This review illustrates the wide range of phenolic compounds present in seaweed and highlight their agricultural importance for improved plant growth and enhanced tolerance against various abiotic and biotic stress factors. The continuous use of synthetic phenolic compounds to improve plant growth whilst minimizing the negative effects of stress conditions is no longer a viable option due to its deleterious effects on human health and the environment. The use of natural phenolic compounds derived from seaweed to improve plant growth and stress tolerance could diminish the use of synthetic chemicals thus limiting the harmful impact on the environment and improve agricultural outputs in a sustainable manner.

It is worth noting that although phenolic compounds have been identified in seaweed, their downstream application in agriculture remains limited. To date, most research has focused on the use of synthetic phenolic compounds instead of natural phenolic compounds to improve plant growth and enhance plant immunity/resilience. Therefore, more research on natural phenolic compounds is encouraged to obtain a holistic understanding of their modes of action for improved plant growth and enhanced stress tolerance especially in economically important food/feed crops.

Author Contributions: Conceptualization, A.K. and O.A.; software, O.A.; resources, O.O.B., A.O.F., D.R.B. and A.G.; writing—original draft preparation, O.A.; writing—review and editing, O.O.B., A.I.D., A.G., A.O.F., D.R.B., M.K. and A.K.; supervision, M.K. and A.K. All authors have read and agreed to the published version of the manuscript.

Funding: This review manuscript received no external funding.

Institutional Review Board Statement: Not applicable.

Informed Consent Statement: Not applicable.

Data Availability Statement: Not applicable.

Acknowledgments: The authors would like to thank the University of the Western Cape and the University of Free State for administrative support.

Conflicts of Interest: The authors declare no conflict of interest.

References

1. Pati, M.P.; Sharma, S.; Nayak, L.; Panda, C.R. Uses of seaweed and its application to human welfare: A review. *Int. J. Pharm. Pharm. Sci.* **2016**, *8*, 12–20. [CrossRef]
2. Xunmeng, L.; Wang, K.; Zhang, S.; Feng, M. Distribution and Flora of Seaweed Beds in the Coastal Waters of China. *Sustainability* **2021**, *13*, 3009. [CrossRef]
3. Battacharyya, D.; Babgohari, M.Z.; Rathor, P.; Prithiviraj, B. Seaweed extracts as biostimulants in horticulture. *Sci. Hortic.* **2015**, *196*, 39–48. [CrossRef]
4. Mouritsen, O.G.; Rhatigan, P.; Pérez-Lloréns, J.L. The rise of seaweed gastronomy: Phycogastronomy. *Bot. Mar.* **2019**, *62*, 195–209. [CrossRef]

5. Lomartire, S.; Marques, J.C.; Gonçalves, A.M. An overview to the health benefits of seaweeds consumption. *Mar. Drugs* **2021**, *19*, 341. [CrossRef]
6. Joshi, S.; Kumari, R.; Upasani, V.N. Applications of algae in cosmetics: An overview. *Int. J. Innov. Res. Sci. Eng. Technol* **2018**, *7*, 1269.
7. Stoyneva-Gärtner, M.; Uzunov, B.; Gärtner, G. Enigmatic microalgae from aeroterrestrial and extreme habitats in cosmetics: The potential of the untapped natural sources. *Cosmetics* **2020**, *7*, 27. [CrossRef]
8. Pereira, L.; Cotas, J. Historical use of seaweed as an agricultural fertilizer in the European Atlantic area. In *Seaweeds as Plant Fertilizer, Agricultural Biostimulants and Animal Fodder*; CRC Press: Boca Raton, FL, USA, 2019; pp. 1–22.
9. Baweja, P.; Kumar, S.; Sahoo, D.; Levine, I. Biology of seaweeds. In *Seaweed in Health and Disease Prevention*; Elsevier: Amsterdam, The Netherlands, 2016; pp. 41–106.
10. Stirk, W.A.; Rengasamy, K.R.; Kulkarni, M.G.; van Staden, J. Plant biostimulants from seaweed: An overview. In *The Chemical Biology of Plant Biostimulants*; John Wiley and Sons Ltd.: Hoboken, NJ, USA, 2020; pp. 31–55.
11. Al, M.A.; Akhtar, A.; Rahman, M.F.; Kamal, A.H.M.; Karim, N.U.; Hassan, M.L. Habitat structure and diversity patterns of seaweeds in the coastal waters of Saint Martin's Island, Bay of Bengal, Bangladesh. *Reg. Stud. Mar. Sci.* **2020**, *33*, 100959.
12. Kumar, Y.; Tarafdar, A.; Badgujar, P.C. Seaweed as a source of natural antioxidants: Therapeutic activity and food applications. *J. Food Qual.* **2021**, *2021*, 5753391. [CrossRef]
13. Koche, D.; Shirsat, R.; Kawale, M. An overerview of major classes of phytochemicals: Their types and role in disease prevention. *Hislopia J.* **2016**, *9*, 976–2124.
14. Cotas, L.; Leandro, A.; Monteiro, P.; Pacheco, D.; Figueirinha, A.; Goncalves, A.M.M.; da Silva, G.J.; Pereira, L. Seaweed Phenolics: From Extraction to Applications. *Mar. Drugs* **2020**, *18*, 384. [CrossRef]
15. Luo, J.; Si, H.; Jia, Z.; Liu, D. Dietary anti-aging polyphenols and potential mechanisms. *Antioxidants* **2021**, *10*, 283. [CrossRef]
16. Alba, J.D.M.; Hernández, S.I.M.; López, G.C.V.; Flores, M.A. Anti-inflammatory effect of caffeic acid in an experimental model of pulpitis in guinea pigs. *Rev. Asoc. Dent. Mex.* **2016**, *73*, 250–254.
17. Gao, Q.; Li, Y.; Li, Y.; Zhang, Z.; Liang, Y. Antioxidant and prooxidant activities of phenolic acids commonly existed in vegetables and their relationship with structures. *Food Sci. Technol.* **2022**, *42*, e07622. [CrossRef]
18. Rosa, L.d.S.; Jordão, N.A.; Soares, N.d.C.P.; Mesquita, J.F.d.; Monteiro, M.; Teodoro, A.J. Pharmacokinetic, antiproliferative and apoptotic effects of phenolic acids in human colon adenocarcinoma cells using in vitro and in silico approaches. *Molecules* **2018**, *23*, 2569. [CrossRef] [PubMed]
19. Birošová, L.; Mikulášová, M.; Vaverková, Š. Antimutagenic effect of phenolic acids. *Biomed. Pap.* **2005**, *149*, 489–491. [CrossRef]
20. Samar, J.; Butt, G.Y.; Shah, A.A.; Shah, A.N.; Ali, S.; Jan, B.L.; Abdelsalam, N.R.; Hussaan, M. Phycochemical and Biological Activities From Different Extracts of *Padina antillarum* (Kützing) Piccone. *Front. Plant Sci.* **2022**, *13*, 929368. [CrossRef]
21. Soumya, K.; James, J.; Archana, T.; Dhanya, A.; Shahid, A.; Sudheesh, S. Cytotoxic and antigenotoxic properties of phenolic compound isolated from the fruit of *Terminalia chebula* on HeLa cell. *Beni Suef Univ. J. Basic Appl. Sci.* **2019**, *8*, 15. [CrossRef]
22. Alvarado-Martinez, Z.; Bravo, P.; Kennedy, N.-F.; Krishna, M.; Hussain, S.; Young, A.C.; Biswas, D. Antimicrobial and antivirulence impacts of phenolics on *Salmonella enterica* serovar Typhimurium. *Antibiotics* **2020**, *9*, 668. [CrossRef]
23. Lomartire, S.; Cotas, J.; Pacheco, D.; Marques, J.C.; Pereira, L.; Gonçalves, A.M. Environmental impact on seaweed phenolic production and activity: An important step for compound exploitation. *Mar. Drugs* **2021**, *19*, 245. [CrossRef]
24. Jimenez-Lopez, C.; Pereira, A.G.; Lourenço-Lopes, C.; Garcia-Oliveira, P.; Cassani, L.; Fraga-Corral, M.; Prieto, M.; Simal-Gandara, J. Main bioactive phenolic compounds in marine algae and their mechanisms of action supporting potential health benefits. *Food Chem.* **2021**, *341*, 128262. [CrossRef] [PubMed]
25. Gall, E.A.; Lelchat, F.; Hupel, M.; Jégou, C.; Stiger-Pouvreau, V. Extraction and purification of phlorotannins from brown algae. In *Natural Products from Marine Algae*; Springer: Berlin/Heidelberg, Germany, 2015; pp. 131–143.
26. Shoubaky, G.A.E.; Abdel-Daim, M.M.; Mansour, M.H.; Salem, E.A. Isolation and identification of a flavone apigenin from marine red alga *Acanthophora spicifera* with antinociceptive and anti-Inflammatory activities. *J. Exp. Neurosci.* **2016**, *10*, JEN-S25096. [CrossRef] [PubMed]
27. Agregan, R.; Munekata, P.E.; Franco, D.; Dominguez, R.; Carballo, J.; Lorenzo, J.M. Phenolic compounds from three brown seaweed species using LC-DAD–ESI-MS/MS. *Food Res. Int.* **2017**, *99*, 979–985. [CrossRef]
28. Rajasulochana, P.; Krishnamoorthy, P.; Dhamotharan, R. Isolation, identification of bromophenol compound and antibacterial activity of Kappaphycus sp. *Int. J. Pharm. Biol. Sci.* **2012**, *3*, 173–186.
29. Sumayya, S.; Murugan, K. Antioxidant potentialities of marine red algae *Gracillaria dura*: A search. *Pharma Innov. J.* **2019**, *8*, 1157–1161.
30. Albuquerque, B.R.; Heleno, S.A.; Oliveira, M.B.P.; Barros, L.; Ferreira, I.C. Phenolic compounds: Current industrial applications, limitations and future challenges. *Food Funct.* **2021**, *12*, 14–29. [CrossRef]
31. Singh, P.; Sharma, A.; Sharma, J. Stress physiology in plants. In *Reforms in Agriculture and Rural Development Under COVID-19 Pandemic*; Society of Human Resources and Innovation: Agra, India, 2020; pp. 175–186.
32. Gull, A.; Lone, A.A.; Wani, N.U.I. Biotic and abiotic stresses in plants. In *Abiotic and Biotic Stress in Plants*; IntechOpen: London, UK, 2019; pp. 1–19.
33. Anami, B.S.; Malvade, N.N.; Palaiah, S. Classification of yield affecting biotic and abiotic paddy crop stresses using field images. *Inf. Processing Agric.* **2020**, *7*, 272–285. [CrossRef]

34. Teshome, D.T.; Zharare, G.E.; Naidoo, S. The threat of the combined effect of biotic and abiotic stress factors in forestry under a changing climate. *Front. Plant Sci.* **2020**, *11*, 601009. [CrossRef]
35. Nkomo, M.A. The Role of *p*-Coumaric Acid on Physiological and Biochemical Response of Chia Seedling Under Salt Stress. Ph.D. Thesis, University of the Western Cape, Cape Town, South Africa, 2020.
36. Nicolopoulou-Stamati, P.; Maipas, S.; Kotampasi, C.; Stamatis, P.; Hens, L. Chemical pesticides and human health: The urgent need for a new concept in agriculture. *Front. Public Health* **2016**, *4*, 148. [CrossRef]
37. Tudi, M.; Daniel Ruan, H.; Wang, L.; Lyu, J.; Sadler, R.; Connell, D.; Chu, C.; Phung, D.T. Agriculture development, pesticide application and its impact on the environment. *Int. J. Environ. Res. Public Health* **2021**, *18*, 1112. [CrossRef]
38. Zhao, B.; He, D.; Wang, L. Advances in Fusarium drug resistance research. *J. Glob. Antimicrob. Resist.* **2021**, *24*, 215–219. [CrossRef] [PubMed]
39. Khan, F.; Jeong, G.-J.; Khan, M.S.A.; Tabassum, N.; Kim, Y.-M. Seaweed-Derived Phlorotannins: A Review of Multiple Biological Roles and Action Mechanisms. *Mar. Drugs* **2022**, *20*, 384. [CrossRef] [PubMed]
40. Singh, A.; Gupta, R.; Pandey, R. Exogenous application of rutin and gallic acid regulate antioxidants and alleviate reactive oxygen generation in *Oryza sativa* L. *Physiol. Mol. Biol. Plants* **2017**, *23*, 301–309. [CrossRef] [PubMed]
41. Jones, S.A. Exogenous Caffeic Acid Alters Molecular Responses in *Salvia hispanica* L. Master's Thesis, University of the Western Cape, Cape Town, South Africa, 2016.
42. Nguyen, T.Q. Application of Exogenous Phenolics for Drought Tolerance Improvement in Rice (*Oryza sativa* L.). Master's Thesis, Hiroshima University, Hiroshima, Japan, 2018.
43. Kumar, Y.; Singhal, S.; Tarafdar, A.; Pharande, A.; Ganesan, M.; Badgujar, P.C. Ultrasound assisted extraction of selected edible macroalgae: Effect on antioxidant activity and quantitative assessment of polyphenols by liquid chromatography with tandem mass spectrometry (LC-MS/MS). *Algal Res.* **2020**, *52*, 102114. [CrossRef]
44. Fabrowska, J.; Ibañez, E.; Łęska, B.; Herrero, M. Supercritical fluid extraction as a tool to valorize underexploited freshwater green algae. *Algal Res.* **2016**, *19*, 237–245. [CrossRef]
45. Generalić Mekinić, I.; Skroza, D.; Šimat, V.; Hamed, I.; Čagalj, M.; Popović Perković, Z. Phenolic content of brown algae (Pheophyceae) species: Extraction, identification, and quantification. *Biomolecules* **2019**, *9*, 244. [CrossRef]
46. Gomes, L.; Monteiro, P.; Cotas, J.; Gonçalves, A.M.; Fernandes, C.; Gonçalves, T.; Pereira, L. Seaweeds' pigments and phenolic compounds with antimicrobial potential. *Biomol. Concepts* **2022**, *13*, 89–102. [CrossRef]
47. Kisiriko, M.; Anastasiadi, M.; Terry, L.A.; Yasri, A.; Beale, M.H.; Ward, J.L. Phenolics from medicinal and aromatic plants: Characterisation and potential as biostimulants and bioprotectants. *Molecules* **2021**, *26*, 6343. [CrossRef]
48. Neoh, Y.Y.; Matanjun, P.; Lee, J.S. Effects of Various Drying Processes on Malaysian Brown Seaweed, *Sargassum polycystum* Pertaining to Antioxidants Content and Activity. *Trans. Sci. Technol.* **2021**, *8*, 25–37.
49. Alara, O.R.; Abdurahman, N.H.; Ukaegbu, C.I. Extraction of phenolic compounds: A review. *Curr. Res. Food Sci.* **2021**, *4*, 200–214. [CrossRef]
50. Yuan, Y.; Zhang, J.; Fan, J.; Clark, J.; Shen, P.; Li, Y.; Zhang, C. Microwave assisted extraction of phenolic compounds from four economic brown macroalgae species and evaluation of their antioxidant activities and inhibitory effects on α-amylase, α-glucosidase, pancreatic lipase and tyrosinase. *Food Res. Int.* **2018**, *113*, 288–297. [CrossRef] [PubMed]
51. Magnusson, M.; Yuen, A.K.; Zhang, R.; Wright, J.T.; Taylor, R.B.; Maschmeyer, T.; de Nys, R. A comparative assessment of microwave assisted (MAE) and conventional solid-liquid (SLE) techniques for the extraction of phloroglucinol from brown seaweed. *Algal Res.* **2017**, *23*, 28–36. [CrossRef]
52. Flórez-Fernández, N.; Muñoz, M.J.G. Ultrasound-assisted extraction of bioactive carbohydrates. In *Water Extraction of Bioactive Compounds*; Elsevier: Amsterdam, The Netherlands, 2017; pp. 317–331.
53. Michalak, I.; Dmytryk, A.; Wieczorek, P.P.; Rój, E.; Łęska, B.; Górka, B.; Messyasz, B.; Lipok, J.; Mikulewicz, M.; Wilk, R. Supercritical algal extracts: A source of biologically active compounds from nature. *J. Chem.* **2015**, *2015*, 597140. [CrossRef]
54. Cikoš, A.-M.; Jokić, S.; Šubarić, D.; Jerković, I. Overview on the application of modern methods for the extraction of bioactive compounds from marine macroalgae. *Mar. Drugs* **2018**, *16*, 348. [CrossRef] [PubMed]
55. Castro-López, C.; Rojas, R.; Sánchez-Alejo, E.J.; Niño-Medina, G.; Martínez-Ávila, G.C. Phenolic compounds recovery from grape fruit and by-products: An overview of extraction methods. In *Grape And Wine Biotechnology*; InTech: Vienna, Austria, 2016; pp. 103–123.
56. Getachew, A.T.; Jacobsen, C.; Holdt, S.L. Emerging technologies for the extraction of marine phenolics: Opportunities and challenges. *Mar. Drugs* **2020**, *18*, 389. [CrossRef]
57. Zaid, A.; Mohammad, F.; Wani, S.H.; Siddique, K.M. Salicylic acid enhances nickel stress tolerance by up-regulating antioxidant defense and glyoxalase systems in mustard plants. *Ecotoxicol. Environ. Saf.* **2019**, *180*, 575–587. [CrossRef]
58. Saidi, I.; Guesmi, F.; Kharbech, O.; Hfaiedh, N.; Djebali, W. Gallic acid improves the antioxidant ability against cadmium toxicity: Impact on leaf lipid composition of sunflower (*Helianthus annuus*) seedlings. *Ecotoxicol. Environ. Saf.* **2021**, *210*, 111906. [CrossRef]
59. Rengasamy, K.R.; Kulkarni, M.G.; Stirk, W.A.; Van Staden, J. Eckol-a new plant growth stimulant from the brown seaweed *Ecklonia maxima*. *J. Appl. Phycol.* **2015**, *27*, 581–587. [CrossRef]
60. Ali, A.S.; Elozeiri, A.A. Metabolic processes during seed germination. In *Advances in Seed Biology*; InTech: Vienna, Austria, 2017; pp. 141–166.
61. Joshi, R. Role of enzymes in seed germination. *Int. J. Creat. Res. Thoughts* **2018**, *6*, 1481–1485.

62. Aremu, A.O.; Masondo, N.A.; Rengasamy, K.R.; Amoo, S.O.; Gruz, J.; Bíba, O.; Šubrtová, M.; Pěnčík, A.; Novák, O.; Doležal, K. Physiological role of phenolic biostimulants isolated from brown seaweed *Ecklonia maxima* on plant growth and development. *Planta* **2015**, *241*, 1313–1324. [CrossRef]
63. Rengasamy, K.R.; Kulkarni, M.G.; Pendota, S.C.; Van Staden, J. Enhancing growth, phytochemical constituents and aphid resistance capacity in cabbage with foliar application of eckol—A biologically active phenolic molecule from brown seaweed. *New Biotechnol.* **2016**, *33*, 273–279. [CrossRef] [PubMed]
64. Briand, X.; Salamagne, S.C. Use of Phlorotannins as a Stimulant for Mychorrhizal and Rhizobial Symbioses. US Patent Application 15/754,741, 13 September 2018.
65. Kulkarni, M.G.; Rengasamy, K.R.; Pendota, S.C.; Gruz, J.; Plačková, L.; Novák, O.; Doležal, K.; Van Staden, J. Bioactive molecules derived from smoke and seaweed *Ecklonia maxima* showing phytohormone-like activity in *Spinacia oleracea* L. *New Biotechnol.* **2019**, *48*, 83–89. [CrossRef] [PubMed]
66. Kumar, M.; Tak, Y.; Potkule, J.; Choyal, P.; Tomar, M.; Meena, N.L.; Kaur, C. Phenolics as plant protective companion against abiotic stress. In *Plant Phenolics in Sustainable Agriculture*; Springer: Berlin/Heidelberg, Germany, 2020; pp. 277–308.
67. Ficke, A.; Cowger, C.; Bergstrom, G.; Brodal, G. Understanding yield loss and pathogen biology to improve disease management: Septoria nodorum blotch-a case study in wheat. *Plant Dis.* **2018**, *102*, 696–707. [CrossRef] [PubMed]
68. Bashir, S.S.; Hussain, A.; Hussain, S.J.; Wani, O.A.; Zahid Nabi, S.; Dar, N.A.; Baloch, F.S.; Mansoor, S. Plant drought stress tolerance: Understanding its physiological, biochemical and molecular mechanisms. *Biotechnol. Biotechnol. Equip.* **2021**, *35*, 1912–1925. [CrossRef]
69. Sun, W.-J.; Nie, Y.-X.; Gao, Y.; Dai, A.-H.; Bai, J.-G. Exogenous cinnamic acid regulates antioxidant enzyme activity and reduces lipid peroxidation in drought-stressed cucumber leaves. *Acta Physiol. Plant.* **2012**, *34*, 641–655. [CrossRef]
70. Quan, N.T.; Xuan, T.D. Foliar application of vanillic and *p*-hydroxybenzoic acids enhanced drought tolerance and formation of phytoalexin momilactones in rice. *Arch. Agron. Soil Sci.* **2018**, *64*, 1831–1846. [CrossRef]
71. Maghraby, Y.R.; Farag, M.A.; Kontominas, M.; Shakour, Z.T.; Ramadan, A.R. Nanoencapsulated Extract of a Red Seaweed (Rhodophyta) Species as a Promising Source of Natural Antioxidants. *ACS Omega* **2022**, *7*, 6539–6548. [CrossRef]
72. Sumayya, S.; Lubaina, A.; Murugan, K. Phytochemical, HPLC and FTIR Analysis of Methanolic Extract from *Gracilaria dura* (C Agardh) J Agardh. *J. Drug Deliv. Ther.* **2020**, *10*, 114–118. [CrossRef]
73. Kumar, S.; Li, G.; Yang, J.; Huang, X.; Ji, Q.; Liu, Z.; Ke, W.; Hou, H. Effect of salt stress on growth, physiological parameters, and ionic concentration of water dropwort (*Oenanthe javanica*) cultivars. *Front. Plant Sci.* **2021**, *12*, 660409. [CrossRef]
74. Babich, O.; Sukhikh, S.; Larina, V.; Kalashnikova, O.; Kashirskikh, E.; Prosekov, A.; Noskova, S.; Ivanova, S.; Fendri, I.; Smaoui, S. Algae: Study of Edible and Biologically Active Fractions, Their Properties and Applications. *Plants* **2022**, *11*, 780. [CrossRef]
75. Parvin, K.; Nahar, K.; Hasanuzzaman, M.; Bhuyan, M.B.; Mohsin, S.M.; Fujita, M. Exogenous vanillic acid enhances salt tolerance of tomato: Insight into plant antioxidant defense and glyoxalase systems. *Plant Physiol. Biochem.* **2020**, *150*, 109–120. [CrossRef]
76. Ozfidan-Konakci, C.; Yildiztugay, E.; Kucukoduk, M. Upregulation of antioxidant enzymes by exogenous gallic acid contributes to the amelioration in *Oryza sativa* roots exposed to salt and osmotic stress. *Environ. Sci. Pollut. Res.* **2015**, *22*, 1487–1498. [CrossRef] [PubMed]
77. Ofori, S.A.; Cobbina, S.J.; Obiri, S. Climate change, land, water, and food security: Perspectives From Sub-Saharan Africa. *Front. Sustain. Food Syst.* **2021**, *5*, 680924. [CrossRef]
78. Hasanuzzaman, M.; Nahar, K.; Alam, M.M.; Roychowdhury, R.; Fujita, M. Physiological, biochemical, and molecular mechanisms of heat stress tolerance in plants. *Int. J. Mol. Sci.* **2013**, *14*, 9643–9684. [CrossRef]
79. Cheng, Z.-Y.; Sun, L.; Wang, X.-J.; Sun, R.; An, Y.-Q.; An, B.-L.; Zhu, M.-X.; Zhao, C.-F.; Bai, J.-G. Ferulic acid pretreatment alleviates heat stress in blueberry seedlings by inducing antioxidant enzymes, proline, and soluble sugars. *Biol. Plant.* **2018**, *62*, 534–542. [CrossRef]
80. Zhizhong, Z.; Lan, M.; Han, X.; Wu, J.; Wang-Pruski, G. Response of ornamental pepper to high-temperature stress and role of exogenous salicylic acid in mitigating high temperature. *J. Plant Growth Regul.* **2020**, *39*, 133–146.
81. Tiwari, S.; Lata, C. Heavy metal stress, signaling, and tolerance due to plant-associated microbes: An overview. *Front. Plant Sci.* **2018**, *9*, 452. [CrossRef] [PubMed]
82. Chakraborty, K.; Joseph, D. Antioxidant potential and phenolic compounds of brown seaweeds Turbinaria conoides and Turbinaria ornata (class: Phaeophyceae). *J. Aquat. Food Prod. Technol.* **2016**, *25*, 1249–1265. [CrossRef]
83. Shuping, D.; Eloff, J.N. The use of plants to protect plants and food against fungal pathogens: A review. *Afr. J. Tradit. Complementary Altern. Med.* **2017**, *14*, 120–127. [CrossRef]
84. Li, L.; Zhu, X.-M.; Zhang, Y.-R.; Cai, Y.-Y.; Wang, J.-Y.; Liu, M.-Y.; Wang, J.-Y.; Bao, J.-D.; Lin, F.-C. Research on the Molecular Interaction Mechanism between Plants and Pathogenic Fungi. *Int. J. Mol. Sci.* **2022**, *23*, 4658. [CrossRef]
85. Zhong, B.; Robinson, N.A.; Warner, R.D.; Barrow, C.J.; Dunshea, F.R.; Suleria, H.A. Lc-esi-qtof-ms/ms characterization of seaweed phenolics and their antioxidant potential. *Mar. Drugs* **2020**, *18*, 331. [CrossRef]
86. Nehela, Y.; Taha, N.A.; Elzaawely, A.A.; Xuan, T.D.; Amin, M.; Ahmed, M.E.; El-Nagar, A. Benzoic acid and its hydroxylated derivatives suppress early blight of tomato (*Alternaria solani*) via the induction of salicylic acid biosynthesis and enzymatic and nonenzymatic antioxidant defense machinery. *J. Fungi* **2021**, *7*, 663. [CrossRef]
87. Chavoushi, M.; Najafi, F.; Salimi, A.; Angaji, S. Improvement in drought stress tolerance of safflower during vegetative growth by exogenous application of salicylic acid and sodium nitroprusside. *Ind. Crops Prod.* **2019**, *134*, 168–176. [CrossRef]

88. An, Y.; Sun, L.; Wang, X.; Sun, R.; Cheng, Z.; Zhu, Z.; Yan, G.; Li, Y.; Bai, J. Vanillic acid mitigates dehydration stress responses in blueberry plants. *Russ. J. Plant Physiol.* **2019**, *66*, 806–817. [CrossRef]
89. Saleh, A.M.; Madany, M. Coumarin pretreatment alleviates salinity stress in wheat seedlings. *Plant Physiol. Biochem.* **2015**, *88*, 27–35. [CrossRef]
90. Ahanger, M.A.; Aziz, U.; Alsahli, A.A.; Alyemeni, M.N.; Ahmad, P. Influence of exogenous salicylic acid and nitric oxide on growth, photosynthesis, and ascorbate-glutathione cycle in salt stressed *Vigna angularis*. *Biomolecules* **2019**, *10*, 42. [CrossRef]
91. Mohamed, G. The Effects of Gallic Acid on the Membrane Proteome and Antioxidant System of Wheat Plants under Salt Stress. Master's Thesis, University of the Western Cape, Cape Town, South Africa, 2020.
92. Mekawy, A.M.M.; Abdelaziz, M.N.; Ueda, A. Apigenin pretreatment enhances growth and salinity tolerance of rice seedlings. *Plant Physiol. Biochem.* **2018**, *130*, 94–104. [CrossRef]
93. Kang, Y.; Liu, J.; Yang, L.; Li, N.; Wang, Y.; Ao, T.; Chen, W. Foliar application of flavonoids (rutin) regulates phytoremediation efficiency of *Amaranthus hypochondriacus* L. by altering the permeability of cell membranes and immobilizing excess Cd in the cell wall. *J. Hazard. Mater.* **2022**, *425*, 127875.
94. Zhang, X.; Ran, W.; Li, X.; Zhang, J.; Ye, M.; Lin, S.; Liu, M.; Sun, X. Exogenous Application of Gallic Acid Induces the Direct Defense of Tea Plant Against *Ectropis obliqua* Caterpillars. *Front. Plant Sci.* **2022**, *13*, 833489. [CrossRef] [PubMed]
95. Yousif, D.Y. Effects sprayed solution of salicylic acid to prevent of wilt disease caused by *Fusarium oxysporium*. *J. Phys. Conf. Ser.* **2018**, *1003*, 012001. [CrossRef]
96. Li, S.; Pi, J.; Zhu, H.; Yang, L.; Zhang, X.; Ding, W. Caffeic Acid in Tobacco Root Exudate Defends Tobacco Plants From Infection by *Ralstonia solanacearum*. *Front. Plant Sci.* **2021**, *12*, 690586. [CrossRef] [PubMed]
97. Xi, D.; Li, X.; Gao, L.; Zhang, Z.; Zhu, Y.; Zhu, H. Application of exogenous salicylic acid reduces disease severity of Plasmodiophora brassicae in pakchoi (*Brassica campestris* ssp. chinensis Makino). *PLoS ONE* **2021**, *16*, e0248648. [CrossRef] [PubMed]
98. Islam, M.T.; Lee, B.-R.; Das, P.R.; Jung, H.-i.; Kim, T.-H. Characterization of *p*-Coumaric acid-induced soluble and cell wall-bound phenolic metabolites in relation to disease resistance to *Xanthomonas campestris* pv. campestris in Chinese cabbage. *Plant Physiol. Biochem.* **2018**, *125*, 172–177. [CrossRef] [PubMed]
99. Leonberger, K.; Jackson, K.; Smith, R.; Ward Gauthier, N. Plant diseases. In *Kentucky Master Gardner Manual*; University of Kentucky: Lexington, KY, USA, 2016.
100. Mondal, K.K. Phytopathogenic Bacteria and Plant Diseases by BS Thind. *Indian Phytopathol.* **2020**, *73*, 587. [CrossRef]
101. Rajauria, G. Optimization and validation of reverse phase HPLC method for qualitative and quantitative assessment of polyphenols in seaweed. *J. Pharm. Biomed. Anal.* **2018**, *148*, 230–237. [CrossRef] [PubMed]
102. Singhal, P.; Nabi, S.U.; Yadav, M.K.; Dubey, A. Mixed infection of plant viruses: Diagnostics, interactions and impact on host. *J. Plant Dis. Prot.* **2021**, *128*, 353–368. [CrossRef]
103. Mishra, J.; Srivastava, R.; Trivedi, P.K.; Verma, P.C. Effect of virus infection on the secondary metabolite production and phytohormone biosynthesis in plants. *3 Biotech* **2020**, *10*, 547. [CrossRef]
104. Zhang, K.; Wang, Y.; Sun, W.; Han, K.; Yang, M.; Si, Z.; Li, G.; Qiao, Y. Effects of exogenous salicylic acid on the resistance response of wild soybean plants (*Glycine soja*) infected with Soybean mosaic virus. *Can. J. Plant Pathol.* **2020**, *42*, 84–93. [CrossRef]
105. Gossner, M.M.; Beenken, L.; Arend, K.; Begerow, D.; Peršoh, D. Insect herbivory facilitates the establishment of an invasive plant pathogen. *ISME Commun.* **2021**, *1*, 6. [CrossRef]
106. Olate-Gallegos, C.; Barriga, A.; Vergara, C.; Fredes, C.; García, P.; Giménez, B.; Robert, P. Identification of polyphenols from chilean brown seaweeds extracts by LC-DAD-ESI-MS/MS. *J. Aquat. Food Prod. Technol.* **2019**, *28*, 375–391. [CrossRef]
107. Tanna, B.; Brahmbhatt, H.R.; Mishra, A. Phenolic, flavonoid, and amino acid compositions reveal that selected tropical seaweeds have the potential to be functional food ingredients. *J. Food Process. Preserv.* **2019**, *43*, e14266. [CrossRef]
108. Jan, R.; Khan, M.A.; Asaf, S.; Lee, I.-J.; Kim, K.-M. Overexpression of OsF3H modulates WBPH stress by alteration of phenylpropanoid pathway at a transcriptomic and metabolomic level in *Oryza sativa*. *Sci. Rep.* **2020**, *10*, 14685. [CrossRef] [PubMed]

life

Article

Effect of Salinity and Plant Growth Promoters on Secondary Metabolism and Growth of Milk Thistle Ecotypes

Noreen Zahra [1,2,*], Abdul Wahid [1], Muhammad Bilal Hafeez [3], Irfana Lalarukh [2], Aaliya Batool [1], Muhammad Uzair [4], Mohamed A. El-Sheikh [5], Saleh Alansi [5] and Prashant Kaushik [6]

1 Department of Botany, University of Agriculture, Faisalabad 38040, Pakistan
2 Department of Botany, Government College for Women University, Faisalabad 38040, Pakistan
3 Department of Agronomy, University of Agriculture, Faisalabad 38040, Pakistan
4 Department of Plant Breeding and Genetics, University of Agriculture, Faisalabad 38040, Pakistan
5 Botany and Microbiology Department, King Saud University, Riyadh 11451, Saudi Arabia
6 Instituto de Conservación y Mejora de la Agrodiversidad Valenciana, Universitat Politècnica de València, 46022 Valencia, Spain
* Correspondence: noreenzahra59@yahoo.com

Simple Summary: The present study shed light on the effect of salinity on the plant growth and secondary metabolites of medicinally important milk thistle plant ecotypes. At the same time, we also studied the effect of external supplementation with ascorbic acid, thiourea, and moringa leaf extract on improving growth-related attributes and secondary metabolites under salinity stress. Various parameters were studied related to stress alleviation. Ascorbic acid, followed by moringa leaf extract, was the most effective in improving growth under salt stress conditions. The present study demonstrated that milk thistle could withstand moderate doses of salt stress, while externally supplemented media improved all the growth parameters by increasing the accumulation of secondary metabolites.

Abstract: Milk thistle (*Silybum marianum* (L.)) is a wild medicinal herbal plant that is widely used in folk medicine due to its high content of secondary metabolites (SMs) and silymarin; however, the data regarding the response of milk thistle to salinity are still scarce and scanty. The present study evaluated the effect of salinity on a geographically diverse population of milk thistle and on the role of medium supplementation (MS) with ascorbic acid, thiourea, and moringa leaf extract in improving the SMs and growth-related attributes under salinity stress (SS). For germination, a 120 mM level of salinity was applied in the soil during the seedling stage. After salinity development, predetermined levels of the following compounds were used for MS: thiourea (250 μM), moringa leaf extract (3%), and ascorbic acid (500 μM). The data regarding growth attributes showed that SS impaired plant growth and development and increased SM production, including alkaloids, anthocyanin, and saponins. Moreover, ascorbic acid, followed by moringa leaf extract, was the most effective in improving growth by virtue of increased SMs, especially under salt stress conditions. The present study demonstrated that milk thistle could withstand moderate doses of SS, while MS improved all the growth parameters by increasing the accumulation of SMs.

Keywords: milk thistle; secondary metabolites; ecotypes; salinity; growth attributes

Citation: Zahra, N.; Wahid, A.; Hafeez, M.B.; Lalarukh, I.; Batool, A.; Uzair, M.; El-Sheikh, M.A.; Alansi, S.; Kaushik, P. Effect of Salinity and Plant Growth Promoters on Secondary Metabolism and Growth of Milk Thistle Ecotypes. *Life* **2022**, *12*, 1530. https://doi.org/10.3390/life12101530

Academic Editors: Hakim Manghwar and Wajid Zaman

Received: 27 August 2022
Accepted: 25 September 2022
Published: 30 September 2022

Publisher's Note: MDPI stays neutral with regard to jurisdictional claims in published maps and institutional affiliations.

Copyright: © 2022 by the authors. Licensee MDPI, Basel, Switzerland. This article is an open access article distributed under the terms and conditions of the Creative Commons Attribution (CC BY) license (https://creativecommons.org/licenses/by/4.0/).

1. Introduction

The cultivation of milk thistle has been increasing all over the world due to its uses in pharmaceutical industries [1]. The seeds of milk thistle contain bioactive compounds such as silychristin, silydianin, isosilybin, silybin, quercetin, apigenin, naringin, dihydrokaempferol, taxifolin, chrysoeriol, flavonolignans, eriodyctiol, and kaempferol [2–4]. It is used in pharmacologically relevant actions against different diseases, such as mushroom poisoning, liver injury due to drugs toxicities, and viral hepatitis, as reviewed by [5]. However, little attention has been paid to evaluating the effect of abiotic stresses on its morphology, growth, physio-chemical mechanisms, yield potential, and medium-supplementation-induced accumulation.

Globally, climate fluctuations are a foremost hazard to global food security for >7.8 billion people as of 2020 [6] and increase salinization [7,8], waterlogging [9], drought [9], and extreme temperatures [10,11]. Around the world, salinity stress (SS) affects over 20% of irrigated agricultural land, which is why it is considered as one of the key challenges for agricultural researchers [12–16]. Salinity stress causes osmotic and ionic stress, results in ionic homeostasis, which decreases growth and yield, and causes the premature senescence of leaves [17,18]. Salinity was found to reduce shoot and root length, and the relative growth rate of shoots and roots in wheat [19]. Similarly, under SS, leaf area, root length, fresh root weight, yield, and total soluble solids were found to be decreased in cherry tomato [20]. In pea, plant height, leaf area, plant fresh and dry weight, and relative water content were found to be decreased under SS [21]. Under SS, dry weight, shoot and root length, shoot and root numbers, leaf area, and relative water content in pomegranate were found to be negatively affected [22]. Under SS, the relative water content and secondary metabolites (total phenolic and total flavonoid contents) were decreased in sesame [23]. However, a maintained sugar production and the production of secondary metabolites could provide defense to ensure a more successful acclimation to SS [24–26].

Medium supplementation (MS) for plant growth was reported to promote the confronting of salinity stress and the overcoming of yield losses in all crops [27–31]. Medium supplementation with 24-brassinosteroid increased the canopy diameter, the length of branches, the diameter of branches, the total number of branches per tree, the number of branches producing fruit per tree, the total number of fruit setting per tree, and the total number of ripe fruit load per tree as compared with the control [32]. Medium supplementation with paclobutrazol increased grain yield, grain weight, and main panicle length in quinoa under SS [33]. Medium supplementation with uniconazole significantly increased plant height, stem, the length of spikes, and top dry weight in barley under SS [34]. Ascorbic acid is one of the plant growth promoters known as vitamin C (water-soluble) that deploys many physio-chemical modulations to provide tolerance against salinity [35]. Ascorbic acid treatment under SS increased shoot and root length and their fresh and dry weight, as well as nutrient elements, while it decreased lipid peroxidation [36]. Ascorbic acid application increased shoot height, leaf number, and tuberous root diameter under SS [37]. Thiourea is a plant growth promoter that scavenges ROS by modulating numerous essential functions under a plethora of abiotic stresses [38]. Exogenous treatment with thiourea enhanced shoot and root length, shoot and root fresh and dry weight, chlorophyll content, potassium, and zinc under SS [39]. Plant bio-stimulants such as moringa leaf extract trigger growth and increase crops' economic yield because they are enriched with amino acids, growth hormones, antioxidants, vitamins, and mineral nutrients [40–42]. Moringa leaf extract application ameliorated SS by improving root and shoot fresh and dry weight, root and shoot length, potassium, calcium, and phosphorus [43]. The purpose of the current trial was to evaluate the efficacy of different plant growth promoters in modulating SMs and growth indicators in milk thistle ecotypes under saline stress conditions.

2. Materials and Methods

2.1. Experimental Details

A two-year pot trail was performed at "Botanical Garden, University of Agriculture, Faisalabad" to investigate the tolerance potential of different milk thistle ecotypes under SS. Different plant growth promoters were used to ameliorate the adversities of salt stress and for regulating silymarin biosynthesis. Plants of three ecotypes (Faisalabad, Gujranwala, and Quetta) were taken from varied geographic regions and sown in Faisalabad in an ecotype agronomic environment. The achenes of the F1 generation obtained from three ecotypes were sown on 17th November (2017–2018) under control conditions. The salinity level (120 mM) [44] was applied on 11th December, and different plant growth promoters were applied via MS. Plant growth promoters, such as thiourea (250 μM) [45], moringa leaf extract (3%) [46], and ascorbic acid (500 μM) [47], were applied at predetermined levels after inducing SS during the seedling stage through irrigation (single time point). Flowering started on 10th, 20th, and 25th February in the Faisalabad, Gujranwala, and Quetta ecotypes, respectively, in 2017, while in 2018, the flowering dates were 12th and 20th January for the Faisalabad and Gujranwala ecotypes, respectively, and 30th January for the Quetta ecotype. After harvest, all the morpho-physiological analyses were performed using standard protocols.

2.2. Growth and Yield Indicators

The careful harvesting of plants was performed on 11th April in both experimental years, and the samples were thoroughly rinsed with tap water. The shoot and root length was measured with a foot scale. After that, the plants shoots were detached from the roots, and the fresh weight (g) of shoots and roots were noted using an electric weight balance. The other indicators, such as the number of roots plant^{-1}, the number of leaves plant^{-1}, the leaf area (mm^2), the root diameter (mm^2), and the number of spines leaf^{-1}, were also evaluated.

2.3. Determination of Secondary Metabolites

After sampling, fresh plants were instantly put in an ice bath and preserved at −40 °C in a laboratory for performing various biochemical analyses. Half of the cut shoots were moved to paper envelopes and dried for one week at 70 °C (Memert, Schwabach, Germany).

2.4. Anthocyanin Determination

The anthocyanin contents were estimated by following STRACK and WRAY [48]. For this, fresh plant samples (0.1 g) were homogenized in acidified methanol (2.5 mL). After that, samples were heated for 1 h (50 °C) and filtered. The optical density was noted at 535 nm.

2.5. Total Alkaloids Determination

The method by Singh and Sahu [49] was used to measure the total alkaloids contents. Plant material (0.1 g) was homogenized in 1 mL of methanol and then diluted with distilled water. For 1 mL of running sample, 0.5 mL of acetic acid and 0.01 M sodium meta periodate were added for each sample. The mixture was boiled, and after that, 0.01 M 3-methyl-2-benzothiazol solution was added to each test tube. The samples were cooled for 20 min in a water bath, and absorbance was recorded at 470 nm.

2.6. Saponin Estimation

For measuring the saponin contents, 0.1 g plant samples were homogenized until they were converted into powder form and then soaked in DH_2O: ethanol in a one-to-one ratio. For 0.5 mL of extract, 5 mL of H_2SO_4 (72%) and 0.5 mL vanillin (10%) were added, while the sample mixtures were put on ice. The samples were moved to a water bath at 60 °C for 10 min. Absorbance was noted at 535 nm.

2.7. Statistical Analysis

The design of the trial was a Completely Randomized Factorial Design (CRD) with three replications. The main and interactive effects among SS, ecotypes, and MS under non-saline conditions were assessed using numerous response variables with the analysis of variance (ANOVA) technique at the 5% probability level. The ANOVA was performed using statistical software "Statistax8.1". The trial data were also dealt with using a principal component analysis (PCA) using R-Stat to assess the existing relationships with original variables.

3. Results

3.1. Plant Length

Significant ($p < 0.01$) differences were noted among all the factors for plant height (root and shoot length) in 2017. Under control conditions, the maximum shoot length were observed in the control, thiourea and moringa leaf extract treatment in the Faisalabad, Gujranwala, and Quetta ecotypes, respectively. Moreover, data displayed that under SS, the trends of the maximum increment in this parameter were found for the control, thiourea, and ascorbic acid treatments in the Faisalabad, Gujranwala, and Quetta ecotypes, respectively. Overall, Quetta performed well under control conditions; however, Gujranwala displayed the maximum shoot height under SS (Figure 1a). In 2018, MS with ascorbic acid, followed by moringa leaf extract, showed profound results for shoot length under non-saline conditions in the Faisalabad ecotype, while the effects of thiourea and moringa leaf extract were the most pronounced in the Gujranwala and Quetta ecotypes, respectively. Under SS, the effect of moringa leaf extract was the significantly increased the shoot length in the Faisalabad ecotype, while ascorbic acid was more effective in the Gujranwala and Quetta ecotypes than other plant growth promoters in 2018.

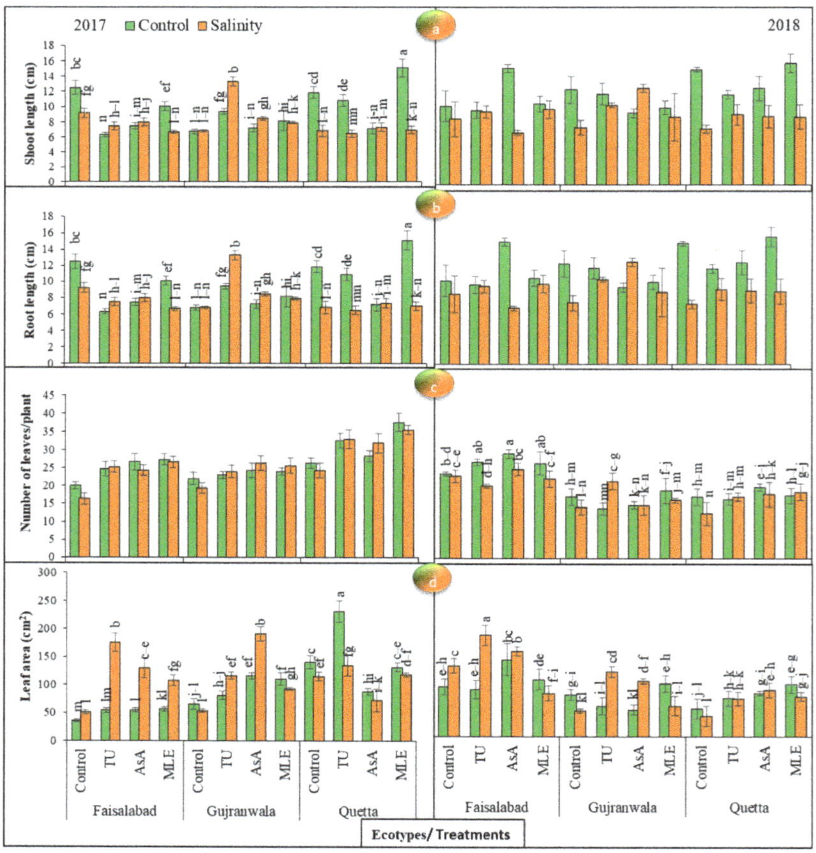

Figure 1. (**a**) Effects of MS with different plant growth promoters on shoot length, (**b**) root length, (**c**) number of leaves/plant, and (**d**) leaf area in milk thistle in 2017 and 2018. Different letters represent significant differences at the $p < 0.05$.

Considering the root length of control plants, the trends of the maximum root length following control, thiourea, and moringa leaf extract treatments were observed in the Faisalabad, Gujranwala, and Quetta ecotypes, respectively, in 2017. Moreover, while under SS conditions, MS with ascorbic acid performed better in the Faisalabad and Quetta ecotypes, and the effectiveness of thiourea was at its peak in Gujranwala. On the other hand, in 2018, an increasing trend for root length was observed in the order Quetta > Faisalabad > Gujranwala ecotypes. Under normal conditions, ascorbic acid showed the maximum increment in Faisalabad, while thiourea and moringa leaf extract showed the maximum increment in the Gujranwala and Quetta ecotypes, respectively. Under SS, moringa leaf extract showed the maximum increment in Faisalabad, while ascorbic acid and thiourea showed the maximum increments in the Gujranwala and Quetta ecotypes, respectively. Overall, Quetta performed well under control conditions; however, Gujranwala displayed the maximum shoot height under SS (Figure 1b).

3.2. Number of Leaves

Significant ($p < 0.01$) differences were recorded among all the factors for the number of leaves in 2018. In 2017, the trend of the maximum number of leaves was observed when moringa leaf extract was applied in all ecotypes under control conditions, while under SS conditions, MS with moringa leaf extract performed better in the Faisalabad and Quetta ecotypes, and the effectiveness of ascorbic acid was at its peak in Gujranwala. In 2018, the trends of the maximum number of leaves were observed following treatment with ascorbic acid in the Faisalabad and Quetta ecotypes, while moringa leaf extract treatment performed better in the Gujranwala ecotype under control conditions; on the other hand, under SS, the effect of ascorbic acid on the Faisalabad ecotype and the effects of thiourea and moringa leaf extract on the Gujranwala and Quetta ecotypes were more pronounced than those of MS with other plant growth promoters (Figure 1c).

3.3. Leaf Area

Significant ($p < 0.01$) differences were noted among all the factors for leaf area in both years. Under control conditions, the trends of the maximum leaf area were observed following treatments with moringa leaf extract and thiourea in Faisalabad, and in the Gujranwala and Quetta ecotypes, respectively, regardless of SS treatment in 2017, while under SS, the trends of leaf-area increment were observed following thiourea treatment in the Faisalabad and Gujranwala ecotypes and moringa leaf extract treatment in the Quetta ecotype. Moreover, ecotypic variation revealed the greatest leaf area in the Quetta ecotype. Furthermore, according to data recorded in 2018, the trend of the increment in this trait in the Faisalabad ecotype was observed with ascorbic acid, while in the Gujranwala and Quetta ecotypes, it was observed with moringa leaf extract. Under SS, this increasing trend was observed following thiourea treatment in the Faisalabad and Gujranwala ecotypes and following ascorbic acid treatment in the Quetta ecotype. Moreover, ecotypic variation revealed the greatest leaf area in the Quetta ecotype (Figure 1d).

3.4. Fresh Weight

Plant fresh root and shoot weight displayed significant ($p < 0.01$) differences among all the factors in both trial years. Under non-saline conditions, the maximum increment in the Faisalabad and Quetta ecotypes was obtained by applying moringa leaf extract treatment; however, in Gujranwala, it was obtained with ascorbic acid treatment. Moreover, under SS, the maximum change in this attribute was obtained via MS with ascorbic acid in the Faisalabad and Gujranwala ecotypes, while in the Quetta ecotype, it was noted with moringa leaf extract treatment in 2017. MS with moringa leaf extract was the most effective in improving shoot fresh weight in all ecotypes under normal circumstances, while under SS, the effect of ascorbic acid treatment was the most pronounced in the Faisalabad and Quetta ecotypes, and in Gujranwala, thiourea-supplemented plants had the greatest fresh weight in 2018 (Figure 2a).

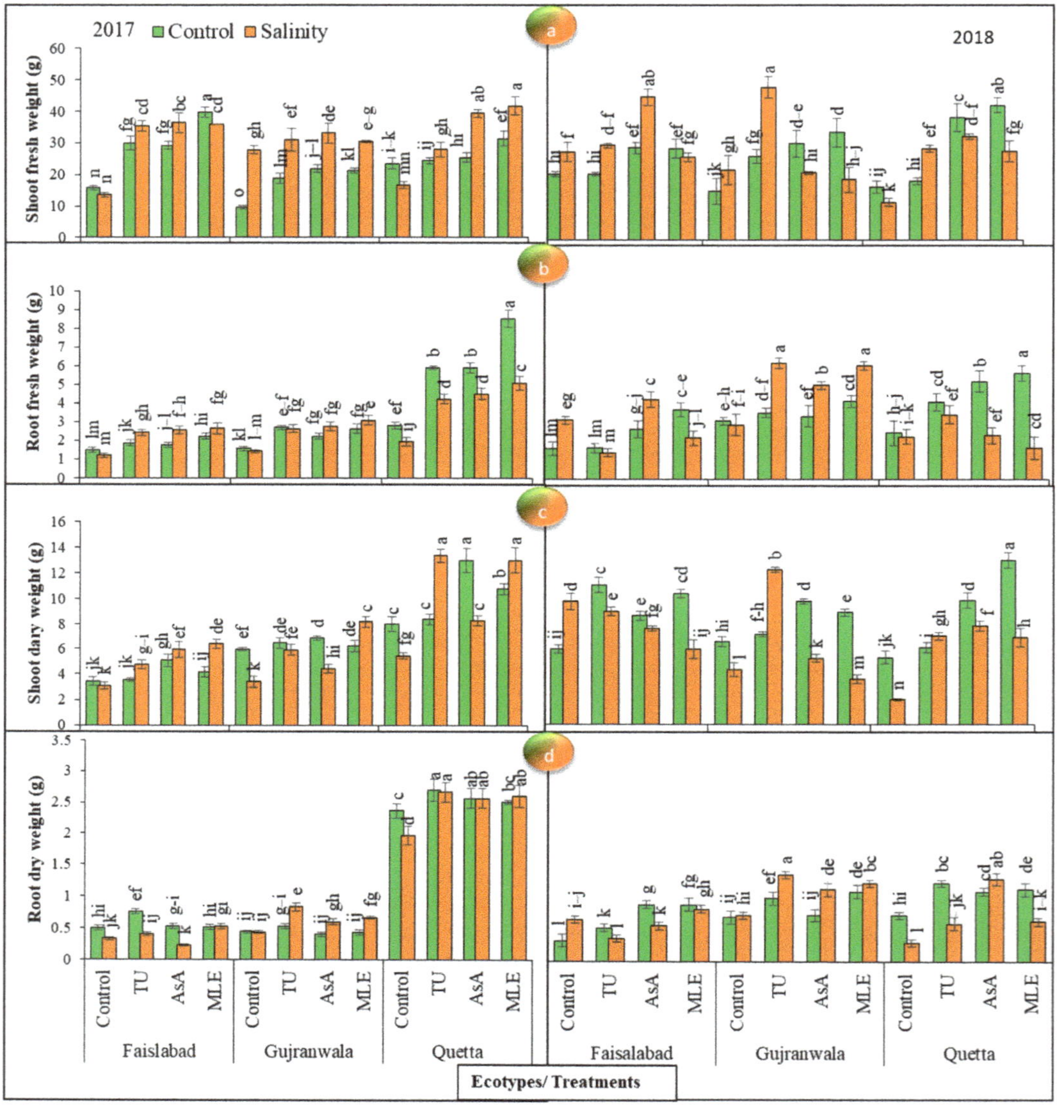

Figure 2. (**a**) Effects of MS with different plant growth promoters on shoot fresh weight, (**b**) root fresh weight, (**c**) shoot dry weight, and (**d**) root dry weight in milk thistle in 2017 and 2018. Different letters represent significant differences at the $p < 0.05$.

In 2017, the Faisalabad, Gujranwala, and Quetta ecotypes treated with moringa leaf extract had the greatest root fresh weight under control and SS conditions. Overall, under SS and normal conditions, the Quetta ecotype showed the maximum root fresh weight. In 2018, the maximum increment in this trait was obtained via MS with moringa leaf extract in all ecotypes, while under SS conditions, it was noted with ascorbic acid MS in Faisalabad, moringa leaf extract treatment = thiourea treatment in Gujranwala, and moringa leaf extract treatment in the Quetta ecotype (Figure 2b).

3.5. Dry Weight

Statistical data for plant dry weight (of shoots and roots) revealed significant ($p < 0.01$) differences among all the factors in both years. The data further revealed that ascorbic acid MS was effective in enhancing shoot dry weight in the first year of the experiment under normal conditions, while under SS conditions, the effects of moringa leaf extract MS were the most evident in all ecotypes. Additionally, the maximum shoot dry weight was verified in Quetta, followed by Gujranwala and Faisalabad, in 2017, while in 2018, under normal conditions, treatments with thiourea, ascorbic acid, and moringa leaf extract were more effective in the Faisalabad, Gujranwala, and Quetta ecotypes, respectively, than other treatments. Moreover, under SS conditions, the greatest shoot dry weight was observed with thiourea application in the Faisalabad and Gujranwala ecotypes, while ascorbic acid was effective in the Quetta ecotype. Furthermore, the greatest dry weight was noted in the Quetta ecotype regardless of the SS conditions in 2017–2018 (Figure 2c).

Data recorded for root dry weight depicted that thiourea MS showed the most pronounced results for dry weight under SS and control conditions in 2017. In 2018, the maximum increment obtained via MS with moringa leaf extract was observed in the Faisalabad and Gujranwala ecotypes, while thiourea treatment caused the maximum increment in the Quetta ecotype under control conditions. On the other hand, under SS, it was found that moringa leaf extract treatment in Faisalabad, thiourea treatment in Gujranwala, and ascorbic acid treatment in the Quetta ecotype showed the maximum root dry weight (Figure 2d). In conclusion, the data revealed that MS with plant growth promoters was quite efficient in accumulating the root dry weight in milk in all ecotypes under control and saline conditions, and the greatest dry weight was confirmed in the Quetta ecotype.

3.6. Number of Roots

The statistical data for the number of roots revealed significant ($p < 0.01$) differences among all the factors in both years. In 2017, an increasing trend in the Faisalabad and Quetta ecotypes was noted following moringa leaf extract treatment, while in the Gujranwala ecotype, this was observed following ascorbic acid treatment under control conditions. On the other hand, under saline conditions, the maximum number of roots was observed in all ecotypes via MS with moringa leaf extract. In 2018, an increasing trend under control conditions was observed following MS with ascorbic acid in the Faisalabad ecotype, while in the Gujranwala and Quetta ecotypes, it was observed following thiourea treatment. On the other hand, under SS, this increasing trend was observed following moringa leaf extract, thiourea, and ascorbic acid treatments in the Faisalabad, Gujranwala, and Quetta ecotypes, respectively (Figure 3a).

3.7. Number of Spines

The statistical data obtained for the number of spines displayed significant ($p < 0.01$) differences among all these factors and their interaction in both years. Under non-saline conditions, in 2017, the maximum increment in the number of spines was obtained via ascorbic acid, thiourea, and control treatments in the Faisalabad, Gujranwala, and Quetta ecotypes, respectively, while under SS conditions, the maximum increment in this parameter was obtained via MS with ascorbic acid in the Faisalabad and Gujranwala ecotypes and following control treatment in the Quetta ecotype. Furthermore, a higher number of spines was present on Quetta leaves than on the leaves of other ecotypes. In 2018, the maximum number of spines was observed following MS with ascorbic acid, which was the most effective in the Faisalabad ecotype, while thiourea was the most effective in the Gujranwala and Quetta ecotypes under normal circumstances; on the other hand, under SS, the effects of thiourea, ascorbic acid, and control treatments were the most efficient in the Faisalabad, Gujranwala, and Quetta ecotypes, respectively, and thiourea-supplemented plants had the highest number of spines in 2018. Interestingly, SS plants showed a greater number of spines than control plants of the Faisalabad and Gujranwala ecotypes, while an antagonistic effect was observed in the Quetta ecotype (Figure 3b).

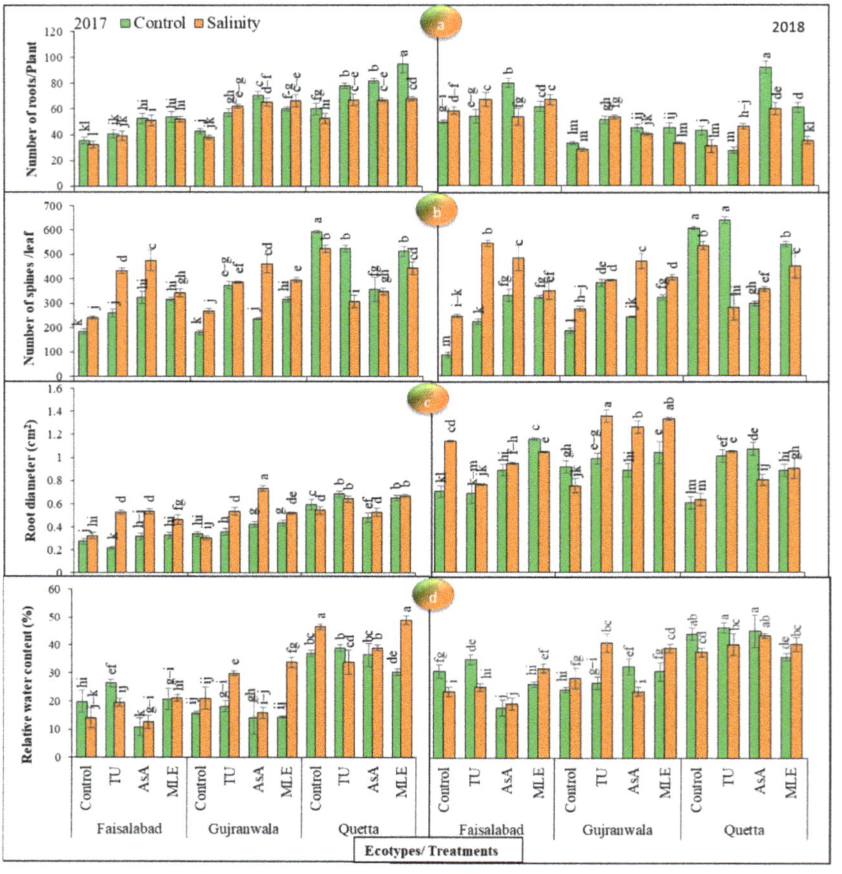

Figure 3. (**a**) Effects of MS with different plant growth promoters on number of roots/plant, (**b**) number of spines/leaf, (**c**) root diameter, and (**d**) relative water content in milk thistle in 2017 and 2018. Different letters represent significant differences at the $p < 0.05$.

3.8. Root Diameter

The results obtained for root diameter showed substantial ($p < 0.01$) variations among all the factors in both years. Under control conditions, the maximum change in 2017 was obtained via MS with moringa leaf extract, ascorbic acid, and thiourea in the Faisalabad, Gujranwala and Quetta ecotypes, respectively, while under SS, the change in this trait was observed following ascorbic acid treatment in all ecotypes. In 2018, the maximum increase in root diameter in non-stressed plants was obtained via MS with moringa leaf extract in the Faisalabad and Gujranwala ecotypes and via ascorbic acid treatment in the Quetta ecotype. However, under SS conditions, control treatment was effective in the Faisalabad ecotype, while thiourea treatment was effective in Gujranwala and Quetta. Furthermore, a greater root diameter was recorded in 2018 than in the previous year. Additionally, the greatest root diameter was measured in Gujranwala in 2017 and 2018 (Figure 3c).

3.9. Relative Water Content

The results found for the RWC exposed significant ($p < 0.01$) differences among all the factors in both years. Under normal conditions, in 2017, the increasing trend was at its maximum following MS with thiourea in all ecotypes as compared with other plant growth promoters. Meanwhile, under SS conditions, moringa leaf extract-supplied plants showed the maximum water content in all ecotypes. Considering the differences in the ecotypes revealed that the Quetta ecotype was efficient in accumulating higher relative water contents (Figure 3d).

The data noted in 2018 exhibited that the increasing trend in the Faisalabad and Quetta ecotypes was at its highest following MS with thiourea, while in Gujranwala, the application of ascorbic acid was the most effective under control conditions. However, under SS conditions, the maximum RWC was observed following MS with moringa leaf extract, thiourea, and ascorbic acid in the Faisalabad, Gujranwala, and Quetta ecotypes, respectively. The increasing trend with respect to the ecotypes was observed as: Quetta > Gujranwala > Faisalabad. Overall, the RWC increased with MS in the different ecotypes, while SS reduced the water content in the two trial studies.

3.10. Saponin Content

The graphical data of saponin contents in both shoots and roots showed significant ($p < 0.01$) differences among all the factors in 2017. Under non-saline conditions, in 2017, the increasing trend in the Faisalabad and Quetta ecotypes with respect to shoot saponin was at its maximum with the soil addition of ascorbic acid, while in Gujranwala, this increase was at its maximum with the addition of thiourea under SS and control conditions. In 2018, ascorbic acid gave distinctive results in comparison with other plant growth promoters in the Faisalabad and Quetta ecotypes, while MS with thiourea was effective for increasing this trait at the maximum level in the Gujranwala ecotype under control and stress conditions (Figure 4a). The ecotypic differences showed the highest shoot saponin content in the Faisalabad ecotype in 2017, while in 2018, the maximum content was found in the Quetta ecotype regardless of SS. Regarding the root saponin content in control plants in 2017, ascorbic acid MS showed the highest value in Faisalabad, while moringa leaf extract MS showed distinctive results in the Gujranwala and Quetta ecotypes as compared with other plant growth promoters under control conditions. On the other hand, under SS, thiourea MS was the most effective in increasing the root saponin content in the Faisalabad and Gujranwala ecotypes, but in Quetta, moringa leaf extract was effective. In 2018, under control conditions, ascorbic acid gave distinctive results in comparison with other plant growth promoters in the Faisalabad and Quetta ecotypes, while thiourea MS was effective in increasing this trait at the maximum level in the Gujranwala ecotype. On the other hand, under SS, in 2018, under control conditions, ascorbic acid gave distinctive results in comparison with other plant growth promoters in the Faisalabad and Quetta ecotypes, while thiourea MS was effective in improving this trait at the maximum level in the Gujranwala ecotype. The data relating to this trait exposed that the lowest saponin content was recorded in Quetta, while the highest one was noted in Faisalabad, regardless of treatment differences (Figure 4b). Furthermore, SS maximally increased the saponin content with MS.

3.11. Anthocyanin Content

The results obtained for anthocyanin contents showed significant ($p < 0.01$) differences among all the factors in 2017. Regarding the shoot anthocyanin content in 2017 under control conditions, the trends of the change in this trait were found to be maximum with thiourea MS in the Faisalabad and Quetta ecotypes, while the effect of plant growth promoters was lower in the Gujranwala ecotype. On the other hand, under SS, similar results were observed with respect to all ecotypes except for the Gujranwala ecotype, in which thiourea was the most effective in increasing anthocyanin. The data further revealed that in 2018, ascorbic acid MS showed the maximum increase in this trait under control and stress conditions. Moreover, the ecotypic variation showed an increasing trend for this trait as follows: Quetta > Gujranwala > Faisalabad in 2017. In 2018, it was: Quetta > Faisalabad > Gujranwala (Figure 5a).

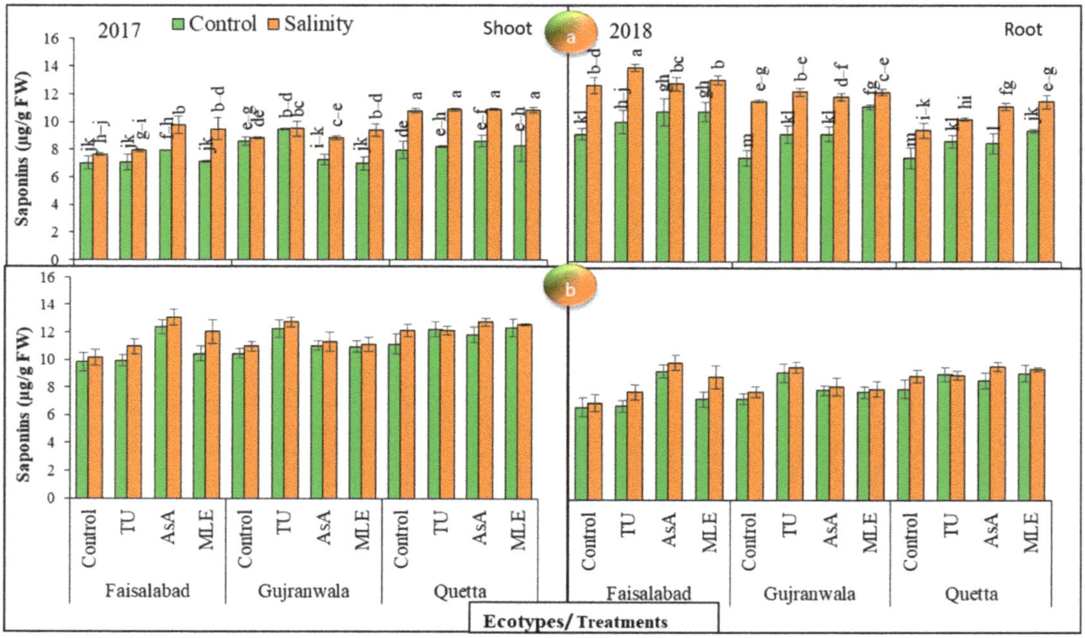

Figure 4. (**a**) Effects of MS with different plant growth promoters on shoot saponin contents and (**b**) root saponin contents in milk thistle in 2017 and 2018. Different letters represent significant differences at the $p < 0.05$.

The data noted for root anthocyanin contents exhibited that MS with ascorbic acid was the most effective in increasing the anthocyanin content in both years, regardless of ecotypic variations and SS. The trends of the change in this trait with respect to the ecotypes was found to be Quetta > Faisalabad > Gujranwala (Figure 5b).

In conclusion, the results exhibited that SS improved the anthocyanin content in both trial years, and MS application increased the anthocyanin content to a significant extent, regardless of SS and the differences in the ecotypes. Moreover, shoots and the 2017 year were superior in synthesizing more anthocyanin with respect to roots and the 2018 year, respectively.

3.12. Alkaloids Content

The alkaloids contents demonstrated significant ($p < 0.01$) differences among all these factors, which were also highly significant in 2017, for both root and shoot parts. Under control conditions, regarding the shoot alkaloids content in 2017, ascorbic acid MS showed the highest value in the Faisalabad and Quetta ecotypes, while thiourea MS = ascorbic acid MS in terms of effectiveness in the Gujranwala ecotype. On the other hand, under SS, ascorbic acid showed the highest value in Faisalabad, and moringa leaf extract = ascorbic acid in the Quetta ecotype, while thiourea MS = ascorbic acid MS in terms of showing the maximum results in Gujranwala. The data further revealed that in 2018, the effect of ascorbic acid was the most pronounced on the Faisalabad and Quetta ecotypes, while thiourea had the greatest effect on increasing this trait in the Gujranwala ecotype under control conditions. On the other hand, under SS, ascorbic acid MS was effective in the Faisalabad and Gujranwala ecotypes, while in the Quetta ecotype, moringa leaf extract MS gave the maximum increase as compared with other plant growth promoters (Figure 6a).

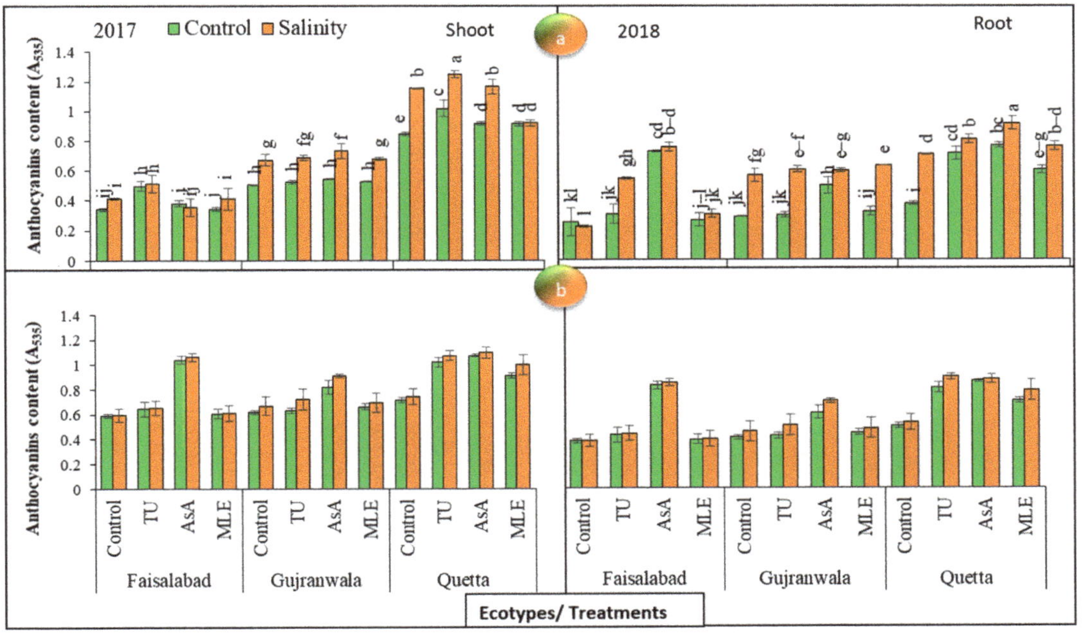

Figure 5. (**a**) Effects of MS with different plant growth promoters on shoot anthocyanin contents and (**b**) root anthocyanin contents in milk thistle in 2017 and 2018. Different letters represent significant differences at the $p < 0.05$.

The data noted for root alkaloids contents depicted that ascorbic acid MS had the greatest effect on the Faisalabad ecotype in improving this attribute, while in the Gujranwala and Quetta ecotypes, moringa leaf extract MS was the most effective under SS and control stress conditions in 2017. On the other hand, in 2018, the maximum root alkaloids content in the Faisalabad ecotype was obtained via MS with moringa leaf extract, in the Gujranwala ecotype via MS with thiourea, and in Quetta via MS with ascorbic acid under non-saline conditions, while under SS conditions, ascorbic acid MS was the most effective in increasing the alkaloids content in the Faisalabad and Quetta ecotypes, and in Gujranwala, thiourea was the most effective. In contrast, the root alkaloids content was the highest in Faisalabad, followed by Quetta and Gujranwala, in 2017, while in 2018, it was Quetta > Gujranwala > Faisalabad. The alkaloids content was higher in shoots than in roots (Figure 6b).

3.13. Principle Component Analysis

For the year 2017, both principal components 1 and 2 explained all the variation, i.e., 100% of the data. On average, the Quetta ecotype performed best for all the traits under study. It performed negatively for root saponins and root alkaloids but positively for all the other traits. However, the Faisalabad ecotype performed best for shoot fresh weight and root saponins but negatively for all the other traits. Similarly, the Gujranwala ecotype performed best for root alkaloids but negatively for all the other traits. In addition, the contribution of all parameters to the total variation is shown by the intensity of the vectors' color (Figure 7a).

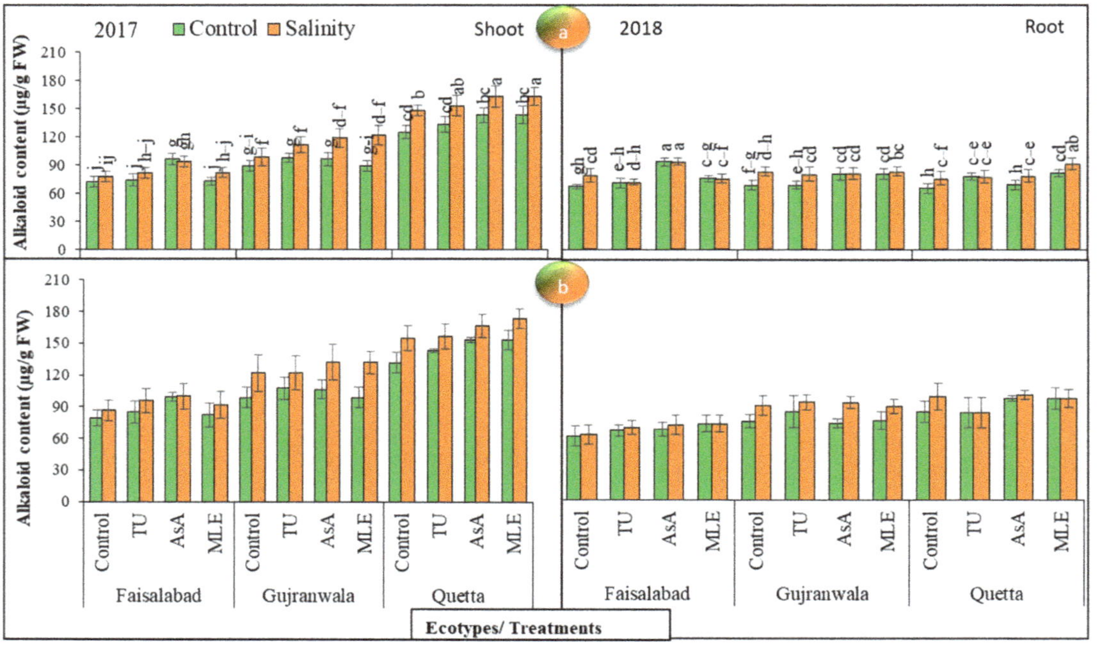

Figure 6. (**a**) Effects of MS with different plant growth promoters on shoot alkaloids contents and (**b**) root alkaloids contents in milk thistle in 2017 and 2018. Different letters represent significant differences at the $p < 0.05$.

For the year 2018, both principal components 1 and 2 explained all the variation, i.e., 100% of the data. On average, the Quetta ecotype performed best for all the traits. It performed negatively for both fresh and dry weight (shoots + roots) but positively for all the other traits. However, the Faisalabad ecotype performed best for shoot weight (fresh + dry) but negatively for all the other traits. Similarly, the Gujranwala ecotype performed best for root weight (fresh + dry) but negatively for all the other traits. In addition, the contribution of all parameters to the total variation is shown by the intensity of the vectors' color (Figure 7b).

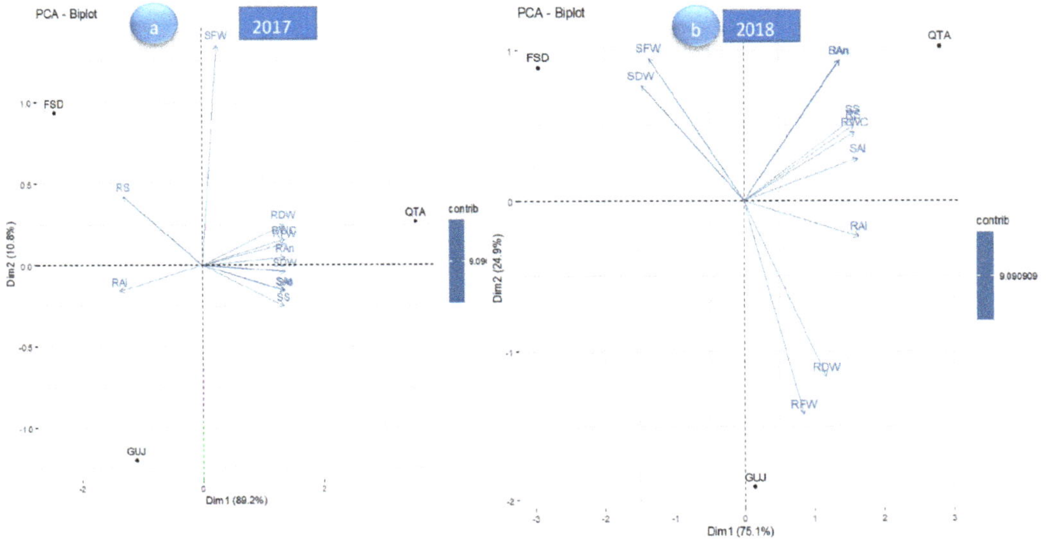

Figure 7. Principal component biplot of three ecotypes for all the traits under study in years (**a**) 2017 and (**b**) 2018. RFW = root fresh weight; SFW = shoot fresh weight; RDW = root dry weight; SDW = shoot dry weight; RAl = root alkaloids; SAl = shoot alkaloids; RAn = root anthocyanins; SAn = shoot anthocyanins; RS = root saponins; SS = shoot saponins; RWC = relative water content.

4. Discussion

The enzymes concerning nitrogen and carbon metabolism are significantly altered under salinity stress, thereby resulting in a lower production of SMs and hampering plant growth [50]. By observing milk thistle performance, it was found that the shoot and root length in milk thistle planted in saline soil was shorter (Figure 1a,b) and that leaf color was slightly changed to yellowish-green as compared with control conditions. Overall, the data showed that salt stress reduced the vegetative growth of milk thistle, such as the number of leaves (Figure 1c), leaf area (Figure 1d), shoot and root fresh and dry weight (Figure 2), the number of roots (Figure 3a), and the RWC (Figure 3d). The present study's thiourea results confirmed the studies of [51,52]; they also observed a significant reduction in plant length, the number of leaves, and root and shoot fresh weight. In another case [53], it was noted that plant height increased in *Alhagi pseudoalhagi* (a leguminous plant) at 5 dS/m, while it decreased at 10 and 20 dS/m. In another study of thiourea, [54] confirmed milk thistle growth reduction at 9 dS/m. However, in this study, increases in the number of spines and root diameter were seen, which confirmed the anticipatory role of salt stress (Figure 3b,c). According to [55], SS increased root area by up to 20% in *Brassica napus*, which indicated the spontaneous response of plants consisting in the uptake of more nutrients and water under stressful conditions. Contrarily, [56] observed a root-surface-area reduction in wheat under SS. Additionally, the application of plant growth promoters significantly increased all growth-related attributes regardless of salinity treatment, and an increasing trend was observed as follows: ascorbic acid > moringa leaf extract > thiourea. Our studies were inconsistent with the findings obtained by [57,58]; they also observed higher growth rate and RWC in maize with the application of the above-mentioned plant growth promoters. They suggested that ascorbic acid and thiourea improved growth-related attributes by prompting the photosynthetic capacity. Moringa leaf extract is also a rich source of ascorbate, which was found to promote a shielding effect against oxidative stress and to improve photosynthetic efficiency [59]. On the other hand, ecotypic variations showed the highest growth rate in Quetta, followed by Gujranwala and Faisalabad.

According to the observations in [60], SMs increased in safflower under saline stress conditions at 5 to 15 dS/m. In this study, a higher SM content in terms of anthocyanin was observed both under SS and in plants supplied with plant growth promoters than in unstressed plants (Figure 5). The authors of [61] reported that the anthocyanin content increased in response to salinity, while it decreased in salt-sensitive genotypes. According to [62], the anthocyanin content increased in wheat genotypes under salt stress conditions. In another study, ref. [63] reported higher anthocyanin contents in tomato and cabbage under SS. In addition, higher total alkaloid and saponin (Figures 4 and 6) contents were observed under salt stress conditions, and MS with plant growth promoters further increased their concentrations, regardless of ecotypic variations. The authors of [64] also deduced that the alkaloids content increased in *Chelidinium majus* L. in response to drought and SS. They presupposed that this increase might have been due to the increased enzymatic activities of stylopine synthase, which takes part in alkaloid biosynthesis. In contrast, ref. [65] found a decreased content of alkaloids in *Catharanthus roseus* (L.) at 5 to 15 dS/m salinity. In another case, ref. [66] observed higher alkaloid and saponin contents in soybean under SS. The effect of ascorbic acid, followed by moringa leaf extract and thiourea, was the most pronounced in enhancing the SM contents in all milk thistle ecotypes. Our findings are corroborated by the results of [67]; they also observed higher SM production in common bean with the application of ascorbic acid (200 or 400 mg L^{-1}) under SS conditions, which may have been directly or indirectly linked with its antioxidative properties. Ecotypic variations showed higher SM contents in Quetta ecotypes than in other ecotypes, regardless of SS and MS with plant growth promoters. In addition, a higher alkaloids content was recorded in 2018 than in 2017, while saponin and anthocyanin contents were higher in 2017 under net house conditions. Therefore, using agronomic practices such as MS with plant growth promoters, especially ascorbic acid in Quetta ecotype, can enhance SS tolerance and could be used as a key for improving the SMs required for sustainable, low-input production in the SS environments of the world.

5. Conclusions

SS imprinted a negative effect on all growth-related attributes of milk thistle except for the number of spines and root diameter, which were surprisingly increased under SS treatment. The harmful effects of SS could be ameliorated via MS with plant growth promoters ascorbic acid, thiourea, and moringa leaf extract in milk thistle plants. Additionally, salt stress increased the synthesis of SMs, as confirmed by the increases in total alkaloids, saponin, and anthocyanin contents in SS-exposed plants, while their production was further enhanced via MS with plant growth promoters. Regarding ecotypic variations, the Quetta ecotype performed better not only under control but also SS conditions. Overall, a higher growth rate was observed in 2018 than in 2017. In a nutshell, an adequate supply of plant growth promoters played the anticipated role in improving the salinity tolerance of milk thistle plants by increasing the production of SMs.

Author Contributions: Conceptualization, N.Z. and A.W.; methodology, N.Z. and M.B.H.; validation, I.L., A.B. and M.U.; formal analysis, M.A.E.-S.; investigation, S.A.; resources, N.Z.; data curation, A.W.; writing—original draft preparation, N.Z.; writing—review and editing, P.K.; visualization, A.W.; supervision, N.Z.; project administration, N.Z. All authors have read and agreed to the published version of the manuscript.

Funding: This research study received no external funding.

Institutional Review Board Statement: Not applicable.

Informed Consent Statement: Not applicable.

Data Availability Statement: The data analyzed for this paper can be requested from the corresponding author.

Acknowledgments: The authors would like to extend their sincere appreciation to Researchers Supporting Project Number RSP-2022/182, King Saud University, Riyadh, Saudi Arabia.

Conflicts of Interest: The authors declare no conflict of interest.

References

1. Shah, M.; Jan, H.; Drouet, S.; Tungmunnithum, D.; Shirazi, J.H.; Hano, C.; Abbasi, B.H. Chitosan Elicitation Impacts Flavonolignan Biosynthesis in *Silybum marianum* (L.) Gaertn Cell Suspension and Enhances Antioxidant and Anti-Inflammatory Activities of Cell Extracts. *Molecules* **2021**, *26*, 791. [CrossRef] [PubMed]
2. Kim, N.-C.; Graf, T.N.; Sparacino, C.M.; Wani, M.C.; Wall, M.E. Complete isolation and characterization of silybins and isosilybins from milk thistle (*Silybum marianum*). *Org. Biomol. Chem.* **2003**, *1*, 1684–1689. [CrossRef] [PubMed]
3. Denev, P.; Ognyanov, M.; Georgiev, Y.; Teneva, D.; Klisurova, D.; Yanakieva, I.Z. Chemical composition and antioxidant activity of partially defatted milk thistle (*Silybum marianum* L.) seeds. *Bulg. Chem. Commun.* **2020**, *52*, 182–187.
4. Choe, U.; Whent, M.; Luo, Y.; Yu, L. Total phenolic content, free radical scavenging capacity, and anti-cancer activity of silymarin. *J. Food Bioact.* **2020**, *10*, 53–64. [CrossRef]
5. Abenavoli, L.; Izzo, A.A.; Milić, N.; Cicala, C.; Santini, A.; Capasso, R. Milk thistle (*Silybum marianum*): A concise overview on its chemistry, pharmacological, and nutraceutical uses in liver diseases. *Phytother. Res.* **2018**, *32*, 2202–2213. [CrossRef] [PubMed]
6. Kogo, B.K.; Kumar, L.; Koech, R. Climate change and variability in Kenya: A review of impacts on agriculture and food security. *Environ. Develop. Sustain.* **2020**, *23*, 23–43. [CrossRef]
7. Hafeez, M.B.; Raza, A.; Zahra, N.; Shaukat, K.; Akram, M.Z.; Iqbal, S.; Basra, S.M.A. Gene regulation in halophytes in conferring salt tolerance. In *Handbook of Bioremediation*; Elsevier: Amsterdam, The Netherlands, 2021; pp. 341–370.
8. Kaya, C.; Ashraf, M.; Alyemeni, M.N.; Ahmad, P. The role of endogenous nitric oxide in salicylic acid-induced up-regulation of ascorbate-glutathione cycle involved in salinity tolerance of pepper (*Capsicum annuum* L.) plants. *Plant Physiol. Biochem.* **2020**, *147*, 10–20. [CrossRef]
9. Zahra, N.; Hafeez, M.B.; Shaukat, K.; Wahid, A.; Hussain, S.; Naseer, R.; Raza, A.; Iqbal, S.; Farooq, M. Hypoxia and Anoxia Stress: Plant responses and tolerance mechanisms. *J. Agron. Crop Sci.* **2021**, *207*, 249–284. [CrossRef]
10. Mubarik, M.S.; Khan, S.H.; Sajjad, M.; Raza, A.; Hafeez, M.B.; Yasmeen, T.; Rizwan, M.; Ali, S.; Arif, M.S. A manipulative interplay between positive and negative regulators of phytohormones: A way forward for improving drought tolerance in plants. *Physiol. Plant.* **2021**, *172*, 1269–1290. [CrossRef]
11. Batool, S.; Khan, S.; Basra, S.M.; Hussain, M.; Saddiq, M.S.; Iqbal, S.; Irshad, S.; Hafeez, M. Impact of natural and synthetic plant stimulants on Moringa seedlings grown under low-temperature conditions. *Int. Lett. Nat. Sci.* **2019**, *76*, 50–59. [CrossRef]
12. Akhter, M.S.; Noreen, S.; Mahmood, S.; Ashraf, M.; Alsahli, A.A.; Ahmad, P. Influence of salinity stress on PSII in barley (*Hordeum vulgare* L.) genotypes, probed by chlorophyll-a fluorescence. *J. King Saud Univ. Sci.* **2021**, *33*, 101239. [CrossRef]
13. Ahmad, P.; Venema, K.; Corpas, F.J. Unravelling salt tolerance mechanisms in plants: From Lab to Field. *Plant Physiol. Biochem.* **2022**, *176*, 31–33. [CrossRef] [PubMed]
14. Ahammed, G.J.; Li, Y.; Li, X.; Han, W.-Y.; Chen, S. Epigallocatechin-3-gallate alleviates salinity-retarded seed germination and oxidative stress in tomato. *J. Plant Growth Regul.* **2018**, *37*, 1349–1356. [CrossRef]
15. Shah, T.; Latif, S.; Saeed, F.; Ali, I.; Ullah, S.; Alsahli, A.A.; Jan, S.; Ahmad, P. Seed priming with titanium dioxide nanoparticles enhances seed vigor, leaf water status, and antioxidant enzyme activities in maize (*Zea mays* L.) under salinity stress. *J King Saud Uni. Sci.* **2020**, *33*, 101207. [CrossRef]
16. Ahanger, M.A.; Mir, R.A.; Alyemeni, M.N.; Ahmad, P. Combined effects of brassinosteroid and kinetin mitigates salinity stress in tomato through the modulation of antioxidant and osmolyte metabolism. *Plant Physiol. Biochem.* **2020**, *147*, 31–42. [CrossRef] [PubMed]
17. Zahra, N.; Raza, Z.A.; Mahmood, S. Effect of salinity stress on various growth and physiological attributes of two contrasting maize genotypes. *Brazil. Arch. Biol. Technol.* **2020**, *63*, e20200072. [CrossRef]
18. Arif, Y.; Singh, P.; Siddiqui, H.; Bajguz, A.; Hayat, S. Salinity induced physiological and biochemical changes in plants: An omic approach towards salt stress tolerance. *Plant Physiol. Biochem.* **2020**, *156*, 64–77. [CrossRef] [PubMed]
19. Saddiq, M.S.; Iqbal, S.; Hafeez, M.B.; Ibrahim, A.M.; Raza, A.; Fatima, E.M.; Baloch, H.; Woodrow, P.; Ciarmiello, L.F. Effect of salinity stress on physiological changes in winter and spring wheat. *Agronomy* **2021**, *11*, 1193. [CrossRef]
20. El-Beltagi, H.S.; Ahmad, I.; Basit, A.; El-Lateef, A.; Hany, M.; Yasir, M.; Tanveer Shah, S.; Ullah, I.; Elsayed Mohamed Mohamed, M.; Ali, I. Effect of azospirillum and azotobacter species on the performance of cherry tomato under different salinity levels. *Gesunde Pflanz.* **2022**, *74*, 487–499. [CrossRef]
21. Ismail, L.M.; Soliman, M.I.; Abd El-Aziz, M.H.; Abdel-Aziz, H.M. Impact of silica ions and nano silica on growth and productivity of pea plants under salinity stress. *Plants* **2022**, *11*, 494. [CrossRef]
22. Liu, C.; Zhao, X.; Yan, J.; Yuan, Z.; Gu, M. Effects of salt stress on growth, photosynthesis, and mineral nutrients of 18 pomegranate (*Punica granatum*) cultivars. *Agronomy* **2020**, *10*, 27. [CrossRef]
23. Khademian, R.; Asghari, B.; Sedaghati, B.; Yaghoubian, Y. Plant beneficial rhizospheric microorganisms (PBRMs) mitigate deleterious effects of salinity in sesame (*Sesamum indicum* L.): Physio-biochemical properties, fatty acids composition and secondary metabolites content. *Ind. Crops Prod.* **2019**, *136*, 129–139. [CrossRef]

24. MacNeill, G.J.; Mehrpouyan, S.; Minow, M.A.; Patterson, J.A.; Tetlow, I.J.; Emes, M.J.; Raines, C. Starch as a source, starch as a sink: The bifunctional role of starch in carbon allocation. *J. Exp. Bot.* **2017**, *68*, 4433–4453. [CrossRef]
25. Hasanuzzaman, M.; Bhuyan, M.; Zulfiqar, F.; Raza, A.; Mohsin, S.M.; Mahmud, J.A.; Fujita, M.; Fotopoulos, V. Reactive oxygen species and antioxidant defense in plants under abiotic stress: Revisiting the crucial role of a universal defense regulator. *Antioxidants* **2020**, *9*, 681. [CrossRef] [PubMed]
26. Khaliq, A.; Zia-ul-Haq, M.; Ali, F.; Aslam, F.; Matloob, A.; Navab, A.; Hussain, S. Salinity tolerance in wheat cultivars is related to enhanced activities of enzymatic antioxidants and reduced lipid peroxidation. *CLEAN–Soil Air Water* **2015**, *43*, 1248–1258. [CrossRef]
27. Kamiab, F. Exogenous melatonin mitigates the salinity damages and improves the growth of pistachio under salinity stress. *J. Plant Nutr.* **2020**, *43*, 1468–1484. [CrossRef]
28. Ishaq, H.; Nawaz, M.; Azeem, M.; Mehwish, M.; Naseem, M.B.B. Ascorbic Acid (Asa) improves Salinity Tolerance in Wheat (*Triticum Aestivum* L.) by Modulating Growth and Physiological Attributes. *J. Bioresour. Manag.* **2021**, *8*, 1. [CrossRef]
29. Ahmadi, F.; Karimi, K.; Struik, P. Effect of exogenous application of methyl jasmonate on physiological and biochemical characteristics of *Brassica napus* L. cv. Talaye under salinity stress. *S. Afr. J. Bot.* **2018**, *115*, 5–11. [CrossRef]
30. Alam, P.; Albalawi, T.H.; Altalayan, F.H.; Bakht, M.A.; Ahanger, M.A.; Raja, V.; Ashraf, M.; Ahmad, P. 24-Epibrassinolide (EBR) confers tolerance against NaCl stress in soybean plants by up-regulating antioxidant system, ascorbate-glutathione cycle, and glyoxalase system. *Biomolecules* **2019**, *9*, 640. [CrossRef]
31. Ahanger, M.A.; Aziz, U.; Alsahli, A.A.; Alyemeni, M.N.; Ahmad, P. Influence of exogenous salicylic acid and nitric oxide on growth, photosynthesis, and ascorbate-glutathione cycle in salt stressed *Vigna angularis*. *Biomolecules* **2020**, *10*, 42. [CrossRef]
32. Zheng, Y.; Xu, B.; Ren, K.; Zhang, Y.; Wu, J. Impact of soil drench and foliar spray of 24-epibrassinolide on the growth, yield, and quality of field-grown *Moringa oleifera* in Southwest China. *J. Plant Growth Regul.* **2017**, *36*, 931–941. [CrossRef]
33. Waqas, M.; Yaning, C.; Iqbal, H.; Shareef, M.; ur Rehman, H.; Iqbal, S.; Mahmood, S. Soil drenching of paclobutrazol: An efficient way to improve quinoa performance under salinity. *Physiol. Plant.* **2019**, *165*, 219–231. [CrossRef] [PubMed]
34. Hussein, M.; Bakheta, M.; Zaki, S. Influence of uniconazole on growth characters, photosynthetic pigments, total carbohydrates and total soluble sugars of *Hordium vulgare* L. plants grown under salinity stress. *Int. J. Sci. Res.* **2014**, *3*, 2208–2213.
35. Akram, N.A.; Shafiq, F.; Ashraf, M. Ascorbic acid-a potential oxidant scavenger and its role in plant development and abiotic stress tolerance. *Front. Plant Sci.* **2017**, *8*, 613. [CrossRef]
36. Wang, Y.-H.; Zhang, G.; Chen, Y.; Gao, J.; Sun, Y.-R.; Sun, M.-F.; Chen, J.-P. Exogenous application of gibberellic acid and ascorbic acid improved tolerance of okra seedlings to NaCl stress. *Acta Physiol. Plant.* **2019**, *41*, 93. [CrossRef]
37. de Sousa Basílio, A.G.; Vieira de Sousa, L.; da Silva, T.I.; de Moura, J.G.; de Melo Gonçalves, A.C.; de Melo Filho, J.S.; Leal, Y.H.; Jardelino Dias, T. Radish (*Raphanus sativus* L.) morphophysiology under salinity stress and ascorbic acid treatments. *Agron. Colomb.* **2018**, *36*, 257–265. [CrossRef]
38. Waqas, M.A.; Kaya, C.; Riaz, A.; Farooq, M.; Nawaz, I.; Wilkes, A.; Li, Y. Potential mechanisms of abiotic stress tolerance in crop plants induced by thiourea. *Front. Plant Sci.* **2019**, *10*, 1336. [CrossRef]
39. Aziz, U.; Qadir, I.; Yasin, G.; Azhar, M.F.; Javed, A.; Akhtar, A. Potential of priming in improving germination, seedling growth and nutrient status of Calotropis procera under salinity. *Pak. J. Bot.* **2021**, *53*, 1953–1958. [CrossRef]
40. Yaseen, A.; Takacsne Hajos, M. Study on moringa tree (*Moringa oleifera* Lam.) leaf extract in organic vegetable production: A review. *Res. Crops* **2020**, *21*, 402–414.
41. Faisal, M.; Iqbal, S.; Basra, S.; Afzal, I.; Saddiq, M.; Bakhtavar, M.; Hafeez, M.; Rehman, H.; Basit, A.; Habib-ur-Rahman, M. Moringa landraces of Pakistan are potential source of premium quality oil. *S. Afr. J. Bot.* **2020**, *129*, 397–403. [CrossRef]
42. Aslam, M.F.; Basra, S.M.; Hafeez, M.B.; Khan, S.; Irshad, S.; Iqbal, S.; Saqqid, M.S.; Akram, M.Z. Inorganic fertilization improves quality and biomass of *Moringa oleifera* L. *Agrofor. Syst.* **2020**, *94*, 975–983. [CrossRef]
43. Ahmed, T.; Abou Elezz, A.; Khalid, M.F. Hydropriming with moringa leaf extract mitigates salt stress in Wheat seedlings. *Agriculture* **2021**, *11*, 1254. [CrossRef]
44. Yaghoubian, I.; Antar, M.; Ghassemi, S.; Modarres-Sanavy, S.A.M.; Smith, D.L. The Effects of Hydro-Priming and Colonization with *Piriformospora indica* and *Azotobacter chroococcum* on Physio-Biochemical Traits, Flavonolignans and Fatty Acids Composition of Milk Thistle (*Silybum marianum*) under Saline Conditions. *Plants* **2022**, *11*, 1281. [CrossRef] [PubMed]
45. Perveen, A.; Wahid, A.; Mahmood, S.; Hussain, I.; Rasheed, R. Possible mechanism of medium-supplemented thiourea in improving growth, gas exchange, and photosynthetic pigments in cadmium-stressed maize (*Zea mays*). *Brazil. J. Bot.* **2015**, *38*, 71–79. [CrossRef]
46. Rashid, N.; Basra, S.M.; Shahbaz, M.; Iqbal, S.; Hafeez, M.B. Foliar applied moringa leaf extract induces terminal heat tolerance in quinoa. *Int. J. Agric. Biol.* **2018**, *20*, 157–164.
47. Peñas, E.; Limón, R.I.; Martínez-Villaluenga, C.; Restani, P.; Pihlanto, A.; Frias, J. Impact of elicitation on antioxidant and potential antihypertensive properties of lentil sprouts. *Plant Foods Human Nutr.* **2015**, *70*, 401–407. [CrossRef]
48. STRACK, D.; WRAY, V. Anthocyanins. In *Methods in Plant Biochemistry*; Elsevier: Amsterdam, The Netherlands, 1989; Volume 1, pp. 325–356.
49. Singh, D.; Sahu, A. Spectrophotometric determination of caffeine and theophylline in pure alkaloids and its application in pharmaceutical formulations. *Anal. Biochem.* **2006**, *349*, 176–180. [CrossRef]

50. Marino, D.; González, E.M.; Arrese-Igor, C. Drought effects on carbon and nitrogen metabolism of pea nodules can be mimicked by paraquat: Evidence for the occurrence of two regulation pathways under oxidative stresses. *J. Exp. Bot.* **2006**, *57*, 665–673. [CrossRef]
51. Su, Y.; Guo, A.; Huang, Y.; Wang, Y.; Hua, J. GhCIPK6a increases salt tolerance in transgenic upland cotton by involving in ROS scavenging and MAPK signaling pathways. *BMC Plant Biol.* **2020**, *20*, 421. [CrossRef]
52. Chartzoulakis, K.; Klapaki, G. Response of two greenhouse pepper hybrids to NaCl salinity during different growth stages. *Sci. Horti.* **2000**, *86*, 247–260. [CrossRef]
53. Kurban, H.; Saneoka, H.; Nehira, K.; Adilla, R.; Premachandra, G.S.; Fujita, K. Effect of salinity on growth, photosynthesis and mineral composition in leguminous plant *Alhagi pseudoalhagi* (Bieb.). *Soil Sci. Plant Nutr.* **1999**, *45*, 851–862. [CrossRef]
54. Ghavami, N.; Ramin, A. Grain yield and active substances of milk thistle as affected by soil salinity. *Comm. Soil Sci. Plant Anal.* **2008**, *39*, 2608–2618. [CrossRef]
55. Arif, M.R.; Islam, M.T.; Robin, A.H.K. Salinity stress alters root morphology and root hair traits in *Brassica napus*. *Plants* **2019**, *8*, 192. [CrossRef] [PubMed]
56. Robin, A.H.K.; Matthew, C.; Uddin, M.J.; Bayazid, K.N. Salinity-induced reduction in root surface area and changes in major root and shoot traits at the phytomer level in wheat. *J. Exp. Bot.* **2016**, *67*, 3719–3729. [CrossRef] [PubMed]
57. Sahu, M.; Solanki, N.; Dashora, L. Effects of thiourea, thiamine and ascorbic acid on growth and yield of maize (*Zea mays* L.). *J. Agron. Crop Sci.* **1993**, *171*, 65–69. [CrossRef]
58. Waqas, M.A.; Khan, I.; Akhter, M.J.; Noor, M.A.; Ashraf, U. Exogenous application of plant growth regulators (PGRs) induces chilling tolerance in short-duration hybrid maize. *Environ. Sci. Pollut. Res.* **2017**, *24*, 11459–11471. [CrossRef]
59. Khan, A.; Ashraf, M. Exogenously applied ascorbic acid alleviates salt-induced oxidative stress in wheat. *Environ. Exp. Bot.* **2008**, *63*, 224–231.
60. Gengmao, Z.; Yu, H.; Xing, S.; Shihui, L.; Quanmei, S.; Changhai, W. Salinity stress increases secondary metabolites and enzyme activity in safflower. *Ind. Crops Prod.* **2015**, *64*, 175–181. [CrossRef]
61. Liang, W.; Ma, X.; Wan, P.; Liu, L. Plant salt-tolerance mechanism: A review. *Biochem. Biophys. Res. Comm.* **2018**, *495*, 286–291. [CrossRef]
62. Mbarki, S.; Sytar, O.; Zivcak, M.; Abdelly, C.; Cerda, A.; Brestic, M. Anthocyanins of Coloured wheat genotypes in specific response to SalStress. *Molecules* **2018**, *23*, 1518. [CrossRef]
63. Eryılmaz, F. The relationships between salt stress and anthocyanin content in higher plants. *Biotechnol. Biotechnol. Equip.* **2006**, *20*, 47–52. [CrossRef]
64. Yahyazadeh, M.; Meinen, R.; Hänsch, R.; Abouzeid, S.; Selmar, D. Impact of drought and salt stress on the biosynthesis of alkaloids in *Chelidonium majus* L. *Phytochemistry* **2018**, *152*, 204–212. [CrossRef] [PubMed]
65. Idrees, M.; Naeem, M.; Aftab, T.; Khan, M.M.A. Salicylic acid mitigates salinity stress by improving antioxidant defence system and enhances vincristine and vinblastine alkaloids production in periwinkle [*Catharanthus roseus* (L.) G. Don]. *Acta Physiol. Plant.* **2011**, *33*, 987–999. [CrossRef]
66. Radhakrishnan, R.; Leelapriya, T.; Kumari, B.D.R. Effects of pulsed magnetic field treatment of soybean seeds on calli growth, cell damage, and biochemical changes under salt stress. *Bioelectromagnetics* **2012**, *33*, 670–681. [CrossRef] [PubMed]
67. Gaafar, A.A.; Ali, S.I.; El-Shawadfy, M.A.; Salama, Z.A.; Sekara, A.; Ulrichs, C.; Abdelhamid, M.T. Ascorbic acid induces the increase of secondary metabolites, antioxidant activity, growth, and productivity of the common bean under water stress conditions. *Plants* **2020**, *9*, 627. [CrossRef]

life

MDPI

Article

Influence of Plant Growth Regulators and Artificial Light on the Growth and Accumulation of Inulin of Dedifferentiated Chicory (*Cichorium intybus* L.) Callus Cells

Rima N. Kirakosyan [1,*], Anton V. Sumin [1], Anna A. Polupanova [1], Maria G. Pankova [1], Irina S. Degtyareva [1], Nikolay N. Sleptsov [2] and Quyet V. Khuat [1,3]

1 Department of Biotechnology, Russian State Agrarian University—Moscow Timiryazev Agricultural Academy, Timiryazevskaya Str., 49, 127550 Moscow, Russia
2 Department of Plant Physiology, Russian State Agrarian University—Moscow Timiryazev Agricultural Academy, Timiryazevskaya Str., 49, 127550 Moscow, Russia
3 Biology and Agricultural Engineering Faculty, Hanoi Pedagogical University 2, Nguyen Van Linh, Phuc Yen 15000, Vietnam
* Correspondence: mia41291@mail.ru; Tel.: +7-(985)-460-66-65

Abstract: Chicory (*Chicorium intybus* L.) is a perennial herb of the family *Asteraceae*, widely distributed in Asia and Europe, commonly used industrially as a raw material for extracting inulin because of a high content of inulin and biologically active compounds. Light conditions and plant growth regulators (PGRs) are two of many factors that affect the growth and inulin content of chicory callus. The aim of this work is to study the effect of PGRs and light conditions on proliferation and accumulation of inulin of chicory callus in vitro. In this study, we used semi-solid MS medium supplemented with different auxins (including Indole-3-acetic acid (IAA), naphthylacetic acid (NAA), and 2,4-dichlorophenoxyacetic acid (2,4-D)) at a concentration of 5.5–9.5 mg/L in combination with 2.0 mg/L 6 benzylaminopurine (BA) to determine induction and proliferation of callus. The increasing value of callus fresh weight was used to assess the growth of the callus in treatments. The results showed that a steady increase in callus fresh weight and inulin content in callus cells was obtained when they were cultured on MS medium supplemented with a combination of 2.0 mg/L BA with 7.5 mg/L IAA in lighting conditions with radiation equalized by the flux density of photosynthetic photons and ratios of radiation levels in the region of FR—far red > R—red. Increasing demand for organic inulin sources in production practice can be met by our finding.

Keywords: artificial light; auxins; chicory; callus cells; inulin; plant growth regulators

Citation: Kirakosyan, R.N.; Sumin, A.V.; Polupanova, A.A.; Pankova, M.G.; Degtyareva, I.S.; Sleptsov, N.N.; Khuat, Q.V. Influence of Plant Growth Regulators and Artificial Light on the Growth and Accumulation of Inulin of Dedifferentiated Chicory (*Cichorium intybus* L.) Callus Cells. *Life* **2022**, *12*, 1524. https://doi.org/10.3390/life12101524

Academic Editors: Hakim Manghwar and Wajid Zaman

Received: 30 August 2022
Accepted: 27 September 2022
Published: 29 September 2022

Publisher's Note: MDPI stays neutral with regard to jurisdictional claims in published maps and institutional affiliations.

Copyright: © 2022 by the authors. Licensee MDPI, Basel, Switzerland. This article is an open access article distributed under the terms and conditions of the Creative Commons Attribution (CC BY) license (https://creativecommons.org/licenses/by/4.0/).

1. Introduction

The production of high-quality drugs, characterized by safety and high efficiency, is one of the priority areas for the development of the pharmaceutical industry. The production of such drugs is based on the use of plant materials, in particular medicinal plants. It plays an important role in expanding the range of medicinal products [1–3]. According to the World Health Organization (WHO) forecast and the WHO Traditional Medicine Strategy 2014–2023, in 15–20 years, the share of herbal medicines in the total range of medicines may increase to 60% [4].

The interest of researchers in medicinal plants is constantly growing, since they are a source of biologically active substances that can be widely used in the food industry as well. Today, special attention is paid to the production of food for dietary and functional purposes, which include, for example, dietary fiber, antioxidants, prebiotics, etc. [5]. One of the effective prebiotics is inulin, which is industrially most often extracted from chicory (*Cichorium intybus* L.) [6]. Inulin is a heterogeneous collection of fructose polymers. It consists of chain-terminating glucosyl moieties and a repetitive fructosyl moiety, which are linked by β(2,1) bonds [7]. The degree of polymerization (DP) of standard inulin ranges from 2 to 60 [8,9]. Because of the β(2,1) linkages, inulin is not digested by enzymes in the human alimentary system, contributing to its functional properties: reduced calorie value, dietary fiber, and prebiotic effects [10].

Chicory is a perennial herb of the family *Asteraceae*, widely distributed in Asia and Europe [11,12], commonly used industrially as a raw material for extracting inulin because of a high content of inulin and biologically active compounds. All parts of this plant possess great medicinal importance due to the presence of a number of medicinally important compounds such as alkaloids, inulin, sesquiterpene lactones, coumarins, vitamins, chlorophyll pigments, unsaturated sterols, flavonoids, saponins and tannins [13–16]. According to Meehye and Shin (1996) [17], fresh chicory typically contains 68% inulin, 14% sucrose, 5% cellulose, 6% protein, 4% ash, and 3% other compounds, while dried chicory contains approximately 98% inulin and 2% other compounds. Chicory has been traditionally used for the treatment of fever, diarrhea, jaundice, and gallstones [18,19]. Several recent studies have been also reported that chicory has a potent hepatoprotective, antioxidant, hypoglycemic, diuretic, anti-testicular toxicity, and immunomodulatory effects [20–24]. Moreover, its roots are often used as a coffee substitute [25], particularly in India [26] and South Africa.

Nowadays, the use of biotechnology methods not only allows to multiply and obtain high-quality planting material but also creates in vitro cell cultures of medicinal plants with an increased content of biologically active substances [27,28]. Several previous reports have demonstrated the regenerative ability of chicory from different plant parts (including leaf explant [29–35], and petiole explant [36,37] by using different hormonal combinations. In most of the reported studies, leaf explants were used most in micropropagation through callus cultures, and combinations of BA with IAA or NAA at different concentrations were used for the initial induction and proliferation of the callus [31–33,35]. In addition, light is known to be an effective abiotic elicitor that influences plant photosynthesis process, development, and morphogenesis [38–40]. Light plays a vital role in regulating primary as well as secondary metabolism to help achieve optimum growth [41,42]. Multiple studies have reported the direct stimulation of secondary metabolites production in the presence of monochromatic lights, especially red light [43–45]. There are currently no similar reports on chicory species.

Based on the foregoing, the aim of this work is to study the effect of plant growth regulators (PGRs) and light conditions on proliferation and accumulation of inulin of Chicory (*Cichorium intybus* L.) callus in vitro.

2. Materials and Methods

2.1. Plant Material Prepareation

The work was carried out at the Department of Biotechnology of the Russian State Agrarian University—Moscow Agricultural Academy named after K. A. Timiryazev (Moscow, Russia). The objects of the study were leaf segments obtained from in vitro seedlings of Chicory (*Cichorium intybus* L.), cultivar Petrovsky.

Seeds were sterilized with a 0.1% mercuric chloride ($HgCl_2$) solution for 9 min, followed by rinsing them three times with sterile distilled water following the protocol described in the literature data [35].

All work on sterilization of seeds, introduction into culture in vitro and further work on the study of callogenesis and morphogenesis were carried out in aseptic conditions of laminar hood flow (BIOBASE BBS—H1800(X)). After surface disinfection, seeds were cultivated on a PGR-free nutrient medium of Murashige and Skoog (MS) [46]. The pH of the medium was adjusted to 5.6–5.8 before being autoclaved at 121 °C and 1.1 atm for 20 min. The cultures were maintained in a culture room at 25 ± 2 °C during a long-day photoperiod (16 h of light: 8 h of dark) with cool white fluorescent light (2000–2500 lux). Five seeds were cultured per Petri dish and ten dishes per treatment.

Seedlings obtained on the 7th day from the time the seeds germinated were subcultured to MS medium supplemented with 1.0 mg/L 6–benzylaminopurine (BA) and 0.1 mg/L Indole–3–acetic acid (IAA) for in vitro plantlet development.

2.2. Cultivation of Chicory Callus under Different PGR Composition

In order to induce callus, leaf segments (5 × 5 mm) isolated from 30 days old in vitro plantlets were cultivated in MS medium supplemented with cytokinin (2.0 mg/L BA) and various auxins (including IAA, naphthylacetic acid (NAA), and 2,4-dichlorophenoxyacetic acid (2,4-D)) at a concentration of 5.5–9.5 mg/L. Ten leaf explants were cultured per Petri dish and fifteen dishes per treatment.

Every 4 weeks, the callus was transferred onto a fresh medium. Five calluses were cultured per Petri dish and thirty dishes per treatment.

The consistency and color of the callus were taken into account. The increasing value of callus fresh weight was used to assess the growth of the callus. Biomass of callus was weighed on scales (AND GR-202) in a laminar flow hood.

2.3. Cultivation of Chicory Callus under Light-Culture Conditions

Chicory leaf segments obtained from 30 days old in vitro plantlets were cultivated on semisolid MS medium supplemented with 2.0 mg/L BA in combination with 7.5 mg /L NAA or 7.5 mg /L IAA. Ten leaf explants were cultured per Petri dish and fifteen dishes per treatment. Petri dishes were placed in light proof grow tents (Urban Grower 60 × 60 × 200 cm (Gorshkoff, Russia)) with radiation equalized by the flux density of photosynthetic photons and different ratios of radiation levels in the region of 660 nm (R—red) and 730 nm (FR—far red). R/FR ratio options:

(1) R/FR = 1, photosynthetic photon flux density (PPFD) = 142 (±10) μmol/m^2s;
(2) R/FR = 1/2, PPFD = 142 (±10) μmol/m^2s;
(3) R/FR = 2, PPFD = 142 (±10) μmol/m^2s;
(4) Control: white linear fluorescent lamp (OSRAM AG 4000K), PPFD 40 μmol/m^2s.

The emission spectra were measured with a PLA-20 instrument (Everfine, Hangzhou, China).

The increasing value of callus fresh weight was used to assess the growth of the callus. The callus was weighed in a laminar flow hood by scales at the beginning and the end of the subculture.

2.4. Determination of Inulin by Spectrophotometry

For one analysis, a dry sample of 600 mg callus was taken, placed in a volumetric flask, poured in 9.5 mL of 90% ethanol, and kept for 30 min in a water bath (UCHEN) at 80 °C, periodically stirring its content. As a control, the same volume of water was added to the sample of callus.

After cooling, 0.05 mL of 25% NaOH solution was added to both flasks (sample and control) and then the contents of them were brought to 10 mL with ethanol at the sample flask and water at the control flask. The contents of the flasks were left for 10–20 min after which centrifugation was carried out at 4000–6000 rpm for 3–5 min. After that, empty 10 mL volumetric flasks were taken and 0.05 mL of centrifugates of both extracts (sample and control), as well as a solution of 3.0 mg/mL fructose (standard) and water (control), were transferred into them. In the next step, the resulting volume was added to 1.0 mL of reagent (2.0 mg/mL resorcinol + 96% ethanol and concentrated hydrochloric acid (HCl) in equal volumes). The flasks were kept in a water bath for 35 min and then cooled. The volume was brought to 10 mL with water and stirred.

The optical density of the analyzed samples and standard solution was measured on an SF-104 spectrophotometer at a wavelength of 480 nm. The content of fructose-containing sugars in the samples was calculated using the formula:

$$X = \frac{A_p \times C_s \times 10}{A_s \times m_p} \quad (1)$$

where A_p and A_s are the optical density of the test sample and the standard solution, respectively; C_s is the concentration of a standard fructose solution (3.0 mg/mL); m_p is the weight of the weighed portion of the analyzed sample, g.

The result obtained for an aqueous extract reflects the total content of water-soluble carbohydrates. The result for the ethanol extract reflects the content of low molecular weight fructosides. The difference between the two measures gives the inulin content. [47,48].

2.5. Statistical Analysis of Experimental Data

The experiments were arranged completely randomly and repeated three times. Mean values of all data were calculated using Microsoft Excel 2013 (Microsoft Corporation, Redmond, WA, USA). Analysis of Variance (ANOVA) was performed in AGROS software (version 2.11, Russian State Agrarian University—Moscow Timiryazev Agricultural Academy, Moscow, Russia, 1999–2001) and means were compared using Fisher's Least Significant Difference (LSD) test at a significance level of $p \leq 0.05$.

3. Results

3.1. Influence of the PGR Composition of the Nutrient Medium on Chicory Callus Growth

Some patterns have been established in chicory callus formation, as a result of the studies carried out. On days 7 to 10, the proliferation of cells was observed at the sites of cut and damage in all treatments. However, it should be noted that the applied different concentrations of various auxins (IAA, NAA, and 2,4-D) in combination with 2.0 mg/L BA had a significant effect on the growth, texture, and color of callus.

Cultivation of leaf explants on semisolid MS medium supplemented with different concentrations of IAA in combination with 2.0 mg/L BA led to the callus formation of a bright yellow color, semisolid, and with the formation of meristematic foci (Figure 1a). The onset of callogenesis was noted on the third day from the beginning of culture. With NAA, the chicory callus was dense and white or light yellow (Figure 1b). The onset of callogenesis was observed on days 7–10. When leaf explants were cultivated on a medium containing 2,4-D, a different result was observed. In this variant, the intensity of callogenesis was minimal. The formed callus was brown, of loose consistency, and died during cultivation. Therefore, the culture medium containing 2,4-D was not used in further experiments (Figure 1c).

Figure 1. Formation callus on semisolid MS medium supplemented with different concentrations of various auxins (IAA, NAA, and 2,4-D) in combination with 2.0 mg/L BA: (**a**) 7.5 mg/L IAA; (**b**) 7.5 mg/L NAA; (**c**) 5.5 mg/L 2,4-D. Callus after 28 days of culture (the first subculture). Scale bars = 1 cm.

The increasing value of callus fresh weight was used to assess the growth of the callus. This indicator was evaluated at fourth and fifth subcultures. The growth of callus was determined depending on the investigated auxin and its concentration (Table 1).

Table 1. Effect of different concentrations of various auxins (IAA and NAA) in combination with 2.0 mg/L BA on the growth of callus at different subcultures.

Auxin	Auxin Concentration, mg/L				
	5.5	6.5	7.5	8.5	9.5
			4th subculture		
IAA	0.77 ± 0.03 d	0.48 ± 0.03 b	0.22 ± 0.01 a	0.54 ± 0.03 bc	0.77 ± 0.03 d
NAA	0.78 ± 0.03 d	1.56 ± 0.41 ef	1.08 ± 0.38 de	1.32 ± 0.51 def	0.66 ± 0.03 c
			5th subculture		
IAA	0.92 ± 0.04 cde	0.84 ± 0.04 cd	0.63 ± 0.03 a	0.69 ± 0.03 ab	0.91 ± 0.04 cde
NAA	0.60 ± 0.03 a	0.80 ± 0.04 c	1.24 ± 0.58 def	1.24 ± 0.55 def	0.90 ± 0.04 cde

Mean (callus fresh weight, g) ± standard error (SE); At each subculture, means followed by a different letter are significantly different at an alpha level of 0.05 according to the Fisher's Least Significant Difference (LSD) test.

The stable growth of callus was observed when NAA was added to the nutrient medium at a concentration of 7.5 and 8.5 mg/L at the fifth subculture (reach 1.24 g) and 6.5 mg/L at the fourth subculture (reach 1.56 g). In the treatment with IAA (at the concentration of 8.5 mg/L in combination with 2.0 mg/L BA), the increasing value of callus fresh weight is about two times less than in the treatment with NAA at the same concentration. However, it should be noted that, on a medium containing NAA, the proliferative activity decreased with an increase in the number of subcultures. A stable increase in callus cells was observed between fourth and fifth subculture during all the IAA treatments. In addition, in these treatments in the callus at the end of the cultivation cycle, the formation of multiple meristematic foci was noted.

The difference in the growth of callus during subculturing was revealed. When using IAA, the greatest difference in growth between the fourth and fifth subcultures was at a concentration of 7.5 mg/L and amounted to 65%. At the same time, the greatest difference in growth between NAA transfers was at the level of 7.5 mg/L and amounted to 12.9%. Combined with fresh weight, this was the best result for this PGR. Therefore, it was decided to use these PGR concentrations for further experiments.

3.2. Influence of Light Quality and PGR Composition on the Morphology of the Chicory Callus

One of the regulatory factors of morphogenesis is the intensity and quality of light. The paper studied the influence of ratios of red LED lamps (R = 660 nm) and far-red spectrum (FR = 730 nm) on the formation of callus from leaf segments of chicory seedlings. Various auxins (IAA or NAA) at the concentration of 7.5 mg/L in combination with 2.0 mg/L BA were used in all variants of nutrient media. It was found that under different light growth conditions, callus of different size, density, and color was formed.

Growing callus under the conditions of a light regime FR > R on semisolid MS medium supplemented with 7.5 mg/L NAA in combination with 2.0 mg/L BA resulted in the formation of a loose, highly hydrated callus, and easily disintegrating into individual cell aggregates. In this case, the obtained callus was yellow and there was an anthocyanin coloration that appeared towards the center (Figure 2a). On semisolid MS medium supplemented with 7.5 mg /L IAA in combination with 2.0 mg/L BA, the obtained callus was green in color, of a dense type, and formed many meristematic foci over the entire surface of the callus (Figure 2b).

Figure 2. Callus obtained by cultivation on semisolid MS medium supplemented with 2.0 mg/L BA in combination with: (**a**) FR > R mode, 7.5 mg/L NAA; (**b**) FR > R mode, 7.5 mg/L IAA; (**c**) FR = R mode, 7.5 mg/L NAA; (**d**) FR = R mode, 7.5 mg/L IAA; (**e**) FR < R mode, 7.5 mg/L NAA; (**f**) FR < R mode, 7.5 mg/L IAA. Callus after 28 days of culture (the first subculture). Scale bars = 1 cm.

When callus was grown under the conditions of the FR = R light regime on semisolid MS medium supplemented with 7.5 mg /L NAA in combination with 2.0 mg/L BA, a loose callus type, poorly hydrated, and easily disintegrating into individual cell aggregates, was formed. In this case, the obtained callus was light yellow or white without anthocyanin patches (Figure 2c). On semisolid MS medium supplemented with 7.5 mg /L IAA in combination with 2.0 mg/L BA, the obtained callus was light green color with an anthocyanin coloration, of a dense type, and formed many meristematic foci over the entire surface of the callus (Figure 2d). However, the formed adventive shoots had a lanceolate leaf blade, which is atypical for plantlets with normal morphology.

When callus was grown under the conditions of the light regime FR < R semisolid MS medium supplemented with 7.5 mg /L NAA in combination with 2.0 mg/L BA, a loose callus type, poorly hydrated, and easily disintegrating into individual cell aggregates, was formed. In this case, the obtained callus was light yellow or white without anthocyanin patches (Figure 2e). On semisolid MS medium supplemented with 7.5 mg /L IAA in combination with 2.0 mg/L BA, callus of a bright green color with an anthocyanin coloration, of a medium density type, and formed many meristematic foci over the entire surface of the callus (Figure 2f). However, the adventive shoots formed were hyperhydrated.

3.3. Influence of the PGR Composition of the Nutrient Medium on the Accumulation of Inulin in Chicory Callus

The results indicated that the ability of callus cells to accumulate inulin depends on the type of auxins used. The presence of 2.0 mg/L BA in combination with 7.5 mg/L IAA in the nutrient medium led to the accumulation of inulin in the callus five times higher than 2.0 mg/L BA in combination with 7.5 mg/L NAA (Table 2). The ratio was consistent on both the fourth and fifth subcultures. The increased content of inulin in callus obtained on semisolid MS medium supplemented with 2.0 mg/L BA in combination with 7.5 mg/L IAA can be explained by the appearance of meristematic foci, in contrast to the medium with NAA, on which non-morphogenic callus was formed.

Table 2. Effect of different auxins (2.0 mg/L BA in combination with 7.5 mg/L NAA or 7.5 mg/L IAA) on the inulin content (%) in chicory callus.

Auxin	Extraction Medium	Mean Optical Density	Fructose-Containing Sugars, %	Inulin Content, %
		4th subculture		
NAA	Alcohol	0.0387	4.2928	1.15 a
	Water	0.0491	5.4464	
IAA	Alcohol	0.0435	4.8215	5.04 b
	Water	0.0889	9.8612	
		5th subculture		
NAA	Alcohol	0.0407	4.3788	1.23 a
	Water	0.0463	5.6002	
IAA	Alcohol	0.0465	4.8441	5.27 b
	Water	0.0859	9.6449	

Means ± SE; At each subculture, means followed by a different letter are significantly different at an alpha level of 0.05 according to the Fisher's Least Significant Difference (LSD) test. Values of inulin content were arcsin $\sqrt{X}$ transformed prior to statistical analysis.

3.4. Influence of Light Quality and PGR Composition on the Accumulation of Inulin in Chicory Callus

With regard to the influence of the spectral composition of light on the accumulation of inulin in chicory callus cells, there was a clear dependence of the accumulation of inulin on the light quality. On semisolid MS medium supplemented with 7.5 mg /L IAA in combination with 2.0 mg/L BA, the responsiveness of the callus to changes in light treatment was observed. With an increase in PPFD by a mean of 100 μmol/m^{2}s, all treatments (PPFD = 142 ± 10 μmol/m^{2}s) differed from the control (fluorescent lamp, PPFD = 40 μmol/m^{2}s) by a mean of 28%. The FR = R and FR < R treatments did not differ from each other during all subcultures. There was no difference between FR > R and FR<R as well. The effects of both the 4thand 5th subcultures were the same (Figure 3).

During the cultivation on semisolid MS medium supplemented with 7.5 mg/L NAA in combination with 2.0 mg/L BA, because the obtained callus was non-morphogenic, their weak responsiveness to lighting was observed in all treatments and control. At the fourth subculture, the R = FR and R < FR treatments were not significantly different from each other but different from the control and the FR > R treatment (Figure 3a). There was a significant difference in all treatments with control at the fifth subculture (Figure 3b). There were no differences between FR = R, FR < R. The treatment with FR > R at 5th subculture had the highest value of inulin content by a mean of 35% higher than the control (fluorescent lamp).

The high ability of callus cells to synthesize and accumulate inulin was obtained by growing cells on a nutrient medium containing IAA. Probably, the high biosynthetic potential of cells for inulin synthesis is due to the fact that it was under these conditions that a well-proliferating and highly morphogenic callus was formed.

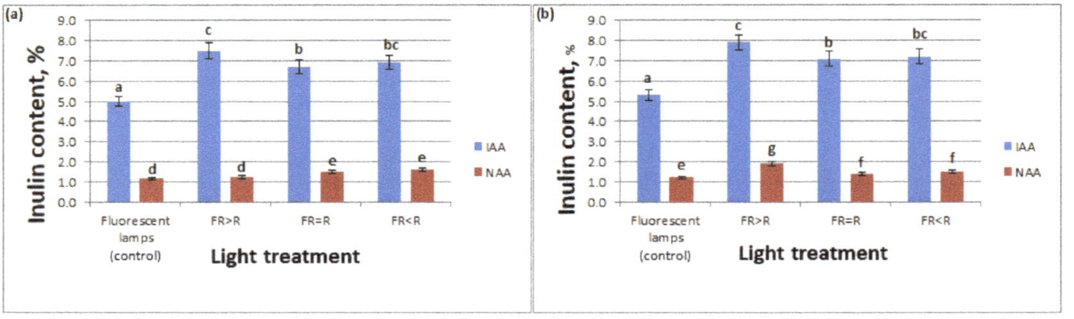

Figure 3. Combined effect of light quality (FR > R, FR = R, FR < R) and plant growth regulators (2.0 mg /L BA in combination with 7.5 mg/L NAA) on inulin content of chicory callus at: (**a**) Fourth subculture; (**b**) Fifth subculture. Means (percentage of inulin content) followed by a different letter are significantly different at an alpha level of 0.05 according to the Fisher's Least Significant Difference (LSD) test. For each treatment n = 150.

4. Discussion

Currently, the most promising area for the development of the food industry is the production of functional and dietary food products. These products include products containing dietary fiber, antioxidants, prebiotics, etc. One of the effective prebiotics is inulin, which is found in chicory, which determines the importance of this crop for the food and pharmaceutical industries.

The use of biotechnology methods not only allows to multiply and obtain high-quality planting material but also creates in vitro cell cultures of medicinal plants with an increased content of biologically active substances [27,28]. The biosynthetic potential of cultured cells in vitro depends on various factors, which include, in particular: the mineral and hormonal composition of the nutrient medium as well as the use of various elicitors (physical and chemical nature) and illumination conditions (light intensity, spectral composition of light). A change in each of these factors can lead to a change in primary and secondary metabolism, and the quality of the obtained medicinal plant raw materials also depends on illumination conditions [38–45].

In most of the reported studies, leaf explants were used most in micropropagation through callus cultures, and combination of BA with IAA or NAA at different concentrations was most suitable for the initial induction and proliferation of chicory callus [31–33,35]. In our studies, the most optimal conditions for callogenesis were the presence of IAA or NAA in combination with 2.0 mg/L BA in the nutrient medium at a concentration of 7.5 and 8.5 mg/L. The use of 2,4-D in any of the treatments did not lead to the formation of a well-proliferating callus. In all treatments, the formation of weakly growing, non-viable tissue was observed, the death of which was observed already in the middle of the first subculture. Cultivation of callus cells on a nutrient medium supplemented with IAA or NAA led to different responses to morphogenesis. Differences in the effect of various auxins on the content of inulin, apparently, were due to the nature of the auxin. IAA is a natural auxin and NAA is a synthetic. IAA is preferred for the formation of morphogenic callus. When using NAA, the formation of a non-morphogenic callus was observed. In our study on medium with IAA, where the morphogenic callus formed, a greater amount of inulin was observed (Table 2). According to other researchers, the content of inulin in callus obtained on a medium with NAA is relatively low. However, the content of inulin in plant leaves in vitro can reach up to 5% dry weight [49]. In our case, it can be assumed that the high content of inulin in the callus on the medium with IAA is due to the formation of meristematic foci over the entire surface of the callus. Its further increase can be controlled by changing the spectral composition and quality of light. During cultivation on a medium supplemented with NAA, the callus was non-morphogenic; therefore, a weak responsiveness to lighting conditions was observed.

It is well known that one of the important factors necessary for the growth, development and productivity of plants is the intensity and spectral composition of light. In conditions of insufficient supply of sunlight, the process of photosynthesis is disrupted, the growth, development, productivity, and resistance of plants are reduced [50–52].

In the literature, there are numerous works devoted to the study of the intensity and spectral composition of light for growth and development, photosynthesis and plant productivity [38–45]. Light also acts as an effective regulator that controls plant morphogenesis during their development in vitro [38–40]. Various authors have found that cultivation of cell cultures under conditions with the use of red light leads to the stimulation of growth, both of the aerial part and of the roots, in comparison with cultivation in white and blue light [53,54]. In addition, it has been shown that red light enhances the synthesis of carbohydrates in the leaves, while blue light enhances the synthesis of proteins [40,53]. Our studies have shown that the use of different lighting conditions (FR > R, FR = R, FR < R) have a significantly stimulating effect on the accumulation of inulin in callus. Moreover, the maximum value of inulin in callus (7.55–7.95%) was obtained in the treatment 7.5 mg/L IAA with FR > R. According to the data obtained, the best combinations of studied factors for callus growth were 7.5 mg/L NAA with FR > R and 7.5 mg/L IAA with FR < R.

5. Conclusions

The cultivation of callus cells is one of the uses of the PGRs. The nature and concentration of PGRs in culture media has an effect on morphogenesis and secondary metabolite biosynthesis. The studies performed allowed us to conclude that in order to obtain a well-proliferating, non-morphogenic callus, the presence of NAA in the nutrient medium is necessary, and IAA to obtain regenerated plants from callus. Such plants can be initial material for the selection of new forms of chicory.

The light treatment had a significant effect on the inulin content in callus. The FR > R treatment with PPFD = 142 ($\pm$10) μmol/m^{2}s increases the content of inulin in callus cells. On the other hand, the ability of callus cells to accumulate inulin also depends on the type of auxin used. The increased content of inulin in callus obtained in the medium with IAA can be explained by the appearance of meristematic foci, in contrast to the medium with NAA, on which non-morphogenic callus was formed.

Author Contributions: Conceptualization, R.N.K. and N.N.S.; methodology, R.N.K., A.A.P., M.G.P., I.S.D., A.V.S. and Q.V.K.; software, Q.V.K.; validation, R.N.K. and A.V.S.; formal analysis, R.N.K., A.A.P., M.G.P., I.S.D. and A.V.S.; investigation, R.N.K. and A.V.S.; resources, R.N.K. and A.V.S.; data curation, Q.V.K.; writing—original draft preparation, R.N.K.; writing—review and editing, R.N.K., A.V.S. and N.N.S.; visualization, R.N.K., A.V.S. and N.N.S.; supervision, R.N.K. and A.V.S.; project administration, R.N.K. and N.N.S.; funding acquisition, R.N.K. All authors have read and agreed to the published version of the manuscript.

Funding: The article was made with support of the Ministry of Science and Higher Education of the Russian Federation in accordance with agreement № 075-15-2020-905 date 16 November 2020 on providing a grant in the form of subsidies from the Federal budget of Russian Federation. The grant was provided for state support for the creation and development of a World-class Scientific Center "Agrotechnologies for the Future".

Institutional Review Board Statement: Not applicable.

Informed Consent Statement: Not applicable.

Data Availability Statement: Not applicable.

Conflicts of Interest: The authors declare no conflict of interest.

References

1. McClelland, J.W.; Allen, J.C.; Zakir, S. Bio-medicinal effect of sweet potato in people with diabetes. *J. Am. Diet. Assoc.* **2007**, *8*, A104. [CrossRef]
2. Ikanone, C.E.O.; Oyekan, P.O. Effect of boiling and frying on the total carbohydrate, vitamin c and mineral contents of Irish (*Solanun tuberosum*) and sweet (*Ipomea batatas*) potato tubers. *Niger. Food J.* **2014**, *32*, 33–39. [CrossRef]
3. Mohammad, K.A. A comprehensive review of sweet potato (*Ipomoea batatas* Lam): Revisiting the associated health benefits. *Trends Food Sci. Technol.* **2021**, *115*, 512–529.
4. World Health Organization. *WHO Traditional Medicine Strategy: 2014–2023*; World Health Organization: Geneva, Switzerland, 2013.
5. Tutelyan, V.A.; Sukhanov, B.P.; Kochetkova, A.A.; Sheveleva, S.A.; Smirnova, E.A. Russian Regulations on Nutraceuticals, Functional Foods, and Foods for Special Dietary Uses. In *Nutraceutical and Functional Food Regulations in the United States and around the World*; Academic Press: Cambridge, MA, USA, 2019; pp. 399–416.
6. Marcel, B.R. Inulin-type fructans: Functional food ingredients. *J. Nutr.* **2007**, *137*, 2493S–2502S.
7. Barclay, T.; Ginic-Markovic, M.; Cooper, P.; Petrovsky, N. Inulin—A versatile polysaccharide with multiple pharmaceutical and food chemical uses. *J. Excip. Food Chem.* **2016**, *1*, 1132.
8. Kathy, R.N. Inulin and oligofructose: What are they? *J. Nutr.* **1999**, *129*, 1402S–1406S.
9. Kalyani, N.K.; Kharb, S.; Thompkinson, D.K. Inulin dietary fiber with functional and health attributes—A review. *Food Rev. Int.* **2010**, *26*, 189–203. [CrossRef]
10. Boeckner, L.S.; Marilynn, I.S.; Bryan, C.T. Inulin: A review of nutritional and health implications. *Adv. Food Nutr. Res.* **2001**, *43*, 1–63.
11. Bais, H.P.; Ravishankar, G.A. *Cichorium intybus* L. cultivation, processing, utility, value addition and biotechnology, with an emphasis on current status and future prospects. *J. Sci. Food Agric.* **2001**, *81*, 467–484. [CrossRef]
12. Abbas, Z.K.; Saggu, S.; Sakeran, M.I.; Zidan, N.; Rehman, H.; Ansari, A.A. Phytochemical, antioxidant and mineral composition of hydroalcoholic extract of chicory (*Cichorium intybus* L.) leaves. *Saudi J. Biol. Sci.* **2015**, *22*, 322–326. [CrossRef]
13. Molan, A.L.; Duncan, A.J.; Barryand, T.N.; McNabb, W.C. Effect of condensed tannins and sesquiterpene lactones extracted from chicory on the motility of larvae of deer lungworm and gastrointestinal nematodes. *Parasitol. Int.* **2003**, *52*, 209–218. [CrossRef]
14. Nandagopal, S.; Ranjitha, B.D. Phytochemical and antibacterial studies of chicory (*Cichorium intybus* L.)—A multipurpose medicinal plant. *Adv. Biol. Res.* **2007**, *1*, 17–21.
15. Muthusamy, V.S.; Anand, S.; Sangeetha, K.N.; Sujatha, S.; Arun, B.; Lakshami, B.S. Tannins present in Cichorium intybus enhance glucose uptake and inhibit adipogenesis in 3T3-L1 adipocytes through PTP1B inhibition. *Chem. Biol. Interact.* **2008**, *174*, 69–78. [CrossRef]
16. Atta, A.H.; Elkoly, T.A.; Mouneir, S.M.; Kamel, G.; Alwabel, N.A.; Zaher, S. Hepatoprotective effect of methanolic extracts of *Zingiber officinale* and *Cichorium intybus*. *Indian J. Pharm. Sci.* **2010**, *72*, 564–570. [CrossRef]
17. Meehye, K.; Shin, H.K. The water-soluble extract of chicory reduces glucose uptake from the perfused jejunum in rats. *J. Nutr.* **1996**, *126*, 2236–2242.
18. Afzal, S.; Afzal, N.; Awan, M.R.; Khan, T.S.; Gilani, A.; Khanum, R.; Tariq, S. Ethno-botanical studies from Northern Pakistan. *J. Ayub Med. Coll. Abbottabad* **2009**, *21*, 52–57.
19. Abbasi, A.M.; Khan, M.A.; Ahmad, M.; Zafar, M.; Khan, H.; Muhammad, N.; Sultana, S. Medicinal plants used for the treatment of jaundice and hepatitis based on socio-economic documentation. *Afr. J. Biotechnol.* **2009**, *8*, 1643–1650.
20. Jamshidzadeha, A.; Khoshnooda, J.M.; Dehghanib, Z.; Niknaha, H. Hepatoprotective activity of *Cichorium intybus* L. leaves extract against carbon tetrachloride induced toxicity. *Iran. J. Pharm. Res.* **2006**, *1*, 41–46.
21. Hassan, H.A. The prophylactic role of some edible wild plants against nitrosamine precursor's experimentally-induced testicular toxicity in male albino rats. *J. Egypt. Soc. Toxicol.* **2008**, *38*, 1–11.
22. Nayeemunnisa, A. Alloxan diabetes-induced oxidative stress and impairment of oxidative defense system in rat brain: Neuroprotective effects of *Cichorium intybus* L. *Int. J. Diabetes Metabol.* **2009**, *17*, 105–109.
23. Mulabagal, V.; Wang, H.; Ngouajio, M.; Nair, M.G. Characterization and quantification of health beneficial anthocyanins in leaf chicory (*Cichorium intybus*) varieties. *Eur. Food Res. Technol.* **2009**, *230*, 47–53. [CrossRef]
24. Hassan, H.A.; Yousef, M.I. Ameliorating effect of chicory (*Cichorium intybus* L.)-supplemented diet against nitrosamine precursors-induced liver injury and oxidative stress in male rats. *Food Chem. Toxicol.* **2010**, *48*, 2163–2169. [CrossRef] [PubMed]
25. Taylor, R.L. *Weeds of Roadsides and Waste Ground in New Zealand*; The Caxton Press: Christchurch, New Zealand, 1981.
26. Arya, P.S.; Saini, S.S. Kalpa Sel 1′chicory. An ideal flavouring agent for coffee. *Indian Hortic.* **1984**, *18*, 55–56.
27. Parsons, J.L.; Cameron, S.I.; Cory, S.; Harris, C.S.; Smith, M.L. Echinacea biotechnology: Advances, commercialization and future considerations. *Pharm. Biol.* **2018**, *56*, 485–494. [CrossRef] [PubMed]
28. Poothong, S.; Morré, J.; Maier, C.S.; Reed, B.M. Metabolic changes and improved growth in micropropagated red raspberry "Indian summer" are tied to improved mineral nutrition. *In Vitro Cell. Dev. Biol. Plant.* **2017**, *53*, 579–590. [CrossRef]
29. Toponi, M. Action combining kinetin and acid Indole acetic on the neoformation of organs by fragments of leaves of endive (*Cichorium intybus* L.) grown in vitro. *C. R. Acad. Sci. Paris* **1963**, *257*, 3030–3033.

30. Park, E.J.; Lim, H.T. Establishment of an efficient in vitro plant regeneration system in Chicory (*Cichorium intybus* L). *Acta Hortic.* **1999**, *483*, 367–370. [CrossRef]
31. Velayutham, P.; Kumari, B.D.; Baskaran, P. An efficient in vitro plant regeneration system for *Cichorium intybus* L.—An important medicinal plant. *J. Agric. Technol.* **2006**, *2*, 287–298.
32. Rehman, R.U.; Israr, M.; Srivastava, P.S.; Bansal, K.C.; Abdin, M.Z. In vitro regeneration of witloof chicory (*Cichorium intybus* L.) from leaf explants and accumulation of esculin. *In Vitro Cell. Dev. Biol.* **2003**, *39*, 142–146. [CrossRef]
33. Yucesan, B.; Turker, A.U.; Gurel, E. TDZ-induced high frequency plant regeneration through multiple shoot formation in witloof chicory (*Cichorium intybus* L.). *Plant Cell Tissue Organ Cult.* **2007**, *91*, 243–250. [CrossRef]
34. Rafsanjani, S.M.O.; Alvari, A.; Mohammad, A.; Abdin, M.; Hejazi, M. In vitro propagation of *Cichorium intybus* L. and quantification of enhanced secondary metabolite (esculin). *Recent Pat. Biotechnol.* **2011**, *5*, 227–234. [CrossRef] [PubMed]
35. Dakshayini, K.; Rao, C.V.; Karun, A.; Bhavyashree, U.; Ujwal, P. High-frequency plant regeneration and histological analysis of callus in Cichorium intybus: An important medicinal plant. *J. Phytol.* **2016**, *8*, 7–12. [CrossRef]
36. Wagner, G.M.; Eneva, T. Positive effect of cefotaxime on plant regeneration from *Cichorium intybus* L. leaf material. *Landbauforsch Voelkenrode* **1996**, *46*, 166–168.
37. Cadalen, T.; Morchen, M.; Blassiau, C. Development of SSR markers and construction of a consensus genetic map for chicory (*Cichorium intybus* L). *Mol. Breed.* **2010**, *25*, 699–722. [CrossRef]
38. Tariq, U.; Ali, M.; Abbasi, B.H. Morphogenic and biochemical variations under different spectral lights in callus cultures of *Artemisia absinthium* L. *J. Photochem. Photobiol. B Biol.* **2014**, *130*, 264–271. [CrossRef]
39. Adil, M.; Ren, X.; Jeong, B.R. Light elicited growth, antioxidant enzymes activities and production of medicinal compounds in callus culture of *Cnidium officinale* Makino. *J. Photochem. Photobiol. B Biol.* **2019**, *196*, e111509. [CrossRef]
40. Shulgina, A.A.; Kalashnikova, E.A.; Tarakanov, I.G.; Kirakosyan, R.N.; Cherednichenko, M.Y.; Polivanova, O.B.; Baranova, E.N.; Khaliluev, M.R. Influence of light conditions and medium composition on morphophysiological characteristics of Stevia rebaudiana Bertoni in vitro and in vivo. *Horticulturae* **2021**, *7*, 195. [CrossRef]
41. Shohael, A.; Ali, M.; Yu, K.; Hahn, E.; Islam, R.; Paek, K. Effect of light on oxidative stress, secondary metabolites and induction of antioxidant enzymes in *Eleutherococcus senticosus* somatic embryos in bioreactor. *Process Biochem.* **2006**, *41*, 1179–1185. [CrossRef]
42. Liu, C.; Zhao, Y.; Wang, Y. Artemisinin: Current state and perspectives for biotechnological production of an antimalarial drug. *Appl. Microbiol. Biotechnol.* **2006**, *72*, 11–20. [CrossRef]
43. Zhong, J.; Seki, T.; Kinoshita, S.; Yoshida, T. Effect of light irradiation on anthocyanin production by suspended culture of *Perilla frutescens*. *Biotechnol. Bioeng.* **1991**, *38*, 653–658. [CrossRef]
44. Kreuzaler, F.; Hahlbrock, K. Flavonoid glycosides from illuminated cell suspension cultures of *Petroselinum hortense*. *Phytochemistry* **1973**, *12*, 1149–1152. [CrossRef]
45. Liu, C.; Guo, C.; Wang, Y.; Ouyang, F. Effect of light irradiation on hairy root growth and artemisinin biosynthesis of *Artemisia annua* L. *Process Biochem.* **2002**, *38*, 581–585. [CrossRef]
46. Murashige, T.; Skoog, F. A revised medium for rapid growth and bioassays with tobacco tissue cultures. *Physiol. Plant.* **1962**, *15*, 473–497. [CrossRef]
47. Kasyan, I.; Kasyan, A.K. Optimisatsiya spektrofotometrichescogo sposoba opredeleniya inulina v klubnyah topinambura (*Helianthus tuberosus* L.). *Likars'ke Rosl. Dosvidu Minulogo Novitnih Tekhnol.* **2019**, 121–124. (In Russian) [CrossRef]
48. Pencheva, D.; Petkova, N.; Denev, P. Determination of inulin in dough products. *Sci. Work. UFT* **2012**, *59*, 339–344.
49. Kumari, B.D.R.; Velautham, P.; Anitha, S. A comparative study on inulin and esculin content of in vitro and in vivo plants of chicory (*Cichorium intybus* L. Cv Lucknow local). *Adv. Biol. Res.* **2007**, *1*, 22–25.
50. Clapa, D.; Fira, A.; Joshee, N. An efficient ex vitro rooting and acclimatization method for horticultural plants using float hydroculture. *Hortscience* **2013**, *48*, 1159–1167. [CrossRef]
51. Wozny, A.; Miler, N. LEDs application in *ex vitro* rooting and acclimatization of chrysanthemum (*Chrysanthemum* x *grandiflorum/Ramat./Kitam*). *Electron. J. Pol. Agric. Univ.* **2016**, *19*, 1–8.
52. Velayutham, P.; Kumari, B.D.R. Influence of photoperiod on in vitro flowering in *Cichorium intybus* L. *Indian J. Plant Physiol.* **2003**, *218*, 90–93.
53. Khurshid, R.; Ullah, M.A.; Tungmunnithum, D.; Drouet, S.; Shah, M.; Zaeem, A.; Abbasi, B.H. Lights triggered differential accumulation of antioxidant and antidiabetic secondary metabolites in callus culture of *Eclipta alba* L. *PLoS ONE* **2020**, *15*, e0233963. [CrossRef]
54. Hung, C.D.; Hong, C.H.; Kim, S.K.; Lee, K.H.; Park, J.Y.; Nam, M.W.; Lee, H.I. LED light for in vitro and ex vitro efficient growth of economically important highbush blueberry (*Vaccinium corymbosum* L.). *Acta Physiol. Plant.* **2016**, *38*, 1–9. [CrossRef]

life

MDPI

Article

Genomic Analysis of *LEA* Genes in *Carica papaya* and Insight into Lineage-Specific Family Evolution in Brassicales

Zhi Zou *, Jingyuan Guo, Yujiao Zheng, Yanhua Xiao and Anping Guo *

Hainan Key Laboratory for Biosafety Monitoring and Molecular Breeding in Off-Season Reproduction Regions, Institute of Tropical Biosciences and Biotechnology, Sanya Research Institute of Chinese Academy of Tropical Agricultural Sciences, Haikou 571101, China
* Correspondence: zouzhi@itbb.org.cn (Z.Z.); guoanping@itbb.org.cn (A.G.)

Abstract: Late embryogenesis abundant (LEA) proteins comprise a diverse superfamily involved in plant development and stress responses. This study presents a first genome-wide analysis of *LEA* genes in papaya (*Carica papaya* L., Caricaceae), an economically important tree fruit crop widely cultivated in the tropics and subtropics. A total of 28 members were identified from the papaya genome, which belong to eight families with defined Pfam domains, i.e., LEA_1 (3), LEA_2 (4), LEA_3 (5), LEA_4 (5), LEA_5 (2), LEA_6 (2), DHN (4), and SMP (3). The family numbers are comparable to those present in *Ricinus communis* (Euphorbiaceae, 28) and *Moringa oleifera* (Moringaceae, 29), but relatively less than that found in *Moringa oleifera* (Cleomaceae, 39) and *Arabidopsis thaliana* (Brassicaceae, 51), implying lineage-specific evolution in Brassicales. Indeed, best-reciprocal-hit-based sequence comparison and synteny analysis revealed the presence of 29 orthogroups, and significant gene expansion in *Tarenaya* and *Arabidopsis* was mainly contributed by whole-genome duplications that occurred sometime after their split with the papaya. Though a role of transposed duplication was also observed, tandem duplication was shown to be a key contributor in gene expansion of most species examined. Further comparative analyses of exon-intron structures and protein motifs supported fast evolution of this special superfamily, especially in *Arabidopsis*. Transcriptional profiling revealed diverse expression patterns of *CpLEA* genes over various tissues and different stages of developmental fruit. Moreover, the transcript level of most genes appeared to be significantly regulated by drought, cold, and salt stresses, corresponding to the presence of *cis*-acting elements associated with stress response in their promoter regions. These findings not only improve our knowledge on lineage-specific family evolution in Brassicales, but also provide valuable information for further functional analysis of *LEA* genes in papaya.

Keywords: papaya (*Carica papaya*); brassicales; late embryogenesis abundant protein; orthogroup; abiotic stress; expression profile

Citation: Zou, Z.; Guo, J.; Zheng, Y.; Xiao, Y.; Guo, A. Genomic Analysis of *LEA* Genes in *Carica papaya* and Insight into Lineage-Specific Family Evolution in Brassicales. *Life* **2022**, *12*, 1453. https://doi.org/10.3390/life12091453

Academic Editors: Hakim Manghwar and Wajid Zaman

Received: 1 August 2022
Accepted: 15 September 2022
Published: 19 September 2022

Publisher's Note: MDPI stays neutral with regard to jurisdictional claims in published maps and institutional affiliations.

Copyright: © 2022 by the authors. Licensee MDPI, Basel, Switzerland. This article is an open access article distributed under the terms and conditions of the Creative Commons Attribution (CC BY) license (https://creativecommons.org/licenses/by/4.0/).

1. Introduction

Late embryogenesis abundant (LEA) proteins comprise a large and diverse superfamily that is widely involved in plant development as well as stress responses [1–3]. Since their first discovery as accumulating late in cotton (*Gossypium hirsutum*) embryogenesis [4–6], over the past four decades, LEA proteins have been found in a wide range of plants as well as bacteria, fungi, and animals [1,7]. According to sequence similarity and particular Pfam domains present, LEAs can be classified into eight main families, i.e., LEA_1 (Pfam accession number PF03760), LEA_2 (PF03168), LEA_3 (PF03242), LEA_4 (PF02987), LEA_5 (PF00477), LEA_6 (PF10714), DHN (dehydrin, PF00257), and SMP (seed maturation protein, PF04927) [3,8,9]. In the model plant arabidopsis (*Arabidopsis thaliana*), the presence of 51 LEA-encoding genes was reported, whereby two members (i.e., AtEM10 and AtEM17) comprise one more family named AtM without significant protein domains [2,10]. Generally, LEA proteins are extremely hydrophilic; however, some members in the LEA_2 family were shown to be hydrophobic and even have a three-dimensional structure [11]. Increasing evidence shows that the accumulation of LEA proteins is not only found in seeds, but also different vegetative tissues especially under stress conditions, e.g., high temperature, low temperature, drought, and salt [2,3,12,13]. Moreover, improved stress tolerance was also observed after overexpressing *LEA* genes in *Escherichia coli*, yeast (*Saccharomyces cerevisiae*), and several model plants such as tobacco (*Nicotiana tabacum*), arabidopsis, and rice (*Oryza sativa*) [14–17]. Although the exact mechanism has not been clarified, LEA proteins are able to stabilize other proteins and membrane structures during water stress [16,18].

Papaya (*Carica papaya* L., $2n$ = 18) is an important tree fruit crop that belongs to the Caricaceae family within the order Brassicales, which also includes arabidopsis as a representative in Brassicaceae, spider flower (*Tarenaya hassleriana*) in Cleomaceae, and horseradish tree (*Moringa oleifera*) in Moringaceae. Compared with the occurrence of two recent whole-genome duplications (WGDs) in both spider flower and arabidopsis, papaya and horseradish tree did not experience any additional WGD after the ancient so-called γ WGD shared by all core eudicots [19–22]. Although originated in Central America, the high nutritional value with significant vitamins and minerals in papaya fruits has prompted its wide cultivation in tropics and subtropics, e.g., India, Nigeria, Brazil, Mexico, Indonesia, and China [23]. In contrast to the considerable drought tolerance of wild relatives, commercial papaya cultivars are highly susceptible to cold and drought stresses [24,25], which frequently occur in subtropical regions such as south China. Therefore, exploring genes involved in stress responses and breeding resistant varieties in these areas are of particular importance. By taking advantage of available genome and transcriptome datasets, in this study, we would like to report a genome-wide analysis of *LEA* genes in papaya, which includes gene locations, exon-intron structures, sequence characteristics, evolutionary relationships, and *cis*-acting elements in the promoter regions, as well as gene expression patterns with a focus on fruit development and stress responses. These findings provide a global view of *CpLEA* genes that can facilitate further functional studies, and the comparative analysis with arabidopsis, spider flower, horseradish tree, and castor bean (*Ricinus communis*) contributes to our knowledge on the lineage-specific evolution of this special superfamily in Brassicales.

2. Materials and Methods

2.1. Data Retrieval and Identification of LEA Genes in Papaya, Horseradish Tree, and Spider Flower

LEA genes reported in arabidopsis and castor bean (see Table S1) were retrieved from Araport11 (https://www.arabidopsis.org/, accessed on 18 June 2022) and Phytozome v13 (https://phytozome-next.jgi.doe.gov/, accessed on 18 June 2022), respectively. Their protein sequences were used to identify homologs from papaya, horseradish tree, and spider flower, whose genome sequences were accessed from Phytozome v13, NCBI (http://www.ncbi.nlm.nih.gov/, accessed on 18 June 2022), and NGDC (https://ngdc.cncb.ac.cn/, accessed on 18 June 2022). The *E*-value of the tBLASTn search [26] was set to 1×10^{-5}, and gene models of candidates were curated with available mRNAs as described before [27]. The presence of certain Pfam domains was confirmed using MOTIF Search (https://www.genome.jp/tools/motif/, accessed on 18 June 2022). Systematic names were assigned with two italic letters denoting the source organism and family name followed by a progressive number of their locations on chromosomes (Chrs) or scaffolds (Scfs).

2.2. Synteny Analysis and Gene Expansion Patterns

Homolog pairs were identified using the all-to-all BLASTP method (*E*-value cutoff 1×10^{-10}) and syntenic blocks were inferred using MCScanX (BLAST hits $\geq$ 5) [26,28]. Tandem repeats were defined when two paralogs were consecutive in a genome; WGD repeats were considered when duplicated genes were located in syntenic blocks of duplicated chromosomes, and transposed repeats were identified using the DupGen_finder pipeline as previously described [29]. Orthologs between different species were determined using the Best Reciprocal Hit (BRH) method [30], as well as information from synteny analysis; and orthogroups (OGs) were assigned only when they were present in at least two species examined.

2.3. Exon-Intron Structure, Phylogenetic Analysis, and Structural Characterization

The exon-intron structure was analyzed using GSDS 2.0 [31] by aligning the coding sequence (CDS) to the corresponding genomic sequence. The molecular weight (MW), theoretical isoelectric point (pI), and grand average of hydropathy (GRAVY) were calculated using ProtParam (http://web.expasy.org/protparam/, accessed on 18 June 2022), and protein subcellular localization was predicted using WoLF PSORT (http://www.genscript.com/wolf-psort.html, accessed on 18 June 2022). Multiple sequence alignment and phylogenetic reconstruction were performed using MEGA6 [32] with MUSCLE and the maximum likelihood method (bootstrap: 1000 replicates), respectively. Conserved motifs in LEA proteins were identified using MEME (v 5.4.1) [33]: any number of repetitions; maximum number of motifs, 20; minimum sites, 2; and, the optimum width of each motif, between 6 and 100 residues.

2.4. Promoter Analysis

PLACE (http://www.dna.affrc.go.jp/PLACE/, accessed on 18 June 2022) was used to examine the presence of two stress-related *cis*-acting elements (i.e., abscisic acid response (ABRE, ACGTG) and low temperature response (LTRE, CCGAC)) in the 2000-bp promoter region of *CpLEA* genes.

2.5. Plant Materials, RNA-seq, and Gene Expression Analysis

Gene expression profiles were analyzed on the basis of RNA sequencing (RNA-seq) samples as shown in Table S2. Various tissues, i.e., root, apical bud, leaf, petiole, leaf vein, male flower, female flower, fruit, peel, and seed, were collected from one-year-old hermaphrodite plants of the cultivar Zhongbai that were planted in 2019 at the Wenchang experimental base, Institute of Tropical Biosciences and Biotechnology, Chinese Academy of Tropical Agricultural Sciences (Wenchang, Hainan, China: 19°32′15.39″ N, 110°45′47.26″ E). Routine management was performed, and three groups of more than five trees were used. As for cold and salt stresses, eight-week-old plantlets were used and treatments of 4 °C low temperature (i.e., 0, 7, 21, and 40 h) and 300 mmol/L NaCl (i.e., 0, 10, 15, and 20 d) were applied. To ensure the consistency of materials, only the second leaf from the top of a plantlet was collected and at least 10 leaves were pooled for total RNA isolation and subsequent Illumina RNA-seq as previously described [34,35]. As for drought stress, watering was withheld from three-month-old plants for 0, 10, and 20 d; and samples of roots, leaves, and phloem sap were sequenced as previously described [36]. Quality control and read mapping were carried out using Trimmomatic [37] and TopHat (v2.0.8) [38], respectively. The gene expression level was represented using FKPM (fragments per kilobase of exon per million fragments mapped) [39], and differentially expressed genes were determined using RSEM (v1.2.27) [40] with default parameters.

3. Results

3.1. Identification, Chromosome Localization, and Synteny Analysis of 28 LEA Genes in Papaya

Thus far, three genome assemblies have been reported in papaya, i.e., two for a virus-resistant transgenic variety SunUp, and one for its progenitor Sunset [20,41]. Whereas the ASGPBv0.4 assembly of SunUp is fragmented in 17,766 scaffolds [20], two recently available assemblies for SunUp and Sunset are chromosomal-level genomes [41], providing a good chance for comparative genomics analysis. Since the *LEA* genes identified in two chromosomal-level genomes are exactly the same, only results from the Sunset genome, as well as the ASGPBv0.4 assembly, were presented in Table 1, where an ortholog (i.e., sunset04G0003920/evm.TU.supercontig_6.122) of *AtLEA13/-43* was not included due to the absence of a significant LEA_4 domain. Based on the presence of Pfam domains in deduced proteins, 28 identified *CpLEA* genes were assigned into eight out of nine families as described in arabidopsis (only excluding the AtM family), and each family contains two to five members, respectively, i.e., *CpLEA1-1* to *-3*, *CpLEA2-1* to *-4*, *CpLEA3-1* to *-5*, *CpLEA4-1* to *-5*, *CpLEA5-1* to *-2*, *CpLEA6-1* to *-2*, *CpDHN1* to *-4*, and *CpSMP1* to *-3* (Table 1). Gene localization analysis indicated that they are not randomly distributed across eight out of nine chromosomes (excluding Chr9), varying from one (i.e., Chr7) to nine (i.e., Chr5) genes. Notably, several hotspots were observed, and a good example is the top of Chr5, which contains the maximum of seven genes (Figure 1). Correspondingly, eight duplicate pairs were identified, which include two tandem repeats (*CpLEA2-4/-3/-2*) and three transposed repeats (*CpLEA2-1/-4*, *CpLEA3-4/-5*, and *CpSMP1/-3*) (Table S1); on the contrary, synteny analysis revealed that the other three duplicate pairs are located in syntenic blocks and thus were defined as WGD repeats, i.e., *CpLEA3-1/-3*, *CpLEA5-1/-2*, and *CpSMP1/-2*. Among them, *CpLEA2-4/-3/-2/-1* as well as *CpLEA3-3* are located in the top region of Chr5, though *CpLEA3-4* is located in the bottom region (Figure 1). Whereas the protein identity between tandem repeats CpLEA2-3 and CpLEA2-4 is relatively low (about 29.0%), *CpLEA2-2* and *CpLEA2-3* exhibit 51.1% and 47.2% sequence identity at the nucleotide or protein level, respectively. Moreover, the first 483-bp sequences (counting from the initiation codon) of these two genes even harbor a relatively high sequence identity of 88.4%, and the low sequence identity of the full CDS was shown to result from the divergence of 3′ sequences (Figure S1).

Table 1. *LEA* genes identified in papaya.

Family	Gene Name	Locus		AS	Deduced Protein				
		Sunset	ASGPBv0.4		AA	MW (kDa)	pI	GRAVY	Loc
LEA_1	*CpLEA1-1*	sunset05G0006380	evm.TU.supercontig_18.65	-	160	17.01	9.65	−0.755	Nucl
	CpLEA1-2	sunset05G0013060	evm.TU.supercontig_41.41	-	102	11.41	7.03	−0.908	Mito
	CpLEA1-3	sunset08G0019430	evm.TU.supercontig_85.72	Yes	158	16.07	8.83	−0.878	Mito
LEA_2	*CpLEA2-1*	sunset05G0003590	evm.TU.supercontig_9.242	Yes	316	35.12	4.69	−0.384	Cyto
	CpLEA2-2	sunset05G0009060	evm.TU.supercontig_11.66	Yes	305	34.10	5.38	−0.243	Chlo
	CpLEA2-3	sunset05G0009070	evm.TU.supercontig_11.68	Yes	185	20.23	5.65	−0.056	Chlo
	CpLEA2-4	sunset05G0009080	evm.TU.supercontig_11.69	Yes	151	16.16	4.75	0.094	Cyto
LEA_3	*CpLEA3-1*	sunset03G0023320	evm.TU.supercontig_16.192	-	103	11.21	10.07	−0.472	Chlo
	CpLEA3-2	sunset04G0017920	evm.TU.supercontig_25.184	Yes	98	10.94	9.52	−0.526	Chlo
	CpLEA3-3	sunset05G0003680	evm.TU.supercontig_9.251	Yes	99	10.61	9.89	−0.531	Cyto
	CpLEA3-4	sunset05G0018090	evm.TU.supercontig_2471.1	Yes	95	10.62	9.66	−0.997	Mito
	CpLEA3-5	sunset06G0002130	evm.TU.supercontig_200.7	-	104	11.78	9.69	−0.839	Cyto
LEA_4	*CpLEA4-1*	sunset01G0016400	evm.TU.supercontig_66.6	-	590	66.20	8.91	−0.515	Extr
	CpLEA4-2	sunset03G0025310	evm.TU.supercontig_209.19	-	581	61.45	5.20	−0.864	Nucl
	CpLEA4-3	sunset05G0000220	evm.TU.supercontig_146.20	-	193	21.63	5.21	−1.053	Extr
	CpLEA4-4	sunset07G0004690	evm.TU.supercontig_464.2	-	222	24.57	8.95	−1.333	Chlo
	CpLEA4-5	sunset08G0016230	evm.TU.supercontig_5.110	Yes	280	30.34	6.17	−1.360	Nucl
LEA_5	*CpLEA5-1*	sunset02G0011780	evm.TU.supercontig_19.160	-	89	9.64	5.51	−1.319	Cyto
	CpLEA5-2	sunset08G0009640	evm.TU.supercontig_2485.2	-	111	12.10	5.51	−1.338	Nucl
LEA_6	*CpLEA6-1*	sunset01G0017510	evm.TU.supercontig_88.61	-	97	10.42	5.56	−0.705	Nucl
	CpLEA6-2	sunset04G0003310	evm.TU.supercontig_6.54	-	78	8.77	5.22	−1.573	Nucl
DHN	*CpDHN1*	sunset01G0014930	evm.TU.supercontig_26.225	Yes	211	24.10	5.05	−1.584	Nucl
	CpDHN2	sunset04G0004410	evm.TU.supercontig_6.176	-	137	14.76	9.45	−1.222	Nucl
	CpDHN3	sunset06G0003520	evm.TU.supercontig_106.3	Yes	167	17.93	5.94	−1.265	Nucl
	CpDHN4	sunset06G0021280	evm.TU.supercontig_161.14	Yes	93	10.50	6.62	−1.984	Nucl
SMP	*CpSMP1*	sunset03G0005590	evm.TU.supercontig_58.99	-	262	26.70	4.70	−0.270	Chlo
	CpSMP2	sunset03G0027120	evm.TU.supercontig_487.3	-	267	27.97	4.56	−0.246	Cyto
	CpSMP3	sunset06G0024460	evm.TU.contig_34050.2	-	244	25.13	6.44	−0.359	Nucl

AA, Amino acid; AS, Alternative splicing; Chlo, Chloroplast; Cyto, Cytoplasmic; Extr, Extracellular; GRAVY, Grand average of hydropathicity; Mito, Mitochondria; MW, Molecular weight; Nucl, Nuclear; pI, Isoelectric point; Loc, Subcellular localization.

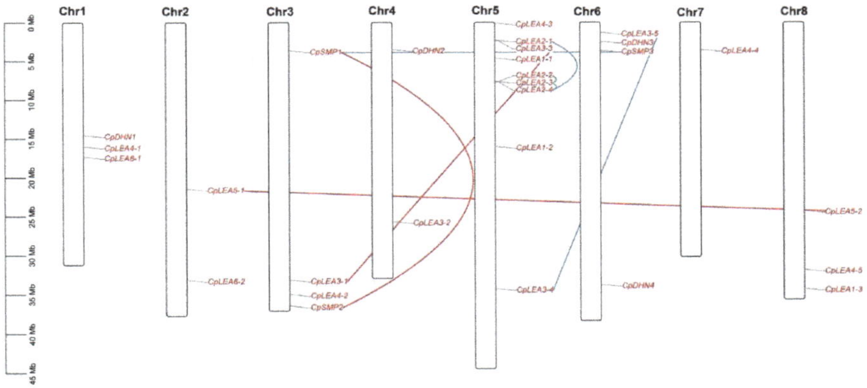

Figure 1. Chromosomal locations and duplication events of 28 *CpLEA* genes. Chromosome serial numbers are indicated at the top of each chromosome. *CpLEA2-4/-3/-2* are clustered as tandem repeats (lines in green); *CpLEA2-1/-4*, *CpLEA3-4/-5*, and *CpSMP1/3* are transposed repeats (lines in blue); and *CpLEA3-1/-3*, *CpLEA5-1/-2*, and *CpSMP1/2* are WGD repeats (lines in red) that are located in syntenic blocks.

3.2. Identification of LEA Genes in Horseradish Tree and Spider Flower and Definition of Orthogroups

The finding of almost half the amount of *LEA* genes in papaya relative to those in arabidopsis impelled us to investigate the lineage-specific evolution of the *LEA* superfamily in different families of Brassicales, i.e., Caricaceae, Moringaceae, Cleomaceae, and Brassicaceae. For this purpose, *LEA* genes were also identified from horseradish tree and spider flower, whose genome sequences have recently been accessible [21,22]. As shown in Table S1, 29 *LEA* genes identified in the horseradish tree are comparable to 28 present in papaya, as well as castor bean (an Euphorbiaceae plant also not having experienced any recent WGD), relatively less than 39 found in spider flower, and considerably less than 51 reported in arabidopsis, implying lineage-specific gene contraction and expansion. The species-specific distribution of *LEA* genes in nine defined gene families is summarized in Figure 2. Notably, no AtM homolog was found beyond arabidopsis.

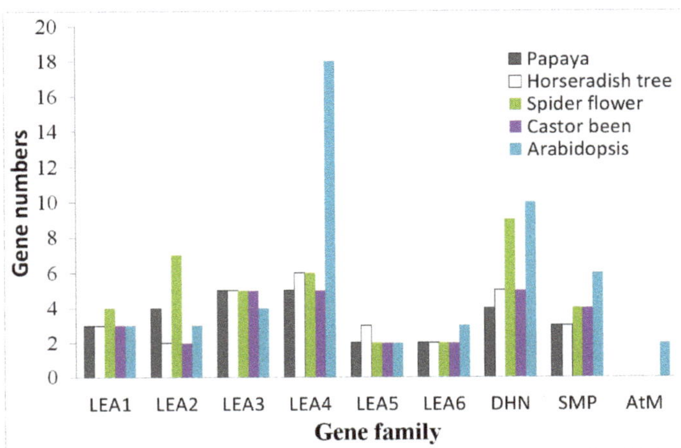

Figure 2. Distribution of papaya, horseradish tree, spider flower, arabidopsis, and castor been *LEA* genes in nine defined gene families.

To gain insights into species-specific evolution patterns, we further conducted BRH-based homology analysis between different species, resulting in 29 orthogroups that are present in more than one species compared (Table 2). In total, 28 *CpLEA* genes belong to 27 orthogroups, and each orthogroup includes one, with the exception of LEA2b containing two. As for two other orthogroups, DHNe is only present in horseradish tree and spider flower, whereas LEA4f is widely found, though a papaya homolog (see above) has lost the LEA_4 domain. Among three species without a recent WGD, i.e., papaya, horseradish tree, and castor bean, nearly one-to-one orthologous relationships were observed, though no member was identified in castor bean for LEA2b, DHNe, or LEA4a. Notably, a LEA4a homolog is actually found in castor bean, i.e., 30074.t000080; however, no significant LEA_4 domain was identified, supporting species-specific divergence. Like papaya, orthogroups that include more than one member were also found in horseradish tree and castor bean, i.e., MoLEA5-2/-3 in LEA5b, RcDHN2/-3 in DHNb, and RcSMP1/-2 in SMPb, all of which were characterized as tandem repeats (Table S1). On the contrary, orthologous relationships between papaya and spider flower/arabidopsis are relatively complex, including one-to-one, one-to-two, one-to-three, and two-to-four. In spider flower, the majority (84.6%) of duplicate pairs within an orthogroup were characterized as WGD repeats, which is relatively more than the 69.2% found in arabidopsis. Moreover, the duplication mode of the remaining duplicate pairs is also different, i.e., dispersed duplication in spider flower and tandem duplication in arabidopsis, respectively (Table S1).

Table 2. 29 Orthogroups identified in this study.

Family	Orthogroup	Papaya	Horseradish Tree	Spider Flower	Castor Been	Arabidopsis
LEA_1	LEA1a	CpLEA1-1	MoLEA1-1	ThLEA1-1 ThLEA1-2	RcLEA1-2	AtLEA6 AtLEA18
	LEA1b	CpLEA1-2	MoLEA1-2	-	RcLEA1-1	-
	LEA1c	CpLEA1-3	MoLEA1-3	ThLEA1-3 ThLEA1-4	RcLEA1-3	AtLEA46
LEA_2	LEA2a	CpLEA2-1	MoLEA2-1	ThLEA2-1 ThLEA2-2	RcLEA2-2	AtLEA26
	LEA2b	CpLEA2-2 CpLEA2-3	-	ThLEA2-3 ThLEA2-4 ThLEA2-5 ThLEA2-6	-	-
	LEA2c	CpLEA2-4	MoLEA2-2	ThLEA2-7	RcLEA2-1	AtLEA1 AtLEA27
LEA_3	LEA3a	CpLEA3-1	MoLEA3-1	ThLEA3-1	RcLEA3-5	AtLEA41
	LEA3b	CpLEA3-2	MoLEA3-2	ThLEA3-2	RcLEA3-4	AtLEA37
	LEA3c	CpLEA3-3	MoLEA3-3	ThLEA3-3 ThLEA3-4	RcLEA3-1	AtLEA2 AtLEA38
	LEA3d	CpLEA3-4	MoLEA3-4	ThLEA3-5	RcLEA3-2	-
	LEA3e	CpLEA3-5	MoLEA3-5	-	RcLEA3-3	-
LEA_4	LEA4a	CpLEA4-1	MoLEA4-1	ThLEA4-1	-	AtLEA9
	LEA4b	CpLEA4-2	MoLEA4-2	ThLEA4-2	RcLEA4-2	AtLEA25
	LEA4c	CpLEA4-3	MoLEA4-3	-	RcLEA4-4	AtLEA30
	LEA4d	CpLEA4-4	MoLEA4-4	ThLEA4-3	RcLEA4-3	AtLEA42 AtLEA48
	LEA4e	CpLEA4-5	MoLEA4-5	ThLEA4-4	RcLEA4-5	AtLEA19 AtLEA36
	LEA4f	-	MoLEA4-6	ThLEA4-5 ThLEA4-6	RcLEA4-1	AtLEA13 AtLEA43

Table 2. *Cont.*

Family	Orthogroup	Papaya	Horseradish Tree	Spider Flower	Castor Been	Arabidopsis
LEA_5	LEA5a	CpLEA5-1	MoLEA5-1	ThLEA5-1	RcLEA5-1	AtLEA20
	LEA5b	CpLEA5-2	MoLEA5-2 MoLEA5-3	ThLEA5-2	RcLEA5-2	AtLEA35
LEA_6	LEA6a	CpLEA6-1	MoLEA6-1	ThLEA6-1	RcLEA6-1	AtLEA17
	LEA6b	CpLEA6-2	MoLEA6-2	ThLEA6-2	RcLEA6-2	AtLEA15 AtLEA16
DHN	DHNa	CpDHN1	MoDHN1	ThDHN1 ThDHN2	RcDHN1	AtLEA4 AtLEA5 AtLEA10
	DHNb	CpDHN2	MoDHN2	ThDHN3	RcDHN2 RcDHN3	AtLEA33 AtLEA34 AtLEA51
	DHNc	CpDHN3	MoDHN3	ThDHN4 ThDHN5	RcDHN4	AtLEA14 AtLEA45
	DHNd	CpDHN4	MoDHN4	ThDHN6 ThDHN7 ThDHN8	RcDHN5	AtLEA8
	DHNe	-	MoDHN5	ThDHN9	-	-
SMP	SMPa	CpSMP1	MoSMP1	ThSMP1	RcSMP3	AtLEA31 AtLEA32
	SMPb	CpSMP2	MoSMP2	ThSMP2	RcSMP1 RcSMP2	AtLEA3
	SMPc	CpSMP3	MoSMP3	ThSMP3 ThSMP4	RcSMP4	AtLEA47

Compared with other species examined, 27.5% of *AtLEA* genes seem to be arabidopsis-specific. To uncover their evolution patterns in Brassicaceae, we further traced their orthologs in representative Brassicaceae plants whose genome sequences are available in Phytozome v13, i.e., *A. lyrata*, *A. halleri*, *Capsella rubella*, *C. grandiflora*, *Eutrema salsugineum*, *Brassica oleracea*, and *B. rapa*. As expected, all of them have orthologs in at least one out of seven species examined, though species-specific evolution was observed (Table S3).

3.3. Exon-Intron Structure, Phylogenetic Analysis, and Structural Characterization

To learn more about the divergence between papaya and arabidopsis, we performed phylogenetic analysis of LEA proteins according to families, and further compared their gene structures and protein motifs. As observed in arabidopsis, *CpLEA* genes feature few introns, varying from zero to two in the coding region, accounting for 14.3%, 75%, and 10.7% of total genes, respectively. Notably, an additional intron was also found in 5′ or 3′ untranslated regions (UTR) of *CpLEA2-1* and *CpDHN4*, though no intron is present in the coding region of *CpDHN4* (Figure 3). Moreover, 12 out of 25 intron-containing *CpLEA* genes appeared to have alternative splicing (AS) isoforms, and the proportion of 48% is relatively more than the 39.5% found in arabidopsis (Table S1). For convenience, the most expressed transcript was selected for further analyses. The deduced protein length of *CpLEA* genes varies from 78 to 590 amino acids (AA), and molecular weight (MW) and isoelectric point (pI) values range from 8.77 to 66.20 kDa, or from 4.56 to 10.07, respectively. Except for CpLEA2-4, the GRAVY value of other CpLEA proteins is less than 0, implying their hydrophilic feature. These proteins were predicted to target mitochondria, chloroplast, nuclear, cytoplasmic as well as extracellular genes (Table 1). A further MEME search resulted in 20 conserved motifs, which were shown to significantly distribute over different families (Figures 3 and S2).

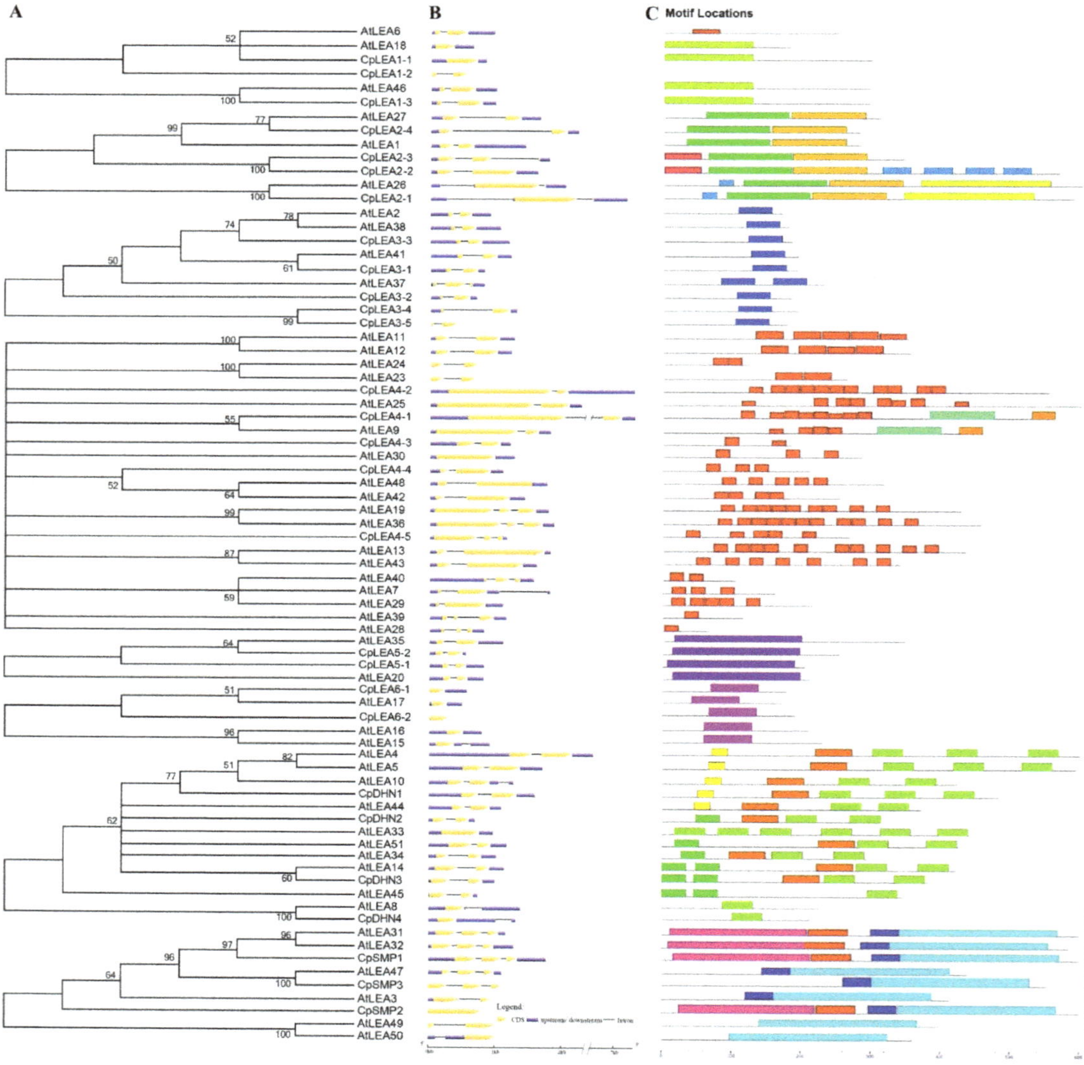

Figure 3. Phylogenetic analysis, gene structure, and motif distribution of papaya and arabidopsis *LEA* genes. (**A**) Phylogenetic analysis of eight families of Cp/AtLEA proteins; (**B**) Exon-intron structures of *Cp/AtLEA* genes; (**C**) Distribution of 20 conserved motifs. Multiple sequence alignments were conducted using MUSCLE and unrooted phylogenetic trees were constructed using MEGA6 (maximum likelihood method; bootstrap, 1000 replicates; shown are bootstrap values at nodes supported by a posterior probability of ≥50%). Motifs were identified using MEME.

3.3.1. LEA_1

The LEA_1 family is also known as D-113 [42]. In papaya, this family includes three members, which is equal to that of arabidopsis (Figure 2). However, their gene origin is not exactly the same. In fact, these genes belong to three phylogenetic groups or orthogroups, i.e., LEA1a, LEA1b, and LEA1c (Figure 2 and Table 2). Among them, *AtLEA18* was characterized as a paralog of *AtLEA6* that were resulted from the α WGD [43]. Whereas the majority of members in this family contain one intron, *CpLEA1-1* and *AtLEA18* in LEA1a are intronless (Figure 3), gene-specific loss of an intron can be speculated. Most proteins in this family were shown to harbor Motif 20, which was characterized as the LEA_1 domain. By contrast, despite the presence of the LEA_1 domain in CpLEA1-2 and AtLEA6 as supported by a MOTIF Search, no motif was detected in CpLEA1-2 due to the parameter of 20 motifs set in this study, whereas AtLEA6 was shown to harbor Motif 1, which was characterized as a LEA_4-like domain, supporting their sequence divergence (Figure 3). The length of three CpLEA1s varies from 102 to 160 AA, and the average of 140 AA is relatively longer than the 130 AA observed in arabidopsis. Correspondingly, the MW value varies from 11.41 to 17.01 kDa, and the average of 14.83 kDa is relatively larger than 13.85 kDa in arabidopsis (Table 1). Nevertheless, the pI value in two species appeared to be greater than 7.0, implying their basic feature.

3.3.2. LEA_2

This family is also known as LEA14 or D-95 [42]. The four members found in papaya are relatively more than the three present in arabidopsis (Figure 2). Similar to LEA_1, the LEA_2 family also includes three orthogroups, i.e., LEA2a, LEA2b, and LEA2c (Table 2). In contrast to *AtLEA1* and *AtLEA27* that are repeats derived from the β WGD [43], *CpLEA2-1* was characterized as a transposed repeat of *CpLEA2-4*, which also resulted in *CpLEA2-3* via tandem duplication; and *CpLEA2-2* is a more recent tandem repeat of *CpLEA2-3* (Figure 1 and Table S1). Most genes in this family harbor a single intron in the coding region; however, *CpLEA2-3* contains two instead and the gain of the second intron can be speculated. Moreover, one more intron was also observed in the 5′ UTR of both *CpLEA2-1* and *AtLEA26*, implying their early origin. All members in this family include Motif 6 and Motif 5, which were characterized as the LEA_2 or LEA_3-like domain, respectively. Moreover, both CpLEA2-1 and AtLEA26 harbor two additional motifs, i.e., Motif 13 and Motif 10, where the latter was characterized as the LEA_2 domain; both CpLEA2-2 and CpLEA2-3 include Motif 16, while CpLEA2-2 also contains eight copies of Motif 13 (Figure 3). The length of CpLEA2s varies from 151 to 316 AA, and the average of 239 AA is relatively longer than 214 AA in arabidopsis. Correspondingly, the MW value varies from 16.16 to 35.12 kDa, and the average of 26.40 kDa is relatively larger than 23.48 kDa in arabidopsis. Nevertheless, the pI value in these two species varies from 4.53 to 5.65 (Table 1), suggesting that they are acidic.

3.3.3. LEA_3

This family is also known as LEA5 or D-73 [42], and the five members present in papaya are relatively more than the four present in arabidopsis (Figure 2), which can be assigned into five orthogroups, i.e., LEA3a, LEA3b, LEA3c, LEA3d, and LEA3e (Table 1). Among them, *AtLEA38* and *AtLEA41* are repeats of *AtLEA2* and were derived from the α or γ WGD, respectively [43]; *CpLEA3-1* may also be derived from *CpLEA3-3* via the γ WGD, whereas *CpLEA3-4* was characterized as a transposed repeat of *CpLEA3-5*, which only exhibit 33.3% sequence identity at the protein level. This family features one intron; however, *AtLEA37* has gained an additional intron in the coding region. All members in this family harbor a single motif (i.e., Motif 7), which was characterized as the LEA_3 domain (Figure 3). The length of CpLEA3s varies from 95 to 104 AA, and the average of 100 AA is relatively shorter than 104 AA in arabidopsis. Correspondingly, the MW value varies from 10.61 to 11.78 kDa, and the average of 11.03 kDa is slightly smaller than

11.38 kDa in arabidopsis. The pI value in two species varies from 9.39 to 10.07 (Table 1), indicating that they are basic.

3.3.4. LEA_4

This family is also known as D-7 or D-29 [42], which contains the most number of 6 or 18 members in papaya and arabidopsis, respectively (Figure 2). This family was shown to be highly diverse, including six main orthogroups and six Brassicaceae-specific groups, i.e., LEA4a, LEA4b, LEA4c, LEA4d, LEA4e, LEA4f, AtLEA7/-29, AtLEA11/-12, AtLEA23/-24, AtLEA28, AtLEA39, and AtLEA40 (Tables 2 and S3). Among them, *AtLEA42/-48*, *AtLEA19/-36*, *AtLEA13/-43*, and *AtLEA7/-29* are duplicates that resulted from the α WGD [43], *AtLEA11/-12* and *AtLEA7/-40* are transposed repeats, and *AtLEA23/-24* are tandem repeats (Table S1). The intron number also varies from zero to two, and the copy number of the widely distributed Motif 1, which was characterized as the LEA_4 domain, varies from one to eleven. Additionally, both CpLEA4-1 and AtLEA9 harbor two more motifs, i.e., Motif 12 and Motif 18, where the former was characterized as a domain of unknown function (DUF4149, PF13664) (Figure 3). The length of CpLEA4s varies from 193 to 590 AA, and the average of 358 AA is considerably longer than 280 AA in arabidopsis. Correspondingly, the MW value varies from 23.63 to 61.45 kDa, and the average of 39.33 kDa is relatively smaller than 30.37 kDa in arabidopsis. Unlike most families, the pI value in both species is highly diverse, varying from 4.82 to 9.71 (Table 1).

3.3.5. LEA_5

This family is also known as D-19 or EM [42], which includes two members in both papaya and arabidopsis, comprising two orthogroups, i.e., LEA5a and LEA5b (Figure 2 and Table 2). Whereas CpLEA5-1 and -2 were characterized as WGD repeats, AtLEA20 and -35 are dispersed repeats (Table S1), implying possible chromosome rearrangement after papaya-arabidopsis divergence. All members in this family feature a single intron and harbor Motif 4 that was characterized as the LEA_5 domain (Figure 3). Nevertheless, the sequence length of LEA5b is relatively longer than LEA5a (i.e., 89–92 vs. 111–152) due to fragment insertion. The MW value of CpLEA5-1 and CpLEA5-2 is 9.64 or 12.10 kDa, respectively, and the average of 10.87 kDa is relatively smaller than 13.27 kDa in arabidopsis. The pI value in two species varies from 5.51 to 6.75 (Table 1), suggesting that they are acidic.

3.3.6. LEA_6

This family is also known as PvLEA18 [44], which harbors two or three members in papaya and arabidopsis, respectively (Figure 2). It is composed of two orthogroups, i.e., LEA6a and LEA6b (Table 2), where *AtLEA15* and *AtLEA16* in LEA6b are tandem repeats (Table S1). Although most genes are intronless, *AtLEA15* was shown to gain one intron in the 3′ UTR. The unique motif identified in this family (i.e., Motif 15) was characterized as the LEA_6 domain (Figure 3). CpLEA6-1 and CpLEA6-2 are 97 or 78 AA in length, respectively, and the average of 88 AA is slightly longer at 83 AA in arabidopsis, whereas the average MW value of 9.60 kDa in papaya is relatively larger than 8.71 kDa in arabidopsis. The pI value in these two species varies from 4.46 to 5.56 (Table 1), implying that they are acidic.

3.3.7. DHN

This family is also known as D-11 [42], and the 4 members found in papaya is considerably less than the 10 present in arabidopsis (Figure 2). These genes constitute five orthogroups and one Brassicaceae-specific group, i.e., DHNa, DHNb, DHNc, DHNd, DHNe, and AtLEA44 (Tables 2 and S3). Among them, *AtLEA4/-5* and *AtLEA33/-34* are tandem repeats (Table S1), where *AtLEA4/-10*, *AtLEA14/-45*, and *AtLEA33/-51* are duplicates that were derived from the α WGD [43]. Most members in this family harbor one intron in the coding region; however, *AtLEA33* has lost the corresponding intron present in its paralogs (i.e., *AtLEA34* and *AtLEA51*). By contrast, one conserved intron was found in the 3′ UTR of both *CpDHN4* and *AtLEA8*, though the intron retention was observed in one alternative splicing isoform of *CpDHN4*, supporting species-specific evolution. All members in this family include Motif 3, which was characterized as the DHN domain (or more precisely as the K-segment), and the motif copies vary from one to six. One copy of Motif 9, which was also characterized as the DHN domain (or more precisely as the S-segment), is widely found with the exception of CpDHN4, AtLEA8, AtLEA33, and AtLEA45. Further sequence alignment revealed the presence of the S-segment at the C-terminal of both CpDHN4 and AtLEA8, and one to three copies of the Y-segment at the N-terminal of CpDHN2, CpDHN3, AtLEA14, AtLEA34, AtLEA45, and AtLEA51. Based on the presence and order of these conserved domains, all five architectures (i.e., K_n, SK_n, K_nS, Y_nK_n, and Y_nSK_n) were found in arabidopsis, while only SK_n, K_nS, and Y_nSK_n were identified in papaya (Figure S3). Additionally, members in DHNa as well as AtLEA44 also harbor Motif 19 (Figures 3 and S3), whose function has not been described yet. The length of CpDHNs varies from 93 to 211 AA, and the average of 152 AA is relatively shorter than 181 AA in arabidopsis. Correspondingly, the MW value varies from 10.50 to 24.10 kDa, and the average of 16.82 kDa is relatively smaller than 19.76 kDa in arabidopsis. Like the LEA_4 family, the pI value in both species is also diverse, varying from 4.74 to 9.38 (Table 1).

3.3.8. SMP

This family is also known as D-34 [42,45], and the three members identified in papaya is considerably less than the six present in arabidopsis (Figure 2). They comprise three orthogroups and one Brassicaceae-specific group, i.e., SMPa, SMPb, SMPc, and AtLEA49/-50 (Tables 2 and S3). Among them, *AtLEA31/-32* and *AtLEA49/-50* are tandem repeats, *AtLEA3/31* are transposed repeats, and *CpSMP2/-3* were characterized as WGD and transposed repeats of *CpSMP1*, respectively. Members in SMPa and SMPc feature two introns, whereas other group members have no or a single one instead. Despite the close evolutionary relationship between *AtLEA49* and *AtLEA50*, they include one intron in the coding region or 5′ UTR, respectively, implying fast evolution and sequence divergence. All members in this family include Motif 2, which was characterized as the SMP domain. Moreover, Motif 14 is also present in members of SMPa, SMPb, and SMPc, whereas two more motifs (i.e., Motif 8 and Motif 17) were also found in members of SMPa and SMPb (Figure 3). Noteworthy, Motif 8 was also characterized as the SMP domain, implying possible fragment duplication or gene fusion. The length of CpSMPs varies from 244 to 267 AA, and the average of 258 AA is relatively longer than 204 AA in arabidopsis. Correspondingly, the MW value varies from 25.13 to 27.97 kDa, and the average of 26.60 kDa is relatively larger than 21.15 kDa in arabidopsis. The pI value in the two species varies from 4.56 to 6.44 (Table 1), indicating that they are acidic.

3.4. ABRE and LTRE cis-Acting Elements Present in the Promoter Region of CpLEA Genes

LTRE, also known as DRE (drought responsive) or CRT (C-repeat), is a key *cis*-acting element for CBF/DREB1 transcription factors, whereas ABRE is a key element involved in ABA signaling [46,47]. Previous studies showed that these two elements are overrepresented in the promoter region of *AtLEA* genes and are associated with ABA, cold and/or drought responses [3]. To reveal possible response patterns of *CpLEA* genes to stresses, we examined the presence of ABRE and LTRE elements in the 2,000-bp promoter regions. Results showed that 89.3% of *CpLEA* genes contain 1 to 10 copies of the ABRE element, only excluding *CpLEA2-3*, *CpLEA3-5*, and *CpLEA6-1*, while 67.9% of them contain 1 to 4 copies of the LTRE element, excluding *CpLEA1-2*, *CpLEA2-2*, *CpLEA3-4*, *CpLEA3-5*, *CpLEA5-2*, *CpLEA6-1*, *CpDHN3*, *CpSMP1*, and *CpSMP3* (Figure 4). The proportion is similar to the 82.0% and 69.0% reported for *AtLEA* genes, respectively [3].

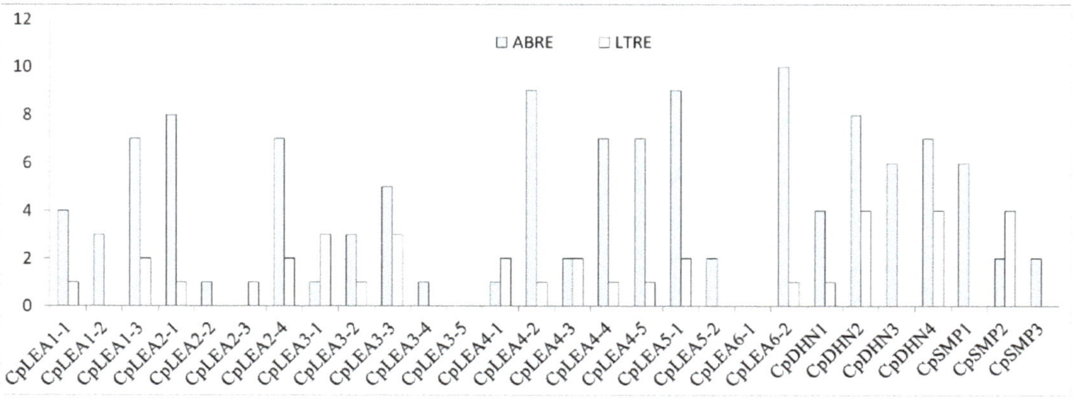

Figure 4. ABRE and LTRE *cis*-acting elements present in 2000-bp promoter regions of *CpLEA* genes.

3.5. Tissue-Specific Expression Profiles of CpLEA Genes

Although some LEA proteins have been reported to be regulated by posttranslational modifications (e.g., phosphorylation), cellular trafficking, homo- and heteromerization [18,48–50], and transcriptional regulation still represent a key mechanism to perform their functions. For this purpose, we first performed global expression profiling of *CpLEA* genes in various tissues.

As shown in Figure 5, our transcriptional profiling supported the expression of all *CpLEA* genes in at least one of 11 tissues examined in this study, i.e., root, apical bud, leaf, petiole, leaf vein, phloem sap, male flower, female flower, fruit, peel, and seed, though the transcript level was highly diverse. As expected, *CpLEA* genes were most expressed in the seed, but considerably less expressed in the leaf and root, which is consistent with the cluster analysis. In total, 22 out of 28 *CpLEA* genes (75.9%) possessed a FKPM value >1 in the seed, which is relatively more than the 15 in the petiole, 15 in the vein, 13 in the root, 12 in the bud, 11 in the fruit, 11 in the peel, 10 in the leaf, 10 in the female flower, 9 in the male flower, and 7 in the sap. Five genes, i.e., *CpLEA3-3*, *CpDHN4*, *CpDHN1*, *CpLEA2-1*, and *CpLEA2-4*, appeared to constitutively express in these tissues, whereas other genes were tissue-specific. As for a certain tissues, several key genes were also identified: *CpLEA3-3* represents the most expressed gene in most tested tissues, whereas *CpLEA1-3* and *CpDHN1* represent the most expressed genes in the seed or bud/fruit, respectively; *CpDHN4* represents the second most expressed gene in the male flower, female flower, petiole, vein, and peel, whereas *CpLEA3-3*, *CpDHN1*, *CpDHN3*, and *CpLEA2-2* represent the second most expressed genes in the bud/fruit, root/leaf, seed, or sap, respectively. According to tissue-specific expression patterns, *CpLEA* genes can be divided into five main clusters: Cluster I includes the most of the 13 genes that are predominantly expressed in the seed; Cluster II includes *CpLEA3-5* (preferentially expressed in fruit), *CpLEA3-2* (preferentially expressed in vein), *CpSMP2* (preferentially expressed in seed), and other four rarely expressed genes; Cluster III includes *CpLEA3-3*, *CpDHN1*, and *CpDHN4*, which are constitutively expressed; Cluster IV includes *CpLEA2-2*, *CpLEA2-3*, and *CpLEA2-4*, which are typically expressed in sap; and Cluster V includes the constitutively expressed *CpLEA2-1*, as well as *CpLEA3-1*, which is preferentially expressed in fruit (Figure 5).

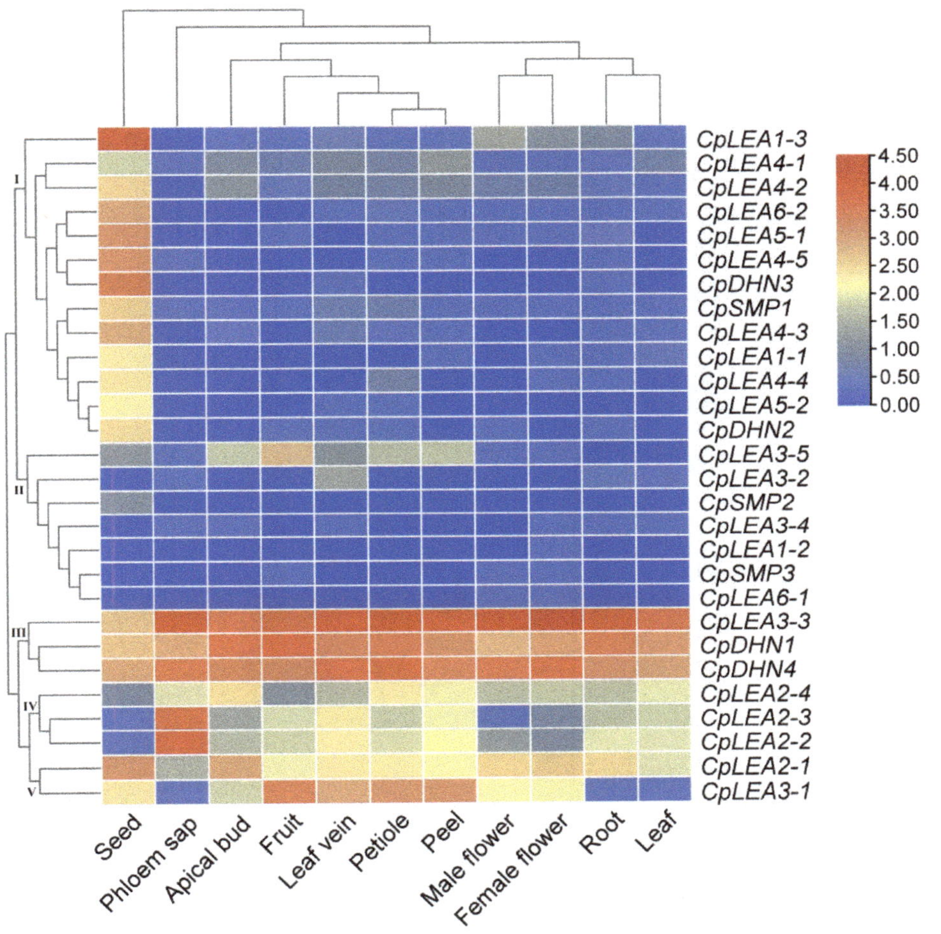

Figure 5. Tissue-specific expression profiles of *CpLEA* genes. Color scale represents FKPM normalized $\log_{10}$ transformed counts, where blue indicates low expression and red indicates high expression.

3.6. Expression Patterns of CpLEA Genes during Fruit Development

To learn more about the expression pattern of *CpLEA* genes during fruit development, six typical stages were investigated, i.e., 30 days post-anthesis (30 DPA), 150 DPA, and stages 1–4 of fruit flesh from immature to ripe, i.e., S1, S2, S3, and S4, as previously defined [51]. Unlike rapid accumulation of *LEA* genes during the late stage of seed development as described in other species, *CpLEA* genes were shown to be the most expressed in the early stages of fruit development, but considerably less expressed in mature fresh fruit. Based on the expression patterns of 15 genes with the FKPM value >1 in at least one of the stages tested, these genes could be grouped into four clusters: Cluster I includes *CpLEA3-1*, *CpLEA3-3*, *CpDHN1*, and *CpDHN4*, which were highly abundant in all stages; Cluster II includes *CpLEA1-3*, *CpLEA3-2*, *CpLEA3-4*, and *CpLEA5-1*, which were rarely or lowly expressed in a few stages; Cluster III includes *CpLEA4-2* and *CpSMP1*, which were lowly expressed in most stages; Cluster IV includes *CpLEA2-1*, *CpLEA2-2*, *CpLEA2-3*, *CpLEA2-4*, and *CpLEA3-5*, which were moderately expressed in most stages (Figure 6).

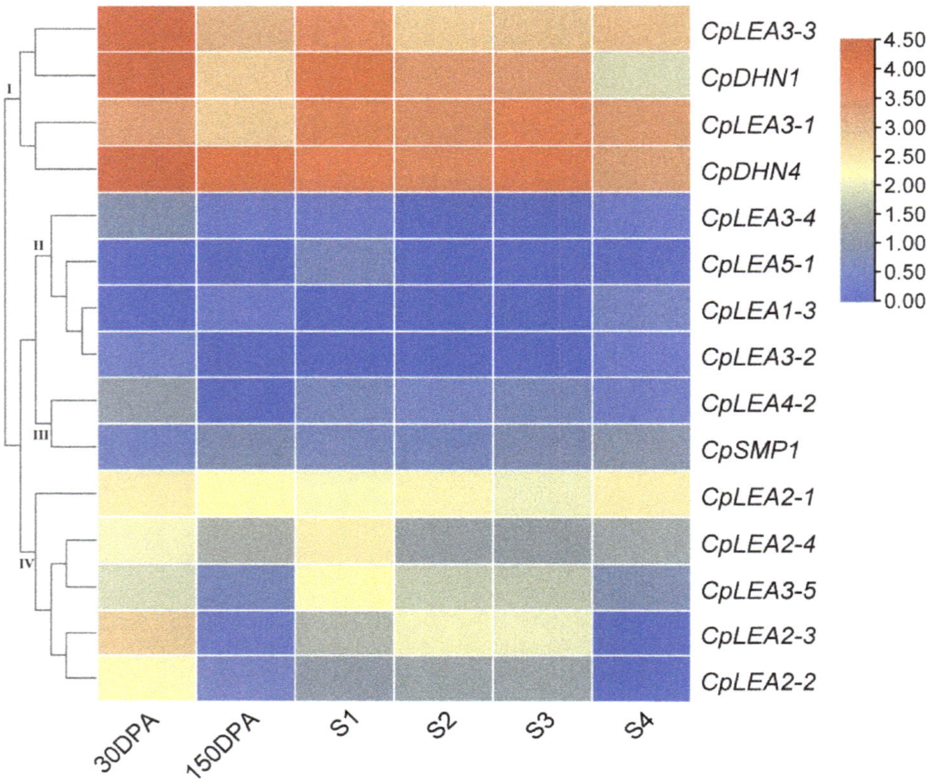

Figure 6. Expression profiles of *CpLEA* genes during fruit development. Color scale represents FKPM normalized $\log_{10}$ transformed counts, where blue indicates low expression and red indicates high expression. (DPA, days post-anthesis; S, stage of developmental fruit).

3.7. Expression Patterns of CpLEA Genes under Drought, Cold and Salt Stresses

The response of *CpLEA* genes to mild (10 d) and severe (20 d) drought was investigated based on transcriptomes of the roots, leaves, and phloem sap [36]. As shown in Figure 7, a total of 15 *CpLEA* genes were differentially expressed in at least one tissue per treatment, and the majority of them (86.7%) were shown to be significantly up-regulated. As for the root, six genes, i.e., *CpLEA1-3*, *CpLEA4-2*, *CpLEA4-3*, *CpLEA4-5*, *CpLEA5-1*, and *CpSMP1*, were up-regulated under both conditions; *CpLEA3-2* was up-regulated by mild drought, whereas *CpLEA4-1* and *CpDHN4* were up-regulated by severe drought; by contrast, *CpDHN1* was down-regulated by severe drought. As for the leaf, in contrast to the down-regulation of *CpDHN1*, four genes, i.e., *CpLEA1-3*, *CpLEA4-2*, *CpLEA4-5*, and *CpDHN4*, were up-regulated by both treatments; *CpSMP1* was up-regulated only by mild drought, whereas *CpLEA2-1*, *CpLEA3-3*, *CpLEA4-3*, *CpLEA5-1*, and *CpLEA6-2* were up-regulated only by severe drought; *CpLEA2-4* and *CpLEA4-1* were down-regulated by mild and severe drought, respectively. As for the sap, only one gene (i.e., *CpLEA2-1*) was up-regulated by severe drought (Figure 7).

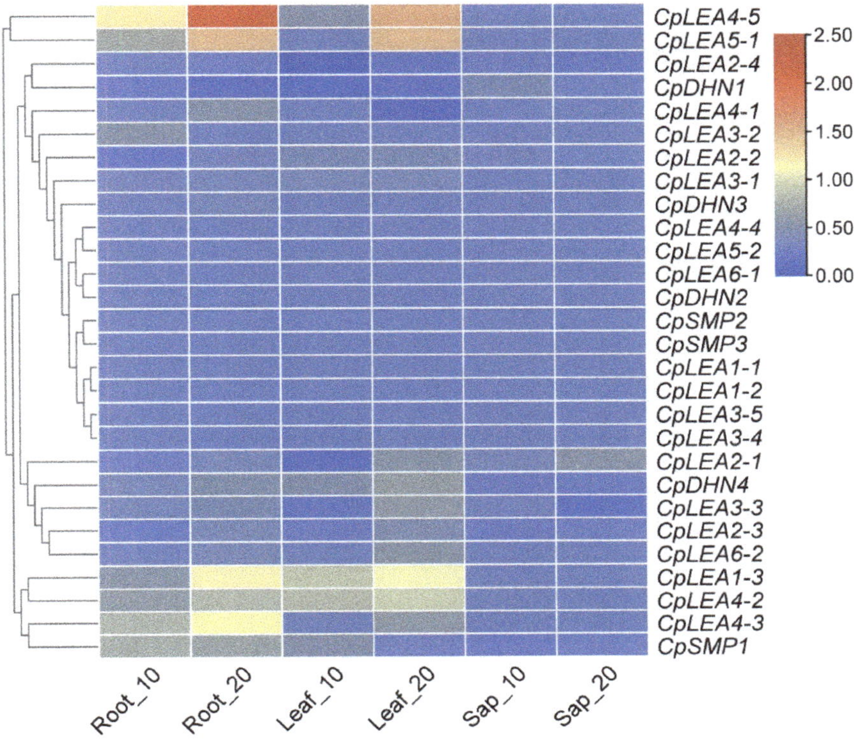

Figure 7. Expression profiles of *CpLEA* genes upon drought stress. The FKPM value of all genes in controls was normalized to one, and the color scale represents normalized $\log_{10}$ transformed fold changes, where blue indicates low expression and red indicates high expression.

To study the response of *CpLEA* genes to cold and salt stresses, eight-week-old plantlets were subjected to 4 °C chilling or 300 mM/L NaCl treatment, and the leaf transcriptome was characterized at 0–40 h or 0–20 d post treatment, respectively. Among 18 *CpLEA* genes with a FKPM value >1, 16 genes were shown to be significantly regulated: six (i.e., *CpLEA1-3*, *CpLEA2-4*, *CpLEA4-1*, *CpLEA4-2*, *CpLEA6-2*, and *CpDHN4*) are shared by cold and salt stresses, whereas five (i.e., *CpLEA2-1*, *CpLEA2-2*, *CpLEA2-3*, *CpLEA3-3*, and *CpDHN1*) and five (i.e., *CpLEA1-1*, *CpLEA3-1*, *CpLEA3-4*, *CpLEA4-3*, and *CpLEA4-5*) are cold- or salt-specific, respectively. Similar to drought stress, most genes were up-regulated, accounting for about 68.8% of total *LEA* genes, though some of them (i.e., *CpLEA1-3*, *CpLEA2-4*, and *CpLEA4-1*) were occasionally down-regulated at a certain time point. As for cold stress, five regulated genes are shared by three time points, including four up-regulated (i.e., *CpLEA2-4*, *CpLEA3-3*, *CpLEA4-2*, and *CpDHN4*) and one down-regulated (i.e., *CpLEA2-3*); *CpLEA2-2* was down-regulated at two former time points, whereas *CpLEA1-3* and *CpDHN1* were up-regulated at the latter two time points; *CpLEA2-1* and *CpLEA6-2* were down-regulated at 7 or 40 h post-treatment, respectively; *CpLEA4-1* was down-regulated at 7 h but up-regulated at 40 h post-treatment. As for salt stress, *CpLEA4-2* and *CpLEA4-5* were up-regulated at three time points, whereas *CpLEA1-3* was down-regulated at 10 d but up-regulated at the latter two time points; *CpLEA1-1*, *CpLEA4-1*, and *CpLEA4-3* were up-regulated at the latter two time points, whereas *CpLEA3-4* was down-regulated at the same time points; *CpDHN4* was up-regulated at 10 d post-treatment, whereas *CpLEA2-4* and *CpLEA6-2* were down-regulated at the same time point; *CpLEA3-1* was up-regulated at 10 and 20 d post-treatment (Figure 8).

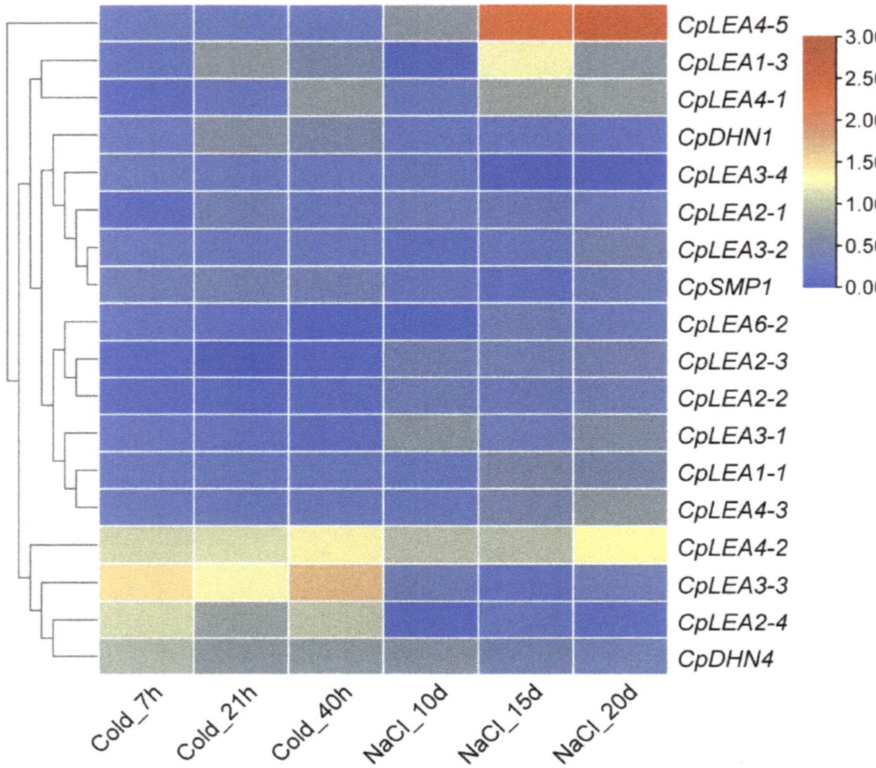

Figure 8. Expression profiles of *CpLEA* genes upon cold and salt stresses. The FKPM value of all genes in the controls was normalized to one, and the color scale represents normalized $\log_{10}$ transformed fold changes, where blue indicates low expression and red indicates high expression.

4. Discussion

4.1. Small Number but High Diversity of LEA Genes in Papaya

Although first identified for their accumulation in the later stages of seed development, LEA proteins have been found in a wide range of plant tissues, as well as different types of organisms [1,7,21,45]. In contrast to a single or few members present in algae, rapid expansion of the *LEA* superfamily was observed in terrestrial plants, which was shown to be essential for survival under water stress [9,52]. Rapid gene expansion is usually accompanied by WGDs, which are widespread and play an important role in the radiation of flowering plants [53]. In eudicots, studies established that the γ whole genome triplication event occurred at 117 million years ago (Mya), sometime before the diversification of core eudicots [54]. After that, arabidopsis, a Brassicaceae plant within the order Brassicales, was proven to experience two additional whole genome doubling events, i.e., β and α, occurred within a window of 61–65 and 23–50 Mya, respectively [19,55]. As a result, a high number of 51 *LEA* genes are present in arabidopsis, including seven dispersed repeats as well as 21 repeats that resulted from γ WGD (1), β WGD (1), α WGD (9), tandem duplication (7), and transposed duplication (4) (Table S1).

In this study, a first genome-wide identification of *LEA* genes was conducted in an important tropical fruit tree of the Caricaceae family, papaya, as well as another two Brassicales plants, i.e., horseradish tree and spider flower. Horseradish tree is an important multipurpose shrub with medicinal and nutritional properties and the ability to grow in the low water conditions of the Moringaceae family, whereas spider flower belongs to a phylogenetic outgroup of the Brassicaceae sister family Cleomaceae [21,22]. Like castor bean (Euphorbiaceae), the papaya and horseradish tree did not experience any additional WGD after the γ WGD. By contrast, the spider flower shared the β WGD but further experienced one genome triplication that is independent of the Brassicaceae-specific α WGD as described in arabidopsis [19–22,56]. As expected, a relatively small number of 28 or 29 *LEA* genes were found in the papaya and horseradish tree, respectively, which are comparable to 28 reported in castor bean, but relatively less than the 39 and 51 present in spider flower and arabidopsis, respectively, reflecting the occurrence of lineage-specific WGDs in the latter after their divergence [3,8,19,21].

LEA genes identified in this study belong to eight out of nine families as described in arabidopsis, i.e., LEA_1, LEA_2, LEA_3, LEA_4, LEA_5, LEA_6, DHN, and SMP [3]. As for the AtM family, which includes two tandem repeats in arabidopsis, it is more likely to be Brassicaceae-specific, because it is widely present in Brassicaceae plants (Table S3) but has not yet been identified in other species [3,8,9,12,13], including species examined in this study. Nevertheless, 28 *CpLEA* genes represent 27 out of 29 orthogroups based on sequence comparison of the above five species, though a LEA4f homolog has lost the corresponding LEA_4 domain. Moreover, no orthologs were identified for CpLEA1-2, CpLEA2-2, CpLEA2-3, CpLEA3-4, or CpLEA3-5 in arabidopsis, though their counterparts are present in at least one of three other species examined.

4.2. Comparative Genomics Analysis Reveals Lineage-Specific Evolution of the LEA Superfamily in Brassicales

Orthology defines genes in different organisms that evolved from a common ancestral gene via speciation, which may perform similar functions [57]. Characterization of 29 orthogroups in five representative species allows us to infer lineage-specific evolution in Brassicales. Notably, a nearly one-to-one orthologous relationship was observed between the papaya/horseradish tree and castor bean, though they belong to different plant families, implying that few *LEA* genes have been lost in either the papaya or horseradish tree after the split with the castor bean. By contrast, tandem duplication plays a predominant role in gene expansion within an orthogroup, i.e., *RcDHN2/-3* in DHNb, and *RcSMP1/-2* in SMPb, *CpLEA2-2/-3* in LEA2b, and *MoLEA5-2/-3* in LEA5b. As for the spider flower, which experienced two WGDs (including the β WGD shared by Brassicaceae plants) after the split with papaya at approximately 72 Mya [21,58], duplicate pairs are mainly contributed by WGD (12), followed by dispersed duplication (3) and transposed duplication (1) (Table S1). The transposed duplication is shared by all five species examined, whereas WGD repeats appear to be spider flower-specific. By contrast, *AtLEA2/-41* and *AtLEA1/-27* were characterized as γ and β WGD-derived repeats, respectively [22], supporting species-specific evolution following WGDs. Nevertheless, since the spider flower-specific WGD is a triplication event, theoretically, it should have given rise to three gene copies from a single ancestral gene. However, in most cases, only one or two copies are maintained. Unlike the spider flower, tandem duplication also plays a key role in gene expansion in arabidopsis.

Further comparative analysis of exon-intron structures and protein motifs revealed frequent gain and/or loss of certain introns/motifs, which includes the loss of the second intron in *CpLEA2-3* relative to *CpLEA2-2*. In fact, compared with papaya, such an occurrence is relatively more prevalent in arabidopsis, which is consistent with a relatively faster evolution of annual than perennial shrubs [59]. Nevertheless, family-specific Pfam domains are highly conserved. It is worth noting that CpLEA2-1 and AtLEA26 contain two LEA_2 domains relative to a single one present in other LEA_2 family members, implying a possible fragment repetition. From an evolutionary perspective, further characterization of these species-specific genes is of particular interest.

4.3. Diverse Expression Patterns of CpLEA Genes and a Role in Fruit Development and Abiotic Stress Responses

As reported in other species, our transcriptional profiling revealed diverse expression patterns of *CpLEA* genes in 11 tissues, as well as six typical stages of fruit development examined in this study. In contrast to the constitutive expression of a few members, e.g., *CpLEA2-1*, *CpLEA2-4*, *CpLEA3-3*, *CpDHN1*, and *CpDHN4*, most *CpLEA* genes appeared to preferentially express in a few tissues, especially in seed. However, except for *CpLEA1-3* and *CpSMP1* that preferentially accumulated in mature fruits, the expression patterns of most *CpLEA* genes differ from that observed in seeds, which undergo a dehydration process [2–6,21]. The high abundance of *CpDHN4*, *CpDHN1*, *CpLEA3-1*, and *CpLEA3-3* in fruits implies their possible important role in this special tissue.

Analyzing promoter sequences of *CpLEA* genes revealed the presence of a high number of ABRE and LTRE *cis*-acting elements, implying their possible involvement in stress responses. As expected, the transcript levels of most *CpLEA* were shown to be significantly regulated by the cold, drought, and high salt conditions examined in this study. Among three genes (i.e., *CpLEA2-3*, *CpLEA3-5*, and *CpLEA6-1*) without ABRE elements in their promoters, none of them were regulated by drought as well as salt, though *CpLEA2-3* was down-regulated by cold, which is consistent with the presence of one copy of the LTRE element in its promoter. Among nine genes (i.e., *CpLEA1-2*, *CpLEA2-2*, *CpLEA3-4*, *CpLEA3-5*, *CpLEA5-2*, *CpLEA6-1*, *CpDHN3*, *CpSMP1*, and *CpSMP3*) without LTRE elements, only *CpLEA2-2* was shown to be down-regulated by cold, while *CpLEA3-4* and *CpSMP1* were regulated by salt or drought, respectively. Among 20 genes containing both ABRE and LTRE *cis*-acting elements, most of them (85.0%) were regulated by at least one of the three stresses tested, only excluding *CpLEA4-4*, *CpDHN2*, and *CpSMP2*, which were preferentially expressed in seed but lowly expressed in the leaf, root and sap examined in this study. Among these 17 regulated genes, all of them were up-regulated by at least one treatment in at least one of three examined tissues: nine genes (i.e., *CpLEA1-1*, *CpLEA3-1*, *CpLEA3-2*, *CpLEA3-3*, *CpLEA4-2*, *CpLEA4-3*, *CpLEA4-5*, *CpLEA5-1*, and *CpDHN4*) exhibit a single up-regulated pattern; *CpLEA2-4*, the unique gene regulated in sap, was up-regulated by drought but down-regulated by cold in leaf; *CpDHN1*, a cold-induced gene, was down-regulated by drought in both the root and leaf; *CpLEA2-4* was up-regulated by cold but down-regulated by both drought and NaCl in the leaf; *CpLEA6-2* was down-regulated by both cold and NaCl but up-regulated by drought in the leaf; *CpLEA4-5*, a NaCl-induced gene, was down-regulated in leaf but up-regulated in root upon drought stress; by contrast, an initial decline followed by a steady increasing trend was observed. Regulation by stresses has been frequently reported in arabidopsis, rice, cassava (*Manihot esculenta*), and other species [2,3,12,13]. In arabidopsis, a study revealed that 54.5% of genes highly expressed in non-seed tissues were induced more than threefold by various stresses, mainly by cold, drought and salt [3]. For example, *AtLEA18*, the ortholog of *CpLEA1-1*, was also induced by salt; *AtLEA41*, the ortholog of *CpLEA3-1*, was induced by ABA, cold, and salt; *AtLEA46*, the ortholog of *CpLEA1-3*, was induced by ABA, cold, drought, and salt [2,3]. Thereby, similar functions could be speculated.

5. Conclusions

This study presents the first genome-wide identification of *LEA* genes in papaya as well another two Brassicales plants, horseradish tree and spider flower; resulting in 28, 29, and 39 members, respectively. These genes belong to eight out of nine families as described in arabidopsis, i.e., LEA_1, LEA_2, LEA_3, LEA_4, LEA_5, LEA_6, DHN, and SMP. Further comparison of *LEA* genes in papaya, horseradish tree, spider flower, castor bean, and arabidopsis reveals lineage-specific evolution in Brassicales, and significant expansion in spider flower and arabidopsis was mainly contributed by WGDs sometime after their split with papaya. Analysis of exon-intron structures and protein motifs supported the fast evolution of this special family, especially in arabidopsis. Moreover, global expression profiles of *CpLEA* genes were comprehensively analyzed, which revealed tissue-specific expression patterns and key roles in fruit development and stress responses. Taken together, these findings provide valuable information for further functional analysis of *LEA* genes in papaya and other species.

Supplementary Materials: The following supporting information can be downloaded at: https://www.mdpi.com/article/10.3390/life12091453/s1, Figure S1: Nucleotide and protein sequence alignments of *CpLEA2-2* and *CpLEA2-3*; Figure S2: Sequence logos of 20 motifs identified in this study; Figure S3: Alignment of DHN proteins in papaya and arabidopsis; Table S1: Detailed information of *LEA* genes present in papaya, horseradish tree, spider flower, castor bean, and arabidopsis; Table S2: Detailed information of transcriptome data used in this study; Table S3: Orthologs in representative Brassicaceae plants for 14 arabidopsis-specific *LEA* genes identified in this study.

Author Contributions: Z.Z.: methodology, data curation and writing—original draft; Z.Z., J.G., Y.Z. and Y.X.: data curation; Z.Z., J.G., Y.Z. and Y.X.: conceptualization and methodology; Z.Z., Y.Z. and Y.X.: software; Z.Z. and A.G.: formal analysis and preparation of materials; Z.Z.: conceptualization, data curation and funding acquisition. All authors have read and agreed to the published version of the manuscript.

Funding: This work was supported by the Natural Science Foundation of Hainan province (320RC705), the National Natural Science Foundation of China (31971688), and the Central Public-interest Scientific Institution Basal Research Fund for Chinese Academy of Tropical Agricultural Sciences (1630052022001).

Conflicts of Interest: The authors declare no conflict of interest.

References

1. Battaglia, M.; Olvera-Carrillo, Y.; Garciarrubio, A.; Campos, F.; Covarrubias, A.A. The enigmatic LEA proteins and other hydrophilins. *Plant. Physiol.* **2008**, *148*, 6–24. [CrossRef] [PubMed]
2. Bies-Etheve, N.; Gaubier-Comella, P.; Debures, A.; Lasserre, E.; Jobet, E.; Raynal, M.; Cooke, R.; Delseny, M. Inventory, evolution and expression profiling diversity of the LEA (late embryogenesis abundant) protein gene family in *Arabidopsis thaliana*. *Plant. Mol. Biol.* **2008**, *67*, 107–124. [CrossRef] [PubMed]
3. Hundertmark, M.; Hincha, D.K. LEA (late embryogenesis abundant) proteins and their encoding genes in *Arabidopsis thaliana*. *BMC Genom.* **2008**, *9*, 118. [CrossRef] [PubMed]
4. Dure, L.; Chlan, C. Developmental biochemistry of cottonseed embryogenesis and germination. XII. Purification and properties of principal storage proteins. *Plant. Physiol.* **1981**, *68*, 180–186. [CrossRef] [PubMed]
5. Dure, L.; Galau, G.A. Developmental biochemistry of cottonseed embryogenesis and germination. XIII. Regulation of biosynthesis of principal storage proteins. *Plant. Physiol.* **1981**, *68*, 187–194. [CrossRef]
6. Dure, L.; Greenway, S.C.; Galau, G.A. Developmental biochemistry of cottonseed embryogenesis and germination: Changing messenger ribonucleic acid populations as shown by in vitro and in vivo protein synthesis. *Biochemistry* **1981**, *20*, 4162–4168. [CrossRef]
7. Hand, S.C.; Menze, M.A.; Toner, M.; Boswell, L.; Moore, D. LEA proteins during water stress: Not just for plants anymore. *Annu. Rev. Physiol.* **2011**, *73*, 115–134. [CrossRef]
8. Zou, Z.; Huang, Q.X.; An, F. Genome-wide identification, classification and phylogenetic analysis of *LEA* gene family in castor bean (*Ricinus communis* L.). *Chin. J. Oil Crop. Sci.* **2013**, *35*, 637–643.
9. Artur, M.A.S.; Zhao, T.; Ligterink, W.; Schranz, E.; Hilhorst, H.W. Dissecting the genomic diversification of late embryogenesis abundant (LEA) protein gene families in plants. *Genome Biol. Evol.* **2019**, *11*, 459–471. [CrossRef]
10. Raynal, M.; Guilleminot, J.; Gueguen, C.; Cooke, R.; Delseny, M.; Gruber, V. Structure, organization and expression of two closely related novel Lea (late-embryogenesis abundant) genes in *Arabidopsis thaliana*. *Plant. Mol. Biol.* **1999**, *40*, 153–165. [CrossRef]
11. Singh, S.; Cornilescu, C.C.; Tyler, R.C.; Cornilescu, G.; Tonelli, M.; Lee, M.S.; Markley, J.L. Solution structure of a late embryogenesis abundant protein (LEA14) from *Arabidopsis thaliana*, a cellular stress-related protein. *Protein Sci.* **2005**, *14*, 2601–2609. [CrossRef] [PubMed]
12. Wang, X.S.; Zhu, H.B.; Jin, G.L.; Liu, H.L.; Wu, W.R.; Zhu, J. Genome-scale identification and analysis of *LEA* genes in rice (*Oryza sativa* L.). *Plant. Sci.* **2007**, *172*, 414–420. [CrossRef]
13. Wu, C.; Hu, W.; Yan, Y.; Tie, W.; Ding, Z.; Guo, J.; He, G. The late embryogenesis abundant protein family in cassava (*Manihot esculenta* Crantz): Genome-wide characterization and expression during abiotic stress. *Molecules* **2018**, *23*, 1196. [CrossRef]
14. Salleh, F.M.; Evans, K.; Goodall, B.; Machin, H.; Mowla, S.B.; Mur, L.A.; Runions, J.; Theodoulou, F.L.; Foyer, C.H.; Rogers, H.J. A novel function for a redox-related LEA protein (SAG21/AtLEA5) in root development and biotic stress responses. *Plant. Cell Environ.* **2012**, *35*, 418–429. [CrossRef] [PubMed]
15. Dang, N.X.; Popova, A.V.; Hundertmark, M.; Hincha, D.K. Functional characterization of selected LEA proteins from *Arabidopsis thaliana* in yeast and in vitro. *Planta* **2014**, *240*, 325–336. [CrossRef] [PubMed]
16. Zhang, X.; Lu, S.; Jiang, C.; Wang, Y.; Lv, B.; Shen, J.; Ming, F. *RcLEA*, a late embryogenesis abundant protein gene isolated from *Rosa chinensis*, confers tolerance to *Escherichia coli* and *Arabidopsis thaliana* and stabilizes enzyme activity under diverse stresses. *Plant. Mol. Biol.* **2014**, *85*, 333–347. [CrossRef]

17. Xiang, D.J.; Man, L.L.; Zhang, C.L.; Li, Z.G.; Zheng, G.C. A new Em-like protein from *Lactuca sativa*, LsEm1, enhances drought and salt stress tolerance in *Escherichia coli* and rice. *Protoplasma* **2018**, *255*, 1089–1106. [CrossRef]
18. Hernández-Sánchez, I.E.; Maruri-López, I.; Molphe-Balch, E.P.; Becerra-Flora, A.; Jaimes-Miranda, F.; Jiménez-Bremont, J.F. Evidence for in vivo interactions between dehydrins and the aquaporin AtPIP2B. *Biochem. Bioph. Res. Commun.* **2019**, *510*, 545–550. [CrossRef]
19. Bowers, J.E.; Chapman, B.A.; Rong, J.; Paterson, A.H. Unravelling angiosperm genome evolution by phylogenetic analysis of chromosomal duplication events. *Nature* **2003**, *422*, 433–438. [CrossRef]
20. Ming, R.; Hou, S.; Feng, Y.; Yu, Q.; Dionne-Laporte, A.; Saw, J.H.; Senin, P.; Wang, W.; Ly, B.V.; Lewis, K.L.; et al. The draft genome of the transgenic tropical fruit tree papaya (*Carica papaya* Linnaeus). *Nature* **2008**, *452*, 991–996. [CrossRef]
21. Cheng, S.; van den Bergh, E.; Zeng, P.; Zhong, X.; Xu, J.; Liu, X.; Hofberger, J.; de Bruijn, S.; Bhide, A.S.; Kuelahoglu, C.; et al. The *Tarenaya hassleriana* genome provides insight into reproductive trait and genome evolution of crucifers. *Plant. Cell* **2013**, *25*, 2813–2830. [CrossRef] [PubMed]
22. Shyamli, P.S.; Pradhan, S.; Panda, M.; Parida, A. De novo whole-genome assembly of *Moringa oleifera* helps identify genes regulating drought stress tolerance. *Front. Plant. Sci.* **2021**, *12*, 766999. [CrossRef] [PubMed]
23. Ming, R.; Moore, P. *Genetics and Genomics of Papaya. Plant Genetics and Genomics: Crops and Models*; Springer: Cham, Switzerland, 2014; Volume 10. [CrossRef]
24. Allan, P. *Carica papaya* responses under cool subtropical growth conditions. *Acta Hortic.* **2002**, *575*, 757–763. [CrossRef]
25. Mahouachi, J.; Socorro, A.; Talon, M. Responses of papaya seedlings (*Carica papaya* L.) to water stress and re-hydration: Growth, photosynthesis and mineral nutrient imbalance. *Plant. Soil.* **2006**, *281*, 137–146. [CrossRef]
26. Altschul, S.F.; Madden, T.L.; Schäffer, A.A.; Zhang, J.; Zhang, Z.; Miller, W.; Lipman, D.J. Gapped BLAST and PSI-BLAST: A new generation of protein database search programs. *Nucleic Acids Res.* **1997**, *25*, 3389–3402. [CrossRef]
27. Zou, Z.; Yang, L.; Gong, J.; Mo, Y.; Wang, J.; Cao, J.; An, F.; Xie, G. Genome-wide identification of *Jatropha curcas* aquaporin genes and the comparative analysis provides insights into the gene family expansion and evolution in *Hevea brasiliensis*. *Front. Plant. Sci.* **2016**, *7*, 395. [CrossRef]
28. Wang, Y.; Tang, H.; DeBarry, J.D.; Tan, X.; Li, J.; Wang, X.; Lee, T.H.; Jin, H.; Marler, B.; Guo, H.; et al. MCScanX: A toolkit for detection and evolutionary analysis of gene synteny and collinearity. *Nucleic Acids Res.* **2012**, *40*, e49. [CrossRef]
29. Qiao, X.; Li, Q.; Yin, H.; Qi, K.; Li, L.; Wang, R.; Zhang, S.; Paterson, A.H. Gene duplication and evolution in recurring polyploidization-diploidization cycles in plants. *Genome Biol.* **2019**, *20*, 38. [CrossRef]
30. Moreno-Hagelsieb, G.; Latimer, K. Choosing BLAST options for better detection of orthologs as reciprocal best hits. *Bioinformatics* **2008**, *24*, 319–324. [CrossRef]
31. Hu, B.; Jin, J.; Guo, A.Y.; Zhang, H.; Luo, J.; Gao, G. GSDS 2.0: An upgraded gene feature visualization server. *Bioinformatics* **2015**, *31*, 1296–1297. [CrossRef] [PubMed]
32. Tamura, K.; Stecher, G.; Peterson, D.; Filipski, A.; Kumar, S. MEGA6: Molecular Evolutionary Genetics Analysis version 6.0. *Mol. Biol. Evol.* **2013**, *30*, 2725–2729. [CrossRef] [PubMed]
33. Bailey, T.L.; Boden, M.; Buske, F.A.; Frith, M.; Grant, C.E.; Clementi, L.; Ren, J.; Li, W.W.; Noble, W.S. MEME SUITE: Tools for motif discovery and searching. *Nucleic Acids Res.* **2009**, *37*, W202–W208. [CrossRef] [PubMed]
34. Zou, Z.; Gong, J.; An, F.; Xie, G.; Wang, J.; Mo, Y.; Yang, L. Genome-wide identification of rubber tree (*Hevea brasiliensis* Muell. Arg.) aquaporin genes and their response to ethephon stimulation in the laticifer, a rubber-producing tissue. *BMC Genom.* **2015**, *16*, 1001. [CrossRef] [PubMed]
35. Zou, Z.; Yang, J.H.; Zhang, X.C. Insights into genes encoding respiratory burst oxidase homologs (RBOHs) in rubber tree (*Hevea brasiliensis* Muell. Arg.). *Ind. Crop. Prod.* **2019**, *128*, 126–139. [CrossRef]
36. Gamboa-Tuz, S.D.; Pereira-Santana, A.; Zamora-Briseño, J.A.; Castano, E.; Espadas-Gil, F.; Ayala-Sumuano, J.T.; Keb-Llanes, M.Á.; Sanchez-Teyer, F.; Rodríguez-Zapata, L.C. Transcriptomics and co-expression networks reveal tissue-specific responses and regulatory hubs under mild and severe drought in papaya (*Carica papaya* L.). *Sci. Rep.* **2018**, *8*, 14539. [CrossRef] [PubMed]
37. Bolger, A.M.; Lohse, M.; Usadel, B. Trimmomatic: A flexible trimmer for Illumina sequence data. *Bioinformatics* **2014**, *30*, 2114–2120. [CrossRef]
38. Trapnell, C.; Williams, B.A.; Pertea, G.; Mortazavi, A.; Kwan, G.; Van Baren, M.J.; Salzberg, S.L.; Wold, B.J.; Pachter, L. Transcript assembly and quantification by RNA-Seq reveals unannotated transcripts and isoform switching during cell differentiation. *Nat. Biotechnol.* **2010**, *28*, 511–515. [CrossRef]
39. Langmead, B.; Salzberg, S.L. Fast gapped-read alignment with Bowtie 2. *Nat. Methods* **2012**, *9*, 357–359. [CrossRef]
40. Li, B.; Dewey, C.N. RSEM: Accurate transcript quantification from RNA-Seq data with or without a reference genome. *BMC Bioinform.* **2011**, *12*, 323. [CrossRef]
41. Yue, J.; VanBuren, R.; Liu, J.; Fang, J.; Zhang, X.; Liao, Z.; Wai, C.M.; Xu, X.; Chen, S.; Zhang, S.; et al. SunUp and Sunset genomes revealed impact of particle bombardment mediated transformation and domestication history in papaya. *Nat. Genet.* **2022**, *54*, 715–724. [CrossRef]
42. Dure, L. A repeating 11-mer amino acid motif and plant desiccation. *Plant. J.* **1993**, *3*, 363–369. [CrossRef]
43. Wang, Y.; Tan, X.; Paterson, A.H. Different patterns of gene structure divergence following gene duplication in *Arabidopsis*. *BMC Genom.* **2013**, *14*, 652. [CrossRef] [PubMed]

44. Colmenero-Flores, J.M.; Moreno, L.P.; Smith, C.E.; Covarrubias, A.A. Pvlea18, a member of a new late-embryogenesis-abundant protein family that accumulates during water stress and in the growing region of well-irrigated bean seedlings. *Plant. Physiol.* **1999**, *120*, 93–103. [CrossRef] [PubMed]
45. Zou, Z.; Zhao, Y.; Zhang, L.; Xiao, Y.; Guo, A. Analysis of *Cyperus esculentus SMP* family genes reveals lineage-specific evolution and seed desiccation-like transcript accumulation during tuber maturation. *Ind. Crop. Prod.* **2022**, *187*, 115382. [CrossRef]
46. Bartels, D.; Sunkar, R. Drought and salt tolerance in plants. *Crit. Rev. Plant. Sci.* **2005**, *24*, 23–58. [CrossRef]
47. Yamaguchi-Shinozaki, K.; Shinozaki, K. Organization of *cis*-acting regulatory elements in osmotic- and cold-stress-responsive promoters. *Trends Plant. Sci.* **2005**, *10*, 88–94. [CrossRef] [PubMed]
48. Nylander, M.; Svensson, J.; Palva, E.T.; Welin, B.V. Stress-induced accumulation and tissue-specific localization of dehydrins in *Arabidopsis thaliana. Plant. Mol. Biol.* **2001**, *45*, 263–279. [CrossRef]
49. Alsheikh, M.K.; Svensson, J.T.; Randall, S.K. Phosphorylation regulated ion-binding is a property shared by the acidic subclass dehydrins. *Plant. Cell Environ.* **2005**, *28*, 1114–1122. [CrossRef]
50. Candat, A.; Paszkiewicz, G.; Neveu, M.; Gautier, R.; Logan, D.C.; Avelange-Macherel, M.H.; Macherel, D. The ubiquitous distribution of late embryogenesis abundant proteins across cell compartments in *Arabidopsis* offers tailored protection against abiotic stress. *Plant. Cell* **2014**, *26*, 3148–3166. [CrossRef]
51. Lü, P.; Yu, S.; Zhu, N.; Chen, Y.R.; Zhou, B.; Pan, Y.; Tzeng, D.; Fabi, J.P.; Argyris, J.; Garcia-Mas, J.; et al. Genome encode analyses reveal the basis of convergent evolution of fleshy fruit ripening. *Nat. Plants.* **2018**, *4*, 784–791. [CrossRef]
52. Rensing, S.A.; Lang, D.; Zimmer, A.D.; Terry, A.; Salamov, A.; Shapiro, H.; Nishiyama, T.; Perroud, P.F.; Lindquist, E.A.; Kamisugi, Y.; et al. The *Physcomitrella* genome reveals evolutionary insights into the conquest of land by plants. *Science* **2008**, *319*, 64–69. [CrossRef] [PubMed]
53. Van de Peer, Y.; Fawcett, J.A.; Proost, S.; Sterck, L.; Vandepoele, K. The flowering world: A tale of duplications. *Trends Plant. Sci.* **2009**, *14*, 680–688. [CrossRef] [PubMed]
54. Jiao, Y.; Leebens-Mack, J.; Ayyampalayam, S.; Bowers, J.E.; McKain, M.R.; McNeal, J.; Rolf, M.; Ruzicka, D.R.; Wafula, E.; Wickett, N.J.; et al. A genome triplication associated with early diversification of the core eudicots. *Genome Biol.* **2012**, *13*, R3. [CrossRef] [PubMed]
55. Vanneste, K.; Baele, G.; Maere, S.; Van de Peer, Y. Analysis of 41 plant genomes supports a wave of successful genome duplications in association with the Cretaceous-Paleogene boundary. *Genome Res.* **2014**, *24*, 1334–1347. [CrossRef] [PubMed]
56. Chan, A.P.; Crabtree, J.; Zhao, Q.; Lorenzi, H.; Orvis, J.; Puiu, D.; Melake-Berhan, A.; Jones, K.M.; Redman, J.; Chen, G. Draft genome sequence of the oilseed species *Ricinus communis. Nat. Biotechnol.* **2010**, *28*, 951–956. [CrossRef]
57. Koonin, E.V. Orthologs, paralogs, and evolutionary genomics. *Annu. Rev. Genet.* **2005**, *39*, 309–338. [CrossRef] [PubMed]
58. Carvalho, F.A.; Renner, S.S. The Phylogeny of the Caricaceae. In *Genetics and Genomics of Papaya. Plant Genetics and Genomics: Crops and Models*; Ming, R., Moore, P., Eds.; Springer: Cham, Switzerland, 2014.
59. Luo, M.C.; You, F.M.; Li, P.; Wang, J.R.; Zhu, T.; Dandekar, A.M.; Leslie, C.A.; Aradhya, M.; McGuire, P.E.; Dvorak, J. Synteny analysis in Rosids with a walnut physical map reveals slow genome evolution in long-lived woody perennials. *BMC Genom.* **2015**, *16*, 707. [CrossRef]

life

Review

Role of WRKY Transcription Factors in Regulation of Abiotic Stress Responses in Cotton

Xiaoqiang Guo [1,*], Abid Ullah [2,*], Dorota Siuta [3], Bożena Kukfisz [4] and Shehzad Iqbal [5]

1 College of Life Science and Technology, Longdong University, Qingyang 745000, China
2 Department of Botany, Post Graduate College Dargai, Malakand 23060, Khyber Pakhtunkhwa, Pakistan
3 Faculty of Process and Environmental Engineering, Lodz University of Technology, Wolczanska Str. 213, 90-924 Lodz, Poland
4 Faculty of Security Engineering and Civil Protection, The Main School of Fire Service, 01-629 Warsaw, Poland
5 College of Plant Sciences and Technology, Huazhong Agricultural University, Wuhan 430070, China
* Correspondence: guo.xiaoq@163.com (X.G.); abid.ullah@uom.edu.pk (A.U.)

Abstract: Environmental factors are the major constraints in sustainable agriculture. WRKY proteins are a large family of transcription factors (TFs) that regulate various developmental processes and stress responses in plants, including cotton. On the basis of *Gossypium raimondii* genome sequencing, WRKY TFs have been identified in cotton and characterized for their functions in abiotic stress responses. WRKY members of cotton play a significant role in the regulation of abiotic stresses, i.e., drought, salt, and extreme temperatures. These TFs either activate or repress various signaling pathways such as abscisic acid, jasmonic acid, salicylic acid, mitogen-activated protein kinases (MAPK), and the scavenging of reactive oxygen species. WRKY-associated genes in cotton have been genetically engineered in *Arabidopsis*, *Nicotiana*, and *Gossypium* successfully, which subsequently enhanced tolerance in corresponding plants against abiotic stresses. Although a few review reports are available for WRKY TFs, there is no critical report available on the WRKY TFs of cotton. Hereby, the role of cotton WRKY TFs in environmental stress responses is studied to enhance the understanding of abiotic stress response and further improve in cotton plants.

Keywords: abiotic stresses; cotton; hormones; signaling pathway; WRKY

Citation: Guo, X.; Ullah, A.; Siuta, D.; Kukfisz, B.; Iqbal, S. Role of WRKY Transcription Factors in Regulation of Abiotic Stress Responses in Cotton. *Life* **2022**, *12*, 1410. https://doi.org/10.3390/life12091410

Academic Editors: Hakim Manghwar, Wajid Zaman and Othmane Merah

Received: 12 August 2022
Accepted: 6 September 2022
Published: 9 September 2022

Publisher's Note: MDPI stays neutral with regard to jurisdictional claims in published maps and institutional affiliations.

Copyright: © 2022 by the authors. Licensee MDPI, Basel, Switzerland. This article is an open access article distributed under the terms and conditions of the Creative Commons Attribution (CC BY) license (https://creativecommons.org/licenses/by/4.0/).

1. Introduction

As an industrial crop, cotton is cultivated almost all over the world due to its crucial role in the economy of a country [1]. Cotton is generally divided into two types, i.e., wild and cultivated cotton. The cultivated cotton species are *Gossypium hirsutum* (*G. Hirsutum*), *G. herbaceum*, *G. arboreum*, and *G. barbadense*. Among the cultivated species, *G. hirsutum*, sometimes also called "American, Mexican or upland cotton", is the most cultivated cotton species globally. Despite cultivated cotton species, there are about 46 wild types of cotton species that majorly belong to Australia and Mexico [2]. The importance of cotton fiber as a product is due to its massive use in the textile industry [3,4]. Regarding world cotton production, 80% comes from India, China, the United States of America (USA), Pakistan, Brazil, Turkey, and Uzbekistan. Cotton production is crucial to the economy of several developing countries such as Pakistan, India, and China [5,6]. However, global warming created several types of abiotic stresses, which limit cotton production worldwide.

The harsh environmental conditions, including drought, salinity, extreme temperatures, and high concentration of heavy metals, result in low crop yield and consequently reduce the economy of a country [7,8]. Although cotton is slightly resistant (glycophytic) to environmental stresses compared with rice, wheat, and maize; however, extreme environmental conditions are still affecting cotton production and fiber quality [9]. Due to the immobile nature of plants, they tackle the abiotic factors in the same environment by inducing different biological pathways and the production of stress hormones [10].

Adaptations to these environmental stresses are critical for the life cycle of cotton plants and also for their successive generations. Therefore, numerous adaptations have been developed by cotton plants over the long course of evolution to perceive, transduce, and respond to environmental stimuli by several morph-physiological, cellular, and molecular processes [11,12]. In response to environmental stresses, extensive molecular reprogramming starts in cotton plants at both the transcriptional and post-transcriptional levels. The principal gene regulators, i.e., transcriptional factors (TFs), trigger several stress-responsive genes to mitigate the effect of abiotic stresses in cotton [13–15]. These transcription factors include MYB, WRKY, ERF, NAC, and bZIP, which are involved in stress responses and development [9]. WRKY TFs are among the major regulators in cotton, which need to be studied for further improvements in cotton. Hereby, the aim of the present study is to highlight the role of WRKY TFs in cotton against abiotic stresses. In addition, it also explains several molecular signaling pathways associated with WRKY TFs in cotton under abiotic stresses.

2. WRKY Transcription Factors

Among the TFs, numerous WRKY TFs have been evaluated in plants for their crucial role in the regulation of stress responses. In addition to stress responses, their role in senescence and development has also been reported [11,13]. The WRKY family is considered one of the largest families of TFs in plants, including cotton. The first member of WRKY transcription factors was studied in sweet potato. Subsequently, several WRKY members have also been identified in other plants, including 74 members in *Arabidopsis*, 109 in rice, 197 in soybean, and 71 in pepper [16,17]. *Arabidopsis* is considered a model plant, and its WRKY proteins are well classified due to their small genome size. WRKY proteins are classified according to the number of WRKY domains and the differences present in the Zinc finger motif. Overall, the members of WRKY proteins are classified into three major groups, i.e., Group I, II, and III, based on the number and diversity of WRKY domains. Group I proteins contain two WRKY domains, while the rest of the two groups contain one in each [18]. On the other hand, Group II and III proteins are differentiated from each other due to the structural differences of the zinc fingers motif. In this way, Group II proteins have a C_2H_2 zinc finger motif, while Group III proteins possess a C_2HXC zinc finger motif [19–21]. Moreover, Group II proteins can be further classified into five subgroups (IIa-IIe) based on the phylogenetic analysis of WRKY. Apart from the WRKY domain and zincfinger motif, most members of WRKY transcription factors have nuclear localization signals, a serine/threonine-rich region, leucine zippers, kinase domains, glutamine-rich region, proline-rich region, and other structures. These various structures confer different transcriptional regulatory functions on WRKY TFs [22].

The name WRKY was assigned to these transcription factors after a conserved region of 60 amino acids called the WRKY domain. The domain is characterized by a highly conserved heptapeptide motif (WRKYGQK) at the N-terminal and a zinc finger-like motif at the C-terminal [23,24]. In order to respond to both internal (development) and external stimuli (stresses), members of the WRKY proteins bind to the W-box (TGACC (A/T)) in the promoter region of its target genes and regulate the expression of downstream genes responsible for the development and/or stress [19,25]. This triggering is usually auto-regulated by the WRKY proteins themselves or by another WRKY transcription factor [11]. During normal conditions, the WRKY genes regulate numerous important biological functions related to the developmental processes in cotton and other plants. For example, a comparative transcriptomic study of cotton species during somatic embryogenesis revealed that 4.8% of WRKY TF encoding genes were detected in somatic embryogenesis [26]. Similarly, *GhWRKY15* is involved in the regulation of root and stem development [17]. In addition, WRKY TFs regulate several physiological processes associated with stress response in cotton either by activating or inhibiting the transcription of physiological processes.

3. Cotton WRKY Transcription Factors

The sequencing of the *Gossypiumraimondii* genome is provided the opportunity to conduct genome-wide identification of WRKY genes in cotton. Like other plants, i.e., *A. thaliana*, *Triticum aestivum*, *Oryza sativa*, and *Cicer arietinum* [27–30], genome-wide identification of cotton WRKY genes has also been assessed [31–33]. Both the whole genome sequence scaffolds of two drafts of the D_5 genome [1,31,34] and express sequence tags (ESTs) from four cotton species, Cai et al. [35] detected 120 candidate WRKY genes. These WRKY genes were based on the sequence information from Paterson et al. [33,35]. Of these TFs, 103 homologous WRKY genes were also found based on the sequence data of Wang et al. [1]. In addition, 3668 ESTs, including 70, 148, 519, and 2935 ESTs from *G. arboreum*, *G. barbadense*, *G. raimondii*, and *G. hirsutum*, respectively, were found to match these WRKY members with at least one EST hit. When the cotton WRKY genes were compared with *Arabidopsis* sequences present in the online database of TAIR (http://www.arabidopsis.org/), 105 WRKY homologs were also found in *Arabidopsis* [35]. These candidate WRKY genes were unevenly distributed on 13 chromosomes in *G. raimondii* from Paterson et al. [34]. Based on the phylogenetic analysis, WRKY members were divided into three major groups, Group I contained 20 members, Group II had 88, and Group III contained 12 members. Group II genes were classified further into five subgroups, groups IIa, b, c, d, and e, which contained 7, 16, 37, 15, and 13 members, respectively [34].

In a genome-wide identification of the WRKY gene family in *G. raimondii*, a total of 116 WRKY genes were found from the complete genome sequence. Members of the WRKY family have distributed unevenly on the chromosomes of *G. raimondii*, chromosome 7 and 5 contained the largest (16.04%) and fewest (2.83%) number of GrWRKY genes, respectively. When the GrWRKY protein domain structures were compared with AtWRKY proteins, variation in the WRKY domain structure was observed [31]. In another study, the authors of the reference [32] identified 109 and 112 WRKY genes in *G. arborium* and *G. raimondii*, respectively. According to the physical mapping, WRKY genes in *G. arborium* were not present on the same chromosome of *G. raimondii*, which revealed that there is a great chromosomal rearrangement in the diploid cotton genome. These studies revealed that there are more than 100 WRKY genes in the *G. arborium* or *G. raimondii* genome. In addition to general identification of the WRKY gene family, drought, salt, heat, cold, and alkalinity responsive members of the WRKY gene family have also been found [31,32,35,36], which shows its crucial role in abiotic stresses. These TFs enhance the stress-induced gene responsiveness and hence the overall stress tolerance. The expression pattern of a large number of GhWRKY genes was high during leaf senescence and fiber development which shows its prominent role in leaf senescence and fiber development of diploid cotton [32,35,37]. Similarly, WRKY genes *GhWRKY27* and *GhWRKY42* have been reported to induce leaf senescence and anther development in transgenic plants [38–40]. The higher expression levels of WRKY transcription factors in different tissues such as root, stem, leaf, petal, and anther reveal their role in the respective tissue [35].

Numerous cotton WRKY genes have been characterized in model plants for their crucial role in modulating stress responses. In a case, overexpression of *GhWRKY15* increased resistance against the infection of viruses and fungi in tobacco [17], while overexpressing *GhWRKY34* and *GhWRKY41* enhanced salt tolerance in *Arabidopsis* and drought tolerance in tobacco, respectively [41,42].

4. Functions of WRKY TFs in Cotton against Abiotic Stresses

The effects of environmental stresses and response mechanisms of cotton are discussed below. It is important to understand the response mechanism of WRKY TFs signaling pathways in manipulating the cotton genome against abiotic stresses.

4.1. Drought and Heat Stresses

According to the United States Department of Agriculture (USDA), cotton production is expected to reduce by drought stress in the USA and other countries [9]. Similarly, in

Pakistan, during the last five years, cotton production declined by 23.7% to an average of 7.42 M bales against the average production of 9.72 M bales from the previous 5 years. In addition to other reasons, Harsh weather is responsible for this decline in cotton production [43,44]. In general, drought conditions restrict plant growth and yield by affecting seed germination, plant height, leaf area index, canopy, fiber quality, and root development [45]. Specifically, it decreases the photosynthetic rate, stomata conduction, transpiration rate, carboxylation efficiency, and water potential of cotton leaves significantly during drought stress [46]. Recently, Ibrahim et al. [47] reported the negative effects of drought stress on cotton plants. According to their study, 31.1% lint yield and 44.4% fresh plant weight of *Gossypium hirsutum* (Zhongmian 41) were reduced under drought stress. In addition, the chlorophyll content and photosynthetic rate were also recorded lower under drought compared with control plants. However, abscisic acid (ABA), indole acetic acid (IAA), superoxide dismutase (SOD), H_2O_2, callose, and proline contents were reported to be higher. The effects of drought stress on cotton and its coping strategies have been extensively reviewed in our previous report [9,46]. Plants have devolved numerous morphological, cellular, and molecular adaptations to cope with drought stress. In the case of biochemical adaptations, cotton WRKY TFs are the key regulators in reducing the effects of drought and heat stresses and consequently lead to various morpho-physiological changes crucial for drought and heat tolerance in cotton.

WRKY transcription factors in cotton have been reported widely for their prominent role in the regulation of drought stress [33,39,48,49]. During an investigation, 34 IId WRKY genes were identified in the *G. hirsutum* genome. Among these, 10 genes were distinctly expressed under drought and salt stresses. The highly expressed gene, Gh_A11G1801, was silenced by Virus-Induced Gene Silencing (VIGS) technology in cotton plants. The VIGS resulted in cotton seedlings showing enhanced sensitivity to drought stress. In addition, these plants had higher malondialdehyde (MDA) content and lower catalase (CAT) content [33]. Similarly, the expression of group III WRKY genes in cotton was assessed under abiotic stresses. Expression patterns of *GhWRKY7*, *50*, *59*, *60*, and *102* were significantly upregulated under ABA, mannitol, and salt treatments. It reveals that these genes may regulate drought and/or salt stress-responsive pathways, i.e., the ABA signaling pathway in cotton plants. In addition, *GhWRKY7* and *GhWRKY7102* genes were markedly expressed in roots under high concentration (200 mmol L^{-1}) of mannitol and NaCl showing their crucial role in root improvement [31]. Moreover, the *GhWRKY1-like* transcription factor was identified in *G. hirsutum* as a drought tolerance regulator. The overexpression of *GhWRKY1-like* in Arabidopsis improved drought tolerance by manipulating ABA synthesis and interaction with several cis-elements [50]. An extensive root system is often counted as favorable for drought tolerance in cotton and other plants. A transcriptomic study of cotton roots under drought stress revealed that *GhWRKY75* is involved in root development [51].

4.2. Salt Stress

Salinity is a global problem that affects approximately 20% of irrigated land and reduces crop yields remarkably [52]. As a glycophyte, cotton is tolerant to mild salt stress; however, high salt concentration affects cotton plants in various ways. It causes oxidative stress and ion toxicity and affects nutrient uptake, increases water deficiency, alters metabolic processes, membrane disorganization, and genotoxicity, and reduces cell division and expansion. Consequently, these factors affect cotton growth, development, productivity, and fiber quality [53]. Cotton plants respond to salinity stress by various morpho-physiological and biochemical changes, where several pathways work together at the transcriptional and post-transcriptional levels. At the transcriptional level, WRKY transcription factors are the key mediators in the salt tolerance of cotton.

With the release of the cotton genome sequence, genome-wide analysis of WRKY family genes has been carried out in *G. arboreum*, *G. raimondii*, and *G. aridum*. Numerous studies have revealed the importance of specific WRKYs in the transcriptional regulation of salt-related genes in cotton [54]. In a transcriptomic analysis of wild-type-salt-tolerant cotton species, 109 GarWRKY genes were identified [55]. Overexpression of a cotton WRKY gene, *GhWRKY25* enhanced salt tolerance in *Nicotiana benthamiana* [56]. Similarly, overexpressing *GhWRKY39-1 Nicotiana benthamiana* plants showed increased salt and oxidative stress tolerance. In addition, overexpression of *GhWRKY39-1* increased the transcriptional level of antioxidant enzyme-associated genes [57]. Moreover, the overexpression of another cotton WRKY gene, *GhWRKY6-like,* markedly enhanced salt tolerance in *Arabidopsis.* On the other hand, the silencing of the *GhWRKY6-like* gene in cotton through VIGS technology increased the sensitivity of cotton plants to drought and salt stresses [58].

4.3. Cold Stress

Cold stress is among the major abiotic stresses that limit cotton growth, productivity, and fiber quality [59–61]. Plants have devolved sophisticated mechanisms involving altered physiological and biochemical processes to cope with cold stress. The coping strategies that plants develop to tolerate harsh conditions are diverse among plants. These strategies often start changes to protect plants in the first instance, followed by cold acclimation, enhancing plant survival under cold stress [62]. Most of these processes are regulated by TFs that trigger the expression of stress-responsive genes. WRKY transcription factors are among those TFs which mediate cold tolerance in cotton plants.

The expression of WRKY TFs during cold treatment in cotton has been widely reported, which exhibited that WRKY TFs regulate cold-stress response. In a transcriptomic analysis of cotton, 10 WRKY TFs were differentially expressed (upregulated) under cold stress [63]. In a previous study, the expression of *GhWRKY41* was significantly induced by cold (4 °C), heat (37 °C), salt, and drought stresses [41]. Similarly, the *GhWRKY15* expression level was also increased after the treatment of the cold. However, the overexpressing *GhWRKY15* and *GhWRKY41* tobacco lines were only checked for drought, salt, and other stresses [17].

The response of cotton plants to low temperatures is not only limited to the transcription network. As a principal stress hormone, ABA promotes phospholipid metabolism and generates second messengers in cold stress [64]. In a study, cold stress slightly enhanced endogenous ABA levels in plants [65]. While the role of WRKY TFs in the ABA pathway has been discussed in Section 5.2, it reveals that WRKY TFs indirectly regulate responses to cold stress.

4.4. Other Abiotic Stresses

Stresses are varied from place to place and time to time; similarly, plants cope with each stress in their own correspondence. In addition to drought, heat, salt, and cold stresses, other abiotic stresses such as waterlog, alkalinity, nutrient deficiency, wounding, and heavy metals stresses have also been reported for cotton. These stresses have also been found to be regulated by WRKY TFs [22,66]. Members of the WRKY gene family were induced in response to waterlog stress. For example, microarray data were used and investigated 50 GhWRKY gene expression patterns in roots and leaves under waterlog stress. As a result, the expression level of several genes was found higher in the root than the in the leaves, which suggests that most of the GhWRKY genes responding to waterlog stress are present in cotton roots. In addition, numerous GhWRKY genes were also induced by pH stress in cotton. The expression level of GhWRKY genes in waterlog and pH stresses suggest the participation of WRKY genes in regulating these stresses [31].

Phosphorus is one of the essential plant nutrients for normal growth and development; however, plants uptake phosphorus in the form of Phosphate [67]. Cotton WRKY gene, *GbWRKY1* regulated Phosphate deficiency in cotton. Overexpressing *GbWRKY1 Arabidopsis* plants reduced Phosphorus deficiency symptoms, accumulated high levels of total phosphorus, increased lateral root development, and Phosphatase activity. These results speculated the positive role of *GbWRKY1* in Phosphate starvation and its involvement in the modulation of Phosphate homeostasis and participation in Phosphate allocation and remobilization [68].

Wounding stress is also counted as one of the abiotic stresses caused by snow, rain, strong wind, herbivores, and insect attack. Wounding tissues are attractive and easy sites for pathogen (bacteria and virus) infection [69]. It activates both local and systematic response mechanisms in plants. The systematic response mechanism includes transcriptional and post-transcriptional changes [70]. In order to validate the role of WRKY TFs in wounding, a cotton WRKY gene, the *GhWRKY15* expression level was markedly high after 2 h of wounding stress [17]. In addition, the transcript level of the cotton WRKY gene, *GhWRKY40*, was increased in cotton upon wounding, while its overexpression enhanced the wounding tolerance in *Nicotiana benthamiana* [71].

Heavy metals are one of the environmental stresses affecting both plants and animals negatively [72]. Upon heavy metal (Cd) stress, several transcription factors, including WRKY, were upregulated in rice [73]. For cotton, although it has been studied successfully in the phytoremediation of heavy metals [74], there is no such report available on the WRKY TF's involvement in the regulation of heavy metals stress response in cotton. There may have functions of cotton WRKY TFs in the regulation of heavy metal stress or phytoremediation of heavy metals. Thus, it is suggested to study WRKY TFs in cotton under heavy metals stress and phytoremediation.

5. Signaling Pathways Associated with WRKY TFs

Cotton has evolved several strategies to cope with abiotic and biotic stresses because they cannot escape from stresses. Numerous phytohormones, Reactive oxygen species (ROS) scavenging, and kinases signaling pathways play a crucial role in signals transmission of abiotic stresses. Phytohormones, ABA, Jasmonate (JA), Auxin (IAA), Salicylic acid (SA), Ethylene (ET), Brassinosteroids (BR), Gibberellin (GA), and Cytokinin (CK) play a key role in regulating cotton plant response against pathogens and abiotic stresses [7,46]. Especially, ABA regulates plant responses against abiotic stresses [15]. These signaling pathways are presented in Figure 1, where several pathways respond to abiotic stresses in a coordinated form.

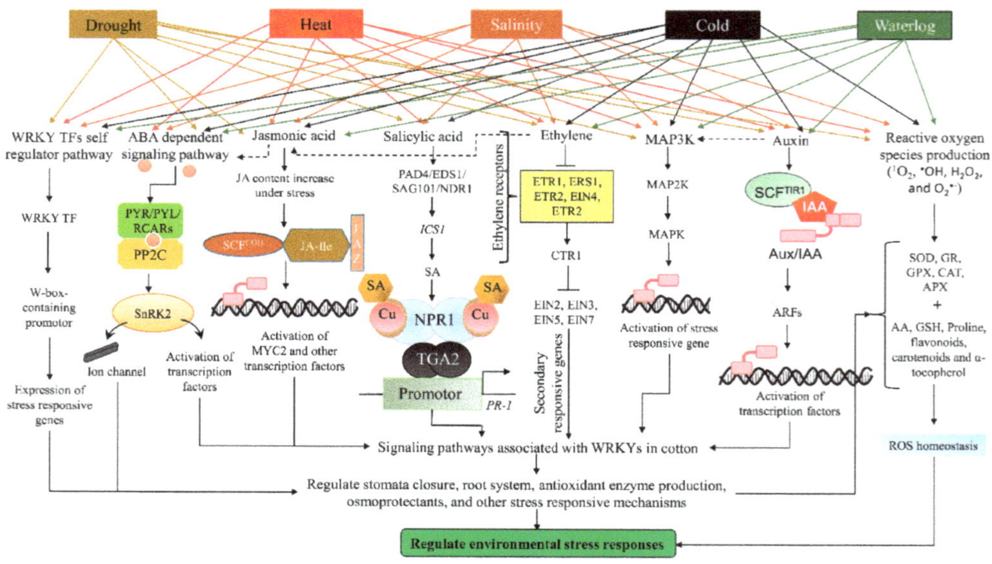

Figure 1. Association of WRKY TFs with stress-responsive signaling pathways during abiotic stresses in cotton.

5.1. Self-Regulatory Pathway

WRKY TFs are involved in stress responses by auto-regulation (self-regulation) or cross-regulation. In auto-regulation, WRKY protein binds to a W-box-containing promotor and auto-regulates the expression of stress-related genes. Apart from self-regulation, in cross-regulation, the expression of stress-related genes is regulated by another WRKY TF [75]. The evidence is available for the model plant, i.e., *Arabidopsis*, where AtWRKY18, 40, and 60 interacted with themselves and with each other under stress conditions [11,22]. In addition, group III WRKY proteins, AtWRKY30, 53, 54, and 70, were also interacting with themselves and with each other [76]. Similarly, GhWRKY91 is directly bound to the W-box promotor of GhWRKY17 and cross-regulated GhWRKY17 expression in cotton under drought stress. GhWRKY17 is involved in ABA signaling and ROS production [77]. It revealed that cotton has a cross-regulatory pathway of WRKY under drought stress. The self/cross-regulatory WRKY signaling pathway is shown in the schematic diagram of overall signaling pathways in cotton under stress conditions (Figure 1).

5.2. Abscisic Acid

Abscisic acid is a major stress-responsive hormone that plays a crucial role in stress signaling pathways. When plants are exposed to different environmental stresses such as extreme temperatures, drought, and salinity, plant growth is modulated by coordination between several plant hormones, proteins, and regulatory factors [78]. ABA plays an essential role in inducing various responses, such as the closing of the stomatal aperture and the expression of stress-responsive genes under abiotic stresses. Abscisic acid responses to stress conditions are divided into slow and rapid responses [79]. The slow response includes the expression of target genes, while a rapid response is mediated by ion channels. Abscisic acid signaling cascade consists of three steps regulatory process, including receptors, protein kinases, and targets, i.e., transcription factors and ion channels [80]. The perception of ABA through ABA receptors (membrane-bound or soluble) initiated the ABA-mediated signaling cascade under stress conditions. The major component of ABA perception and signaling consists of soluble cytoplasmic PYrabactin Resistance (PYR)/PYrabactin Resistance such as (PYL)/Regulatory Component of ABA Receptors (RCAR) proteins. Protein Phosphatase 2C (PP2Cs) is the negative regulator in the ABA signaling pathway, i.e., it inhibits the SNF1-related kinases (SnRK2). Under stress conditions, ABA rapidly accumulates and binds to the ABA receptors (PYR/PYL/RCARs), which in turn bind and inactivate PP2C and lead to the auto-activation of SnRK2 [81–83]. The activated SnRK2 further phosphorylates the downstream targets in the form of TFs and ion channels. The ABA signaling pathway in response to environmental stresses has been presented schematically in Figure 1. Recently, several studies revealed that WRKY TFs are positive regulators of ABA-induced stomatal closure and hence abiotic stress responses, i.e., drought, heat, salt, etc. Further, the role of WRKY TFs in the ABA signaling pathway has been reviewed by the authors of the reference [75] in detail.

WRKY TFs in cotton are key elements in the ABA-mediated signaling network under environmental stresses. Numerous cotton WRKY genes have been testified to be induced by ABA treatment, and the same genes enhanced/reduced abiotic stress tolerance in cotton. In a case study, the overexpression of the cotton WRKY gene, *GhWRKY41*, improved drought and salt tolerance in *Nicotiana benthamiana* by an ABA-dependent signaling pathway. It has been reported widely that ABA regulates stomatal movement. Hereby, stomatal aperture and conductivity were remarkably reduced by ABA treatment. In addition, *GhWRKY41* is highly expressed in stomata. Moreover, the genes associated with ABA signaling were also upregulated during drought or salt stress. Among these, the *SnRK2* gene was also induced significantly by drought or salt treatment, while SnRK2 is directly involved in the ABA signaling pathway [41]. In contrast, *GhWRKY17* increased plant sensitivity to drought stress by reducing the ABA content and expression level of ABA-associated genes. In addition, the accumulation of ROS and the production of antioxidant enzymes were also enhanced in overexpressing *GhWRKY17* plants [48]. These studies revealed the essential role of ABA in cotton and its association with WRKY genes during abiotic stresses.

5.3. Jasmonic Acid, Ethylene, and Salicylic Acid

Jasmonic acid (JA) and its active derivatives, jasmonates, serve as important signaling molecules in the regulation of stress responses and play an important role in mediating the expression of defense-associated genes [84]. In addition, JA is also involved in plant growth and development, such as root growth, tendril coiling, fruit ripening, and viable pollen production [85]. On the other hand, SA is another phytohormone that plays an essential role in plant growth, development, and stress response mechanisms [86]. Similarly, ethylene is a naturally occurring gaseous hormone with multiple actions, including growth, fruit ripening, senescence, flowering, seed germination, and response to both biotic and environmental stresses [46,87]. The JA, SA, and ethylene signaling pathways during stress conditions are presented in Figure 1. WRKY TFs have a crucial role in the regulation of these phytohormones under environmental stresses.

WRKY TFs in cotton are also involved in regulating jasmonic acid, ethylene, and salicylic acid signaling pathways [17,88]. Overexpressing *GarWRKY5* increased salt tolerance in transgenic *Arabidopsis* by regulating ROS scavenging, jasmonic acid, and salicylic acid pathways [55]. The overexpressing *GhWRKY40* enhanced wounding tolerance in *Nicotiana benthamiana*. The transcript level of *GhWRKY40* was increased by the stress hormones methyl jasmonate, salicylic acid, and ethylene [71]. The expression pattern of several GhWRKY genes (*GhWRKY5*, *GhWRKY7*, *GhWRKY27*, *GhWRKY31*, *GhWRKY50*, *GhWRKY56*, *GhWRKY59*, *GhWRKY60*, and *GhWRKY102*) were upregulated by one or more treatments of SA, JA, Ethylene, ABA, mannitol, and NaCl [31]. In another case study, *GhWRKY15* was significantly induced by the application of JA and SA [17].

5.4. Scavenging of ROS

Reactive oxygen species consist of singlet oxygen ($^1O_2$), hydrogen peroxide (H_2O_2), superoxide anion (O_2^-), and hydroxyl radical (OH) [89]. Production of ROS enhances under abiotic and biotic stresses beyond the threshold level and causes ROS-related injuries in plants [90]. When the concentration of ROS in a cell reaches beyond the threshold level, it causes oxidative damage to DNA, RNA, proteins, and membranes. Eventually, it may cause program cell death if the oxidation did not control on time [9]. However, plants evolved a complicated scavenging system to control the excess amount of ROS in cells. The plant scavenging system against ROS consists of enzymatic [e.g., SOD, catalase (CAT), guaiacol peroxidase, glutathione reductase, and ascorbate peroxidase] and non-enzymatic mechanisms [e.g., ascorbic acid, glutathione (GSH), proline, flavonoids, carotenoids, and α-tocopherol] [91–93]. Like other plants, cotton has also developed a complicated scavenging system to cope with abiotic stresses and relieve the effects of oxidative stress. While WRKY TFs have been studied for their prominent role in mediating ROS scavenging pathways in cotton, the evaluation of a WRKY TF gene, *GhWRKY6-like,* showed an elevated response to oxidative stress. Overexpression of *GhWRKY6-like* reduced the H_2O_2 and MDA content in *Arabidopsis* under osmotic stress [58]. Overproduction of ROS in the cell increased the MDA level [94] and is used as a marker for the destructive effects of ROS under stress conditions [92]. In addition, SOD and POD activities were also recorded higher in overexpressing *GhWRKY6-like* transgenic lines under salt and osmotic stresses. Proline played an essential role in oxidative stress responses and was also accumulated higher in overexpressing lines than wild-type (WT) plants under salt and osmotic stresses. They suggested that *GhWRKY6-like* promoted the expression of marker genes associated with the ABA signaling pathway and other stress-responsive genes, thereby enhancing drought and salt tolerance. The enhanced tolerance might be due to the activation of the ABA signaling pathway and improved scavenging system for ROS [58]. The increased antioxidant enzyme activities under salt and drought conditions reflect the reduced MDA level in overexpressing plants suggesting that *GhWRKY6-like* is involved in the scavenging of ROS. In contrast, another WRKY gene, *GhWRKY17*, reduced salt and oxidative stress tolerance in *Nicotiana benthamiana*. Under salt and drought stresses, the overexpressing plants had higher H_2O_2 and O_2^- content compared with WT plants, indicating that ROS levels were increased by the ectopic expression of *GhWRKY17* in tobacco plants. On the other hand, antioxidant enzyme activities in overexpressing plants subjected to drought stress exhibited a decreased level of SOD, POD, APX, and CAT content. In addition, proline content was also recorded to be lower in transgenic lines than in WT plants [48]. The reduced antioxidant enzyme activities under salt and drought conditions reflect oxidative damage in overexpressing plants. These results revealed that *GhWRKY17* is involved in the ROS signaling pathway negatively. These contrasting studies reveal that WRKY transcription factors regulate the ROS signaling pathway in both positive and negative manner in plants under environmental stresses. Thus, we suggest determining the role of WRKY genes before up- or downregulation in a plant and whether it positively or negatively regulates the ROS signaling pathway. Upregulate the expression of WRKY genes that positively regulate the ROS scavenging signaling pathway and downregulate the expression if it is negatively involved in regulating the ROS signaling pathway.

5.5. Kinases (MAPK)

Mitogen-activated protein kinase (MAPK) cascade is one of the principal pathways mediating plant responses to environmental stresses. The MAPK cascade is involved in transferring extracellular signals to the nucleus for appropriate cellular response [95]. This cascade is minimally consisting of three kinases, a MAPKKK (MAP3K), a MAPKK (MAP2K), and a MAPK, which activate each other in a consecutive manner via phosphorylation (Figure 1) [96]. These MAPKs play a significant role in cellular signaling by transferring information from sensors to responders. Mitogen-activated protein kinase cascades are involved in plant responses to water deficit, extreme temperatures, salinity, wounding, and pathogens [97,98]. WRKY TFs in cotton have been studied in the regulation of MAPK activation.

In a genome-wide identification of the MAPKKK gene family in cotton, the authors of the reference [99] identified 157 GhMAPKKKs. Various cis-elements were identified in the promotor regions of GhMAP3Ks related to stress responses. The transcription levels of maximum genes were significantly altered under abiotic stresses. In addition, the expression of some GhMP3Ks genes, i.e., *GhMEKK1012, 24, 31, 36, 38, 40, 45, GhRAF2, 3, 4, 7, 8, 21, 49,* and *GhRAF78* were induced under abiotic stresses in cotton. Silencing of *GhMEKK12* and *GhRAF4* through VIGS technology increased drought sensitivity in cotton seedlings. Drought-related physiological parameters, i.e., relative water content, proline, SOD, and POD contents, were reduced in the gene-silenced seedlings, while the stomatal aperture and MDA contents were enhanced. This study revealed the importance of protein kinases in cotton under abiotic stresses. An earlier study revealed that the cotton WRKY TF gene, *GhWRKY59* played a significant role in the ABA-independent GhMAPK cascade. A GhMAPK cascade consisting of GhMAPKKK15 (MAP3K)-GhMKK4 (MAP2K)-GhMPK6 (MAPK) was identified in cotton by Li et al. [100]. GhWRKY59 regulates MAPK activation and GhMAPKKK expression through feedback. GhWRKY59 binds to the GhDREB2 promoter and regulates the expression of downstream drought-responsible genes. Moreover, it regulates GhMAP3K15 expression positively by establishing a feedback loop. Overexpression of *GhWRKY59* in *Arabidopsis* increased drought tolerance in transgenic plants. A novel phosphorylation loop (GhMAP3K15-GhMKK4-GhMPK6-GhWRKY59) has been found in cotton that regulates the GhDREB2-mediated and ABA-independent drought response, which shows the WRKY-associated MAPK cascade in cotton [100].

The current advancement in the field of plant signaling under stress conditions has revealed that a single factor does not trigger the overall response mechanism in cotton plants. Mostly, these pathways activate each other and work in the form of an integrative signaling mechanism [101]. Under single or multiple stresses, the coordinated response mechanism of cotton in the form of stress signaling pathways associated with WRKY, either directly or indirectly, has been elaborated in the schematic diagram (Figure 1). The tolerance mechanism of cotton to several environmental stresses still needs to be understood due to the complex nature of their response.

6. Cotton WRKY TFs Genetically Engineered Plants

Genetically engineered plants could be a way to mitigate the effects of abiotic stresses. Generally, genetic modification in crop plants comes through the introduction of a beneficial foreign gene or silencing of the expression of an endogenous gene [102]. In the last few decades, genetic manipulation has been started in several crop plants to develop stress-resistant varieties [11]. Several crops have been genetically modified successfully with increased crop yield, stress-resistant, and improved food quality [103–105]. Although concerns persist in humans about the use of genetically modified crops, 526 transgenic events have been approved to date in more than 30 crops for cultivation in various countries of the world. Among them, *Zea maiys* L. (maize) accounts for the highest number of events, i.e., 238, followed by cotton, i.e., 67, and then others [105,106]. Cotton has also been genetically modified to enhance tolerance against abiotic stresses by overexpressing stress-related genes and/or silencing interested genes. Of these, cotton WRKY genes have also been overexpressed and/or silenced in other plants, mostly in *Arabidopsis* and tobacco. Thus far, all the genetically modified cotton WRKY genes have been enlisted in Table 1. Previous research work revealed that WRKY TFs regulate several mechanisms related to biotic and abiotic stresses. It includes root development, stomatal aperture regulation, phytohormone signaling pathways, the generation of antioxidants and osmoprotectant substances, accumulation of metabolites, and stress-associated gene expression [11,22,66,107]. For instance, overexpression of cotton WRKY gene, *GarWRKY5*, enhanced salt tolerance in *Arabidopsis*. Overexpressing lines accumulated higher levels of antioxidant enzymes, i.e., SOD and POD. In addition, transgenic lines had longer roots than wild type. These characteristics in overexpressing transgenic lines improved salt tolerance, while silencing of *GarWRKY5* in cotton increased plant sensitivity to salt stress [55]. In contrast, the ectopic expression of a GhWRKY gene, *GhWRKY17*, enhanced salt and drought sensitivity in *Nicotiana benthamiana*. The overexpression of *GhWRKY17* reduced root length, seed germination, stomatal closure, chlorophyll content, and expression pattern of ABA-signaling-pathway-associated genes. In addition, increased water loss rate, electrolyte leakage, and accumulation of O_2^-, H_2O_2, and MDA contents were examined in overexpressing tobacco which led to enhanced salt and drought sensitivity [48].

Most of the cotton WRKY genes were characterized in *Arabidopsis* and tobacco (Table 1), while a few WRKY genes have also been studied in cotton through virus-induced gene silencing technology. However, there is no report available on the overexpression of WRKY genes in cotton, although it has been overexpressed in other crops. For example, overexpressing *TaWRKY2*, a WRKY transcription factor, significantly enhanced grain yield and drought tolerance in transgenic wheat [108]. On the other side, no one tested the yield parameters, which is the most important part for the farmers and industry (Table 1). Till now, we did not find any report on cotton WRKY genes overexpressing in cotton, although several WRKY family genes have been overexpressed in Arabidopsis and tobacco. Thus, it is strongly suggested to select and overexpress the WRKY genes (as summarized in Table 1) in cotton. In addition, it would be highly appreciated if two or more genes regulating different parameters (stress-tolerant, increased yield, high fiber quality, etc.) would be engineered in the cotton.

Table 1. Cotton WRKY genes regulated abiotic stresses.

WRKY Gene (Sub Group)	Stress	Cellular Localization	Expression in Plant	Traits Regulated	References
GbWRKY1 (I)	Reduced Phosphorus starvation, and tolerance to drought and salt stresses	Nucleus	Overexpressed in Arabidopsis and cotton	Overexpressing lines reduced the accumulation of anthocyanin and enhanced the activity of phosphatase, MDA content, ion leakage and root inhibition	[68,109]
GhWRKY3 (I)	Expressed under drought, salt and low temperature	Nucleus	NA	NA	[110]
GarWRKY5 (III)	Enhanced salt tolerance	NA	Silenced in cotton and overexpressed in Arabidopsis	Overexpressing lines exhibited higher activities of SOD, POD and enhanced root length	[55]
GhWRKY6	Regulated drought and salt stresses	NA	Silenced in cotton and overexpressed in Arabidopsis	Overexpressing plants showed shorter root length, larger stomatal aperture, increased H_2O_2 and MDA level, reduced proline accumulation and participated in ABA signaling pathway	[111]
GhWRKY6-like	Improved salt and drought tolerance	Nucleus	Overexpressed in Arabidopsis and silenced in cotton	In overexpressing lines, MDA and H_2O_2 content were reduced and proline, SOD and POD contents were increased. Expression pattern of ABA signaling pathway genes was also reported higher in overexpressing lines.	[58]
GhWRKY15 (IId)	Increased resistant against wounding Viral and fungal infection	Nucleus	Overexpressed in Tobacco	Increased POD, APX and expression of stress related genes	[12]
GhWRKY17 (IId)	Enhanced drought and salt sensitivity	Nucleus	Overexpressed in *Nicotiana benthamiana*	Reduced root length, seed germination, stomatal closure, chlorophyll content and expression pattern of marker genes involved in ABA signaling pathway Increased water loss rate, electrolyte leakage and accumulation of O_2^-, H_2O_2, MDA content	[48]
GhWRKY25 (I)	Enhanced salt tolerance but reduced drought tolerance and plant defense against fungal pathogen	Nucleus	Overexpressed in *Nicotiana benthamiana*	Increased MDA, O_2^-, and H_2O_2 content Decreased SOD, POD and CAT activities, inhibited root length under drought stress	[56]
GhWRKY27a (III)	Reduced drought tolerance	Nucleus	Overexpressed in *Nicotiana benthamiana* and silenced in cotton	Overexpressing lines exhibited short roots, closed stomata, high content O_2^-, and H_2O_2, and low level of drought related genes expression	[49]
GhWRKY33 (III)	Reduced drought tolerance	Nucleus	Overexpressed in Arabidopsis	Inhibited seed germination, early seedling growth and root length, reduced sensitivity to ABA	[24]
GhWRKY34 (III)	Enhanced salt tolerance	Nucleus	Overexpressed in Arabidopsis	Seed germination, root length, chlorophyll content and expression pattern of stress related genes was higher	[42]
GhWRKY39 (IId)	Enhanced resistance to pathogen infection and salinity	Nucleus	Overexpressed in *Nicotiana benthamiana*	Reduced hydrogen peroxide accumulation and increased the level of APX, CAT, GST and SOD	[112]
GhWRKY39-1 (IId)	Enhanced resistance to pathogen infection and salinity	Nucleus	Overexpressed in *Nicotiana benthamiana*	Enhanced root length and expression pattern of SOD, GST, APX, and CAT. Decreased H_2O_2 content	[57]
GhWRKY40 (IIa)	Enhanced tolerance against wounding	Nucleus	Overexpressed in *Nicotiana benthamiana*	Decreased level of O_2^- and H_2O_2 Transcript level of JA and SA associated genes in overexpressing plants was decreased	[71]
GhWRKY41 (III)	Enhanced drought and salinity tolerance	Nucleus	Overexpressed in *Nicotiana benthamiana*	Reduced MDA, H_2O_2 content and ABA dependent stomatal opening. Accumulated increased level of SOD, POD and CAT	[41]
GhWRKY42 (IId)	Induced drought and salinity stress	Nucleus	Overexpressed in Arabidopsis	Increased senescence	[33]
GhWRKY68 (IIc)	Reduced drought and salt tolerance	Nucleus	Overexpressed in *Nicotiana benthamiana*	Stomatal opening, O_2^-, H_2O_2, and MDA content were increased. Reduced total chlorophyll, CAT, SOD and POD content, root length, expression pattern of ABA-dependent pathway genes	[113]
GhWRKY91 (IIe)	Enhanced drought tolerance and delayed senescence	NA	Overexpressed in Arabidopsis	Expression pattern of marker genes involved in drought tolerance was significantly higher.	[77]

Footnote: ABA = Abscisic acid, APX = Ascrobate peroxidase, CAT = Catalase, GST = Glutathione peroxidase, MDA = Malondialdehyde, NA = Not applicable, POD = Peroxidase, SOD = Superoxide dismutase.

7. Concluding Remarks, Complications, and Recommendations

Cotton is an important industrial crop, and its production has been hampered badly due to various factors in most of the world. Consequently, reduced the economic growth of several developing countries such as Pakistan, India, China, Brazil, Turkey, etc. [9]. In this regard, research is on the way to improving tolerance in cotton against abiotic stresses. Transcription factors are the major regulators of biological processes involved in abiotic stresses [114]. Among the transcription factors, WRKYs are the largest family of transcription factors regulating several stress-related pathways, including ABA, JA, SA, MAPK, and the scavenging of ROS. Understanding stress-responsive signaling pathways and the consequent manipulation of these could be a gateway for the resistant cotton variety. WRKY transcription factors and their related genes have been expressed in other plants (mostly in Arabidopsis and tobacco), showing significant responses against abiotic stresses. However, the role of those genes/TF may not show the same results in cotton due to its tetra genomic (AD genome) nature. On the other hand, producing transgenic cotton plants require a long time to obtain homozygous transgenic seeds, which disheartens scientists if the resulting transgenic cotton does not produce the same results shown in model plants. Still, various transgenic cotton plants have been produced after a long struggle by scholars. The transgenic cotton plants showed better resistance than existing varieties. Even if successful genetically modified cotton plants are produced, research should still continue to improve and combat future challenges. The available reports on cotton WRKY transcription factors reveal that these TFs regulate abiotic stresses positively and/or negatively.

Author Contributions: X.G. wrote the paper, A.U. designed the study, D.S. helped in writing the paper, B.K. critically analyzed the paper, and S.I. drew the figure. All authors have read and agreed to the published version of the manuscript.

Funding: The authors highly acknowledge the Faculty of Process and Environmental Engineering, Lodz University of Technology, Wolczanska Str. 213, 90-924 Lodz, Poland, for providing financial support to the current study.

Conflicts of Interest: The authors declare no conflict of interest.

References

1. Wang, K.; Wang, Z.; Li, F.; Ye, W.; Wang, J.; Song, G.; Yue, Z.; Cong, L.; Shang, H.; Zhu, S.; et al. The draft genome of a diploid cotton *Gossypium raimondii*. *Nat. Genet.* **2012**, *4410*, 1098–1103. [CrossRef] [PubMed]
2. Chen, Z.J.; Scheffler, B.E.; Dennis, E.; Triplett, B.A.; Zhang, T.; Guo, W.; Chen, X.; Stelly, D.M.; Rabinowicz, P.D.; Town, C.D.; et al. Toward sequencing cotton Gossypium genomes. *Plant Physiol.* **2007**, *1454*, 1303–1310. [CrossRef] [PubMed]
3. Egbuta, M.A.; McIntosh, S.; Waters, D.L.; Vancov, T.; Liu, L. Biological Importance of Cotton By-Products Relative to Chemical Constituents of the Cotton Plant. *Molecules* **2017**, *221*, 93. [CrossRef]
4. Zhang, T.; Hu, Y.; Jiang, W.; Fang, L.; Guan, X.; Chen, J.; Zhang, J.; Saski, C.A.; Scheffler, B.E.; Stelly, D.M.; et al. Sequencing of allotetraploid cotton *Gossypium hirsutum* L. acc. TM-1 provides a resource for fiber improvement. *Nat. Biotechnol.* **2015**, *335*, 531–537. [CrossRef]
5. Shi, Y.; Liu, A.; Li, J.; Zhang, J.; Zhang, B.; Ge, Q.; Jamshed, M.; Lu, Q.; Li, S.; Xiang, X.; et al. Dissecting the genetic basis of fiber quality and yield traits in interspecific backcross populations of Gossypium hirsutum × Gossypium barbadense. *Mol. Genet. Genom.* **2019**, *294*, 1385–1402. [CrossRef]
6. Statista. 2022. Available online: https://www.statista.com/statistics/263055/cotton-production-worldwide-by-top-countries/ (accessed on 12 February 2022).
7. He, M.; He, C.Q.; Ding, N.Z. Abiotic Stresses: General Defences of Land Plants and Chances for Engineering Multi stress Tolerance. *Front. Plant Sci.* **2018**, *9*, 1771. [CrossRef]
8. Khan, A.; Tan, D.K.Y.; Afridi, M.Z.; Luo, H.; Tung, S.; Ajab, M.; Fahad, S. Nitrogen fertility and abiotic stresses management in cotton crop: A review. *Environ. Sci. Pollut. Res.* **2017**, *24*, 14551–14566. [CrossRef]
9. Ullah, A.; Sun, H.; Yang, X.; Zhang, X. Drought coping strategies in cotton: Increased crop per drop. *Plant Biotechnol. J.* **2017**, *153*, 271–284. [CrossRef]
10. Lata, R.; Chowdhury, S.; Gond, S.K.; White, J.F., Jr. Induction of abiotic stress tolerance in plants by endophytic microbes. *Lett. Appl. Microbiol.* **2018**, *664*, 268–276. [CrossRef]
11. Banerjee, A.; Roychoudhury, A. WRKY proteins: Signaling and regulation of expression during abiotic stress responses. *Scient. World J.* **2015**, *2015*, 807560. [CrossRef]

12. Gull, A.; Lone, A.A.; Wani, N.U.I. Biotic and Abiotic Stresses in Plants. In *Abiotic and Biotic Stress in Plants*; IntechOpen: London, UK, 2019.
13. Abdelraheem, A.; Esmaeili, N.; O'Connell, M.; Zhang, J. Progress and perspective on drought and salt stress tolerance in cotton. *Ind. Crops Prod.* **2019**, *130*, 118–129. [CrossRef]
14. Roy, S. 2016. Function of MYB domain transcription factors in abiotic stress and epigenetic control of stress response in plant genome. *Plant Signal. Behav.* **2016**, *111*, e1117723. [CrossRef] [PubMed]
15. Wei, Y.; Xu, Y.; Lu, P.; Wang, X.; Li, Z.; Cai, X.; Zhou, Z.; Wang, Y.; Zhang, Z.; Lin, Z.; et al. Salt stress responsiveness of a wild cotton species *Gossypium klotzschianum* based on transcriptomic analysis. *PLoS ONE* **2015**, *125*, e0178313. [CrossRef]
16. Chen, F.; Hu, Y.; Vannozzi, A.; Wu, K.; Cai, H.; Qin, Y.; Mullis, A.; Lin, Z.; Zhang, L. The WRKY transcription factor family in model plants and crops. *Crit. Rev. Plant Sci.* **2017**, *365*, 311–335. [CrossRef]
17. Yu, F.; Huaxia, Y.; Lu, W.; Wu, C.; Cao, X.; Guo, X. *GhWRKY15*, a member of the WRKY transcription factor family identified from cotton *Gossypium hirsutum* L., is involved in disease resistance and plant development. *BMC Plant Biol.* **2012**, *121*, 144–161. [CrossRef]
18. Wani, S.H.; Anand, S.; Singh, B.; Bohra, A.; Joshi, R. WRKY transcription factors and plant defense responses: Latest discoveries and future prospects. *Plant Cell Rep.* **2021**, *40*, 1071–1085. [CrossRef]
19. Bakshi, M.; Oelmüller, R. WRKY transcription factors: Jack of many trades in plants. *Plant Signal. Behav.* **2014**, *92*, e27700. [CrossRef]
20. Song, H.; Wang, P.; Nan, Z.; Wang, X. The WRKY Transcription Factor Genes in Lotus japonicus. *Int. J. Genom.* **2014**, *2014*, 420128.
21. Xie, T.; Chen, C.; Li, C.; Liu, J.; Liu, C.; He, Y. Genome-wide investigation of WRKY gene family in pineapple: Evolution and expression profiles during development and stress. *BMC Genom.* **2018**, *191*, 490. [CrossRef]
22. Jiang, J.; Ma, S.; Ye, N.; Jiang, M.; Cao, J.; Zhang, J. WRKY transcription factors in plant responses to stresses. *J. Integr. Plant Biol.* **2017**, *592*, 86–101. [CrossRef]
23. Rushton, P.J.; Somssich, I.E.; Ringler, P.; Shen, Q.J. WRKY transcription factors. *Trends Plant Sci.* **2010**, *155*, 247–258. [CrossRef] [PubMed]
24. Wang, N.N.; Xu, S.W.; Sun, Y.L.; Liu, D.; Zhou, L.; Li, Y.; Li, X.B. The cotton WRKY transcription factor *GhWRKY33* reduces transgenic Arabidopsis resistance to drought stress. *Sci. Rep.* **2019**, *91*, 724. [CrossRef] [PubMed]
25. Schluttenhofer, C.; Yuan, L. Regulation of specialized metabolism by WRKY transcription factors. *Plant Physiol.* **2015**, *1672*, 295–306. [CrossRef] [PubMed]
26. Sun, R.; Tian, R.; Ma, D.; Wang, S.; Lium, C. Comparative transcriptome study provides insights into acquisition of embryogenic ability in upland cotton during somatic embryogenesis. *J. Cotton Res.* **2018**, *11*, 9. [CrossRef]
27. Ross, C.A.; Liu, Y.; Shen, Q.X.J. The WRKY gene family in rice *Oryza sativa*. *J. Integr. Plant Biol.* **2007**, *496*, 827–842. [CrossRef]
28. Song, Y.; Gao, J. Genome-wide analysis of WRKY gene family in *Arabidopsis lyrata* and comparison with *Arabidopsis thaliana* and *Populus trichocarpa*. *Chin. Sci. Bull.* **2014**, *598*, 754–765. [CrossRef]
29. Huang, X.; Li, K.; Xu, X.; Yao, Z.; Jin, C.; Zhang, S. Genome-wide analysis of WRKY transcription factors in white pear *Pyrus bretschneideri* reveals evolution and patterns under drought stress. *BMC Genom.* **2015**, *161*, 1104. [CrossRef]
30. Waqas, M.; Azhar, M.T.; Rana, I.A.; Azeem, F.; Ali, M.A.; Nawaz, M.A.; Chung, G.; Atif, R.M. Genome-wide identification and expression analyses of WRKY transcription factor family members from chickpea *Cicer arietinum* L. reveal their role in abiotic stress-responses. *Genes Genom.* **2019**, *414*, 467–481. [CrossRef]
31. Dou, L.; Zhang, X.; Pang, C.; Song, M.; Wei, H.; Fan, S.; Yu, S. Genome-wide analysis of the WRKY gene family in cotton. *Mol. Genet. Genom.* **2014**, *2896*, 1103–1121. [CrossRef]
32. Ding, M.; Chen, J.; Jiang, Y.; Lin, L.; Cao, Y.; Wang, M.; Zhang, Y.; Rong, J.; Ye, W. Genome-wide investigation and transcriptome analysis of the WRKY gene family in Gossypium. *Mol. Genom.* **2014**, *290*, 151–171. [CrossRef]
33. Gu, L.; Wang, H.; Wei, H.; Sun, H.; Li, L.; Chen, P.; Elasad, M.; Su, Z.; Zhang, C.; Ma, L.; et al. Identification, Expression, and Functional Analysis of the Group IId WRKY Subfamily in Upland Cotton *Gossypium hirsutum* L. *Front. Plant Sci.* **2018**, *9*, 1684. [CrossRef] [PubMed]
34. Paterson, A.H.; Wendel, J.F.; Gundlach, H.; Guo, H.; Jenkins, J.; Jin, D.; Llewellyn, D.; Showmaker, K.C.; Shu, S.; Udall, J.; et al. Repeated polyploidization of Gossypium genomes and the evolution of spinnable cotton fibres. *Nature* **2012**, *4927429*, 423–427. [CrossRef] [PubMed]
35. Cai, C.; Niu, E.; Du, H.; Zhao, L.; Feng, Y.; Guo, W. Genome-wide analysis of the WRKY transcription factor gene family in *Gossypium raimondii* and the expression of orthologs in cultivated tetraploid cotton. *Crop J.* **2014**, *2*, 87–101. [CrossRef]
36. Fan, X.; Guo, Q.; Xu, P.; Gong, Y.; Shu, H.; Yang, Y.; Ni, W.; Zhang, X.; Shen, X. Transcriptome-wide identification of salt-responsive members of the WRKY gene family in *Gossypium aridum*. *PLoS ONE* **2015**, *10*, e0216148.
37. Hande, A.S.; Katageri, I.S.; Jadhav, M.P.; Adiger, S.; Gamanagatti, S.; Padmalatha, K.V.; Dhandapani, G.; Kanakachari, M.; Kumar, P.A.; Reddy, V.S. Transcript profiling of genes expressed during fibre development in diploid cotton *Gossypium arboreum* L. *BMC Genom.* **2017**, *181*, 675. [CrossRef]
38. Gu, L.; Wei, H.; Wang, H.; Su, J.; Yu, S. Characterization and functional analysis of *GhWRKY42*, a group IId WRKY gene, in upland cotton *Gossypium hirsutum* L. *BMC Genet.* **2018**, *191*, 48. [CrossRef]

39. Gu, L.; Dou, L.; Guo, Y.; Wang, H.; Li, L.; Wang, C.; Chen, P.; Elasad, M.; Su, Z.; Zhang, C.; et al. The WRKY transcription factor *GhWRKY27* coordinates the senescence regulatory pathway in upland cotton *Gossypium hirsutum* L. *BMC Plant Biol.* **2019**, *191*, 116. [CrossRef]
40. Wang, Y.; Li, Y.; He, S.P.; Gao, Y.; Wang, N.N.; Lu, R.; Li, X.B. A cotton *Gossypium hirsutum* WRKY transcription factor *GhWRKY22* participates in regulating anther/pollen development. *Plant Physiol. Biochem.* **2019**, *141*, 231–239. [CrossRef]
41. Chu, X.; Wang, C.; Chen, X.; Lu, W.; Li, H.; Wang, X.; Hao, L.; Guo, X. The Cotton WRKY Gene *GhWRKY41* Positively Regulates Salt and Drought Stress Tolerance in Transgenic *Nicotiana benthamiana*. *PLoS ONE* **2015**, *1011*, e0143022. [CrossRef]
42. Zhou, L.; Wang, N.; Gong, S.; Lu, R.; Li, Y.; Li, X. Overexpression of a cotton *Gossypium hirsutum* WRKY gene, *GhWRKY34*, in Arabidopsis enhances salt–tolerance of the transgenic plants. *Plant Physiol. Biochem.* **2015**, *96*, 311–320. [CrossRef]
43. Index Mundi. 2020. Available online: https://www.indexmundi.com/agriculture/?country=pk&commodity=cotton&graph=production (accessed on 12 April 2020).
44. Shakeel, A. Dawn Newspaper. 2020. Available online: https://www.dawn.com/news/1529289 (accessed on 12 June 2020).
45. Loka, D.A.; Oosterhuis, D.M. *Water Stress and Reproductive Development in Cotton*; Department of Crop, Soil, and Environmental Sciences University of Arkansas: Fayetteville, AR, USA, 2012.
46. Ullah, A.; Manghwar, H.; Shaban, M.; Khan, A.H.; Akbar, A.; Ali, U.; Ali, E.; Fahad, S. Phytohormones enhanced drought tolerance in plants: A coping strategy. *Environ. Sci. Pollut. Res.* **2018**, *2533*, 33103–33118. [CrossRef] [PubMed]
47. Ibrahim, W.; Zhu, Y.M.; Chen, Y.; Qiu, C.W.; Zhu, S.; Wu, F. Genotypic differences in leaf secondary metabolism, plant hormones and yield under alone and combined stress of drought and salinity in cotton genotypes. *Physiol. Plant.* **2019**, *1652*, 343–355. [CrossRef] [PubMed]
48. Yan, H.; Jia, H.; Chen, X.; Hao, L.; An, H.; Guo, X. The cotton WRKY transcription factor *GhWRKY17* functions in drought and salt stress in transgenic *Nicotiana benthamiana* through ABA signaling and the modulation of reactive oxygen species production. *Plant Cell Physiol.* **2014**, *5512*, 2060–2076. [CrossRef] [PubMed]
49. Yan, Y.; Jia, H.; Wang, F.; Wang, C.; Liu, S.; Guo, X. Overexpression of *GhWRKY27a* reduces tolerance to drought stress and resistance to *Rhizoctonia solani* infection in transgenic *Nicotiana benthamiana*. *Front. Physiol.* **2015**, *6*, 265. [CrossRef]
50. Hu, Q.; Ao, C.; Wang, X.; Wu, Y.; Du, X. GhWRKY1-like, a WRKY transcription factor, mediates drought tolerance in Arabidopsis via modulating ABA biosynthesis. *BMC Plan Biol.* **2021**, *21*, 458. [CrossRef]
51. Ranjan, A.; Sawant, S. Genome-wide transcriptomic comparison of cotton *Gossypium herbaceum* leaf and root under drought stress. *3 Biotech* **2015**, *54*, 585–596. [CrossRef]
52. Negrao, S.; Schmockel, S.M.; Tester, M. Evaluating physiological responses of plants to salinity stress. *Ann. Bot.* **2017**, *1191*, 1–11. [CrossRef] [PubMed]
53. Sharif, I.; Aleem, S.; Farooq, J.; Rizwan, M.; Younas, A.; Sarwar, G.; Chohan, S.M. Salinity stress in cotton: Effects, mechanism of tolerance and its management strategies. *Physiol. Mol. Biol. Plants* **2019**, *254*, 807–820. [CrossRef]
54. Ashraf, J.; Zuo, D.; Wang, Q.; Malik, W.; Zhang, Y.; Abid, M.A.; Cheng, H.; Yang, Q.; Song, G. Recent insights into cotton functional genomics: Progress and future perspectives. *Plant Biotechnol. J.* **2018**, *163*, 699–713. [CrossRef]
55. Guo, Q.; Zhao, L.; Fan, X.; Xu, P.; Xu, Z.; Zhang, X.; Meng, S.; Shen, X. Transcription Factor *GarWRKY5* Is Involved in Salt Stress Response in Diploid Cotton Species *Gossypium aridum* L. *Int. J. Mol. Sci.* **2021**, *2021*, 5244. [CrossRef]
56. Liu, X.; Song, Y.; Xing, F.; Wang, N.; Wen, F.; Zhu, C. *GhWRKY25*, a group I WRKY gene from cotton, confers differential tolerance to abiotic and biotic stresses in transgenic *Nicotiana benthamiana*. *Protoplasma* **2016**, *2535*, 1265–1281. [CrossRef]
57. Shi, W.; Hao, L.; Li, J.; Liu, D.; Guo, X.; Li, H. The *Gossypium hirsutum* WRKY gene *GhWRKY39–1* promotes pathogen infection defense responses and mediates salt stress tolerance in transgenic *Nicotiana benthamiana*. *Plant Cell Rep.* **2014**, *33*, 483–498. [CrossRef]
58. Ullah, A.; Sun, H.; Yang, X.; Zhang, X. A novel cotton WRKY gene, *GhWRKY6*—Like, improves salt tolerance by activating the ABA signaling pathway and scavenging of reactive oxygen species. *Physiol. Plant.* **2018**, *1624*, 439–454. [CrossRef]
59. Sanghera, G.S.; Wani, S.H.; Hussain, W.; Singh, N.B. Engineering cold stress tolerance in crop plants. *Curr. Genom.* **2011**, *121*, 30. [CrossRef]
60. Chinnusamy, V.; Zhu, J.K.; Sunkar, R. Gene Regulation during Cold Stress Acclimation in Plants. In *Plant Stress Tolerance*; Humana Press: Totowa, NJ, USA, 2011; pp. 39–55.
61. Yuan, P.; Yang, T.; Poovaiah, B.W. Calcium Signaling-Mediated Plant Response to Cold Stress. *Int. J. Mol. Sci.* **2018**, *1912*, 3896. [CrossRef]
62. Vyse, K.; Pagter, M.; Zuther, E.; Hincha, D.K. Deacclimation after cold acclimation-a crucial, but widely neglected part of plant winter survival. *J. Exp. Bot.* **2019**, *7018*, 4595–4604. [CrossRef]
63. Zhu, Y.N.; Shi, D.Q.; Ruan, M.B.; Zhang, L.L.; Meng, Z.H.; Liu, J.; Yang, W.C. Transcriptome analysis reveals crosstalk of responsive genes to multiple abiotic stresses in cotton *Gossypium hirsutum* L. *PLoS ONE* **2013**, *811*, e80218. [CrossRef] [PubMed]
64. Barrero-Sicilia, C.; Silvestre, S.; Haslam, R.P.; Michaelson, L.V. Lipid remodelling: Unravelling the response to cold stress in Arabidopsis and its extremophile relative *Eutrema salsugineum*. *Plant Sci.* **2017**, *263*, 194–200. [CrossRef]
65. Pareek, A.; Khurana, A.K.; Sharma, A.; Kumar, R. An overview of signaling regulons during cold stress tolerance in plants. *Curr. Genom.* **2017**, *186*, 498–511. [CrossRef]
66. Phukan, U.J.; Jeena, G.S.; Shukla, R.K. 2016 WRKY transcription factors: Molecular regulation and stress responses in plants. *Front. Plant Sci.* **2016**, *7*, 760. [CrossRef]

67. Hasan, M.M.; Hasan, M.M.; da Silva, J.A.T.; Li, X. Regulation of phosphorus uptake and utilization: Transitioning from current knowledge to practical strategies. *Cell. Mol. Biol. Lett.* **2016**, *211*, 7. [CrossRef] [PubMed]
68. Xu, L.; Jin, L.; Long, L.; Liu, L.; He, X.; Gao, W.; Zhu, L.; Zhang, X. Overexpression of *GbWRKY1* positively regulates the Pi starvation response by alteration of auxin sensitivity in Arabidopsis. *Plant Cell Rep.* **2012**, *3112*, 2177–2188. [CrossRef]
69. León, J.; Rojo, E.; Sánchez-Serrano, J.J. Wound signalling in plants. *J. Exp. Bot.* **2001**, *52*, 1–9. [CrossRef]
70. Savatin, D.V.; Gramegna, G.; Modesti, V.; Cervone, F. Wounding in the plant tissue: The defense of a dangerous passage. *Front. Plant Sci.* **2014**, *5*, 470. [CrossRef]
71. Wang, X.; Yan, Y.; Li, Y.; Chu, X.; Wu, C.; Guo, X. *GhWRKY40*, a multiple stress–responsive cotton WRKY gene, plays an important role in the wounding response and enhances susceptibility to *Ralstonia solanacearum* infection in transgenic *Nicotiana benthamiana*. *PLoS ONE* **2014**, *94*, e93577. [CrossRef]
72. Ullah, A.; Heng, S.; Munis, M.F.H.; Fahad, S.; Yang, X. Phytoremediation of heavy metals assisted by plant growth promoting PGP bacteria: A review. *Environ. Exp. Bot.* **2015**, *117*, 28–40. [CrossRef]
73. Ogawa, I.; Nakanishi, H.; Mori, S.; Nishizawa, N.K. Time course analysis of gene regulation under cadmium stress in rice. *Plant Soil* **2009**, *325*, 97. [CrossRef]
74. Kaur, R.; Bhatti, S.S.; Singh, S.; Singh, J.; Singh, S. Phytoremediation of heavy metals using cotton plant: A field analysis. Bull. Environ. *Contam. Toxicol.* **2018**, *1015*, 637–643. [CrossRef]
75. Rushton, D.L.; Tripathi, P.; Rabara, R.C.; Lin, J.; Ringler, P.; Boken, A.K.; Langum, T.J.; Smidt, L.; Boomsma, D.D.; Emme, N.J.; et al. WRKY transcription factors: Key components in abscisic acid signalling. *Plant Biotechnol. J.* **2012**, *10*, 2–11. [CrossRef]
76. Besseau, S.; Li, J.; Palva, E.T. WRKY54 and WRKY70 co-operate as negative regulators of leaf senescence in *Arabidopsis thaliana*. *J. Exp. Bot.* **2012**, *637*, 2667–2679. [CrossRef]
77. Gu, L.; Ma, Q.; Zhang, C.; Wang, C.; Wei, H.; Wang, H.; Yu, S. The Cotton GhWRKY91 Transcription Factor Mediates Leaf Senescence and Responses to Drought Stress in Transgenic *Arabidopsis thaliana*. *Front. Plant Sci.* **2019**, *10*, 1352. [CrossRef]
78. Xiong, L.; Schumaker, K.S.; Zhu, J.K. Cell signaling during cold, drought, and salt stress. *Plant Cell* **2002**, *14* (Suppl. S1), S165–S183. [CrossRef]
79. Vishwakarma, K.; Upadhyay, N.; Kumar, N.; Yadav, G.; Singh, J.; Mishra, R.K.; Kumar, V.; Verma, R.; Upadhyay, R.G.; Pandey, M.; et al. Abscisic acid signaling and abiotic stress tolerance in plants: A review on current knowledge and future prospects. *Front. Plant Sci.* **2017**, *8*, 161. [CrossRef]
80. Wasilewska, A.; Vlad, F.; Sirichandra, C.; Redko, Y.; Jammes, F.; Valon, C.; Frey, N.F.D.; Leung, J. An update on abscisic acid signaling in plants and more. *Mol. Plant* **2008**, *12*, 198–217. [CrossRef]
81. Ilyas, M.; Nisar, M.; Khan, N.; Hazrat, A.; Khan, A.H.; Hayat, K.; Fahad, S.; Khan, A.; Ullah, A. Drought Tolerance Strategies in Plants: A Mechanistic Approach. *J. Plant Growth Regul.* **2020**, *40*, 926–944. [CrossRef]
82. Kim, T.H.; Böhmer, M.; Hu, H.; Nishimura, N.; Schroeder, J.I. Guard cell signal transduction network: Advances in understanding abscisic acid, CO_2, and Ca^{2+} signaling. *Annu. Rev. Plant Biol.* **2010**, *61*, 561–591. [CrossRef]
83. Yu, F.; Lou, L.; Tian, M.; Li, Q.; Ding, Y.; Cao, X.; Wu, Y.; Belda-Palazon, B.; Rodriguez, P.L.; Yang, S.; et al. ESCRT-I component VPS23A affects ABA signaling by recognizing ABA receptors for endosomal degradation. *Mol. Plant* **2016**, *912*, 1570–1582. [CrossRef]
84. Fujita, M.; Fujita, Y.; Noutoshi, Y.; Takahashi, F.; Narusaka, Y.; Shinozaki, K.Y.; Shinozaki, K. Crosstalk between abiotic and biotic stress responses: A current view from the points of convergence in the stress signaling networks. *Curr. Opin. Plant Biol.* **2006**, *9*, 4436–4442. [CrossRef]
85. Sun, H.; Chen, L.; Li, J.; Hu, M.; Ullah, A.; He, X.; Yang, X.; Zhang, X. The JASMONATE ZIM-domain gene family mediates JA signaling and stress response in cotton. *Plant Cell Physiol.* **2017**, *5812*, 2139–2154. [CrossRef]
86. Khan, M.I.R.; Fatma, M.; Per, T.S.; Anjum, N.A.; Khan, N.A. Salicylic acid-induced abiotic stress tolerance and underlying mechanisms in plants. *Front. Plant Sci.* **2015**, *6*, 462. [CrossRef]
87. Iqbal, N.; Khan, N.A.; Ferrante, A.; Trivellini, A.; Francini, A.; Khan, M.I.R. Ethylene role in plant growth, development and senescence: Interaction with other phytohormones. *Front. Plant Sci.* **2017**, *8*, 475. [CrossRef] [PubMed]
88. Xiong, X.P.; Sun, S.C.; Zhang, X.Y.; Li, Y.J.; Liu, F.; Zhu, Q.H.; Xue, F.; Sun, J. *GhWRKY70D13* regulates resistance to *Verticillium dahliae* in Cotton through the Ethylene and Jasmonic Acid signaling pathways. *Front. Plant Sci.* **2020**, *11*, 69. [CrossRef] [PubMed]
89. Waszczak, C.; Carmody, M.; Kangasjärvi, J. Reactive oxygen species in plant signaling. *Annu. Rev. Plant Biol.* **2018**, *69*, 209–236. [CrossRef]
90. Choudhury, F.K.; Rivero, R.M.; Blumwald, E.; Mittler, R. Reactive oxygen species, abiotic stress and stress combination. *Plant J.* **2017**, *905*, 856–867. [CrossRef]
91. Noctor, G.; Mhamdi, A.; Foyer, C.H. The roles of reactive oxygen metabolism in drought: Not so cut and dried. *Plant Physiol.* **2014**, *1644*, 1636–1648. [CrossRef] [PubMed]
92. Sharma, P.; Jha, A.B.; Dubey, R.S.; Pessarakli, M. Reactive oxygen species, oxidative damage, and antioxidative defense mechanism in plants under stressful conditions. *J. Bot.* **2012**, *2012*, 217037. [CrossRef]
93. Zandalinas, S.I.; Fichman, Y.; Devireddy, A.R.; Sengupta, S.; Azad, R.K.; Mittler, R. Systemic signaling during abiotic stress combination in plants. *Proc. Natl. Acad. Sci. USA* **2020**, *24*, 14551–14566. [CrossRef]
94. Huang, H.; Ullah, F.; Zhou, D.X.; Yi, M.; Zhao, Y. Mechanisms of ROS Regulation of Plant Development and Stress Responses. *Front. Plant Sci.* **2019**, *10*, 800. [CrossRef]

95. Zhu, J.K. Abiotic stress signaling and responses in plants. *Cell* **2016**, *167*, 313–324. [CrossRef]
96. de Zelicourt, A.; Colcombet, J.; Hirt, H. 2016 The role of MAPK modules and ABA during abiotic stress signaling. *Trends Plant Sci.* **2016**, *218*, 677–685. [CrossRef]
97. Danquah, A.; de Zelicourt, A.; Colcombet, J.; Hirt, H. The role of ABA and MAPK signaling pathways in plant abiotic stress responses. *Biotechnol. Adv.* **2014**, *321*, 40–52. [CrossRef] [PubMed]
98. Sinha, A.K.; Jaggi, M.; Raghuram, B.; Tuteja, N. Mitogen–activated protein kinase signaling in plants under abiotic stress. *Plant Signal. Behav.* **2011**, *62*, 196–203. [CrossRef] [PubMed]
99. Zhang, J.; Wang, X.; Wang, Y.; Chen, Y.; Luo, J.; Li, D.; Li, X. Genome–wide identification and functional characterization of cotton (*Gossypium hirsutum*) MAPKKK gene family in response to drought stress. *BMC Plant Biol.* **2020**, *20*, 217. [CrossRef] [PubMed]
100. Li, F.; Li, M.; Wang, P.; Cox Jr, K.L.; Duan, L.; Dever, J.K.; Shan, L.; Li, Z.; He, P. Regulation of cotton *Gossypium hirsutum* drought responses by mitogen-activated protein MAP kinase cascade-mediated phosphorylation of *GhWRKY59*. *New Phytol.* **2017**, *2154*, 1462–1475. [CrossRef]
101. Kaleem, F.; Shabir, G.; Aslam, K.; Rasul, S.; Manzoor, H.; Shah, S.M.; Khan, A.R. An overview of the genetics of plant response to salt stress: Present status and the way forward. *Appl. Biochem. Biotechnol.* **2018**, *1862*, 306–334. [CrossRef]
102. Takeda, S.; Matsuoka, M. Genetic approaches to crop improvement: Responding to environmental and population changes. *Nat. Rev. Genet.* **2008**, *96*, 444–457. [CrossRef]
103. Bailey-Serres, J.; Parker, J.E.; Ainsworth, E.A.; Oldroyd, G.E.; Schroeder, J.I. Genetic strategies for improving crop yields. *Nature* **2019**, *5757781*, 109–118. [CrossRef]
104. Newell-McGloughlin, M. Nutritionally improved agricultural crops. *Plant Physiol.* **2008**, *1473*, 939–953. [CrossRef]
105. Kumar, K.; Gambhir, G.; Dass, A.; Tripathi, A.K.; Singh, A.; Jha, A.K.; Yadava, P.; Choudhary, M.; Rakshit, S. Genetically modified crops: Current status and future prospects. *Planta* **2020**, *251*, 91. [CrossRef]
106. ISAA. 2020. Available online: http://www.isaaa.org/gmapprovaldatabase/eventslist/default.asp (accessed on 12 March 2020).
107. Chen, L.; Song, Y.; Li, S.; Zhang, L.; Zou, C.; Yu, D. The role of WRKY transcription factors in plant abiotic stresses. *Biochim. Biophys. Acta Gene Regul. Mech.* **2012**, *18192*, 120–128. [CrossRef]
108. Gao, H.; Wang, Y.; Xu, P.; Zhang, Z. Overexpression of a WRKY transcription factor TaWRKY2 enhances drought stress tolerance in transgenic wheat. *Front. Plant Sci.* **2018**, *9*, 997. [CrossRef] [PubMed]
109. Luo, X.; Li, C.; He, X.; Zhang, X.; Zhu, L. ABA signaling is negatively regulated by *GbWRKY1* through JAZ1 and ABI1 to affect salt and drought tolerance. *Plant Cell Rep.* **2020**, *392*, 181–194. [CrossRef] [PubMed]
110. Guo, R.; Yu, F.; Gao, Z.; An, H.; Cao, X.; Guo, X. GhWRKY3, a novel cotton (*Gossypium hirsutum* L.) WRKY gene, is involved in diverse stress responses. *Mol. Biol. Rep.* **2011**, *38*, 49–58. [CrossRef] [PubMed]
111. Li, Z.; Li, L.; Zhou, K.; Zhang, Y.; Han, X.; Din, Y.; Ge, X.; Qin, W.; Wang, P.; Li, F.; et al. *GhWRKY6* acts as a negative regulator in both transgenic Arabidopsis and cotton during drought and salt stress. *Front. Genet.* **2019**, *10*, 392. [CrossRef]
112. Shi, W.; Liu, D.; Hao, L.; Wu, C.A.; Guo, X.; Li, H. *GhWRKY39*, a member of the WRKY transcription factor family in cotton, has a positive role in disease resistance and salt stress tolerance. *Plant Cell Tissue Organ Cult.* **2014**, *1181*, 17–32. [CrossRef]
113. Jia, H.; Wang, C.; Wang, F.; Liu, S.; Li, G.; Guo, X. GhWRKY68 reduces resistance to salt and drought in transgenic Nicotiana benthamiana. *PLoS ONE* **2015**, *10*, e0120646.
114. Ayaz, A.; Huang, H.; Zheng, M.; Zaman, W.; Li, D.; Saqib, S.; Zhao, H.; Lü, S. Molecular Cloning and Functional Analysis of GmLACS2-3 Reveals Its Involvement in Cutin and Suberin Biosynthesis along with Abiotic Stress Tolerance. *Int. J. Mol. Sci.* **2021**, *22*, 9175. [CrossRef]

life

Article

Genome-Wide Identification and Characterization of Heat Shock Protein 20 Genes in Maize

Huanhuan Qi [1,†], Xiaoke Chen [1,†], Sen Luo [1], Hongzeng Fan [1], Jinghua Guo [1], Xuehai Zhang [2], Yinggen Ke [1], Pingfang Yang [1] and Feng Yu [1,*]

[1] State Key Laboratory of Biocatalysis and Enzyme Engineering, School of Life Sciences, Hubei University, Wuhan 430062, China
[2] National Key Laboratory of Wheat and Maize Crop Science, Henan Agricultural University, Zhengzhou 450002, China
* Correspondence: yufeng@hubu.edu.cn
† These authors contributed equally to this work.

Abstract: Maize is an important cereal crop worldwide and is sensitive to abiotic stresses in fluctuant environments that seriously affect its growth, yield, and quality. The small heat shock protein (*HSP20*) plays a crucial role in protecting plants from abiotic stress. However, little is known about *HSP20* in maize (*ZmHSP20*). In this study, 44 *ZmHSP20s* were identified, which were unequally distributed over 10 chromosomes, and 6 pairs of *ZmHSP20s* were tandemly presented. The gene structure of *ZmHSP20s* was highly conserved, with 95% (42) of the genes having no more than one intron. The analysis of the cis-element in *ZmHSP20s* promoter demonstrated large amounts of elements related to hormonal and abiotic stress responses, including abscisic acid (ABA), high temperature, and hypoxia. The ZmHSP20s protein had more than two conserved motifs that were predictably localized in the cytoplasm, nucleus, endoplasmic reticulum, peroxisome, mitochondria, and plasma. Phylogenetic analysis using HSP20s in *Arabidopsis*, rice, maize, and *Solanum tuberosum* indicated that ZmHSP20s were classified into 11 categories, of which each category had unique subcellular localization. Approximately 80% (35) of *ZmHSP20* were upregulated under heat stress at the maize seedling stage, whereas the opposite expression profiling of 10 genes under 37 and 48 °C was detected. A total of 20 genes were randomly selected to investigate their expression under treatments of ABA, gibberellin (GA), ethylene, low temperature, drought, and waterlogging, and the results displayed that more than half of these genes were downregulated while *ZmHSP20-3*, *ZmHSP20-7*, *ZmHSP20-24*, and *ZmHSP20-44* were upregulated under 1 h treatment of ethylene. A yeast-one-hybrid experiment was conducted to analyze the binding of four heat stress transcription factors (ZmHSFs) with eight of the *ZmHSP20s* promoter sequences, in which ZmHSF3, ZmHSF13, and ZmHSF17 can bind to most of these selected *ZmHSP20s* promoters. Our results provided a valuable resource for studying *HSP20s* function and offering candidates for genetic improvement under abiotic stress.

Keywords: heat shock protein 20; maize; abiotic stress; yeast-one-hybrid

Citation: Qi, H.; Chen, X.; Luo, S.; Fan, H.; Guo, J.; Zhang, X.; Ke, Y.; Yang, P.; Yu, F. Genome-Wide Identification and Characterization of Heat Shock Protein 20 Genes in Maize. *Life* **2022**, *12*, 1397. https://doi.org/10.3390/life12091397

Academic Editors: Wajid Zaman and Hakim Manghwar

Received: 21 August 2022
Accepted: 5 September 2022
Published: 8 September 2022

Publisher's Note: MDPI stays neutral with regard to jurisdictional claims in published maps and institutional affiliations.

Copyright: © 2022 by the authors. Licensee MDPI, Basel, Switzerland. This article is an open access article distributed under the terms and conditions of the Creative Commons Attribution (CC BY) license (https://creativecommons.org/licenses/by/4.0/).

1. Introduction

In the changing environment, numerous adverse stress conditions such as drought, salinity, heat, cold, and chemicals, nematodes, insects, and rodents were imposed on plants, which significantly influence their growth and development [1]. These abiotic stresses can cause damage to plant cells and cause secondary damage, such as osmotic and oxidative stress [2,3]. Plants have a series of elaborate mechanisms in response to environmental changes compared to animals, including maintaining cell membrane stability [4], capturing reactive oxygen species (ROS), synthesizing antioxidants, osmotic accumulation, and osmotic regulation, inducing some enzymes in response to stress, and enhancing the transcription and signaling of partners [5], to adapt morphologically and physiologically [6]. Abiotic stresses in plants are often interrelated and lead to physiological, morphological, cellular, and molecular changes [1], and two or more abiotic stresses are often more lethal than single stress [7].

Heat shock protein (*HSP*) in *Drosophila melanogaster* was primarily discovered under exposure to high-temperature stress [8]. The response to stresses on the molecular level was found in all organisms, especially the sudden changes in genotypic expression resulting in an increase in the synthesis of *HSP* proteins [9–11]. The *HSPs* are characterized by the presence of a carboxyl terminus called a heat shock domain [12]. Under environmental stress conditions, plants reduce the synthesis of normal proteins and facilitate the transcription and translation of *HSPs* [13]. The expression of *HSPs* is mediated by the binding of heat stress transcription factors (HSFs) to heat shock element (HSE) sequences that are located in the promoter region of *HSPs* [2]. The heat shock promoter is characterized by a conserved palindromic element with the consensus motif "nGAAnnTTCn". This HSE motif or its various variants have different effects on the interaction of HSFs with HSE [14]. *HSPs* can be divided into five classes according to their molecular weight and sequence homology, including *HSP100*, *HSP90*, *HSP70*, *HSP60*, and Small *HSPs* (*HSP20*) [15,16], in which *HSP20s* are 12–25 kDa polypeptides. Most *HSP20s* occur together in oligomers with 12 subunits. Plants have many types of *HSP20s*, and some species have more than 40 types of *HSP20s* [15]. The structure of *HSP20s* presents remarkable diversity, but all *HSP20s* share a common α-crystalline domain (ACD) that allows them to be recognized [17], reflecting their fitness with diversity stresses. *HSP20s* are widely present in plants and help to protect plant cells against protein breakdown and maintain their functional conformation [18]. *HSP20s* can also act as an ATP-independent molecular chaperone to capture substrate proteins denatured by stress [19], preventing the irreversible denaturation of substrates [20]. The feature facilitates the refolding of denatured proteins and improves plant performance in adapting to environmental stress.

Most *HSP20s* are not expressed under normal conditions but can be omnipresent in various biotic and abiotic stresses [21]. *HSP20s* are thus considered a component of cell protein quality control to defend against stresses and coordinate defensive signaling cascades by participating in the build-up of various resistance proteins. *HSP20s* are also involved in plant embryogenesis, germination, and fruit development. For example, *HSP21* in tomato participates in the accumulation of carotenoids during ripening [22]. *HSP20* plays an important role in abiotic stress and has been identified in various plants. *PtHSP17.8* of *Populus trichocarpa* enhances heat and salt tolerance by maintaining ROS homeostasis and collaboration [23]. Overexpression of maize *HSP16.9* in tobacco can increase heat tolerance and oxidation resistance [24]. The expression of *HSP22.8* in watermelon is reduced under abscisic acid (ABA) stress and salt stress [25]. *HSP17.7* in tomato can maintain intracellular Ca^{2+} homeostasis and improve cold tolerance [26]. Moreover, most *HSP20s* in apple were upregulated under heat stress [27].

Maize (*Zea mays*) is an essential staple crop in Latin America, Asia, and sub-Saharan Africa, mainly for human consumption and animal feed production [28]. Aside from its agronomic importance, maize has been a key model for fundamental research for almost a century [29]. However, the *HSP20* gene family in maize (*ZmHSP20s*) has not been fully researched [30,31]. In this study, we systematically identified and characterized *ZmHSP20s* in maize genome, which included the gene structure, conserved motif, cis-element in the promoter, and phylogenetic relationship. We also analyzed the expression level of *ZmHSP20s* under hormone treatments and abiotic stresses, especially for expression levels under high temperatures. Moreover, the possible interactions between ZmHSP20s and ZmHSFs were experimentally verified. These results provide valuable resources for investigating the function of *HSP20s* in plants.

2. Materials and Methods

2.1. Plant Growth and Treatment

Seeds of the maize inbred line B73 were planted in a greenhouse with a controlled temperature (~25 °C/22 °C, day/light cycle), a 14 h/10 h light/dark cycle, and 60% average humidity. As previously described, the treatments were imposed on seedlings at the second leaf stage [32]. For the high-temperature treatment, the seedlings were transferred to an artificial climate chamber at 37, 42, and 48 °C, and the leaves were collected after 4 h of stress. The seedling leaves under 10 °C, drought stress after 1, 2, and 4 h treatments, and waterlogged roots were also collected. For hormone treatments, 100 μM of ethylene (ET), 100 mM of ABA, and 100 mM of gibberellin (GA) were applied to treat the seedlings, and the leaves after 1, 2, and 4 h treatments were sampled. The leaves and roots of seedlings growing under 25 °C conditions were collected as the control. For each sample, more than six seedlings were mixed and immediately frozen at −80 °C.

2.2. RNA Extraction and Quantitative Reverse Transcription PCR (qRT-PCR)

Total RNA was isolated using TRIZOL reagent (Invitrogen, Gaithersburg, MD, USA) and treated with RNase-free DNase (Invitrogen). Purified RNA was used to synthesize single-stranded cDNA using recombinant M-MLV reverse transcriptase (Invitrogen). Quantitative reverse transcription PCR (qRT-PCR) was performed using gene-specific primers (Table S1) and a 2 × iTaq^{TM} Universal SYBR Green Super Mix (Bio-Rad, Hercules, CA, USA). *ZmActin1* (*GRMZM2G126010*) was used as an internal control for the normalization of expression data. Relative expression levels were calculated using the $2^{-\Delta\Delta CT}$ (cycle threshold) method [33]. PCR involved an initial denaturation step at 95 °C for 5 min, followed by 40 cycles of 15 s at 95 °C, 10 s at 58 °C, and 20 s at 72 °C. The primers used for qRT-PCR were designed using online software Primer3Plus (https://www.primer3plus.com/ (accessed on 1 June 2022)).

2.3. Identification of ZmHSP20s

Two approaches were applied to identify the *ZmHSP20s* family genes in maize. The conserved *ZmHSP20* domain (PF00011) from the Pfam database [34] was used to query the maize B73 proteome (RefGen_v4) [35] with the *ZmHSP20* HMM using the HMMER3.0 package [36], and the *ZmHSP20* proteins were collected based on the E-value < 1×10^{-5}. Moreover, the protein sequences of *HSP20* family members in *Arabidopsis* and rice [37] were downloaded from TAIR [38] and the MSU Rice Genome Annotation Project [39] databases, respectively. These protein sequences from these *Arabidopsis* and rice *HSP20* were used as queries to search against the maize proteome with an E-value < 1×10^{-5} based on a local BLASTP program with the default parameters. The proteins from these two approaches were collected and redundant sequences were manually eliminated. The Pfam [34] and SMART [40] databases were utilized to confirm the conserved domain in the identified proteins. The molecular weight (MW) and isoelectric point (pI) of ZmHSP20s were computed with the online ExPASy tool [41]. Four online tools (Predotar [42], WOLF PSORT (https://www.genscript.com/wolf-psort.html (accessed on 6 June 2022)), TargetP [43], and CELLO [44]) were used to predict the subcellular localization. Some subcellular localizations of ZmHSP20s that cannot be predicted using software will be predicted by affinities with other species.

2.4. Analysis of Gene Structure, Chromosome Distribution, Duplication, Collinearity, and Conserved Motif

DNA, coding sequences (CDSs), and protein sequences of ZmHSP20 family genes and their corresponding physical location in the maize B73 reference genome (RefGen_v4) were downloaded from the MaizeGDB database. The gene structures were drawn and displayed by Gene Structure Display Server (GSDS) [45] using DNA and CDS sequences of each gene. The online program of Multiple Em for Motif Elicitation (MEME, V5.0.3, https://meme-suite.org/meme/doc/meme.html, accessed on 6 June 2022) was applied to predict the potential motifs with default parameters. The MG2C (MapGene2Chromosome V2, http://mg2c.iask.in/mg2c_v2.0/, accessed on 8 June 2022) software was used to display the physical location of each gene in its corresponding position. According to the manual, the *ZmHSP20s* gene collinearity analysis within the maize genome was conducted using MCScanX software with default parameters [46].

2.5. Phylogenetic Analysis

To illuminate the evolutionary relationship of ZmHSP20s, the phylogenetic tree was constructed using representative HSP20s protein sequences of *Arabidopsis thaliana* (AtHSP20s) [47], rice (OsHSP20s) [48], *Solanum tuberosum* (StHSP20s) [49], and 44 ZmHSP20s. After being aligned using ClustalW [50], the aligned sequences were imported into MEGA11 [51] to construct an unrooted neighbor-joining phylogenetic tree (NJ) using 1000 bootstrap repetitions. The phylogenetic tree was modified using the online software iTOL [52].

2.6. Predicting the Cis-Regulatory Elements

The 1.5 kb sequences of the promoter of *ZmHSP20* genes were obtained from the EnsemblPlants database and were then uploaded to the website of PlantCare [53] to predict the cis-regulatory DNA elements. The elements related to stress response and hormones were selected and displayed through Tbtools [54].

2.7. Prediction of the Interaction between ZmHSP20s and ZmHSFs

Protein sequences for *HSF* family members in maize (ZmHSFs) [55] were downloaded from maizeGDB [56], which were uploaded onto STRING database [57] to predict the interaction with ZmHSP20s. The interaction networks were drawn through Cytoscape_v3.9.1 [58]. The promoter sequences of *ZmHSP20s* were uploaded onto PlantRegMap [59] to predict the binding of ZmHSFs.

2.8. Yeast One- and Two-Hybrid Assays

A full-length CDS of *ZmHSFs* was cloned into vector pGADT7-Rec2 and the 1.5 kb promoter sequence of *ZmHSP20s* was cloned into vector pHIS2 using a CloneExpressII One Step Cloning Kit (Vazyme, Nanjing, China) with the corresponding primers (Table S1). Recombinant vectors were co-transfected into yeast competent *AH109*. Transformants were cultured on SD/-Leu-Trp and were then placed on SD/-Leu-Trp-His with a special concentration of 3-amino-1,2,4-triazole (3-AT). For the yeast-two-hybrid experiment, the full-length CDS of *ZmHSFs* was cloned into vector pGADT7 while the full-length CDS of *ZmHSP20s* was cloned into vector pGBKT7, and the transformants were screened on SD/-Leu-Trp-His-Ade.

3. Results

3.1. The Characters of ZmHSP20 Gene Members

A total of 44 members of *ZmHSP20s* in maize were finally identified through BLASP and HMMER programs, which were referred to as *ZmHSP20-1* to *ZmHSP20-44* based on their location in chromosomes (Table S2). *ZmHSP20s* locate across 10 chromosomes, and chromosomes 1 (11) and 3 (7) had the largest member of *ZmHSP20s* while chromosomes 4 (*ZmHSP20-22* and *ZmHSP20-23*), 7 (*ZmHSP20-32* and *ZmHSP20-33*), and 10 (*ZmHSP20-44*) had the smallest member of *ZmHSP20s*. The isoelectric point ranged from 4.75 (*ZmHSP20-30*) to 11.66 (*ZmHSP20-17*) and the molecular weight (MW) ranged from 13.98 to 62.73 kilodalton (Kd), most of which were around 20 Kd. *ZmHSP20-8* (62.73 Kd) and *ZmHSP20-21* (46.21 Kd) had higher apparent MWs although both proteins had a conserved domain of HSP20 (Table S2). The subcellular localization of ZmHSP20s demonstrated that most of these proteins localized in the cytoplasmic region, while some proteins localized in nuclear, mitochondrial, endoplasmic reticulum (ER), plastid, and peroxisome (Po) regions. Four gene clusters of *ZmHSP20s* in chromosomes 1 (2 clusters), 3 (1 cluster), and 9 (1 cluster) were identified, of which cluster 1 contained three genes (*ZmHSP20-2* to *ZmHSP20-4*), cluster 2 contained three genes (*ZmHSP20-5* to *ZmHSP20-7*), cluster 3 contained five genes (*ZmHSP20-16* to *ZmHSP20-20*), and cluster 4 contained three genes (*ZmHSP20-40* to

ZmHSP20-42) (Figure 1). Moreover, six pairs of *ZmHSP20s* exhibited collinearity, which included *ZmHSP20-8* and *ZmHSP20-11*, *ZmHSP20-11* and *ZmHSP20-24*, *ZmHSP20-12* and *ZmHSP20-44*, *ZmHSP20-16* and *ZmHSP20-34*, *ZmHSP20-22* and *ZmHSP20-39*, and *ZmHSP20-30* and *ZmHSP20-36*.

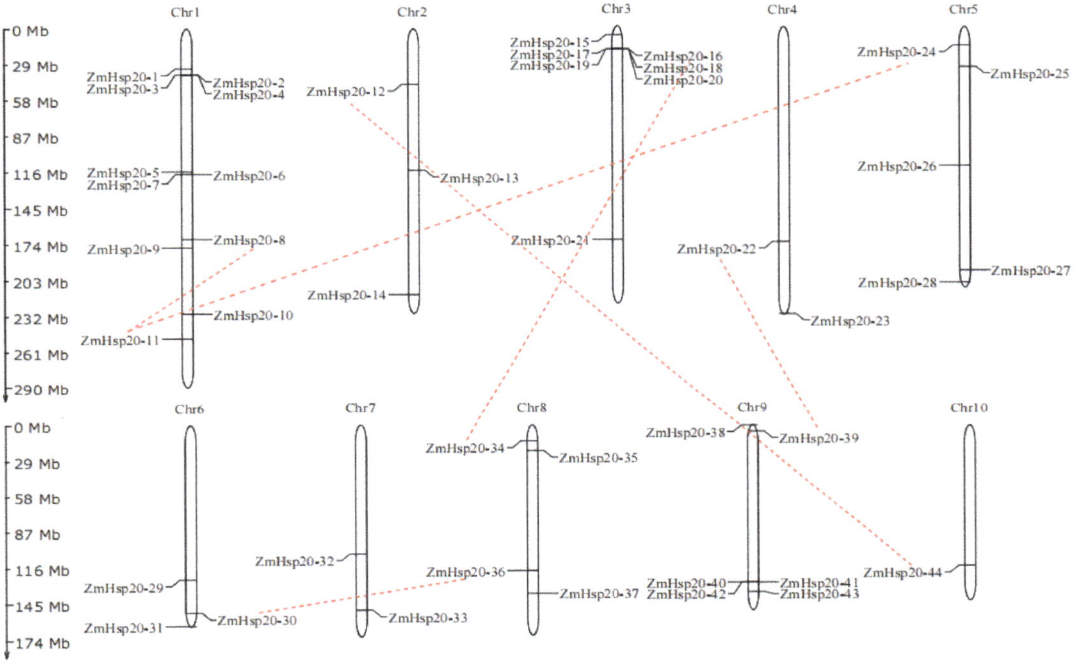

Figure 1. Genome-wide distribution of *ZmHSP20* genes on maize chromosomes. The chromosomal location of each *ZmHSP20* gene is annotated with the gene name. Chromosome numbers are indicated at the top of each bar. The *ZmHSP20* genes present on duplicated chromosomal segments are connected by red dashed lines.

Of 44 *ZmHSP20s*, 22 members had no intron, 20 members had only 1 intron, and 1 member (*ZmHSP20-21*) had 6 introns (Figure 2). One gene, *ZmHSP20-13*, had an ultra-long intron. Seventeen *ZmHSP20s* did not predict the 5′-UTR and 3′-UTR regions. To explore the potential regulatory and function of *ZmHSP20* genes, the cis-acting elements of the *ZmHSP20s* promoter region involved in hormone stimulus and stress response were analyzed (Figure 3). Ten elements involved in hormone stimulus were detected, including the ABA response element (ABRE), auxin response element (TGA-element, AuxRR-core), GA response element (TATC-box, GARE-motif, and P-box), MeJA response element (CGTCA-motif and TGACG-motif), SA response element (TCA-element and SARE). The cis-elements of drought responsiveness (DRE), anaerobic responsiveness (ARE and GC-motif), low-temperature responsiveness (LTR), wound responsiveness (WUN-motif), and light responsiveness (G-box) were also identified. The number of cis-elements ranged from 4 (*ZmHSP20-43*) to 39 (*ZmHSP20-21*). *ZmHSP20-3* had 26 cis-elements, of which 16 cis-elements were related to hormone and abiotic stresses, including ABA, auxin, GA, MeJA, drought, low-temperature, and light. The G-box occupied the most genes, which appeared in the promoter regions of 40 *ZmHSP20* genes, except *ZmHSP20-7*, *ZmHSP20-22*, *ZmHSP20-28*, and *ZmHSP20-42*. ABRE presented in 36 genes, of which *ZmHSP20-19* contained 10, and *ZmHSP20-10* and *ZmHSP20-21* contained 8, respectively. In particular, GC-motif appeared 6 times in *ZmHSP20-33* but no more 2 in the other genes. Moreover, 23 of *ZmHSP20s* contained the MBS element, 10 of *ZmHSP20s* contained the TGA element, and 9 of *ZmHSP20* contained the TATC_box. These results indicated that *ZmHSP20s* were involved in multiple hormonal and abiotic responses.

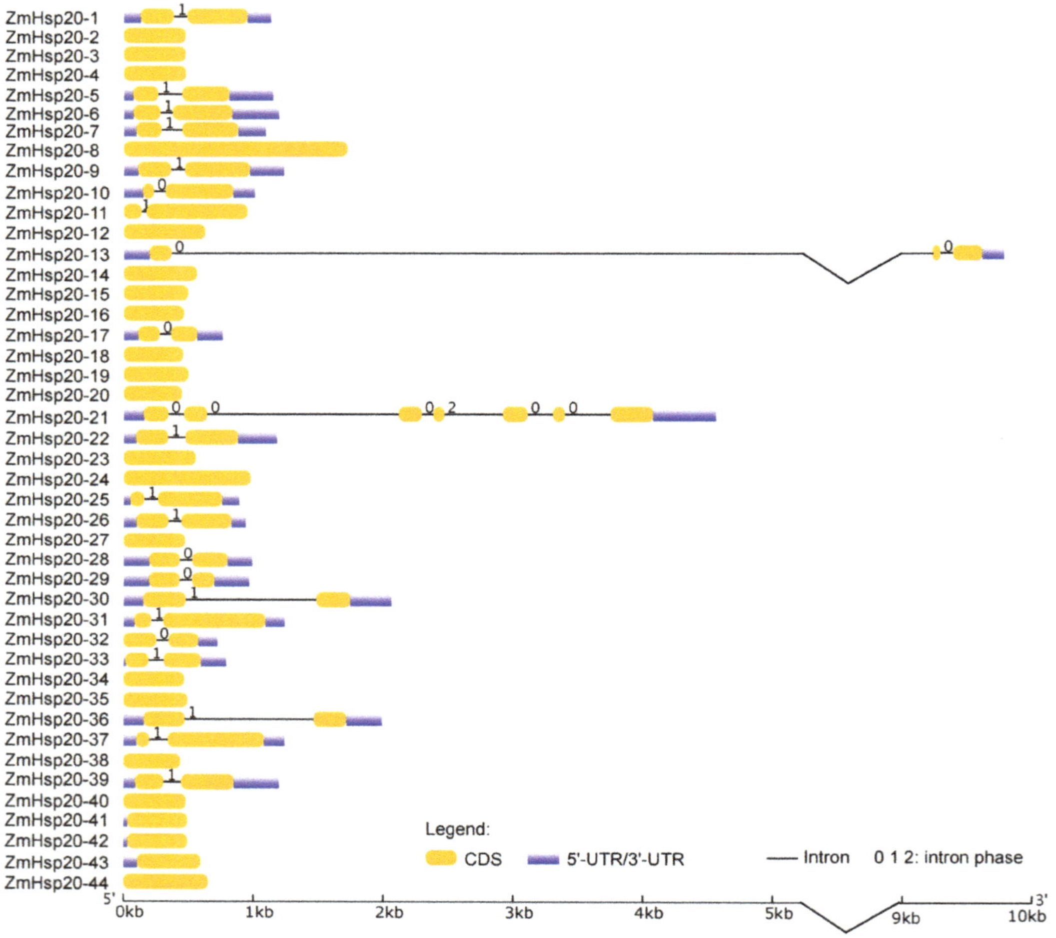

Figure 2. The gene structure of *ZmHsp20s*. The CDSs are displayed with yellow rectangles. The introns are displayed with black lines. Purple rectangles represent UTR. CDS, coding sequence; UTR, untranslated region.

Figure 3. Characters of cis-elements in promoter regions of *ZmHSP20s*. Cis-elements related to hormone responsiveness are represented as cylindrical and cis-elements related to abiotic stress responsiveness are represented as a wedge. ABRE was the response to ABA; ARE and GC-motif were the response to anaerobic conditions; CGTCA-motif was the response to MeJA; DRE and MBS were the response to drought; G-box was light response, GARE-motif and TATC-box were the response to GA; LTR was the response to low temperature; TC-rich repeat was the response to defense and stress; and TGA-element was the response to auxin.

3.2. Conserved Function of ZmHSP20s

The conserved motifs in ZmHSP20s proteins were analyzed using MEME (Figure 4). A total of five motifs were identified, of which Motif 1 was detected in all ZmHSP20s proteins, and more than two motifs in one protein were identified (Figure 4A). Motif 1, Motif 3, and Motif 4 were distributed on most of the proteins while Motif 5 was only found on 8 members, including ZmHSP20-02, ZmHSP20-03, ZmHSP20-04, ZmHSP20-16, ZmHSP20-17, ZmHSP20-18, ZmHSP20-34, and ZmHSP20-43. These ZmHSP20s were divided into two subgroups based on whether they contained the Motif 5 at the N-terminal. Interestingly, members in group 1 (containing Motif 5) were localized to the cytoplasm and had no intron except ZmHSP20-17 had 1 intron. In particular, ZmHSP20-17 in group 1 lacked Motif 3 compared with other members. The length of these conserved motifs varied from 15 to 29 amino acids (Figure 4B). The GO enrichment analysis of 44 ZmHSP20 genes was conducted, of which 35 genes were enriched (Figure S1). The significant GO terms mainly included the response to hydrogen peroxide, response to hydrogen peroxide, response to reactive oxygen species, response to heat, response to osmotic stress, response to stimulus, and protein oligomerization, indicating the important roles in abiotic stress.

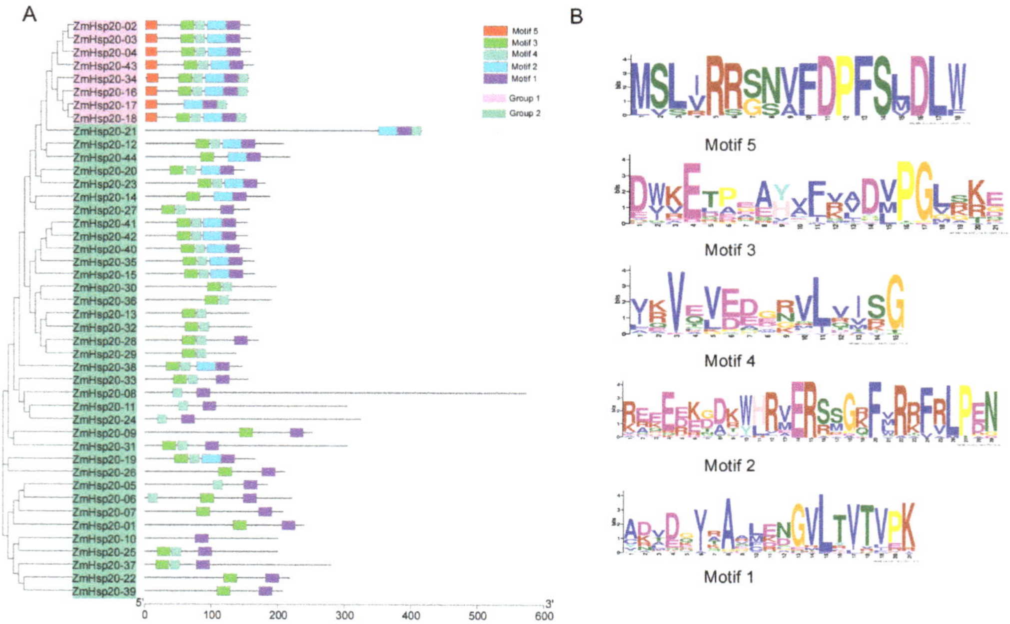

Figure 4. Analysis of the conserved motif in the ZmHSP20s protein. (**A**) Conserved motifs in ZmHSP20 proteins. The phylogenetic tree of ZmHSP20s was constructed with amino acid sequences using MEGA11 software. Different motifs are presented in different colors. ZmHSP20s were classified into group 1 and group 2 based on the presence or absence of Motif 5. (**B**) Motif sequences were predicted in the ZmHSP20s protein. The overall height of the amino acid stacks plotted on the y-axis indicates the sequence conservation at a given position, while the height of individual symbols within a stack indicates the relative frequency of a nucleotide base at that position.

To explore the evolutionary relationship of *HSP20s* in plants, 44 of *ZmHSP20s*, 18 of *AtHSP20s*, 18 of *OsHSP20s*, and 35 of *StHSP20s* were subjected to construction of a phylogenetic tree, which was divided into 11 categories according to a previous classification [47,60,61] (Figure 5). These proteins were predicted to localize in 6 organelles, including the cytoplasm and nucleus (C), endoplasmic reticulum (ER), peroxisome (Po), mitochondria (M), and plasma (P). The proteins in the categories of CI, CII, CIII, CV, CVI, and CVII were mainly localized in the cytoplasm and nucleus, proteins in the category of MI and MII were mainly localized in the mitochondria, while proteins in the category of ER, Po, and P were mainly localized in the endoplasmic reticulum, peroxisome, and plasma, respectively. The CI category had the largest number of members, and most of the members in category CII belonged to maize and rice. The category of CV and Po had only four members, with one member of each species. The P category had only three ZmHSP20s (ZmHSP20-5, ZmHSP20-6, and ZmHSP20-7), and the CVII category had three members (AtHSP14.7, StHSP20-27, and StHSP20-28) from dicotyledonous plants, and the CVI category had four members from maize, *Solanum tuberosum*, and *Arabidopsis*. The phylogenetic relationship indicated the conservation and difference in HSP20s in plant evolution.

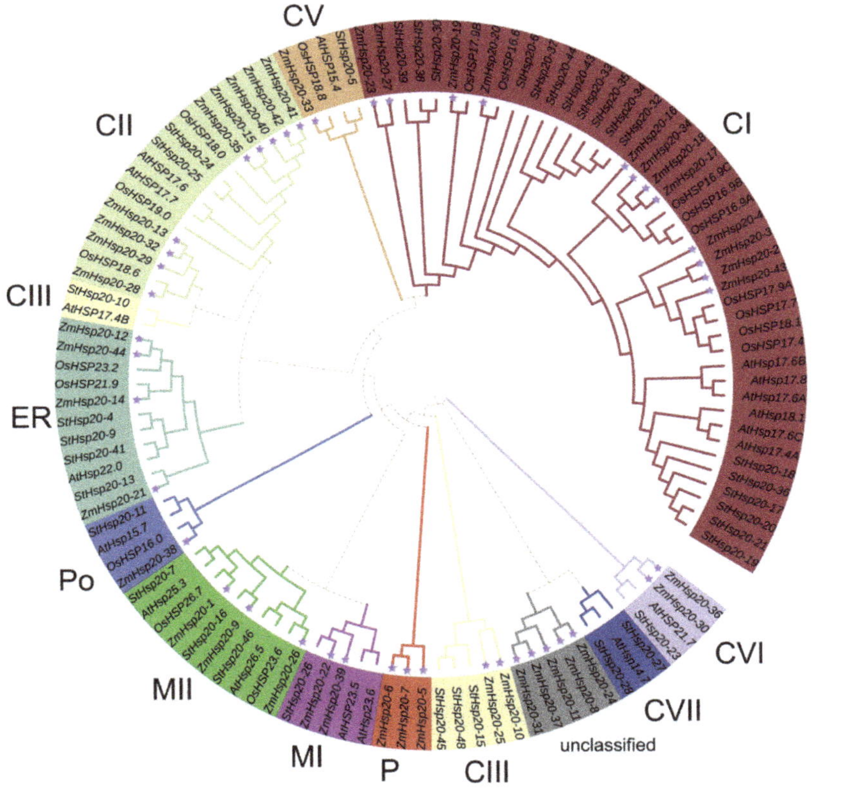

Figure 5. Phylogenetic tree of HSP20 proteins of rice (Os), *Arabidopsis* (At), *Solanum tuberosum* (St), and maize (Zm) using MEGA11 software based on the NJ method. Eleven subfamilies with different colors were classified and unclassified ZmHSP20s are labeled with grey.

3.3. High Temperature Strongly Induced the Expression of ZmHSP20s

To investigate the response of *ZmHSP20s* to high temperature, qRT-PCR was applied to analyze the expression level of 44 *ZmHSP20s* under 37, 42, and 48 °C stresses (Figure 6). Of 44 genes, 31 genes were upregulated after heat stress, while 12 genes such as *ZmHSP20-3*, *ZmHSP20-16*, *ZmHSP20-17*, *ZmHSP20-18*, *ZmHSP20-34*, and *ZmHSP20-43* were increasingly induced under three temperature gradients (Figure 6A). The highest upregulation of *ZmHSP20s* was under 42 °C stress, which was more than 1000-fold compared with the normal condition (25 °C). Only 23 genes were upregulated under 48 °C stress, of which one gene, *ZmHSP20-24*, was only upregulated (116-fold) at this temperature point. One gene, *ZmHSP20-38*, was only upregulated (32-fold) under 37 °C stress. The interaction network of ZmHSP20s showed that only 30 genes interacted with each other (Figure 6B,C). Except for *ZmHSP20-24*, these 14 *ZmHSP20s* that were not in the network were not upregulated by heat stress. We further compared the expression level of *ZmHSP20s* under 37 and 48 °C stresses, and nine genes such as *ZmHSP20-20*, *ZmHSP20-24*, *ZmHSP20-28*, and *ZmHSP20-36 to ZmHSP20-39* displayed opposite expression profiling under 37 and 48 °C stresses (Figure 6B,C). Moreover, a significantly higher expression level of *ZmHSP20s* under 37 and 42 °C stresses than under 48 °C stress was detected (Figure S2), implying the differential expression of *ZmHSP20s* under different degrees of heat stress.

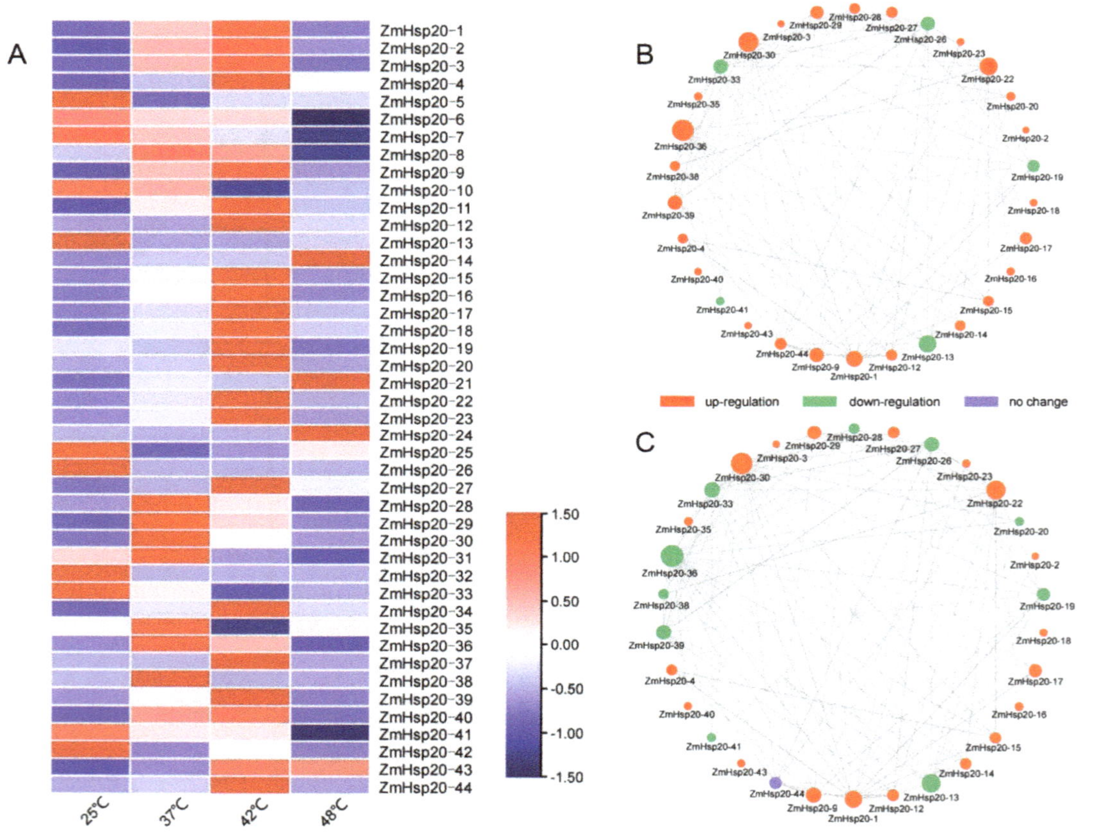

Figure 6. The expression level of *ZmHSP20s* under heat stress. (**A**) Heatmap showing the expression levels of *ZmHSP20s* at 25, 37, 42, and 48 °C based on qRT-PCR. (**B**) The PPI map of ZmHSP20s was drawn and the genes upregulated at 37 °C are shown in red, and the downregulated genes are shown in green. PPI, protein–protein interaction. (**C**) The PPI map of ZmHSP20s was drawn and the genes upregulated at 48 °C are shown in red, and the downregulated genes are shown in green.

3.4. Differential Expression of ZmHSP20s under Hormonal Stimuli and Abiotic Stresses

Given that a large number of cis-elements related to hormone and abiotic response occurred in the promoter region of *ZmHSP20s*, 20 *ZmHSP20s* were randomly selected to analyze their expression level under three treatments of hormone (ABA, ethylene, and GA) and three abiotic stresses (cold, drought, and waterlogging) (Figure 7). Under the ABA treatment, *ZmHSP20-40* was apparently upregulated, while *ZmHSP20-3*, *ZmHSP20-10*, and *ZmHSP20-30* had minor changes. All four genes (*ZmHSP20-3*, *ZmHSP20-7*, *ZmHSP20-24*, and *ZmHSP20-44*) were upregulated under 1 h of ethylene treatment, whereas *ZmHSP20-3* and *ZmHSP20-24* were downregulated under 2 and 4 h of treatment. *ZmHSP20-4* had more than 10-fold induction after GA treatment, while *ZmHSP20-27* was reduced by GA. Cold stress strongly restricted the expression of four genes (*ZmHSP20-12*, *ZmHSP20-25*, *ZmHSP20-37*, and *ZmHSP20-38*), and the expression restriction of *ZmHSP20-4*, *ZmHSP20-6*, *ZmHSP20-33*, and *ZmHSP20-38* was different under waterlogging stress. Furthermore, *ZmHSP20-18* were upregulated after 2 and 4 h of drought stress, whereas *ZmHSP20-37* was strongly limited. These results indicated that the members of *ZmHSP20s* play different roles in different stimuli.

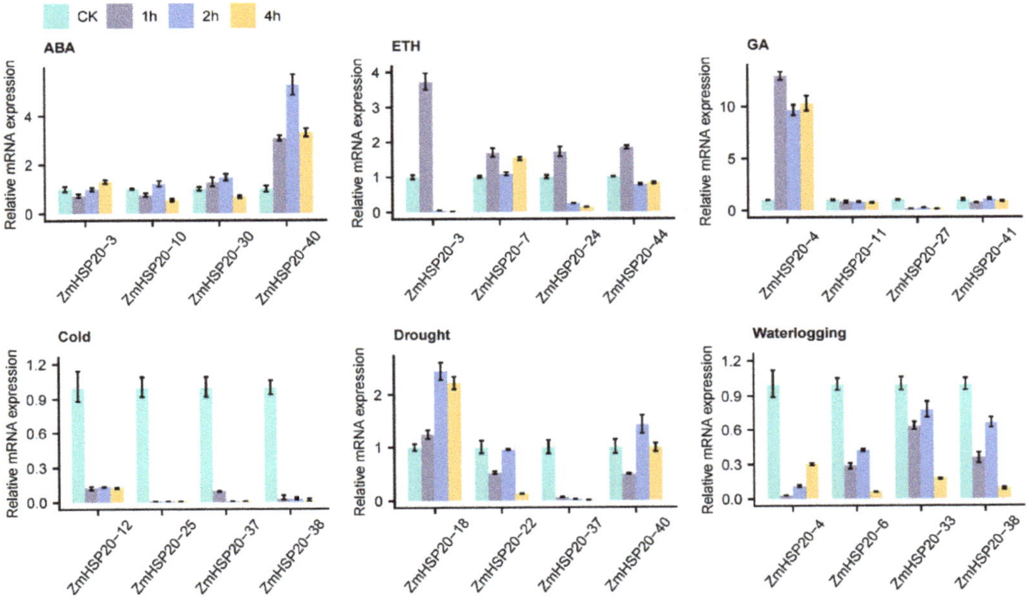

Figure 7. Expression level of *ZmHSP20s* under hormone stimulus and abiotic stresses. The height of each column indicates the mean value of three technical replicates. ABA, abscisic acid; ETH, ethylene; GA, gibberellin; 1, 2, and 4 h indicate the time of treatment.

3.5. Interaction of the ZmHSP20s with the ZmHSFs

The protein interaction between ZmHSP20s and ZmHSFs was predicted using Strings [57] and it was found that 7 of the ZmHSP20s interacted with 14 of the ZmHSFs (Figure S3). To verify their interaction at the protein level, six *ZmHSFs* CDSs (*ZmHSH02*, *ZmHSH10*, *ZmHSH15*, *ZmHSH17*, *ZmHSH24*, and *ZmHSH25*) were inserted into the *pGADT7* vector, and six *ZmHSPs* CDSs (*ZmHSP20-1*, *ZmHSP20-9*, *ZmHSP20-26*, *ZmHSP20-38*, *ZmHSP20-41*, and *ZmHSP20-44*) were inserted into the *pGBKT7* vector (Table S1). Yeast-two-hybrid experiments detected no interaction between ZmHSFs and ZmHSP20s (Figure S4). The predicted binding of ZmHSFs with the promoter sequence of *ZmHSP20s* in PlantRegMap [59] showed that four ZmHSFs (ZmHSF3, ZmHSF6, ZmHSF13, and ZmHSF17) can bind to 32 of the *ZmHSP20s* promoters (Table S3). Yeast-one-hybrid was applied to verify the binding of four ZmHSFs with the promoter sequence of eight *ZmHSP20s* (*ZmHSP20-1*, *ZmHSP20-12*, *ZmHSP20-14*, *ZmHSP20-20*, *ZmHSP20-26*, *ZmHSP20-27*, *ZmHSP20-31*, and *ZmHSP20-44*) (Figure 8). The ZmHSF3 and ZmHSF13 can interact with *ZmHSP20-1*, *ZmHSP20-12*, *ZmHSP20-14*, *ZmHSP20-20*, *ZmHSP20-26*, *ZmHSP20-31*, and *ZmHSP20-44*, ZmHSF6 can interact with *ZmHSP20-1*, *ZmHSP20-14*, *ZmHSP20-20*, and *ZmHSP20-44*, and ZmHSF17 can interact with *ZmHSP20-12*, *ZmHSP20-20*, and *ZmHSP20-31*. The differential strength of interactions between *ZmHSPs* and ZmHSFs was also observed, which included the strong interaction between ZmHSF3, ZmHSF13, and ZmHSF17 with the promoter of *ZmHSP20-20*, *ZmHSP20-26*, and *ZmHSP20-31*.

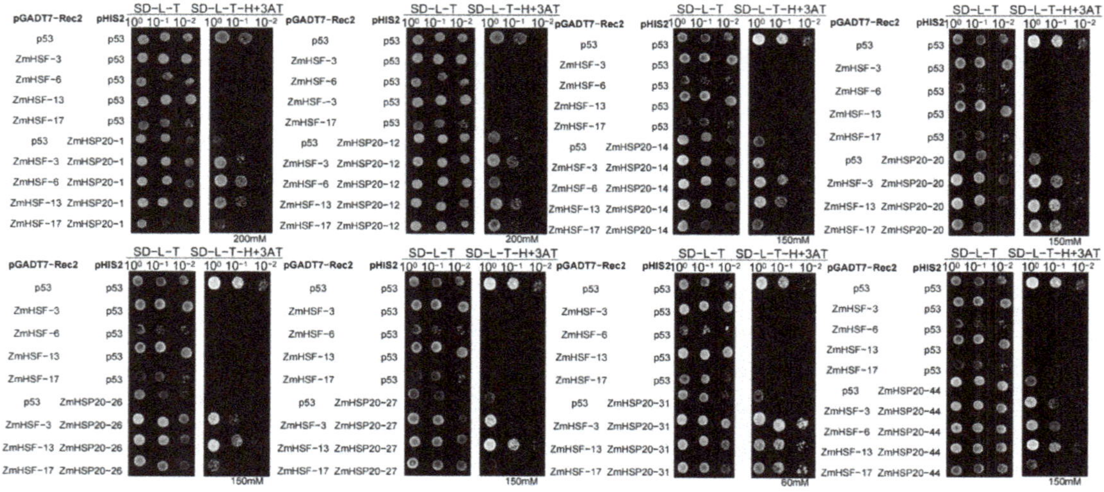

Figure 8. The binding of ZmHSFs with the promoter of *ZmHSP20s* using the yeast-one-hybrid experiment. The p53 represents the positive control; SD, synthetic dropout medium; L, leucine; T, tryptophan; H, histidine; 3-AT, 3-amino-1,2,4-triazole.

4. Discussion

Abiotic stress hurts crop development and yield and is a major barrier to meeting food demand worldwide. Plants have different strategies for coping with different types of stress. *HSPs* were induced in almost all stresses [2], and each member of the *HSPs* group has a unique roles [62]. *HSP20s* is a subfamily of *HSPs* groups, which is also called small *HSPs*. The expression levels of *HSP20s* were regulated by heat, salt, and powdery mildew in barley (*Hordeum vulgare* L.) [63], and the expression of *Lilium davidii HSP16.45* in *Arabidopsis thaliana* enhanced the latter cell activity in heat, salt, and oxidative stress [64], indicating that *HSP20s* play essential roles in biotic and abiotic stresses. In the present study, a total of 44 *ZmHSP20s* were identified (Table S2), and four clusters in three chromosomes were detected (Figure 1). The gene structure and amino acid sequence were conserved among 44 members (Figures 2 and 4), and six pairs of genes were collinear, of which these characters were also detected in tomatoes and apples [65,66]. The analysis of the phylogenetic relationship in maize, rice, *Arabidopsis*, and potato demonstrated that the specific subcellular localization of each category was presented, indicating the specific function of *HSP20s* in each category. Some evolution-related categories such as P and CVI were also identified, which may play vital roles in maize and dicotyledonous plants, respectively. Moreover, the member of *OsHSP20s* was not detected in the CVI category, implying the possible association with the aquatic environment.

Gene expression was strongly affected by environmental stimuli, which was regulated through multiple factors such as cis-elements and trans-factors. The protein of trans-factors can bind to the cis-elements in the promoter to activate or inhibit the expression of targets. The cis-elements in the promoter of one gene can reflect its potential expression profiling. The *HSP20s* participate in diverse biotic and abiotic stresses [49], which implied that some cis-elements related to stresses may be located in the promoter of *HSP20s*. Using the online tool PlantCare [53], the cis-elements in the promoter of 44 *ZmHSP20s* were identified (Figure 3), and large amounts of elements associated with hormone and abiotic stress were detected in all genes, indicating that *ZmHSP20s* are also tightly associated with abiotic stresses. To verify these results, qRT-PCR was conducted to analyze the response of hormone and abiotic stresses (Figure 7). All selected genes responded to hormone stimuli (ABA, GA, and ethylene) and abiotic stresses (hypoxia, low temperature, and drought), of which all four genes increased their expression after 1 h of ethylene, suggesting their possible roles in ethylene-mediated signals. The expression of four *ZmHSP20s* were restricted under cold stress, similar to previous transcriptome analysis [67]. Under given conditions, some *ZmHSP20s* were upregulated while some *ZmHSP20s* were inhibited, demonstrating their differential function in response to stresses.

Heat stress seriously affects growth, development, and yield, which frequently occurs with the increasing global climate. The expression of *HSP20s* was activated, and yielded proteins can avoid protein degradation [13,18], which usually play roles in molecular chaperone, retaining suitable conformations [19,20]. The GO analysis of 44 *ZmHSP20s* displayed that these genes are mainly involved in stresses such as high temperature, osmosis, and salt stress. They also involved in protein assembly, folding, and membrane composition (Figure S1), which were also discovered in rice [68], indicating the conserved characters of *HSP20s* in plants under heat stress. Transcriptome analysis of maize seedling leaves revealed that *ZmHSP20* were obviously upregulated under heat stress [67], and qRT-PCR analysis of *ZmHSP20* under 37, 42, and 48 °C stresses showed that approximately 80% of *ZmHSP20* were upregulated (Figure 6), implying the essential roles of *ZmHSP20s* under heat stress. Moreover, the differential expression profiling of *ZmHSP20s* under 37 and 48 °C conditions indicated their diverse roles. Specifically, the genes in cluster 3 such as *ZmHSP20-16*, *ZmHSP20-17*, and *ZmHSP20-20* were significantly upregulated (more than 1000-fold) under heat stress, which would be a potential target for genetic improvement of heat stress. Moreover, the induced expression of *ZmHSP20s* under heat stress depended on the binding of ZmHSFs proteins in their promoter regions (Figure 8), but not on protein–protein interaction between ZmHSFs and ZmHSP20s (Figure S4), suggesting the molecular mechanism of ZmHSP20s in response to heat stress.

Supplementary Materials: The following supporting information can be downloaded at: https://www.mdpi.com/article/10.3390/life12091397/s1, Figure S1. Functional analysis of 44 *ZmHSP20s* based on Gene Ontology. Figure S2. Boxplots showing the expression levels of *ZmHSP20s* under heat stress. Expression levels of *ZmHSP20s* increased under 37 °C, 42 °C, and 48 °C conditions, comparing with 25 °C. Figure S3. Predicted Interaction between ZmHSP20s and ZmHSFs protein using STRING database. Figure S4. The interaction between ZmHSFs and ZmHSP20s based on yeast-two-hybrid experiments. SD, synthetic dropout medium; Leu, leucine; Trp, tryptophan; His, histidine; Ade, adenine. Table S1. The primer using in this study. Table S2. Information of 44 ZmHSP20s, including Gene, Gene ID, position start, position end, chromosome, isoelectric point (PI), molecular weight (MW), and predicted subcellular localization. Table S3. Predicted gene fragments for ZmHSP20 and HSF interaction.

Author Contributions: Conceptualization, methodology, formal analysis, and writing—original draft preparation, H.Q.; formal analysis, software, visualization, and writing—original draft preparation, X.C.; Investigation and validation, S.L., J.G. and Y.K.; formal analysis and software, H.F. and X.Z.; conceptualization, supervision, and funding acquisition, F.Y. and P.Y. All authors have read and agreed to the published version of the manuscript.

Funding: This research was supported by the open fund of the National Key Laboratory of Wheat and Maize Crop Science (30500888) and the National Natural Science Foundation of China (Project 31801369).

Conflicts of Interest: The authors declare no conflict of interest.

References

1. Levitt, J. *Responses of Plants to Environmental Stresses*, 2nd ed.; Physiological Ecology; Academic Press: New York, NY, USA, 1980.
2. Al-Whaibi, M.H. Plant Heat-Shock Proteins: A Mini Review. *J. King Saud Univ. Sci.* **2011**, *23*, 139–150. [CrossRef]
3. Wang, W.; Vinocur, B.; Altman, A. Plant Responses to Drought, Salinity and Extreme Temperatures: Towards Genetic Engineering for Stress Tolerance. *Planta* **2003**, *218*, 1–14. [CrossRef] [PubMed]
4. Nakamoto, H.; Vígh, L. The Small Heat Shock Proteins and Their Clients. *Cell. Mol. Life Sci.* **2007**, *64*, 294–306. [CrossRef] [PubMed]
5. Wahid, A.; Gelani, S.; Ashraf, M.; Foolad, M. Heat Tolerance in Plants: An Overview. *Environ. Exp. Bot.* **2007**, *61*, 199–223. [CrossRef]

6. Shao, H.B.; Guo, Q.J.; Chu, L.Y.; Zhao, X.N.; Su, Z.L.; Hu, Y.C.; Cheng, J.F. Understanding Molecular Mechanism of Higher Plant Plasticity under Abiotic Stress. *Colloids Surf. B* **2007**, *54*, 37–45.
7. Mittler, R. Abiotic Stress, the Field Environment and Stress Combination. *Trends Plant Sci.* **2006**, *11*, 15–19. [CrossRef]
8. Ritossa, F. A New Puffing Pattern Induced by Temperature Shock and DNP in Drosophila. *Experientia* **1962**, *18*, 571–573. [CrossRef]
9. Lindquist, S.; Craig, E.A. The Heat-Shock Proteins. *Annu. Rev. Genet.* **1988**, *22*, 631–677. [CrossRef]
10. Gupta, S.C.; Sharma, A.; Mishra, M.; Mishra, R.K.; Chowdhuri, D.K. Heat Shock Proteins in Toxicology: How Close and How Far? *Life Sci.* **2010**, *86*, 377–384. [CrossRef]
11. De Maio, A. Heat Shock Proteins: Facts, Thoughts, and Dreams. *Shock* **1999**, *11*, 1–12. [CrossRef]
12. Helm, K.W.; LaFayette, P.R.; Nagao, R.T.; Key, J.L.; Vierling, E. Localization of Small Heat Shock Proteins to the Higher Plant Endomembrane System. *Mol. Cell. Biol.* **1993**, *13*, 238–247.
13. Wu, J.; Gao, T.; Hu, J.; Zhao, L.; Yu, C.; Ma, F. Research Advances in Function and Regulation Mechanisms of Plant Small Heat Shock Proteins (SHSPs) under Environmental Stresses. *Sci. Total Environ.* **2022**, *825*, 154054. [CrossRef]
14. Pelham, H.R.B. A Regulatory Upstream Promoter Element in the Drosophila Hsp70 Heat-Shock Gene. *Cell* **1982**, *30*, 517–528. [CrossRef]
15. Waters, E.R.; Vierling, E. Plant Small Heat Shock Proteins—Evolutionary and Functional Diversity. *New Phytol.* **2020**, *227*, 24–37. [CrossRef] [PubMed]
16. Waters, E.R. The Evolution, Function, Structure, and Expression of the Plant SHSPs. *J. Exp. Bot.* **2013**, *64*, 391–403. [CrossRef] [PubMed]
17. Banerjee, A.; Roychoudhury, A. Small Heat Shock Proteins. In *Plant Metabolites and Regulation Under Environmental Stress*; Elsevier: Amsterdam, The Netherlands, 2018; pp. 367–376.
18. Ferguson, D.L.; Guikema, J.A.; Paulsen, G.M. Ubiquitin Pool Modulation and Protein Degradation in Wheat Roots during High Temperature Stress. *Plant Physiol.* **1990**, *92*, 740–746. [CrossRef]
19. Miernyk, J.A. Protein Folding in the Plant Cell. *Plant Physiol.* **1999**, *121*, 695–703. [CrossRef]
20. Sun, W.; Van Montagu, M.; Verbruggen, N. Small Heat Shock Proteins and Stress Tolerance in Plants. *Biochim. Biophys. Acta Gene Struct. Expr.* **2002**, *1577*, 1–9. [CrossRef]
21. Morimoto, R.I. Cells in Stress: Transcriptional Activation of Heat Shock Genes. *Science* **1993**, *259*, 1409–1410. [CrossRef]
22. Neta-Sharir, I.; Isaacson, T.; Lurie, S.; Weiss, D. Dual Role for Tomato Heat Shock Protein 21: Protecting Photosystem II from Oxidative Stress and Promoting Color Changes during Fruit Maturation. *Plant Cell* **2005**, *17*, 1829–1838. [CrossRef]
23. Li, J.; Zhang, J.; Jia, H.; Li, Y.; Xu, X.; Wang, L.; Lu, M. The Populus Trichocarpa PtHSP17.8 Involved in Heat and Salt Stress Tolerances. *Plant Cell Rep.* **2016**, *35*, 1587–1599. [CrossRef] [PubMed]
24. Sun, L.; Liu, Y.; Kong, X.; Zhang, D.; Pan, J.; Zhou, Y.; Wang, L.; Li, D.; Yang, X. ZmHSP16.9, a Cytosolic Class I Small Heat Shock Protein in Maize (*Zea mays*), Confers Heat Tolerance in Transgenic Tobacco. *Plant Cell Rep.* **2012**, *31*, 1473–1484. [CrossRef]
25. He, Y.; Yao, Y.; Li, L.; Li, Y.; Gao, J.; Fan, M. A Heat-Shock 20 Protein Isolated from Watermelon (ClHSP22.8) Negatively Regulates the Response of Arabidopsis to Salt Stress via Multiple Signaling Pathways. *PeerJ* **2021**, *9*, e10524. [CrossRef] [PubMed]
26. Zhang, N.; Zhao, H.; Shi, J.; Wu, Y.; Jiang, J. Functional Characterization of Class I SlHSP17.7 Gene Responsible for Tomato Cold-Stress Tolerance. *Plant Sci.* **2020**, *298*, 110568. [CrossRef]
27. Guo, L.M.; Li, J.; He, J.; Liu, H.; Zhang, H.M. A Class I Cytosolic HSP20 of Rice Enhances Heat and Salt Tolerance in Different Organisms. *Sci. Rep.* **2020**, *10*, 1383. [CrossRef] [PubMed]
28. Mahuku, G.; Lockhart, B.E.; Wanjala, B.; Jones, M.W.; Kimunye, J.N.; Stewart, L.R.; Cassone, B.J.; Sevgan, S.; Nyasani, J.O.; Kusia, E.; et al. Maize Lethal Necrosis (MLN), an Emerging Threat to Maize-Based Food Security in Sub-Saharan Africa. *Phytopathology* **2015**, *105*, 956–965. [CrossRef]
29. Strable, J.; Scanlon, M.J. Maize (*Zea mays*): A Model Organism for Basic and Applied Research in Plant Biology. *Cold Spring Harb. Protoc.* **2009**, *2009*, pdb.emo132. [CrossRef]
30. Han, Z.; Ku, L.; Zhang, Z.; Zhang, J.; Guo, S.; Liu, H.; Zhao, R.; Ren, Z.; Zhang, L.; Su, H.; et al. QTLs for Seed Vigor-Related Traits Identified in Maize Seeds Germinated under Artificial Aging Conditions. *PLoS ONE* **2014**, *9*, e92535.
31. Xing, L.-M.; Lyu, W.Z.; Lei, W.; Liang, Y.-H.; Lu, Y.; Chen, J.-Y. Response of HSP20 Genes to Artificial Aging Treatment in Maize Embryo. *Zuo Wu Xue Bao* **2018**, *44*, 1733. [CrossRef]
32. Yu, F.; Liang, K.; Fang, T.; Zhao, H.; Han, X.; Cai, M.; Qiu, F. A Group VII Ethylene Response Factor Gene, ZmEREB180, Coordinates Waterlogging Tolerance in Maize Seedlings. *Plant Biotechnol. J.* **2019**, *17*, 2286–2298. [CrossRef]
33. Livak, K.J.; Schmittgen, T.D. Analysis of Relative Gene Expression Data Using Real-Time Quantitative PCR and the $2^{-\Delta\Delta CT}$ Method. *Methods* **2001**, *25*, 402–408. [CrossRef]
34. Punta, M.; Coggill, P.C.; Eberhardt, R.Y.; Mistry, J.; Tate, J.; Boursnell, C.; Pang, N.; Forslund, K.; Ceric, G.; Clements, J.; et al. The Pfam Protein Families Database. *Nucleic Acids Res.* **2012**, *40*, D290–D301. [CrossRef] [PubMed]
35. Jiao, Y.; Peluso, P.; Shi, J.; Liang, T.; Stitzer, M.C.; Wang, B.; Campbell, M.S.; Stein, J.C.; Wei, X.; Chin, C.S.; et al. Improved Maize Reference Genome with Single-Molecule Technologies. *Nature* **2017**, *546*, 524–527. [CrossRef] [PubMed]
36. Eddy, S.R. Hidden Markov Models. *Curr. Opin. Struct. Biol.* **1996**, *6*, 361–365. [CrossRef]
37. Chen, X.; Lin, S.; Liu, Q.; Huang, J.; Zhang, W.; Lin, J.; Wang, Y.; Ke, Y.; He, H. Expression and Interaction of Small Heat Shock Proteins (SHsps) in Rice in Response to Heat Stress. *Biochim. Biophys. Acta Proteins Proteom.* **2014**, *1844*, 818–828. [CrossRef]

38. Lamesch, P.; Berardini, T.Z.; Li, D.; Swarbreck, D.; Wilks, C.; Sasidharan, R.; Muller, R.; Dreher, K.; Alexander, D.L.; Garcia-Hernandez, M.; et al. The Arabidopsis Information Resource (TAIR): Improved Gene Annotation and New Tools. *Nucleic Acids Res.* **2012**, *40*, D1202–D1210. [CrossRef]
39. Kawahara, Y.; de la Bastide, M.; Hamilton, J.P.; Kanamori, H.; McCombie, W.R.; Ouyang, S.; Schwartz, D.C.; Tanaka, T.; Wu, J.; Zhou, S.; et al. Improvement of the Oryza Sativa Nipponbare Reference Genome Using next Generation Sequence and Optical Map Data. *Rice* **2013**, *6*, 4. [CrossRef]
40. Letunic, I.; Khedkar, S.; Bork, P. SMART: Recent Updates, New Developments and Status in 2020. *Nucleic Acids Res.* **2021**, *49*, D458–D460. [CrossRef]
41. Ison, J.; Kalas, M.; Jonassen, I.; Bolser, D.; Uludag, M.; McWilliam, H.; Malone, J.; Lopez, R.; Pettifer, S.; Rice, P. EDAM: An Ontology of Bioinformatics Operations, Types of Data and Identifiers, Topics and Formats. *Bioinformatics* **2013**, *29*, 1325–1332. [CrossRef]
42. Small, I.; Peeters, N.; Legeai, F.; Lurin, C. Predotar: A Tool for Rapidly Screening Proteomes ForN-Terminal Targeting Sequences. *Proteomics* **2004**, *4*, 1581–1590. [CrossRef]
43. Almagro Armenteros, J.J.; Salvatore, M.; Emanuelsson, O.; Winther, O.; von Heijne, G.; Elofsson, A.; Nielsen, H. Detecting Sequence Signals in Targeting Peptides Using Deep Learning. *Life Sci. Alliance* **2019**, *2*, e201900429. [CrossRef] [PubMed]
44. Yu, C.S.; Lin, C.J.; Hwang, J.K. Predicting Subcellular Localization of Proteins for Gram-Negative Bacteria by Support Vector Machines Based on n -Peptide Compositions. *Protein Sci.* **2004**, *13*, 1402–1406. [CrossRef] [PubMed]
45. Hu, B.; Jin, J.; Guo, A.-Y.; Zhang, H.; Luo, J.; Gao, G. GSDS 2.0: An Upgraded Gene Feature Visualization Server. *Bioinformatics* **2015**, *31*, 1296–1297. [CrossRef] [PubMed]
46. Wang, Y.; Tang, H.; DeBarry, J.D.; Tan, X.; Li, J.; Wang, X.; Lee, T.-H.; Jin, H.; Marler, B.; Guo, H.; et al. MCScanX: A Toolkit for Detection and Evolutionary Analysis of Gene Synteny and Collinearity. *Nucleic Acids Res.* **2012**, *40*, e49. [CrossRef]
47. Siddique, M.; Gernhard, S.; von Koskull-Döring, P.; Vierling, E.; Scharf, K.-D. The Plant SHSP Superfamily: Five New Members in Arabidopsis Thaliana with Unexpected Properties. *Cell Stress Chaperones* **2008**, *13*, 183–197. [CrossRef]
48. Ouyang, Y.; Chen, J.; Xie, W.; Wang, L.; Zhang, Q. Comprehensive Sequence and Expression Profile Analysis of Hsp20 Gene Family in Rice. *Plant Mol. Biol.* **2009**, *70*, 341–357. [CrossRef]
49. Zhao, P.; Wang, D.; Wang, R.; Kong, N.; Zhang, C.; Yang, C.; Wu, W.; Ma, H.; Chen, Q. Genome-Wide Analysis of the Potato Hsp20 Gene Family: Identification, Genomic Organization and Expression Profiles in Response to Heat Stress. *BMC Genom.* **2018**, *19*, 61. [CrossRef]
50. Li, K.-B. ClustalW-MPI: ClustalW Analysis Using Distributed and Parallel Computing. *Bioinformatics* **2003**, *19*, 1585–1586. [CrossRef]
51. Tamura, K.; Stecher, G.; Kumar, S. MEGA11: Molecular Evolutionary Genetics Analysis Version 11. *Mol. Biol. Evol.* **2021**, *38*, 3022–3027. [CrossRef]
52. Letunic, I.; Bork, P. Interactive Tree Of Life (ITOL) v5: An Online Tool for Phylogenetic Tree Display and Annotation. *Nucleic Acids Res.* **2021**, *49*, W293–W296. [CrossRef]
53. Lescot, M. PlantCARE, a Database of Plant Cis-Acting Regulatory Elements and a Portal to Tools for in Silico Analysis of Promoter Sequences. *Nucleic Acids Res.* **2002**, *30*, 325–327. [CrossRef] [PubMed]
54. Chen, C.; Chen, H.; Zhang, Y.; Thomas, H.R.; Frank, M.H.; He, Y.; Xia, R. TBtools: An Integrative Toolkit Developed for Interactive Analyses of Big Biological Data. *Mol. Plant* **2020**, *13*, 1194–1202. [CrossRef]
55. Zhang, H.; Li, G.; Fu, C.; Duan, S.; Hu, D.; Guo, X. Genome-Wide Identification, Transcriptome Analysis and Alternative Splicing Events of Hsf Family Genes in Maize. *Sci. Rep.* **2020**, *10*, 8073. [CrossRef]
56. Woodhouse, M.R.; Cannon, E.K.; Portwood, J.L.; Harper, L.C.; Gardiner, J.M.; Schaeffer, M.L.; Andorf, C.M. A Pan-Genomic Approach to Genome Databases Using Maize as a Model System. *BMC Plant Biol.* **2021**, *21*, 385. [CrossRef] [PubMed]
57. Szklarczyk, D.; Gable, A.L.; Nastou, K.C.; Lyon, D.; Kirsch, R.; Pyysalo, S.; Doncheva, N.T.; Legeay, M.; Fang, T.; Bork, P.; et al. The STRING Database in 2021: Customizable Protein–Protein Networks, and Functional Characterization of User-Uploaded Gene/Measurement Sets. *Nucleic Acids Res.* **2021**, *49*, D605–D612. [CrossRef]
58. Shannon, P.; Markiel, A.; Ozier, O.; Baliga, N.S.; Wang, J.T.; Ramage, D.; Amin, N.; Schwikowski, B.; Ideker, T. Cytoscape: A Software Environment for Integrated Models of Biomolecular Interaction Networks. *Genome Res.* **2003**, *13*, 2498–2504. [CrossRef] [PubMed]
59. Tian, F.; Yang, D.C.; Meng, Y.Q.; Jin, J.; Gao, G. PlantRegMap: Charting Functional Regulatory Maps in Plants. *Nucleic Acids Res.* **2019**, *48*, gkz1020. [CrossRef]
60. Vierling, E.; Harris, L.M.; Chen, Q. The Major Low-Molecular-Weight Heat Shock Protein in Chloroplasts Shows Antigenic Conservation among Diverse Higher Plant Species. *Mol. Cell. Biol.* **1989**, *9*, 461–468.
61. Sarkar, N.K.; Kim, Y.-K.; Grover, A. 099 Rice SHsp Genes: Genomic Organization and Expression Profiling under Stress and Development. *BMC Genom.* **2009**, *10*, 393. [CrossRef]
62. Panaretou, B.; Zhai, C. The Heat Shock Proteins: Their Roles as Multi-Component Machines for Protein Folding. *Fungal Biol. Rev.* **2008**, *22*, 110–119. [CrossRef]
63. Li, J.; Liu, X. Genome-Wide Identification and Expression Profile Analysis of the Hsp20 Gene Family in Barley (*Hordeum vulgare* L.). *PeerJ* **2019**, *7*, e6832. [CrossRef] [PubMed]

64. Mu, C.; Zhang, S.; Yu, G.; Chen, N.; Li, X.; Liu, H. Overexpression of Small Heat Shock Protein LimHSP16.45 in Arabidopsis Enhances Tolerance to Abiotic Stresses. *PLoS ONE* **2013**, *8*, e82264. [CrossRef] [PubMed]
65. Yao, F.; Song, C.; Wang, H.; Song, S.; Jiao, J.; Wang, M.; Zheng, X.; Bai, T. Genome-Wide Characterization of the HSP20 Gene Family Identifies Potential Members Involved in Temperature Stress Response in Apple. *Front. Genet.* **2020**, *11*, 609184. [CrossRef] [PubMed]
66. Yu, J.; Cheng, Y.; Feng, K.; Ruan, M.; Ye, Q.; Wang, R.; Li, Z.; Zhou, G.; Yao, Z.; Yang, Y.; et al. Genome-Wide Identification and Expression Profiling of Tomato Hsp20 Gene Family in Response to Biotic and Abiotic Stresses. *Front. Plant Sci.* **2016**, *7*, 1215. [CrossRef]
67. Li, Y.; Wang, X.; Li, Y.; Zhang, Y.; Gou, Z.; Qi, X.; Zhang, J. Transcriptomic Analysis Revealed the Common and Divergent Responses of Maize Seedling Leaves to Cold and Heat Stresses. *Genes* **2020**, *11*, 881. [CrossRef]
68. Sarkar, N.K.; Kim, Y.K.; Grover, A. Coexpression Network Analysis Associated with Call of Rice Seedlings for Encountering Heat Stress. *Plant Mol. Biol.* **2014**, *84*, 125–143. [CrossRef]

life

MDPI

Article

Sorghum Allelopathy: Alternative Weed Management Strategy and Its Impact on Mung Bean Productivity and Soil Rhizosphere Properties

Raza Ullah [1,2], Zubair Aslam [2], Houneida Attia [3], Khawar Sultan [4], Khalid H. Alamer [5], Muhammad Zeeshan Mansha [6], Ashwaq T. Althobaiti [3], Najla Amin T. Al Kashgry [3], Badreyah Algethami [3] and Qamar uz Zaman [4,*]

1 Department of Environmental Sciences, University of Okara, Punjab 56300, Pakistan
2 Department of Agronomy, University of Agriculture Faisalabad, Punjab 38040, Pakistan
3 Department of Biology, College of Science, Taif University, P.O. Box 11099, Taif 21944, Saudi Arabia
4 Department of Environmental Sciences, The University of Lahore-Lahore, Punjab 54590, Pakistan
5 Biological Sciences Department, Faculty of Science and Arts, King Abdulaziz University, P.O. Box 80200, Rabigh 21911, Saudi Arabia
6 College of Agriculture, Bahauddin Zakariya University, Bahadur Sub Campus, Layyah, Punjab 31200, Pakistan
* Correspondence: qamar.zaman1@envs.uol.edu.pk

Citation: Ullah, R.; Aslam, Z.; Attia, H.; Sultan, K.; Alamer, K.H.; Mansha, M.Z.; Althobaiti, A.T.; Al Kashgry, N.A.T.; Algethami, B.; Zaman, Q.u. Sorghum Allelopathy: Alternative Weed Management Strategy and Its Impact on Mung Bean Productivity and Soil Rhizosphere Properties. *Life* **2022**, *12*, 1359. https://doi.org/10.3390/life12091359

Academic Editors: Kousuke Hanada, Hakim Manghwar and Wajid Zaman

Received: 2 August 2022
Accepted: 29 August 2022
Published: 31 August 2022

Publisher's Note: MDPI stays neutral with regard to jurisdictional claims in published maps and institutional affiliations.

Copyright: © 2022 by the authors. Licensee MDPI, Basel, Switzerland. This article is an open access article distributed under the terms and conditions of the Creative Commons Attribution (CC BY) license (https://creativecommons.org/licenses/by/4.0/).

Simple Summary: Plants are subjected to a variety of biotic and abiotic stresses, which affect the rhizospheric attributes and limit agricultural crop productivity. To meet the food and energy demands of the future, several diverse approaches are used for achieving more stress-tolerant and climate-flexible crops for sustainable yields. Several organic and inorganic amendments are used to ameliorate these stresses. Crop-mediated modification (crop residues and allelopathic extracts) has great effects on weed management, improving rhizospheric attributes, and ultimately producing the best quality yield. Sorghum crop residues and their allelopathic extract can be used as a nutrient resource to enhance soil and crop productivity through their application. Sorghum-mediated crop modification will support the soil health and environmental sustainability, providing insight into the improvement of crop productivity. This study will help policymakers in modelling and enhancing sustainable crop production.

Abstract: The reduction of herbicide use and herbicide-resistant weeds through allelopathy can be a sustainable strategy to combat the concerns of environmental degradation. Allelopathic crop residues carry great potential both as weed suppressers and soil quality enhancers. The influence of sorghum crop residues and water extracts on the weed population, soil enzyme activities, the microbial community, and mung bean crop productivity was investigated in a two-year experiment at the Student Research Farm, University of Agriculture Faisalabad. The experimental treatments comprised two levels of sorghum water extract (10 and 20 L ha^{-1}) and two residue application rates (4 and 6 t ha^{-1}), and no sorghum water extract and residues were used as the control. The results indicated that the incorporation of sorghum water extract and residue resulted in significant changes in weed dynamics and the soil quality indices. Significant reduction in weed density (62%) and in the dry weight of weeds (65%) was observed in T_5. After the harvest, better soil quality indices in terms of the microbial population (72–90%) and microbial activity (32–50%) were observed in the rhizosphere (0–15 cm) by the same treatment. After cropping, improved soil properties in terms of available potassium, available phosphorus soil organic matter, and total nitrogen were higher after the treatment of residue was incorporated, i.e., 52–65%, 29–45%, 62–84%, and 59–91%, respectively. In the case of soil enzymes, alkaline phosphatase and dehydrogenase levels in the soil were 35–41% and 52–77% higher, respectively. However, residue incorporation at 6 t ha^{-1} had the greatest effect in improving the soil quality indices, mung bean productivity, and reduction of weed density. In conclusion, the incorporation of 6 t ha^{-1} sorghum residues may be opted to improve soil quality indices, suppress weeds, harvest a better seed yield (37%), and achieve higher profitability (306 $ ha^{-1}) by weed suppression, yield, and rhizospheric properties of spring-planted mung beans. This strategy can provide a probable substitute for instigating sustainable weed control and significant improvement of soil properties in the mung bean crop, which can be a part of eco-friendly and sustainable agriculture.

Keywords: crop residues; profitability; soil fertility; soil biology; allelopathy

1. Introduction

The human population will be 8.6 billion by 2030, as estimated by the United Nations, and is projected to be 9.8 billion by 2050 [1]. To feed this many humans, there must be an increase of 40% to 70% in food production [2] to feed the growing number of people by the year 2050. Food production increased by about 146% between 1961 and 2000, while the agricultural land for crops increased by only 8% [3]. This milestone was reached by an intensive use of topsoil nutrients and agrochemicals, rendering the soil exhausted and polluted [4]. Geographically, a large part of Pakistan is located in a dry-land environment where 80% of the land is classified as arid to semi-arid and only 8% is humid [5]. More than 60% of Pakistan's population depends on dry land to support their life and income, mainly through agriculture and pastoral activities [6]. Sustainable land management and crop production systems are necessary for developing countries to achieve stable production in the food supply system [5,6].

Weeds are another major concern in agriculture, and a large number of herbicides have been manufactured and used in soil [7]. The rhizosphere contains millions of weed seeds that grow when they get suitable growing conditions, otherwise remaining dormant [7]. Of the total pesticides manufactured around the world, 15% are herbicides [8]. Although the use of herbicides in Pakistan is higher because of the availability of labor for manual weeding, that is changing fast due to the introduction of mechanized systems in agriculture, and the tendency of applying chemical herbicides is on the rise with time [7].

Furthermore, the rise in the application of herbicides for managing weeds is a serious environmental risk to organisms and the planet. According to government statistics, Pakistan has seen an increase in the use of the pesticide glyphosate. In 2015, over 1100 tons of glyphosate were imported from other countries. In 2020, this number increased to 2000 tons, with importers including both domestic and foreign pesticide manufacturers [9]. Excessive herbicide use may result in a shift with the emergence of herbicide-resistant weeds and related health problems, which have changed the research interest to find alternative tools for managing weeds [10]. Finding more sustainable alternative options for weed control that will decrease the dependence on traditional farming practices, including synthetic herbicides, need time [11].

Allelopathy is the best substitute compared to synthetic herbicides, as allelochemicals have no residual toxic effects; however, many allelochemicals have limited efficacy and specificity [12,13]. Allelopathic crops carry great potential for the development of cultivars that are weed-suppressive. The application of residue of allelopathic crops suppresses weeds and improves soil health and crop production [14]. Crop residue is not only an excellent source of nutrients, but a significant source of organic material applied to soils as it improves soil health by increasing nutrient and water-holding potential [13]. Many crop plants including sorghum have been widely used for allelopathy. Besides these allelopathic properties, sorghum ranks among the top three important grains, as its industrial demand is increasing particularly in the food, beverage, and livestock feed industries [14]. Recently, sorghum has also gained interest as a new-generation bioenergy crop because of its multiple uses and wide adaptability to varied agroclimatic conditions [15]. Crops such as sunflower, sorghum, wheat, rice, rye, barley, maize, cucurbits, and alfalfa all show strong allelopathic potential. Among them, sorghum is the most investigated crop concerning its allelopathic potential [16]. The application of crushed sorghum mulch significantly decreased the total weed dry weight (26–56%) with a yield increase of 6–17% in wheat crops [17,18]. In cotton and maize, the use of sorghum surface mulch substantially decreased the weed population with a significant crop yield increase [11,16,19]. Species-specific compounds found in root exudates have important ecological consequences on soil health (macro- and microbiota) and plant health. Symbiotic relationships are supported and soil qualities, such as chemical and physical properties, are altered by the exudation of diverse substances [20]. The concentrations and classifications of these metabolites have a direct effect on ion absorption. For example, N and K absorption is boosted by a modest concentration of dibutyl phthalate and diphenylamine [16,20].

In the rhizospheric biome, there are millions of bacteria that have positive interactions with the plants and promote their growth and survival, while only a few are found to be pathogenic to plants [21]. The beneficial bacteria stimulate plant growth, make nutrients available to plants, suppress the growth of pathogens, and improve the soil structure, thus playing an essential role in sustainable crop production [3,7,11]. These beneficial microbes in the rhizosphere decompose the added residues which results in the improvement of soil health such as soil organic C sequestration, microbial biomass C, activity of soil biota [22], increase in soil organic matter, reduction in the fertilizer cost, and weed control, which ultimately results in better production [23].

A wide range of factors, including soil health, production costs, net revenue per acre, crop yields, gross income per acre, individual farm income, and many more can be improved through sustainable agriculture [16]. Organic and sustainable farming practices move us one acre closer to sustainability, or at the very least one acre less likely to cause harm [24]. Some studies have documented weed control by using sorghum water extract and sequential plantation of sorghum [7,11,16,24]. However, not much is known about the possible changes of such uses of allelopathic interventions on the soil–plant environment and microbial diversity. In this study, we hypothesized that sorghum water extracts and residues may suppress weeds while improving soil health and mung bean productivity, which is the only viable option available for meeting the growing demand for food in developing countries. The precise objective of the experiment was to find out the impact of sorghum water extracts and the residues upon soil enzymatic and chemical activities, weed dynamics, the population of microbes, and mung bean productivity by the modulation of physiological parameters.

2. Materials and Methods

2.1. Experimental Site, Climate and Soil Sampling

The experiment in this study was conducted at the Student Research Farm, Department of Agronomy (University of Agriculture Faisalabad), Pakistan (Latitude: 31°26′ N and Longitude: 73°06′ E; Altitude: 184.4 mASL). As per the classification system, the soil belongs to the Lyallpur series. According to the US Department of Agriculture classification system, the soil type is arid sol-fine-silty, hyperthermic Ustalfic, mixed, and Haplargid. According to the Food and Agriculture Organization classification system, it is classified as Haplic Yermosols soil type.

The soil samples from mung bean plants' rhizosphere were sampled 20 days after planting and the following harvest. Before the experiment began, a composite sample of ten soil samples (0–15 cm) was taken from the experimental site. A second composite sample was then taken from the same field after the crop had been harvested to determine the enzymatic characteristics and microbial counts of the soil samples. The soil samples were subjected to air drying, grinding, and sieving (using a 2 mm sieve), and all parameters except for microbial culture and dehydrogenase activity were examined. Soil samples were kept at 4 °C for both dehydrogenase and microbiological analysis. Soil properties, nutrient dynamics, soil enzyme activities, and microbial populations of the experimental soil were measured before sowing by following standard protocols as depicted in Table 1.

Table 1. Soil properties, nutrient dynamics, soil enzyme activities, and microbial populations of the experimental soil before sowing.

Soil Indices	Experiment Year 1	Experiment Year 2
pH	7.85	7.79
Electrical Conductivity (dS m^{-1})	1.11	1.19
Total Soil Organic Matter (%)	0.53	0.61
Available Phosphorous (mg kg^{-1})	6.74	6.95
Available Potassium (mg kg^{-1})	123.00	131.00
Total Soil Nitrogen (g kg^{-1})	0.24	0.29
Bacteria (cfu/g $\times 10^5$)	35.00	45.00
Fungi (cfu/g $\times 10^4$)	5.00	8.00
Microbial Activity (mg CO_2-C kg^{-1} d^{-1})	3.05	3.14
Alkaline Phosphatase Activity (µg NP g^{-1} soil h^{-1})	135.00	143.00
Dehydrogenase Activity (µg TPFg^{-1} soil h^{-1})	21.00	25.00

The field data related to weather parameters for the whole period of crop growth and management were taken from the Meteorological Observatory, Department of Agronomy, University of Agriculture, Faisalabad, Pakistan and are given in Table 2.

Table 2. Weather indices of an experimental site for the study period.

Months	Weather Indices	Experiment Year 1	Experiment Year 2
March	Maximum Temperature (°C)	25	25
	Minimum Temperature (°C)	14	14
	Rain Fall (mm)	42	68
	Relative Humidity (%)	60	64
April	Maximum Temperature (°C)	32	33
	Minimum Temperature (°C)	19	21
	Rain Fall (mm)	28	33
	Relative Humidity (%)	52	44
May	Maximum Temperature (°C)	37	39
	Minimum Temperature (°C)	24	25
	Rain Fall (mm)	41	17
	Relative Humidity (%)	33	28
June	Maximum Temperature (°C)	41	38
	Minimum Temperature (°C)	28	26
	Rain Fall (mm)	7	12
	Relative Humidity (%)	34	39
July	Maximum Temperature (°C)	37	35
	Minimum Temperature (°C)	28	27
	Rain Fall (mm)	58	128
	Relative Humidity (%)	54	61

2.2. Experimental Treatments and Design

The field experiment in this study was designed with the treatments: control (plots with no crop residues or extract application), sorghum water extract at 10 L ha^{-1}, sorghum water extract at 20 L ha^{-1}, sorghum residues at 4 t ha^{-1}, and sorghum residues at 6 t ha^{-1}. The field experiment design involved a randomized complete block design (RCBD) with three replications. The size (area) of each plot for each treatment was measured to be 15 m^2 (3.0 m $\times$ 5.0 m).

2.3. Crop Management

The experimental site was ploughed twice using a tractor-drawn cultivator and then planked. Flat wooden planks were utilized for breaking up clods, and a laser leveler was used for leveling soil. Wheat was the fore-crop for mung beans. The seed of mung bean cultivar NM-92 was collected from the National Institute of Agriculture and Biology (NIAB), Faisalabad. It was planted on 15 March 2014 and repeated on 20 March 2015. A recommended rate of mung bean seeds (25 kg ha^{-1}) was used to maintain the plant population in 30 cm apart rows using a hand drill. Urea, diammonium phosphate, and sulphate of potash were applied at the rate of 3 kg N, 58 kg P_2O_5, and 63 kg K_2O ha^{-1} for the nutrient requirement. The recommended dose of P and K and 1/3rd of N was applied at the time of sowing, and the remaining N was applied with the first and second irrigation using the top-dressing method. After ten days of sowing, the first irrigation was applied (7.5 cm), while subsequent irrigation was applied upon crop requirement. To control the termites and pod borers, insecticides Fipronil and Emamectin Benzoate were applied at the rate of 1.73% *w/w* ha^{-1} and 5.03% *w/w* ha^{-1}, respectively. The crop harvesting was done on the 10th and 15th of July during both years.

2.4. Sorghum Crop Water Extracts Preparation

Sorghum plant residue samples were obtained from the Student Research Farm (University of Agriculture Faisalabad). The plant samples were harvested at maturity, shade dried, and cut into pieces (<3 cm in size) by using the electric fodder cutter. These shredded pieces of sorghum residues were then placed in distilled water for 24 h with a ratio of 1:10 (*w/v*%) and the filtrate liquid obtained from it was used as fresh [25]. This sorghum water extract was used as 10 & 20 L ha^{-1} by spraying (300 L ha^{-1}) with the help of a T-jet nozzle using a knapsack sprayer 5 days after the sowing (3–5 leaf stage) of the mung bean plants. A total volume of sprayed liquid was determined by the calibration method.

2.5. Sorghum Crop Residues Preparation

Samples of sorghum plant residues were also collected from Student Research Farm (University of Agriculture Faisalabad). Plants were harvested at maturity, shade dried, and then cut into tiny pieces in sizes less than 3 cm with the help of a machine (electric fodder). These dried pieces of the crop were then applied to the soil before the sowing as per treatments of 4 and 6 t ha^{-1} in the experiment.

2.6. Data Analysis

For the measurement of weeds-related, soil-related and yield-related attributes, the following protocols were followed.

2.6.1. Soil Attributes, Microbial Population, and Soil Enzymatic Activities

Soil electrical conductivity (EC) and the pH of the soil were determined by following the protocols of Ryan et al. [26]. Water/soil suspension was utilized at a ratio of 2:1 to measure soil EC and pH. The value of EC and pH was measured using a Jenway Model 4510 digital conductivity meter and a Kent Eil 7015 pH meter. The protocols of Blake and Hartge [27] and Vomocil [28] were followed to determine the total porosity of soil (TP) and soil bulk density (BD), respectively. Similarly, methods developed by Bremner and Mulvaney [29], Walkley and Black [30], Olsen and Sommers [31], and Helmke and Sparks [32] were used for the calculation of total nitrogen (TN), available potassium (K), available phosphorus (P), and SOM (soil organic matter). The microbial populations of soil samples were measured by the method of spiral plating serial dilutions of each sample on agar plates [33]. A total number of culturable bacteria populations was measured on R2A (half-strength) agar plates following the methods described by Janssen et al. [34] and Wu et al. [35]. The culturable fungi in samples were determined using the plating method in Rose Bengal media of potato dextrose agar [36] and the colony counts were carried out after enough time (48 h) for culturing. The microbial activity as indicated by CO_2 evolution was measured by acid-base titration procedure and reported as mg CO_2-C kg^{-1} d^{-1}. Soil dehydrogenase enzymatic activity was measured by the procedures described by Min et al. [37] and was reported as µg TPF g^{-1} 12 h^{-1}. Alkaline phosphatase activity was determined spectrophotometrically by following the method described by Tabatabai and Bremner, [38] and was reported as μg p-nitrophenol g^{-1} h^{-1} in this study.

2.6.2. Weeds Dynamics

Weed dynamics such as the total number of weeds (0.25 m^{-2}), fresh weight (g 0.25 m^{-2}), and dry weight (g 0.25 m^{-2}) were observed and recorded from each plot for 30 days after sowing by randomly selecting two quadrates (50 cm × 50 cm). First, weeds were counted one by one and then cut at levels above the ground surface. For determining the dry weight, weeds were dried under the sunlight for 48 h and then dried in an electric oven for 72 h, maintaining temperatures at 70 °C. The dry weight of samples was recorded by using an electric balance after attaining the constant weight.

2.6.3. Yield Attributes

The yield components such as the number of pods per plant, number of seeds per pod, and 1000-seed weight were observed and recorded by the methods described by Rab et al. [39]. Mung bean crop samples were harvested at maturity and threshed manually for the separation of seeds from straw. The seed yield of each experimental unit was recorded and expressed as kg ha^{-1} in the experiment.

2.6.4. Statistical Analysis

Statistical analysis of data (both years) was carried out using Statistix 8.1. For the comparison of treatments, the LSD (least significance difference) test at 5% probability was applied.

3. Results

3.1. Weeds Dynamics

Horse purslane (*Trianthema portulacastrum*) and purple nutsedge (*Cyperus rotundus*) both were dominant in each experimental unit during both years of study. This study indicated that the density and dry weight of horse purslane significantly differed with various allelopathic weed management strategies. Total weed density and dry weight also significantly differed with various allelopathic weed management strategies. However, the dry weight of purple nutsedge was non-significant among various allelopathic weed management strategies (Table 3). The effect of time was significant for all weed parameters except the dry weight of purple nutsedge (Table 3). The interaction of allelopathic weed management strategies and the year was significant for total weed density but non-significant for the dry weight of the weeds, as well as for density and dry weight of the horse purslane and purple nutsedge (Table 3).

The lowest horse purslane (13) and purple nutsedge (3) densities were recorded with sorghum residues at 6 t ha^{-1} compared to the control (40 and 10, respectively). The lowest values were observed in the control (Table 3). Total weed density and dry weight decreased over time and the minimum values were observed during the 2nd year (Table 2). In the case of horse purslane, dry weight (14 g/0.25 m^{-2}) was observed with sorghum residues at 6 t ha^{-1}, followed by sorghum residues at 4 t ha^{-1} (Table 3). The maximum value of dry weight (48 g/0.25 m^{-2}) was observed in the control (Table 3). In the case of total weed density, the interactive effect of allelopathic weed management strategies and the year showed a statistically significant effect. The minimum total weed density (18.42) was recorded with 6 t ha^{-1} sorghum residues during the 2nd year, as compared to the control (56.55). The lowest total weed dry weight (56.23 and 19.97 g/0.25 m^{-2}, respectively) was observed with sorghum residues at 6 tons ha^{-1} and the maximum total weed density (56.85) was recorded in the control (Table 3).

Table 3. Effect of sorghum water extracts and residues on weed dynamics in mung bean crops.

Treatments	Year 1	Year 2	Mean [(a)] (T)	Year 1	Year 2	Mean (T)	Year 1	Year 2	Mean (T)
	Horse Purslane Density (0.25 m^{-2})			Horse Purslane Fresh Weight (g/0.25 m^2)			Horse Purslane Dry Weight (g/0.25 m^2)		
T_1	41	40	40 A	156	147	152 A	49	47	48 A
T_2	41	32	36 B	135	126	130 B	40	43	41 B
T_3	38	27	32 C	118	99	108 C	37	32	34 C
T_4	22	16	19 D	82	65	73 D	26	21	23 D
T_5	13	12	13 E	48	43	46 E	15	13	14 E
Mean [(b)] (Y)	31 A	26 B		104 A	100 B		33 A	31 B	
LSD ($p \leq 0.05$)		*T = 3.76; Y = 2.38*			*T = 14.19; Y = 3.15*			*T = 4.51; Y = 1.85*	
	Purple Nutsedge Density (0.25 m^{-2})			Purple Nutsedge Fresh Weight (g/0.25 m^2)			Purple Nutsedge Dry Weight (g/0.25 m^2)		
T_1	10	10	10 A	13	13	13	4	4	4
T_2	9	7	8 B	10	6	8	3	2	3
T_3	7	7	7 C	6	6	6	2	2	2
T_4	6	5	5 D	3	3	3	1	1	1
T_5	3	2	3 E	3	3	3	1	1	1
Mean (Y)	7 A	6 B		7	6		2	2	
LSD ($p \leq 0.05$)		*T = 1.09; Y = 0.68*			*NS*			*NS*	
	Total Weed Density (0.25 m^{-2})			Total Weed Fresh Weight (g/0.25 m^2)			Total Weed Dry Weight(g/0.25 m^2)		
T_1	57.15 a	56.55 a	56.85 A	175.18	165.94	170.56 A	58.05	55.12	56.58 A
T_2	52.93 ab	49.61 bc	51.27 B	147.90	141.61	144.76 B	49.39	47.39	48.39 B
T_3	45.77 c	36.55 d	41.16 C	130.28	112.08	121.18 C	43.79	38.02	40.91 C
T_4	35.98 d	25.78 e	30.88 D	88.16	78.70	83.93 D	31.37	26.15	28.76 D
T_5	24.35 e	18.42 f	21.39 E	59.78	52.68	56.23 E	20.78	19.16	19.97 E
Mean (Y)	43.12 A	37.50 B		120.26 A	110.20 B		39.69 A	38.15 B	
LSD ($p \leq 0.05$)		*T = 3.81; Y = 2.41; T × Y = 5.39*			*T = 14.44; Y = 8.13*			*T = 4.59; Y = 0.55*	

Figures of interaction and main effects sharing the same case letter do not differ significantly ($p \leq 0.05$) by the least significant difference test; likewise, the figures of main effects and interaction without lettering do not differ significantly ($p \leq 0.05$) by the least significant difference test; T_1 = Control (plots with no crop residues or extract application); T_2 = Sorghum water extract at 10 L ha^{-1}; T_3 = Sorghum water extract at 20 L ha^{-1}; T_4 = Sorghum residues at 4 t ha^{-1}; T_5 = Sorghum residues @ 6 t ha^{-1}; [(a)] T = treatments; [(b)] Y = year.

3.2. Yield and Yield Parameters

Yield and yield parameters differed significantly among the various allelopathic weed management strategies (Table 4). Likewise, the year effect was significant for the weight of 1000 seeds, biological yield, harvest index, and yield, but non-significant for No. of pods per plant and No. of seeds per pod (Table 4). The interaction of allelopathic weed management strategies and the year was significant only for yield (Table 4). However, the interaction was non-significant for No. of pods per plant, No. of seed per pod, the weight of 1000 seeds, biological yield, and harvest index (Table 4). The results indicated that the maximum values of No. of pods per plant (24.6), No. of seed per pod (9.9), weight of 1000 seeds (55.33 g), biological yield (4106 kg ha^{-1}), harvest index (26.01%), and yield (1019.3 kg ha^{-1}) were recorded with sorghum residues at 6 tons ha^{-1}. The minimum values of No. of pods per plant (14.6), No. of seeds per pod (5.9), weight of 1000 seeds (50.25 g), biological yield (3206 kg ha^{-1}), harvest index (22.74%), and yield (744.3 kg ha^{-1}) were observed in the control (Table 4). A linear increase in the No. of pods per plant, No. of seeds per pod, weight of 1000 seeds, biological yield, harvest index, and yield was observed over time and all the above observations had a significant increase in values during the 2nd year of the study (Table 4). In the present study, all treatments gave higher net returns as compared with the control during both the years of study. Among all treatments, sorghum residue at 6 tons ha^{-1} gave maximum economical returns during both years, while a minimum net benefit was obtained from the control (Table 5).

Table 4. Effect of sorghum water extracts and residues on yield and yield components of mung bean plants.

Treatments	Year 1	Year 2	Mean [(a)] (T)	Year 1	Year 2	Mean (T)	Year 1	Year 2	Mean (T)
	Final Emergence Count per Plot			**Plant Height at Maturity (cm)**			**Number of Nodules per Plant**		
T_1	557	558	558	40.7	41.2	40.9 C	5	5	5 C
T_2	558	560	559	42.5	42.7	42.6 C	7	8	8 B
T_3	560	562	561	42.6	43.7	43.1 C	7	8	8 B
T_4	561	562	562	45.5	46.6	46.0 B	9	9	9 AB
T_5	563	565	564	47.7	49.0	48.3 A	10	11	11 A
Mean [(b)] (Y)	560	561		44.0	44.4		8	8	
LSD ($p \leq 0.05$)		*NS*			*T = 2.1*			*T = 1.79*	
	No. of Pods per Plant			**No. of Seeds per Pod**			**Weight of 1000-Seeds (g)**		
T_1	13.76	15.33	14.55 D	5.43	6.37	5.90 E	49.95	50.54	50.25 E
T_2	17.00	19.09	18.05 C	6.55	7.80	7.17 D	52.58	54.03	53.31 D
T_3	19.45	21.19	20.32 BC	6.95	8.02	7.48 C	53.25	54.90	54.08 C
T_4	20.03	23.99	22.32 AB	7.07	9.01	8.04 B	53.76	55.66	54.71 B
T_5	23.55	25.72	24.63 A	9.24	10.61	9.92 A	54.49	56.16	55.33 A
Mean (Y)	17.96	21.86		7.05	8.36		52.81 B	54.26 A	
LSD ($p \leq 0.05$)		*T = 2.62*			*T = 0.25*			*T = 0.59; Y = 1.45*	
	Biological Yield (kg ha^{-1})			**Harvest Index (%)**			**Yield (kg ha^{-1})**		
T_1	3196	3216	3206 E	22.62	22.85	22.74 C	741.9 e	746.7 e	744.3 E
T_2	3351	3410	3380 D	22.48	23.07	22.78 C	789.2 d	811.5 d	800.4 D
T_3	3525	3587	3556 C	23.95	24.15	24.05 B	844.1 c	867.6 c	855.8 C
T_4	3660	3670	3665 B	24.26	24.74	24.50 B	931.2 b	934.2 b	932.7 B
T_5	3970	4242	4106 A	25.67	26.35	26.01 A	1009.1 a	1029.4 a	1019.3 A
Mean (Y)	3540 B	3625 A		23.80 B	24.23 A		863.70 B	877.29 A	
LSD ($p \leq 0.05$)		*T = 105.07; Y = 75.92*			*T = 0.39; Y = 0.41*			*T = 21.97; Y = 11.45; T × Y = 31.07*	

Figures of interaction and main effects sharing the same case letter do not differ significantly ($p \leq 0.05$) by the least significant difference test; likewise, the figures of main effects and interaction without lettering do not differ significantly ($p \leq 0.05$) by the least significant difference test; T_1 = Control (plots with no crop residues or extract application); T_2 = Sorghum water extract at 10 L ha^{-1}; T_3 = Sorghum water extract at 20 L ha^{-1}; T_4 = Sorghum residues at 4 t ha^{-1}; T_5 = Sorghum residues at 6 t ha^{-1}; [(a)] T = treatments; [(b)] Y = year.

Table 5. Economics of mung bean crops grown in various allelopathic weed management strategies.

Treatments	Yield (kg ha^{-1})	Adjusted Yield (kg ha^{-1})	Gross Income [(d)] \$ ha^{-1}	Total Cost \$ ha^{-1}	Net Benefits \$ ha^{-1}	Benefit–Cost Ratio
[(a)] Control	744	670	750	615	135	0.22
[(b)] SWE at 10 L ha^{-1}	800	720	806	628	179	0.29
SWE at 20 L ha^{-1}	856	770	863	633	230	0.36
[(c)] SR at 4 tons ha^{-1}	933	840	940	688	252	0.37
SR at 6 tons ha^{-1}	1019	917	1027	721	306	0.42
Remarks			\$44.67/40 kg			

[(a)] Control = (plots with no crop residues or extract application); [(b)] SWE = sorghum water extract; [(c)] SR = sorghum residues; [(d)] \$ = US dollar.

3.3. Rhizosphere Soil Microbial Population and Enzymes Activity

Microbiological and biochemical indicators are also used as full indicators of soil health. They are more susceptible than physical and chemical attributes to changes imposed on the environment. Microbiological indicators such as the population of bacteria, fungi, and microbial activity at 20 days after sowing and harvesting differed significantly among various allelopathic weed management strategies (Table 6). The year effect was also significant for all the above parameters (Table 6). The interactive effect of allelopathic weed management strategies and the year was significant for the population of fungi but non-significant for the population of bacteria at 20 days after sowing and at harvesting (Table 6). Biochemical indicators like soil enzymes (alkaline phosphatase and dehydrogenase) differed significantly among various allelopathic weed management strategies at harvest (Table 6).

Table 6. Effect of sorghum water extracts and residues on microbial population, microbial activity, and soil enzymatic activity in the rhizosphere of mung bean crops.

Treatments	Year 1	Year 2	Mean [(a)] (T)	Year 1	Year 2	Mean (T)	Year 1	Year 2	Mean (T)	Year 1	Year 2	Mean (T)
	Bacteria (cfu/g × 10^5) 20 [(c)] DAS			Fungi (cfu/g × 10^4) 20 DAS			Microbial Activity (mg CO_2-C kg^{-1} d^{-1}) 20 DAS			Alkaline Phosphatase (µg NP g^{-1} Soil h^{-1})		
T_1	43	44	43 D	7 d	8 d	8 C	3.63	3.77	3.70 C	135.47 e	135.50 e	135.48 C
T_2	46	48	47 CD	8 d	8 d	8 C	3.65	3.78	3.71 C	135.48 e	135.55 e	135.52 C
T_3	48	49	48 C	8 d	9 d	9 C	3.72	3.81	3.77 C	135.78 e	135.85 e	135.82 C
T_4	64	74	69 B	15 c	20 b	18 B	4.88	5.05	4.97 B	167.26 d	173.46 c	170.36 B
T_5	74	83	79 A	21 b	25 a	23 A	5.45	5.58	5.52 A	185.24 b	196.22 a	190.73 A
Mean [(b)] (Y)	55 B	60 A		12 B	14 A		4.26 B	4.40 A		151.85 B	155.32 A	
LSD ($p \leq 0.05$)	*T = 4.72; Y = 2.98*			*T = 1.55; Y = 0.98; T × Y = 2.20*			*T = 0.20; Y = 0.13*			*T = 3.85; Y = 2.43; T × Y = 5.44*		
	Bacteria (cfu/g × 10^5) [(d)] AH			Fungi (cfu/g × 10^4) AH			Microbial Activity (mg CO_2-C kg^{-1} d^{-1}) AH			Dehydrogenase (µg TPFg^{-1} Soil h^{-1})		
T_1	21	22	22 C	5 f	6 f	6 D	2.99	3.13	3.06 C	22.68 d	23.33 d	23.00 C
T_2	23	24	24 C	6 f	7 f	7 CD	3.05	3.15	3.10 C	23.09 d	23.59 d	23.34 C
T_3	24	25	25 C	6 f	9 e	8 C	3.08	3.17	3.13 C	23.33 d	24.19 d	23.76 C
T_4	35	40	38 B	11 d	17 b	14 B	3.99	4.11	4.05 B	32.00 c	38.00 b	35.00 B
T_5	40	44	42 A	14 c	20 a	17 A	4.50	4.65	4.58 A	37.33 b	44.00 a	40.67 A
Mean (Y)	29 B	31 A		9 B	12 A		3.52 B	3.64 A		27.69 B	30.62 A	
LSD ($p \leq 0.05$)	*T = 2.21; Y = 1.40*			*T = 1.21; Y = 0.76; T × Y = 1.71*			*T = 0.28; Y = 0.10*			*T = 2.18; Y = 1.37; T × Y = 3.08*		

Figures of interaction and main effects sharing the same case letter do not differ significantly ($p \leq 0.05$) by the least significant difference test; likewise, the figures of main effects and interaction without lettering do not differ significantly ($p \leq 0.05$) by the least significant difference test; T_1 = Control (plots with no crop residues or extract application); T_2 = Sorghum water extract at 10 L ha^{-1}; T_3 = Sorghum water extract at 20 L ha^{-1}; T_4 = Sorghum residues at 4 t ha^{-1}; T_5 = Sorghum residues at 6 t ha^{-1}; [(a)] T = treatments; [(b)] Y = year; [(c)] DAS = days after sowing; [(d)] AH = after harvesting.

Interaction (allelopathic weed management strategies × year) was significant for the fungal population. The highest fungal population (25 cfu/g × 10^4 and 20 cfu/g × 10^4, respectively) was recorded with the application of sorghum residues at 6 t ha^{-1} at both stages, i.e., 20 days after sowing and at the harvesting of the second year of the experiment. However, the highest bacterial population (79 cfu/g × 10^4 and 42 cfu/g × 10^4, respectively) and microbial activity (5.52 mg CO_2-C kg^{-1} d^{-1} and 4.58 mg CO_2-C kg^{-1} d^{-1}, respectively) were recorded with the application of sorghum residues at 6 t ha^{-1} at both stages, i.e., 20 days after sowing and at harvesting. The lowest microbial activity (3.70 mg CO_2-C kg^{-1} d^{-1}) and the lowest populations of both bacteria (22 cfu/g × 10^5) and fungi (6 cfu/g × 10^4) were observed in the control (Table 6). A linear increase in the bacterial population was observed over time at 20 days after sowing and at harvesting, and the highest bacterial population was observed during the second year (Table 6). In the case of soil enzymes, the interactive effect of the allelopathic weed management strategies and the year resulted in a significant effect on the activity of both enzymes alkaline phosphatase and dehydrogenase. The highest value (196.22 µg NP g^{-1} soil h^{-1} and 44.00 µg TPFg^{-1} soil h^{-1}, respectively) was observed with the application of sorghum residues at 6 t ha^{-1} during the second year, which was followed with the same treatment in the first year. The lowest value (135.50 µg NP g^{-1} soil h^{-1} and 23.33 µg TPFg^{-1} soil h^{-1}, respectively) was recorded in the control (Table 6).

3.4. Rhizosphere Soil Properties and Nutrient Dynamics

At the end of the experiment, the physical indicators of soil health like soil porosity and bulk density significantly differed among various allelopathic weed management strategies (Table 7). The year effect was also statistically significant for the soil's physical indicators but the interaction (allelopathic weed management strategies × year) was non-significant (Table 7). Chemical indicators of soil health such as EC (electrical conductivity), SOM (soil organic matter), N (nitrogen), available K (potassium), and P (phosphorus) significantly differed among various allelopathic weeds management strategies (Table 7). The year effect was also statistically significant for all soil chemical indicators except soil pH and available K. The interaction (allelopathic weed management strategies × year) was statistically significant for SOM, N, and available P. However, for soil pH, EC and available K interactions were non-significant (Table 7). The lowest bulk density (1.30 g cm^{-3}) and the highest soil porosity (50.22%) were observed in treatments when sorghum residues at 6 tons ha^{-1} were applied, as compared to the control, while the lowest bulk density (1.23 g cm^{-3}) and highest soil porosity (51.79%) were observed in second year of the experiment (Table 7). In case of SOM, N, and available P, the highest values (1.37%, 0.45 g kg^{-1}, 10.31 mg kg^{-1}, respectively) were observed during the second year when sorghum residues at 6 tons ha^{-1} were applied as compared to the control (0.69%, 0.21 g kg^{-1}, 6.77 mg kg^{-1}, respectively). Among all allelopathic weed management strategies, the statistically highest values of soil EC (1.34 dS m^{-1}) and available K (200.83 mg kg^{-1}) were obtained with the application of sorghum residues at 6 tons ha^{-1}. The statistically lowest values for all parameters given above were observed in the control, which was statistically similar to sorghum water extracts at 10 and 20 L ha^{-1} (Table 7). A linear increase in soil EC and available K was observed over time, and these parameters (soil EC and available K) had the highest values during the second year of the experiment (Table 7). In the case of soil pH, a decreasing trend was observed. The lowest soil pH (7.28) was observed with the application of sorghum residues at 6 tons ha^{-1} and the highest soil pH (7.73) was observed in the control, which was statistically similar to the sorghum water extract at 10 and 20 L ha^{-1} (Table 7).

Table 7. Effect of sorghum water extracts and residues on soil properties and nutrient dynamics in the rhizosphere of mung bean crops at harvest.

Treatments	2014	2015	Mean [(a)] (T)	2014	2015	Mean (T)	2014	2015	Mean (T)	2014	2015	Mean (T)
	Soil Bulk Density (g cm^{-3})			**Total Soil Porosity (%)**			**Soil pH**			**Soil EC (dS m^{-1})**		
T_1	1.48	1.46	1.47 A	42.82	43.73	43.27 C	7.75	7.72	7.73 A	1.07	1.11	1.09 C
T_2	1.47	1.46	1.46 A	43.53	44.10	43.82 C	7.75	7.70	7.73 A	1.11	1.14	1.12 C
T_3	1.47	1.46	1.46 A	43.80	44.11	43.96 C	7.74	7.69	7.72 A	1.12	1.14	1.13 C
T_4	1.41	1.29	1.35 B	48.50	50.39	49.44 B	7.44	7.41	7.43 B	1.23	1.27	1.25 B
T_5	1.38	1.23	1.30 C	48.66	51.79	50.22 A	7.38	7.18	7.28 C	1.32	1.36	1.34 A
Mean [(b)] (Y)	1.44 A	1.38 B		45.46 B	46.83 A		7.61 A	7.54 B		1.17 B	1.20 A	
LSD ($p \leq 0.05$)	*T = 0.04; Y = 0.06*			*T = 0.75; Y = 0.92*			*T = 0.12; Y = 0.04*			*T = 0.07; Y = 0.02*		
	Total Soil Organic Matter (%)			**Total Soi Nitrogen (g kg^{-1})**			**Available Potassium (mg kg^{-1})**			**Available Phosphorous (mg kg^{-1})**		
T_1	0.67 d	0.69 d	0.68 C	0.22 d	0.21 d	0.22 C	121.45	121.93	121.69 C	6.74 d	6.77 d	6.76 C
T_2	0.68 d	0.69 d	0.69 C	0.22 d	0.21 d	0.22 C	121.52	121.64	121.58 C	6.77 d	6.77 d	6.78 C
T_3	0.69 d	0.71 d	0.70 C	0.22 d	0.21 d	0.22 C	121.52	122.00	122.76 C	6.77 d	6.80 d	6.79 C
T_4	0.96 c	1.24 ab	1.10 B	0.32 c	0.38 b	0.35 B	178.85	190.00	184.43 B	8.09 c	9.28 b	8.69 B
T_5	1.12 b	1.37 a	1.25 A	0.38 b	0.45 a	0.42 A	195.00	206.65	200.83 A	9.25 b	10.31 a	9.78 A
Mean (Y)	0.82 B	0.94 A		0.27 B	0.29 A		147.67	152.45		7.53 B	7.99 A	
LSD ($p \leq 0.05$)	*T = 0.10; Y = 0.06; T × Y = 0.14*			*T = 0.03; Y = 0.01; T × Y = 0.03*			*T = 8.43*			*T = 0.41; Y = 0.26; T × Y = 0.58*		

Figures of interaction and main effects sharing the same case letter do not differ significantly ($p \leq 0.05$) by the least significant difference test; likewise, the figures of main effects and interaction without lettering do not differ significantly ($p \leq 0.05$) by the least significant difference test; T_1 = Control (plots with no crop residues or extract application); T_2 = Sorghum water extract at 10 L ha^{-1}; T_3 = Sorghum water extract @ 20 L ha^{-1}; T_4 = Sorghum residues at 4 t ha^{-1}; T_5 = Sorghum residues at 6 t ha^{-1}; [(a)] T = treatments; [(b)] Y = year.

4. Discussion

Incorporating allelopathic crop residues is a green approach to manage weeds in field crops. Our results showed significant weed suppression potential with the incorporation of sorghum residues and water extract. This approach had a maximum reduction in weed density, fresh weight, and dry weight of weed species in mung bean crops (Table 3). This reduction was due to the release of phenolic compounds, including phenolic acids (Dhurrin, p-hydroxybenzaldehyde, sorgoleone, vanillic acid, p-hydroxybenzoic acid, p-hydroxybenzaldehyde, p-coumaric acid, and ferulic acid) with a wide spectrum of biological activities, including allelopathy [40,41]. In the case of field crops, sorghum had the highest allelopathic potential, which has been reported by many researchers [16,17,42]. Inhibitory activity of sorghum allelochemicals on grassy and broad-leaved weeds has been reported [17]. Cheema and Khaliq [18] investigated that 35–49% of weed density and weed biomass was reduced by using water extract of mature sorghum crop plants as compared with the control group. The sorghum residue treatments showed the highest suppression of weeds compared to sorghum water extracts treatments (Table 3); adding sorghum at 2–6 Mg ha^{-1} to the soil reduced the weed biomass by 40–50%. Crop residues may change the weed frequency and distribution, and may cause the suppression of weeds [14,43]. Zaji and Majd [44] showed that the fresh weight and dry weight of different weed biota, viz., red root pigweed (*Amaranthus retroflexus*), palmer amaranth (*Amaranthus palmeri*), black nightshade or wonder berry (*Solanum nigrum*), and curled dock (*Rumex crispus*) were decreased severely by the impact of canola crop residues. The growth suppression of dominant weed biota in this experiment might have been observed due to the physical resistance by sorghum residues' incorporation or the release of chemicals from these residues [7]. Allelochemicals released through different parts of plants are dependent on many factors, i.e., applied crop family, size and dose of mulching, decomposition rate, moisture contents, the texture of the soil, and soil microbiota [45,46]. Weed suppression level is directly related to the dose of allelopathic products [47,48]. The higher the amount of plant material used for mulch, the greater the total amount of allelochemicals present in the mulch and released, leading to a higher concentration of allelochemicals into the soil [49–51]. Generally, by incorporating a higher amount of crop residues, greater weed suppression was observed. A two-year field experiment was conducted by Alsaadawi et al. [52] who stated that sorghum residue incorporation significantly reduced the weed number and produced a higher yield of broad bean than weedy check.

In our study, more than a 37% increase in mung bean yield was achieved through effective allelopathic weed management strategies (Table 4). This increase in crop yield might be due to the improvement of soil properties and reduced weed competition during the critical periods of crop growth. The effective reduction of weeds also increases the obtainability of resources such as light, moisture, nutrients, and yield gap [7,53]. Research on wheat residue application in the Mediterranean environment by Stagnari et al. [54] concluded that the conservation of soil moisture was improved, especially during the critical growth period of the test crop. The residues which are completely decomposed into the soil not only provide allelochemicals, but also participate in nutrition for crop plants. They provide nitrogen by releasing it into the rhizosphere soil of the tested crop plant. The application of sorghum residues as biological weed management helps in the mineralization of nitrogen and enhances nitrogen availability in the rhizosphere [7,14,50]. However, at later stages of crop growth, the obtainability of nitrogen was improved by mineralization, so this sustained supply of nitrogen was a nonstop source of nutrition for test crops as well as next crops. Therefore, the incorporation of sorghum residues improved soil properties, viz., moisture retention; restored physical properties; enhanced nutrient cycling and microbial activity due to the presence of phenolic compounds [55–57]; and suppressed weeds due to the physical hindrance by residues, reduced light penetration, and the suppressing ability of allelochemicals, which, released from these plant residues, harvested better seed yield and achieved higher profitability in spring-planted mung bean [16,45,46,58].

Our results indicated that using sorghum residues as allelopathic weed management strategies in mung bean crops improved the microbial population and enzymatic activities of the soil (Table 6). Microbial abundance and soil enzymes are biological soil activities and important indicators of soil quality [59–61]. The incorporation of different crop residues in the soil modified the bio-chemical attributes, i.e., soil microbial population and soil enzymatic activity [62]. Soil enzymes and microbiota play a key role in the availability of nutrients. The dehydrogenase enzyme is important for the oxidation of soil organic matter (SOM), transferring the hydrogen and electrons from substrates to acceptors. The activity of soil enzymes, viz., dehydrogenase and phosphatase, depends on the type of residues incorporated in the soil. It also depends on the moisture content and the temperature of the soil. It affects the activity of dehydrogenase by changing the oxidation-reduction status of soil [63,64]. Incorporation of crop residues, viz., tobacco and sunflower in the soil increased the activities of most of the soil enzymes, while the residues of tomato crop only increased the activity of amylase and phosphodiesterase [16,65]. In Akola, Maharashtra, Ravankar et al. [66] reported that the incubation of soil with 1% organic residues, stalks, straw, stubble, stovers, trash, and husks of various field crops showed a wide variation in the rate of decomposition, C:N ratio, and the microbial population at different intervals. Fungal, bacterial, and actinomycetes populations increased after 30 days of incubation. Bacteria were predominant over fungi and actinomycetes.

The incorporation of sorghum residues not only had a positive effect in the case of reducing the weed population and biomass, but also improved nodulation and nitrogen fixation processes, as well as the physical, chemical, and nutritional statuses of field soils. Our results indicate that increased quantities of crop residues have decreased the bulk density and increased the total porosity of the soil over time (Table 7). Soil porosity is directly related to the soil bulk density because as soil bulk density decreases, the soil porosity increases [67]. In the case of soil properties, sorghum residues as an allelopathic weed management strategy improved the SOM, N, available K, and P in the soil (Table 7). Crop residues are good sources of nutrients and are the primary source of organic material added to the soil [68]. They increase the nutrient availability and water-holding capacity of the soil [69]. Moisture retention is the main benefit of residue incorporation. It is caused by a decrease in runoff and evaporation of water from the soil [70,71]. The improvement in nutrient accumulation (especially P and K) might be attributed to the enhanced moisture retention within the soils [16,72]. Improved moisture availability due to residue incorporation also indicated that the soil's water-holding capacity was improved and the soil moisture was available for longer times to support plant growth [73]. This increase in moisture retention properties might decrease the irrigational requirements of the crops, which should be investigated in future studies. In one study, Raut et al. [74] stated that incorporation of sunflower straw at 4 t ha^{-1} and RDF at (125% N + 100% P) in green gram recorded significantly higher soil N, K, and P content in green gram–sunflower sequence. As a result, the incorporation of sorghum residues increased the soil's physical characteristics, microbial activity, and nutrient cycling [55,56,75,76]. The decreased chance of light penetration and the potential of allelochemicals emitted from the plant debris to limit growth also repressed weeds [46,65,77]. The spring-planted mung bean crop was more profitable and produced a higher yield of seeds as a result of all the aforementioned operations.

5. Conclusions

Due to their direct mode of action on the soil surface, herbicides are hazardous to both plants and soil microbes. Weeds and soil quality were significantly impacted by the sorghum crop's allelopathy. In our study, the differential ability to suppress weeds was observed among various sorghum residue and water extract application treatments. A high suppression of weed density, fresh weight, and dry weight was observed when sorghum residues were incorporated into the soil at 6 t ha^{-1}. The residues favorably affected the soil properties, viz., microbial populations, activity, and soil enzymes. The improvement in soil properties and the suppression of weeds harvested a better seed yield and achieved higher profitability in spring-planted mung bean. In short, sorghum residue implementation may provide better weed control along with enhancing the soil health and seed yield of spring-planted mung bean. Different multidisciplinary approaches that incorporate sorghum crops for strategic weed control might be potential alternatives that can also serve as lead compounds for herbicide discovery programs. Future studies should also focus on the interactions of micronutrients in the soil environment under multiple allelopathic weed management techniques in the field. Additionally, there are still relevant issues to look into regarding nitrogen and weed management using different allelopathic strategies and observing the allelopathic effect between the crop residue and the crop that is applied.

Author Contributions: Conceptualization, Z.A.; data curation, R.U. and Q.u.Z.; formal analysis, R.U., M.Z.M., H.A. and B.A.; methodology, Z.A., R.U. and Q.u.Z.; project administration, Z.A.; resources, Z.A.; software, R.U., K.H.A. and A.T.A.; supervision, Z.A.; visualization, Z.A.; writing—original draft, R.U., Q.u.Z., K.S., and Z.A.; writing—review and editing, N.A.T.A.K. and Q.u.Z. All authors have read and agreed to the published version of the manuscript.

Funding: The financial support from the Higher Education Commission Pakistan (HEC) under project no. 20-2114/NRPU/R&D/12/4188 is highly acknowledged.

Institutional Review Board Statement: Not applicable.

Informed Consent Statement: Not applicable.

Data Availability Statement: Not applicable.

Conflicts of Interest: The authors declare no conflict of interest.

References

1. Islam, S.M.F.; Karim, Z. World's demand for food and water: The consequences of climate change. *Desal. Chall. Opport.* **2019**, 57–84. [CrossRef]
2. World Resources Institute. *Creating a Sustainable Food Future. Report 2013–2014: Interim Findings*; World Resources Institute: Washington, DC, USA, 2014.
3. Tian, X.; Engel, B.A.; Qian, H.; Hua, E.; Sun, S.; Wang, Y. Will reaching the maximum achievable yield potential meet future global food demand? *J. Clean. Prod.* **2021**, *294*, 126285. [CrossRef]
4. Liu, Y.; Liu, X.; Liu, Z. Effects of climate change on paddy expansion and potential adaption strategies for sustainable agriculture development across Northeast China. *Appl. Geogr.* **2022**, *141*, 102667. [CrossRef]
5. Khan, M.T.; Rafi, M.A.; Sultana, R.; Munir, A.; Ahmad, S. Diversity and Bio-Geography of Subfamily Eumeninae (Vespidae: Hymenoptera) in Sindh, Pakistan. *Pak. J. Zool.* **2022**, *54*, 1729. [CrossRef]
6. dos Reis, J.C.; Rodrigues, G.S.; de Barros, I.; Rodrigues, R.D.A.R.; Garrett, R.D.; Valentim, J.F.; Smukler, S. Integrated crop-livestock systems: A sustainable land-use alternative for food production in the Brazilian Cerrado and Amazon. *J. Clean. Prod.* **2021**, *283*, 124580. [CrossRef]
7. Farooq, N.; Abbas, T.; Tanveer, A.; Jabran, K. Allelopathy for weed management. *Coevol. Sec. Metab.* **2020**, *1152*, 505–519.
8. Sharma, A.; Kumar, V.; Shahzad, B.; Tanveer, M.; Sidhu, G.P.S.; Handa, N.; Thukral, A.K. Worldwide pesticide usage and its impacts on ecosystem. *SN Appl. Sci.* **2019**, *1*, 1446. [CrossRef]
9. Nadeem, M.A.; Abbas, T.; Tanveer, A.; Maqbool, R.; Zohaib, A.; Shehzad, M.A. Glyphosate hormesis in broad-leaved weeds: A challenge for weed management. *Arch. Agron. Soil Sci.* **2017**, *63*, 344–351. [CrossRef]
10. Peerzada, A.M.; O'Donnell, C.; Adkins, S. Optimizing Herbicide Use in Herbicide-Tolerant Crops: Challenges, Opportunities, and Recommendations. *Agron. Crops* **2019**, 283–316. [CrossRef]
11. Farooq, M.; Jabran, K.; Cheema, Z.A.; Wahid, A.; Siddique, K.H.M. Role of allelopathy in agricultural pest management. *Pest Manag. Sci.* **2011**, *67*, 494–506. [CrossRef]
12. Macias, F.A. New approaches in allelopathy, challenge for the new millenium. *Third World Congr. Allelopath. Abstr.* **2002**, *38*, 227–233.
13. Islam, A.M.; Yeasmin, S.; Qasem, J.R.S.; Juraimi, A.S.; Anwar, M.P. Allelopathy of medicinal plants: Current status and future prospects in weed management. *Agric. Sci.* **2018**, *9*, 1569–1588. [CrossRef]
14. Khaliq, A.; Matloob, A.; Hussain, A.; Hussain, S.; Aslam, F.; Zamir, S.I.; Chattha, M.U. Wheat residue management options affect productivity, weed growth and soil properties in direct-seeded fine aromatic rice. *Clean Soil Air Water* **2015**, *43*, 1259–1265. [CrossRef]
15. Qi, G.; Li, N.; Sun, X.S.; Wang, D. Overview of sorghum industrial utilization. *Sorghum State Art Future Perspect.* **2019**, *58*, 463–476.
16. Hussain, M.I.; Danish, S.; Sánchez-Moreiras, A.M.; Vicente, Ó.; Jabran, K.; Chaudhry, U.K.; Reigosa, M.J. Unraveling sorghum allelopathy in agriculture: Concepts and implications. *Plants* **2021**, *10*, 1795. [CrossRef] [PubMed]

17. Farooq, M.; Khan, I.; Nawaz, A.; Cheema, M.A.; Siddique, K.H. Using sorghum to suppress weeds in autumn planted maize. *Crop Prot.* **2020**, *133*, 105162. [CrossRef]
18. Cheema, Z.A.; Khaliq, A. Use of sorghum allelopathic properties to control weeds in irrigated wheat in a semi-arid region of Punjab. *Agric. Ecosyst. Environ.* **2000**, *79*, 105–112. [CrossRef]
19. Cheema, Z.A.; Khaliq, A.; Akhtar, S. Use of sorghum water extract as a natural weed inhibitor in spring mung bean. *Int. J. Agric. Biol.* **2001**, *3*, 515–518.
20. More, S.S.; Shinde, S.E.; Kasture, M.C. Root exudates a key factor for soil and plant: An overview. *Pharma Innov. J.* **2020**, *8*, 449–459.
21. Kumar, A.; Dubey, A. Rhizosphere microbiome: Engineering bacterial competitiveness for enhancing crop production. *J. Adv. Res.* **2020**, *24*, 337–352. [CrossRef]
22. Komal, N.; Zaman, Q.U.; Yasin, G.; Nazir, S.; Ashraf, K.; Waqas, M.; Ahmad, M.; Batool, A.; Talib, I.; Chen, Y. Carbon Storage Potential of Agroforestry System near Brick Kilns in Irrigated Agro-Ecosystem. *Agriculture* **2022**, *12*, 295. [CrossRef]
23. Bao, Y.; Dolfing, J.; Guo, Z.; Chen, R.; Wu, M.; Li, Z.; Feng, Y. Important ecophysiological roles of non-dominant Actinobacteria in plant residue decomposition, especially in less fertile soils. *Microbiome* **2021**, *9*, 84. [CrossRef] [PubMed]
24. Ferdous, Z.; Zulfiqar, F.; Datta, A.; Hasan, A.K.; Sarker, A. Potential and challenges of organic agriculture in Bangladesh: A review. *J. Crop Improv.* **2021**, *35*, 403–426. [CrossRef]
25. Scavo, A.; Abbate, C.; Mauromicale, G. Plant allelochemicals: Agronomic, nutritional and ecological relevance in the soil system. *Plant Soil.* **2019**, *442*, 23–48.
26. Ryan, J.; Estefan, G.; Rashid, A. *Soil and Plant Analysis: Laboratory Manual; International Center for Agricultural Research in Dry Areas (ICARDA): Aleppo, Syria*; Nacional Agricultural Research Centre: Islamabad, Pakistan, 2001; p. 172.
27. Blake, G.R.; Hartge, K.H. Bulk density. In *Methods of Soil Analysis: Part 1. Physical and Mineroalogical Methods*, 2nd ed.; Agronomy Monograph No. 9; Lute, A., Ed.; American Society of Agronomy: Madison, WI, USA, 1986; pp. 363–382.
28. Vomocil, J.A. Porosity. In *Methods of Soil Analysis*; Blake, C.A., Ed.; American Society of Agronomy: Madison, WI, USA, 1965; pp. 299–314.
29. Bremner, J.M.; Mulvaney, C.S. Total nitrogen. In *Methods of Soil Analysis*; Page, A.L., Miller, R.H., Keeny, D.R., Eds.; American Society of Agronomy and Soil Science Society of America: Madison, WI, USA, 1982; pp. 1119–1123.
30. Walkley, A.; Black, I.A. An examination of Degtjareff method for determining soil organic matter and aproposed modification of the chromic acid titration method. *Soil Sci.* **1934**, *37*, 29–38. [CrossRef]
31. Olsen, S.O.; Sommers, I.E. Phosphorus. In *Methods of Soil Analysis*, 2nd ed.; Chemical and Microbial Properties: Part 2; Page, A.L., Ed.; American Society of Agronomy: Madison, WI, USA, 1982; pp. 403–430.
32. Helmke, P.A.; Sparks, D.L. Lithium, sodium and potassium, rubidium and cesium. In *Methods of Soil Analysis*; American Society of Agronomy: Madison, WI, USA, 1996; pp. 551–575.
33. Aslam, Z.; Yasir, M.; Jeon, C.O.; Chung, Y.R. Lysobacter oryzae sp. nov., isolated from the rhizosphere of rice (*Oryza sativa* L.) managed under no-tillage practice. *Int. J. Syst. Evol. Microbiol.* **2008**, *59*, 675–680. [CrossRef]
34. Janssen, P.H.; Yates, P.S.; Grinton, B.E.; Taylor, P.M.; Sait, M. Improved culturability of soil bacteria and isolation in pure culture of novel members of the divisions Acidobacteria, Actinobacteria, Proteobacteria, and Verrucomicrobia. *Appl. Environ. Microbiol.* **2002**, *68*, 2391–2396. [CrossRef]
35. Wu, W.X.; Ye, Q.F.; Min, H.; Duan, X.J.; Jin, W.M. Bt-transgenic rice straw affects the culturable micro biota and dehydrogenase and phosphatase activities in a flooded paddy soil. *Soil Biol. Biochem.* **2004**, *36*, 289–295.
36. Martin, J.P. Use of acid, rose bengal and streptomycin in the plate method for enumerating soil fungi. *Soil Sci.* **1950**, *69*, 215–232. [CrossRef]
37. Min, H.; Ye, Y.F.; Chen, Z.Y.; Wu, W.X.; Du, Y.F. Effects of butachlor on microbial populations and enzyme activities in paddy soil. *J. Environ. Sci. Health* **2001**, *36*, 581–595. [CrossRef]
38. Tabatabai, M.A.; Bremner, J.M. Use of p-nitrophenyl phosphate for assay of soil phosphatase activity. *Soil Biol. Biochem.* **1969**, *1*, 301–307. [CrossRef]
39. Rab, A.; Khan, M.R.; Haq, S.U.; Zahid, S.; Asim, M.; Afridi, M.Z.; Munsif, F. Impact of biochar on mungbean yield and yield components. *Pure Appl. Biol.* **2016**, *5*, 632–640. [CrossRef]
40. Won, O.J.; Uddin, M.R.; Park, K.W.; Pyon, J.Y.; Park, S.U. Phenolic compounds in sorghum leaf extracts and their effects on weed control. *Allelopath. J.* **2013**, *31*, 147.
41. Ullah, R.; Aslam, Z.; Maitah, M.; Zaman, Q.; Bashir, S.; Hassan, W.; Chen, Z. Sustainable weed control and enhancing nutrient use efficiency in crops through Brassica (*Brassica compestris* L.) allelopathy. *Sustainability* **2020**, *12*, 5763.
42. Cheema, Z.A.; Khaliq, A.; Abbas, M.; Farooq, M. Allelopathic potential of sorghum (*Sorghum bicolor* L. Moench) cultivars for weed management. *Allelopath. J.* **2007**, *20*, 167.
43. Essien, B.; Essien, J.; Nwite, J.; Eke, K.; Anaele, U.; Ogbu, J. Effect of organic mulch materials on maize performance and weed growth in the derived savanna of South Eastern Nigeria. *Nigeria Agric. J.* **2009**, *40*, 1–9. [CrossRef]
44. Zaji, B.; Majd, A. Allelopathic potential of canola (*Brassica napus* L.) residues on weed suppression and yield response of maize (*Zea mays* L.). In Proceedings of the International Conference on Chemical, Ecology and Environmental Sciences (IICCEES), Pattaya, Thailand, 17–18 December 2011; pp. 457–460.

45. Kamara, A.; Akobundu, I.; Chikoye, D.; Jutzi, S. Selective control of weeds in an arable crop by mulches from some multipurpose trees in south western Nigeria. *Agrofor. Sys.* **2000**, *50*, 17–26.
46. Khaliq, A.; Hussain, S.; Matloob, A.; Tanveer, A.; Aslam, F. Swine cress (*Cronopus didymus* L. Sm.) residues inhibit rice emergence and early seedling growth. *Phillipine Agric. Sci.* **2014**, *96*, 419–425.
47. Khaliq, A.; Matloob, A.; Farooq, M.; Mushtaq, M.N.; Khan, M.B. Effect of crop residues applied isolated or in combination on the germination and seedling growth of horse purslane (*Trianthema portulacastrum*). *Planta Daninha* **2011**, *29*, 121–128.
48. Khaliq, A.; Matloob, A.; Irshad, M.S.; Tanveer, A.; Zamir, M.S.I. Organic weed management in maize through integration of allelopathic crop residues. *Pak. J. Weed Sci. Res.* **2010**, *16*, 409–420.
49. Khanh, T.D.; Chung, M.I.; Xuan, T.D.; Tawata, S. The exploitation of crop allelopathy in sustainable agricultural production. *J. Agron. Crop Sci.* **2005**, *191*, 172–184.
50. Jabran, K.; Mahajan, G.; Sardana, V.; Chauhan, B.S. Allelopathy for weed control in agricultural systems. *Crop Protect.* **2015**, *72*, 57–65.
51. Shehzad, T.; Okuno, K. Genetic analysis of QTLs controlling allelopathic characteristics in sorghum. *PLoS ONE* **2020**, *15*, 235896.
52. Alsaadawi, I.S.; Khaliq, A.; Lahmod, N.R.; Matloob, A. Weed management in broad bean (*Vicia faba* L.) through allelopathic *Sorghum bicolor* (L.) Moench residues and reduced rate of a pre plant herbicide. *Allelopath. J.* **2011**, *32*, 203–212.
53. Kruidhof, H.; Bastiaans, L.; Kropff, M. Ecological weed management by cover cropping: Effects on weed growth in autumn and weed establishment in spring. *Weed Res.* **2008**, *48*, 492–502.
54. Stagnari, F.; Galieni, A.; Speca, S.; Cafiero, G.; Pisante, M. Effects of straw mulch on growth and yield of durum wheat during transition to conservation agriculture in Mediterranean environment. *Field Crops Res.* **2014**, *167*, 51–63.
55. Alam, M.K.; Islam, M.M.; Salahin, N.; Hasanuzzaman, M. Effect of tillage practices on soil properties and crop productivity in wheat mungbean rice cropping system under subtropical climatic conditions. *Sci. World J.* **2014**, *1*, 437283.
56. Adugna, A.; Abegaz, A. Effects of land use changes on the dynamics of selected soil properties in northeast Wellega, Ethiopia. *Soil* **2016**, *2*, 63–70.
57. Marchiosi, R.; dos Santos, W.D.; Constantin, R.P.; de Lima, R.B.; Soares, A.R.; Finger-Teixeira, A.; Mota, T.R.; de Oliveira, D.M.; Foletto-Felipe, M.D.P.; Abrahão, J. Biosynthesis and metabolic actions of simple phenolic acids in plants. *Phytochem. Rev.* **2020**, *19*, 865–906.
58. Raza, T.; Khan, M.Y.; Nadeem, S.M.; Imran, S.; Qureshi, K.N.; Mushtaq, M.N.; Eash, N.S. Biological management of selected weeds of wheat through co-application of allelopathic rhizobacteria and sorghum extract. *Biol. Control* **2021**, *164*, 104775.
59. Nawaz, A.; Lal, R.; Shrestha, R.K.; Farooq, M. Mulching affects soil properties and greenhouse gases emissions under long term no-till and plough till systems in Alfisol of central Ohio. *Land Develop. Degrad.* **2016**, *28*, 673–681.
60. Vilkiene, M.; Mockeviciene, I.; Karcauskiene, D.; Suproniene, S.; Doyeni, M.O.; Ambrazaitiene, D. Biological indicators of soil quality under different tillage systems in retisol. *Sustainability* **2021**, *13*, 9624. [CrossRef]
61. Duke, O.S. Proving Allelopathy in crop-weed interactions. *Weed Sci.* **2015**, *63*, 121–132. [CrossRef]
62. Rathore, S.S.S.; Shekhawat, K. Crop Residue Recycling for Improving Crop Productivity and Soil Health. In *Handbook of Research on Green Technologies for Sustainable Management of Agricultural Resources*; IGI Global: New Delhi, India, 2022; pp. 290–308.
63. Farhangi-Abriz, S.; Ghassemi-Golezani, K.; Torabian, S. A short-term study of soil microbial activities and soybean productivity under tillage systems with low soil organic matter. *Appl. Soil Ecol.* **2021**, *168*, 104122. [CrossRef]
64. Głąb, L.; Sowiński, J.; Bough, R.; Dayan, F.E. Allelopathic potential of sorghum (*Sorghum bicolor* (L.) Moench) in weed control: A comprehensive review. *Adv. Agron.* **2017**, *145*, 43–95.
65. Burezq, H.; Davidson, M.K. Ecological Intensification for Soil Management: Biochar—A Natural Solution for Soil from Agricultural Residues. In *Sustainable Intensification for Agroecosystem Services and Management*; Springer: Singapore, 2021; pp. 403–455.
66. Ravankar, H.N.; Patil, R.; Puranik, R.B. Decomposition of different organic residues in soil. *PKV Res. J.* **2000**, *24*, 23–25.
67. Anda, P. The reciprocal effect between soil water content and the soil bulk density on the growth and yield of onion (*Allium cepa* L.). *J. Appl. Agric. Sci. Technol.* **2021**, *5*, 84–94.
68. Andrews, E.M.; Kassama, S.; Smith, E.E.; Brown, P.H.; Khalsa, S.D.S. A review of potassium-rich crop residues used as organic matter amendments in tree crop agroecosystems. *Agriculture* **2021**, *11*, 580.
69. Krishna, G.A.; Misra, A.K.; Hati, K.M.; Bandyopadhyay, K.K.; Ghosh, P.K.; Mohanty, M. Rice residue management options and effects on soil properties and crop productivity. *Food Agric. Environ.* **2004**, *2*, 224–231.
70. Verhulst, N.; Nelissen, V.; Jespers, N.; Haven, H.; Sayre, K.D.; Raes, D.; Deckers, J.; Govaerts, B. Soil water content, maize yield and its stability as affected by tillage and crop residue management in rainfed semi-arid highlands. *Plant Soil* **2011**, *344*, 73–85. [CrossRef]
71. Nazir, S.; Zaman, Q.U.; Abbasi, A.; Komal, N.; Riaz, U.; Ashraf, K.; Ahmad, N.; Agerwal, S.; Chen, Y. Bioresource Nutrient Recycling in the Rice–Wheat Cropping System: Cornerstone of Organic Agriculture. *Plants* **2021**, *10*, 2323. [CrossRef]
72. Zhou, J.; Xu, D.; Xue, C. Study of comprehensive utilization efficiency of returning rice straw to field. *Chin. Agric. Sci. Bull.* **2002**, *4*, 7–10.
73. Jin, Y.Q.; Du, D.J.; Gao, H.J.; Chang, J.; Zhang, L.G. Effects of maize straw returning on water dynamics and water use efficiency of winter wheat in lime concretion black soil. *J. Triticeae Crops* **2013**, *33*, 1–7.
74. Raut, V.U.; Bhowate, R.T.; Waghmare, A.G. Effect of crop residues on nutrient contents in green gram-sunflower cropping sequence. *Green Farm* **2010**, *1*, 14–19.

75. Saqib, S.; Uddin, S.; Zaman, W.; Ullah, F.; Ayaz, A.; Asghar, M.; Rehman, S.; Munis, M.F.H.; Chaudhary, H.J. Characterization and phytostimulatory activity of bacteria isolated from tomato (*Lycopersicon esculentum* Mill.) rhizosphere. *Microb. Pathog.* **2021**, *140*, 103966.
76. Abbas, T.; Ahmad, A.; Kamal, A.; Nawaz, M.Y.; Jamil, M.A.; Saeed, T.; Ateeq, M. Ways to use allelopathic potential for weed management: A review. *Int. J. Food Sci. Agric.* **2021**, *5*, 492–498. [CrossRef]
77. Hussain, M.I.; Reigosa, M.J. Secondary metabolites, ferulic acid and *p*-hydroxybenzoic acid induced toxic effects on photosynthetic process in *Rumex acetosa* L. *Biomolecules* **2021**, *11*, 233. [CrossRef]

life

MDPI

Article

Genome-Wide Identification and Expression Analyses of the Chitinase Gene Family in Response to White Mold and Drought Stress in Soybean (*Glycine max*)

Peiyun Lv [1], Chunting Zhang [1], Ping Xie [1], Xinyu Yang [1], Mohamed A. El-Sheikh [2], Daniel Ingo Hefft [3], Parvaiz Ahmad [4,*], Tuanjie Zhao [1,*] and Javaid Akhter Bhat [1,*]

1 National Center for Soybean Improvement, State Key Laboratory of Crop Genetics and Germplasm Enhancement, Nanjing Agricultural University, Nanjing 210095, China
2 Botany and Microbiology Department, College of Science, King Saud University, Riyadh 11451, Saudi Arabia
3 School of Chemical Engineering, Edgbaston Campus, University of Birmingham, Birmingham B15 2TT, UK
4 Department of Botany, GDC, Pulwama 192301, Jammu and Kashmir, India
* Correspondence: parvaizbot@yahoo.com (P.A.); tjzhao@njau.edu.cn (T.Z.); javid.akhter69@gmail.com (J.A.B.)

Abstract: Chitinases are enzymes catalyzing the hydrolysis of chitin that are present on the cell wall of fungal pathogens. Here, we identified and characterized the chitinase gene family in cultivated soybean (*Glycine max* L.) across the whole genome. A total of 38 chitinase genes were identified in the whole genome of soybean. Phylogenetic analysis of these chitinases classified them into five separate clusters, I–V. From a broader view, the I–V classes of chitinases are basically divided into two mega-groups (X and Y), and these two big groups have evolved independently. In addition, the chitinases were unevenly and randomly distributed in 17 of the total 20 chromosomes of soybean, and the majority of these chitinase genes contained few introns ($\leq$2). Synteny and duplication analysis showed the major role of tandem duplication in the expansion of the chitinase gene family in soybean. Promoter analysis identified multiple cis-regulatory elements involved in the biotic and abiotic stress response in the upstream regions (1.5 kb) of chitinase genes. Furthermore, qRT-PCR analysis showed that pathogenic and drought stress treatment significantly induces the up-regulation of chitinase genes belonging to specific classes at different time intervals, which further verifies their function in the plant stress response. Hence, both in silico and qRT-PCR analysis revealed the important role of the chitinases in multiple plant defense responses. However, there is a need for extensive research efforts to elucidate the detailed function of chitinase in various plant stresses. In conclusion, our investigation is a detailed and systematic report of whole genome characterization of the chitinase family in soybean.

Keywords: *Glycine max* L.; PR proteins; chitinase; genome-wide; plant stresses

Citation: Lv, P.; Zhang, C.; Xie, P.; Yang, X.; El-Sheikh, M.A.; Hefft, D.I.; Ahmad, P.; Zhao, T.; Bhat, J.A. Genome-Wide Identification and Expression Analyses of the Chitinase Gene Family in Response to White Mold and Drought Stress in Soybean (*Glycine max*). *Life* **2022**, *12*, 1340. https://doi.org/10.3390/life12091340

Academic Editors: Hakim Manghwar and Wajid Zaman

Received: 2 August 2022
Accepted: 23 August 2022
Published: 29 August 2022

Publisher's Note: MDPI stays neutral with regard to jurisdictional claims in published maps and institutional affiliations.

Copyright: © 2022 by the authors. Licensee MDPI, Basel, Switzerland. This article is an open access article distributed under the terms and conditions of the Creative Commons Attribution (CC BY) license (https://creativecommons.org/licenses/by/4.0/).

1. Introduction

Plants, being immobile, are often subjected to different environmental stresses that lead to a decrease in plant growth and productivity [1,2]. However, to combat these external threats, plants have developed well established defense mechanisms. For example, a small group of heterogeneous proteins called pathogenesis-related (PR) proteins are produced following the attack of disease pathogens, and these proteins plays critical role in inducing plants' potential to resist pathogen attack [3,4]. Many studies have documented the accumulation and activation of these proteins under multiple abiotic stresses, and thus they are recognized as part of multiple defense systems. Up to now, many families of PR proteins have been characterized [3]; among them, the PR3 family consist of chitinases enzymes that inhibit fungal growth by degrading heterogenous polysaccharide (chitin), a major component of the fungi cell wall [4]. Under normal conditions, these proteins are expressed at basal level; however, pathogen attack or abiotic stress such as drought increases their expression considerably, resulting in systemic acquired resistance (SAR) [4].

Chitinases are ubiquitous in nature and are found in living organisms across different kingdoms of life [5]. The proteins are categorized into two glycosyl hydrolases (GH) families, GH18 & GH19, based on the presence of specific catalytic domains [6]. In addition, by considering the different characteristics of chitinases such as structure, catalytic reaction, phylogenetic relationship and specificity to inhibitors, etc., these chitinases represent five distinct classes (classes I–V) [4]. The members of the GH19 family are specifically found in plants only; however, GH18 family members are widely distributed across different kingdoms, including plants. A lack of chitin in the plant cell wall and other tissue parts makes chitinase an important component of the plant defense system. Chitinase has been documented to the control positive feedback cycle in the plant defense system [7]. This pathway is used by plants in the regulation of plant defense reactions against fungal pathogens [8]. Hence, the chitinases are important targets for enhancing plant growth, especially under environment stresses [9]. To this end, recent studies have also documented the role of chitinase in abiotic situations such as salinity and water deficit conditions [10–12].

Soybean (*Glycine max* L. Merr.), an important legume crop, possesses high levels of edible oil and protein in its seed [13]. However, many environment stresses, including both biotic and abiotic conditions, have a negative influence on soybean growth and yield, and the frequency of these stress events has increased due to the changing global climate [14]. Among the biotic stresses, pathogenic diseases such as white mold (caused by *Sclerotinia sclerotiorum*) are a major stress affecting the growth, yield and quality of soybean [15,16]. White mold disease is documented as the fourth major cause of yield losses in soybean [17]. Lack of information about the genes regulating disease resistance is the major hindrance to developing pathogenic-resistant cultivars [18], and the phenotypic evaluation of disease scoring in the field is also technically challenging. Development of resistant cultivars against pathogens requires the identification of underlying genes. The gene family of chitinase has been identified in multiple species, and research studies have confirmed its role against the invasion of fungal pathogens [3,19]; for example, transgenic lines of chitinase genes possess increased resistance to pathogens of fungal origin [3,5]. To this end, chitinases are documented to modulates abiotic stress responses, such as to drought in various plant species [6,11,20]. However, until now, the gene family has not been identified and characterized at the whole genome level in *Glycine max* L. Nevertheless, there are research studies that have used chitinase genes from other organisms to develop transgenic soybean lines [21].

Until now, almost negligible efforts have been made to characterize and identify the chitinase gene family in soybean at the whole genome scale. However, the availability of the whole genome sequence of crop plants is allowing characterization of the whole gene families in plants. In this context, the whole genome sequence of the soybean plant is freely available in public databases (SoyBase and Phytozome); hence, in the current investigation, we identify and characterize the chitinase gene family at the genome-wide scale in soybean. In addition, we also studied the response of the identified chitinase genes under pathogenic attack and drought stress, to confirm their role in plant defense.

2. Materials and Methods

2.1. Identifying Chitinase Genes in Soybean

For chitinase gene family identification in soybean, the whole genome sequences of soybean were downloaded from the Phytozome database (https://phytozome-next.jgi.doe.gov/ (accessed on 11 November 2019)), using the *Glycine max Wm82.a2.v1*. This genome sequence was used to develop the protein local database of soybean, using Bioedit ver 7.2 software. Moreover, the 24 known chitinase genes of *Arabidopsis thaliana* freely available at the TAIR database (https://www.arabidopsis.org/ (accessed on 11 November 2019)) were used as a query sequence to identify putative orthologs in soybean, using BLASTp [22]. The e-value $<10^{-5}$ and bit scores >100 were the fitted parameters used to pick out high scoring pairs (HSPs). Redundant hits possessing highest similarity were eliminated to select the unique sequences. To confirm the Glyco_hydro_18 or Glyco_hydro_19 conserved domains, we submitted all identified unique sequences to NCBI-The Conserved Domain Database (https://www.ncbi.nlm.nih.gov/cdd/?term=) (accessed on 17 November 2019).

2.2. Phylogenetic Analysis and Multiple Sequence Alignment

Protein sequences of chitinases were aligned using the CLUSTALW function present in MEGA 7.0 [23]. The neighbor-joining method and a bootstrap value of 1000 were used to develop the phylogenetic tree. Chitinases of cultivated soybean (*Glycine max* L.) plus 24 chitinases of *Arabidopsis thaliana* were utilized to develop the phylogenetic tree. Grouping of the chitinases were based on the different chitinase classes (I–V) of *A. thaliana*. Finally, using EvolView (https://www.evolgenius.info//evolview/#login (accessed on 2 December 2019)), the evolutionary trees were developed.

2.3. Structure Analysis and Chromosomal Location of Chitinase Genes

The ProtParam database (https://web.expasy.org/protparam/ (accessed on 7 December 2019)), an online program for determining physical protein properties such the molecular weight (MW), length of protein and isoelectric points (pI), was utilized in tge present study for chitinase proteins [24]. The genomic and coding sequence of all chitinases genes were collected from an online database (Phytozome); and gene structures (i.e., exon-intron structures) analysis was performed using the online Gene Structure Display Server tool (http://gsds.gao-lab.org/ (accessed on 7 December 2019)). Chromosomal location information of individual genes of chitinase was obtained from the Phytozome database (https://phytozome-next.jgi.doe.gov/ (accessed on 10 December 2019)); and chromosomal maps were developed with MapChat software (www.https://mapchat.ca/ (accessed on 15 December 2019)).

2.4. Promoter Analysis and Three-Dimensional (3D) Structure of Chitinase Genes

The PlantCARE Database (https://bioinformatics.psb.ugent.be/webtools/plantcare/html/ (accessed on 19 December 2019)) was utilized for analysis of cis-regulatory elements in the promotor region (upstream region of 1.5 kb) of chitinase [3].

PHYRE2 server software (http://www.sbg.bio.ic.ac.uk/phyre2/html/page.cgi?id=help (accessed on 19 December 2019)) was used for generating three-dimensional (3D) models, and the thresholds were kept as alignment coverage >65% and confidence = 100%. The transmembrane helix and topology of chitinases proteins were predicted by the MEMSAT-SVM prediction method, available at the PSIPRED online site (https://bio.tools/memsat-svm (accessed on 20 December 2019)).

2.5. Synteny and Duplication Analysis

The syntenic information about *Glycine max* and *Arabidopsis thaliana* was downloaded from the Phytozome database (https://phytozome-next.jgi.doe.gov/ (accessed on 23 December 2019)). Using the comparison of inter-genomic, the mapping of chitinase genes were performed, and TBtools software (https://bio.tools/tbtools (accessed on 24 December 2019)) was used to draw a syntenic diagram. By using the criteria of physical positions of chitinase genes in the genome of cultivated soybean, we identified the tandem duplications. Tandem duplication genes are considered as those that are separated by not more than one intervening gene.

2.6. Plant Materials and Culture

To sterilize the seeds of soybean (*W82*), we initial used ethanol (70% *v*/*v*) for 1 min, and after this for 6 min these seeds were bleached (10%); this was followed by sowing them in a 10 cm diameter pot containing vermiculite and nutritive soil at 1:1 (*v*/*v*) mixture. The soybean seedlings were raised in a growth chamber by maintaining the controlled conditions followed by Aleem et al. [25]. After every four days, seedlings were supplied with water in half-length Hoagland solution. The V3 stage of the seedlings were selected for the stress treatments, i.e., fungus inoculation and osmotic stress treatment.

2.7. Pathogenic and Drought Treatments

The white mold pathogen of soybean (*Sclerotinia sclerotiorum)* was cultured by following the detailed procedure described by Hoffman et al. [26]. The drop-mycelium method was used for the inoculation of *Sclerotinia sclerotiorum* to soybean leaves, using four replications [27]. The experiments were conducted in controlled conditions at the Soybean Research Institute, Nanjing Agricultural University, China. The *S. sclerotiorum* isolate 105 HT was provided by the Department of Plant Protection, Nanjing Agricultural University and used in disease evaluation. Procedures for the controlled evaluation of white mold diseases in soybean were followed, as described by Chen and Wang, [27]. For about three to four days, potato dextrose agar (PDA) medium was used to grow the sclerotia (sterilized), and fresh stock was maintained by re-culturing the sclerotia. Small pieces of mycelia were put into the liquid broth of potato dextrose, and homogenization of the potato dextrose broth was performed in a G10 Gyrotory shaker (Edison, NJ) at 200 rpm for four nights. A household blender was used to homogenize the suspension of mycelia for maintaining mycelium uniformity immediately before the inoculation. A battery-operated hand sprayer was used to spray a suspension of blended mycelia at ~4.6 ml/plant on the plant leaves, and this spray was used at the V3 growth stage. The inoculated plants were placed in controlled chambers, maintaining near 100% humidity inside the chambers. A control was also used, that was not inoculated with the pathogen.

Seedlings were randomly grouped in four replicates for the osmatic treatments. Three replicates were subjected to drought stress and treated using 20% PEG-6000, whereas the fourth one was used as control, and not subjected to drought treatment. Collection of fresh and healthy leaf tissues was carried out for both control and treated plants (in case of both disease and drought stress) at time intervals of 6, 12, 24 and 48 h post-inoculation (hpi)/post-treatment for the extraction of RNA, and were rapidly flash frozen in liquid nitrogen and stored at −80 °C.

2.8. qRT-PCR Analysis

Total RNA was extracted from the leaf tissue (100 mg) that was collected from soybean plants using a PureLink RNA Mini Kit (Ambion Life Technologies, 5791 Van Allen Way Carlsbad, CA, USA). A nanodrop spectrophotometer (Thermo Scientific, Wilmington, DE, USA) was used for checking the quality and quantity of RNA. The protocol used for cDNA synthesis was same as followed by us in the previous study of Sharmin et al. [27]. The primers used in the qRT-PCR analysis are listed in the Table S1. The qRT-PCR reaction was performed as initial annealing at 95 °C for 5 min, followed by 40 cycles as 94 °C for 30 s, 60 °C for 30 s, and 72 °C for 30 s. The reaction mixture and replication used is as per our previous study [28,29].

In our experiment, we used the actin gene as an internal control, and relative expression of each gene was estimated by the Delta Ct method [30]. The $p < 0.05$ was used to check the level of significance.

2.9. Statistics

In our experiments we used replicates of three, and every replicate was repeated three times. Student's *t*-test was used to check for significance differences in gene expression of chitinases. In all experiments, the difference among the groups is reported as statistically significant (* $p < 0.05$) or extremely significant (** $p < 0.01$).

3. Results

3.1. Chitinase Genes Identified in the Glycine max Genome

Soybean whole-genome sequence availability has allowed the characterization of novel gene families in these crop plants, but it requires already known orthologs query genes from the model plants. Therefore, by using the known sequence of 24 chitinase genes of *A. thaliana* as a query, we identified the 38 chitinase genes in cultivated soybean (Table 1). These sequences were further subjected to functional annotation using the Conserved Domain Database (CDD), and the results revealed that the predicted protein sequence of these genes possess either the Glyco_hydro 18 or Glyco_hydro 19 domain (Table 1). These domains are the key component needed by the chitinase enzymes to hydrolyze the chitin; therefore, this confirmed their role as chitinase enzymes. Protein sequences containing the catalytic domain of Glyco_hydro 18 are members of either Class III or V, whereas those possessing Glyco_hydro 19 are the members of any of the three different classes, Class I, II or IV. Interestingly, out of the 38 identified chitinases in soybean, 25 possess Glyco_hydro 18, while only 13 harbored the Glyco_hydro 19 domain (Table 1).

3.2. Phylogenetic Analysis and Chromosomal Location of Chitinase in Glycine max

The protein sequence of the 38 chitinases of soybean, along with the 24 known chitinases from *A. thaliana,* were utilized for developing an unrooted maximum likelihood phylogenetic tree (Figure 1). Based on the phylogenetic relationship, chitinases are classified into five different groups representing five classes of chitinases, I, II, III, IV and V (Figure 1). Each class of chitinase is grouped into separate cluster. Broadly, chitinases are grouped into two mega-groups. All the chitinases of classes I, II and IV, comprising the GH19 family, are clustered into mega-group 1, while mega-group 2 possesses the chitinases of the GH18 family. Naming of chitinases for *Glycine max* is based on their known ortholog of *A. thaliana*, which shows three, seven, nineteen, three and six chitinases of class I, class II, class III, class IV and class V, respectively.

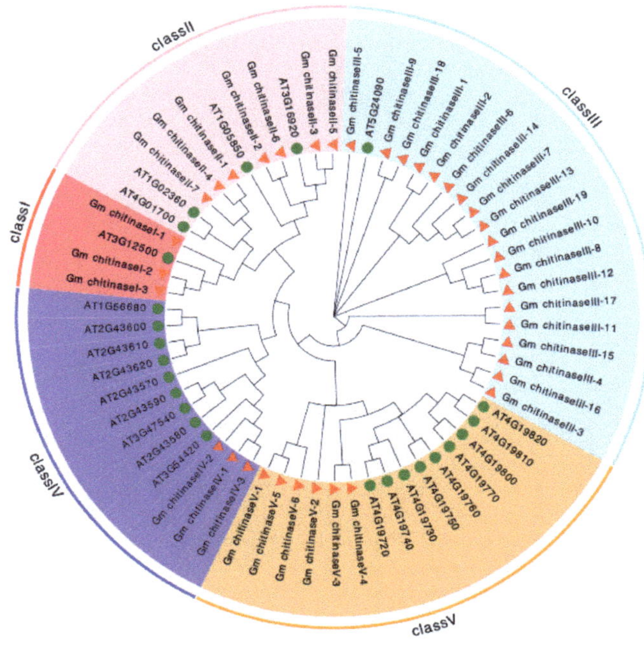

Figure 1. Phylogenetic analysis and chromosomal distribution of chitinase genes identified in the soybean genome.

Table 1. Genome-wide identification and distribution of chitinase genes in soybean.

S. No.	Name	Gene ID	Class	Protein Length(aa)	Mol. Wt.(Da)	PI(pH)	Instability Index	GRAVY	*Arabidopsis* Ortholog Locus	*Arabidopsis* Locus Description
1	Gm_chitinaseI-1	Glyma.01G160100	I	275	30,182.55	5.34	38.39	−0.331	AT3G12500	BASIC CHITINASE, PR3
2	Gm_chitinaseI-2	Glyma.02G042500	I	320	34,341.3	7.40	29.21	−0.404	AT3G12500	BASIC CHITINASE, PR3
3	Gm_chitinaseI-3	Glyma.16G119200	I	317	34,445.61	8.10	38.56	−0.350	AT3G12500	BASIC CHITINASE, PR3
4	Gm_chitinaseII-1	Glyma.02G007400	II	281	31,229.36	8.83	49.95	−0.290	AT1G02360	Chitinase family protein
5	Gm_chitinaseII-2	Glyma.08G259200	II	326	36,029.92	5.83	34.73	−0.180	AT1G05850	CHITINASE-LIKE protein 1
6	Gm_chitinaseII-3	Glyma.09G038500	II	317	34,709.20	7.01	34.06	−0.266	AT3G16920	Encodes a chitinase-like protein
7	Gm_chitinaseII-4	Glyma.10G138400	II	245	27,411.72	8.66	42.22	−0.578	AT1G02360	Chitinase family protein
8	Gm_chitinaseII-5	Glyma.15G143600	II	318	34,889.50	6.97	34.32	−0.254	AT3G16920	Encodes a chitinase-like protein
9	Gm_chitinaseII-6	Glyma.18G283400	II	329	36,375.35	5.91	37.68	−0.176	AT1G05850	CHITINASE-LIKE protein 1
10	Gm_chitinaseII-7	Glyma.19G221800	II	272	29,960.98	6.80	36.74	−0.181	AT4G01700	Chitinase family protein
11	Gm_chitinaseIII-1	Glyma.01G055200	III	296	31,735.72	5.39	35.54	−0.113	AT5G24090	Chitinase A (class III)
12	Gm_chitinaseIII-2	Glyma.02G113600	III	296	31,687.55	5.18	34.54	−0.106	AT5G24090	Chitinase A (class III)
13	Gm_chitinaseIII-3	Glyma.03G254300	III	303	32,588.90	8.97	38.48	−0.231	AT5G24090	Chitinase A (class III)
14	Gm_chitinaseIII-4	Glyma.05G075000	III	298	32,643.35	9.41	38.72	−0.115	AT5G24090	Chitinase A (class III)
15	Gm_chitinaseIII-5	Glyma.07G061600	III	289	31,297.28	6.31	23.17	−0.068	AT5G24090	Chitinase A (class III)
16	Gm_chitinaseIII-6	Glyma.08G299700	III	300	32,004.46	8.08	38.61	0.024	AT5G24090	Chitinase A (class III)
17	Gm_chitinaseIII-7	Glyma.08G300300	III	245	25,864.39	4.87	35.41	0.106	AT5G24090	Chitinase A (class III)
18	Gm_chitinaseIII-8	Glyma.09G126200	III	292	30,880.19	4.07	33.02	−0.016	AT5G24090	Chitinase A (class III)
19	Gm_chitinaseIII-9	Glyma.10G227700	III	304	32,429.85	7.58	39.30	0.049	AT5G24090	Chitinase A (class III)

Table 1. *Cont.*

S. No.	Name	Gene ID	Class	Protein Length(aa)	Mol. Wt.(Da)	PI(pH)	Instability Index	GRAVY	*Arabidopsis* Ortholog Locus	*Arabidopsis* Locus Description
20	Gm_chitinaseIII-10	Glyma.12G156600	III	298	31,508.36	5.51	30.94	−0.050	AT5G24090	Chitinase A (class III)
21	Gm_chitinaseIII-11	Glyma.15G015100	III	820	91,012.67	6.31	35.59	−0.141	AT5G24090	Chitinase A (class III)
22	Gm_chitinaseIII-12	Glyma.16G173000	III	297	31,768.55	5.01	34.00	−0.043	AT5G24090	Chitinase A (class III)
23	Gm_chitinaseIII-13	Glyma.18G120200	III	295	31,225.32	5.87	35.97	0.095	AT5G24090	Chitinase A (class III)
24	Gm_chitinaseIII-14	Glyma.18G120700	III	295	31,266.46	7.50	32.96	0.045	AT5G24090	Chitinase A (class III)
25	Gm_chitinaseIII-15	Glyma.19G076200	III	316	34,753.54	9.42	37.34	−0.238	AT5G24090	Chitinase A (class III)
26	Gm_chitinaseIII-16	Glyma.19G251900	III	148	16,384.63	8.91	37.51	−0.124	AT5G24090	Chitinase A (class III)
27	Gm_chitinaseIII-17	Glyma.20G035400	III	800	88,944.66	7.93	42.27	−0.170	AT5G24090	Chitinase A (class III)
28	Gm_chitinaseIII-18	Glyma.20G164600	III	301	32,393.00	9.34	41.78	−0.017	AT5G24090	Chitinase A (class III)
29	Gm_chitinaseIII-19	Glyma.20G164900	III	299	32,114.60	4.27	38.89	−0.092	AT5G24090	Chitinase A (class III)
30	Gm_chitinaseIV-1	Glyma.11G124500	IV	235	25,871.79	4.90	34.87	−0.261	AT3G54420	CHITINASE CLASS IV
31	Gm_chitinaseIV-2	Glyma.12G049200	IV	280	30,569.11	4.94	26.76	−0.276	AT3G54420	CHITINASE CLASS IV
32	Gm_chitinaseIV-3	Glyma.13G346700	IV	274	29,829.05	5.02	28.58	−0.301	AT3G54420	CHITINASE CLASS IV
33	Gm_chitinaseV-1	Glyma.13G155800	V	379	41,065.40	4.78	16.18	0.141	AT4G19800	Glycoside hydrolase, family 18
34	Gm_chitinaseV-2	Glyma.15G206400	V	762	86,075.18	6.40	39.94	−0.167	AT4G19800	Glycoside hydrolase, family 18
35	Gm_chitinaseV-3	Glyma.15G206800	V	365	40,085.06	8.92	34.77	−0.304	AT4G19810	CLASS V CHITINASE
36	Gm_chitinaseV-4	Glyma.17G076100	V	374	41,252.20	8.79	33.08	−0.102	AT4G19810	CLASS V CHITINASE
37	Gm_chitinaseV-5	Glyma.17G103500	V	377	41,059.92	9.11	18.08	0.158	AT4G19800	CLASS V CHITINASE
38	Gm_chitinaseV-6	Glyma.17G217000	V	384	43,291.89	6.14	32.33	−0.228	AT4G19810	CLASS V CHITINASE

By analyzing the distribution of the chitinase genes on the different chromosomes in soybean, we identified that all of the 38 chitinase genes are distributed on 17 of the total of 20 soybean chromosomes (Figure S1). Distribution of these chitinase genes was random and uneven across the soybean genome. For example, Chr.15 possess four genes, whereas Chr.04, Chr.06 and Chr.14 possess no chitinase gene; however, the remaining chromosomes contain one to three genes. Hence, the results of current study showed that *Glycine max* chitinases are not evenly distributed in the soybean genome.

3.3. Structural Analysis of Chitinase Genes in Glycine max

Exon–intron analysis of soybean chitinase genes was carried out by comparing the genomic and coding sequence of each gene (Figure 2). Structural analysis showed that most of the genes of same chitinase class possess almost the same number of exons or introns. For instance, all the three chitinases of class I have two introns; similarly, chitinase genes of class IV and class V contain one intron, except *Gm_chitinaseV-2* of class V, that possesses six introns. Moreover, out of seven chitinases of class II, four have two introns, two have one intron and one has three introns. However, the 19 chitinase genes of class III are very diverse in terms of intron number, which varies from 0–6 introns; for example, eight of them contains zero introns, another eight possess one intron, and the remaining one has three, one has two and two have six introns. Overall, structural analysis revealed that soybean chitinases showed significant variation in exon and intron numbers, and this ultimately leads to differences in the length of different chitinases and their physio-chemical properties (Table 1).

To understand the role and response of the chitinases in plant growth and multiple plant stresses, 1.5 kb upstream promoter sequences of ten randomly selected chitinase genes (two each from classes I, II, III, IV and V) were utilized for cis-regulatory element identification (Figure S2; Table 2). Our results showed the presence of multiple cis-elements regulating the response against biotic and abiotic stresses. For example, biotic stress responsive elements were observed as EIRE (fungal elicitor responsive elements), Box-W, TCA-element (SA-responsive element), CGTCA-motif and TGACG-motif (JA responsive element) and TC-rich repeats (ATTTTC). Similarly, abiotic stress response cis-elements were identified in the chitinase promoter genes such as LTRE motif (TGG/ACC GAC), involved in cold/chilling response, MBS/MYB motif (TAACTG) for water-deficit, HSE motif (CNNGAANNTTCNNG), involved in heat stress, WUN-motif, involved in wound response and ABREs motif (ACGT), regulated by expression of ABA. To this end, many elements showing responsiveness for hormones are also identified, such as gibberellin- (P-box and GARE-motif), ethylene- (ERA) and auxin-responsive elements (TGA) (Table 2). The presence of these elements in the chitinase promoters suggests their regulatory role in multiple abiotic and biotic stresses.

3.4. Molecular Modeling of Chitinases in G. max

Dynamic and energetic information regarding the chitin binding domain of the chitinase proteins can be determined by using the bioinformatic approach of molecular modeling. This information is very laborious and expensive to obtain, as well as taking a long time. The PHYRE2 server, freely available online, was used to construct 3D models for chitinases of I–V classes, and this analysis provides a better understanding about the structural properties of chitinase genes in soybean (Figure 3). The following parameters were used to generate the 3D model of chitinase proteins: confidence >90% and residue coverage of 72–98. These predicted 3D protein structures can serve as the preliminary basis to understand the function of chitinase genes at the molecular level. Our results revealed that, except class II members, all of the chitinases have a N-terminal signal peptide that possesses a different number of amino acids; however, all the five classes of chitinases possess pore linings with varying amino acid numbers. A signal peptide at the N-terminal guides chitinase proteins to their proper location, and after they reaches their destination, the signal peptide is cleaved off. In addition, results showed the cytoplasmic nature of all chitinases, and extra-cellular mode of action (Figure 3).

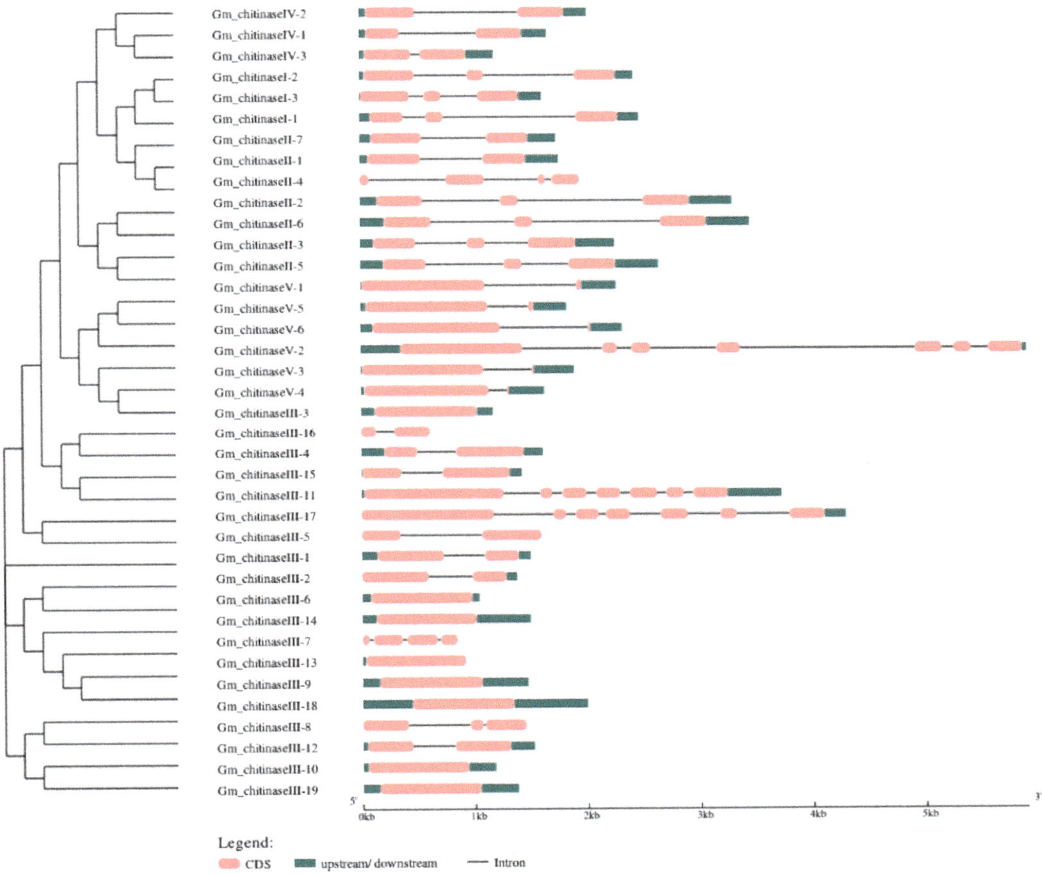

Figure 2. Exon–intron analysis of chitinase genes of soybean. Graphic representation of the gene models of 38 *GmChis* genes identified from *Glycine max*. L genome revealed presence of varied numbers of introns. Exons are shown as red boxes and introns are shown as black lines.

Table 2. Putative cis-regulatory elements in BjPR1 promoter sequence, identified by PlantCARE and PLACE promoter databases.

Cis-Acting Element	Function	Sequence
ABRE	ABA-dependent expression	ACGTG/AACCCGG
ABRE3a	ABA-dependent expression	TACGTG
ABRE4	ABA-dependent expression	CACGTA/CACGTA
AuxRE	part of an auxin-responsive element	TGTCTCAATAAG
CGTCA-motif	JA responsive element	CGTCA
GARE-motif	gibberellin-responsive element	TCTGTTG
GT1-motif	pathogen and salt response	GGTTAA/GTGTGTGAA
LTR	cis-acting element involved in low-temperature responsiveness	CCGAAA
MBS	drought stress	CAACTG
MYB	drought stress	CAACCA/CAACAG/TAACCA
MYB-like sequence	drought stress	TAACCA
MYC	early response to drought and ABA induction	CAATTG/CATGTG/CATTTG
P-box	gibberellin-responsive element	CCTTTTG
TATC-box	cis-acting element involved in gibberellin-responsiveness	TATCCCA
TCA-element	cis-acting element involved in salicylic acid responsiveness	CCATCTTTTT/TCAGAAGAGG
TC-rich repeats	cis-acting element involved in defense and stress responsiveness	ATTCTCTAAC
TGACG-motif	cis-acting regulatory element involved in MeJA-responsiveness	TGACG
TGA-element	auxin-responsive element	AACGAC
W-box	activation of defense and wounding-related genes	TTGACC
WUN-motif	wound response	AAATTACT/TTATTACAT

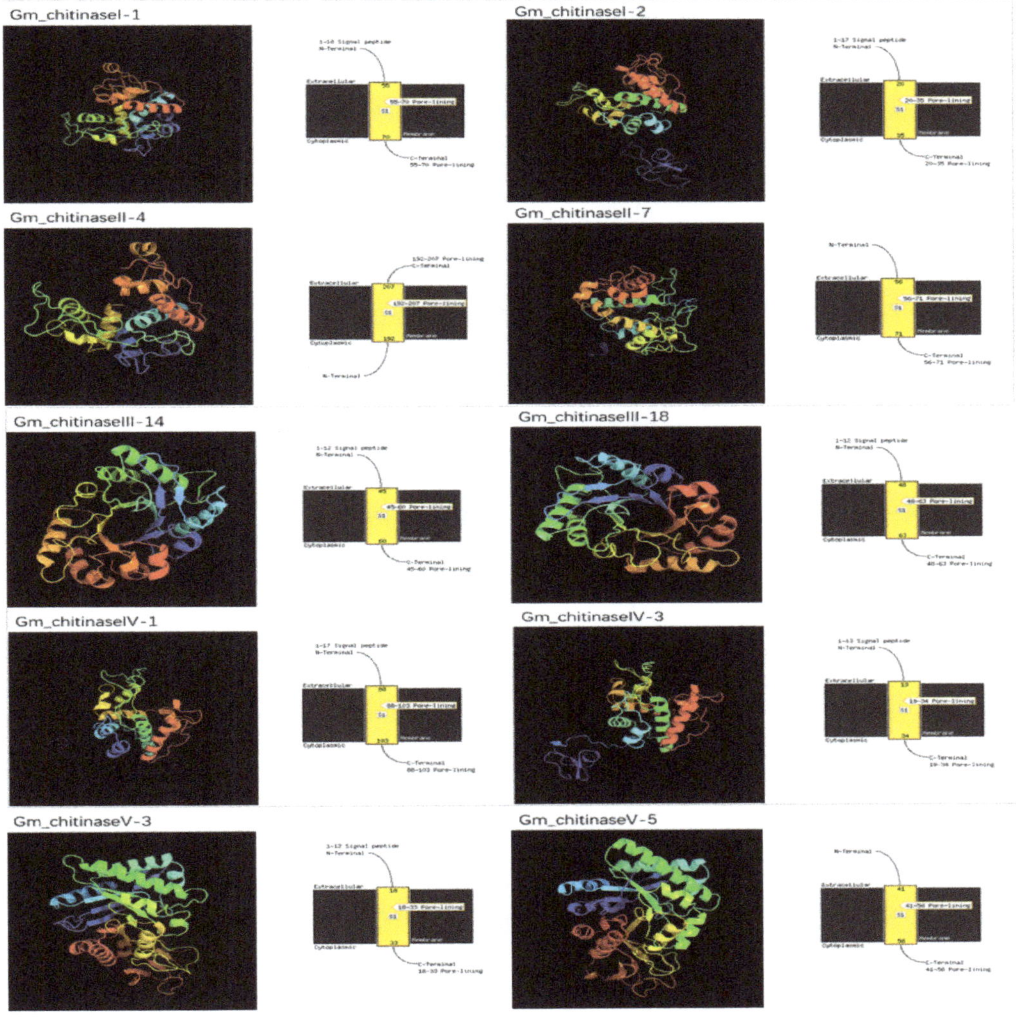

Figure 3. Predicted 3D structures and transmembrane helix (TM) of 10 randomly selected soybean chitinase proteins, two from each class I–V, from top to bottom.

3.5. Synteny Analysis of Chitinases

Soybean crops have encountered different duplication events, such as one WGD and WGT events, during their evolution [31]; these events give rise to many copies of different soybean genes, and a highly duplicated genome [32]. Hence, it is expected that each *Arabidopsis thaliana* chitinase gene might have multiple copies in the soybean genome. In this context, we identified only 38 chitinase orthologs from the 24 chitinase genes of *Arabidopsis thaliana*. It is interesting these 38 chitinase genes represent the orthologs of only nine chitinase genes of Arabidopsis, i.e. *At_chitinaseI-1*, *At_chitinaseII-1*, *At_chitinaseII-2*, *At_chitinaseII-3*, *At_chitinaseII-4*, *At_chitinaseIII-1*, *At_chitinaseIV-9*, *At_chitinaseV-7* and *At_chitinaseV-8*. The remaining 15 chitinase genes of *Arabidopsis thaliana* do not have any orthologs in the soybean genome, perhaps because these chitinase genes have been lost during the evolution of the soybean genome. The highest number of 19 ortholog genes was observed for *Arabidopsis* At_chitinaseIII-1 in the soybean genome, followed by three genes each for At_chitinaseI-1, At_chitinaseIV-9, At_chitinaseV-7 and At_chitinaseV-8 and two genes each for At_chitinaseII-1, At_chitinaseII-2, At_chitinaseII-3. At_chitinaseII-4 has the lowest, one ortholog gene, in the soybean genome. The Circos and synteny analysis showed that both tandem duplication and segmental duplication are involved in the expansion of the chitinase gene family in the soybean (Figure 4).

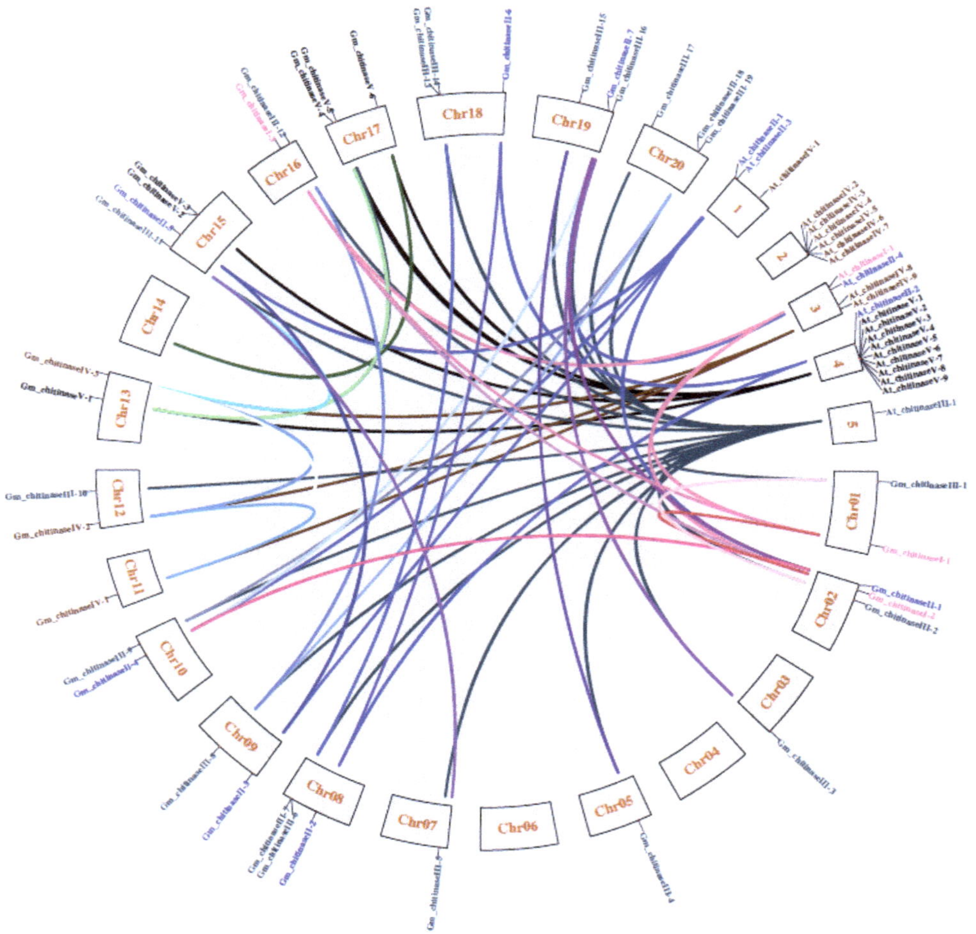

Figure 4. Syntenic relationships of among *A. thaliana and G. max* L. chitinase genes are indicated in different colors. Synteny relationships were lined by Circos (http://circos.ca/ (accessed on 23 December 2019)).

3.6. Transcriptional Analysis of Chitinase Genes in Response to White Mold and Drought Stress

Research evidence has revealed the regulatory role of chitinases in biotic stress such as antifungal disease resistance [3,10,33], and abiotic stress such as drought [11,34–36]. In addition, the role of chitinases in modulating plant growth and productivity has been also reported [37]. Hence, the current investigation examined changes in the expression of the genes in response to white mold fungal pathogen (*Sclerotinia sclerotiorum*) and drought stress (Figure 5). In this regard, we randomly selected two *GmChis* genes from each of five different classes (I–V) of chitinases identified in the soybean to determine their expression pattern in response to pathogen infection and drought stress. Our results revealed that chitinase of different classes showed a considerably varied response under both pathogen and drought stresses. For example, the chitinases belonging to class I and class III were significantly up-regulated (6-fold to 10-fold) at different intervals following pathogen infection. In contrast, the chitinases of class-II, class IV and class V did not show any significant response under the pathogen treatment. Under drought stress, only the chitinase of class V showed significantly higher up-regulation (up to a 16-fold increase in expression) at all the four time intervals (6 h, 12 h, 24 h and 48 h) following the stress treatment. Chitinases of the remaining four classes did not significantly change under drought stress. This suggests a diverse and specific role of different chitinase genes of soybean in the regulation of biotic and abiotic stresses. Hence, research efforts are needed to functionally elucidate the role of chitinase genes in the regulation of different biotic and abiotic stresses in soybean.

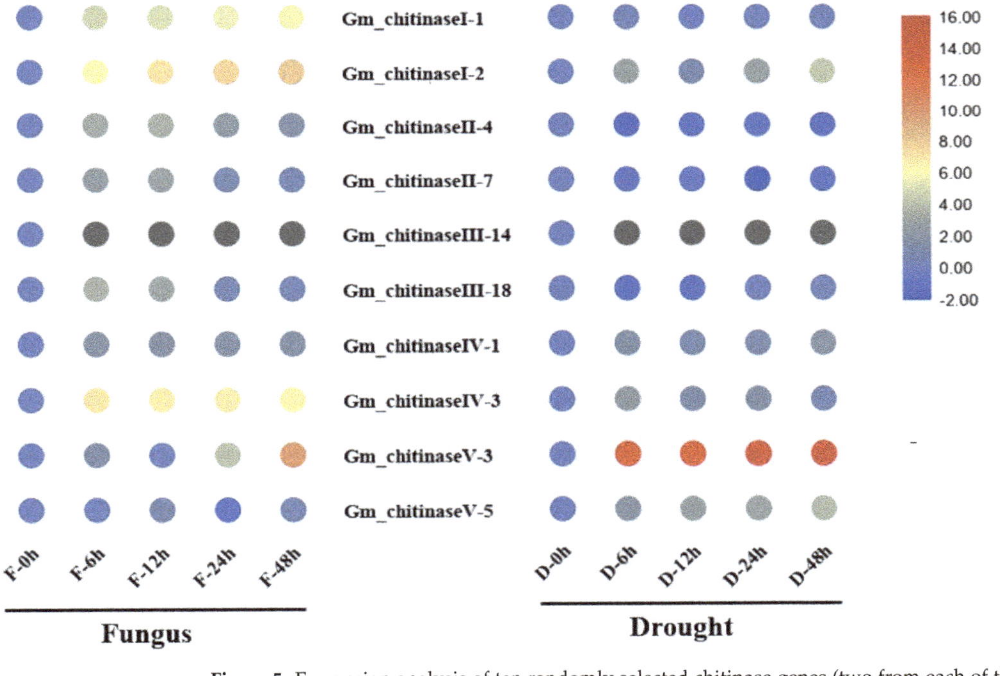

Figure 5. Expression analysis of ten randomly selected chitinase genes (two from each of the five classes) of the soybean ate 6, 12, 24 and 48 h after the inoculation of white mold pathogen (*Sclerotinia sclerotiorum)* and drought stress treatment. Three biological replicates were used to calculate error bars, based on standard error.

4. Discussion

Plants, being immobile, often encounter various environmental stresses, leading to negative effects on the plants' growth [1]. Plants possess well established defense mechanism to alleviate these stresses. For example, PR proteins are a diverse range of proteins produced by the plants in response to stress, and chitinases are one class of PR proteins that are ubiquitously found in prokaryotes and eukaryotes, including plants [3,38]. Chitinases regulate plant growth and development under biotic (such as fungal pathogens) and abiotic stresses [3,5]. Research investigation has confirmed the important role of chitinases in plant defense, but there is a need to identify and elucidate the function of these genes for their potential use in crop improvement. To date, little is known about chitinases in cultivated soybean (*Glycine max* L.), and no systematic investigation has been carried in soybean. Hence, we undertook a comprehensive and systematic investigation to identify and characterize the chitinase gene family across the whole soybean genome. We identified 38 chitinase genes in the soybean genome, and this number was relatively higher than previously reported in *A. thaliana* [5]. This can be explained as follows: the soybean genome is complex, and in its evolutionary history it has gone through the events of WGD and WGT, ~13–130 million years ago, which might have created multiple gene copies [31]. However, soybean possesses a very similar number of chitinases to what has previously been reported in rice (37), grape (38), *B. rapa* (33) and cucumber (28) [39–42]. In contrast, soybean possesses a lower number of chitinase genes than *Gossypium hirsutum* (92), *Gossypium barbadense* (116), *E. grandis* (67) and *C. sativa* (79) [3,5,43]. This can be attributed to the large genome size and more duplication events present in these species, compared to soybean [3]. Moreover, chitinase genes in *Glycine max* L. are unevenly and randomly distributed in 17 of the 20 chromosomes (Figure S1). Chen et al. [41] also reported the distribution of 33 chitinases genes on eight of the 10 total chromosomes. Similar findings were observed in rice [39] and *P. trichocarpa* [44].

Based on the phylogenetic relationship, soybean chitinase, along with the known chitinases of *Arabidopsis thaliana,* are classified into five separate clusters, and these five clusters represent five chitinase classes, i.e., I, II, III, IV and V in soybean. From a broader viewpoint, these five clusters are basically separated into two mega-clusters ("mega-cluster 1" & "mega-cluster 2"). The GH19 family chitinases that include class I, II and IV are grouped in "mega-group 1", and "mega-group 2" possess the chitinases of the GH18 family (class III and V). However, GH19 and GH18 are distinct from each other, as well as having an independent history of evolution [3]. For example, chitinases of the GH18 family possess the catalytic domains triosepho-sphateisomerise (TIM barrel) with highly conserved motif (DxDxE), and these chitinases function in hydrolytic reactions, whereas chitinases of the GH19 family contains alpha-helices and catalyze single displacement [45–47]. The chitinase classes of I and II are grouped close to each other, because class II has originated from class I via chitin-binding domain insertion [48]. In addition, the two mega-clusters can be easily identified based on their domain; for example, "mega-cluster 1" chitinases are characterized by the Glyco_hydro_19 domain, whereas "mega-cluster 2" possess the Glyco_hydro_18 domain. Chitinases of "mega-cluster 2" are present in diverse living organisms, such as microorganisms, animals and plants; in contrast, the chitinases of "mega-cluster 1" are uniquely found in plants [44]. However, our results showed that the soybean genome possesses a lower number of GH19 chitinases (13) than GH18 chitinases (25). Similar differences in the contribution of GH18 and GH19 genes to the chitinase family has been also previously reported in *B. rapa* [41], *Musa acuminata* [49] and *Zea mays* [50], etc.

Stress-related genes have been observed to contain a smaller number of introns, relative to other genes that possess no role in plant stress response (Jeffares et al. 2008). Hence, our study showed that, out of 38 chitinase genes identified in the soybean, 36 possess three or fewer introns, and confirmed the above conception. Similar findings were recently reported by Mir et al. [3], who also reported fewer introns in the chitinase genes of *B. juncea* and *C. sativa*. Moreover, many authors have reported lower intron numbers in different stress-related genes such as the LEA family [51], leucine-rich repeat (LRR) family [52] and the trehalose-6-phosphate synthase gene family [53]. Genes that possess a higher number of introns need a longer time for transcription, hence the product of these genes is not available immediately for cellular function. In contrast, genes with reduced intron numbers are quickly transcribed, and are thus rapidly available for defense response [54]. In this context, the reduced number of introns in the soybean chitinase genes allows them to react quickly and respond to stress conditions immediately.

In order to understand chitinase functioning in the various stress responses, we scanned the 1.5 kb upstream promoter regions of chitinase genes for cis-elements. The bioinformatic analysis revealed the presence of multiple cis-regulatory elements, either in one or more copies, in the upstream promoter regions. The biotic stress regulatory cis-elements present in the promoter region include SA motifs, TC-rich repeats, JA motifs and fungal responsive elements. Hence, this suggests a function of chitinase in modulating the stress response in plants. The ABA-dependent pathway activates the genes involved in the abiotic stress response in plants, and it requires the presence of single or multiple copies of ABREs motifs. In addition, these genes are activated independently via binding of different DREBPs groups to DRE motifs (TAC CGA CAT) [55]. To this end, the MYB and MBS cis-elements identified in the upstream region are drought-inducibility elements/motifs, suggesting role of the chitinase in drought stress [56]. Additionally, cold/chilling responsive cis-elements (LTRE) were also identified [57], and HSEs are the important cis-elements present in the heat shock protein genes (HSPs) regulating the heat stress response in plants [58]. Moreover, the presence of ERA, GARE- motif, P-box and TGA-element in the chitinase promoters suggests their regulatory influence by plant hormones. The motifs of SA and JA are present in many stress-related genes and regulate stress tolerance in plants [4]. Similar to our findings, these motifs (existing in one or more copies) were also previously reported in chitinase genes and other PR genes in different plants, such as *B. juncea* and *C. sativa* [3,4], and thus our results provide preliminary evidence for the functioning of chitinase genes in multiple plant stresses. Therefore, cis-regulatory element analysis showed that soybean chitinase might be involved in modulating both biotic and abiotic stress tolerance in soybean.

Widening of gene families occurs through different types of duplication events, such as WGD/WGT, segmental and tandem duplications [59]. The two and one WGD and WGT events experienced by soybean genome in its evolution have produced many copies of soybean genes and led to the genome's complexity [31,32]. However, all *A. thaliana* chitinase genes do not have homologous genes in the soybean genome; only nine chitinase genes of *Arabidopsis* possess homologs in the soybean genome. The remaining 15 chitinase genes of *Arabidopsis* do not have any orthologs in the soybean genome, perhaps because these chitinase genes have been lost during the evolution of the soybean genome. Interestingly, *At_chitinaseIII-1* has 19 chitinase orthologs in the soybean genome, and they represent mostly tandem duplications, but a few are segmental duplications. Four genes, *At_chitinaseI-1*, *At_chitinaseIV-9*, *At_chitinaseV-7* and *At_chitinaseV-8*, revealed triplication, and this has evolved through tandem duplication. The remaining three genes, *At_chitinaseII-1*, *At_chitinaseII-2* and *At_chitinaseII-3*, showed duplication, and this has also evolved through tandem duplications, and *At_chitinaseII-4* has only a single copy in the soybean genome. Hence, the widening of the soybean chitinase gene family has mainly resulted from tandem duplications (Figure 4). Our results suggest that *Arabidopsis* chitinase genes might have been conserved before speciation, but have been lost during the evolution of the soybean genome as well as during artificial selection. Similar to our findings, the homologs of *Arabidopsis* chitinase has been lost in other plant species as well. For example, 10 *Arabidopsis* chitinase genes do not have orthologs, and are lost in *B. rapa* [41]. Similar findings were observed in *B. Juncea* and *C. sativa* by Mir et al. [3]. In addition, the WGD and WGT events leading to the loss of genes in soybean have been reported for other gene families, such as cytokinin oxidase/dehydrogenase (CKX) genes [60], nucleotide binding site (NBS)-encoding genes [61] and MKK and MPK genes [62]. These results suggest that expansion or elimination of some *Arabidopsis* chitinase genes in the soybean genome might have occurred due to functional differentiation of these genes under diverse environmental stresses. The soybean probably has retained a sufficient number of chitinase genes during its evolution to respond to external stress properly.

In plants, PR proteins modulate the plant defense system to provide protection against various environmental stresses. Hence, the PR-3 family of PR proteins represents the chitinases class [9], and expression of PR-3 proteins has been demonstrated to be induced by both biotic and abiotic stresses [5,11]. Therefore, our results revealed that chitinases belonging to specific classes were significantly induced under white mold fungal pathogen and drought stress treatments. For example, the chitinases belonging to class I and class III were significantly up-regulated (6-fold to 10-fold) at different intervals following pathogen infection. In contrast, the chitinases of class II, class IV and class V did not show any significant response under the pathogen treatment, which is similar to reports of different studies in various plants [63–65]. Moreover, in the cotton plant, the expression of chitinase genes was induced by inoculation of a pathogen (*Verticillium dahlia*) and significantly reached peak level 24 h following inoculation [5]. Under drought stress, only the chitinases of class V showed significantly higher up-regulation (up to a 16-fold increase in expression) at all the four time intervals (6 h, 12 h, 24 h and 48 h) following the stress treatment. Chitinases of the remaining four classes did not undergo significant changes under drought stress. In agreement with our report, chitinase expression induced by drought stress has also been reported in *Arabidopsis thaliana* [11] and *Crocus sativus* [20]. Hence, the above findings suggest an important role of chitinase genes in controlling multiple plant stress (diseases and abiotic) responses in soybean plants. Therefore, the more research efforts are required to elucidate the detailed function and mechanism involved in chitinase-mediated regulation of plant defense.

5. Conclusions

The current investigation provides a comprehensive and systematic report of the chitinase gene family at the whole genome scale in soybean. Here, we detected 38 chitinase genes in the soybean genome, and these genes were randomly and unevenly distributed on the soybean chromosomes. Phylogenetic analysis grouped these chitinase genes into five distinct clusters representing five classes of chitinase (I, II, III, IV and V). In addition, synteny and duplication analysis revealed that tandem duplication has played the major role in widening the family of chitinase genes in soybean, while segmental duplication has the smallest role. Promoter analysis showed multiple cis-regulatory elements related to biotic and abiotic stresses in the upstream region of the chitinase genes, suggesting their role in plant defense response against multiple stresses. Moreover, gene expression analysis revealed that pathogenic and drought stress treatments significantly induce the up-regulation of chitinase genes belonging to specific classes at different time intervals, which further confirmed their role in plant stress response. Overall, our study provides evidence about the role of the chitinases in multiple plant stress responses in soybean. However, there is a need for future research efforts to validate the specific or general functions of different chitinases against different biotic and abiotic stresses. Therefore, extensive research efforts are required to elucidate the detailed mechanism involved in chitinase-mediated modulation for different plant stresses, for their potential use in soybean improvement.

Supplementary Materials: The following supporting information can be downloaded at: https://www.mdpi.com/article/10.3390/life12091340/s1, Figure S1: Diagram showing the distribution of the 38 chitinase genes identified among the different 17 of the total 20 chromosomes of the soybean; Figure S2: In silico analysis of Chitinase gene promoters of *G. max* L. Promoter cis-elements of 10 chitinase genes (two genes from each five classes of chitinases identified in soybean) in response to biotic, abiotic and hormonal stresses are shown in different shapes and colors along with their respective positions from the start codon ATG; Table S1: List of primers used in the qRT-PCR analysis of the selected chitinase genes of soybean.

Author Contributions: Conceptualization, J.A.B., T.Z. and P.A.; methodology, M.A.E.-S.; software, P.A.; validation, P.L., C.Z. and P.X.; formal analysis, M.A.E.-S.; investigation, P.L.; resources, P.L., P.X.; data curation, P.L., C.Z. and X.Y., P.L.; writing—original draft preparation, J.A.B.; writing—review and editing, J.A.B., P.A.; D.I.H., T.Z.; visualization, J.A.B.; supervision, P.A.; project administration, J.A.B., T.Z.; funding acquisition, D.I.H.; M.A.E.-S. All authors have read and agreed to the published version of the manuscript.

Funding: This work was financially supported by the National Natural Science Foundation of China (Grant Nos. 31871646 and 31571695), the MOE Program for Changjiang Scholars and Innovative Research Team in University (PCSIRT_17R55), the Fundamental Research Funds for the Central Universities (KYT201801), and the Jiangsu Collaborative Innovation Center for Modern Crop Production (JCIC-MCP) Program.

Institutional Review Board Statement: Not applicable.

Informed Consent Statement: Not applicable.

Data Availability Statement: Data is contained within the article and Supplementary Materials.

Acknowledgments: The authors would like to extend their sincere appreciation to the Researchers Supporting Project Number (RSP-2022/182), King Saud University, Riyadh, Saudi Arabia. The authors specifically thank Nanjing Agricultural University for providing the research facilities.

Conflicts of Interest: The authors declare no conflict of interest.

References

1. Bhat, J.A.; Shivaraj, S.M.; Singh, P.; Navadagi, D.B.; Tripathi, D.K.; Dash, P.K.; Solanke, A.U.; Sonah, H.; Deshmukh, R. Role of silicon in mitigation of heavy metal stresses in crop plants. *Plants* **2019**, *8*, 71. [CrossRef] [PubMed]
2. Ali, S.; Ganai, B.A.; Kamili, A.N.; Bhat, A.A.; Mir, Z.A.; Bhat, J.A.; Tyagi, A.; Islam, S.T.; Mushtaq, M.; Yadav, P.; et al. Pathogenesis-related proteins and peptides as promising tools for engineering plants with multiple stress tolerance. *Microbiol. Res.* **2018**, *212*, 29–37. [CrossRef] [PubMed]
3. Mir, Z.A.; Ali, S.; Shivaraj, S.M.; Bhat, J.A.; Singh, A.; Yadav, P.; Rawat, S.; Paplao, P.K.; Grover, A. Genome-wide identification and characterization of Chitinase gene family in *Brassica juncea* and *Camelina sativa* in response to *Alternaria brassicae*. *Genomics* **2020**, *112*, 749–763. [CrossRef]
4. Ali, S.; Mir, Z.A.; Tyagi, A.; Bhat, J.A.; Chandrashekar, N.; Papolu, P.K.; Rawat, S.; Grover, A. Identification and comparative analysis of *Brassica juncea* pathogenesis-related genes in response to hormonal, biotic and abiotic stresses. *Acta Physiol. Plant.* **2017**, *39*, 268. [CrossRef]
5. Xu, J.; Xu, X.; Tian, L.; Wang, G.; Zhang, X.; Wang, X.; Guo, W. Discovery and identification of candidate genes from the chitinase gene family for *Verticillium dahliae* resistance in cotton. *Sci. Rep.* **2016**, *6*, 29022. [CrossRef]
6. Cao, J.; Tan, X. Comprehensive analysis of the chitinase family genes in tomato (*Solanum lycopersicum*). *Plants* **2019**, *8*, 52. [CrossRef]
7. Wang, X.; He, N.; Zeng, Q.; Xiang, Z. Identification and expression analyses of chitinase genes in mulberry (*Morus* L.). *Plant Omics* **2015**, *8*, 183.
8. Wan, J.; Zhang, X.C.; Neece, D.; Ramonell, K.M.; Clough, S.; Kim, S.Y.; Stacey, M.G.; Stacey, G. A LysM receptor-like kinase plays a critical role in chitin signaling and fungal resistance in *Arabidopsis*. *Plant Cell* **2008**, *20*, 471–481. [CrossRef]
9. Van Loon, L.C.; Van Strien, E.A. The families of pathogenesis-related proteins, their activities, and comparative analysis of PR-1 type proteins. *Physiol. Mol. Plant Pathol.* **1999**, *55*, 85–97. [CrossRef]
10. Dana, M.D.; Pintor-Toro, J.A.; Cubero, B. Transgenic tobacco plants overexpressing chitinases of fungal origin show enhanced resistance to biotic and abiotic stress agents. *Plant Physiol.* **2006**, *142*, 722–730. [CrossRef]
11. Takenaka, Y.; Nakano, S.; Tamoi, M.; Sakuda, S.; Fukamizo, T. Chitinase gene expression in response to environmental stresses in *Arabidopsis thaliana*: Chitinase inhibitor allosamidin enhances stress tolerance. *Biosci. Biotechnol. Biochem.* **2009**, *73*, 1066–1071. [CrossRef] [PubMed]
12. Liu, T.; Guo, X.; Bu, Y.; Zhou, Y.; Duan, Y.; Yang, Q. Structural and biochemical insights into an insect gut-specific chitinase with antifungal activity. *Insect Biochem. Mol. Biol.* **2020**, *119*, 103326. [CrossRef] [PubMed]
13. Hina, A.; Cao, Y.; Song, S.; Li, S.; Sharmin, R.A.; Elattar, M.A.; Bhat, J.A.; Zhao, T. High-resolution mapping in two RIL populations refines major "QTL Hotspot" regions for seed size and shape in soybean (*Glycine max* L.). *Int. J. Mol. Sci.* **2020**, *21*, 1040.
14. Bhat, J.A.; Deshmukh, R.; Zhao, T.; Patil, G.; Deokar, A.; Shinde, S.; Chaudhary, J. Harnessing High-throughput Phenotyping and Genotyping for Enhanced Drought Tolerance in Crop Plants. *J. Biotechnol.* **2020**, *324*, 248–260. [PubMed]
15. Wutzki, C.R.; de Souza Jaccoud Filho, D.; Neto, A.B.; Tullio, H.E.; Juliatti, F.C.; do Nascimento, A.J. Reduction of white mold level on soybean by fungicide management strategies. *Biosci. J.* **2016**, *32*, 642–651. [CrossRef]
16. Kandel, Y.R.; Mueller, D.S.; Legleiter, T.; Johnson, W.G.; Young, B.G.; Wise, K.A. Impact of fluopyram fungicide and preemergence herbicides on soybean injury, population, sudden death syndrome, and yield. *Crop Prot.* **2018**, *106*, 103–109. [CrossRef]
17. Koenning, S.R.; Wrather, J.A. Suppression of soybean yield potential in the continental United States by plant diseases from 2006 to 2009. *Plant Health Prog.* **2010**, *11*, 5. [CrossRef]
18. Peltier, A.J.; Bradley, C.A.; Chilvers, M.I.; Malvick, D.K.; Mueller, D.S.; Wise, K.A.; Esker, P.D. Biology, yield loss and control of Sclerotinia stem rot of soybean. *J. Integr. Pest Manag.* **2012**, *3*, B1–B7. [CrossRef]
19. Cao, H.; Li, X.; Dong, X. Generation of broad-spectrum disease resistance by overexpression of an essential regulatory gene in systemic acquired resistance. *Proc. Natl. Acad. Sci. USA* **1998**, *95*, 6531–6536. [CrossRef]
20. Békésiová, B.; Hraška, Š.; Libantová, J.; Moravčíková, J.; Matušíková, I. Heavy-metal stress induced accumulation of chitinase isoforms in plants. *Mol. Biol. Rep.* **2008**, *35*, 579–588. [CrossRef]

21. Yang, X.; Yang, J.; Li, H.; Niu, L.; Xing, G.; Zhang, Y.; Xu, W.; Zhao, Q.; Li, Q.; Dong, Y. Overexpression of the chitinase gene CmCH1 from *Coniothyrium minitans* renders enhanced resistance to Sclerotinia sclerotiorum in soybean. *Transgenic Res.* **2020**, *29*, 187–198. [CrossRef] [PubMed]
22. Ayaz, A.; Saqib, S.; Huang, H.; Zaman, W.; Lü, S.; Zhao, H. Genome-wide comparative analysis of long-chain acyl-CoA synthetases (LACSs) gene family: A focus on identification, evolution and expression profiling related to lipid synthesis. *Plant Physiol. Biochem.* **2021**, *161*, 1–11. [CrossRef] [PubMed]
23. Kumar, S.; Nei, M.; Dudley, J.; Tamura, K. MEGA: A biologist-centric software for evolutionary analysis of DNA and protein sequences. *Brief. Bioinform.* **2008**, *9*, 299–306. [CrossRef]
24. Gasteiger, E.; Gattiker, A.; Hoogland, C.; Ivanyi, I.; Appel, R.D.; Bairoch, A. ExPASy: The proteomics server for in-depth protein knowledge and analysis. *Nucleic Acids Res.* **2003**, *31*, 3784–3788. [CrossRef] [PubMed]
25. Aleem, M.; Riaz, A.; Raza, Q.; Aleem, M.; Aslam, M.; Kong, K.; Atif, R.M.; Kashif, M.; Bhat, J.A.; Zhao, T. Genome-wide characterization and functional analysis of class III peroxidase gene family in soybean reveal regulatory roles of GsPOD40 in drought tolerance. *Genomics* **2022**, *114*, 45–60. [CrossRef] [PubMed]
26. Hoffman, M.L.; Owen, M.D.; Buhler, D.D. Effects of crop and weed management on density and vertical distribution of weed seeds in soil. *Agron. J.* **1998**, *90*, 793–799. [CrossRef]
27. Chen, Y.; Wang, D. Two convenient methods to evaluate soybean for resistance to *Sclerotinia sclerotiorum*. *Plant Dis.* **2005**, *89*, 1268–1272. [CrossRef]
28. Sharmin, R.A.; Bhuiyan, M.R.; Lv, W.; Yu, Z.; Chang, F.; Kong, J.; Bhat, J.A.; Zhao, T. RNA-Seq based transcriptomic analysis revealed genes associated with seed-flooding tolerance in wild soybean (*Glycine soja* Sieb. & Zucc.). *Environ. Exp. Bot.* **2020**, *171*, 103906.
29. Ayaz, A.; Huang, H.; Zheng, M.; Zaman, W.; Li, D.; Saqib, S.; Zhao, H.; Lü, S. Molecular cloning and functional analysis of GmLACS2-3 reveals its involvement in cutin and suberin biosynthesis along with abiotic stress tolerance. *Int. J. Mol. Sci.* **2021**, *22*, 9175. [CrossRef]
30. Livak, K.J.; Schmittgen, T.D. Analysis of relative gene expression data using real-time quantitative PCR and the $2^{-\Delta\Delta CT}$ method. *Methods* **2001**, *25*, 402–408. [CrossRef]
31. Severin, A.J.; Cannon, S.B.; Graham, M.M.; Grant, D.; Shoemaker, R.C. Changes in twelve homoeologous genomic regions in soybean following three rounds of polyploidy. *Plant Cell* **2011**, *23*, 3129–3136. [CrossRef]
32. Schmutz, J.; Cannon, S.B.; Schlueter, J.; Ma, J.; Mitros, T.; Nelson, W.; Hyten, D.L.; Song, Q.; Thelen, J.J.; Cheng, J.; et al. Genome sequence of the palaeopolyploid soybean. *Nature* **2010**, *463*, 178–183. [CrossRef] [PubMed]
33. Ahmad, M.Z.; Hussain, I.; Muhammad, A.; Ali, S.; Ali, G.M.; Roomi, S.; Zia, M.A.; Ijaz, A. Factor affecting Agrobacterium-mediated transformation of rice chitinase gene in *Solanum tuberosum* L. *Afr. J. Biotechnol.* **2012**, *11*, 9716–9723.
34. Liu, K.; Ding, H.; Yu, Y.; Chen, B. A cold-adapted chitinase-producing bacterium from antarctica and its potential in biocontrol of plant pathogenic fungi. *Mar. Drugs* **2019**, *17*, 695. [CrossRef] [PubMed]
35. Zhou, N.; An, Y.; Gui, Z.; Xu, S.; He, X.; Gao, J.; Zeng, D.; Gan, D.; Xu, W. Identification and expression analysis of chitinase genes in *Zizania latifolia* in response to abiotic stress. *Sci. Hortic.* **2020**, *261*, 108952. [CrossRef]
36. Cheng, S.P.; Lee, J.J.; Chang, Y.C.; Lin, C.H.; Li, Y.S.; Liu, C.L. Overexpression of chitinase-3-like protein 1 is associated with structural recurrence in patients with differentiated thyroid cancer. *J. Pathol.* **2020**, *252*, 114–124. [CrossRef] [PubMed]
37. Grover, A. Plant chitinases: Genetic diversity and physiological roles. *Crit. Rev. Plant Sci.* **2012**, *31*, 57–73. [CrossRef]
38. Hamid, R.; Khan, M.A.; Ahmad, M.; Ahmad, M.M.; Abdin, M.Z.; Musarrat, J.; Javed, S. Chitinases: An update. *J. Pharm. Bioallied Sci.* **2013**, *5*, 21.
39. Xu, F.; Fan, C.; He, Y. Chitinases in *Oryza sativa* ssp. japonica and Arabidopsis thaliana. *J. Genet. Genom.* **2007**, *34*, 138–150. [CrossRef]
40. Zheng, T.; Zhang, K.; Sadeghnezhad, E.; Jiu, S.; Zhu, X.; Dong, T.; Liu, Z.; Guan, L.; Jia, H.; Fang, J. Chitinase family genes in grape differentially expressed in a manner specific to fruit species in response to *Botrytis cinerea*. *Mol. Biol. Rep.* **2020**, *47*, 7349–7363. [CrossRef]
41. Chen, J.; Piao, Y.; Liu, Y.; Li, X.; Piao, Z. Genome-wide identification and expression analysis of chitinase gene family in *Brassica rapa* reveals its role in clubroot resistance. *Plant Sci.* **2018**, *270*, 257–267. [CrossRef] [PubMed]
42. Bartholomew, E.S.; Black, K.; Feng, Z.; Liu, W.; Shan, N.; Zhang, X.; Wu, L.; Bailey, L.; Zhu, N.; Qi, C.; et al. Comprehensive analysis of the chitinase gene family in cucumber (*Cucumis sativus* L.): From gene identification and evolution to expression in response to Fusarium oxysporum. *Int. J. Mol. Sci.* **2019**, *20*, 5309. [CrossRef] [PubMed]
43. Tobias, P.A.; Christie, N.; Naidoo, S.; Guest, D.I.; Külheim, C. Identification of the *Eucalyptus grandis* chitinase gene family and expression characterization under different biotic stress challenges. *Tree Physiol.* **2017**, *37*, 565–582. [CrossRef]
44. Jiang, C.; Huang, R.F.; Song, J.L.; Huang, M.R.; Xu, L.A. Genomewide analysis of the chitinase gene family in *Populus trichocarpa*. *J. Genet.* **2013**, *92*, 121–125. [CrossRef]
45. van Aalten, D.M.; Komander, D.; Synstad, B.; Gåseidnes, S.; Peter, M.G.; Eijsink, V.G. Structural insights into the catalytic mechanism of a family 18 exo-chitinase. *Proc. Natl. Acad. Sci. USA* **2001**, *98*, 8979–8984. [CrossRef]
46. Vaaje-Kolstad, G.; Vasella, A.; Peter, M.G.; Netter, C.; Houston, D.R.; Westereng, B.; Synstad, B.; Eijsink, V.G.; van Aalten, D.M. Interactions of a family 18 chitinase with the designed inhibitor HM508 and its degradation product, chitobiono-δ-lactone. *J. Biol. Chem.* **2004**, *279*, 3612–3619. [CrossRef]

47. Hoell, I.A.; Vaaje-Kolstad, G.; Eijsink, V.G. Structure and function of enzymes acting on chitin and chitosan. *Biotechnol. Genet. Eng. Rev.* **2010**, *27*, 331–366. [CrossRef]
48. Araki, T.; Torikata, T. Structural classification of plant Chitinases: Two subclasses in class I and class II Chitinase. *Biosci. Biotechnol. Biochem.* **1995**, *59*, 336–338. [CrossRef]
49. Backiyarani, S.; Uma, S.; Nithya, S.; Chandrasekar, A.; Saraswathi, M.S.; Thangavelu, R.; Mayilvaganan, M.; Sundararaju, P.; Singh, N.K. Genome-wide analysis and differential expression of chitinases in banana against root lesion nematode (*Pratylenchus coffeae*) and eumusa leaf spot (*Mycosphaerella eumusae*) pathogens. *Appl. Biochem. Biotechnol.* **2015**, *175*, 3585–3598. [CrossRef]
50. Hawkins, L.K.; Mylroie, J.E.; Oliveira, D.A.; Smith, J.S.; Ozkan, S.; Windham, G.L.; Williams, W.P.; Warburton, M.L. Characterization of the maize chitinase genes and their effect on *Aspergillus flavus* and aflatoxin accumulation resistance. *PLoS ONE* **2015**, *10*, e0126185. [CrossRef]
51. Liang, Y.; Xiong, Z.; Zheng, J.; Xu, D.; Zhu, Z.; Xiang, J.; Gan, J.; Raboanatahiry, N.; Yin, Y.; Li, M. Genome-wide identification, structural analysis and new insights into late embryogenesis abundant (LEA) gene family formation pattern in *Brassica napus*. *Sci. Rep.* **2016**, *6*, 24265. [CrossRef] [PubMed]
52. Zhou, F.; Guo, Y.; Qiu, L.J. Genome-wide identification and evolutionary analysis of leucine-rich repeat receptor-like protein kinase genes in soybean. *BMC Plant Biol.* **2016**, *16*, 58. [CrossRef] [PubMed]
53. Xie, D.W.; Wang, X.N.; Fu, L.S.; Sun, J.; Zheng, W.; Li, Z.F. Identification of the trehalose-6-phosphate synthase gene family in winter wheat and expression analysis under conditions of freezing stress. *J. Genet.* **2015**, *94*, 55–65. [CrossRef] [PubMed]
54. Jeffares, D.C.; Penkett, C.J.; Bähler, J. Rapidly regulated genes are intron poor. *Trends Genet.* **2008**, *24*, 375–378. [CrossRef]
55. Basu, S.; Roychoudhury, A. Expression profiling of abiotic stress-inducible genes in response to multiple stresses in rice (*Oryza sativa* L.) varieties with contrasting level of stress tolerance. *BioMed Res. Int.* **2014**, *2014*, 706890. [CrossRef]
56. Smita, S.; Katiyar, A.; Chinnusamy, V.; Pandey, D.M.; Bansal, K.C. Transcriptional regulatory network analysis of MYB transcription factor family genes in rice. *Front. Plant Sci.* **2015**, *6*, 1157. [CrossRef]
57. Brown, T.A.; Campbell, L.A.; Lehman, C.L.; Grisham, J.R.; Mancill, R.B. Current and lifetime comorbidity of the DSM-IV anxiety and mood disorders in a large clinical sample. *J. Abnorm. Psychol.* **2001**, *110*, 585. [CrossRef]
58. Guo, M.; Liu, J.H.; Ma, X.; Luo, D.X.; Gong, Z.H.; Lu, M.H. The plant heat stress transcription factors (HSFs): Structure, regulation, and function in response to abiotic stresses. *Front. Plant Sci.* **2016**, *7*, 114. [CrossRef]
59. Freeling, M. Bias in plant gene content following different sorts of duplication: Tandem, whole-genome, segmental, or by transposition. *Annu. Rev. Plant Biol.* **2009**, *60*, 433–453. [CrossRef]
60. Hai, N.N.; Chuong, N.N.; Tu, N.H.C.; Kisiala, A.; Hoang, X.L.T.; Thao, N.P. Role and regulation of cytokinins in plant response to drought stress. *Plants* **2020**, *9*, 422. [CrossRef]
61. Zhang, Y.M.; Shao, Z.Q.; Wang, Q.; Hang, Y.Y.; Xue, J.Y.; Wang, B.; Chen, J.Q. Uncovering the dynamic evolution of nucleotide-binding site-leucine-rich repeat (NBS-LRR) genes in Brassicaceae. *J. Integr. Plant Biol.* **2016**, *58*, 165–177. [CrossRef] [PubMed]
62. Jiang, M.; Chu, Z. Comparative analysis of plant MKK gene family reveals novel expansion mechanism of the members and sheds new light on functional conservation. *BMC Genom.* **2018**, *19*, 407. [CrossRef] [PubMed]
63. Rawat, S.; Ali, S.; Mittra, B.; Grover, A. Expression analysis of chitinase upon challenge inoculation to *Alternaria* wounding and defense inducers in *Brassica juncea*. *Biotechnol. Rep.* **2017**, *13*, 72–79. [CrossRef] [PubMed]
64. Rasmussen, U.; Bojsen, K.; Collinge, D.B. Cloning and characterization of a pathogen-induced chitinase in *Brassica napus*. *Plant Mol. Biol.* **1992**, *20*, 277–287. [CrossRef]
65. Mukherjee, A.K.; Lev, S.; Gepstein, S.; Horwitz, B.A. A compatible interaction of *Alternaria brassicicola* with *Arabidopsis thaliana* ecotype DiG: Evidence for a specific transcriptional signature. *BMC Plant Biol.* **2009**, *9*, 31. [CrossRef]

life

MDPI

Article

Functional Characterization of *MaZIP4*, a Gene Regulating Copper Stress Tolerance in Mulberry (*Morus atropurpurea* R.)

Yisu Shi †, Qiaonan Zhang †, Lei Wang †, Qiuxia Du, Michael Ackah, Peng Guo, Danyan Zheng, Mengmeng Wu and Weiguo Zhao *

Key Laboratory of Silkworm and Mulberry Genetic Improvement, Ministry of Agriculture, School of Biology and Technology, Jiangsu University of Science and Technology, Zhenjiang 212018, China
* Correspondence: wgzsri@126.com
† These authors contributed equally to this work.

Abstract: *ZIP4* (zinc transporter 4) plays important roles in transporting Cu^{2+} ions in plants, which may contribute to the maintenance of plant metal homeostasis in growth, plant development and normal physiological metabolism. However, *ZIP4* transporters have not been described in mulberry and the exact function of *ZIP4* transporters in regulating the homeostasis of Cu in mulberry remains unclear. In this study, a new *ZIP4* gene (*MaZIP4*) was isolated and cloned from *Morus atropurpurea* R. Phylogenetic analysis of amino sequences suggested that the amino-acid sequence of the *MaZIP4* protein shows high homology with other *ZIP4* proteins of *Morus notabilis*, *Trema orientale*, *Ziziphus jujube* and *Cannabis sativa*. In addition, a *MaZIP4* silenced line was successfully constructed using virus-induced gene silencing (VIGS). The analysis of *MaZIP4* expression by quantitative real-time PCR in mulberry showed that the level of *MaZIP4* expression increased with increasing Cu concentration until the Cu concentration reached 800 ppm. Relative to the blank (WT) and the negative controls, malondialdehyde (MDA) levels increased significantly and rose with increasing Cu concentration in the *MaZIP4* silenced line, whereas the soluble protein and proline content, superoxide dismutase (SOD) and peroxidase (POD) activities of these transgenic plants were lower. These results indicated that *MaZIP4* may play an important role in the resistance of mulberry to Cu stress.

Keywords: phylogenetic; virus-induced gene silencing; transgenic lines; physiological and biochemical analysis

Citation: Shi, Y.; Zhang, Q.; Wang, L.; Du, Q.; Ackah, M.; Guo, P.; Zheng, D.; Wu, M.; Zhao, W. Functional Characterization of *MaZIP4*, a Gene Regulating Copper Stress Tolerance in Mulberry (*Morus atropurpurea* R.). *Life* **2022**, *12*, 1311. https://doi.org/10.3390/life12091311

Academic Editors: Wajid Zaman and Hakim Manghwar

Received: 4 August 2022
Accepted: 23 August 2022
Published: 26 August 2022

Publisher's Note: MDPI stays neutral with regard to jurisdictional claims in published maps and institutional affiliations.

Copyright: © 2022 by the authors. Licensee MDPI, Basel, Switzerland. This article is an open access article distributed under the terms and conditions of the Creative Commons Attribution (CC BY) license (https://creativecommons.org/licenses/by/4.0/).

1. Introduction

Although copper (Cu) is an essential nutrient for plant growth and development, above certain physiological levels it can be toxic. It participates in numerous physiological processes and is an essential cofactor for many metalloproteins [1]. For example, Cu activates many enzymes in plants that are involved in lignin synthesis, and it is the key to the formation of chlorophyll for photosynthesis [2–4]. In addition, it is essential in plant respiration and assists in the plant metabolism of carbohydrates and proteins [2]. Another important function of Cu is to promote the development of flower organs and to intensify flower coloring [5]. Cu is one of the micronutrients needed by plants and the ideal range for Cu in the tissue is 20 times lower than that of iron [6]. Excess Cu may have a negative impact on plant growth and quality [4], and impairs leaf Cu concentration, gas exchange and protein profiles [7]. Excess Cu in the growth medium can restrict tap root growth by burning the root tips and thereby promoting lateral root growth [8]. High concentrations of Cu can compete with plant uptake of Ca, Mg, K, Zn and Fe, initially resulting in greener new growth than normal, but later to the exhibition of iron or other micronutrient deficiencies [8]. The continued exposure to excess Cu toxicity can reduce aerial branching and lead to further reductions in plant health. Plants must therefore have evolved appropriate strategies to maintain Cu homeostasis in response to different environmental Cu levels.

Such strategies must prevent the accumulation of the metal in its reactive form within detoxification pathways and ensure that the element is properly transported for storage or for the biosynthesis of target metalloproteins. Although there has been substantial recent research on the mechanisms involved in Cu acquisition and transport into and within cells in yeast and other eukaryotic organisms, including *Arabidopsis thaliana* [9], these processes in plants remain incompletely understood in plant systems. Nevertheless, several families of heavy metal transporters involved in the maintenance of intracellular heavy metal homeostasis in plants have been identified [2,10].

ZIP transporters are divalent metal transporters, which have been identified in a variety of plant species, especially dicots such as *Arabidopsis* and soybean [10]. They are responsible for transporting various metal cations into the cytoplasm, such as Zn^{2+}, Mn^{2+}, Fe^{2+}/Fe^{3+}, Cd^{2+}, Co^{2+}, Ni^{2+} and Cu^{2+}. It has been reported that ZIP proteins are involved in cellular uptake of Zn^{2+}. *ZIP* transporters contain eight transmembrane domains and a histidine-rich variable loop between TM3 and TM4 [11]. There are 14 additional members of the *ZIP* family in *Arabidopsis* [12]. It has been shown that *AtZIP2* and *AtZIP4* can supplement the growth defects of Cu and Zn transport mutants in yeast [13,14]. Expression of the two genes is upregulated in *Arabidopsis* in deficiency of Cu and Zn, but not Fe [15]. In addition, *AtZIP2* has been proposed to participate in Cu acquisition by *Arabidopsis* roots. In the model legume *Medicago truncatula* L., six cDNA encoding *ZIP* family members have been identified, and their ability to complement yeast metal-absorption mutants have been tested [16]. Furthermore, according to the differential expression analysis of mRNA in response to different transition metal concentrations, a role for ZIP proteins in the maintenance of metal homeostasis was proposed [17]. In summary, the differential preferences of ZIP family members for divalent metals suggest that ZIP2 and ZIP4 proteins may play important roles in transporting Cu^{2+} ions. However, their role of these proteins in Cu transport needs to be confirmed and further investigated.

Mulberry (*Morus. atropurpurea* R.) is an economically important perennial tree in China that is widely distributed in the northern temperate regions. As the sole food source of the domesticated silkworm, mulberry cultivation is crucial to the development of sericulture, and has many other important economic and ecological values [18]. Mulberry produces delicious fruits with medicinal value in the treatment of hypertension, oral and dental diseases, diabetes, arthritis and anemia. In addition, the fruit are also used in production of jams, juices, liquors, natural dyes and in the cosmetics industry. Relative to other plant species, mulberry shows high tolerance to multiple abiotic environmental factors in China, such as low temperature, high salinity, waterlogging and high soil concentrations of heavy metal ions [19–21]. Despite this tolerance, an excess of Cu in mulberry has been shown to disturb the cellular redox environment in young leaves, accelerate the rate of leaf senescence and damage roots. In addition, a deficiency in Cu is harmful to mulberry through the aggravation of oxidative stress through an enhanced generation of reactive oxygen species (ROS) and disturbed redox coupling [22]. Several families of heavy metal transporters are thought to be involved in the maintenance of intracellular Cu homeostasis in plants, including ZIP, COPT, P_{IB}-ATPase, ATX, CCS and YSL [2]. ZIP family proteins are thought responsible for the uptake and allocation of many micronutrients including Fe, Mn, Zn and Cu [10,15]. OsZIP1, as a metal efflux transporter, limits the accumulation of excessive Zn, Cu and Cd in rice [23]. AtZIP2 has been implicated in Cu homeostasis and AtZIP4 was demonstrated to be involved in Cu transport in *Arabidopsis* [14]. However, the molecular roles of these heavy metal transporters have not been described in mulberry. Information on the role of the Cu transporters in mulberry and how the expression can be regulated by Cu concentrations may help to cultivate Cu-resistant plants, this expanding the practical planting range of mulberry.

Virus-induced gene silencing (VIGS) is a technique based on RNA interference to construct gene silencing lines, which is widely used to explore gene function [24,25]. Compared with other transgenic techniques, VIGS offers unique advantages in terms of a short cycle time, ease of operation, no need to build stably transformed plants and low costs. In this study, *MaZIP4* was associated with the response to Cu stress in mulberry through comparative transcriptome analysis. *MaZIP4* gene silencing by VIGS technology was subsequently performed to analyze the function of this gene.

2. Materials and Methods

2.1. Sample Collection, RNA Extraction and Sequencing

The mulberry (*M. atropurpurea* R.) materials used in this research were obtained from the sericulture research institute of Chinese Academy of Agricultural Sciences. The seedlings were transplanted in pots containing vermiculite and loamy soil (pH 6.5). seedlings with a length of 20 cm and consistent growth status were selected, three of which were treated with 500 mL of MS culture solution containing either 0 (control), or 200 ppm $CuSO_4$ and grown for a further 20 days. All selected plants were watered every day. Three seedlings were selected from each of the experimental and control groups, representing three biological replicates. Leaf samples were collected from plants 1, 3, 5, 10, 15 and 20 days after Cu treatments. After 20 days of stress incubation, mulberry leaves showed yellowing and shrinkage, and burn marks appear on the leaf edges (Figure 1). For the expression analysis of *MaZIP4*, seedlings were selected as above and three biological replicates were treated with 0, 100, 200, 400 or 800 ppm $CuSO_4$, following by daily watering for a further five days, after which young leaves were collected from each treatment group. In all cases, the collected leaves were immediately frozen in liquid nitrogen and stored at −80 °C for later RNA extraction.

Figure 1. Mulberry leaves collected after 10, 15 and 20 days of stress response at 200 ppm $CuSO_4$.

The total RNA of leaf samples was extracted using a reagent RNAiso Plus (Takara, Shanghai, China). The quality and quantity of total RNA were determined using an Agilent 2100 Bioanalyzer (Agilent Technologies, Palo Alto, CA, USA) and a NanoDrop 1000 spectrophotometer, respectively.

RNA sequence libraries of Cu-treated and untreated leaf samples were prepared at the Novogene biotechnology company in Beijing, China. 1 μg total RNA from each sample was used for the construction of a RNA sequence library using NEBNext®UltraTM RNA Library Prep Kit from Illumina® (NEB, Newburyport Tpke, Rowley, MA, USA) following the manufacturer's specifications and index codes were added to attribute sequences to each sample.

2.2. Transcript Quantification and Differential Expression Analysis

Raw RNA-Seq reads were processed through in-house Perl scripts for trimming adapters, reads containing poly-N as well as low-quality bases from the reads ends. The clean paired-end reads of high quality were aligned to the reference genome for mulberry (https://morus.swu.edu.cn/morusdb/datasets, accessed on 10 November 2017) using Hisat2 v2.0.5 [26]. The mapped reads of each sample were assembled by StringTie (v1.3.3b) (https://github.com/gpertea/stringtie, accessed on 3 August 2022). The FPKM of each gene was calculated based on the read counts mapped to the gene calculated by featureCounts v1.5.0-p3 and the length of the gene. The DESeq2 R package (1.16.1) was used to perform the differential expression analysis between the Cu-treatments and the control groups (three biological replicates per group). p-Values were adjusted using the Benjamini and Hochberg's approach [27] for controlling the false discovery rate in multiple testing and genes with an adjusted p-value < 0.05 and an absolute fold change of ≥ 2 were considered differentially expressed. In order to validate the result, differential expression analysis was performed using the edgeR R package (3.18.1).

2.3. Cloning and Sequence Analysis of the ZIP4 Gene Homolog in Mulberry (MaZIP4)

9 µg of total RNA was used as template for reverse transcription with M-MLV reverse transcriptase (Takara Bio, Beijing, China). The cDNA was used as template for amplification of *MaZIP4* gene using the PCR primers, *MaZIP4*-F: 5′-ATGGCGAATACAAGTTGCCAGAGC-3′ and *MaZIP4*-R: 5′-TCAAGCCCAAATAGCTAATGAAGAC-3′, which were designed by Oligo7 based on the coding sequence of *MaZIP4* gene obtained from transcriptome data prepared above. The purified DNA fragment amplified by PCR was ligated into the pMDTM18-T Vector and amplified by transformation of *E. coli* TOP 10 cells (Takara Bio, Beijing, China). The bacterial solution was sequenced with an automated DNA sequencer (Sangon Biotech, Shanghai, China). DNASTAR was used to integrate the sequence fragments of upstream and downstream to obtain the whole sequence of the cloned *MaZIP4*.

DNAMAN was used to perform the sequences analysis of *MaZIP4*. The online software ExPASy (http://web.expasy.org/protparam/, accessed on 1 July 2020) was used to predict theoretical isoelectric point (pI) and the molecular weight of protein encoded by *MaZIP4*. The molecular modeling of protein encoded by *MaZIP4* was predicted using SWISS-MODEL (http://swissmodel.expasy.org/, accessed on 1 June 2022). Homologous sequences were searched from the NCBI database using BLAST [28]. Alignments of DNA/protein sequences were carried out using BLAST/Protein BLAST and non-rooted phylogenetic tree drawings were carried out using MEGA6.0.

2.4. Expression Analysis of MaZIP4 under Different Degree of Cu Stress

The qRT-PCR method was used to analyze the different expression levels of *MaZIP4* in response to Cu stress [29]. cDNA synthesis used 9 µg of total RNA and M-MLV Reverse Transcriptase (TaKaRa-Bio, China) with oligo (dT)18 primer. The β-actin gene [30] was amplified using the primers: β-actin-F; 5′-AGCAACTGGGATGACATGGAGA-3′ and β-actin-R; 5′-CGACCACTGGCGTAAAGGGA-3′ as the internal reference gene. qRT-PCR was performed on a ABI Quant Studio 6 Flex (Maywood Ave, CA, USA) instrument with gene-specific forward and reverse primers (q*MaZIP*-F:5′-TGCTGCATTATCCTTCCACCA-3′, q*MaZIP*-R: 5′-AAGCAATGGCAGTCCCAA-3′) designed by Oligo7 based on the sequence of the cloned *MaZIP4* gene. 4 µL cDNA was used as the template of qRT-PCR. SYBR Green RT-PCR was performed according to FastStart universal SYBR Green Master Mix Kit (Novoprotein, Nanjing, China) specification in the LightCycler ®96 real-time PCR system (Novoprotein, Nanjing, China). The qRT-PCR conditions were 95 °C for 1 min followed by 35 cycles of 95 °C for 20 s, 57 °C for 20 s and 72 °C for 30 s. PCR specificity was checked by melting curve analysis, and data were calculated using the $2^{-\Delta\Delta Ct}$ method [31]. Standard errors and standard deviations were also calculated simultaneously.

2.5. Functional Analysis of MaZIP4 in Mulberry

2.5.1. Transient Transformation of Mulberry Leaves for *MaZIP4* Repression

According to the obtained sequence of *MaZIP4*, the primers *MaZIP4*-F1 (5′-GGGGTAC CATGGCGAATACAAGTTGCCAGAGC-3′) and *MaZIP4*-R1 (5′-GCTCTAGATCAAGCCC AAATAGCTAATGAAGAC), each including a Kpn I or Xbal I restriction enzyme site at their 5′ end, respectively.) were used for the PCR amplification of *MaZIP4*. The PCR products were then ligated into the pMDTM18-T Vector, which was used to transform E. coli TOP 10 cells. The recombinant plasmid was then isolated and the *MaZIP4* insert isolated after digestion with Kpn I and Xba I and inserted into the similarly digested pTRV2 vector with T4 DNA ligase to construct the pTRV2-*MaZIP4* vector. The pTRV2-*MaZIP4* vector was then transferred into *Agrobacterium tumefaciens* GV3101 using the freeze-thaw method [32]. The primers *MaZIP4*-F1 and *MaZIP4*-R1 were used to PCR amplify the *MaZIP4* insert in the isolated pTRV2-*MaZIP4* vector for sequence verification. pTRV1 and pTRV2 were used as negative controls.

A. tumefaciens transformed with pTRV1, pTRV2 or pTRV2-*MaZIP4* was resuspended in transient transformation buffer (150 mM AS (acetosyringone), 10 mM MES, 10 mM $MgCl_2$). Cultures transformed with pTRV1 and pTRV2-*MaZIP4* were mixed in equal volumes for gene silencing. A similar mixture of pTRV1 and pTRV2 was prepared as a mock control (Young et al.). Each *Agrobacterium* mixture was injected into 25 mulberry seedlings by pressure. A further 25 seedlings were injected with transformation buffer alone as a blank control.

Thirty days after injection, the survival rate of seedlings in each group was greater than 70%, and the survival rates of seedlings in the three groups were 72.0%, 80.0%, and 76.0%, for the positive, mock and blank experimental groups, respectively. To assess the virus multiplication in the injected seedlings, PCR amplification of the tobacco brittle virus capsid protein *CP* gene sequence (Genbank No: Z36974.2) was performed (three replicates). From each treatment group, 16 seedlings with the same growth status were selected, divided into four groups (4 × 4) and cultivated in vermiculite nutrient soil watered with either 100 ppm, 200 ppm, 400 ppm or 800 ppm $CuSO_4$. After 5 days, leaf samples were collected from each treatment (3 × 4 × 4) and immediately frozen in liquid nitrogen and stored at −80 °C for later RNA extraction, Cu concentration determinations and physiological analyses.

2.5.2. Quantification of Gene Expression by qRT-PCR

The qRT-PCR analysis was used to assess the expression level of *MaZIP4* in *MaZIP4*-VIGS plants treated with Cu in different concentrations based on the methods described previously. The relative expression differences of mRNAs were calculated using $2^{-\Delta\Delta Ct}$ method.

2.5.3. Determination of Cu Concentration in Leaves

Leaves from WT, negative control and the *MaZIP4*-VIGS plants under different Cu stresses were washed with deionized water and dried, before digestion in a mixture of concentrated HNO_3:H_2O_2 (3:2) in a microwave oven (Galanz, Shanghai, China). The concentrations of Cu were determined by inductively coupled plasma optical emission spectroscopy (ICP-OES, Avio 560 Max, Syngistix, PerkinElmer, Waltham, MA, USA).

2.5.4. Determination of Physiological and Biochemical Indicators of Cu-Stress

Leaves of seedlings treated with Cu in different concentrations were collected to determine the content of soluble protein, free proline (PRO) and malondialdehyde (MDA), and to measure superoxide dismutase (SOD) and peroxidase (POD) activities as described in Liang et al. [33]. Each measurement was repeated three times.

3. Results

3.1. Gene Expression Analysis and the Cloning of MaZIP4

A total of 54,400,717 and 57,624,783 paired-end reads were obtained after sequencing all the libraries constructed from the Cu-treated and control plants on the Illumia NextSeq 500 platform, respectively (Table 1). After removing the low-quality reads, 53,546,796 and 56,701,537 clean reads were mapped to the reference mulberry genome, corresponding to more than 70% successfully mapped reads (Table 2).

Table 1. Statistical table of sequencing data quality.

Sample Name	Raw Reads	Clean Reads	Clean Bases	Error Rate (%)	Q20 (%)	Q30 (%)	GC Content (%)
Control	57,624,783	56,701,537	8.51G	0.03	96.76	91.62	45.54
Cu stress	54,400,717	53,546,796	8.04G	0.03	96.84	91.78	45.96

Table 2. Statistical table of the comparison between reads and reference genome.

Sample Name	Total Reads	Total Mapped	Multiple Mapped	Uniquely Mapped	Reads Map to '+'	Reads Map to '−'
Control	56,701,537	40,105,596 (70.73%)	1,528,234 (2.69%)	38,577,363 (68.04%)	19,238,895 (33.93%)	19,338,468 (34.11%)
Cu stress	53,546,796	39,001,494 (72.84%)	1,439,133 (2.69%)	37,562,362 (70.15%)	18,764,150 (35.05%)	18,798,212 (35.11%)

The comparison of the transcriptomes of Cu-treated and control plants indicated 5486 differentially expressed transcripts, 3078 of which were up-regulated and 2408 transcripts were down-regulated (Figure S2, Table S1. According to the transcriptome sequencing results, the expression of Zinc transporter 4 (*ZIP4*) gene was significantly upregulated after the Cu-treatment relative to the control group.

The sequence of cloned *ZIP4*-like fragment, named *MaZIP4*, was 1405 bp in length with a full ORF (open reading frame) of 1254 bp (Figures S1 and S3), which was predicted to encode a 417-amino acid protein with a weight of 44.065 KD, the isoelectric point was 6.46. Protein sequence alignment analysis of Blast hits showed that the majority of high scoring hits were from members of the *ZIP* superfamily (Figure S4). Swiss-model software was used to predict the tertiary structure of the MaZIP4 protein (Figure S5), and the predicted results indicated that its spatial architecture was similar to that of ZIP4 proteins from other plants, suggesting that MaZIP4 may have similar functions. The amino-acid sequence alignment of *MaZIP4* protein show high homology with ZIP4 proteins of *Morus notabilis* (>80%), *Trema orientale*, *Ziziphus jujube* and *Cannabis sativa* (70%). This indicated that, although ZIP4 proteins show a high level of conservation among species, all these *ZIP4* proteins had a significantly long variant region (Figure S6).

To further analyze the evolutionary relationship of ZIP4 members among species, a phylogenetic tree was constructed f their amino-acid sequences in mulberry and 19 other related species. The results revealed that *M. atropurpurea* R. showed a close evolutionary distance with *Morus notabilis* and *Trema orientale*, but was furthest from *Arachis hypogaea* and *Prosopis alba* (Figure 2).

3.2. Expression of MaZIP4 under Cu Stress Treatments

To further investigate the role of *MaZIP4* in Cu stress, the level of *MaZIP4* transcription under different concentrations of Cu stress was measured by qRT-PCR. The relative expression level of *MaZIP4* increased to a maximum at 400 ppm (11.35) and then decreased at 800 ppm (9.03), although maintaining a higher level than that of the control (0 ppm). This indicates that *MaZIP4* may have the ability to transport Cu within a certain range of Cu stress (Figure 3).

3.3. MaZIP4 Expression under Cu Stress after MaZIP4 Silencing

To further verify the effect of *MaZIP4*, the construct pTRV2-*MaZIP4* was created for subsequent VIGS (Figure 4). To assess the proliferation of the virus, PCR amplification of the *CP* coat capsid gene was performed on randomly selected young leaf samples from the experimental group (pTRV2-*MaZIP4*), the negative control group (pTRV2), and the blank control group (WT; injected with transformation buffer alone). The results (Figure 5) indicated successful viral replication in the experimental and negative control groups, but not in the uninfected seedlings.

qRT-PCR analysis was used to assess the expression levels of *MaZIP4* in *MaZIP4*-VIGS plants treated with Cu in different concentrations (Figures 4 and 6). Consistent with that observed in WT plants, the expression of *MaZIP4* in *MaZIP4*-VIGS plants increased up to 400 ppm Cu then decreased at 800 ppm Cu. However, the expression levels of *MaZIP4* were consistently reduced to about 50% of that in WT plants, indicating that *MaZIP4* was successfully knocked down in *MaZIP4*-VIGS plants.

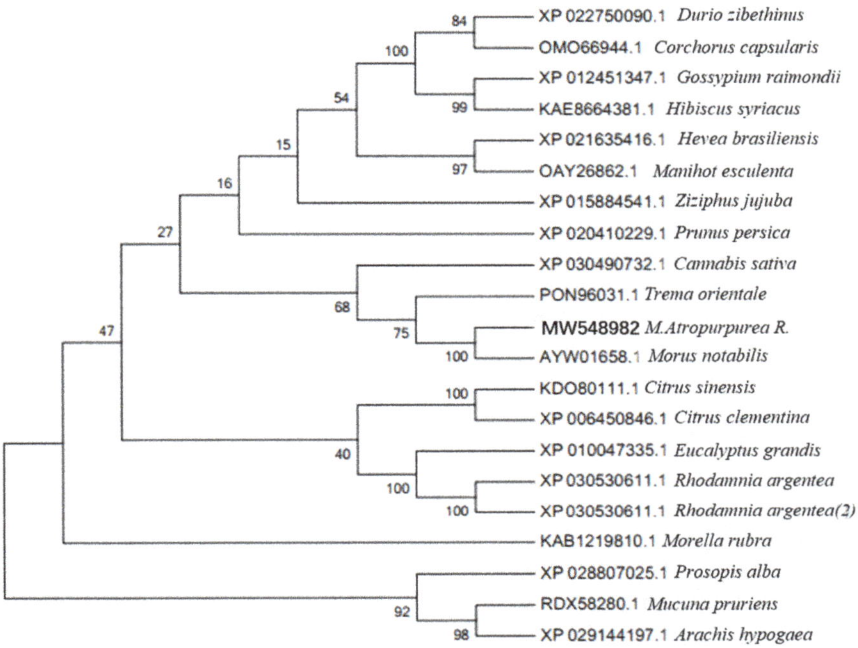

Figure 2. The phylogenetic tree based on the amino acid sequence of the *ZIP4* gene from mulberry (*M. atropurpurea* R.) and other homologous sequences from 19 different species. The protein accessions for the ZIP4 homologs and source species are given to the right of the figure. The evolutionary tree was established in the MEGA 6.0 program using the minimum-evolution test method. The numerals at the branch points indicate bootstrap percentages.

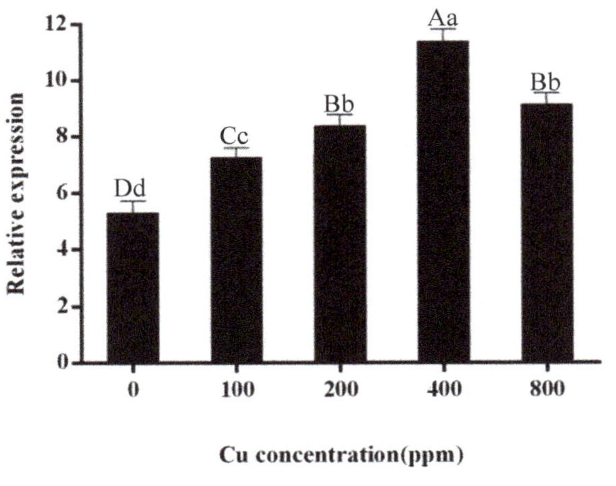

Figure 3. qRT-PCR measurements of the relative expression levels of *MaZIP4* gene in WT under different Cu concentrations. The error bars represent the mean ± SD of three replicates. Bars with different letters indicate a significant difference between expression levels at different concentrations ($p \leq 0.05$) on the basis of Duncan's multiple range test.

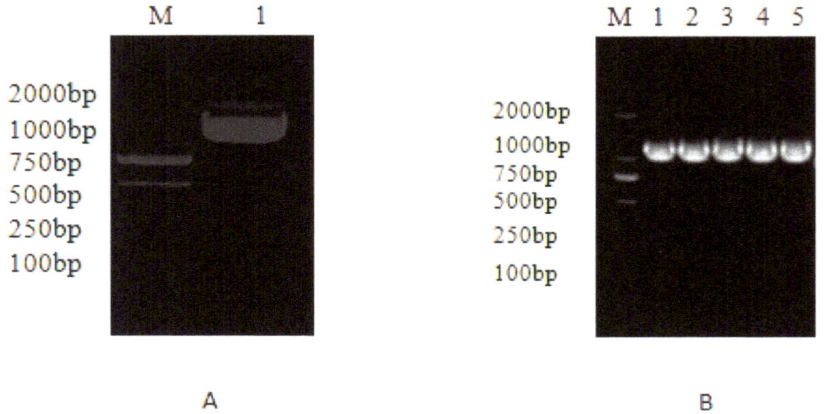

Figure 4. Construction and PCR detection of *MaZIP4* in the recombinant vector pTRV2-*MaZIP4* A: PCR amplification of *MaZIP4*. M: DL2000 bp marker, 1: *MaZIP4* B: PCR amplification of *MaZIP4* from the pTRV2-*MaZIP4* vector, M: molecular weight markers, 1–5: PCR of *MaZIP4*.

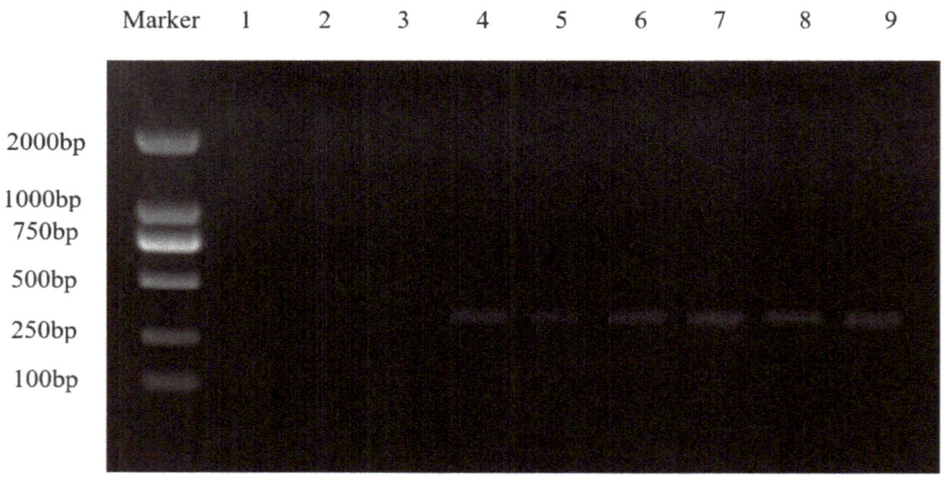

Figure 5. Infected plant virus expression results. Tobacco brittle virus capsid protein CP gene was not expressed in WT lines 1–3, and can be expressed normally in the negative control lines 4–6 and in *MaZIP4*-VIGS seedling lines 7–9. M: molecular weight markers.

3.4. MaZIP4 Silencing Increased Cu Accumulation in Cu Treated Mulberry Leaves

The Cu concentration in leaves was higher in *MaZIP4*-VIGS plants than the blank and negative controls, and increased with the intensity of the Cu stress applied over 0–400 ppm. It is obvious that MaZIP4 gene is likely to be involved in the maintaining the homeostasis of Cu^{2+} in leaves (Figure 7). In addition, with the increase in Cu concentration, mulberry leaves showed higher degrees of yellowing. At 800 ppm Cu, the WT, negative control and *MaZIP4*-VIGS plants all showed obvious leaf yellowing and shrinkage (Figure 8) However, it is clear that relative to the negative and blank controls, the reduced expression of *MaZIP4* in *MaZIP4*-VIGS lines resulted in a significantly greater Cu accumulation in leaves in plants exposed to 400 and 800 ppm Cu, thus supporting a role for *MaZIP4* in Cu homeostasis.

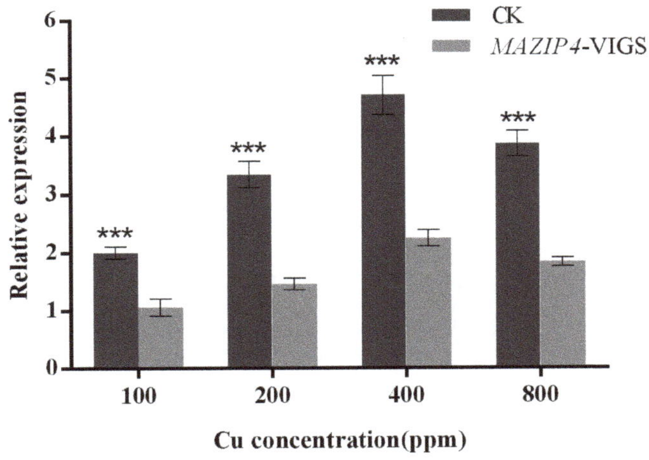

Figure 6. The expression level of *MaZIP4* gene under Cu stress after *MaZIP4* knock-down by VIGS. The error bars represent the mean ± standard deviation (SD) of the three biological replicates. Control means the negative control using the unaltered vector, pTRV2. *MaZIP4*-VIGS refers to the knock-down of *MaZIP4* gene in the experimental group. Asterisks indicate significant differences in expression levels between the control and *MaZIP4*-VIGS (*** $p < 0.001$), based on Duncan's multiple range test.

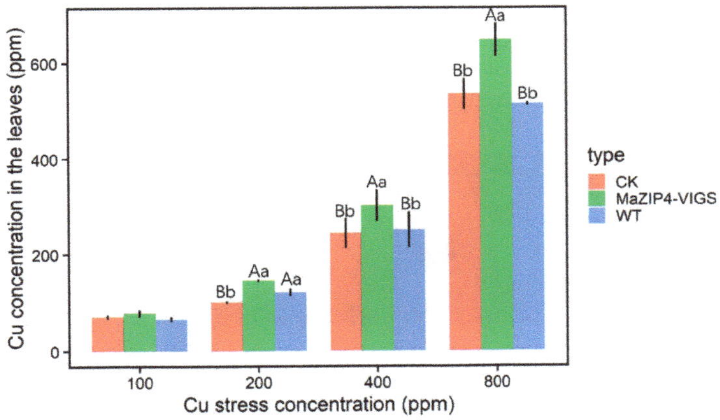

Figure 7. Mean values of Cu concentration in leaves of CK, MaZIP4-VIGS and WT plants under different Cu stress concentrations. Error bars represent the mean ± SD of the three biological replicates. Different lowercase letters above bar represent significant differences ($p < 0.05$, one-way ANOVA). Different capital letters above bars indicate significant differences ($p < 0.001$, one-way ANOVA). The same letter or no letter above bars indicates no significant difference.

3.5. MaZIP4 Silencing Induced Physiological and Biochemical Changes in Mulberry Associated with Cu Stress

Several important physiological and biochemical indicators related to plant stress responses were compared between in the blank control, negative control and *MaZIP4*-VIGS mulberry plants. MDA concentration is commonly used to measure the degree of lipid peroxidation caused by the stress-induced accumulation of reactive oxygen species (ROS) (Figure 9e). The MDA content of *MaZIP4*-VIGS plants, negative control and WT lines gradually increased in all treatment groups with increasing concentrations of Cu stress treatment. Relative to the negative control and the WT, the MDA content in *MaZIP4*-VIGS plants increased rapidly and reached a maximum at 800 ppm Cu (Figure 9e). This indicates that WT levels *MaZIP4* can more effectively inhibit the increase in MDA content in mulberry under Cu stress. The activity of POD and SOD in plants is often considered to be related to the plant stress response. Under abiotic stress, antioxidant enzymes in plants such as POD and SOD promote stress tolerance by reducing stress-induced accumulation of ROS. The POD and SOD activities in WT and the negative control groups were enhanced with increasing concentration of Cu up to 400 ppm and showed a reduced increase at 800 ppm. However, the activity of both enzymes in the *MaZIP4*-VIGS transgenic plants was lower than that of WT and negative control group under the 4 different Cu treatment concentrations tested, and reached their maximum at 400 ppm Cu (Figure 9a,d). These results suggest that the mulberry response to Cu stress involves an increase in POD and SOD activities to help reduce ROS levels. The reduced mobilization of this response in *MaZIP4*-VIGS lines also suggests that *MaZIP4* expression is required for the positive regulation of this response. PRO is a common osmolyte which widely exists in tissues of plants and accumulates in response to abiotic stress (heat, cold, salinity, drought) and is often positively correlated with plant resistance. In WT, negative control lines and *MaZIP4*-VIGS plants, the PRO content increased with increasing of Cu concentration and reached a maximum at 800 ppm. Relative to the WT and negative control, the PRO content in *MaZIP4*-VIGS plants was the lowest, and showed no correlation with the Cu concentration (Figure 9c). Similar to that observed for POD and SOD activities, these results also indicate that PRO accumulation reflects Cu-stress and that PRO accumulation is positively regulated under such conditions by *MaZIP4*. The content of soluble protein accumulation in leaves in WT, negative control lines and *MaZIP4*-VIGS plants all increased gradually with increasing Cu concentration over the range 0–800 ppm. The soluble protein content of *MaZIP4*-VIGS plants also increased with increasing copper concentration, but in lower levels than that observed in WT and negative control lines at all Cu treatments (Figure 9b). This suggested that soluble proteins may be involved in the copper stress response in mulberry and that *MaZIP4* may be required for this response.

Figure 8. The changes in phenotypes of the blank control (WT), the negative control [2] and *MaZIP4*-VIGS plants under different copper stress concentrations. (**a**) coercive concentration of 100 ppm, (**b**) coercive concentration of 200 ppm, (**c**) coercive concentration of 400 ppm, (**d**) coercive concentration of 800 ppm. And from left to right in one photo are CK, *MaZIP4-VIGS* plants and WT.

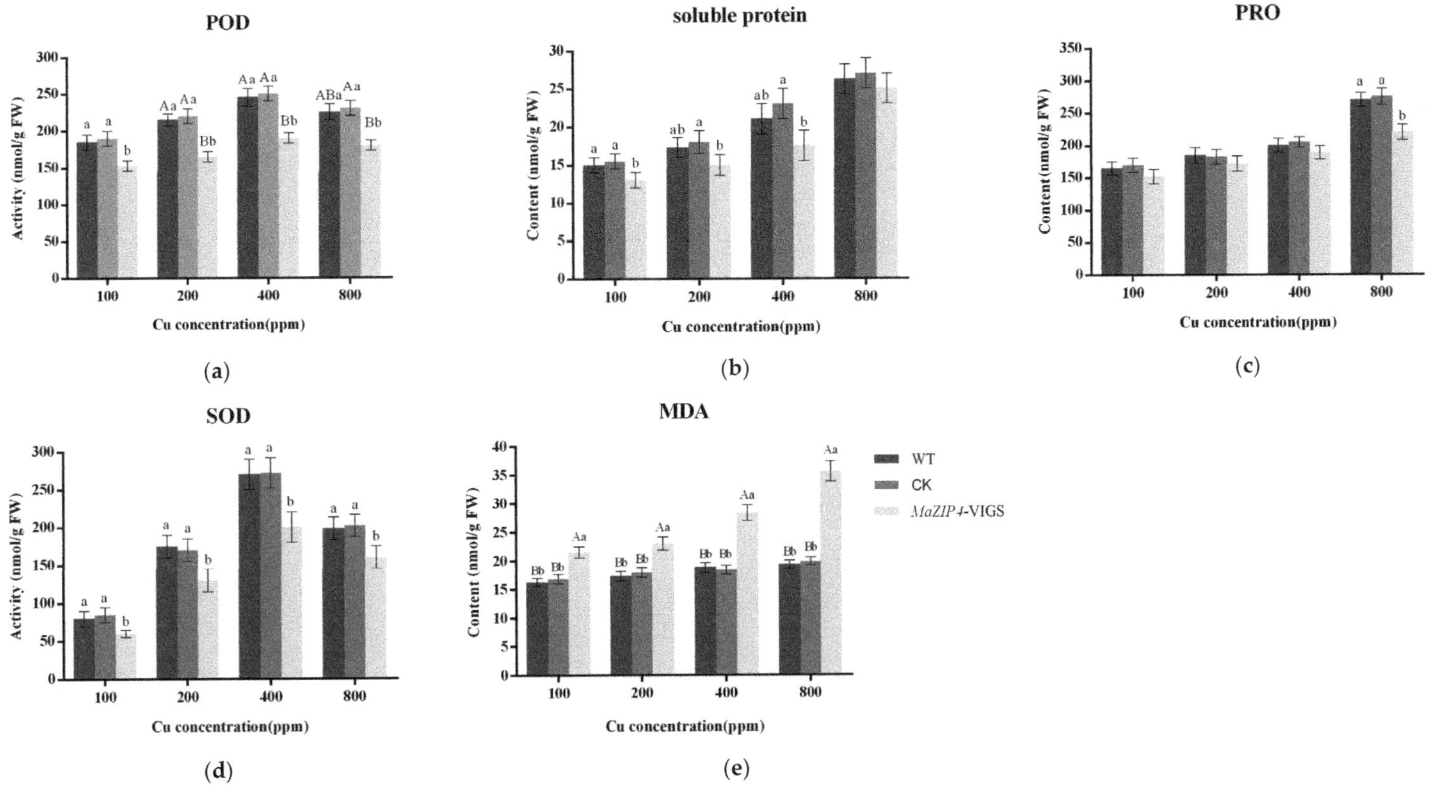

Figure 9. Analysis of contents of soluble protein, PRO, MDA and activities of SOD, POD in mulberry leaves of *MaZIP4*-VIGS lines (light grey bars), negative control (gray bars) and WT (dark grey bars) exposed to various concentrations of Cu. Bars represent mean ± SD ($n = 3$). Different lowercase letters above bar represent significant differences ($p < 0.05$, one-way ANOVA); Different capital letters mean significant differences ($p < 0.001$, one-way ANOVA); The same letter and no letter above bars mean no significant difference. (**a**) POD. (**b**) soluble protein. (**c**) proline. (**d**) SOD. (**e**) MDA.

4. Discussion

The heavy metal Cu is one of essential micronutrients for plant growth and development, but can be toxic in excess. To prevent excess metal toxicity, plants evolved various mechanisms for regulating metal uptake and transport. ZIP family proteins belong to a family of metal transporters involved in the uptake and transportation of Cu, Zn, Ca, Fe and Mn and play vital roles in the maintenance of metal homeostasis in plant tissues [23,34,35]. In our study, a *ZIP4* gene homolog (*MaZIP4*) was identified and cloned for the first time in *M. atropurpurea* R. We further characterized the *MaZIP4* gene to explore its potential function in the mulberry response to Cu stress.

A comparative transcriptome analysis indicated that 3078 transcripts were significantly up-regulated in mulberry in response to treatment with 200 ppm $CuSO_4$, including *MaZIP4*, suggesting The *MaZIP4* gene is 1405 bp in length with a full ORF of 1245 bp and encodes a 417 amino acid protein. Homology and evolutionary analyses indicated that the MaZIP4 showed more than 67% sequence identity with ZIPs of 19 other plant species. The high degree of sequence structural similarities between *MaZIP4* and other ZIP proteins suggests they share similar biological functions. The expression of some zinc-regulated ZIP transporters is also up-regulated under Cu stress, and ZIPs have long been considered responsible for the uptake and allocation of Cu in *Arabidopsis* [14,35]. Therefore, we questioned whether *MaZIP4* might be involved in Cu^{2+} uptake and transport in Mulberry. In support of this, qRT-PCR analyses indicated that with the increase of Cu stress up to 400 ppm, the expression level of *MaZIP4* was significantly enhanced. However, with further increases in Cu (400–800 pm), the expression of *MaZIP4* was decreased, suggesting that the regulatory effect of the gene on Cu stress was limited to a certain range. To further test the function of *MaZIP4*, pTRV2-*MaZIP4* VIGS lines of mulberry (*MaZIP4*-VIGS). qRT-PCR analysis showed the expression level of *MaZIP4* in the transgenic lines was reduced to 50% of the negative control and WT lines under different levels of Cu stress. We next wished to compare physiological and biochemical indicators of the stress response in *MaZIP4*-VIGS, WT and negative control lines subjected to different degrees of Cu stress. When subjected to abiotic stress, plants produce reactive oxygen species (ROS) [36], leading to the peroxidation of membrane lipids which is often assayed indirectly by measuring MDA. MDA is a widely used marker of environmental stress in plants [37]. The content of MDA and the activity of antioxidant enzymes is expected to increase initially and then decrease due to excessive Cu [38–40], which is consistent with our findings. Zhou et al. (2015) studied the changes of physiological and biochemical reactions of forest trees under different levels of Pb stress, and showed that the activities of superoxide dismutase (SOD), peroxidase (POD) and MDA contents in trees were significantly increased under Pb stress [41]. SODs catalyze the dismutation of superoxide into molecular oxygen and hydrogen peroxide which is subsequently catalyzed to H_2O by peroxidase enzymes (POD), and are often used as an indicator of abiotic stress tolerance [42]. The upregulation of SODs and POD has been reported in many plant species including mulberry under abiotic stresses, including drought, salt, and heavy metals [43–48]. The content of PRO is often considered a biomarker of stress because it accumulates under adverse environmental conditions. The accumulation of soluble protein is generally considered a signal of metabolic disorder but not simply a response to stress. Although the change of physiological and biochemical state is not the only way for plants to cope with the environment stress, plants with higher SOD, soluble protein, PRO, POD and lower MDA usually display higher stress resistance [49,50]. Our results showed that the activities of SOD, POD and MDA, soluble protein, PRO contents in mulberry were significantly increased under Cu stress, indicating that Cu stress may stimulate the mobilization of the stress response in mulberry to reduce heavy metal damages. However, while the response to Cu stress in transient transgenic lines for the reduced expression of *MaZIP4* showed higher values for MDA content, these lines also showed lower activities of POD and SOD activities, as well as PRO and soluble protein contents relative to WT and negative control lines. This indicates that the reduced expression of *MaZIP4* in response to Cu stress occurred with the repression of aspects of the plant response to stress. These results indicate that *MaZIP4* may directly contribute to mulberry resistance to Cu stress through its role as a Cu transporter, but that it may also indirectly contribute to the response to Cu stress via a regulatory role in the mobilization of the general stress response. However, further studies are required to elucidate the underlying mechanisms involved in the mitigating role of *MaZIP4* in mulberry under Cu stress.

5. Conclusions

Transcriptome sequencing yielded a new *ZIP4* gene (*MaZIP4*), which was isolated and cloned from *M. atropurpurea* Roxb. This is also the first time a zinc transporter 4 member has been cloned from *M. atropurpurea* Roxb. Phylogenetic analysis indicated that MaZIP4 shows high homology with ZIP4 proteins of *Morus notabilis*, *Trema orientale*, *Ziziphus jujube* and *Cannabis sativa*. The expression of *MaZIP4* in mulberry was increased under a certain degree of Cu stress, which may indicate that *MaZIP4* gene may be involved with the transport of Cu ions in mulberry to maintain Cu homeostasis. After *MaZIP4* knock-down by VIGS, the expression of *MaZIP4* was reduced to about 50% of WT plants (Young et al., 2010). Physiological analyses of *MaZIP4*-VIGS land negative control lines revealed that the partial repression of *MaZIP4* resulted in an increased leaf content of MDA, whereas the contents of soluble protein and proline, as well as the activities of SOD and POD), were decreased. These results indicated that *MaZIP4* could enhance Cu tolerance through its function as a Cu transporter, but also as a positive regulator of the general stress response in mulberry. Our study provides preliminary evidence that *MaZIP4* may have a critical role in regulating Cu^{2+} homeostasis in mulberry and lays the foundation for future studies into the mechanism underlying the plant response to Cu stress.

Supplementary Materials: The following supporting information can be downloaded at: https://www.mdpi.com/article/10.3390/life12091311/s1, Figure S1: MaZIP4 gene amplification of mulberry: PCR in vitro amplification was performed using the designed primers MaZIP4-F and MaZIP4-R, and gene product band of about 1200 bp in length was obtained. M: DL2000 DNA molecular marker 1: Product of RT-PCR; Figure S2. Volcano map of differential genes; Figure S3. *MaZIP4* gene cDNA sequence and translated amino acid sequence; Figure S4. Prediction of amino acid sequence encoded by *MaZIP4* gene; Figure S5. *MaZIP4* gene protein tertiary structure; Figure S6. *MaZIP4* encode multiple sequence alignments of amino acids; Table S1: Screening of differential genes and their corresponding primers.

Author Contributions: Y.S. conceived and designed the project, assembled the genomes, analyzed the data and wrote the original manuscript; Q.Z. designed the experiments, analyzed portions of the data and revised and edited the manuscript; L.W., M.W., D.Z., P.G., Q.D. and M.A. collected leaf samples and extracted chloroplast DNA. W.Z. supervised the experiment and edited the manuscript. Y.S., Q.Z. and L.W. contributed equally to this work. All authors contributed to the editing of the final manuscript. All authors have read and agreed to the published version of the manuscript.

Funding: This work was supported by the earmarked fund for CARS-18, Guangxi innovation-driven development project (AA19182012-2), National Key R&D Program of China (2021YFE0111100), Zhenjiang Science and Technology support project (GJ2021015).

Data Availability Statement: The data presented in this study are available in supplementary material here.

Conflicts of Interest: The authors declare no conflict of interest.

Abbreviations

TM3: transmembrane domains III; TM4: transmembrane domains IV; VIGS: virus-induced gene silencing; FPKM: The fragments per kilobase of transcript per million mapped reads; ORF: open reading frame; MDA: malondialdehyde; PRO: proline; SOD: superoxide dismutase; POD: peroxidase; PI: isoelectric point; AS: acetosyringone; CK: the negative control inoculated with empty vector pTRV2; WT: the blank control.

References

1. Burkhead, J.; Reynolds, K.; Abdel-Ghany, S.; Cohu, C. Copper homeostasis. *New Phytol.* **2009**, *182*, 799–816. [CrossRef] [PubMed]
2. Yruela, I. Copper in plants. *Braz. J. Plant Physiol.* **2005**, *17*, 145–156. [CrossRef]
3. Marques, D.M.; Veroneze Júnior, V.; da Silva, A.B.; Mantovani, J.R.; Magalhães, P.C.; de Souza, T.C. Copper Toxicity on Photosynthetic Responses and Root Morphology of *Hymenaea courbaril* L. (Caesalpinioideae). *Water Air Soil Pollut.* **2018**, *229*, 138. [CrossRef]
4. Saleem, M.H.; Kamran, M.; Zhou, Y.; Parveen, A.; Rehman, M.; Ahmar, S.; Malik, Z.; Mustafa, A.; Ahmad Anjum, R.M.; Wang, B.; et al. Appraising growth, oxidative stress and copper phytoextraction potential of flax (*Linum usitatissimum* L.) grown in soil differentially spiked with copper. *J. Environ. Manag.* **2020**, *257*, 109994. [CrossRef]
5. Yan, J.; Chia, J.C.; Sheng, H.; Jung, H.I.; Zavodna, T.O.; Zhang, L.; Huang, R.; Jiao, C.; Craft, E.J.; Fei, Z.; et al. Arabidopsis Pollen Fertility Requires the Transcription Factors CITF1 and SPL7 That Regulate Copper Delivery to Anthers and Jasmonic Acid Synthesis. *Plant Cell* **2017**, *29*, 3012–3029. [CrossRef]
6. Wuana, R.A.; Okieimen, F.E. Heavy Metals in Contaminated Soils: A Review of Sources, Chemistry, Risks and Best Available Strategies for Remediation. *ISRN Ecol.* **2011**, *2011*, 402647. [CrossRef]
7. Huang, W.-L.; Wu, F.-L.; Huang, H.-Y.; Huang, W.-T.; Deng, C.-L.; Yang, L.-T.; Huang, Z.-R.; Chen, L.-S. Excess Copper-Induced Alterations of Protein Profiles and Related Physiological Parameters in Citrus Leaves. *Plants* **2020**, *9*, 291. [CrossRef]
8. Juang, K.-W.; Lo, Y.-C.; Chen, T.-H.; Chen, B.-C. Effects of Copper on Root Morphology, Cations Accumulation, and Oxidative Stress of Grapevine Seedlings. *Bull. Environ. Contam. Toxicol.* **2019**, *102*, 873–879. [CrossRef]
9. Bruno, P.; Stanley, L.; Jean-Francois, H.; Kjell, S.J. Copper Trafficking in Plants and Its Implication on Cell Wall Dynamics. *Front. Plant Sci.* **2016**, *7*, 601.
10. Grotz, N.; Guerinot, M.L. Molecular aspects of Cu, Fe and Zn homeostasis in plants. *Biochim. Biophys. Acta* **2006**, *1763*, 595–608. [CrossRef]
11. Moreau, S.; Thomson, R.M.; Kaiser, B.N.; Trevaskis, B.; Guerinot, M.L.; Udvardi, M.K.; Puppo, A.; Day, D.A. GmZIP1 encodes a symbiosis-specific zinc transporter in soybean. *J. Biol. Chem.* **2002**, *277*, 4738–4746. [CrossRef] [PubMed]
12. Mäser, P.; Thomine, S.; Schroeder, J.I.; Ward, J.M.; Hirschi, K.; Sze, H.; Talke, I.N.; Amtmann, A.; Maathuis, F.J.M.; Sanders, D.; et al. Phylogenetic Relationships within Cation Transporter Families of Arabidopsis. *Plant Physiol.* **2001**, *126*, 1646–1667. [CrossRef] [PubMed]
13. Grotz, N.; Fox, T.; Connolly, E.; Park, W.; Guerinot, M.L.; Eide, D. Identification of a family of zinc transporter genes from *Arabidopsis* that respond to zinc deficiency. *Proc. Natl. Acad. Sci. USA* **1998**, *95*, 7220. [CrossRef] [PubMed]
14. Wintz, H.; Fox, T.; Wu, Y.Y.; Feng, V.; Chen, W.; Chang, H.S.; Zhu, T.; Vulpe, C. Expression Profiles of Arabidopsis thaliana in Mineral Deficiencies Reveal Novel Transporters Involved in Metal Homeostasis. *J. Biol. Chem.* **2003**, *278*, 47644–47653. [CrossRef] [PubMed]
15. Guerinot, M.L. The ZIP family of metal transporters. *Biochim. Biophys. Acta (BBA) Biomembr.* **2000**, *1465*, 190–198. [CrossRef]
16. Burleigh, S.H.; Kristensen, B.K.; Bechmann, I.E. A plasma membrane zinc transporter from Medicago truncatula is up-regulated in roots by Zn fertilization, yet down-regulated by arbuscular mycorrhizal colonization. *Plant Mol. Biol.* **2003**, *52*, 1077–1088. [CrossRef]
17. López-Millán, A.-F.; Ellis, D.R.; Grusak, M.A. Identification and Characterization of Several New Members of the ZIP Family of Metal Ion Transporters in Medicago Truncatula. *Plant Mol. Biol.* **2004**, *54*, 583–596. [CrossRef]
18. Jiang, Y.; Huang, R.; Yan, X.; Jia, C.; Jiang, S.; Long, T. Mulberry for environmental protection. *Pak. J. Bot.* **2017**, *49*, 781–788.
19. Guha, A.; Rasineni, G.; Reddy, A. Drought tolerance in mulberry (*Morus* spp.): A physiological approach with insights into growth dynamics and leaf yield production. *Exp. Agric.* **2010**, *46*, 471–488. [CrossRef]
20. Gai, Y.-P.; Yuan, S.-S.; Zhao, Y.-N.; Zhao, H.-N.; Zhang, H.-L.; Ji, X.-L. A Novel LncRNA, MuLnc1, Associated With Environmental Stress in Mulberry (*Morus multicaulis*). *Front. Plant Sci.* **2018**, *9*, 669. [CrossRef]
21. Vijayan, K. Approaches for enhancing salt tolerance in mulberry (Morus L)—A review. *Plant Omics* **2009**, *2*, 41.
22. Tewari, R.K.; Kumar, P.; Sharma, P.N. Antioxidant responses to enhanced generation of superoxide anion radical and hydrogen peroxide in the copper-stressed mulberry plants. *Planta* **2006**, *223*, 1145–1153. [CrossRef] [PubMed]
23. Liu, X.S.; Feng, S.J.; Zhang, B.Q.; Wang, M.Q.; Cao, H.W.; Rono, J.K.; Chen, X.; Yang, Z.M. OsZIP1 functions as a metal efflux transporter limiting excess zinc, copper and cadmium accumulation in rice. *BMC Plant Biol.* **2019**, *19*, 283. [CrossRef] [PubMed]
24. Zhang, C.; Wu, Z.; Li, Y.; Wu, J. Biogenesis, Function, and Applications of Virus-Derived Small RNAs in Plants. *Front. Microbiol.* **2015**, *6*, 1237. [CrossRef] [PubMed]
25. Geuten, K.; Viaene, T.; Vekemans, D.; Kourmpetli, S.; Drea, S. Analysis of developmental control genes using virus-induced gene silencing. *Methods Mol. Biol.* **2013**, *975*, 61–69. [CrossRef]
26. Zhang, Y.; Park, C.; Bennett, C.; Thornton, M.; Kim, D. Rapid and accurate alignment of nucleotide conversion sequencing reads with HISAT-3N. *Genome Res.* **2021**, *31*, 1290–1295. [CrossRef]
27. Benjamini, Y.; Hochberg, Y. Controlling the False Discovery Rate: A Practical and Powerful Approach to Multiple Testing. *J. R. Stat. Soc. Ser. B* **1995**, *57*, 289–300. [CrossRef]
28. Altschul, S.F.; Gish, W.; Miller, W.; Myers, E.W.; Lipman, D.J. Basic local alignment search tool. *J. Mol. Biol.* **1990**, *215*, 403–410. [CrossRef]

29. Naeem, M.; Shahzad, K.; Saqib, S.; Shahzad, A.; Nasrullah; Younas, M.; Afridi, M.I. The Solanum melongena COP1LIKE manipulates fruit ripening and flowering time in tomato (*Solanum lycopersicum*). *Plant Growth Regul.* **2022**, *96*, 369–382. [CrossRef]
30. Jiang, X.; Yao, F.; Li, X.; Jia, B.; Zhong, G.; Zhang, J.; Zou, X.; Hou, L. Molecular cloning, characterization and expression analysis of the protein arginine N-methyltransferase 1 gene (As-PRMT1) from *Artemia sinica*. *Gene* **2015**, *565*, 122–129. [CrossRef]
31. Schmittgen, T.D.; Livak, K.J. Analyzing real-time PCR data by the comparative CT method. *Nat. Protoc.* **2008**, *3*, 1101–1108. [CrossRef] [PubMed]
32. Weigel, D.; Glazebrook, J. Transformation of agrobacterium using the freeze-thaw method. *CSH Protoc.* **2006**, *2006*. [CrossRef]
33. Liang, Q.Y.; Wu, Y.H.; Wang, K.; Bai, Z.Y.; Liu, Q.L.; Pan, Y.Z.; Zhang, L.; Jiang, B.B. Chrysanthemum WRKY gene DgWRKY5 enhances tolerance to salt stress in transgenic chrysanthemum. *Sci. Rep.* **2017**, *7*, 4799. [CrossRef] [PubMed]
34. Ishimaru, Y.; Suzuki, M.; Kobayashi, T.; Takahashi, M.; Nakanishi, H.; Mori, S.; Nishizawa, N.K. OsZIP4, a novel zinc-regulated zinc transporter in rice. *J. Exp. Bot.* **2005**, *56*, 3207–3214. [CrossRef] [PubMed]
35. Kumar, V.; Pandita, S.; Singh Sidhu, G.P.; Sharma, A.; Khanna, K.; Kaur, P.; Bali, A.S.; Setia, R. Copper bioavailability, uptake, toxicity and tolerance in plants: A comprehensive review. *Chemosphere* **2021**, *262*, 127810. [CrossRef]
36. Zabalza, A.; Gálvez, L.; Marino, D.; Royuela, M.; Arrese-Igor, C.; González, E.M. The application of ascorbate or its immediate precursor, galactono-1,4-lactone, does not affect the response of nitrogen-fixing pea nodules to water stress. *J. Plant Physiol.* **2008**, *165*, 805–812. [CrossRef] [PubMed]
37. Fan, W.; Zhang, M.; Zhang, H.; Zhang, P. Improved Tolerance to Various Abiotic Stresses in Transgenic Sweet Potato (*Ipomoea batatas*) Expressing Spinach Betaine Aldehyde Dehydrogenase. *PLoS ONE* **2012**, *7*, e37344.
38. Rombel-Bryzek, A.; Rajfur, M.; Zhuk, O. The Impact of Copper Ions on Oxidative Stress in Garden Cress *Lepidium sativum*. *Ecol. Chem. Eng. S* **2017**, *24*, 627–636. [CrossRef]
39. Lin, M.Z.; Jin, M.F. Soil Cu contamination destroys the photosynthetic systems and hampers the growth of green vegetables. *Photosynthetica* **2018**, *56*, 1336–1345. [CrossRef]
40. Gong, Q.; Wang, L.; Dai, T.; Zhou, J.; Kang, Q.; Chen, H.; Li, K.; Li, Z. Effects of copper on the growth, antioxidant enzymes and photosynthesis of spinach seedlings. *Ecotoxicol. Environ. Saf.* **2019**, *171*, 771–780. [CrossRef]
41. Zhou, F.; Wang, J.; Yang, N. Growth responses, antioxidant enzyme activities and lead accumulation of Sophora japonica and Platycladus orientalis seedlings under Pb and water stress. *Plant Growth Regul.* **2015**, *75*, 383–389. [CrossRef]
42. Gill, S.S.; Anjum, N.A.; Gill, R.; Yadav, S.; Hasanuzzaman, M.; Fujita, M.; Mishra, P.; Sabat, S.C.; Tuteja, N. Superoxide dismutase—Mentor of abiotic stress tolerance in crop plants. *Environ. Sci. Pollut. Res. Int.* **2015**, *22*, 10375–10394. [CrossRef] [PubMed]
43. Gapińska, M.; Skłodowska, M.; Gabara, B. Effect of short- and long-term salinity on the activities of antioxidative enzymes and lipid peroxidation in tomato roots. *Acta Physiol. Plant.* **2007**, *30*, 11. [CrossRef]
44. Mobin, M.; Khan, N.A. Photosynthetic activity, pigment composition and antioxidative response of two mustard (*Brassica juncea*) cultivars differing in photosynthetic capacity subjected to cadmium stress. *J. Plant Physiol.* **2007**, *164*, 601–610. [CrossRef] [PubMed]
45. Li, Y.; Song, Y.; Shi, G.; Wang, J.; Hou, X. Response of antioxidant activity to excess copper in two cultivars of *Brassica campestris* ssp. chinensis Makino. *Acta Physiol. Plant.* **2009**, *31*, 155–162. [CrossRef]
46. Khan, N.A.; Samiullah; Singh, S.; Nazar, R. Activities of Antioxidative Enzymes, Sulphur Assimilation, Photosynthetic Activity and Growth of Wheat (*Triticum aestivum*) Cultivars Differing in Yield Potential Under Cadmium Stress. *J. Agron. Crop Sci.* **2007**, *193*, 435–444. [CrossRef]
47. Kumar, N.; Ebel, R.; Roberts, P. Antioxidant Isozyme Variability in Different Genotypes of Citrus and Kumquat. *J. Crop Improv.* **2011**, *25*, 86–100. [CrossRef]
48. Guo, T.; Zhang, G.; Zhou, M.; Wu, F.; Chen, J. Effects of aluminum and cadmium toxicity on growth and antioxidant enzyme activities of two barley genotypes with different Al resistance. *Plant Soil* **2004**, *258*, 241–248. [CrossRef]
49. Li, R.; Liu, L.; Dominic, K.; Wang, T.; Fan, T.; Hu, F.; Wang, Y.; Zhang, L.; Li, L.; Zhao, W. Mulberry (*Morus alba*) MmSK gene enhances tolerance to drought stress in transgenic mulberry. *Plant Physiol. Biochem.* **2018**, *132*, 603–611. [CrossRef]
50. Qi, W.; Wang, F.; Ma, L.; Qi, Z.; Liu, S.; Chen, C.; Wu, J.; Wang, P.; Yang, C.; Wu, Y.; et al. Physiological and Biochemical Mechanisms and Cytology of Cold Tolerance in Brassica napus. *Front. Plant Sci.* **2020**, *11*, 1241. [CrossRef]

life

Article

Circadian Rhythm Regulates Reactive Oxygen Species Production and Inhibits Al-Induced Programmed Cell Death in Peanut

Aaron Ntambiyukuri [1], Xia Li [1], Dong Xiao [1,2,3,*], Aiqin Wang [1,2,3], Jie Zhan [1,2,3] and Longfei He [1,2,3,*]

1 National Demonstration Center for Experimental Plant Science Education, College of Agriculture, Guangxi University, Nanning 530004, China
2 Guangxi Key Laboratory for Agro-Environment and Agro-Product Safety, Nanning 530004, China
3 Guangxi Colleges and Universities Key Laboratory of Crop Cultivation and Tillage, Nanning 530004, China
* Correspondence: xiaodong@gxu.edu.cn (D.X.); lfhe@gxu.edu.cn (L.H.)

Abstract: Peanut is among the most important oil crops in the world. In the southern part of China, peanut is highly produced; however, the arable land is acidic. In acidic soils, aluminum (Al) inhibits plant growth and development by changing the properties of the cell wall and causing the disorder of the intracellular metabolic process. Circadian rhythm is an internal mechanism that occurs about every 24 h and enables plants to maintain internal biological processes with a daily cycle. To investigate the effect of photoperiod and Al stress on the Al-induced programmed cell death (PCD), two peanut varieties were treated with 100 μM $AlCl_3$ under three photoperiodic conditions (8/16, SD; 12/12, ND; 16/8 h, LD). The results show that Al toxicity was higher in ZH2 than in 99-1507 and higher under LD than under SD. Root length decreased by 30, 37.5, and 50% in ZH2 and decreased by 26.08, 34.78, and 47.82% in 99-1507 under SD, ND, and LD, respectively, under Al stress. Photoperiod and Al induced cell death and ROS production. MDA content, PME activity, and LOX activity increased under SD, ND, and LD, respectively, under Al stress both in ZH2 and 99-1507. APX, SOD, CAT, and POD activities were higher under SD, ND, and LD, respectively. Al stress increased the level of *AhLHY* expression under SD and ND but decreased it under LD in both ZH2 and 99-1507. Contrastingly, *AhSTS* expression levels increased exponentially and were higher under SD, LD, and ND, respectively, under Al stress. Our results will be a useful platform to research PCD induced by Al and gain new insights into the genetic manipulation of the circadian clock for plant stress response.

Keywords: circadian clock; reactive oxygen species; Al-induced PCD; photoperiodism; peanut

Citation: Ntambiyukuri, A.; Li, X.; Xiao, D.; Wang, A.; Zhan, J.; He, L. Circadian Rhythm Regulates Reactive Oxygen Species Production and Inhibits Al-Induced Programmed Cell Death in Peanut. *Life* **2022**, *12*, 1271. https://doi.org/10.3390/life12081271

Academic Editors: Hakim Manghwar and Wajid Zaman

Received: 2 August 2022
Accepted: 17 August 2022
Published: 19 August 2022

Publisher's Note: MDPI stays neutral with regard to jurisdictional claims in published maps and institutional affiliations.

Copyright: © 2022 by the authors. Licensee MDPI, Basel, Switzerland. This article is an open access article distributed under the terms and conditions of the Creative Commons Attribution (CC BY) license (https://creativecommons.org/licenses/by/4.0/).

1. Introduction

It has been shown that about 50% of the total arable land around the world is acidic [1]. Al can be concentrated in the plant root tips and interfere with areas of plant growth, such as root and shoot growth, decrease biomass production and nutrient imbalance, and alter physiological and metabolic parameters, which will lead to a decrease in crop yield [2]. The presence of Al in the environment affects plants in various ways, including the generation of ROS [3]. Numerous reports have revealed mechanisms such as Al exclusion from the roots and evacuation to the vacuole in response to Al toxicity in rice [4,5], wheat, barley, maize [6,7], and rye and Arabidopsis [8]. QTLs related to Al tolerance have been identified and used to develop Al-tolerant crops such as maize [9,10], Arabidopsis [11], wheat [12], and rice [13] using breeding or molecular approaches. Genes such as MATE, ALMT, ASR, and ABC transporters have been implicated in some plants for resistance to Al [14], for instance, Al-responsive genes in potato [15], rice [16], wheat [17], sorghum [18], rye [19], and sugarcane [20]. Various transcription factors induced by Al stress, such as STOP1 [21,22], STOP2 [23], ART1, ASR5, and STAR1 [24], regulating the other Al-responsive genes to confer Al tolerance, have been reported. There are less reports on the influence of photoperiod on Al stress tolerance in plants; therefore, we need to understand how day length affects Al stress resistance in peanut.

Circadian clock is an intrinsic timekeeping mechanism that synchronizes with the periodic environment through daily entrainment, especially light and temperature to adjust the internal rhythm [25–27]. The circadian clock anticipates daily environmental fluctuations by coordinating diverse physiological and developmental processes in a day-specific manner to enhance plant fitness and survival [28–30]. The circadian clock plays an important role under adverse environmental conditions, modulating biotic and abiotic stress responses [31,32]. However, many aspects, such as circadian rhythm and its role in Al stress, remain unclear.

Programmed cell death (PCD) or apoptosis is a molecular process in which cells that are not needed commit suicide by activating an intracellular death program [33]. Numerous reports suggested that one of the mechanisms of PCD is the elimination of specific cells under developmental or environmental stimuli such as an increase in ROS in the presence of abiotic stress [34–36]. Photoperiod length is detected by a sensing mechanism consisting of chloroplasts and photoreceptors, which transfer the light information to the circadian clock. This photoperiod sensing influences the development of plants, induces abiotic and biotic stress tolerance, and causes photoperiod stress [37]. It has been revealed that short-day entrained plants were more affected than long-day entrained plants, and the minimal light treatment of 12 h could be necessary for inducing PCD [38]. The circadian clock contributes to cellular processes that maintain ROS at physiological levels in diverse organisms such as mouse [39], zebrafish [40], and *Arabidopsis thaliana* [41]. Plants under the long-day period exhibit reduced CCA1/LHY expression and the induction of PCD. Contrastingly, there was no PCD observed after short nights, so the expression of CCA1/LHY was similar to wild-type levels or higher [42], and it has been reported that CCA1 is a key regulator of ROS homeostasis through association with the evening component (EC) in ROS promoter genes [43]. The proper matching of internal circadian timing with environment enhances plant fitness and survival [27]. Circadian stress (perturbation) regimes have detrimental consequences that lead to the failure of ROS removal and induced PCD. However, the regulatory role of the circadian clock and day length on ROS production and the regulation of Al-induced PCD remains unclear.

Antioxidant defense mechanisms keep the formed ROS at a low level [44–47]. CAT is the first peroxisomal antioxidant enzyme that detoxifies cellular H_2O_2 to be characterized [48], and its expression in Arabidopsis is controlled by the circadian clock [41]. It has been revealed that APX and CAT enzymes are involved in the elimination of H_2O_2 and ROS scavenging in the roots of rice [49]. There is no available report on the exact mechanism of the circadian clock regulating Al-induced PCD by regulating the antioxidant system under photoperiod and Al stress. To further understand the contribution of the circadian clock in the regulation of Al-induced PCD, we investigated the effect of different photoperiods and Al stress on root growth, membrane lipid peroxidation, ROS production, enzyme activities, and gene expression in peanuts.

2. Results

2.1. Effects of Photoperiod and Al Stress on Root Elongation and Cell Death

Root growth and cell death were influenced by the interaction of photoperiod and Al stress. Maximum root growth was observed in 99-1507 without Al treatment under SD (short day, 12/12 h) (Figure 1A). Compared to ND, root growth increased by 15% but decreased by 14.5% in ZH2 and increased by 4.34% but decreased by 19.56% in 99-1507 under SD (short day, 8/16 h) and LD (long day, 16/8 h), respectively, without Al treatment. When treated with Al, compared to ND with no Al, root length decreased by 30, 37.5, and 50% in ZH2 and decreased by 26.08, 34.78, and 47.82% in 99-1507 under SD, ND, and LD. Al highly inhibited root growth under LD compared to SD and ND in ZH2 and 99-1507 (Figure 1A). Photoperiod also significantly influenced Al-induced cell death in the root tips of peanut (Figure 1B). Dead cells decreased by 13.79% but increased by 20.68% in ZH2 and decreased by 37.5% but increased by 10.41% in 99-1507 under SD and LD, respectively, without Al treatment. Compared to ND with no Al, death cells increased by 68.96, 79.3, and 96.55% in ZH2 and increased by 30, 46.25, and 62.5% in 99-1507 under SD, ND, and LD, respectively, under Al stress. The inhibition of root growth and cell death induced by Al stress were higher under LD than ND and SD (Figure 1A,B).

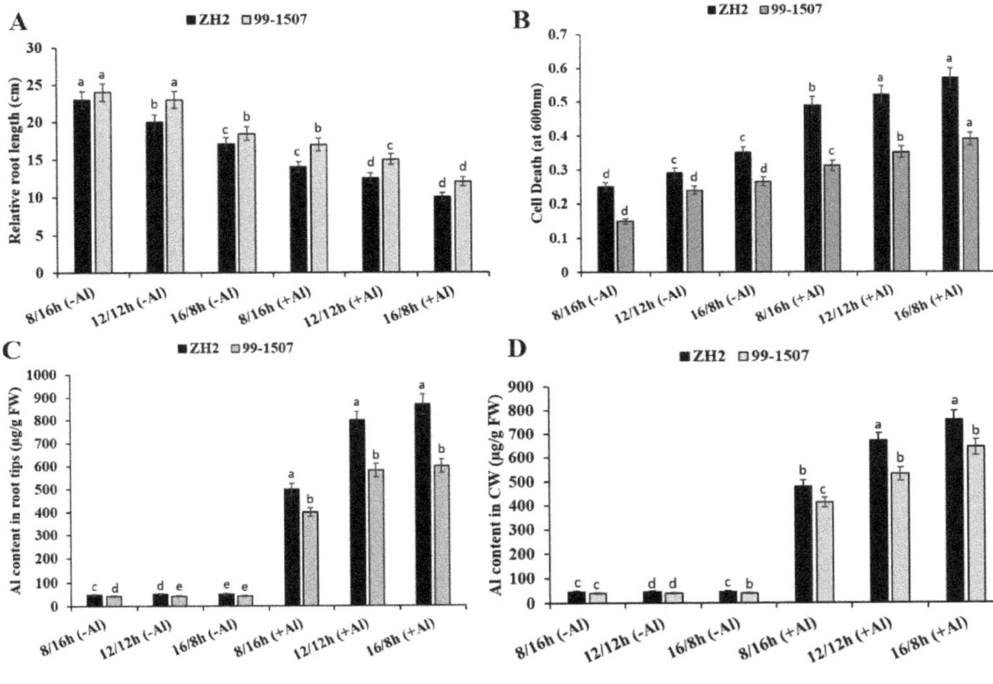

Figure 1. Effect of photoperiodism and Al stress on the root tips in peanut. (**A**) Relative root length (cm); (**B**) Root tip cell death; (**C**) Al content in the root tips; (**D**) Al content in the cell wall. Peanut cultivars (Zh2 and 99-1507) were treated with 100 μM Al, different photoperiods: SD (short day, 8/16 h), ND (normal day, 12/12 h), and LD (long day, 16/8 h). The experiment was carried out in triplicate. The data are presented as means. All the samples were used as fresh weight. Different letters (a–d) assigned to the error bar represent different levels of significance. The results were significant at $p < 0.05$ with different letters.

2.2. *Effects of Photoperiod and Al Stress on Al Accumulation in the Root Tips and CW*

The treatment of ZH2 and 99-1507 under ND, SD, and LD periods with or without Al stress showed that Al content in root tips was highly observed in ZH2 and 99-1507 under LD and SD compared to ND (Figure 1C). However, there was no significant change in Al content in the root tips of ZH2 and 99-1507 under SD, ND, and LD, respectively, without Al treatment (Figure 1C). However, after Al treatment, Al content in the root tips increased exponentially by 10, 16, and 17.4 times in ZH2 under SD, ND, and LD, respectively, and increased by 9.52, 13.8, and 14.28 times in 99-1507 under SD, ND, and LD, respectively. There was no significant change in the Al content observed in CW without Al treatment (Figure 1D). However, with Al treatment, compared to ND with no Al, Al content was 10.66, 14.88, and 16.88 times higher in ZH2 under SD, ND, and LD, respectively, and 10.3, 13.25, and 16.075 times higher in 99-1507 under SD, ND, and LD, respectively. There was a very big difference between the influence of photoperiod alone and photoperiod with Al stress on the content of Al in the root tips and CW, and this was highly observed in ZH2 and 99-1507 under LD and SD when compared to the ND (Figure 1C,D).

2.3. Effects of Photoperiod and Al Stress on MDA Content, PME Activity, and LOX Activity

Photoperiod influenced lipid peroxidation and induced MDA content in the root tips of ZH2 and 99-1507. However, this MDA content was highly increased under photoperiod and Al stress (Figure 2A). Compared to ND, without Al treatment, MDA content decreased by 30.79% but increased by 2.17% in ZH2 and decreased by 66.51% but increased by 4.24% in 99-1507 under SD and LD, respectively. After Al treatment, compared to ND with no Al, MDA content increased by 55.58, 72.75, and 111.71% in ZH2 and increased by 17.57, 35.45, and 65.15% in 99-1507 under SD, ND, and LD, respectively. PME activity was influenced by photoperiod and Al stress (Figure 2B). Compared to ND, without Al treatment, PME activity decreased by 31.96% but increased by 4.53% in ZH2 and decreased by 17.57% but increased by 11.86% in 99-1507 under SD and LD, respectively. Compared to ND with no Al, PME activity increased by 15.95, 17.17, and 20.24% in ZH2 and increased by 22.28, 29.89, and 50.36% in 99-1507 under SD, ND, and LD, respectively, under Al stress. From Figure 2C, LOX activity was induced by photoperiod and Al stress. Without Al treatment, LOX activity decreased by 29.2% but increased by 21.63% in ZH2 and decreased by 20.56% but increased by 16.89% in 99-1507. On the other side, LOX activity increased by 24.64, 49.5, and 91.26% in ZH2 and increased by 22.75, 60.74, and 87.5% in 99-1507 under SD, ND, and LD, respectively, under Al stress.

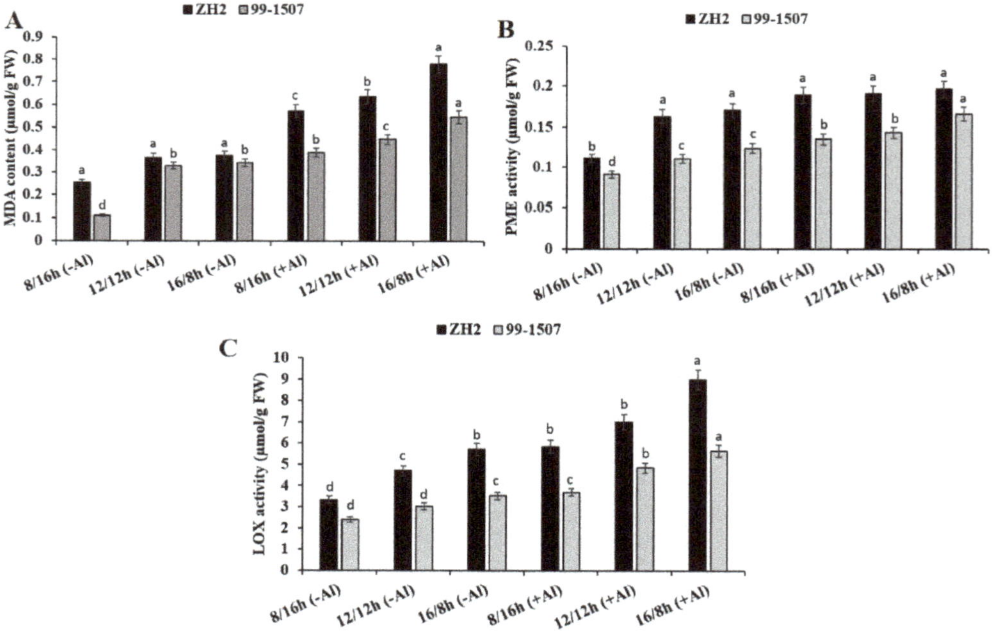

Figure 2. Effect of photoperiodism and Al on MDA content, PME, and LOX activities. (**A**) MDA content; (**B**) PME activity; and (**C**) LOX activity. The root tips of ZH2 and 99-1507 were treated with 100 μM Al under different light/dark periods for 24 h. The experiment was carried out in triplicate to ensure significant results. The data are presented in mean ± standard deviation. All the samples were used as fresh weight. Different letters (a–d) assigned to the error bar represent different levels of significance. The results were significant at $p < 0.05$ with different letters.

2.4. Effects of Photoperiod and Al on ROS Production and Antioxidant Enzyme Activities

To detect the levels of ROS production in peanut, H_2O_2 and $O_2^{\cdot-}$ content was analyzed (Figure 3A,B). Root tips were treated under photoperiod alone or the interaction of photoperiod with Al stress. The results show that compared to ND, without Al treatment, the levels of H_2O_2 decreased by 14.3% but increased by 26.16% in ZH2 and decreased by 28.84% but increased by 22.75% in 99-1507 under SD and LD, respectively. With Al treatment, H_2O_2 increased by 31.53, 35.73, and 75.32% in ZH2 and increased by 33.87, 44.15, and 72% in 99-1507 under SD, ND, and LD, respectively, compared to ND with no Al treatment. The result of $O_2^{\cdot-}$ content is shown in the Figure 3B. $O_2^{\cdot-}$ content without Al, compared to ND, decreased by 31.15% but increased by 17.02% in ZH2 and decreased by 14.84% but increased by 10.95% in 99-1507 under SD and LD, respectively. With Al treatment, $O_2^{\cdot-}$ increased by 32.76, 50.1, and 98.07% in ZH2 and increased by 13.73, 56.17, and 78.91% in 99-1507 under SD, ND, and LD, respectively, compared to ND with no Al. There was a significant effect of photoperiod stress alone or together with Al stress on ROS production, but the level was higher in ZH2 than in 99-1507 and higher under LD than under SD (Figure 3A,B).

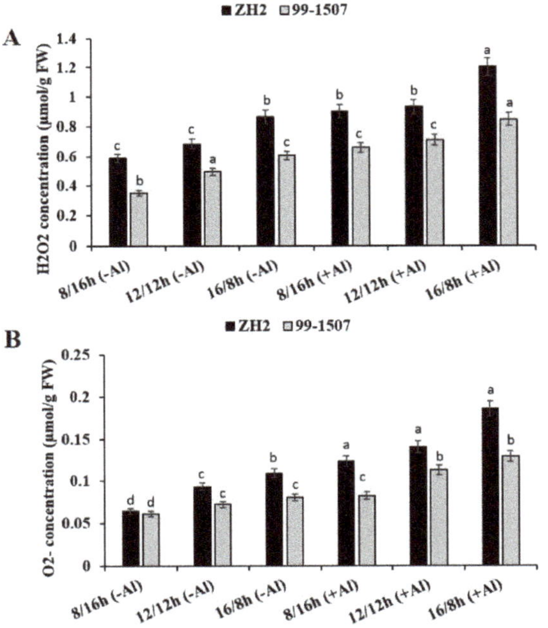

Figure 3. Effect of photoperiodism and Al stress on the production of ROS. (**A**) H_2O_2 production; (**B**) $O_2^{\cdot-}$ production. The experiment was carried out in triplicate to ensure significant results. The data are presented in mean ± standard deviation. Different letters (a–d) assigned to the error bar represent different levels of significance. The results were significant at $p < 0.05$ with different letters.

There was a significant effect of photoperiod or photoperiod interacted with Al stress on antioxidant enzyme activities in ZH2 and 99-1507 cultivars of peanut (Figure 4). APX (Figure 4A), SOD (Figure 4B), CAT (Figure 4C), and POD (Figure 4D) activities were induced under photoperiod or photoperiod–Al stress interaction. Without Al treatment, compared to ND, APX increased by 25.47% but decreased by 13.31% in ZH2 and increased by 6.95% but decreased by 34.8% in 99-1507 under SD and LD, respectively. Compared to ND with no Al, APX increased by 105.91, 61.47, and 61.47% in ZH2 and increased by 50.32, 31.91, and 6.93% in 99-1507 under SD, ND, and LD, respectively, under Al stress.

Compared to ND, without Al stress, SOD activity increased by 0.05% but decreased by 25.8% in ZH2 and increased by 33.13% but decreased by 14.19% in 99-1507 under SD and LD, respectively. However, under Al stress, SOD activity increased by 82.24, 67.63, and 41.09% in ZH2 and increased by 88.48, 85.16, and 56.36% in 99-1507 under SD, ND, and LD, respectively. The highest CAT activity was under SD as opposed to ND or LD (Figure 4C). Without Al treatment, compared to ND, CAT activity increased by 26.83% but decreased by 2.83% in ZH2 and increased by 2.89% but decreased by 8.1% in 99-1507 under SD and LD, respectively. With Al treatment, compared to ND with no Al, CAT activity increased by 76.3, 32.52, and 30.19% in ZH2 and increased by 92.34, 47.82, and 39.76% in 99-1507 under SD, ND, and LD, respectively. We also detected POD activity and found that, without Al treatment, POD activity increased by 21.82% but decreased by 7.86% in ZH2 and increased by 11.15% but decreased by 17.02% in 99-1507, while with Al treatment, POD activity increased by 92.95, 63.1, and 49.92% in ZH2 and increased by 87.79, 73.82, and 63.56% in 99-1507 under SD, ND, and LD, respectively. Compared to photoperiod alone, Al stress significantly increased APX, SOD, CAT, and POD activities (Figure 4).

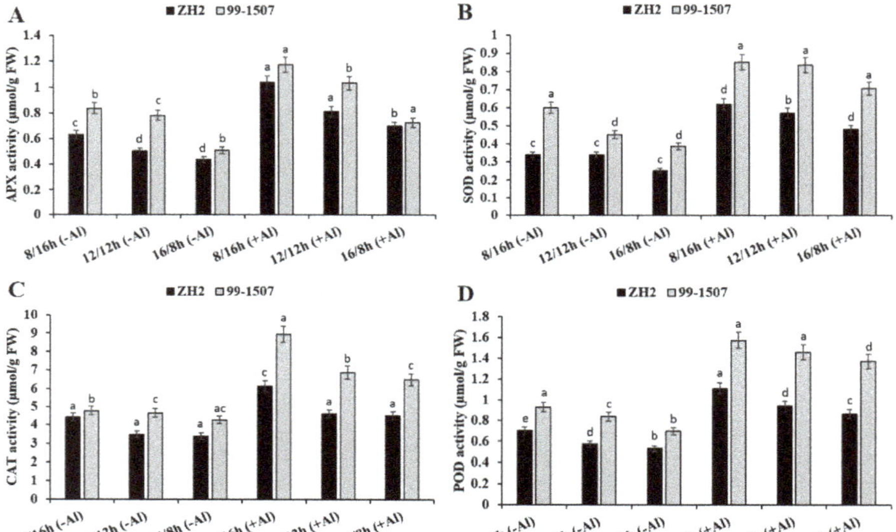

Figure 4. Effect of photoperiodism and Al stress on antioxidant enzyme activities. (**A**) APX activity, (**B**) SOD activity, (**C**) CAT activity, and (**D**) POD activity. The experiment was carried out in triplicate to ensure significant results. The data are presented in mean ± standard deviation. All the samples were used as fresh weight. Different letters (a–d) assigned to the error bar represent different levels of significance. The results were significant at $p < 0.05$ with different letters.

2.5. Effect of Photoperiod on AhLHY and AhSTS Gene Expression under Al Stress

To obtain insight into the potential functional roles of photoperiod in Al stress, two genes, *AhLHY* (AH02G04520.1) and *AhSTS* (AH04G08660.1), were selected based on the transcriptome (PRJ-NA525247) in NCBI to detect their expression levels under Al stress through qRT-PCR (Figure 5). Both the *AhLHY* and *AhSTS* genes were up-regulated under photoperiod alone or interaction of photoperiod with Al treatment in ZH2 and 99-1507 (Figure 5A, B). Photoperiod positively regulated expressed *AhLHY* (Figure 5A), and the expression highly shifted when photoperiod interacted with Al treatment. Without Al treatment, compared to ND, *AhLHY* expression level decreased by 25.92 and 56.7% in ZH2 and decreased by 38.87 and 36.38% in 99-1507 under short day (SD, 8/16 h) and long day

(LD, 16/8 h), respectively. To analyze the level of expression before and after Al treatment, we compared ND without Al to the treatments under Al stress. *AhLHY* expression increased by 9.25 and 51.72% in ZH2 and increased by 62.04 and 10.52% in 99-1507 under ND and SD, respectively. However, it decreased by 30.49% in ZH2 and 23.85% in 99-1507 under LD. The highest expression level was observed in 99-1507 with Al treatment under ND, while the lowest was observed in ZH2 without Al treatment under LD (Figure 5A). *AhSTS* expression under ND was lower than it was under SD or LD, but it was also higher under SD than it was under LD or ND without Al stress (Figure 5B). On the other side, when treated with Al, compared to ND with no Al, *AhSTS* expression increased dramatically by 5.33, 1.88, and 5.16 times in ZH2 and increased by 8.57, 2.97, and 5.16 times in 99-1507 under SD, ND, and LD, respectively. The highest expression level of *AhLHY* expression was observed under ND in 99-1507 and the lowest level was ZH2 under LD. Contrastingly, the highest expression level of *AhSTS* was observed in 99-1507 under SD, and the lowest level was in ZH2 under ND. Compared to photoperiod alone, interaction of photoperiod with Al treatment significantly increased the expression of *AhLHY* and *AhSTS* (Figure 5B).

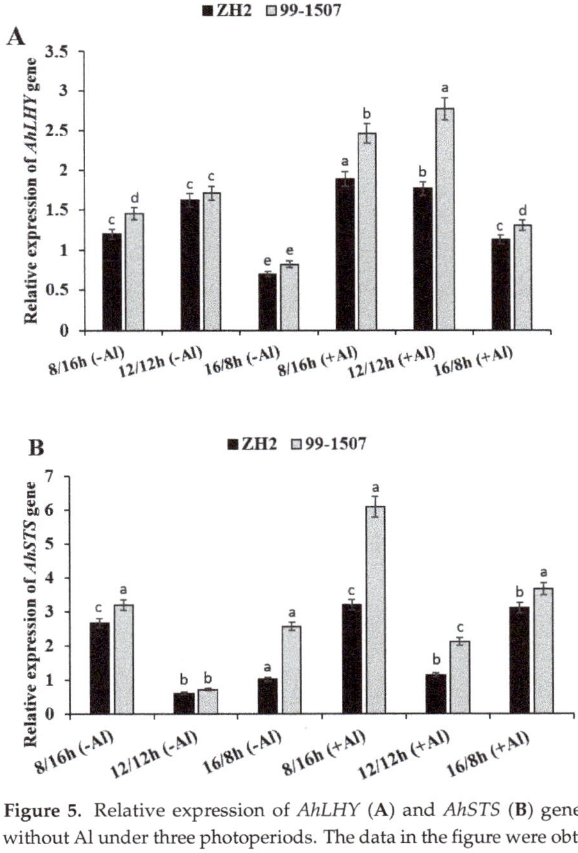

Figure 5. Relative expression of *AhLHY* (**A**) and *AhSTS* (**B**) genes in peanut treated with Al or without Al under three photoperiods. The data in the figure were obtained from three treatments, and analysis of variance (ANOVA) was performed; $p < 0.05$ indicated that the gene differential expression is significant among the three treatments; $n = 3$; Different letters assigned to the error bar represent different levels of significance, error bar is presented as mean ± standard deviation; reference gene was *AhActin*.

3. Discussion

3.1. Effects of Photoperiods on Root Elongation and Cell Death under Al Stress

Environmental changes such as seasonal variation in photoperiod can modulate circadian rhythms, allowing organisms to adjust to the time of the year [50]. Photoperiod stress causes the induction of numerous stress responsive genes, which are indicators of oxidative stress, during the night following the extended light period. Light itself acts as a stressor and, in addition, regulates the outcome of the Al stress response. Our results show that photoperiod influenced root growth. Photoperiod stress alone affected root growth in a way that compared to ND, and root length under SD was higher than under LD (Figure 1A). However, this contradicted the study in radish (long-day plant), which stated that long photoperiod enhanced the formation of radish root [51], but it may be consistent with the study on potato, which revealed that tubers below ground were formed under SD [52]. There was a significant influence of photoperiod and Al stress on root growth; root growth was highly inhibited under LD and SD compared to ND. Numerous reports have discussed the effect of Al stress and revealed that plants exposed to Al stress result in the inhibition of root growth and decrease in crop production [53]. In the present study, we showed that different photoperiods together with Al stress affected root growth and inhibited the growth of root and induced cell death in peanut. Though photoperiodic change regulates various processes in plants, it can also induce numerous stresses [54]. Without Al stress, we observed more dead cells under LD and SD compared to ND (Figure 2). Though there was a significant effect of photoperiod on Al content in root tips and CW under Al stress, there was no significant effect without Al treatment (Figure 1C,D). A recent study revealed that during the night after the prolongation of the light period, stress and cell death marker genes were induced in Arabidopsis [55]. Our results show that root growth was highly inhibited in ZH2 under LD rather than SD, and cell death was also highly induced under LD rather than SD in ZH2. This reveals that peanut is highly sensitive to Al stress under LD.

3.2. Effect of Photoperiod and Al Stress on MDA, PME, and LOX Activity

Photoperiod affected the level of lipid peroxidation; hence, MDA formation was strongly affected by the photoperiodic changes [43]. Our result shows that, compared to ND, both photoperiod alone and photoperiod together with Al stress influenced MDA content in a way that MDA level was higher under LD than under SD (Figure 2A), and this difference was highly observed under Al stress. The highest MDA content was higher in ZH2 than 99-1507. It shows that LD and Al stress lead to membrane lipid peroxidation in peanut, which is more serous in Al-sensitive cultivar and LD. A recent study reported that MDA content was higher under LD in leaves of *Pfaffia glomerata*, leading to higher levels of membrane lipid peroxidation and signaling photooxidative damage [56]. In peanut, increased MDA content is an indicator of membrane lipid peroxidation and abnormal root growth [57].

LOX initiates subsequent biological reactions and activates cellular signaling mechanisms through specific cell surface receptors [58]. LOX activity is involved in catalyzing the formation of H_2O_2 derivatives and activating the lipid peroxidation of membranes. In our study, we found that LOX activity was significantly affected by photoperiod, but this activity level was higher under Al stress (Figure 2C). Compared to ND, LOX was higher under LD than under SD in both ZH2 and 99-1507, but it was also higher in ZH2 than 99-1507. It has been reported that, in potato, LOX activity in morning glory was greatly enhanced and then declined after switching from the light to the dark condition, while the activity did not vary when switching from the dark to the light condition [59].

PME activity reduces pectin methylation [60]. The lower the level of pectin methylation in the cell wall, the greater the Al accumulation was in the cell wall and root tips [61]. In this study, we found that photoperiod influenced PME activity, and its activity level was lower under SD than under LD in both ZH2 and 99-1507 (Figure 2B); this influence was higher under Al stress than photoperiod alone. This increase in PME activity level in the presence of Al stress was as the response to Al stress [57]. In the present study, Al toxicity was higher in ZH2 under LD. This was consistent with the higher Al accumulation measured in ZH2 and LD, demonstrating that the increased PME activity induced by Al and LD accelerated Al toxicity.

Our results consistently suggest that MDA content, PME activity, and LOX activity are directly connected with photoperiod and Al stress.

3.3. Circadian Rhythm Regulates Al-Induced PCD by Controlling ROS Production and Antioxidant Enzyme Activity

Light is an important source of energy and a developmental signal for plants, but it can also cause stress to plants and modulates responses to stress and PCD; excess and fluctuating light result in photoinhibition and ROS accumulation [62]. ROS production plays a critical role in plant development, response to abiotic stresses and immune responses. In the present study, we found that photoperiod could influence ROS production itself but also influence ROS induced by Al stress (Figure 3). Compared to ND with no Al treatment, H_2O_2 and $O_2^{\cdot -}$ production levels were significantly higher under LD than under SD and ND in both ZH2 and 99-1507, but they were greater in ZH2 than in 99-1507 under Al stress (Figure 3A, B). In SD-entrained Arabidopsis, shorter prolongation of the light period causes lower stress levels, is perceived as not harmful and may present a beneficial stress, while higher stress levels by longer prolongations induce a true stress [63]. ROS production in the form of H_2O_2 and $O_2^{\cdot -}$ was significantly affected by photoperiod. However, it was highly affected by photoperiod interacted with Al stress. The lower levels of ROS production in 99-1507 (Al tolerant) under SD rather than LD are evidence that peanut is more resistant to Al stress under SD than under LD (Figure 3A,B). A recent study has revealed that LD species are generally more Al-sensitive than SD species and that the genetic conversion of tomato for the SD growth habit boosts Al tolerance [64].

The day-length sensing mechanisms have been identified to be diverged more between LD plants and SD plants than the circadian clock [65]. In the present study, we found that photoperiod and Al stress induced ROS and activate cellular endogenous antioxidant systems to prevent oxidative stress. The presence of ROS induced various antioxidant enzyme activities, such as APX, SOD, CAT, and POD. There is evidence that ROS play a critical role as the signaling molecules throughout the entire cell death pathway [66]. When we treated ZH2 and 99-1507 peanut cultivars under different photoperiods, compared to ND, APX, SOD, CAT, and POD activities were lower under LD than under SD and antioxidant enzyme activities were higher in 99-1507 than in ZH2 (Figure 4). Further analysis under Al stress revealed that antioxidant enzymes were highly activated. Compared to ND with no Al treatment, APX, SOD, CAT, and POD activities were also higher under SD, ND and LD, respectively, in ZH2 and 99-1507 (Figure 4). In rye, CAT was degraded under light, and the degradation was clearly observed from 16 h after the onset of light [67,68].

3.4. Effect of Photoperiod and Al Stress on AhLHY and AhSTS Gene Expression

LHY/CCA1 regulates photoperiodic flowering, and it has been found that LHY-defective mutants (lhy-7 and lhy-20) exhibit accelerated flowering under both LD and SD [69]. In this study, we observed a remarkable difference between AhLHY gene expression in ZH2 and 99-1507 under ND, SD, and LD either treated with photoperiod alone or together with Al (Figure 5A). Though the influence of photoperiod on AhLHY expression was significant, it was highly significant under the interaction with Al stress. The highest expression level was in 99-1507 under ND and Al stress, but this expression level was also higher under SD than it was under LD (Figure 5A). LHY, CCA1, and TOC1 constitute the core of the circadian clock [69]. We confirmed that the expression pattern of *AhLHY* under photoperiod and Al stress reveals the role of the circadian clock in Al stress control. A previous study revealed that the integration of abiotic stress response into the circadian system provides control over daily plant metabolism [70].

Peanut Stilbene synthase (AhSTS) expression was also influenced by photoperiod, but photoperiod together with Al treatment influenced the expression of *AhSTS* more

(Figure 5B). *AhSTS* was highly expressed in 99-1507 and ZH2 under SD rather than LD and ND, but it was higher in 99-1507 than in ZH2. However, we could not find evidence to prove that this photoperiodic influence is similar to the circadian clock. Therefore, further study is needed to investigate the molecular feature of *AhSTS* under circadian rhythm. Plants may be classified as long-day, short-day, or neutral, and their resistance toward the stress is different. Peanut is a short-day crop [71]. Al toxicity was clearly observed in ZH2 under LD rather than under SD, and it was more tolerant in 99-1507 under SD than under LD. Al induced ROS production, antioxidant enzyme activities, and *AhLHY* and *AhSTS* expression under different photoperiods; this proves the role of photoperiodism in the regulation of Al-induced PCD.

Here, we proved that peanut as a short-day crop is more sensitive to Al stress under LD and more Al-tolerant under SD. Overall, this study reveals that day length plays an important role in determining whether ROS production is enhanced under Al stress. Understanding the physiology of the plant face to Al stress under photoperiod can help to regulate Al stress in plants. However, deeper investigation is needed to truly understand the molecular mechanisms and pathways of circadian clock systems under photoperiod and Al stress in plants.

4. Materials and Methods

4.1. Plant Material and Growth Conditions

Two varieties of peanut, ZH2 (Al-sensitive) and 99-1507 (Al-tolerant), were prescribed as Al-sensitive and Al-resistant, respectively, and were used as plant materials. Plant material and growth condition preparation was conducted following the method of [72] with little modifications. In short, the seeds of peanut were placed into wet perlite sand for 3–4 days at 26 ± 2 °C to induce germination. After germination, the seedlings were 2–3 cm long and were placed in Hoagland nutrient solution for 4 days (d). After 4 d, the seedlings had four leaves and were pretreated with 0.1µM $CaCl_2$ solution at pH 4.2 for 24 h; then, the seedlings were treated with 100 µM $AlCl_3$ for 24 h at three different lighting periods (8/16, 12/12, and 16/8 h light–dark periods). All the treatments were conducted in a controlled environment with 26 ± 2 °C, 70% relative humidity (RH), and light intensity 2000 lux. Al was washed off the root surface prior to the analysis to avoid bias, which may result in the influence of external Al.

4.2. Relative Root Growth, Evan's Blue Staining, and Cell Death Assay

To measure root elongation, Al-treated peanut root tips (main roots) under the circadian rhythms were cut. The data were presented by a histogram chart as relative root elongation. The main root tips were stained with Evan's blue, and the picture was taken by a Canon scanner (DR-S150). For cell death assay, root tips (approximately 1 cm) were stained with 0.5% (*w*/*v*) Evan's blue for 15 min and rinsed in distilled water for 30 min. Stained root tips were immersed in a centrifuge tube containing 4 mL N, N-dimethylformamide for 1 h; then, a solution containing Evan's blue dye, which had leached from dead cells, was measured at 600 nm with spectrophotometer UV/VIS (specord plus 50, Analytik Jena, Konrad-Zuse-Strasse, Germany).

4.3. Al Content in Root Tips and Cell Wall

To assay the total Al content in the root tips, fresh root tips were cut (approximately 1 cm), rinsed in 1 mL of 2 M HCl, and then incubated at 25 °C for 24 h with occasional shaking to ensure that Al was released from the root tips. The upper solution was collected and used to measure the total Al content in root tips. To determine Al content in the CW, root tips were collected, frozen in liquid nitrogen, grounded with mortar and pestle, and then homogenized in 7 mL 75% ethanol and left on ice for 20 min. The homogenized samples were centrifuged at 13,000× *g* for 15 min, and the supernatant was discarded. The pellets were washed with precooled (4 °C) acetone followed by methanol:chloroform (1:1), and then with methanol. After washing, the pellets were dried in an oven for 12 h at 60 °C and suspended in 1 ml 2 M HCl for 24 h at room temperature (RT) with occasional shaking. Al content in the root tips and CW were measured by UV/VIS spectrophotometer (specord plus 50, Germany) at 600 nm.

4.4. Lipid Peroxidation and PME Activity Assay

To analyze the lipid peroxidation (LPO) levels, the malondialdehyde (MDA) content was assayed following the method of [72] with little modifications. In brief, approximately 0.2 g fresh root tips were collected and stored in liquid nitrogen at −80 °C. Thereafter, the root tips were grounded and homogenized in 10 mL 10% trichloroacetic acid (TCA). The homogenate was centrifuged at 13,000× *g* for 10 min. We aliquoted 2 mL of the supernatant mixed with 2 mL 0.6% (*w*/*v*) thiorbarbituric acid in 10% TCA, and incubated in the water bath at 95 °C for 15 min. The ice-cooled mixture was centrifuged at 13,000× *g* for 10 min, and the supernatant was measured at 450, 532, and 600 nm. The MDA content ([C]) was calculated using the following formula:

$$\text{MDA [C] (nM)} = 6.45 \times (\text{A532} - \text{A600}) - 0.56 \times \text{A450}$$

To determine the pectin methylesterase (PME) activity, 0.2 g of the root tips was grounded and homogenized with 2 mL 1 M sodium chloride (NaCl) containing 1% (*w*/*v*) polyvinylpyrrolidone (PVP) then centrifuged at 13,000× *g* for 20 min, 4 °C. The enzyme activity was analyzed by mixing 1 ml 0.5% (*w*/*v*) pectin (pH 7.5), 0.4 mL 0.01% (*w*/*v*) bromothymol blue (pH 7.5), 1.55 mL distilled water (pH 7.5), and 50 μL of the root tip extract; then, the mixture was measured with a spectrophotometer at 620 nm.

4.5. ROS (H_2O_2 and $O_2^{\cdot-}$) Content

The H_2O_2 and $O_2^{\cdot-}$ content in the root tips was detected following the method of [73]. Briefly, the root tips of peanut (0.5 g) were ground with mortar and pestle in liquid nitrogen, then homogenized with 3 mL of cold acetone and centrifuged at 5000× *g* at 4 °C for 10 min. A total of 1 ml of the supernatant was mixed with 0.1 mL of 5% (*w*/*v*) titanic sulfonate ($Ti(SO_4)_2$) and 0.1 mL 25% ammonia, and was centrifuged at 3000× *g* for 10 min at 4 °C; the pellet was resuspended in 4 mL of 2 N sulfuric acid (H2SO4). H_2O_2 content was spectrophotometrically determined at 415 nm.

$O_2^{\cdot-}$ content was measured as described by [73]. Root tips (0.5 g) were homogenized in 2 mL of 65 mM phosphate buffer (pH 7.8), then centrifuged at 5000× *g* for 10 min at 4 °C. The aliquot of 1 mL from the supernatant was mixed with 0.9 mL of 65 mM phosphate buffer (pH 7.8) and 0.1 mL of 10 mM hydroxylamine hydrochloride ($HONH_2{\cdot}HCl$), then the mixture was incubated at 25 °C for 20 min. A total of 1 ml of the mixture was extracted and added to 1 mL of 17 mM anhydrous aminobenzene sulfonic acid ($H_3NC_6H_4SO_3$), as well as 1 mL of 17 mM 1-naphthylamine ($C_{10}H_9N$). The mixture was incubated at 25 °C for 20 min, then 3 ml of n-butanol was added. The $O_2^{\cdot-}$ content was measured at 530 nm.

4.6. Antioxidant Enzyme Activity

To determine the antioxidant enzyme activity, root tips (0.2 g) were ground with mortar and pestle in liquid nitrogen and then homogenized with 50 mM sodium phosphate buffer (pH 7.8) containing 1 mM ethylenediaminetetraacetic acid (EDTA), 2% (*w*/*v*) PVP, 1 mM phenylmethanesulfonyl fluoride (PMSF), 1 mM dithiothreitol (DTT), and 0.05% (*v*/*v*) Triton X-100 and centrifuged the homogenate at 13,000× *g* for 15 min at 4 °C. Lipoxygenase (LOX) activity was assayed following the method of [72]. In total, 50 μL of the supernatant was extracted and mixed with 2.75 mL potassium phosphate buffer (pH 6.5), 0.2 ml 7.5 mM linoleic acid containing 0.25% (*v*/*v*) Tween 20, and then measured at 234 nm. CAT, APX, and SOD activities were assayed following the method of [73]. A total of 0.1 ml of the root tips was extracted with 1.9 ml 0.5 M phosphate buffer pH 7.0, and 1 ml 0.5 mM 30% H_2O_2 solution and CAT activity was spectrophotometrically measured as H_2O_2 decomposition at 240 nm. To detect APX activity, the root tip extract was mixed with 0.25 mM ascorbic acid,

0.5 mM H_2O_2 and measured at 290 nm. SOD activity was determined, as the root tips were homogenized in 1 ml cold 100 mM potassium phosphate buffer pH 7.8 containing 0.1 mM EDTA, 1% PVP, and 0.5% v/v triton x-100, and it was measured at 560 nm, as enzyme amount required us to inhibit the reduction of 50% of NBT at 560 nm.

POD activity was assayed following the "standard operating procedures" of scientific engineering response and analytical services (SERAS, SOP 2035, page: 1–7, date: 11/28/94). Frozen root tips (1 g) were ground using mortar and pestle with liquid nitrogen; then, they were homogenized into ice-cold 0.5 M calcium chloride solution ($CaCl_2$). After the homogenization, the solution was centrifuged at 1000× g for 8 min. The supernatant was extracted into a 10 mL centrifuge tube and kept on ice. The pellet was resuspended with 2.5 mL 0.5 M CaCl2 and centrifuged again (repeated twice). Before POD assay, buffer solutions (A and B) were brought to 25 °C with a water bath. We mixed 1.4 mL of solution A(Phenol and 4-aminoantipyrene), 1.5 mL of solution B (0.5 mL of 30% H_2O_2 and MES or HEPES buffer solution, and the volume was brought to 50 mL to make a 3% H_2O_2 solution with 0.01 M final buffer concentration), and 200 μL roots extract. Then, we measured the absorbance at 510 nm. All activities were measured with a spectrophotometer (Specord 50 plus UV/VIS, Germany).

4.7. Quantitative Real-Time Polymerase Chain Reaction (qRT-PCR) Analysis

Total RNA was extracted using Eastep® Super Total RNA Extraction KitRNeasy LS1040 (Promega, Shanghai, China) following the manufacturer's instructions. The quality of RNA was proved by the ThermoFisher Scientific NanoDrop 2000c, Germany (RNA concentration 1.9–2.2 μg was used). cDNA was synthesized through RNA reverse transcription using PrimeScriptTM RT reagent Kit with gDNA eraser (Perfect Real Time) Cat#RR047A, Takara bio, China. qRT-PCR was performed with CF × 96 TM Real-Time System, Bio-Rad Laboratories, Hong Kong, using SYBR Green qPCR super mix. The actin was used as the normalization control. The following primers were used:

Gene Name	Gene ID	Primer
AhLHY-forward AhLHY-reverse	AH02G04520.1	5′-ATTGACTCTAGTAATCGTCGTA-3′ 5′-CTTTGTGGCAACACCTCT-3′
AhSTS-forward AhSTS-reverse	AH04G08660.1	5′-CCCAAGCGTCAAGAGGTA-3′ 5′-TTGCCCACAAGACTATCCA-3′
Actin-forward	AH03G02610.1	5′-ACCTTCTACAACGAGCTTCGTGTG-3′
Actin-reverse		5′-GAAAGAACAGCCTGAATGGCAAC-3′

4.8. Statistical Analysis

The experiments were independently replicated three times and their mean values were subjected to sata processing and statistical analysis with Excel 2007 and SPSS 12.0 (SPSS software Inc., Chicago, IL, USA). Statistical analysis was performed with Student's paired t-test to test the differences between groups. The data are presented as mean ± standard deviation (SD). p-value ≤ 0.05 was considered statistically significant.

5. Conclusions

Different light–dark conditions showed different levels of Al toxicity, which could reveal the influence of photoperiod in the regulation of Al stress. The induction of ROS by Al stress led to the deterioration of the cell wall, the release of free radicals (H_2O_2 and $O_2^{\cdot -}$), and the acceleration of lipid peroxidation; increased MDA concentration; accelerated PME activity; and positively induced the expression of LHY and STS genes. Prolongation of the light period resulted in photoperiod stress, and it has an important influence on Al stress and tolerance. Peanut showed a strong resistance to Al toxicity under SD, moderate under ND, and low under LD. The molecular mechanism and pathways of *AhLHY* and *AhSTS* in Al tolerance under the circadian rhythms should be studied next, including the metabolic engineering of stilbene biosynthesis as a strategy to directly demonstrate the role of this phytoalexin in plant stress resistance.

Author Contributions: Conceptualization, D.X. and L.H.; Formal analysis, A.N. and X.L.; Funding acquisition, L.H.; Investigation, A.N.; Methodology, A.W., J.Z. and L.H.; Project administration, L.H.; Resources, L.H.; Software, A.N. and X.L.; Supervision, L.H.; Writing—original draft, A.N.; Writing—review & editing, D.X., A.W., J.Z. and L.H. All authors have read and agreed to the published version of the manuscript.

Funding: This work was supported by the National Science Foundation of China (No. 31860334, 31701356, 31776190).

Institutional Review Board Statement: Not applicable.

Informed Consent Statement: Not applicable.

Conflicts of Interest: The authors declare that there is no conflict of interest.

References

1. Panda, S.K.; Baluška, F.; Matsumoto, H. Aluminum stress signaling in plants. *Plant Signal. Behav.* **2009**, *4*, 592–597. [CrossRef] [PubMed]
2. Shetty, R.; Vidya, C.S.-N.; Prakash, N.B.; Lux, A.; Vaculík, M. Aluminum toxicity in plants and its possible mitigation in acid soils by biochar: A review. *Sci. Total Environ.* **2021**, *765*, 142744. [CrossRef] [PubMed]
3. Viehweger, K. How plants cope with heavy metals. *Bot. Stud.* **2014**, *55*, 35. [CrossRef] [PubMed]
4. Yang, Q.; Wang, Y.; Zhang, J.; Shi, W.; Qian, C.; Peng, X. Identification of aluminum-responsive proteins in rice roots by a proteomic approach: Cysteine synthase as a key player in Al response. *PROTEOMICS* **2007**, *7*, 737–749. [CrossRef] [PubMed]
5. Liu, S.; Gao, H.; Wu, X.; Fang, Q.; Chen, L.; Zhao, F.-J.; Huang, C.-F. Isolation and Characterization of an Aluminum-resistant Mutant in Rice. *Rice* **2016**, *9*, 60. [CrossRef] [PubMed]
6. Savić, J.; Stevic, N.; Maksimović, V.; Samardžić, J.; Nikolić, D.; Nikolić, M. Root malate efflux and expression of taalmt1 in serbian winter wheat cultivars differing in Al tolerance. *J. Soil Sci. Plant Nutr.* **2018**, *18*, 90–99. [CrossRef]
7. Szurman-Zubrzycka, M.; Chwiałkowska, K.; Niemira, M.; Kwaśniewski, M.; Nawrot, M.; Gajecka, M.; Larsen, P.B.; Szarejko, I. Aluminum or Low pH—Which Is the Bigger Enemy of Barley? Transcriptome Analysis of Barley Root Meristem Under Al and Low pH Stress. *Front. Genet.* **2021**, *12*, 675260. [CrossRef]
8. Zhou, G.; Delhaize, E.; Zhou, M.; Ryan, P.R. Biotechnological solutions for enhancing the aluminium resistance of crop plants. In *Abiotic Stress in Plants—Mechanisms and Adaptations*; IntechOpen: London, UK, 2011; pp. 119–142.
9. Ninamango-Cárdenas, F.E.; Guimaraes, C.T.; Martins, P.R.; Parentoni, S.N.; Carneiro, N.P.; Lopes, M.A.; Moro, J.R.; Paiva, E. Mapping QTLs for aluminum tolerance in maize. *Euphytica* **2003**, *130*, 223–232. [CrossRef]
10. Mattiello, L.; da Silva, F.R.; Menossi, M. Linking microarray data to QTLs highlights new genes related to Al tolerance in maize. *Plant Sci.* **2012**, *191–192*, 8–15. [CrossRef]
11. Kobayashi, Y.; Koyama, H. QTL analysis of Al tolerance in recombinant inbred lines of *Arabidopsis thaliana*. *Plant Cell Physiol.* **2002**, *43*, 1526–1533. [CrossRef]
12. Farokhzadeh, S.; Fakheri, B.A.; Nezhad, N.M.; Tahmasebi, S.; Mirsoleimani, A. Mapping QTLs of flag leaf morphological and physiological traits related to aluminum tolerance in wheat (*Triticum aestivum* L.). *Physiol. Mol. Biol. Plants* **2019**, *25*, 975–990. [CrossRef] [PubMed]
13. Nguyen, B.D.; Brar, D.S.; Bui, B.C.; Nguyen, T.V.; Pham, L.N.; Nguyen, H.T. Identification and mapping of the QTL for aluminum tolerance introgressed from the new source, ORYZA RUFIPOGON Griff., into indica rice (*Oryza sativa* L.). *Theor. Appl. Genet.* **2003**, *106*, 583–593. [CrossRef]
14. Sade, H.; Meriga, B.; Surapu, V.; Gadi, J.; Sunita, M.S.L.; Suravajhala, P.; Kishor, P.B.K. Toxicity and tolerance of aluminum in plants: Tailoring plants to suit to acid soils. *BioMetals* **2016**, *29*, 187–210. [CrossRef] [PubMed]
15. Li, Y.; He, H.; He, L.-F. Genome-wide analysis of the MATE gene family in potato. *Mol. Biol. Rep.* **2018**, *46*, 403–414. [CrossRef]
16. Augusto, A.R.; Yang, B.; de Oliveira, V.; Felipe, L.; Lauro, B.N.; Schunemann, M.; dos Santos, M.F.; Mariath, J.; Silveiro, A.; Sachetto-Martins, G.; et al. New Insights into Aluminum Tolerance in Rice: The ASR5 Protein Binds the STAR1 Promoter and Other Aluminum-Responsive Genes. *Mol. Plant* **2014**, *7*, 709–721. [CrossRef]
17. Zhaorong, H.; Tian, X.; Wang, F.; Zhang, L.; Xin, M.; Hu, Z.; Yao, Y.; Ni, Z.; Sun, Q.; Peng, H. Characterization of wheat MYB genes responsive to high temperatures. *BMC Plant Biol.* **2017**, *17*, 208. [CrossRef]
18. Caniato, F.F.; Guimarães, C.T.; Schaffert, R.E.; Alves, V.M.C.; Kochian, L.V.; Borém, A.; Klein, P.E.; Magalhaes, J.V. Genetic diversity for aluminum tolerance in sorghum. *Theor. Appl. Genet.* **2007**, *114*, 863–876. [CrossRef]
19. Wakao, S.; Niyogi, K.K. *Chlamydomonas* as a model for reactive oxygen species signaling and thiol redox regulation in the green lineage. *Plant Physiol.* **2021**, *187*, 687–698. [CrossRef]
20. Watt, D.A. Aluminium-responsive genes in sugarcane: Identification and analysis of expression under oxidative stress. *J. Exp. Bot.* **2003**, *54*, 1163–1174. [CrossRef] [PubMed]

21. Zhang, Y.; Zhang, J.; Guo, J.; Zhou, F.; Singh, S.; Xu, X.; Xie, Q.; Yang, Z.; Huang, C.-F. F-box protein RAE1 regulates the stability of the aluminum-resistance transcription factor STOP1 in *Arabidopsis*. *Proc. Natl. Acad. Sci. USA* **2018**, *116*, 319–327. [CrossRef]
22. Fang, Q.; Zhou, F.; Zhang, Y.; Singh, S.; Huang, C.-F. Degradation of STOP1 mediated by the F-box proteins RAH1 and RAE1 balances aluminum resistance and plant growth in *Arabidopsis thaliana*. *Plant J.* **2021**, *106*, 493–506. [CrossRef] [PubMed]
23. Kobayashi, Y.; Ohyama, Y.; Kobayashi, Y.; Ito, H.; Iuchi, S.; Fujita, M.; Zhao, C.-R.; Tanveer, T.; Ganesan, M.; Kobayashi, M.; et al. STOP2 Activates Transcription of Several Genes for Al- and Low pH-Tolerance that Are Regulated by STOP1 in *Arabidopsis*. *Mol. Plant* **2014**, *7*, 311–322. [CrossRef] [PubMed]
24. Huang, D.; Gong, Z.; Chen, X.; Wang, H.; Tan, R.; Mao, Y. Transcriptomic responses to aluminum stress in tea plant leaves. *Sci. Rep.* **2021**, *11*, 5800. [CrossRef] [PubMed]
25. Hardin, P.E.; Panda, S. Circadian timekeeping and output mechanisms in animals. *Curr. Opin. Neurobiol.* **2013**, *23*, 724–731. [CrossRef] [PubMed]
26. Sanchez, A.; Shin, J.; Davis, S.J. Abiotic stress and the plant circadian clock. *Plant Signal. Behav.* **2011**, *6*, 223–231. [CrossRef]
27. Shalit-Kaneh, A.; Kumimoto, R.W.; Filkov, V.; Harmer, S.L. Multiple feedback loops of the Arabidopsis circadian clock provide rhythmic robustness across environmental conditions. *Proc. Natl. Acad. Sci. USA* **2018**, *115*, 7147–7152. [CrossRef]
28. Green, R.M.; Tingay, S.; Wang, Z.-Y.; Tobin, E.M. Circadian Rhythms Confer a Higher Level of Fitness to Arabidopsis Plants. *Plant Physiol.* **2002**, *129*, 576–584. [CrossRef] [PubMed]
29. Li, N.; Zhang, Y.; He, Y.; Wang, Y.; Wang, L. Pseudo Response Regulators Regulate Photoperiodic Hypocotyl Growth by Repressing *PIF4/5* Transcription. *Plant Physiol.* **2020**, *183*, 686–699. [CrossRef] [PubMed]
30. Mase, K.; Tsukagoshi, H. Reactive Oxygen Species Link Gene Regulatory Networks during Arabidopsis Root Development. *Front. Plant Sci.* **2021**, *12*, 660274. [CrossRef]
31. Spoel, S.H.; Van Ooijen, G. Circadian redox signaling in plant immunity and abiotic stress. *Antioxid. Redox Signal.* **2014**, *20*, 3024–3039. [CrossRef]
32. Egrundy, J.; Estoker, C.; Carre, I.A. Circadian regulation of abiotic stress tolerance in plants. *Front. Plant Sci.* **2015**, *6*, 648. [CrossRef]
33. Mishra, A.P.; Salehi, B.; Sharifi-Rad, M.; Pezzani, R.; Kobarfard, F.; Sharifi-Rad, J.; Nigam, M. Programmed Cell Death, from a Cancer Perspective: An Overview. *Mol. Diagn. Ther.* **2018**, *22*, 281–295. [CrossRef] [PubMed]
34. Xiao, D.; Li, X.; Zhou, Y.-Y.; Wei, L.; Keovongkod, C.; He, H.-Y.; Zhan, J.; Wang, A.-Q.; He, L.-F. Transcriptome analysis reveals significant difference in gene expression and pathways between two peanut cultivars under Al stress. *Gene* **2021**, *781*, 145535. [CrossRef] [PubMed]
35. Tong, B.; Shi, Y.; Ntambiyukuri, A.; Li, X.; Zhan, J.; Wang, A.; Xiao, D.; He, L. Integration of Small RNA and Degradome Sequencing Reveals the Regulatory Network of Al-Induced Programmed Cell Death in Peanut. *Int. J. Mol. Sci.* **2021**, *23*, 246. [CrossRef] [PubMed]
36. Das, K.; Roychoudhury, A. Reactive oxygen species (ROS) and response of antioxidants as ROS-scavengers during environmental stress in plants. *Front. Environ. Sci.* **2014**, *2*, 53. [CrossRef]
37. Roeber, V.M.; Schmülling, T.; Cortleven, A. The Photoperiod: Handling and Causing Stress in Plants. *Front. Plant Sci.* **2022**, *12*, 781988. [CrossRef]
38. Nitschke, S.; Cortleven, A.; Iven, T.; Feussner, I.; Havaux, M.; Riefler, M.; Schmülling, T. Circadian Stress Regimes Affect the Circadian Clock and Cause Jasmonic Acid-Dependent Cell Death in Cytokinin-Deficient Arabidopsis Plants. *Plant Cell* **2016**, *28*, 1616–1639. [CrossRef]
39. Wang, T.A.; Yu, Y.V.; Govindaiah, G.; Ye, X.; Artinian, L.; Coleman, T.P.; Sweedler, J.V.; Cox, C.L.; Gillette, M.U. Circadian Rhythm of Redox State Regulates Excitability in Suprachiasmatic Nucleus Neurons. *Science* **2012**, *337*, 839–842. [CrossRef]
40. Hirayama, J.; Cho, S.; Sassone-Corsi, P. Circadian control by the reduction/oxidation pathway: Catalase represses light-dependent clock gene expression in the zebrafish. *Proc. Natl. Acad. Sci. USA* **2007**, *104*, 15747–15752. [CrossRef]
41. Shim, J.S.; Imaizumi, T. Circadian Clock and Photoperiodic Response in *Arabidopsis*: From Seasonal Flowering to Redox Homeostasis. *Biochemistry* **2014**, *54*, 157–170. [CrossRef]
42. Lai, A.G.; Doherty, C.J.; Mueller-Roeber, B.; Kay, S.A.; Schippers, J.H.M.; Dijkwel, P.P. *CIRCADIAN CLOCK-ASSOCIATED 1* regulates ROS homeostasis and oxidative stress responses. *Proc. Natl. Acad. Sci. USA* **2012**, *109*, 17129–17134. [CrossRef] [PubMed]
43. Abuelsoud, W.; Cortleven, A.; Schmülling, T. Photoperiod stress induces an oxidative burst-like response and is associated with increased apoplastic peroxidase and decreased catalase activities. *J. Plant Physiol.* **2020**, *253*, 153252. [CrossRef] [PubMed]
44. Sharma, S.S.; Dietz, K.-J. The relationship between metal toxicity and cellular redox imbalance. *Trends Plant Sci.* **2009**, *14*, 43–50. [CrossRef] [PubMed]
45. Liu, L.-Z.; Gong, Z.-Q.; Zhang, Y.-L.; Li, P.-J. Growth, Cadmium Accumulation and Physiology of Marigold (*Tagetes erecta* L.) as Affected by Arbuscular Mycorrhizal Fungi. *Pedosphere* **2011**, *21*, 319–327. [CrossRef]
46. Liu, Y.-T.; Chen, Z.-S.; Hong, C.-Y. Cadmium-induced physiological response and antioxidant enzyme changes in the novel cadmium accumulator, *Tagetes patula*. *J. Hazard. Mater.* **2011**, *189*, 724–731. [CrossRef]
47. Uraguchi, S.; Watanabe, I.; Yoshitomi, A.; Kiyono, M.; Kuno, K. Characteristics of cadmium accumulation and tolerance in novel Cd-accumulating crops, *Avena strigosa* and *Crotalaria juncea*. *J. Exp. Bot.* **2006**, *57*, 2955–2965. [CrossRef]

48. Mhamdi, A.; Queval, G.; Chaouch, S.; Vanderauwera, S.; Van Breusegem, F.; Noctor, G. Catalase function in plants: A focus on Arabidopsis mutants as stress-mimic models. *J. Exp. Bot.* **2010**, *61*, 4197–4220. [CrossRef]
49. Ahn, H.R.; Kim, Y.-J.; Lim, Y.J.; Duan, S.; Eom, S.H.; Jung, K.-H. Key Genes in the Melatonin Biosynthesis Pathway with Circadian Rhythm Are Associated with Various Abiotic Stresses. *Plants* **2021**, *10*, 129. [CrossRef]
50. Xiang, Y.; Sapir, T.; Rouillard, P.; Ferrand, M.; Jiménez-Gómez, J.M. Interaction between photoperiod and variation in circadian rhythms in tomato. *BMC Plant Biol.* **2022**, *22*, 187. [CrossRef]
51. Guo, R.; Li, W.; Wang, X.; Chen, B.; Huang, Z.; Liu, T.; Chen, X.; XuHan, X.; Lai, Z. Effect of photoperiod on the formation of cherry radish root. *Sci. Hortic.* **2018**, *244*, 193–199. [CrossRef]
52. Osnato, M.; Cota, I.; Nebhnani, P.; Cereijo, U.; Pelaz, S. Photoperiod Control of Plant Growth: Flowering Time Genes beyond Flowering. *Front. Plant Sci.* **2022**, *12*, 805635. [CrossRef] [PubMed]
53. Duressa, D.; Soliman, K.; Chen, D. Identification of Aluminum Responsive Genes in Al-Tolerant Soybean Line PI 416937. *Int. J. Plant Genom.* **2010**, *2010*, 164862. [CrossRef] [PubMed]
54. Riboni, M.; Test, A.R.; Galbiati, M.; Tonelli, C.; Conti, L. Environmental stress and flowering time: The photoperiodic connection. *Plant Signal. Behav.* **2014**, *9*, e29036. [CrossRef] [PubMed]
55. Frank, M.; Cortleven, A.; Novák, O.; Schmülling, T. Root-derived *trans* -zeatin cytokinin protects *Arabidopsis* plants against photoperiod stress. *Plant Cell Environ.* **2020**, *43*, 2637–2649. [CrossRef]
56. Fortini, E.A.; Batista, D.S.; De Castro, K.M.; Silva, T.D.; Felipe, S.H.S.; Correia, L.N.F.; Chagas, K.; Farias, L.M.; Leite, J.P.V.; Otoni, W.C. Photoperiod modulates growth and pigments and 20-hydroxyecdysone accumulation in Brazilian ginseng [*Pfaffia glomerata* (Spreng.) Pedersen] grown in vitro. *Plant Cell Tissue Organ Cult.* **2020**, *142*, 595–611. [CrossRef]
57. He, H.; Oo, T.L.; Huang, W.; He, L.-F.; Gu, M. Nitric oxide acts as an antioxidant and inhibits programmed cell death induced by aluminum in the root tips of peanut (*Arachis hypogaea* L.). *Sci. Rep.* **2019**, *9*, 9516. [CrossRef]
58. Laczko, R.; Csiszar, K. Lysyl Oxidase (LOX): Functional Contributions to Signaling Pathways. *Biomolecules* **2020**, *10*, 1093. [CrossRef]
59. Mashima, R.; Okuyama, T. The role of lipoxygenases in pathophysiology; new insights and future perspectives. *Redox Biol.* **2015**, *6*, 297–310. [CrossRef]
60. Sun, C.; Lu, L.; Yu, Y.; Liu, L.; Hu, Y.; Ye, Y.; Jin, C.; Lin, X. Decreasing methylation of pectin caused by nitric oxide leads to higher aluminium binding in cell walls and greater aluminium sensitivity of wheat roots. *J. Exp. Bot.* **2015**, *67*, 979–989. [CrossRef]
61. Darley, C.P.; Forrester, A.M.; McQueen-Mason, S.J. The molecular basis of plant cell wall extension. *Plant Mol. Biol.* **2001**, *47*, 179–195. [CrossRef]
62. Roeber, V.M.; Bajaj, I.; Rohde, M.; Schmülling, T.; Cortleven, A. Light acts as a stressor and influences abiotic and biotic stress responses in plants. *Plant Cell Environ.* **2020**, *44*, 645–664. [CrossRef] [PubMed]
63. Krasensky-Wrzaczek, J.; Kangasjärvi, J. The role of reactive oxygen species in the integration of temperature and light signals. *J. Exp. Bot.* **2018**, *69*, 3347–3358. [CrossRef] [PubMed]
64. Siqueira, J.A.; Wakin, T.; Batista-Silva, W.; Silva, J.C.F.; Vicente, M.H.; Silva, J.C.; Clarindo, W.R.; Zsögön, A.; Peres, L.E.; De Veylder, L.; et al. A long and stressful day: Photoperiod shapes aluminium tolerance in plants. *J. Hazard. Mater.* **2022**, *432*, 128704. [CrossRef] [PubMed]
65. Song, Y.H.; Ito, S.; Imaizumi, T. Similarities in the circadian clock and photoperiodism in plants. *Curr. Opin. Plant Biol.* **2010**, *13*, 594–603. [CrossRef]
66. Kapoor, D.; Singh, S.; Kumar, V.; Romero, R.; Prasad, R.; Singh, J. Antioxidant enzymes regulation in plants in reference to reactive oxygen species (ROS) and reactive nitrogen species (RNS). *Plant Gene* **2019**, *19*, 100182. [CrossRef]
67. Schmidt, M.; Grief, J.; Feierabend, J. Mode of transcreiptional activation of the catalase (*cat1*) mRNA of rye leaves (*Secale cereale* L.) and its control through blue light and reactive oxygen. *Planta* **2006**, *223*, 835–846. [CrossRef]
68. Hertwig, B.; Streb, P.; Feierabend, J. Light Dependence of Catalase Synthesis and Degradation in Leaves and the Influence of Interfering Stress Conditions. *Plant Physiol.* **1992**, *100*, 1547–1553. [CrossRef]
69. Park, M.-J.; Kwon, Y.-J.; Gil, K.-E.; Park, C.-M. LATE ELONGATED HYPOCOTYL regulates photoperiodic flowering via the circadian clock in *Arabidopsis*. *BMC Plant Biol.* **2016**, *16*, 114. [CrossRef]
70. Markham, K.K.; Greenham, K. Abiotic stress through time. *N. Phytol.* **2021**, *231*, 40–46. [CrossRef]
71. Bagnall, D.; King, R. Response of peanut (*Arachis hypogaea*) to temperature, photoperiod and irradiance 1. Effect on flowering. *Field Crop. Res.* **1991**, *26*, 263–277. [CrossRef]
72. Pan, C.-L.; Yao, S.-C.; Xiong, W.-J.; Luo, S.-Z.; Wang, Y.-L.; Wang, A.-Q.; Xiao, D.; Zhan, J.; He, L.-F. Nitric Oxide Inhibits Al-Induced Programmed Cell Death in Root Tips of Peanut (*Arachis hypogaea* L.) by Affecting Physiological Properties of Antioxidants Systems and Cell Wall. *Front. Physiol.* **2017**, *8*, 1037. [CrossRef] [PubMed]
73. Liu, Y.; Wu, R.; Wan, Q.; Xie, G.; Bi, Y. Glucose-6-Phosphate Dehydrogenase Plays a Pivotal Role in Nitric Oxide-Involved Defense Against Oxidative Stress Under Salt Stress in Red Kidney Bean Roots. *Plant Cell Physiol.* **2007**, *48*, 511–522. [CrossRef] [PubMed]

life

Article

Effect of Smut Infection on the Photosynthetic Physiological Characteristics and Related Defense Enzymes of Sugarcane

Xiupeng Song [1], Fenglian Mo [2], Meixin Yan [1], Xiaoqiu Zhang [1], Baoqing Zhang [1], Xing Huang [1], Dongmei Huang [1], Yangfei Pan [2], Krishan K. Verma [1,*] and Yang-Rui Li [1,2,*]

[1] Sugarcane Research Institute, Guangxi Academy of Agricultural Sciences/Key Laboratory of Sugarcane Biotechnology and Genetic Improvement (Guangxi), Ministry of Agriculture and Rural Affairs/Guangxi Key Laboratory of Sugarcane Genetic Improvement, Nanning 530007, China

[2] College of Agriculture, Guangxi University, Nanning 530004, China

* Correspondence: drvermakishan@gmail.com (K.K.V.); liyr@gxaas.net (Y.-R.L.)

Abstract: Pathogen infection seriously affects plant development and crop productivity, sometimes causing total crop failure. In this study, artificial stab inoculation was used to inoculate sugarcane smut. The changes in leaf gas exchange, chlorophyll fluorescence variables, and related defense enzyme activities were measured in sugarcane cultivar ROC22 after pathogen infection. The results showed that the net photosynthetic rate (Pn), stomatal conductance (gs), and transpiration rate (Tr) downregulated in the first three days after smut infection and upregulated on the fourth day; intercellular CO_2 concentration (Ci) increased in the first three days of smut infection and reduced on the fourth day. The chlorophyll fluorescence parameters, i.e., Fo, Fm, Fv/Fm, Fs, and Fv′/Fm′ decreased at the initial stage of pathogen infection but increased rapidly up to 3 days after smut infection. It can be seen that sugarcane seedlings showed a positive response to pathogen infection. The correlation coefficient relationship between Pn, gs, and Tr reached above 0.800, showing a significant correlation; Ci was positively correlated with Fv′/Fm′ and ΦPSII, reaching above 0.800 and showing a significant correlation; Fo positively correlated with Fv/Fm, Fs, and ETR; Fv /Fm was positively correlated with Fv′/Fm′; Fs significantly correlated with Fv′/Fm′; and Fv′/Fm′ positively correlated with ΦPSII. After inoculation with smut, the related defense enzymes, i.e., POD, SOD, PPO, and PAL, were increased and upregulated; photosynthetic parameters can be associated with an increase in enzymatic activities. The results of this study will help to further study of the response mechanism to smut in the sugarcane growing period and provide a theoretical reference for sugarcane resistance to smut breeding.

Keywords: chlorophyll fluorescence efficiency; photosynthetic responses; enzyme activity; sugarcane; smut

Citation: Song, X.; Mo, F.; Yan, M.; Zhang, X.; Zhang, B.; Huang, X.; Huang, D.; Pan, Y.; Verma, K.K.; Li, Y.-R. Effect of Smut Infection on the Photosynthetic Physiological Characteristics and Related Defense Enzymes of Sugarcane. *Life* **2022**, *12*, 1201. https://doi.org/10.3390/life12081201

Academic Editors: Hakim Manghwar and Wajid Zaman

Received: 8 July 2022
Accepted: 30 July 2022
Published: 8 August 2022

Publisher's Note: MDPI stays neutral with regard to jurisdictional claims in published maps and institutional affiliations.

Copyright: © 2022 by the authors. Licensee MDPI, Basel, Switzerland. This article is an open access article distributed under the terms and conditions of the Creative Commons Attribution (CC BY) license (https://creativecommons.org/licenses/by/4.0/).

1. Introduction

Sugarcane is an important bioenergy crop grown worldwide. Its sugar production accounts for 80% of global sugar production and 92% of China's total sugar production [1,2]. Disease is one of the main factors causing the loss of sugarcane yield and sugar content [3]. A variety of pathogens can infect sugarcane during the growth process. Pathogens can accumulate in the sugarcane germplasm, leading to variability, uneven growth, reduced stem weight, sugar loss, and so on. Sugarcane in different countries has about 150 diseases, of which, smut is the main disease and can cause a 20–50% loss of sugarcane production [4,5]. The most apparent symptom of sugarcane infection in the late stage of smut is the extraction of smut whip at the tail of the cane, but there is no obvious morphological feature in the early stage of infection. Studies have shown that pathogen infection can seriously affect the photosynthetic physiological responses in crops [5,6]. There are various studies on the changes in physiological traits of sugarcane after smut infection [7,8], but there are no reports of the photosynthetic responses of sugarcane seedlings in the early stage of smut infection.

When pathogens infect plants, plant cells produce a series of physiological and biochemical changes to prevent the infection. These changes include early defense reactions after plant susceptibility such as thickening of the corpus callosum, changes in protective enzyme activities, induction and accumulation of disease-related proteins, and hormonal and metabolic disorders [5,9–11]. In these defense reactions, enzymatic activities are most active and closely related to plant disease resistance. Under normal physiological conditions, various enzymes in plants are generally in a dynamic equilibrium state. When pathogens infect plant cells, a specific type of enzyme changes the action of the mechanisms, thereby losing its original equilibrium state and harming the organism [5]. Therefore, it is of great significance to study the resistance of the host to the disease when a particular disease infects it. The enzymes involved in disease resistance include peroxidase (POD), superoxide dismutase (SOD), polyphenol oxidase (PPO), and phenylalanine ammonia-lyase (PAL) [5,12]. After studying the relationship between various physiological and biochemical metabolic reactions of the host and plant disease resistance, it is concluded that SOD, POD, and other related enzymes can resist the damage of reactive oxygen and oxygen free radicals to the cell membrane system, while PAL and PPO can promote the production of various secondary metabolites in plants, thereby preventing the invasion and reproduction of pathogens [13,14].

In the present study, after inoculation with smut pathogen, changes in parameters between sugarcane plant leaves, such as the intercellular CO_2 concentration (Ci), net photosynthetic rate (Pn), transpiration rate (Tr), stomatal conductance (gs), initial fluorescence (Fo), the maximum potential quantum efficiency of photosystem II (Fv/Fm), minimum fluorescence under light (Fo′), non-photochemical quenching coefficient (qNP), maximum light energy conversion efficiency (Fv′/Fm′), steady-state fluorescence (Fs), electron transport rate (ETR), photochemical quenching coefficient (qP), and maximum fluorescence under light (Fm′), were measured. The preliminary analysis of relationships with resistance and the detection of POD, SOD, PPO, and PAL activities after smut infection, the effects of smut infection on photosynthetic physiological changes, and resistance-related enzyme activities and response mechanisms were explored, providing a theoretical basis for further research on the resistance mechanism and disease management of sugarcane to smut.

2. Materials and Methods

2.1. Plant Materials, Growth Conditions, and Treatments

The sugarcane cultivar ROC22 was used in this experiment, the most prevalent variety in China. The experiment was completed in the greenhouse of Guangxi University, Nanning, Guangxi, China. The healthy single bud of the test material was disinfected with hot water treatment and used for sowing in the sand. After the emergence of the seedlings, the healthy seedlings with strong growth and consistency were strictly selected and planted in plastic barrels (40 cm upper diameter, 30 cm lower diameter, and 40 cm height). The controlled moisture content was 70% of the maximum water holding capacity in the field, and the experiment was carried out when the seedlings had 6–7 true leaves. The smut pathogen teliospores were collected from the sugarcane base of the Agricultural College of Guangxi University, heated at 40 °C (1 h), stored in sterilized paper bags, and stored at 4 °C for further use.

The test was carried out by the stab inoculation method, and the pathogen teliospores germination rate was more than 95%. They were collected and diluted with sterile water into a spore suspension with a concentration of 5×10^6 spores/mL using a sterile syringe. The sterile syringe was stabbed four times into the sugarcane at the growing stage, and then a 4-drop suspension (about 50 μL) was added along the leaf sheath. The control group replaced the spore suspension with sterilized ddH_2O. Inoculated once every day, the photosynthetic physiological indexes of the sugarcane (+1) leaves inoculated for 1, 2, 3, and 4 days were uniformly measured up to the 4th day after inoculation. The enzyme activity was measured 1, 3, 5, 7, and 9 days after smut inoculation.

2.2. Measurement of Leaf Gas Exchange and Chlorophyll Fluorescence

Photosynthetic responses were measured from sugarcane cultivar ROC 22 1, 2, 3, and 4 days after smut inoculation. The photosynthetic rate (Pn), stomatal conductance (gs), transpiration rate (Tr), and internal CO_2 concentration (Ci) were observed using a portable photosynthesis system (Li-6400xt, LICOR, Lincoln, NE, USA) from a photosynthetically fully mature leaf (+1). For each treatment and control, a minimum of five ($n = 5$) measurements were recorded between 10:00–11:00 am. The photosynthetic photon flux density ((PPFD) 1000 μmol $m^{-2}s^{-1}$), the leaf chamber temperature (35 °C), and the flow rate (500 μmol s^{-1}) were used while recording photosynthetic leaf gas exchange.

Chlorophyll fluorescence was measured by using an FMS-2 Modulate fluorometer (Hansatech, UK). Dark-adapted leaf (30 min), initial fluorescence (Fo) with weak measurement light, maximum fluorescence (Fm) with saturated pulsed light (9000 μmol $m^{-2}s^{-1}$), and photochemical light (1200 μmol $m^{-2}s^{-1}$) were used to determine the steady-state fluorescence (Fs), maximum (Fm′), and minimum fluorescence (Fo′) under the light. The PSII maximum variable fluorescence (Fv = Fm − Fo), maximum variable fluorescence under light (Fv′ = Fm − Fo′), maximum light energy conversion efficiency (Fv/Fm), actual light energy conversion efficiency (ΦPSII = (Fm′ − Fs)/Fm′), maximum light energy conversion efficiency (Fv′/Fm′) of the PSII reaction center under light adaptation, electron transport rate (ETR), non-photochemical quenching coefficient (qNP), and photochemical quenching coefficient (qP) were calculated by the chlorophyll fluorescence parameters. Each treatment was replicated thrice ($n = 3$).

2.3. Enzyme Extraction

For the determination of antioxidative enzyme activities, 1 g of leaf samples were homogenized in 5 mL of pre-cooled sodium borate buffer (pH 8.8) containing 1 mM EDTA, 5 mM β-mercaptoethanol, and 4% (*w*/*v*) PVP, incubated at 4 °C for 5 min. After incubation, the homogenate was centrifuged (12,000× *g*) for 20 min at 4 °C, and the supernatant was used for the subsequent estimation of POD, SOD, PPO, and PAL activities. The enzyme activities were expressed as U g^{-1} FW.

2.3.1. Determination of Peroxidase and Superoxide Dismutase Activity

The peroxidase (POD) activity was assessed according to Nakano and Asada [15] with slight modifications. The reaction mixture contained 0.1M phosphate buffer (pH 5.8) and 18 mM guaiacol, mixed with the 50 μL of enzyme extract followed by the addition of 2.5% H_2O_2 (*v*/*v*). The absorbance of the mixture was measured at 470 nm by a UV-spectrophotometer. The specific POD activity was calculated by using the below formula and expressed as U g^{-1} FW.

$$\text{POD activity (U g}^{-1}\text{ FW)} = (\Delta A_{470} \times Vt/(W \times Vs \times 0.01 \times t)$$

where ΔA_{470} indicates the time for the change in absorbance, Vt is the total volume of the reaction mixture, W is the sample fresh weight, Vs is the volume of the crude enzyme extract, and t is the reaction time (min).

Superoxide dismutase (SOD) activity was determined in terms of its capacity for 50% inhibition of the photochemical reduction in NBT monitored at 560 nm as previously described by Giannopolitis and Reis [16]. The reaction mixture contained 50 mM phosphate buffer (pH 7.8), 13 mM methionine, 63 μM NBT, 1.3 μM riboflavin, and 0.1 mM EDTA mixed with 0.5 mL of the enzyme solution. The specific SOD activity was calculated by using the below formula and expressed as U g^{-1} FW.

$$\text{SOD activity (U g}^{-1}\text{ FW)} = (\Delta A_{560} \times Vt/(W \times Vs \times 0.05 \times t)$$

where ΔA_{560} is the change in absorbance, Vt is the total volume of the reaction mixture, W is the fresh weight of the sample, Vs is the volume of the crude enzyme, and t is the reaction time (min).

2.3.2. Quantification of Polyphenol Oxidase and Phenylalanine Ammonia-Lyase Activity

Polyphenol oxidase (PPO) activity was assayed as described by Zhang and Shao [17] with minor modifications using a mixture (5 mL) containing 0.1 M sodium phosphate buffer (pH 6.8), 0.02 M catechol, and crude enzyme extract. The enzyme extract was added to start the reaction. A heat-killed crude enzyme was used in the control. The absorbance at 420 nm was observed for 3 min at 30-s intervals and the values per minute were calculated. The results were presented as U g^{-1} FW.

Phenylalanine ammonia-lyase (PAL) was quantified by the procedure described by Aoki et al. [18]. The reaction mixture (3 mL) consisted of 0.02 M L-phenylalanine (0.75 mL), 0.01 M borate buffer (2.15 mL, pH 8.8) and 0.1 mL of crude enzyme extract. Phenylalanine conversion into cinnamic acid was estimated at 290 nm and expressed as U g^{-1} FW. The samples were incubated at 30 °C for 1 h. In the control, the enzyme extract was replaced with 1 mL of borate buffer. The reaction was stopped in an icebox. One activity unit was defined as a change in absorbance of 0.01 at 290 nm.

2.4. *Data Processing*

The analytical data were analyzed using Microsoft Excel and SPSS 15.0 software, and the significance test was performed using Duncan's new complex range method.

3. Results

3.1. *Effect of Smut Infection on Photosynthetic Leaf Gas Exchange*

With the prolongation of the infection time of the smut pathogen, the photosynthesis rate (Pn) of sugarcane seedlings showed a change of "rise-lower-rise", which was significantly decreased when compared with the control on the third day; it showed a substantial increase on the fourth day after smut inoculation. It can be seen that the smut infection attacked the normal photosynthesis of sugarcane seedlings and promoted photosynthetic improvement. This may be the vital reason why it affects the early growth of sugarcane seedlings (Table 1). The changing trend of stomatal conductance (gs) of sugarcane seedlings is consistent with the changing trend of Pn, and there is a positive correlation between Pn and gs. However, compared with the control, the changes in the gs of sugarcane seedlings after pathogen infection were not apparent. It can be seen that stomatal factors do not entirely control the changes in the Pn rate (Table 1).

The intracellular CO_2 concentration (Ci) after smut infection showed a trend of increasing first and then reducing, as compared to control plants (Table 1). The enhancement was the largest after the third day of smut inoculation, significantly different from the control condition. This situation occurred when the difference in gs was not noticeable, and the Pn was decreased. It is speculated that the photorespiration of sugarcane seedlings is enhanced after smut infestation, which causes an increase in the Ci level.

The transpiration rate (Tr) of plants reflects the degree of water loss in the aboveground part, which can be used to observe the ability of plants to regulate water. This index is closely associated with multiple physiological and metabolic pathways and also indirectly reflects the regulation of the plant's photosynthesis. It can be seen from Table 1 that with the prolongation of infection time, the Tr of sugarcane leaves decreased and then increased. The treatment and control trends were consistent, but the difference between the treatment and control did not increase significantly (Table 1).

Table 1. The changes in photosynthetic responses, i.e., net photosynthetic rate (Pn), stomatal conductance (gs), intercellular CO_2 concentration (Ci), rate of transpiration (Tr), and photosynthetic water-use efficiency (WUE) during the inoculation of smut pathogen in sugarcane cv. ROC22.

Photosynthetic Response	Treatment Condition	Days of Smut Inoculation				S	R
		1	2	3	4		
Pn (μmol m^{-2} s^{-1})	CK	14.9 ± 0.72 ^{a}	18.9 ± 0.10 ^{a}	15.4 ± 0.35 ^{a}	19.6 ± 0.62 ^{b}	2.331	0.978
	T	15.9 ± 0.74 ^{a}	17.7 ± 0.18 ^{b}	10.7 ± 1.23 ^{b}	25.2 ± 0.92 ^{a}	7.233	0.837
gs (mol m^{-2} s^{-1})	CK	0.077 ± 0.005 ^{a}	0.133 ± 0.002 ^{a}	0.081 ± 0.003 ^{a}	0.128 ± 0.010 ^{a}	0.031	0.912
	T	0.092 ± 0.008 ^{a}	0.127 ± 0.004 ^{a}	0.078 ± 0.006 ^{a}	0.130 ± 0.029 ^{a}	0.030	0.917
Ci (μmol mol^{-1})	CK	54.93 ± 2.93 ^{a}	108.33 ± 0.67 ^{a}	55.27 ± 8.03 ^{b}	100.53 ± 1.85 ^{a}	31.063	0.863
	T	72.23 ± 11.15 ^{a}	113.33 ± 3.84 ^{a}	131.67 ± 7.69 ^{a}	97.06 ± 7.55 ^{a}	20.155	0.960
Tr (mmol m^{-2} s^{-1})	CK	2.410 ± 0.153 ^{a}	3.610 ± 0.063 ^{a}	2.523 ± 0.121 ^{a}	3.397 ± 0.271 ^{a}	0.623	0.952
	T	2.950 ± 1.182 ^{a}	3.497 ± 0.029 ^{a}	2.367 ± 0.152 ^{a}	3.607 ± 0.711 ^{a}	0.689	0.944
PWUE (μmol CO_2 mmol H_2O^{-1})	CK	6.185 ± 1.021 ^{a}	5.230 ± 0.801 ^{a}	6.104 ± 0.871 ^{a}	5.432 ± 0.832 ^{a}	1.601	0.901
	T	5.391 ± 0.924 ^{a}	5.147 ± 1.023 ^{a}	4.472 ± 0.591 ^{a}	5.315 ± 1.201 ^{a}	1.623	0.867

Note: CK: inoculation with ddH_2O, T: inoculation with smut pathogen. S: standard error, R: correlation coefficient. The superscript letters represent a significant difference between different treatments ($p < 0.05$ LSD), $n = 5$.

Photosynthetic water use efficiency (WUE), the amount of CO_2 fixed by plant consumption per unit of weight water, usually uses Pn/Tr to indicate the level of plant water-use efficiency. When the water supply to the plant is low, the plant generally tends to achieve a higher WUE by adjusting the openness of the stomata while maintaining high Pn; when the environmental water supply is especially insufficient, the plant can reduce the transpiration rate to improve WUE. It can be seen from Table 1 that the smut infection reduced the WUE of sugarcane plants, and the downregulation was most significant on the third day after smut inoculation.

3.2. Effect of Smut Infection on Chlorophyll Fluorescence Variables

Minimal chlorophyll fluorescence (Fo) is the fluorescence yield of the photosystem II (PS II) reaction center when it is completely open. The changes are closely interconnected to the leaf chlorophyll concentration, showing the permanent damage of stress to PSII in the plant leaves. The maximum fluorescence yield (Fm) reflects the electron transfer through the PSII, which is the fluorescence yield when the PSII reaction center is completely closed. The non-photochemical energy dissipation of PSII causes a loss in Fo. If reversible deactivation or destruction occurs in the reaction center of PSII, it will increase Fo. Thus, the intrinsic mechanism of this change can be reflected by changes in the initial fluorescence. After being infected by smut, the Fo of the leaves of sugarcane seedlings showed a trend of increasing first and then decreasing with the prolongation of infection time; Fm showed a trend of reducing rather than increasing (Figure 1). The pathogen infection induced the reversible inactivation or destruction of the PSII reaction center for the first day. The Fo of the sugarcane leaves showed a downward trend with the infestation time, indicating that the energy absorbed by the antenna pigment of the PSII of the sugarcane leaves was dissipated in the form of fluorescence and heat. At the same time, the amount of flow to photochemistry is reduced. It can be seen that the pathogen infection caused a certain degree of damage to the photosynthetic apparatus of the sugarcane leaves.

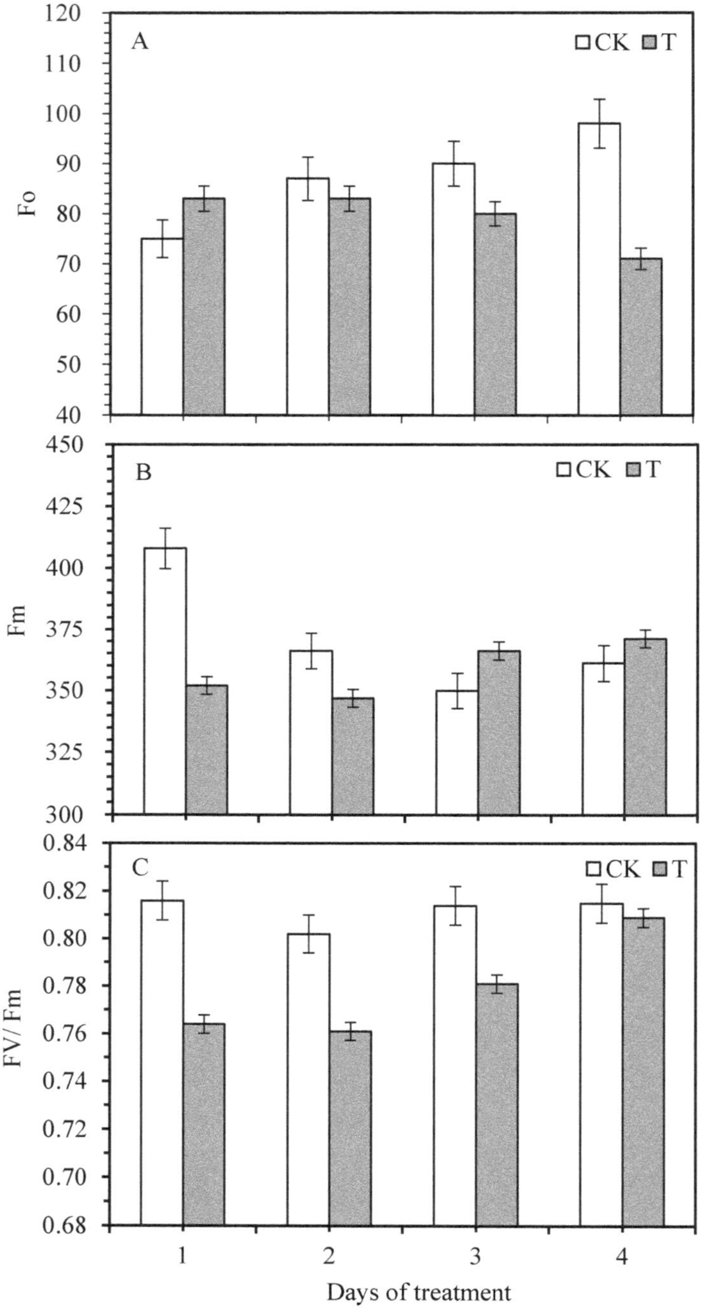

Figure 1. The variation of minimum ((**A**), Fo), maximum ((**B**), Fm), and optimum chlorophyll fluorescence yield of PS II ((**C**), Fv/Fm) of sugarcane cv. ROC22 plant leaves in response to smut inoculation at different time periods. CK—control, T—smut inoculation.

The maximum photochemical quantum yield (Fv/Fm) reflects the conversion efficiency of the original light energy in the PSII reaction center. It is minimal in plant leaves under non-infected conditions, nearly 0.8. It can be seen from Figure 1 that the original light energy conversion efficiency Fv/Fm shows a trend of decreasing first and then increasing gradually with the treatment time. The difference between Fv/Fm and the control sugarcane seedling leaves reached a significant level, indicating that sugarcane was severely affected 3–4 days after smut inoculation. The actual photochemical efficiency (ΦPSII) is the effective quantum yield of PSII and a relative indicator of the rate of photosynthetic electron transport in plants. This indicator can be directly measured under light conditions without a dark-adapted leaf. It can capture the actual primary light energy of PSII when part of the reaction center is closed. It can be seen from the change of PSII actual photochemical efficiency (ΦPSII), with the prolongation of infestation time, that ΦPSII initially decreased then increased (Figure 2).

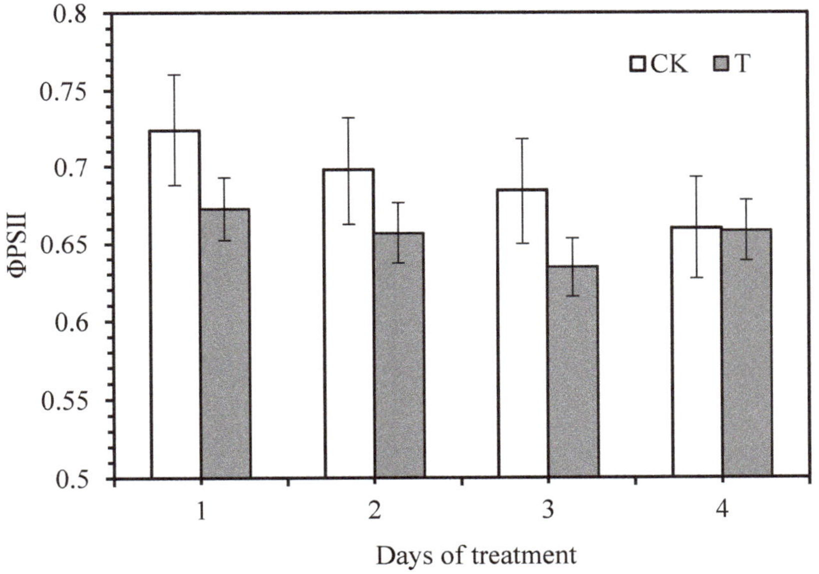

Figure 2. Effect of ΦPSII on sugarcane cv. ROC22 plants during smut inoculation. CK—control, T—smut inoculation.

It can be seen from Figure 3 that the maximum light energy conversion efficiency (Fv′/Fm′) of sugarcane seedlings showed decreasing trend initially and then increased as compared to control plants. The overall performance of Fv′/Fm is an upward trend with the infestation time, and the difference between the treatment and the control reached a significant level. The changing trend of Fv′/Fm′ of the PSII reaction center in sugarcane leaves under light adaptation conditions is consistent with the changing trend of Fv/Fm after dark adaptation. Compared with the control, the steady-state fluorescence (Fs) of the light showed an increasing pattern and then decreased; the maximum decrease was observed when pathogen-infected for 4 days. As the pathogen infection time prolonged, the overall trend gradually reduced.

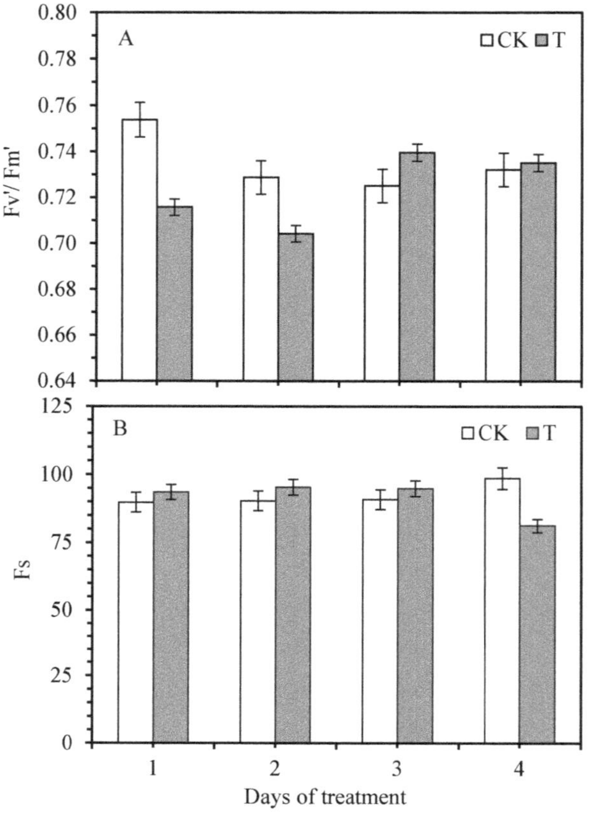

Figure 3. The changes in maximum light energy conversion efficiency ((**A**), Fv′/Fm′) and steady-state fluorescence ((**B**), Fs) of sugarcane cv. ROC22 plant leaves in response to smut inoculation at different time periods. CK—control, T—smut inoculation.

3.3. Changes in Photochemical Quenching Coefficient and Non-Photochemical Quenching Coefficient

Fluorescence quenching includes photochemical (qP) and non-photochemical quenching (qNP). Photochemical quenching can reflect the proportion of light energy absorbed by the PSII in photochemical electron transport and indirectly reflect the degree of opening and closing of the PSII reaction center and the ratio of the QA oxidation state. qNP reflects the proportion of light energy absorbed by the PSII in the form of heat, which also demonstrates the energization of the plant photosynthetic membrane. It can be seen from Table 2 that the photochemical quenching coefficient (qP) of the pathogen decreased and then increased as compared to control plants. The difference between the treatment and control reached a significant level, and the increase in qP was the largest with three days of pathogen infection. qNP is opposite to qP, i.e., with the prolongation of infection time, qNP shows a trend of "rise-lower-rise". Except for the 1-day infestation, the other time treatments were significantly different from the control. The magnitude of qNP reduction was most significant on the third day after being infected by pathogens. This indicates that when sugarcanes were infected by smut, it caused an increase in the proportion of the closed part of the PSII reaction center in the sugarcane leaves and hindered the electron flow in the PSII (oxidation lateral reaction center), which further decreased the quantum yield of electron transport.

Table 2. The variation of qP and qNP characteristics during smut inoculation.

Fluorescence Parameters	Treatment	Days of Smut Inoculation				Loss or Gain (%)
		1	2	3	4	
qP	CK	0.959 ± 0.003 [a]	0.949 ± 0.003 [a]	0.903 ± 0.007 [b]	0.923 ± 0.008 [b]	−0.04
	T	0.948 ± 0.004 [b]	0.930 ± 0.004 [b]	0.956 ± 0.003 [a]	0.948 ± 0.002 [a]	−0.01
qNP	CK	0.253 ± 0.017 [a]	0.230 ± 0.019 [b]	0.321 ± 0.021 [a]	0.137 ± 0.020 [b]	−0.46
	T	0.245 ± 0.011 [a]	0.286 ± 0.011 [a]	0.167 ± 0.009 [b]	0.278 ± 0.017 [a]	0.13

Note: CK: inoculation with ddH_2O, T: inoculation with smut pathogen. Means labeled by different letters are significantly different at $p < 0.05$ using the LSD test.

3.4. Correlation Coefficient between Photosynthesis, Stomatal Conductance, Intercellular CO_2 Concentration, Transpiration Rate, and Water-Use Efficiency

The correlation coefficients of Pn, gs, Ci, Tr, and WUE are shown in Table 3. The Pn of sugarcane seedlings was positively correlated with gs, Tr, and WUE and negatively correlated with Ci. It can be seen that gs, Tr, and WUE are the main factors affecting Pn.

Table 3. The correlation coefficient relationships between photosynthetic parameters.

Variable	Pn	gs	Ci	Tr	WUE
Pn	1.000				
gs	0.926 *	1.000			
Ci	−0.461 **	−0.093	1.000		
Tr	0.984 **	0.978 **	−0.295	1.000	
WUE	0.838 **	0.577 **	−0.864 **	0.727 **	1.000

Note: * and ** indicate significant difference at $p < 0.05$ (r = 0.3291), $p < 0.01$ (r = 0.4238).

3.5. Correlation Coefficient between Leaf Photosynthetic Parameters and Chlorophyll Fluorescence Variables

During smut inoculation, the Ci was significantly positively correlated with Fv′/Fm′ and ΦPSII; the initial fluorescence (Fo) positively correlated with Fv/Fm, Fs, and ETR; Fv/Fm positively correlated with Fv′/Fm; Fs significantly positively correlated with Fv′/Fm′; and Fv′/Fm′ positively correlated with ΦPSII. The results showed a close relationship between the photosynthetic responses and chlorophyll fluorescence variables of sugarcane seedlings in response to smut infestation of sugarcane plants (Table 4).

Table 4. The correlation coefficient of photosynthetic and chlorophyll fluorescence parameters.

Variable	Pn	gs	Ci	Fo	Fv/Fm	Fs	Fv′/Fm′	ΦPSII	qP	qNP	ETR
Pn	1.000										
gs	0.995	1.000									
Ci	0.301	0.283	1.000								
Fo	0.457	0.525	0.554	1.000							
Fv/Fm	−0.484	−0.537	−0.693	−0.984	1.000						
Fs	0.360	0.416	0.714	0.971	−0.990	1.000					
Fv′/Fm′	−0.265	−0.267	−0.978	−0.691	0.804	−0.834	1.000				
ΦPSII	−0.359	−0.345	−0.997	−0.603	0.736	−0.750	0.982	1.000			
qP	−0.757	−0.698	−0.662	−0.181	0.317	−0.241	0.533	0.677	1.000		
qNP	0.062	−0.020	0.706	−0.191	0.012	0.009	−0.554	−0.667	−0.699	1.000	
ETR	0.114	0.211	0.091	0.846	−0.744	0.756	−0.290	−0.137	0.352	−0.634	1.000

$p < 0.05$ (r = 0.3291), $p < 0.01$ (r = 0.4238).

3.6. Effects of Smut Infection on Enzymatic Activities in Sugarcane Plants

The results of POD activity determination after inoculation with smut are shown in Figure 4A. The POD activity was increased after inoculation with smut, and the POD activity showed a bimodal trend; the activity reached peaks I and II, respectively. On the third and seventh days after smut inoculation, peak II was more significant than peak I.

Compared with the control plants, the POD activity of peak I and II of ROC22 increased by 45.9 and 51.9%, respectively, and the difference reached a significant level ($p < 0.01$).

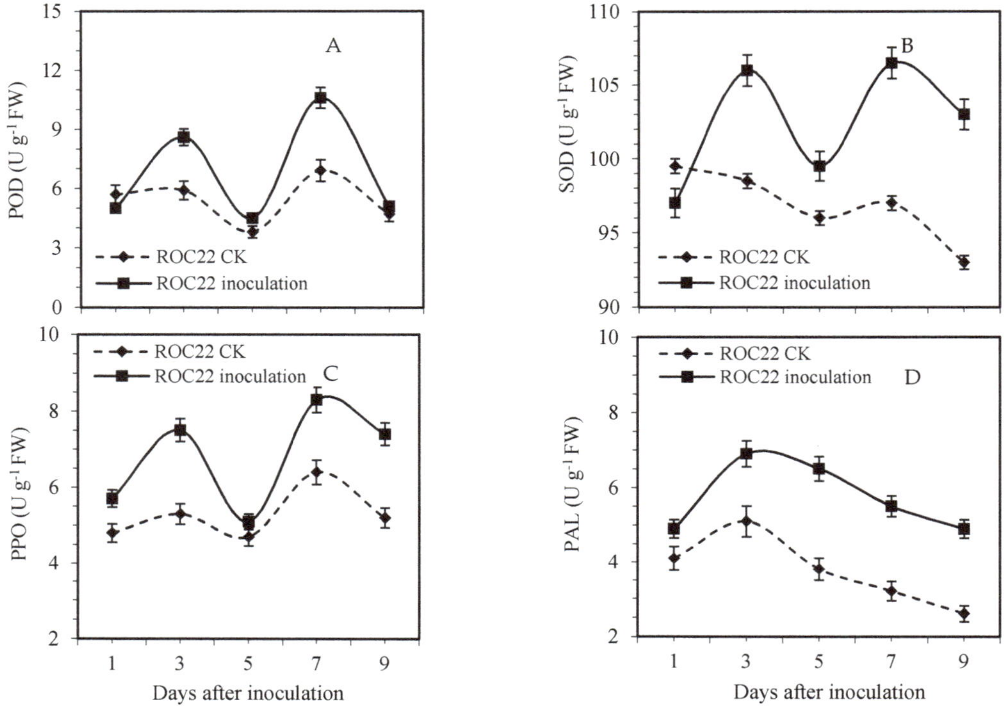

Figure 4. Effect of smut infection on peroxidase ((**A**), POD), superoxide dismutase ((**B**), SOD), polyphenol oxidase ((**C**), PPO), and phenylalanine ammonia-lyase ((**D**), PAL) activities in sugarcane cv. ROC22 plants at specific time intervals.

Figure 4B shows that the SOD activity in sugarcane plant leaves increased after inoculation, and SOD activity peaks were generated or tended to reach the activity peak I and II on the third and seventh days after inoculation, respectively. Compared with the control, ROC22 activity peaks I and II increased by 7.3 and 4.6%, respectively, but the difference was insignificant ($p < 0.05$). It can be seen from Figure 4C, that PPO activity was enhanced after smut inoculation, and reached peak I and peak II activity on the third and seventh days. Peak II was more significant than peak I, respectively. Compared with the control, the activity peak I and II of ROC22 increased by 36.5 and 26.9%, respectively, with a significant difference ($p < 0.05$). As shown in Figure 4D, PAL activity was higher than that of the control during the inoculation period, and increased initially and then decreased, with the peak activity on the third day after inoculation. PAL activity was significantly higher (36.5%) than the control after the third day of smut inoculation.

4. Discussion

The influence of sugarcane smut pathogens on sugarcane seedlings is multi-faceted and multi-layered. Green plants synthesize organic matter and gain energy through photosynthesis. Pathogen stress can affect the activity of enzymes related to photosynthetic electron transport and dark reactions in plants, and it can also cause direct damage to the photosynthetic apparatus system. Therefore, the impact of pathogens on plant photosynthesis is multifaceted [5,19]. In this study, the photosynthesis rate of sugarcane seedlings showed a "rising-lowering-liter" change after the artificial stab inoculation of smut pathogens. The photosynthetic rate was observed to have significant reductions on the second and third days after smut infection, and a more significant increase began to occur on the fourth day. Leaf gas exchange in plants between in vivo and in vitro conditions is mainly done through stomata, so changes in gs affect plant Tr and Pn. Plants can regulate the concentration of CO_2 and the loss of water in plants by changing the opening and closing of their pores or the size of the pores. Therefore, the gs can directly reflect changes in the physiological activity of the plant. The study found that the changes in gs of sugarcane seedlings after smut infection were insignificant. It is considered that there are two main reasons for the decline of Pn; stomatal and non-stomatal factors are mainly affected by the regulation mechanism [2,5,20].

It is speculated that the changes in the Pn of sugarcane seedlings are caused by non-stomatal factors during smut infection. In addition, the study also found that the Ci level in sugarcane plants increased after smut infection. This result may be due to the increase in photorespiration of sugarcane seedlings caused by the infection of the pathogen, decreasing the photosynthetic capacity of leaves, which made the supply capacity of CO_2 exceed the ability of the photosynthetic mechanism to assimilate. In addition, the obstruction of the photosynthetic product transport leads to the accumulation of photosynthetic products in leaves, an important reason for the decrease in Pn. It is speculated that the proliferation of pathogen in the early stage of smut infection inhibits the output of leaf photosynthetic products, which is another possible reason for the decrease in the Pn of sugarcane plants. When the pathogens multiply in the sugarcane plants, they stimulate the excessive Pn of the sugarcane seedlings, eventually leading to excessive nutrient consumption and pre-existing length. Yu et al. [21] found that the trend of Pn and Tr infected with acne scars was consistent with this study. The chlorophyll degradation in leaves is caused by pathogen infection [21].

The change of chlorophyll fluorescence is closely related to the photosynthetic performance. The chlorophyll fluorescence kinetic changes can reflect the effects of stress on the different processes of plant Pn. Therefore, chlorophyll fluorescence parameters can be used to evaluate the function of the plant photosynthetic system and analyze the effects of environmental stress on plants and the extent of damage and the degree of damage in the photosynthetic structure during the adversity. Schnettger et al. [22] suggested that the destruction of the PSII reaction center led to an increase in Fo and a decrease in Fv, Fm, and ΦPSII, which was consistent with the results of the study on sugarcane seedlings 1 and 2 days after smut infection. It can be seen that the infection of smut pathogens caused severe damage to the active center of the PSII in the leaves of sugarcane seedlings, inhibiting the original reaction process and also affecting the photosynthetic electrons from the reaction center of the PSII to the plastids and electron acceptors A and B. In the transmission process, the invasion of pathogens causes photoinhibition in plants. The smut infection causes the decrease in ΦPSII in the sugarcane leaves, indicating that the disease stress reduces the number of electrons involved in CO_2 fixation and the open ratio of the reaction center in PSII, which leads to the weakening of photosynthetic electron transport ability, the obstruction of the dark reaction of sugarcane leaves, and the slow conversion of light energy captured by photosynthetic pigment into chemical energy [5,23].

The ΦPSII and Pn decreases, which will be detrimental to the formation of the final yield of sugarcane. The chlorophyll fluorescence parameters of sugarcane leaves also changed with the duration of pathogen infection. It was found that the smut affected the chlorophyll fluorescence of sugarcane plants. The present findings also found that on the third and fourth day, Fo decreased, and Fv, Fm, and ΦPSII increased after pathogen infection. It can be seen that sugarcane seedlings responded positively by self-regulation in the early stage of smut pathogen infection and reduced their loss.

After smut inoculation, the Fv/Fm value increased initially and then decreased. It is inconsistent with the loss of Fv/Fm value caused by tobacco mosaic virus infection [23]. This may be due to the period of measurement of the initial stage of smut infection, and that sugarcane seedlings have more resistance and regulation to the invasion of the pathogen.

The loss of Fv′/Fm′ of sugarcane seedlings on the first and second day after pathogen infection was more significant than that of Fv/Fm, indicating that the rate and efficiency of the original light energy converted into chemical energy were caused by pathogen infection. The energy of light was not as sufficient as that of control, which may be another reason for the decreased photosynthetic rate caused by the smut infection. qP reflects the share of light energy absorbed by the PSII for photochemical electron transport, which also reflects the reduction state of the PSII primary electron acceptor QA. The larger the qP, the more electron transfer activity of the PSII [24]. The qP value of sugarcane seedlings decreased during the first 2 days after smut infestation and reduced power consumption caused by the acyclic electron transfer process was lower, which also indicated that the reduction degree of QA was higher, the proportion of the open part of the PSII reaction center decreased, and the balance of the closed position of the reaction center increased. Stable charge separation cannot be completed when the PSII reaction center is closed. Therefore, it is impossible to realize the ordered linear transfer process of photosynthetic electrons [25]. The qP value increased rapidly on the third and fourth days after infestation, reflecting that the sugarcane seedlings responded positively to the invasion of smut pathogens and reduced their damage.

The activity of POD and SOD in sugarcane was increased after smut infection. Under normal conditions, the reactive oxygen removal system in the plant, such as SOD, POD, etc., keeps the reactive oxygen metabolism in low dynamic equilibrium. Still, the pathogen infects the host and causes the sudden onset of reactive oxygen species (ROS) in the infected parts of the plants. The accumulation of ROS in sugarcane caused by smut led to increased SOD and POD activity which indicated that the sugarcane could reduce the injury to sugarcane by improving SOD and POD activity. PPO is mainly involved in the oxidation of phenols to form strontium and the polymerization of lignin precursors. PAL is an important and rate-limiting enzyme in the phenylpropanoid metabolic pathway. The roles of these two enzymes, mainly PPO and PAL, promote the production of phenolic compounds with antibiotic properties, killing the host's cells while killing the infected pathogens. PPO and PAL are involved in the synthesis of lignin, while lignin itself is toxic to germs and has an antimicrobial effect. PPO and PAL activities were enhanced after smut infection, indicating involvement in the process of sugarcane response to smut [5,19].

In addition, photoinhibition occurred in the early stage of infection, and photosynthetic capacity increased on the third day, which might be related to increased enzyme activity. Under natural conditions, when plants are subjected to various environmental stresses, the photosynthesis ability of plants is reduced, resulting in inevitably generating excess excitation energy [26]. Plants have developed a series of protective mechanisms during long-term evolution, such as heat dissipation, light respiration, etc., which depend on the xanthophyll cycle. Among them, the Mehler reaction is also considered to play a role in excess light energy dissipation [27,28]. It is speculated that its role may be reflected in two aspects: direct consumption of excess excitation electrons and the establishment of a transmembrane proton gradient to initiate heat dissipation [29]. It showed apparent photoinhibition on the first and second day and recovered on the third day; additionally, the activity of SOD and POD increased, which may be the result of the start of the Mehler reaction. The Mahler reaction is the process by which O_2 is reduced to O_2^- as an excited single electron acceptor [30]. The O_2^- produced by this reaction converts harmful superoxide radicals into H_2O_2 by SOD. Although hydrogen peroxide is still toxic to the body, the body's CAT and POD immediately break it down into completely harmless water. The three enzymes form a complete antioxidant chain [19,31].

Quinone can intervene in the reaction of photosynthesis. PPO only exists in those chloroplasts that produce high-level oxygen and is related to chloroplasts with a high ratio of chlorophyll. While the strong-offset PPO preparation of KCN can significantly improve the release of oxygen in photosynthesis, the broad bean PPO and the PSII protein are co-separated. The first 15 amino acids of the N-terminus of spinach PPO were identical to those of the PSII light-harvesting complex (LHCII) [32]. Therefore, PPO may act as a metal oxidoreductase, regulating the redox level in the cytoplasm, binding to oxygen, and delivering molecular oxygen to regulate the harmful photooxidation reaction rate in the chloroplast, participate in the electron transfer, and function as an energy conversion [33,34].

5. Conclusions

This study can help to understand the effects of sugarcane smut pathogens on the photosynthesis, chlorophyll fluorescence, and related defense enzymes of sugarcane seedlings. The Pn, gs, and Tr rates of sugarcane seedlings decreased on the first three days after smut infection and increased on the fourth day. Intercellular CO_2 concentration increased in the first three days after smut infection and decreased on the fourth day. The chlorophyll fluorescence parameters Fv, Fm, ΦPSII, Fv/Fm, and Fv′/Fm′ decreased at the initial stage of pathogen infection but increased rapidly on the third day after infection. It can be seen that sugarcane seedlings showed a positive response to pathogen infection. Correlation analysis showed that the correlation coefficient between Pn, gs, and Tr reached above 0.800, indicating a significant positive correlation; Ci was significantly positively correlated with Fv′/Fm′ and ΦPSII; Fo positively correlated with Fv/Fm, Fs, and ETR; Fv /Fm significantly positively correlated with Fv′/Fm′; Fs positively correlated with Fv′/Fm′, and Fv′/Fm′ significantly positively correlated with ΦPSII. After inoculation, the related defense enzymes POD, SOD, PPO, and PAL were increased and were associated with the sugarcane response to the smut. The upregulation in photosynthetic capacity may be interconnected to the increase in enzymatic activities.

Author Contributions: X.S. and Y.-R.L. designed the experiment. X.S., F.M., M.Y., X.Z., B.Z., X.H., D.H., Y.P. and K.K.V. performed the literature search, experiment, and data analysis. The original draft was written by X.S. and K.K.V. and Y.-R.L. critically revised the manuscript. All authors have read and agreed to the published version of the manuscript.

Funding: This study was financially supported by the Youth Program of the National Natural Science Foundation of China (31901594), Guangxi Natural Science Foundation (2021GXNSFAA220022), Guangxi characteristic crop experimental station (GTS2022022), Fund for Guangxi Innovation Teams of Modern Agriculture Technology (nycytxgxcxtd-2021-03), Natural Science Foundation of China (31960521), and Fund of Guangxi Academy of Agricultural Sciences (2021YT011/2021YT007).

Institutional Review Board Statement: Not applicable.

Informed Consent Statement: Not applicable.

Data Availability Statement: Not applicable.

Acknowledgments: We are very thankful to the Guangxi Academy of Agricultural Sciences, Nanning, Guangxi, China for providing the necessary facilities.

Conflicts of Interest: The authors declare no conflict of interest.

References

1. Su, Y.; Wang, Z.; Xu, L.; Peng, Q.; Liu, F.; Li, Z.; Que, Y. Early selection for smut resistance in sugarcane using pathogen proliferation and changes in physiological and biochemical indices. *Front. Plant Sci.* **2016**, *7*, e84426. [CrossRef]
2. Verma, K.K.; Song, X.P.; Yadav, G.; Degu, H.D.; Parvaiz, A.; Singh, M.; Huang, H.H.; Mustafa, G.; Xu, L.; Li, Y.R. Impact of agroclimatic variables on proteogenomics in sugar cane (*Saccharum* spp.) plant productivity. *ACS Omega* **2022**, *7*, 22997–23008. [CrossRef] [PubMed]
3. Que, Y.; Su, Y.; Guo, J.; Wu, Q.; Xu, L. A global view of transcriptome dynamics during *Sporisorium scitamineum* challenge in sugarcane by RNA-seq. *PLoS ONE* **2014**, *9*, e106476. [CrossRef] [PubMed]
4. Wang, B. The occurrence status and the research progress of sugarcane disease in China. *Sugar Crops China* **2007**, *3*, 48–51.
5. Rajput, M.A.; Rajput, N.A.; Syed, R.N.; Lodhi, A.M.; Que, Y. Sugarcane smut: Current knowledge and the way forward for management. *J. Fungi* **2021**, *7*, 1095. [CrossRef]
6. Sun, G.; Wang, W. Effects of leaf mosaic infection on photosynthesis and transpiration of papaya leaves. *Acta Phytophylacica Sin.* **1985**, *15*, 230–234.
7. Pinon, D.; de Armas, R.; Vicente, C.; Legaz, M.E. Role of polyamines in the infection of sugarcane buds by *Ustilago scitaminea* spores. *Plant Physiol. Biochem.* **1999**, *37*, 57–64. [CrossRef]

8. De Armas, R.; Santiago, R.; Legaz, M.-E.; Vicente, C. Levels of phenolic compounds and enzyme activity can be used to screen for resistance of sugarcane to smut (*Ustilago scitaminea*). *Australas Plant Path.* **2007**, *36*, 32–38. [CrossRef]
9. Kim, Y.C.; Kim, S.Y.; Choi, D.; Ryu, C.-M.; Park, J.M. Molecular characterization of a pepper C_2 domain-containing SRC2 protein implicated in resistance against host and non-host pathogens and abiotic stresses. *Planta* **2008**, *227*, 1169–1179. [CrossRef]
10. Torres, M.A.; Jones, J.D.; Dangl, J.L. Reactive oxygen species signaling in response to pathogens. *Plant Physiol.* **2006**, *141*, 373–378. [CrossRef]
11. Wawrzynska, A.; Christiansen, K.M.; Lan, Y.; Rodibaugh, N.L.; Innes, R.W. Powdery mildew resistance conferred by loss of the enhanced disease resistance1 protein kinase is suppressed by a missense mutation in keep on going, a regulator of abscisic acid signaling. *Plant Physiol.* **2008**, *148*, 1510–1522. [CrossRef]
12. Hao, L.; Wang, L.; Ma, C.; Yan, Z.F.; Wang, J.B. Research advances in interaction between pathogenic fungi and plant host. *J. Hebei Agric.* **2001**, *5*, 73–77.
13. Girousse, C.; Faucher, M.; Kleinpeter, C.; Bonnemain, J.L. Dissection of the effects of the pea aphid *Acyrthosiphon pisum* feeding on assimilate partitioning in *Medicago sativa*. *New Phytol.* **2003**, *157*, 83–92. [CrossRef] [PubMed]
14. Li, C.X.; Li, P.; Su, Y.F.; Zheng, P.Q.; Zhang, F.Q.; Zhang, Y. Effect of salicylic acid on permeability of plasma membrane and activities of protect enzymes of maize (*Zea mays* L.) seedlings under cadmium stress. *Plant Physiol Comm.* **2006**, *45*, 882–884.
15. Nakano, Y.; Asada, K. Hydrogen peroxide is scavenged by ascorbate-sepecific peroxidase in spinach choloroplasts. *Plant Cell Physiol.* **1981**, *22*, 867–880.
16. Giannopolitis, C.N.; Ries, S.K. Superoxide dismutases: I occurrence in higher plants. *Plant Physiol.* **1977**, *59*, 309–314. [CrossRef] [PubMed]
17. Zhang, X.; Shao, X. Characterisation of polyphenol oxidase and peroxidase and the role in browning of loquat fruit. *Czech J. Food Sci.* **2016**, *33*, 109–117. [CrossRef]
18. Aoki, S.; Araki, C.; Kaneko, K.; Katayama, O. Occurrence of L-phenylalanine ammonia-lyase activity in peach fruit during growth. *Agric. Biol. Chem.* **1971**, *35*, 784–787. [CrossRef]
19. Song, X.P.; Verma, K.K.; Tian, D.D.; Zhang, X.Q.; Liang, Y.J.; Huang, X.; Li, C.L.; Li, Y.R. Exploration of silicon functions to integrate with biotic stress tolerance and crop improvement. *Biol. Res.* **2021**, *54*, 19. [CrossRef] [PubMed]
20. Lal, A.; Ku, M.; Edwards, G.E. Analysis of inhibition of photosynthesis due to water stress in the C_3 species Hordeum vulgare and *Vicia faba*: Electron transport, CO_2 fixation and carboxylation capacity. *Photosynth. Res.* **1996**, *49*, 57–69. [CrossRef]
21. Yu, W.Y.; Pan, T.G.; Ke, Y.Q. Studies on photosynthesis of sweet potato under the stress of sweet potato scab. *Chin. J. Eco-Agric.* **2006**, *14*, 161–164.
22. Schnettger, B.; Critchley, C.; Santore, U.J.; Graf, M.; Krause, G.H. Relationship between photoinhibition of photosynthesis, D1 protein turnover and chloroplast structure: Effects of protein synthesis inhibitors. *Plant Cell Environ.* **1994**, *17*, 55–64. [CrossRef]
23. Song, X.P.; Mo, F.; Verma, K.K.; Wei, J.; Zhang, X.; Yang, L.; Li, Y.R. Effect of sugarcane smut (*Ustilago scitaminea* Syd.) on ultrastructure and biochemical indices of sugarcane. *Biomed. J. Sci. Tech. Res.* **2019**, *17*, 12546–12551.
24. Wang, K.; Xu, C. The effects of water stress on some in vivo chlorophyll: A fluorescence parameters of wheat flag leaves. *Acta Biophys. Sin.* **1997**, *13*, 273–278.
25. Fu, D.Y.; Hong, J.; Chen, J.S.; Wu, J.X. Accumulation of coat protein of turnip mosaic virus in host chloroplasts and its effect on PSII activity. *J. Plant Physiol. Mol. Biol.* **2004**, *30*, 34–40.
26. Feng, Y.; Zhang, Y.; Zhu, C. Relationship between photo-inhibition of photosynthesis and reactive oxygen species in leaves of poplars suffering root osmotic stress. *Chin. J. Appl. Ecol.* **2003**, *14*, 1213–1217.
27. Tao, Z.Y.; Zou, Q. Roles of mehler reaction in dissipating excess light energy in soybean leaves. *Acta Phytophysiol. Sin.* **2001**, *27*, 66–72.
28. Zou, Q.; Xu, C.; Zhao, S.; Meng, Q. The role of SOD in protecting the photosynthetic apparatus of soybean leaves from midday high light stress. *Acta Phytophysiol. Sin.* **1995**, *21*, 397–401.
29. Chen, H.X.; Gao, H.Y.; An, S.Z.; Li, W.J. Dissipation of excess energy in Mehler-peroxidase reactioin in *Rumex* leaves during salt shock. *Photosynthetica* **2004**, *42*, 117–122. [CrossRef]
30. Mehler, A.H. Studies on reactions of illuminated chloroplasts. I. Mechanism of the reduction of oxygen and other Hill reagents. *Arch. Biochem. Biophys.* **1951**, *33*, 65–77. [CrossRef]
31. Song, R.; Ding, Y.; Gong, C.; Xu, G.; Han, X. Research advances in relationship between tobacco resistance and protective enzymes activity. *Chin. Agric. Sci. Bull.* **2007**, *23*, 309–314.
32. Wang, M.L.; Hu, Z.L.; Zhou, M.Q.; Song, Y.C. Advances in research of polyphenol oxidase in plants. *Chin. Bull. Bot.* **2005**, *22*, 215–222.
33. Trebst, A.; Dep, K.B. Polyphenol oxidase and photosynthesis research. *Photosynth. Res.* **1995**, *46*, 414–432. [CrossRef]
34. Huang, M.; Peng, S. Progress on polyphenol oxidases in plants. *J. Guangxi Norm. Univ.* **1998**, *16*, 66–70.

life

MDPI

Article

Drought-Induced Oxidative Stress in Pearl Millet (*Cenchrus americanus* L.) at Seedling Stage: Survival Mechanisms through Alteration of Morphophysiological and Antioxidants Activity

Shuvasish Choudhury [1,*,†], Debojyoti Moulick [1,†], Dibakar Ghosh [2], Mohamed Soliman [3], Adel Alkhedaide [3], Ahmed Gaber [4] and Akbar Hossain [5,*]

1 Plant Stress Biology and Metabolomics Laboratory, Department of Life Science and Bioinformatics, Assam University, Silchar 788011, India; drubha31@gmail.com
2 ICAR-Indian Institute of Water Management, Bhubaneswar 751023, India; dibakar.ghosh@icar.gov.in
3 Clinical Laboratory Sciences Department, Turabah University College, Taif University, Taif 21995, Saudi Arabia; mmsoliman@tu.edu.sa (M.S.); a.khedaide@tu.edu.sa (A.A.)
4 Department of Biology, College of Science, Taif University, Taif 21944, Saudi Arabia; a.gaber@tu.edu.sa
5 Department of Agronomy, Bangladesh Wheat and Maize Research Institute, Dinajpur 5200, Bangladesh
* Correspondence: shuvasish@gmail.com (S.C.); akbar.hossain@bwmri.gov.bd (A.H.)
† These authors contributed equally to this work.

Citation: Choudhury, S.; Moulick, D.; Ghosh, D.; Soliman, M.; Alkhedaide, A.; Gaber, A.; Hossain, A. Drought-Induced Oxidative Stress in Pearl Millet (*Cenchrus americanus* L.) at Seedling Stage: Survival Mechanisms through Alteration of Morphophysiological and Antioxidants Activity. *Life* **2022**, *12*, 1171. https://doi.org/10.3390/life12081171

Academic Editor: Balazs Barna

Received: 7 July 2022
Accepted: 28 July 2022
Published: 31 July 2022

Publisher's Note: MDPI stays neutral with regard to jurisdictional claims in published maps and institutional affiliations.

Copyright: © 2022 by the authors. Licensee MDPI, Basel, Switzerland. This article is an open access article distributed under the terms and conditions of the Creative Commons Attribution (CC BY) license (https://creativecommons.org/licenses/by/4.0/).

Abstract: We report the impact of drought stress on pearl millet during the early seedling stage and its survival mechanism. Drought stress imposed for a period of 7, 14 and 21 days showed considerable changes in morphophysiological attributes, which were evident by a decline in seedling elongation, fresh and dry biomass, and relative water content (RWC) and degradation of chlorophyll pigment. Besides this, visible chlorosis lesions were observed in leaves as compared to the control. As compared to the respective controls, a nearly 60% decline in chlorophyll content was recorded after 14 and 21 days of drought stress. In both root and shoot, drought stress raised the reactive oxygen species (ROS) levels. Both H_2O_2 and $O_2^{\bullet-}$ levels were significantly elevated along with a significant increase in lipid peroxidation in both roots and shoots, which clearly indicated ROS-induced oxidative stress. Concomitant with the increase in ROS levels and malondialdehyde (MDA) content in roots, membrane integrity was also lost, which clearly indicated ROS-induced peroxidation of membrane lipids. The activities of antioxidant enzymes and levels of non-enzymatic antioxidants were significant ($p \leq 0.001$). After 7, 14 and 21 days of drought stress, activities of all the antioxidant enzymes viz., catalase (CAT), guaiacol peroxidase (GPX), superoxide dismutase (SOD) and glutathione reductase (GR) were inhibited, clearly indicating a loss of antioxidant defense machinery. Likewise, the levels of ascorbate (AsA) and reduced glutathione (GSH) levels declined significantly ($p \leq 0.01$). Our results reveal that, being tolerant to arid climatic conditions, pearl millet is highly susceptible to drought stress at the early seedling stage.

Keywords: antioxidants; drought; oxidative stress; pearl millet; redox implications; ROS

1. Introduction

The utmost impact of climate change has resulted in the alteration of global precipitation patterns, either causing the emergence of recurrent droughts or excessive floods in many regions of the world. Today, among the other constraints present in the agro-environmental system (like heavy metals, salinity, etc.) drought is one of the most devastating abiotic stresses that strongly affect agriculture and threaten global food security [1–7]. The emergence of recurrent droughts has resulted in the desertification of agricultural lands and often rendering it unusable for a prolonged period. Further, the incidences of drought may likely increase progressively in the coming years in major food-producing regions of the world as a consequence of climate change, which will have a direct and stringent impact on agriculture and crop productivity [7,8]. The severity of drought on crop productivity depends mainly upon its duration and intensity [5,7,9]. Besides climate change patterns mainly influencing drought worldwide, other factors such as uncontrolled deforestation and related anthropogenic activities also contribute significantly to converting many productive areas into arid and drought-prone zones [7,10]. With the increase in the global population in the last few decades, food security is emerging as a major global crisis and droughts are contributing significantly to it as compared to any other abiotic factors [11,12]. The primary effect of drought stress in plants is the inhibition of seed germination due to the lack of moisture in the soil [13–16]. During drought, the osmotic balance is disturbed, reduction in turgor pressure causes impaired plant growth [14,17–21], reduction in root and shoot biomass, reduced leaf area and affects the overall growth and development [21–24]. Studies have demonstrated that drought stress results in altered mitosis, causing the cessation of cell expansion and elongation, which eventually affects crop yield [14,24].

Abiotic stresses are largely accompanied by an imbalance in cellular redox homeostasis. This arises mainly due to the overproduction of reactive oxygen species (ROS). The ROS are produced mainly due to the failure of antioxidant defense metabolism to counteract ROS overproduction under the environmentally stressed condition in a wide range of crops [25–29]. Drought stress imposes oxidative stress load on the cellular system by inducing such overproduction of ROS [30,31]. ROS entities such as hydrogen peroxide (H_2O_2), superoxide radical ($O_2^{\bullet-}$), singlet oxygen ($^1O_2$) and hydroxyl radical ($OH^{\bullet}$) are highly reactive chemical entities, capable of reacting with cellular components and interfering with normal physiological and metabolic functions, and imbalance the cellular redox homeostasis [32–34]. Exposure to drought stress causes ROS-induced damage in plants; besides, ROS also act as signalling molecules to regulate diverse cellular responses [28,30,31]. Drought responses in plants involve complex traits, which can be achieved either by means of drought avoidance or tolerance [35].

Cereal crops such as pearl millet are widely grown in the arid agro-ecosystem of Africa and Asia. The crop is well known for its resistance against drought and high temperature in comparison to other cereals such as rice, barley, sorghum and wheat [36,37].

The investigation aims to study the impact of drought at the early seedling stage of pearl millet. Besides being well suited to arid conditions, the morphophysiological responses and drought-induced oxidative stress responses at early growth stages are not consistently known in this cereal crop, the basis of redox metabolism in pearl millet is quite fragmented. Thus, to ascertain this, the morpho-physiology, oxidative stress responses and antioxidant metabolism were evaluated after 7, 14 and 21 days of drought stress imposition in 2 days old pearl millet seedlings.

2. Materials and Methods

2.1. Plant Material and Drought Stress Imposition

The seeds of Pearl millet (*Cenchrus americanus* (L.) Morrone] were procured from the local market of Rajasthan, India, and brought to the laboratory in sterile plastic bags. The seeds were surface-sterilized with 0.1% (*w/v*) mercuric chloride ($HgCl_2$) solution, followed by repeated rinsing in sterile deionized water. Sterile seeds were transferred to sterile plastic trays and germinated over moistened paper towels for 24–48 h at 30 ± 2 °C. The germinated seeds were transferred to plastic pots containing a soil mixture composed of sand (50%), sieved clay soil (40%) and vermiculite (10%) and grown for 3 days over growth racks with a 16 h photoperiod at 32 ± 2 °C. After 2 days of growth, drought was imposed by withdrawing water supply for a period of 21 days, while controls sets were moderately watered at every 2 days interval. After 7, 14 and 21 days of drought imposition, the root and shoot were harvested for analysis.

2.2. Growth Responses, Chlorophyll Content and Relative Water Content (RWC)

To study the morphophysiological responses, plant growth responses were recorded in terms of root and shoot elongation, fresh biomass and dry biomass. The RWC was calculated as RWC (%) = $[(FW - DW)/(TW - DW)] \times 100$. The chlorophyll content was measured as per the method suggested by Arnon [38]. The total chlorophyll (Chl_t) content was calculated as Chl_a (mgL^{-1}) = 12.7 A_{663} − 2.69 A_{645}; Chl_b (mgL^{-1}) = 22.9 A_{663} − 4.68 A_{645}. Total chlorophyll was determined as $Chl_t = Chl_a + Chl_b$.

2.3. ROS Production, Lipid Peroxidation and Loss of Plasma Membrane Integrity

The ROS production was measured by determining the levels of hydrogen peroxide (H_2O_2) as per the method of Sagisaka [39] and $O_2^{\bullet -}$ production as per the method suggested by Elstner and Heupel [40]. The lipid peroxidation was measured by measuring the malondialdehyde (MDA) content in pearl millet root and shoot as per the method of Heath and Packer [41]. The loss of plasma membrane integrity was measured spectrophotometrically by assay of Evans Blue (EB) uptake, as suggested by Yamamoto et al. [42].

2.4. Activities of Antioxidant Enzymes

The activities of the antioxidant enzymes in pearl millet root and shoot were determined by measuring the activities of enzymes such as catalase (CAT) [EC 1.11.1.6], guaiacol peroxidase (GPx) [EC 1.11.1.7], superoxide dismutase (SOD) [EC 1.15.1.1] and glutathione reductase (GR) [EC 1.8.1.7]. The CAT and GPX activities were determined as per the method of Chance and Maehly [43]. The SOD activity was determined as suggested by Beauchamp and Fridovich [44], while the GR activity was measured as per the method of Smith et al. [45].

2.5. Determination of Non-Enzymatic Antioxidants

To determine the non-enzymatic antioxidants, the levels of ascorbate (AsA) and total glutathione [GSH] were determined in pearl millet root and shoot [46,47]. For extraction of the metabolites, 0.2 g of fresh tissue samples were grounded to fine powder using liquid nitrogen and homogenised with 5% (*w/v*) slufosalicylic acid. The homogenate was centrifuged at 12,000 g for 10 min at 4 °C and supernatant was obtained. The reaction mixture for AsA was comprised of 2 mL each of 2% (w/v) Na–molybdate (Na_2MoO_4) and 0.15 N sulphuric acid (H_2SO_4), followed by the addition of 1 mL each of 1.5 mM disodium hydrogen phosphate (Na_2HPO_4) and supernatant extract. The mixture was vortexed and incubated at 60 °C for 40 min in a water bath, cooled and centrifuged at 3000 g for 10 min. The absorbance was recorded at 660 nm. The reaction mixture for determination of GSH content was comprised of 0.5 M K–phosphate buffer (pH 7.2) containing ethylenediamine tetraacetic acid (EDTA), 0.6 mM 5,5′-dithiobis-2-nitrobenzoic acid (DTNB), 2 mM nicotinamide adenine dinucleotide phosphate (NADPH) and 1U yeast GR (Type II). The absorbance of the reaction mixture was recorded at 412 nm.

The spectrophotometric measurements were made with a UV–Visible Spectrophotometer (Lambda 35 UV-Vis Spectrophotometer, Perkin Elmer, Waltham, MA, USA).

2.6. Statistical Analysis

The data presented are the mean of three replicates ± SE. All the datasets were statistically analyzed using the Tukey–Kramer multiple comparison test to evaluate the significant difference among the treatments at $p \leq 0.05/0.01$, either using Microsoft Excel 2007 (Microsoft Inc. Redmond, Washington, USA) or INSTAT (Ver 3.0, Chicago, IL, USA).

3. Results

3.1. Morphophysiological Attributes

The impact of drought stress on pearl millet was clearly observed in its morphological attributes such as growth, biomass and relative water content (RWC), along with changes in the total chlorophyll content. These attributes, when compared to the respective controls, showed a relatively significant impact of drought. Though drought stress did not alter the root or shoot growth (elongation) of pearl millet after 7 days of drought stress, a significant reduction in growth was observed after 14 and 21 days of stress imposition as compared to the controls (Figures 1 and 2A,B).

Figure 1. Impact of drought on pearl millet growth after 14 and 21 days of stress. Falling of leaves as a consequence of drought stress with respect to the controls was observed. Note: 14d and 21d indicate 14 and 21 days.

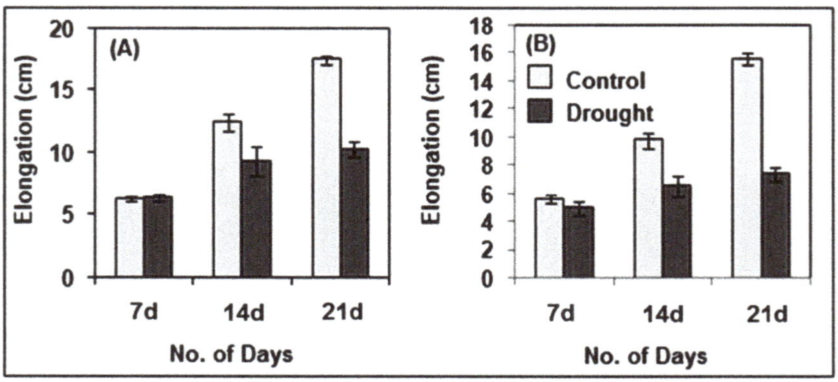

Figure 2. Effect of drought on the root (**A**) and shoot (**B**) elongation in pearl millet seedlings. The data presented are the mean of three replicates (n = 3), ±Standard Error (SE). Note: 7d, 14d and 21d indicate 7, 14 and 21 days.

Along with a decline in root and shoot elongation, several noticeable drought-induced symptoms such as curling of leaves and strong necrotic lesions were also observed (Figure 3A–D). The total chlorophyll content also showed a significant decline after 14 and 21 days of drought stress as compared to the controls, which indicated a possible decline in photosynthetic efficiency as a consequence of drought stress (Figure 3E).

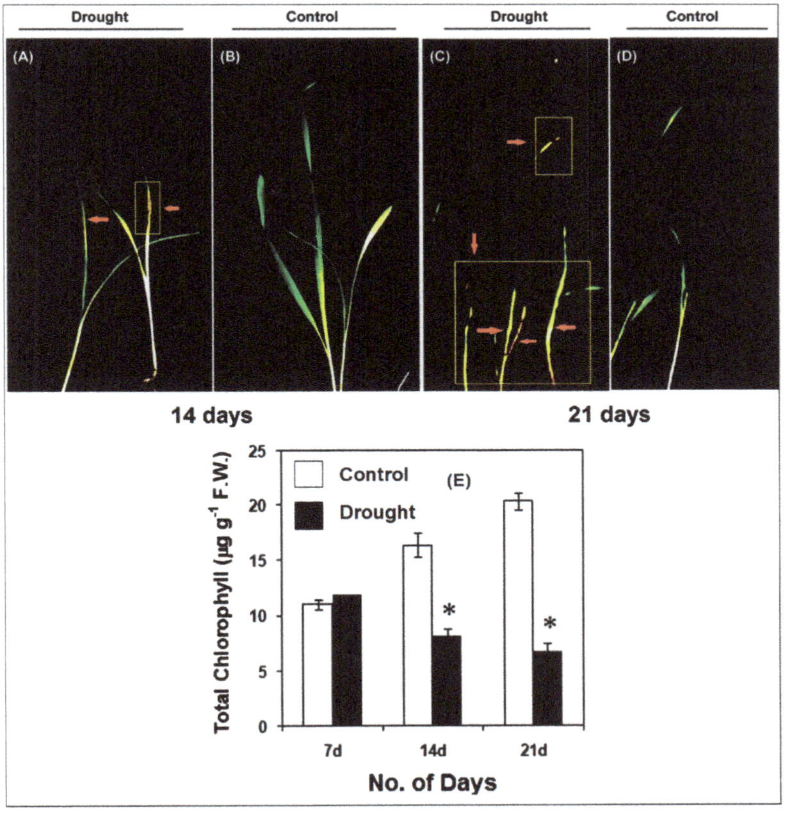

Figure 3. Drought induces changes in pearl millet seedlings after 14 (**A**,**B**) and 21 d (**C**,**D**) treatments, showing leaf curling and chlorotic lesions. (**E**) Shown is a decline in total chlorophyll content in pearl millet after 7, 14 and 21 days of drought stress. The data presented are the mean of three replicates (n = 3), ±Standard Error (SE). Asterisks (*) represent the significant difference at $p \leq 0.01$ with respect to the controls. Note: 7d, 14d and 21d indicate 7, 14 and 21 days.

After 7 days of drought, pearl millet seedlings apparently have similar chlorophyll content as that of the controls seedlings. However, with the drought conditions continued, a gradual and significant ($p \leq 0.01$) decline in the chlorophyll content was observed. After 14 and 21 days of drought, leaves of pearl millet showed almost 2- and 3-fold less chlorophyll content as compared to the respective controls. The fresh and dry biomass of pearl millet seedlings declined significantly ($p \leq 0.01$) after 14 and 21 days of drought stress as compared to the controls (Figure 4A–D). The impact of drought on root and shoot biomass was considerably noticeable after 14 days of drought stress, with a strong decline in biomass after 21 days. Under drought, the RWC was strongly affected in both root and shoot of pearl millet after 14 and 21 days of stress imposition, with practically no changes after 7 days of drought as compared to the controls (Figure 4E,F).

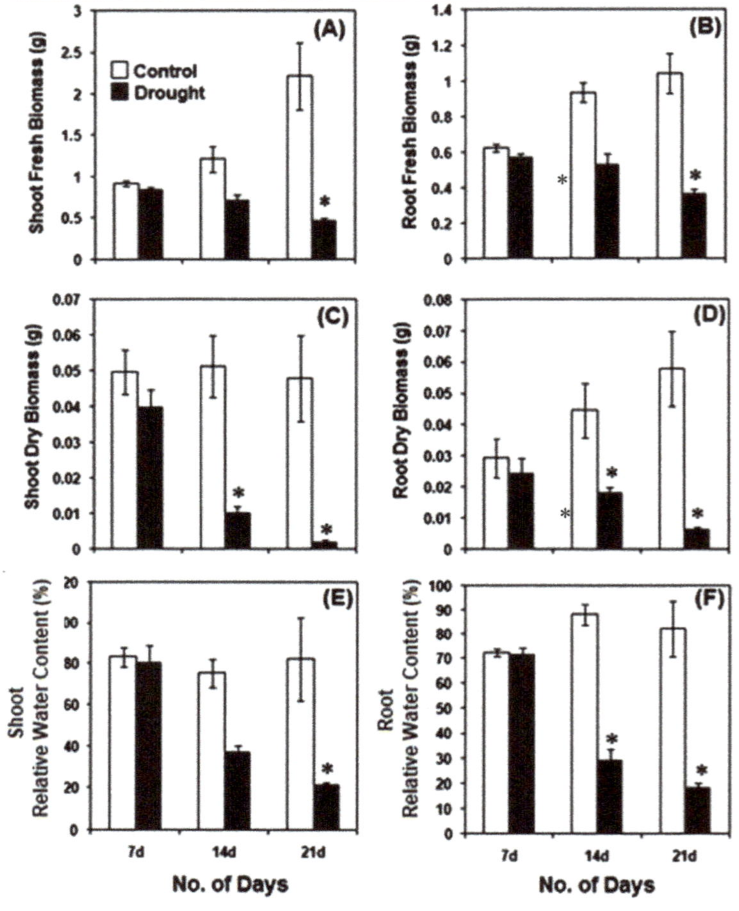

Figure 4. Drought-induced changes in fresh biomass (**A**,**B**), dry biomass (**C**,**D**) and relative water content (**E**,**F**) in pearl millet seedlings after 7, 14 and 21 days of treatment. The data presented are the mean of three replicates ($n = 3$), ±Standard Error (SE). Asterisks (*) represent the significant difference at $p \leq 0.01$ with respect to the controls. Note: 7d, 14d and 21d indicate 7, 14 and 21 days.

3.2. Drought-Induced Biomarkers

Drought (stress)-induced elevated production of ROS such as changes in H_2O_2 and $O_2^{\bullet-}$ (Figure 5A–C), and associated oxidative stress markers such as lipid peroxidation (Figure 6A,B) and loss of plasma membrane integrity in roots (Figure 6C), can be observed in pearl millet seedlings after 7, 14 and 21 days of drought stress imposition. The H_2O_2 content increased significantly ($p \leq 0.01$) in both root and shoot after 14 and 21 days of stress as compared to the controls (Figure 5A,B). After 21 days of stress, H_2O_2 content was also significantly high ($p \leq 0.01$) in both root and shoot as compared to 7 and 14 days of drought stress. Likewise, there was a significant increase ($p \leq 0.01$) in $O_2^{\bullet-}$ levels in shoot and root, respectively, of pearl millet seedlings as compared to the controls after 14 and 21 days of drought stress (Figure 5C,D). When the $O_2^{\bullet-}$ levels during stress periods of 7 and 14 days were compared with those after 21 days, significantly ($p \leq 0.01$) higher levels of the $O_2^{\bullet-}$ content were observed in both shoot and root. The $O_2^{\bullet-}$ content after 14 days was also significantly high ($p \leq 0.01$) as compared to those after 7 days.

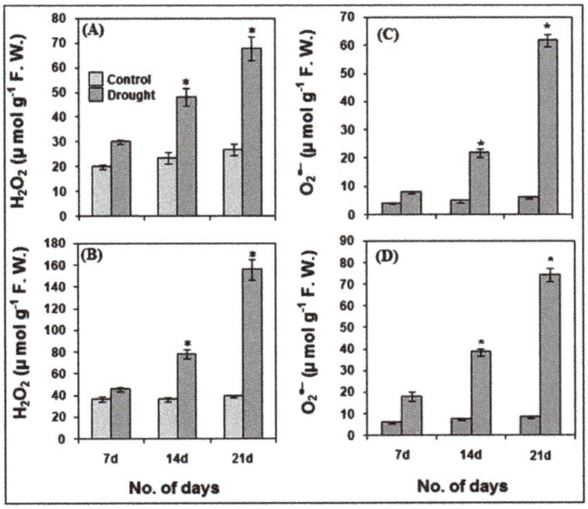

Figure 5. Production of reactive oxygen species (ROS) viz., hydrogen peroxide (H_2O_2) (**A**,**B**) and superoxide radical ($O_2^{\bullet-}$) (**C**,**D**) in pearl millet after 7, 14 and 21 days of drought stress. The data presented are the mean of three replicates ±Standard Error (SE). Asterisks (*) represent the significant difference at $p \leq 0.01$ as compared to the respective controls. Note: 7d, 14d and 21d indicate 7, 14 and 21 days.

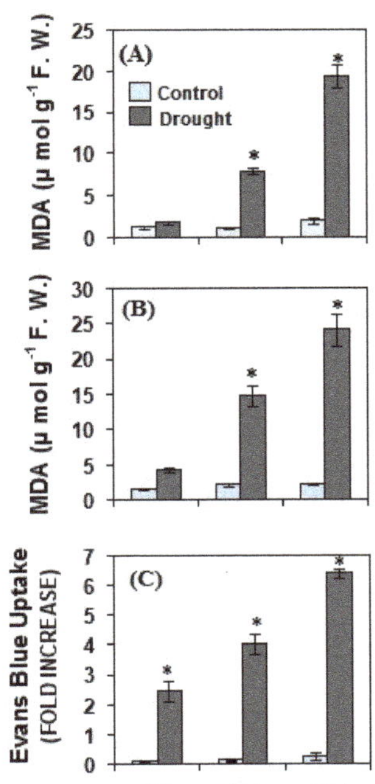

Figure 6. The onset of lipid peroxidation in pearl millet shoot (**A**) and root (**B**) and loss of root plasma membrane integrity (**C**) after drought stress. The data presented are the mean of three replicates ±Standard Error (SE). Asterisks (*) represent the significance level at $p \leq 0.01$ as compared to the controls. Note: 7d, 14d and 21d indicate 7, 14 and 21 days.

As a marker of oxidative stress, the malondialdehyde (MDA) content in pearl millet shoot and root (Figure 6A,B) was determined. A progressive and significant ($p \leq 0.01$) increase in MDA content was observed in both shoot and root after 14 and 21 days of drought stress as compared to the controls. The results were concomitant with the rise in ROS levels, which induced a significantly high level of lipid peroxidation during drought stress. Further, we also assessed the plasma membrane integrity of pearl millet roots (Figure 6C) during drought stress by observing the fold increase in the uptake of Evan's blue dye. Drought stress for the periods of 7, 14 and 21 days showed a significant ($p \leq 0.01$) increase in the uptake of the dye as compared to the controls, which indicated that drought has resulted in a loss of plasma membrane due to ROS-induced peroxidation of plasma membrane lipids.

3.3. Antioxidant Metabolism

Apart from the morpho-physiological perspective, there has been a significant influence of drought stress on antioxidant metabolism (Figures 7 and 8).

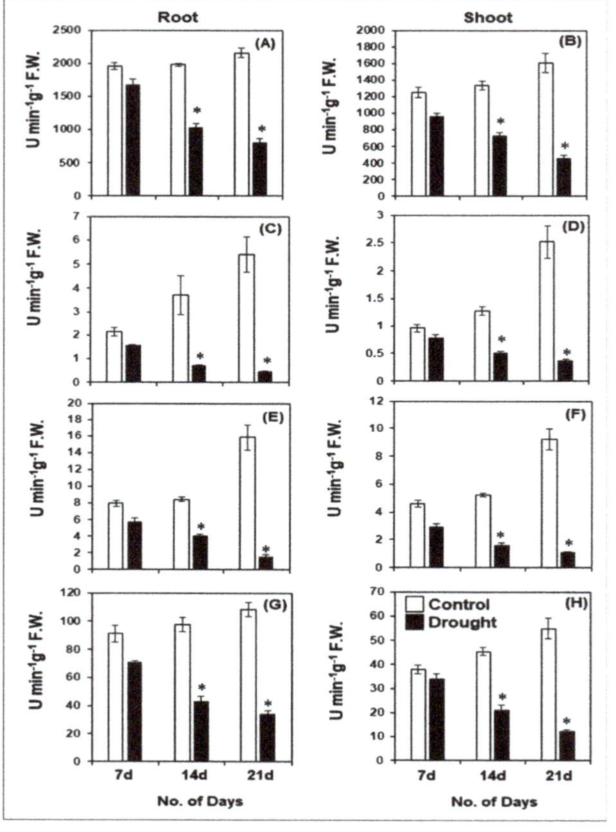

Figure 7. Changes in activities of catalase [CAT] (**A**,**B**), Guaiacol peroxidase [GPx] (**C**,**D**), glutathione reductase [GR] (**E**,**F**) and superoxide dismutase [SOD] (**G**,**H**) in root and shoot of pearl millet seedlings after 7, 14 and 21 days of drought stress. The data presented are the mean of three replicates ±Standard Error (SE). Asterisks (*) represent the significant difference at $p \leq 0.01$ as to the respective controls. Note: 7d, 14d and 21d indicate 7, 14 and 21 days.

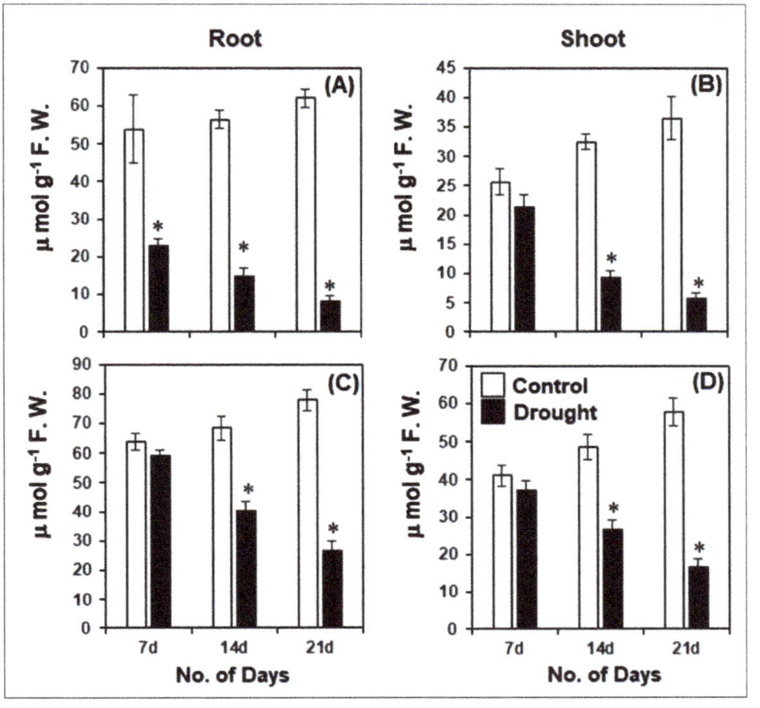

Figure 8. Changes in ascorbate [AsA] (**A**,**B**) and total glutathione [GSH] (**C**,**D**) levels in root and shoot of pearl millet seedlings after 7, 14 and 21 days of drought stress. The data presented are the mean of three replicates ±Standard Error (SE). Asterisks (*) represent the significant difference at $p \leq 0.01$ as to the respective controls. Note: 7d, 14d and 21d indicate 7, 14 and 21 days.

Drought stress for a period of 7, 14 and 21 days significantly ($p \leq 0.01$) affected the activities of enzymatic antioxidants such as CAT, GPx, GR and SOD in both root and shoot (Figure 7A–H). In comparison to their respective controls, the activities of these enzymes were gradually inhibited with the increase in the duration of drought. As compared to the 7 days old drought-stressed seedlings, the CAT activity was inhibited by almost 38% and 52%, respectively, after 14 and 21 days of stress in roots, while in shoot the CAT activity was inhibited by almost 23% and 52% after 14 and 21 days of stress, respectively, as compared to those after 7 days of drought stress. The GPx activity in roots was strongly inhibited by almost 54% and 70% after 14 and 21 days as compared to 7 days of drought stress. A similar trend was also observed in shoots, where GPx activity was inhibited by nearly 33% and 55% after 14 and 21 days, respectively, as compared to 7 days of drought stress. Similarly, the activity of GR in roots was inhibited by 29% and 72% of roots after 14 and 21 days of drought stress, respectively, while GR activity in the shoot was inhibited by 45 and 62% after 14 and 21 days of drought stress, respectively, as compared to the GR activity in root and shoot after 7 days of drought. The SOD activity in pearl millet root was inhibited by almost 39 and 51%, respectively, as compared to those after 7 days of drought stress. In shoots, SOD activity was inhibited by nearly 37 and 64%, respectively, after 14 and 21 days as compared to 7 days of drought stress. The decline in the activity of these enzymes clearly indicated the strong deleterious impact of high levels of ROS in the root and shoot of pearl millet seedlings under drought stress.

The levels of non-enzymatic antioxidants such as AsA and GSH declined significantly ($p \leq 0.01$) in pearl millet root and shoot as compared to the controls (Figure 8A–D). After

21 days, the AsA and GSH levels declined by almost 86% and 65% in respective roots, while nearly 84 and 70% declines in AsA and GSH levels were observed in shoot, respectively, as compared to the controls. Amongst the drought-stressed seedlings, the AsA levels declined by almost 34 and 63% in respective roots after 14 and 21 days as compared to 7 days of drought stress. In the shoot, the AsA levels declined by 56% and 72%, respectively, after 14 and 21 days as compared to 7 days of drought stress. The trend of decline in GSH levels was similar to AsA under drought for 14 and 21 days as compared to 7 days of drought-stressed pearl millet seedlings. In roots, the GSH levels were reduced by nearly 32 and 54% after 14 and 21 days, respectively, as compared to 7 days of drought. In the case of shoot, the GSH levels after 14 and 21 days of drought stress were reduced by 28% and 54%, respectively, as compared to 7 days of drought-stressed pearl millet seedlings.

4. Discussion

As an impact of climate change, the majority of agroclimatic conditions are altered and affected by diverse abiotic stresses. Drought is considered to be the major contributing factor to the decline in crop productivity and yield, thus threatening global food security. The scarcity of water from an agricultural perspective is a global concern and intensified with every passing day. Among grain crops, pearl millet is ranked at the sixth position and is widely cultivated in India, covering an agricultural area of close to 7.0 million ha, with an average annual production of almost 9 million tons [48]. With the ever-increasing impact of drought, especially on cereal crops, in-depth insight into physiological and biochemical responses of pearl millet under drought stress is attempted in the present investigation.

Subjected to drought stress after 7 days of growth under normal conditions till 21 days, pearl millet seedlings showed significant alterations in their morpho-physiology, redox metabolism, oxidative stress and antioxidant metabolism in root and shoot. Compared to the controls, drought-exposed seedlings have significantly less root and shoot elongation, reduction in fresh or dry biomass and highly altered RWC. These parameters were greatly impacted after 14 and 21 days of drought. Under drought conditions, the disposition of these traits exhibited considerable susceptibility of pearl millet to drought stress, and, as such, these can be considered useful datasets for comprehensively improving drought stress adaptation [49]. Further, our findings also showed the pattern of root growth (vertical and lateral), which indicated an effort to penetrate deeper into the soil for the search for available moisture [50]. In the absence of an external water supply under laboratory conditions, the soil drying rate was ominously high, thus affecting the root and progressively shoot RWC. Drought stress also affected the photosynthesis of pearl millet seedlings, as indicated by a significant decline in total chlorophyll content. Photosynthesis is considered an essential metabolic process, which is known to be affected by a variety of stresses, including drought. With a significant reduction in chlorophyll content with practically no phyto-availability of water, the overall photosynthetic process is substantially hampered, leading to low biomass and growth [51–53].

The impact of drought on pearl millet can also be observed by significant changes in redox metabolism and the onset of oxidative stress. Significantly high accumulation of ROS such as H_2O_2 and $O_2^{\bullet-}$ were observed in both root and shoot of pearl millet seedlings under drought stress. High ROS production is accompanied by an increase in MDA content, which clearly indicated oxidative stress. Further, in roots, the increase in the uptake of Evans blue dye indicated the loss of plasma membrane integrity, which is the consequence of membrane lipid peroxidation. ROS are inevitable entities of aerobic life. As a consequence of drought, the rate of photosynthesis decreases and ROS production can increase by several folds, leading to an oxidative burst [30,54,55]. Thus, drought stress disrupts the cellular redox homeostasis by increasing the levels of ROS and enhancing the process of lipid peroxidation, loss of plasma membrane integrity, altered stomatal function, growth retardation, early senescence and ultimately results in poor crop yield [56].

The response of both enzymatic and non–enzymatic antioxidants in pearl millet seedlings under drought stress clearly reflected the loss of cellular redox balance. The activities of all the antioxidant enzymes studied, such as CAT, GPX, GR and SOD, were found to be significantly affected by drought. Likewise, the levels of non-enzymatic antioxidants such as AsA and GSH were significantly lowered in both the root and shoot of pearl millet seedlings. Although antioxidants function to scavenge the ROS and maintain the cellular redox homeostasis, severe drought stress imposed for a period of 21 days at the seedling stage disrupted the antioxidant metabolism in pearl millet. In the absence of CAT and GPx activities in pearl millet due to drought stress, the levels of H_2O_2 increased progressively. Concomitant with the rise of H_2O_2 levels, the $O_2^{\bullet-}$ levels also increased significantly due to the degradation of SOD activity. All these enzymes together serve as frontline antioxidants to scavenge H_2O_2 and $O_2^{\bullet-}$ in order to maintain proper redox balance [27,57]. With an enormous load of ROS, the levels of non-enzymatic antioxidants such as AsA and GSH declined significantly in pearl millet seedlings under drought stress. The sharp decline in the levels of AsA and GSH clearly indicated a comprehensive loss of an antioxidant defense mechanism as the duration of the drought was increased in pearl millet seedlings.

The findings of the present study reflect physiological variations and oxidative stress responses in pearl millet subjected to drought stress at an early seedlings stage. With a minimum requirement of water, pearl millet is largely cultivated in arid agro-ecosystems. Although pearl millet can withstand prolonged water-deficit conditions, at an early seedling stage, drought can turn highly injurious and threatens its survival. An increase in ROS levels and failure of antioxidative metabolism leads to the onset of oxidative stress, causing considerable alterations in cellular homeostasis and disrupting normal cellular functions.

5. Conclusions

Although pearl millet is considered naturally tolerant to semi-arid and arid climatic conditions, complete withdrawal of irrigation leads to a serious deleterious impact on its morphological attributes, redox metabolism and antioxidant defense system, thus disrupting the cellular and functional homeostasis. In this study, drought was simulated for a period of 21 days at an early seedling stage of pearl millet, which resulted in high production of ROS and caused complete deterioration of antioxidant defense metabolism. Our study revealed that, at the early seedling stage, pearl millet is highly susceptible to drought stress.

Author Contributions: Conceptualization: S.C., D.M. and D.G.; methodology: S.C., D.M. and D.G.; software: S.C., A.H. and D.M.; validation: S.C., D.M. and D.G.; formal analysis: S.C., A.H. and D.G.; investigation: S.C., D.M. and D.G.; resources: S.C., D.M. and D.G.; data curation: S.C., D.M., A.H. and D.G.; writing—original draft: S.C., D.M. and D.G.; writing—review and editing: A.A., M.S., A.G. and A.H.; visualization: S.C., D.M. and D.G.; supervision: D.G. and S.C.; project administration: S.C., A.A., M.S., A.G. and A.H.; funding acquisition: A.A., M.S., A.G. and A.H. All authors have read and agreed to the published version of the manuscript.

Funding: The financial support for the current study was funded by Assam University, Silchar, India. The current research was also partially funded by the Taif University Researchers for funding this research with Supporting Project number (TURSP-2020/104), Taif University, Taif, Saudi Arabia.

Institutional Review Board Statement: Not Applicable.

Informed Consent Statement: Not Applicable.

Data Availability Statement: Not Applicable.

Acknowledgments: The authors extend thanks to Assam University, Silchar, India, for providing funding and all necessary facilities to carry out the research work. The authors also extend their appreciation to the Taif University Researchers Supporting Project number (TURSP-2020/104), Taif University, Taif, Saudi Arabia, for supporting the current study.

Conflicts of Interest: The authors declare no conflict of interest.

References

1. Moulick, D.; Samanta, S.; Sarkar, S.; Mukherjee, A.; Pattnaik, B.K.; Saha, S.; Awasthi, J.P.; Bhowmick, S.; Ghosh, D.; Samal, A.C.; et al. Arsenic contamination, impact and mitigation strategies in rice agro-environment: An inclusive insight. *Sci. Total Environ.* **2021**, *800*, 149477. [CrossRef] [PubMed]
2. Moulick, D.; Samanta, S.; Saha, B.; Mazumder, M.K.; Dogra, S.; Panigrahi, K.; Banerjee, S.; Ghosh, D.; Santra, S.C. Salinity Stress Responses in Three Popular Field Crops Belonging to Fabaceae Family: Current Status and Future Prospect. In *The Plant Family Fabaceae*; Springer: Singapore, 2020; pp. 519–541.
3. Moulick, D.; Chowardhara, B.; Panda, S.K. Agroecotoxicological aspect of arsenic (As) and cadmium (Cd) on Field crops and its mitigation: Current status and future prospect. In *Plant-Metal Interactions*; Springer: Cham, Switzerland, 2019; pp. 217–246.
4. Choudhury, S.; Mazumder, M.K.; Moulick, D.; Sharma, P.; Tata, S.K.; Ghosh, D.; Ali, H.M.; Siddiqui, M.H.; Brestic, M.; Skalicky, M.; et al. A Computational Study of the Role of Secondary Metabolites for Mitigation of Acid Soil Stress in Cereals Using Dehydroascorbate and Mono-Dehydroascorbate Reductases. *Antioxidants* **2022**, *11*, 458. [CrossRef] [PubMed]
5. Sahoo, S.; Borgohain, P.; Saha, B.; Moulick, D.; Tanti, B.; Panda, S.K. Seed priming and seedling pre-treatment induced tolerance to drought and salt stress: Recent advances. In *Priming and Pretreatment of Seeds and Seedlings*; Springer: Singapore, 2019; pp. 253–263. [CrossRef]
6. Toker, C.; Canci, H.; Yildirim, T.O.L.G.A. Evaluation of perennial wild Cicer species for drought resistance. *Genet. Res. Crop Evol.* **2007**, *54*, 1781–1786. [CrossRef]
7. Mir, R.R.; Zaman-Allah, M.; Sreenivasulu, N.; Trethowan, R.; Varshney, R.K. Integrated genomics, physiology and breeding approaches for improving drought tolerance in crops. *Theor. Appl. Genet.* **2012**, *125*, 625–645. [CrossRef]
8. Collins, N.C.; Tardieu, F.; Tuberosa, R. Quantitative trait loci and crop performance under abiotic stress: Where do we stand? *Plant Physiol.* **2008**, *147*, 469–486. [CrossRef]
9. Serraj, R.; Hash, C.T.; Rizvi, S.M.H.; Sharma, A.; Yadav, R.S.; Bidinger, F.R. Recent advances in marker-assisted selection for drought tolerance in pearl millet. *Plant Prod. Sci.* **2005**, *8*, 334–337. [CrossRef]
10. Held, I.M.; Delworth, T.L.; Lu, J.; Findell, K.U.; Knutson, T.R. Simulation of Sahel drought in the 20th and 21st centuries. *Proc. Natl. Acad. Sci. USA* **2005**, *102*, 17891–17896. [CrossRef]
11. Hossain, A.; Maitra, S.; Pramanick, B.; Bhutia, K.L.; Ahmad, Z.; Moulick, D.; Syed, M.A.; Shankar, T.; Adeel, M.; Hassan, M.M.; et al. Wild relatives of plants as sources for the development of abiotic stress tolerance in plants. In *Plant Perspectives to Global Climate Changes*; Academic Press: Cambridge, MA, USA, 2022; pp. 471–518.
12. Tuberosa, R.; Salvi, S. Genomics-based approaches to improve drought tolerance of crops. *Trends Plant Sci.* **2006**, *11*, 405–412. [CrossRef]
13. Harris, D.; Tripathi, R.S.; Joshi, A. *On-Farm Seed Priming to Improve Crop Establishment and Yield in Dry Direct-Seeded Rice*; International Research Institute: Manila, Philippines, 2002; pp. 231–240.
14. Farooq, M.; Wahid, A.; Kobayashi, N.S.M.A.; Fujita, D.B.S.M.A.; Basra, S.M.A. Plant drought stress: Effects, mechanisms and management. In *Sustainable Agriculture*; Springer: Dordrecht, The Netherlands, 2009; pp. 153–188.
15. Kaya, M.D.; Okçu, G.; Atak, M.; Cıkılı, Y.; Kolsarıcı, Ö. Seed treatments to overcome salt and drought stress during germination in sunflower (*Helianthus annuus* L.). *Eur. J. Agron.* **2006**, *24*, 291–295. [CrossRef]
16. Okçu, G.; Kaya, M.D.; Atak, M. Effects of salt and drought stresses on germination and seedling growth of pea (*Pisum sativum* L.). *Turk. J. Agric. For.* **2005**, *29*, 237–242.
17. Semida, W.M.; Abdelkhalik, A.; Rady, M.O.A.; Marey, R.A.; El-Mageed, T.A.A. Exogenously applied proline enhances growth and productivity of drought stresses onion by improving photosynthetic efficiency, water use efficiency and up-regulating osmoprotectants. *Sci. Hortic.* **2020**, *272*, 109850. [CrossRef]
18. Lawlor, D.W.; Cornic, G. Photosynthetic carbon assimilation and associated metabolism in relation to water deficits in higher plants. *Plant Cell Environ.* **2002**, *25*, 275–294. [CrossRef] [PubMed]
19. Zhu, J.K. Salt and drought stress signal transduction in plants. *Annu. Rev. Plant Biol.* **2002**, *53*, 247–273. [CrossRef]
20. Zeid, I.M.; Shedeed, Z.A. Response of alfalfa to putrescine treatment under drought stress. *Biol. Plant.* **2006**, *50*, 635–640. [CrossRef]
21. Hossain, A.; Pramanick, B.; Bhutia, K.L.; Ahmad, Z.; Moulick, D.; Maitra, S.; Ahmad, A.; Aftab, T. Emerging roles of osmoprotectant glycine betaine against salt-induced oxidative stress in plants: A major outlook of maize (*Zea mays* L.). In *Frontiers in Plant-Soil Interaction*; Academic Press: Cambridge, MA, USA, 2021; pp. 567–587.
22. Eziz, A.; Yan, Z.; Tian, D.; Han, W.; Tang, Z.; Fang, J. Drought effect on plant biomass allocation: A meta-analysis. *Ecol. Evol.* **2017**, *7*, 11002–11010. [CrossRef] [PubMed]
23. Manickavelu, A.; Nadarajan, N.; Ganesh, S.K.; Gnanamalar, R.P.; Chandra Babu, R. Drought tolerance in rice: Morphological and molecular genetic consideration. *Plant Growth Regul.* **2006**, *50*, 121–138. [CrossRef]
24. Hussain, M.; Malik, M.A.; Farooq, M.; Ashraf, M.Y.; Cheema, M.A. Improving drought tolerance by exogenous application of glycinebetaine and salicylic acid in sunflower. *J. Agron. Crop Sci.* **2008**, *194*, 193–199. [CrossRef]
25. Choudhury, S.; Moulick, D.; Mazumder, M.K. Secondary metabolites protect against metal and metalloid stress in rice: An in silico investigation using dehydroascorbate reductase. *Acta Physiol. Plant.* **2021**, *43*, 1–10. [CrossRef]
26. Choudhury, S.; Sharma, P.; Moulick, D.; Mazumder, M.K. Unrevealing metabolomics for abiotic stress adaptation and tolerance in plants. *J. Crop Sci. Biotechnol.* **2021**, *24*, 479–493. [CrossRef]

27. Scandalios, J.G. The rise of ROS. *Trend Biochem. Sci.* **2002**, *27*, 483–486. [CrossRef]
28. Choudhury, S.; Panda, P.; Sahoo, L.; Panda, S.K. Reactive oxygen species signaling in plants under abiotic stress. *Plant Signal. Behav.* **2013**, *8*, e23681. [CrossRef]
29. Kalia, R.; Sareen, S.; Nagpal, A.; Katnoria, J.; Bhardwaj, R. ROS-induced transcription factors during oxidative stress in plants: A tabulated review. In *Reactive Oxygen Species and Antioxidant Systems in Plants: Role and Regulation under Abiotic Stress*; Springer: Singapore, 2017; pp. 129–158.
30. de Carvalho, M.H.C.; Contour-Ansel, D. (h) GR, beans and drought stress. *Plant Signal. Behav.* **2008**, *3*, 834. [CrossRef]
31. Hu, H.; Xiong, L. Genetic engineering and breeding of drought-resistant crops. *Annu. Rev. Plant Biol.* **2014**, *65*, 715–741. [CrossRef]
32. Verma, G.; Srivastava, D.; Tiwari, P.; Chakrabarty, D. ROS modulation in crop plants under drought stress. In *Reactive Oxygen, Nitrogen and Sulfur Species in Plants: Production, Metabolism, Signaling and Defense Mechanisms*; Wiley-Blackwell: Hoboken, NJ, USA, 2019; pp. 311–336.
33. Xie, X.; He, Z.; Chen, N.; Tang, Z.; Wang, Q.; Cai, Y. The roles of environmental factors in regulation of oxidative stress in plants. *BioMed Res. Int.* **2019**, *2019*, 9732325. [CrossRef]
34. Vranová, E.; Inzé, D.; Van Breusegem, F. Signal transduction during oxidative stress. *J. Exp. Bot.* **2002**, *53*, 1227–1236. [CrossRef]
35. Dorion, S.; Ouellet, J.C.; Rivoal, J. Glutathione metabolism in plants under stress: Beyond reactive oxygen species detoxification. *Metabolites* **2021**, *11*, 641. [CrossRef] [PubMed]
36. Shivhare, R.; Lata, C. Assessment of pearl millet genotypes for drought stress tolerance at early and late seedling stages. *Acta Physiol. Plant.* **2019**, *41*, 1–10. [CrossRef]
37. Yadav, R.S.; Sehgal, D.; Vadez, V. Using genetic mapping and genomics approaches in understanding and improving drought tolerance in pearl millet. *J. Exp. Bot.* **2011**, *62*, 397–408. [CrossRef] [PubMed]
38. Arnon, D.I. Copper enzymes in isolated chloroplasts: Polyphenol oxidase in Beta vulgaris. *Plant Physiol.* **1949**, *24*, 1–15. [CrossRef]
39. Sagisaka, S. The occurrence of peroxide in a perennial plant, Populus gelrica. *Plant Physiol.* **1976**, *57*, 308–309. [CrossRef]
40. Elstner, E.F.; Heupel, A. Inhibition of nitrite formation from hydroxylammoniumchloride: A simple assay for superoxide dismutase. *Anal. Biochem.* **1976**, *70*, 616–620. [CrossRef]
41. Heath, R.L.; Packer, L. Photoperoxidation in isolated chloroplasts: I. Kinetics and stoichiometry of fatty acid peroxidation. *Arch. Biochem. Biophys.* **1968**, *125*, 189–198. [CrossRef]
42. Yamamoto, Y.; Kobayashi, Y.; Matsumoto, H. Lipid peroxidation is an early symptom triggered by aluminum, but not the primary cause of elongation inhibition in pea roots. *Plant Physiol.* **2001**, *125*, 199–208. [CrossRef] [PubMed]
43. Chance, B.; Maehly, A.C. Assay of catalases and peroxidases. *Method Enzym.* **1955**, *2*, 764–778.
44. Beauchamp, C.; Fridovich,, I. Superoxide dismutase: Improved assays and an assay applicable to acrylamide gels. *Anal. Biochem.* **1971**, *44*, 276–287. [CrossRef]
45. Smith, I.K.; Vierheller, T.L.; Thorne, C.A. Assay of glutathione reductase in crude tissue homogenates using 5, 5′-dithiobis (2-nitrobenzoic acid). *Anal. Biochem.* **1988**, *175*, 408–413. [CrossRef]
46. Oser, B.L. *Hawk's Physiological Chemistry*; McGraw-Hills: New York, NY, USA, 1979.
47. Griffith, O.W. Determination of glutathione and glutathione disulfide using glutathione reductase and 2-vinylpyridine. *Anal. Biochem.* **1980**, *106*, 207–212. [CrossRef]
48. Vadez, V.; Hash, T.; Bidinger, F.R.; Kholova, J., II. 1.5 Phenotyping pearl millet for adaptation to drought. *Front. Physiol.* **2012**, *3*, 386. [CrossRef]
49. Slafer, G.A.; Araus, J.L.; ROYO, C.; DEL MORAL, L.F.G. Promising eco-physiological traits for genetic improvement of cereal yields in Mediterranean environments. *Ann. Appl. Biol.* **2005**, *146*, 61–70. [CrossRef]
50. Comas, L.; Becker, S.; Cruz, V.M.V.; Byrne, P.F.; Dierig, D.A. Root traits contributing to plant productivity under drought. *Front. Plant Sci.* **2013**, *4*, 442. [CrossRef]
51. Lu, C.; Zhang, J. Effects of water stress on photosystem II photochemistry and its thermostability in wheat plants. *J. Exp. Bot.* **1999**, *50*, 1199–1206. [CrossRef]
52. Woo, N.S.; Badger, M.R.; Pogson, B.J. A rapid, non-invasive procedure for quantitative assessment of drought survival using chlorophyll fluorescence. *Plant Methods* **2008**, *4*, 1–14. [CrossRef]
53. Pinheiro, C.; Chaves, M.M. Photosynthesis and drought: Can we make metabolic connections from available data? *J. Exp. Bot.* **2011**, *62*, 869–882. [CrossRef]
54. Moran, J.F.; Becana, M.; Iturbe-Ormaetxe, I.; Frechilla, S.; Klucas, R.V.; Aparicio-Tejo, P. Drought induces oxidative stress in pea plants. *Planta* **1994**, *194*, 346–352. [CrossRef]
55. Sgherri, C.L.M.; Navari-Izzo, F. Sunflower seedlings subjected to increasing water deficit stress: Oxidative stress and defence mechanisms. *Physiol. Plant.* **1995**, *93*, 25–30. [CrossRef]
56. Riemann, M.; Dhakarey, R.; Hazman, M.; Miro, B.; Kohli, A.; Nick, P. Exploring jasmonates in the hormonal network of drought and salinity responses. *Front. Plant Sci.* **2015**, *6*, 1077. [CrossRef] [PubMed]
57. Mazumder, M.K.; Moulick, D.; Choudhury, S. Iron (Fe3+)-mediated redox responses and amelioration of oxidative stress in cadmium (Cd2+) stressed mung bean seedlings: A biochemical and computational analysis. *J. Plant Biochem. Biotechnol.* **2021**, *31*, 49–60. [CrossRef]

MDPI
St. Alban-Anlage 66
4052 Basel
Switzerland
www.mdpi.com

Life Editorial Office
E-mail: life@mdpi.com
www.mdpi.com/journal/life

Disclaimer/Publisher's Note: The statements, opinions and data contained in all publications are solely those of the individual author(s) and contributor(s) and not of MDPI and/or the editor(s). MDPI and/or the editor(s) disclaim responsibility for any injury to people or property resulting from any ideas, methods, instructions or products referred to in the content.

www.ingramcontent.com/pod-product-compliance
Lightning Source LLC
Chambersburg PA
CBHW061501220926
51850CB00004B/141